IV International Conference "Supercomputer Technologies of Mathematical Modelling" (SCTeMM'19)

Journal of Physics: Conference Series Volume 1392

Moscow, Russia
19 - 21 June 2019

ISBN: 978-1-7138-1162-6
ISSN: 1742-6588

Printed from e-media with permission by:

Curran Associates, Inc.
57 Morehouse Lane
Red Hook, NY 12571

CURRAN ASSOCIATES INC.
proceedings
.com

Some format issues inherent in the e-media version may also appear in this print version.

Institute of Physics
Dirac House, Temple Back
Bristol BS1 6BE UK

Phone: 44 1 17 929 7481
Fax: 44 1 17 920 0979

techtracking@iop.org

Additional copies of this publication are available from:

Curran Associates, Inc.
57 Morehouse Lane
Red Hook, NY 12571 USA
Phone: 845-758-0400
Fax: 845-758-2633
Email: curran@proceedings.com
Web: www.proceedings.com

TABLE OF CONTENTS

MATHEMATICAL PROBLEMS OF CONTINUUM MECHANICS

SUPERCOMPUTER TECHNOLOGIES IN RESEARCH

COMPUTATIONAL ALGORITHMS FOR SOLVING APPLIED PROBLEMS

HIGH PERFORMANCE COMPUTING SOFTWARE

MULTISCALE MODELS AND METHODS

NUMERICAL METHODS FOR SOLVING INVERSE PROBLEMS

The IV International conference "Supercomputer Technologies of Mathematical Modelling (SCTeMM'19) was organized by Steklov Mathematical Institute of the Russian Academy of Sciences in cooperation with Ammosov North-Eastern Federal University, Ivannikov Institute for System Programming of the Russian Academy of Sciences and Bauman Moscow State Technical University on June 19-21, 2019 in Moscow (Russia).

The purpose of the conference is to share new achievements and create conditions for fruitful cooperation between researchers involved in mathematical modeling, high-performance computing. It is also aimed at discussing current problems of mathematical modeling of continuum mechanics, computational algorithms on multiprocessor systems, scientific computing, and prospects for development of parallel computing technologies.

Conference sections:

- Mathematical problems of continuum mechanics- Professor A.T. Ilyichev;
- Computational algorithms for engineering and scientific research - Professor V.M. Goloviznin;
- High Performance Computing Software- Corresponding Member of the Russian Academy of Sciences A.I. Avetisyan;
- Supercomputer technologies for applied problems - Professor B.M. Glinsky;
- Multiscale finite element method - Professor Y. Efendiev;
- Numerical methods for solving inverse problems - Corresponding Member of the Russian Academy of Sciences S.I. Kabanikhin.

Over 120 participants from around the world attended the conference. 13 plenary reports were presented:

1. Yalchin Efendiev - professor, TAMU, NEFU leading scientist

Multiscale modeling and learning of heterogeneous differential equations;

2. Vladimir Vasilievich Markov - professor, Mathematical Institute, RAS

Multidimensional gas detonation and its application: virtual experiment;

3. Arutyun Ishkhanovich Avetisyan - corresponding member of the RAS, Ivannikov Institute for System Programming, RAS

System programming is a basis for competitive development in digital economy;

4. Yuri Viktorovich Vasilevsky - Corresponding Member of the RAS, Institute of Computational Mathematics, RAS

Two numerical approaches for solving systems of PDEs and related open source software;

5. Sergey Igorevich Kabanikhin - Corresponding Member of the RAS, Institute of Computational Mathematics and Mathematical Geophysics, SB RAS

Multidimensional equation analogues of I.M. Gelfand, B.M. Levitan, V.A. Marchenko and M.G. Krane;

6. Vladimir Viktorovich Shaidurov - Corresponding Member of the RAS, Institute of Computational Mathematics, SB RAS

Numerical calculation methods for problems in mean field games (Mean Field Games);

7. Nikolai Ivanovich Sidnyaev - professor, Bauman Moscow State Technical University

On the theorem of pulses and energy in determination of the rocket engine parameters;

8. Boris Mikhailovich Glinsky - Professor, Institute of Computational Mathematics and Mathematical Geophysics, SB RAS

Supercomputer technologies in solving computationally complex problems;

9. Boris Nikolaevich Chetverushkin - Academician of the RAS, Moscow State University

Computational algorithms for systems with extra-massive parallelism;

10. Vladimir Fedorovich Tishkin - Corresponding Member of the RAS, Institute of Applied Mathematics, RAS

Investigation of DG schemes convergence for viscous terms in Navier-Stokes equations;

11. Vasily Mikhailovich Goloviznin - professor, Moscow State University

Balance - characteristic difference schemes for conservation laws of hyperbolic type;

12. Petr Nikolaevich Vabishchevich – Professor, Nuclear Safety Institute, RAS

Incomplete iterative implicit schemes;

13. Sergey Yuryevich Dobrokhotov - Professor, Ishlinsky Institute for Problems in Mechanics, RAS

Analytical and numerical algorithms of localized solutions calculation in problems of propagation and incursion of long waves on water.

The conference organizers express sincere gratitude to all the members of the Scientific Committee for their valuable contribution to creating a special research atmosphere during the conference and for organizing a review of the reports presented. Special thanks to chairmen of the conference sections for high quality work.

We express special gratitude to administration of Steklov Mathematical Institute of the Russian Academy of Sciences and personal thanks to A.G. Kulikovsky, academician of the Russian Academy of Sciences, Chairman of the Organizing Committee, and Professor A.T. Ilyichev, Deputy Chairman of the Organizing Committee, for excellent condition created for fruitful work of the conference!

Journal of Physics: Conference Series **1392** (2019) 011002 doi:10.1088/1742-6596/1392/1/011002

Peer review statement

All papers published in this volume of *Journal of Physics: Conference Series* have been peer reviewed through processes administered by the proceedings Editors. Reviews were conducted by expert referees to the professional and scientific standards expected of a proceedings journal published by IOP Publishing.

Contact symmetries and conservation laws of the first order of the equation of one-dimensional shallow water over a rough bottom in Lagrange's variables

Alexander V Aksenov and Konstantin P Druzhkov

Lomonosov Moscow State University, Moscow, Russia
Keldysh Institute of Applied Mathematics RAS, Moscow, Russia

E-mail: aksenov@mech.math.msu.su, Konstantin.Druzhkov@gmail.com

Abstract. The systems of equations of one-dimensional shallow water over a rough bottom in Euler's and Lagrange's variables are considered. The problem of group classification of contact symmetries of the shallow water equation in Lagrange's variables is solved. First order conservation laws of the shallow water equation in Lagrange's variables obtained using the Noether's theorem. Bottom profiles, providing additional conservation laws, are given.

1. Introduction

There are different approaches to the construction of conservation laws for equations of mathematical physics [1, 2, 3, 4]. The most well-known method of constructing conservation laws is based on the use of Noether's theorem.

Hydrodynamic conservation laws of one-dimensional shallow water equations over a rough bottom in Euler's variables were obtained in [5, 6].

In present work the problem of group classification of contact symmetries for the equation of shallow water in Lagrangian variables is solved and the first order conservation laws using the Noether's theorem are obtained.

2. Basic equations

In dimensionless variables, the system of one-dimensional shallow-water equations over uneven bottom has the following form [7]

$$\begin{aligned} &u_t + uu_x + \eta_x = 0\,, \\ &\eta_t + ((\eta + h(x))u)_x = 0\,. \end{aligned} \tag{1}$$

Here $h(x)$ is the thickness of the unperturbed layer of the liquid, $u = u(x,t)$ is the depth-average horizontal velocity and $\eta = \eta(x,t)$ is the deviation of the free surface ($\eta + h(x) \geq 0$). The bottom profile is given by the relation $z = -h(x)$ (z is the vertical coordinate).

Using the second equation of the system (1) we introduce a new variable $m = m(x,t)$ and consider the following system of equations

$$\begin{aligned} &u_t + uu_x + \rho_x = h'(x)\,, \\ &m_x = \rho\,, \qquad m_t = -u\rho\,, \end{aligned} \tag{2}$$

where $\rho = \eta + h(x)$. From the second and third equations of the system of equations (2) it follows that due to the relation

$$\frac{dm}{dt} = m_t + u m_x = 0$$

the variable m is Lagrange's variable.

One can get the equation of one-dimensional shallow water in Lagrange's variables by choosing m and t as independent variables [8]

$$x_{tt} - \frac{x_{mm}}{x_m^3} = h'(x)\,. \tag{3}$$

3. Group classification of contact symmetries of the equation of one-dimensional shallow water in Lagrange's variables

A contact symmetry operator of the equation (3) has canonical form [9, 10]

$$Y = W\frac{\partial}{\partial x}\,.$$

Here $W = W(m, t, x, x_m, x_t)$ is characteristic of the symmetry

$$X = \xi^1\frac{\partial}{\partial m} + \xi^2\frac{\partial}{\partial t} + \eta\frac{\partial}{\partial x} + \zeta^1\frac{\partial}{\partial x_m} + \zeta^2\frac{\partial}{\partial x_t}\,, \tag{4}$$

where

$$\xi^1 = -\frac{\partial W}{\partial x_m}\,, \qquad \xi^2 = -\frac{\partial W}{\partial x_t}\,, \qquad \eta = W + \xi^1 x_m + \xi^2 x_t\,,$$
$$\zeta^1 = \frac{\partial W}{\partial m} + x_m\frac{\partial W}{\partial x}\,, \qquad \zeta^2 = \frac{\partial W}{\partial t} + x_t\frac{\partial W}{\partial x}\,.$$

Using the invariance criterion [9, 10], we get the following overdetermined system of defining equations

$$\frac{\partial^2 W}{\partial x_t^2} - x_m^3\frac{\partial^2 W}{\partial x_m^2} = 0\,,$$

$$\frac{\partial^2 W}{\partial m\partial x_t} - x_m^3\frac{\partial^2 W}{\partial t\partial x_m} - x_m^3 x_t\frac{\partial^2 W}{\partial x\partial x_m} + x_m\frac{\partial^2 W}{\partial x\partial x_t} - h'(x)x_m^3\frac{\partial^2 W}{\partial x_m\partial x_t} = 0\,,$$

$$2x_m\frac{\partial^2 W}{\partial m\partial x_m} - 2x_m\frac{\partial^2 W}{\partial t\partial x_t} + 2x_m^2\frac{\partial^2 W}{\partial x\partial x_m} - 2x_m x_t\frac{\partial^2 W}{\partial x\partial x_t} -$$
$$- 2h'(x)x_m^4\frac{\partial^2 W}{\partial x_t^2} - 3\frac{\partial W}{\partial m} - 3x_m\frac{\partial W}{\partial x} = 0\,,$$

$$\frac{\partial^2 W}{\partial m^2} + 2x_m\frac{\partial^2 W}{\partial m\partial x} - 2h'(x)x_m^3\frac{\partial^2 W}{\partial m\partial x_m} - x_m^3\frac{\partial^2 W}{\partial t^2} - 2x_m^3 x_t\frac{\partial^2 W}{\partial t\partial x} +$$
$$+ x_m^2(1 - x_m x_t^2)\frac{\partial^2 W}{\partial x^2} - 2h'(x)x_m^4\frac{\partial^2 W}{\partial x\partial x_m} + (h'(x))^2 x_m^6\frac{\partial^2 W}{\partial x_m^2} +$$
$$+ 3h'(x)x_m^2\frac{\partial W}{\partial m} + 2h'(x)x_m^3\frac{\partial W}{\partial x} - h''(x)x_m^3\left(x_m\frac{\partial W}{\partial x_m} + \frac{\partial W}{\partial x_t} - W\right).$$

This system was tested for compatibility. The following are the results of this investigation for all possible bottom profiles $h = h(x)$.

1. $h = h(x)$ is an arbitrary function. For any bottom profile $h(x)$, the equation (3) has contact symmetries with characteristics

$$W_1 = x_m\,, \qquad W_2 = 2x_t\,.$$

Corresponding symmetry operators are

$$X_1 = -\frac{\partial}{\partial m}\,, \qquad X_2 = -2\frac{\partial}{\partial t}\,.$$

2. $\boldsymbol{h = a_1 x + a_2}$**.** Denote $r = x_t - a_1 t$, $s = 1/x_m$, then characteristics of additional contact symmetries of the equation (3) are

$$W_3 = 2t\,, \qquad W_4 = 48tr^2 + 48ts + 2(9a_1t^2 - 18x)r - \frac{12mr}{s}\,,$$

$$W_5 = 20tr + 6a_1t^2 - 12x - \frac{4m}{s}\,, \qquad W_6 = 2x - 3tr - a_1t^2\,, \qquad W_\infty = \frac{w(r,s)}{s}\,,$$

where $w = w(r,s)$ is an arbitrary solution of the linear equation

$$w_{rr} = s w_{ss}\,.$$

Corresponding symmetry operators are

$$X_3 = 2t\frac{\partial}{\partial x} + 2\frac{\partial}{\partial x_t}\,,$$

$$X_4 = \left(48ts^2 + 12mr\right)\frac{\partial}{\partial m} + \left(\frac{12m}{s} - 96tr - 18a_1t^2 + 36x\right)\frac{\partial}{\partial t} +$$

$$+ \left(96ts + 48tr^2 - 2a_1t(9a_1t^2 - 18x) - 96t(r + a_1t)r + \frac{12m}{s}(r + a_1t)\right)\frac{\partial}{\partial x} +$$

$$+ \left(48s + 12r^2 - 96a_1tr - 2a_1(9a_1t^2 - 18x) + \frac{12a_1m}{s}\right)\frac{\partial}{\partial x_t} - \frac{48r}{s}\frac{\partial}{\partial x_m}\,,$$

$$X_5 = 4m\frac{\partial}{\partial m} - 20t\frac{\partial}{\partial t} - (14a_1t^2 + 12x)\frac{\partial}{\partial x} - \frac{16}{s}\frac{\partial}{\partial x_m} + (8r - 20a_1t)\frac{\partial}{\partial x_t}\,,$$

$$X_6 = 3t\frac{\partial}{\partial t} + 2(x + a_1t^2)\frac{\partial}{\partial x} + \frac{2}{s}\frac{\partial}{\partial x_m} + (3a_1t - r)\frac{\partial}{\partial x_t}\,,$$

$$X_\infty = (sw_s - w)\frac{\partial}{\partial m} - \frac{w_r}{s}\frac{\partial}{\partial t} + \left(w_s - (r + a_1t)\frac{w_r}{s}\right)\frac{\partial}{\partial x} - a_1\frac{w_r}{s}\frac{\partial}{\partial x_t}\,.$$

3. $\boldsymbol{h = a_1x^2/2 + a_2x + a_3}$**.**

3.1. $\boldsymbol{h = a_1x^2/2 + a_2x + a_3,\quad a_1 > 0}$**.** In this case characteristics of additional contact symmetries of the equation (3) are

$$W_3 = \sqrt{a_1}e^{-t\sqrt{a_1}}\,, \qquad W_4 = -\sqrt{a_1}e^{t\sqrt{a_1}}\,, \qquad W_5 = 3a_1mx_m - a_1x - a_2\,.$$

Corresponding symmetry operators are

$$X_3 = \sqrt{a_1}e^{-t\sqrt{a_1}}\frac{\partial}{\partial x} - a_1e^{-t\sqrt{a_1}}\frac{\partial}{\partial x_t}\,, \qquad X_4 = -\sqrt{a_1}e^{t\sqrt{a_1}}\frac{\partial}{\partial x} - a_1e^{t\sqrt{a_1}}\frac{\partial}{\partial x_t}\,,$$

$$X_5 = -3a_1m\frac{\partial}{\partial m} - (a_1x + a_2)\frac{\partial}{\partial x} - a_1x_t\frac{\partial}{\partial x_t} + 2a_1x_m\frac{\partial}{\partial x_m}\,.$$

3.2. $\boldsymbol{h = a_1x^2/2 + a_2x + a_3,\quad a_1 < 0}$**.** In this case characteristics of additional contact symmetries of the equation (3) are

$$W_3 = \sqrt{-a_1}\sin(t\sqrt{-a_1})\,, \qquad W_4 = -\sqrt{-a_1}\cos(t\sqrt{-a_1})\,, \qquad W_5 = 3a_1mx_m - a_1x - a_2\,.$$

Corresponding symmetry operators are

$$X_3 = \sqrt{-a_1}\sin(t\sqrt{-a_1})\frac{\partial}{\partial x} - a_1\cos(t\sqrt{-a_1})\frac{\partial}{\partial x_t}\,,$$

$$X_4 = -\sqrt{-a_1}\cos(t\sqrt{-a_1})\frac{\partial}{\partial x} - a_1\sin(t\sqrt{-a_1})\frac{\partial}{\partial x_t}\,,$$

$$X_5 = -3a_1m\frac{\partial}{\partial m} - (a_1x + a_2)\frac{\partial}{\partial x} - a_1x_t\frac{\partial}{\partial x_t} + 2a_1x_m\frac{\partial}{\partial x_m}\,.$$

4. $h = a_1(x+a_2)^{a_4} + a_3$, $a_1 \neq 0$, $a_4 \neq 0, 1, 2$. In this case characteristic of additional contact symmetry of the equation (3) is

$$W_3 = 3(a_4 - 2)tx_t - 6(a_4 + 1)mx_m + 6(x + a_2)\,.$$

Corresponding symmetry operator is

$$X_3 = -3(a_4 - 2)t\frac{\partial}{\partial t} + 6(a_4 + 1)m\frac{\partial}{\partial m} + 6(x + a_2)\frac{\partial}{\partial x} - 6a_4x_m\frac{\partial}{\partial x_m} + 3a_4x_t\frac{\partial}{\partial x_t}\,.$$

5. $h = a_1 \ln(x + a_2) + a_3$, $a_1 \neq 0$, $x > -a_2$. In this case characteristic of additional contact symmetry of the equation (3) is

$$W_3 = tx_t + mx_m - (x + a_2)\,.$$

Corresponding symmetry operator is

$$X_3 = -t\frac{\partial}{\partial t} - m\frac{\partial}{\partial m} - (x + a_2)\frac{\partial}{\partial x}\,.$$

6. $h = a_1e^{a_2x} + a_3$, $a_1 \neq 0$, $a_2 \neq 0$. In this case characteristic of additional contact symmetry of the equation (3) is

$$W_3 = a_2tx_t - 2a_2mx_m + 2\,.$$

Corresponding symmetry operator is

$$X_3 = -a_2t\frac{\partial}{\partial t} + 2a_2m\frac{\partial}{\partial m} + 2\frac{\partial}{\partial x} - 2a_2x_m\frac{\partial}{\partial x_m} + a_2x_t\frac{\partial}{\partial x_t}\,.$$

4. Conservation laws of the first order

One can show that the equation (3) is an Euler-Lagrange equation for a variational problem with Lagrange function

$$L = \frac{x_t^2}{2} - \frac{1}{2x_m} + h(x)\,.$$

One can derive conservation laws of the equation (3) with Noether's theorem [1, 2]. By conservation laws of the equation (3) we understand relations in divergent form

$$D_m(P) + D_t(Q) = 0$$

that hold modulo (3). Let us recall, that symmetry with operator (4), is variational symmetry if there is a vector field (T_1, T_2), that

$$X(L) + L\big(D_m(\xi^1) + D_t(\xi^2)\big) = D_m(T_1) + D_t(T_2)\,, \tag{5}$$

where D_m D_t are total derivatives.

Using the results of group classification of the equation (3) and the relation (5), we find the variational symmetries and corresponding conservation laws.

1. $h = h(x)$ is an arbitrary function. Symmetries of the equation (3) with operators

$$X_1 = -\frac{\partial}{\partial m}, \qquad X_2 = -2\frac{\partial}{\partial t}$$

are variational symmetries. Characteristics of these symmetries are characteristics of conservation laws with functions

$$P_1 = -\frac{x_t^2}{2} + \frac{1}{x_m} - h(x), \qquad Q_1 = x_tx_m\,,$$

$$P_2 = x_t\Big(\frac{1}{x_m^2} - h^2(x)\Big), \qquad Q_2 = x_t^2 + x_m\Big(\frac{1}{x_m} - h(x)\Big)^2.$$

2. $h = a_1x + a_2$. In this case additional variational symmetries of the equation (3) are

Journal of Physics: Conference Series **1392** (2019) 012001 doi:10.1088/1742-6596/1392/1/012001

symmetries with operators

$$X_3 = 2t\frac{\partial}{\partial x} + 2\frac{\partial}{\partial x_t}\,,$$

$$X_4 =(48ts^2 + 12mr)\frac{\partial}{\partial m} + \Big(\frac{12m}{s} - 96tr - 18a_1t^2 + 36x\Big)\frac{\partial}{\partial t} +$$
$$+ \Big(96ts + 48tr^2 - 2a_1t(9a_1t^2 - 18x) - 96t(r + a_1t)r + \frac{12m}{s}(r + a_1t)\Big)\frac{\partial}{\partial x} +$$
$$+ \Big(48s + 12r^2 - 96a_1tr - 2a_1(9a_1t^2 - 18x) + \frac{12a_1m}{s}\Big)\frac{\partial}{\partial x_t} - \frac{48r}{s}\frac{\partial}{\partial x_m}\,,$$

$$X_5 = 4m\frac{\partial}{\partial m} - 20t\frac{\partial}{\partial t} - (14a_1t^2 + 12x)\frac{\partial}{\partial x} - \frac{16}{s}\frac{\partial}{\partial x_m} + (8r - 20a_1t)\frac{\partial}{\partial x_t}\,,$$

$$X_\infty = (sq_{rs} - q_r)\frac{\partial}{\partial m} - \frac{q_{rr}}{s}\frac{\partial}{\partial t} + \Big(q_{rs} - (r + a_1t)\frac{q_{rr}}{s}\Big)\frac{\partial}{\partial x} - a_1\frac{q_{rr}}{s}\frac{\partial}{\partial x_t}\,,$$

where $p = p(r, s)$, $q = q(r, s)$ is an arbitrary solution of the linear system of equations

$$p_r = sq_s + rq_r\,, \qquad p_s = rq_s + q_r\,. \tag{6}$$

Characteristics of these symmetries are characteristics of conservation laws with functions

$$P_3 = (a_1a_2t^2 - 2a_2x)r + ts^2 - a_2^2t\,, \qquad Q_3 = 2tr + \frac{2a_2x - a_1a_2t^2}{s} + a_1t^2 - 2x,$$

$$P_4 = 2mr^3 + 24tr^2s^2 - (18x - 9a_1t^2)rs^2 - 12mrs + 16ts^3\,,$$

$$Q_4 = 16tr(r^2 + 3s) + (9a_1t^2 - 18x)(r^2 + s) - \frac{6mr^2}{s} - 12m\ln s,$$

$$P_5 = 10trs^2 + 2mr^2 + (3a_1t^2 - 6x)s^2 - 4ms\,, \qquad Q_5 = 10tr^2 + 10ts + (6a_1t^2 - 12x)r - \frac{4mr}{s}\,,$$

$$P_\infty = p(r, s) - rq(r, s), \qquad Q_\infty = \frac{q(r, s)}{s}\,,$$

where $r = x_t - a_1t$, $s = 1/x_m$; pair $p(r, s)$, $q(r, s)$ is an arbitrary solution of the system of equations (6).

3. $h = a_1x^2/2 + a_2x + a_3$.

3.1. $h = a_1x^2/2 + a_2x + a_3$, $a_1 > 0$. In this case additional variational symmetries of the equation (3) are symmetries with operators

$$X_3 = \sqrt{a_1}e^{-t\sqrt{a_1}}\frac{\partial}{\partial x} - a_1e^{-t\sqrt{a_1}}\frac{\partial}{\partial x_t}\,, \qquad X_4 = -\sqrt{a_1}e^{t\sqrt{a_1}}\frac{\partial}{\partial x} - a_1e^{t\sqrt{a_1}}\frac{\partial}{\partial x_t}$$

Characteristics of these symmetries are characteristics of conservation laws with functions

$$P_3 = e^{-t\sqrt{a_1}}\Big(x_thh' + \frac{\sqrt{a_1}}{2}\Big(\frac{1}{x_m^2} - h^2\Big)\Big), \qquad Q_3 = e^{-t\sqrt{a_1}}(\sqrt{a_1}x_t + (1 - x_mh)h'),$$

$$P_4 = e^{t\sqrt{a_1}}\Big(x_thh' - \frac{\sqrt{a_1}}{2}\Big(\frac{1}{x_m^2} - h^2\Big)\Big), \qquad Q_4 = e^{t\sqrt{a_1}}(-\sqrt{a_1}x_t + (1 - x_mh)h').$$

3.2. $h = a_1x^2/2 + a_2x + a_3$, $a_1 < 0$. In this case additional variational symmetries of the equation (3) are symmetries with operators

$$X_3 = \sqrt{-a_1}\sin(t\sqrt{-a_1})\frac{\partial}{\partial x} - a_1\cos(t\sqrt{-a_1})\frac{\partial}{\partial x_t}\,,$$

$$X_4 = -\sqrt{-a_1}\cos(t\sqrt{-a_1})\frac{\partial}{\partial x} - a_1\sin(t\sqrt{-a_1})\frac{\partial}{\partial x_t}\,.$$

Characteristics of these symmetries are characteristics of conservation laws with functions

$$P_3 = \cos(t\sqrt{-a_1}\,)x_t hh' + \frac{\sqrt{-a_1}}{2}\sin(t\sqrt{-a_1}\,)\Big(\frac{1}{x_m^2} - h^2\Big),$$
$$Q_3 = \sin(t\sqrt{-a_1}\,)\sqrt{-a_1}x_t + \cos(t\sqrt{-a_1}\,)(1 - x_m h)h',$$
$$P_4 = \sin(t\sqrt{-a_1}\,)x_t hh' - \frac{\sqrt{-a_1}}{2}\cos(t\sqrt{-a_1}\,)\Big(\frac{1}{x_m^2} - h^2\Big),$$
$$Q_4 = -\cos(t\sqrt{-a_1}\,)\sqrt{-a_1}x_t + \sin(t\sqrt{-a_1}\,)(1 - x_m h)h'.$$

4. $\boldsymbol{h = a_1(x + a_2)^{-4/3} + a_3,\quad a_1 \neq 0,\quad x > -a_2}$**.** In this case additional variational symmetry of the equation (3) is symmetry with operator

$$X_3 = 10t\frac{\partial}{\partial t} - 2m\frac{\partial}{\partial m} + 6(x + a_2)\frac{\partial}{\partial x} + 8x_m\frac{\partial}{\partial x_m} - 4x_t\frac{\partial}{\partial x_t}.$$

Characteristic of this symmetry is characteristic of conservation law with functions

$$P_3 = -\frac{5tx_t}{x_m^2} - mx_t^2 + \frac{3(x + a_2)}{x_m^2} + 2m\Big(\frac{1}{x_m} - h\Big),$$
$$Q_3 = -5t\Big(x_t^2 + \frac{1}{x_m}\Big) + 6(x + a_2)x_t + (10h - 8a_3)t + 2mx_t x_m.$$

One can show that other symmetries of the equation (3) are non-variational.

5. Conclusion

The main results of the work are the solution of a group classification problem for the equation of one-dimensional shallow water in Lagrangian variables and the derivation of all it's first order conservation laws. The paper presents bottom profiles with extensions of contact symmetry operators kernel and conservation laws kernel. These results can be used to build new exact solutions.

Acknowledgments

This research was supported by Russian Science Foundation Grant No 18-11-00238.

References

[1] Noether E 1918 Invariante Variationsprobleme *Nachr. D. König. Gesellsch. D. Wissen. Zu Göttingen, Math.-Phys. Klasse.* 235–257 (English translation: *Transport Theory and Stat. Phys.* 1971 **1** (**3**) 186–207)

[2] Olver P J 1993 *Applications of Lie Groups to Differential Equations* (New York: Springer-Verlag)

[3] *Symmetries and Conservation Laws for Differential Equations of Mathematical Physics* Editors: A M Vinogradov and I S Krasil'schik 1999 **182** (American Mathematical Society)

[4] Bluman G W Cheviakov A F and Anco S C *Applications of Symmetry Methods to Partial Differential Equations* 2010 (New York: Springer-Verlag)

[5] Aksenov A V and Druzhkov K P 2016 Zakony sokhranenija, simmetrii i tochnye reshenija sistemy uravnenij melkoj vody nad nerovnym dnom [Conservation Laws, Symmetries, and Exact Solutions of the System of Equations of Shallow Water over an Irregular Bottom]. *Vestnik Nacionalnogo issledovatelskogo yadernogo universiteta MIFI* **5** (**1**) 38–46

[6] Aksenov A V and Druzhkov K P 2016 Conservation laws and symmetries of the shallow water system above rough bottom *J. Phys.: Conf. Ser.* 2016 **722** (**012001**) 1–7

[7] Stoker J J 1957 *Water Waves. The Mathematical Theory With Applications* (New York: Interscience Publishers)

[8] Cherny G G 1988 *Gas dynamics* (Moscow: Nauka) (in Russian).

[9] Ovsiannikov L V 1982 *Group Analysis of Differential Equations* (New York: Academic Press)

[10] Ibragimov N H 1985 *Transformation Groups Applied to Mathematical Physics* (New York: Springer-Verlag)

Journal of Physics: Conference Series **1392** (2019) 012002 doi:10.1088/1742-6596/1392/1/012002

Equations that Describe Waves in Tubes with Elastic Walls and Analysis of Numerical Methods for Reversible and Weakly Dissipative Systems

I. B. Bakholdin

Keldysh Institute of Applied Mathematics, Miusskaya sq. 4, 125047 Moscow, Russia

E-mail: `ibbakh@yandex.ru`

Abstract. Models of tube with elastic walls are presented: with controlled pressure, filled with incompressible fluid, filled with compressible gas. The non-linear theory of hyperelasticity is applied. The walls of a tube are described with complete membrane model. Equations are solved numerically. General approach for calculation of non-dissipative and weakly dissipative systems is given. Three-layer time and space centered reversible numerical scheme and similar two-layer space reversible numerical schemes with approximation of time derivatives by high-order Runge-Kutta methods are analysed. A method of correction of numerical schemes by inclusion of terms with high-order derivatives is developed. Additional terms with high-order derivatives my be included also in order to calculate non-dissipative and dissipative shocks at the same time.

1. Equations for waves in tubes

Models that describe waves in elastic tubes are of great interest for technical and biological applications. Here equations based on the hyperelastic model and the complete membrane model are analysed. Such equations were derived in [1] and analitically investigated in [2] and [3]. The purpose of this work is the application of the theory of reversible and weakly dissipative shocks [4, 5], the development of numerical methods.

The equations of wave motion of the walls of an elastic incompressible cylindrical tube with fixed internal and external pressure are given by [2]

$$\left(R\sigma_1 z'/\lambda_1^2\right)' - P^* r r' = \rho R\ddot{z}, \quad \left(R\sigma_1 r'/\lambda_1^2\right)' - \sigma_2/\lambda_2 + P^* r z' = \rho R\ddot{r}, \qquad (1)$$

$$\lambda_1 = \sqrt{r'^2 + z'^2}, \quad \lambda_2 = r/R, \quad \lambda_3 = h/H, \quad \sigma_i = \lambda_i W_{\lambda_i} - p, \quad P^* = P/H.$$

Here the prime denotes the derivative with respect to the variable Z, which stands for Lagrange initial longitudinal spatial coordinate, the upper dots denote derivative with respect to the time. Unknowns z and r determine the surface of the tube in the cylindrical system of coordinates, the z axis of this system coincides with the central line of the tube. The parameter P stands for the difference between internal and external pressures,the unknown p is the pressure in the material. The parameter ρ is the density of material per square unit. The thickness of the wall is denoted as h. Here λ_i are the principal stretches and σ_i are the principal Cauchy stresses, W

is the stress function. It is assumed that in the absence of any external forces $z = Z$, $r = R$, $h = H$. Due to the condition of incompressibility,

$$\lambda_1\lambda_2\lambda_3 = 1, \quad \sigma_i = \lambda_i \hat{W}_i, \quad \hat{W}_i = \hat{W}_{\lambda_i}, \quad \hat{W} = W(\lambda_1, \lambda_2, 1/(\lambda_1\lambda_2)).$$

The stress function used in this paper for numerical calculations corresponds to the Gent material

$$W = -\tilde{\mu} J_m \ln \left[1 - (\lambda_1^2 + \lambda_2^2 + \lambda_3^2 - 3)/J_m\right].$$

It is considered that this stress function is applicable for rubber.

There are dispersion branches of longitudinal and transversal elastic waves in this model. Transversal branches do not intersect point of origin in (ω, κ) plane. Hence there are no shock structures of transversal waves.

For the case of a fluid-filled tube with arbitrary pressure we must include the equations [3]

$$\dot{r}z' - r'\dot{z} + vr' + \frac{1}{2}rv' = 0, \tag{2}$$

$$\rho_f(\dot{v}z' - v'\dot{z} + vv') + P' = 0. \tag{3}$$

Here v is the fluid velocity, ρ_f is the fluid density. For the calculations, the equations were transformed to the form with excluded pressure [8]

$$\left[\rho_f^* z' - \left(\frac{r'\rho R^2}{rz'^2\xi}\right)'\right]\dot{v} - \left[\frac{r'\rho R^2}{rz'^2\xi} + \left(\frac{\rho R^2}{2z'^2\xi}\right)'\right]\dot{v}' - \frac{\rho R^2}{2z'^2\xi}\dot{v}'' = \rho_f^*(v'q - vv') + \left[-P_b^* + \frac{\mathcal{P}}{\xi}\right]' - \{c_{v4}v''''\},$$

$$\dot{r}z' - r'q + vr' + \frac{1}{2}rv' = 0 + \{c_{r6}r''''''z'\}, \quad \dot{z} = q - \{c_{z4}z''''\}, \quad \rho R\dot{q} = \left(R\sigma_1\frac{z'}{\lambda_1^2}\right)' - (P^* + P_b^*)rr',$$

$$P^* + P_b^* = \frac{P + P_b}{H} = \frac{-1}{\xi}\left[\frac{r'\rho R}{rz'^2}\dot{v} + \frac{\rho R}{2z'^2}\dot{v}' + \mathcal{P}\right], \qquad \xi = 1 + \frac{r'^2}{z'^2},$$

$$\mathcal{P} = \frac{1}{rz'}\left[\left(R\sigma_1\frac{r'}{\lambda_1^2}\right)' - \frac{r'}{z'}\left(R\sigma_1\frac{z'}{\lambda_1^2}\right)' - \frac{\sigma_2}{\lambda_2}\right] - \frac{\rho R}{rz'^2}(q-v)\left(\frac{z'q - vr' - \frac{1}{2}rv'}{z'}\right)' - \frac{\rho R}{rz'^3}\left(r'q - vr' - \frac{1}{2}rv'\right)\left(\frac{1}{2}v' - q'\right).$$

Here terms in braces may be included for effectiveness of calculations (see below).

There are branches of longitudinal and transversal elastic waves in this model. All branches intersect point of origin.

The analysis shows that for the case of the gas-filled tube [8] the equation (2) must be replaced by

$$(\dot{\rho}_f z' - \rho_f'\dot{z})r^2 + 2\rho_f r(\dot{r}z' - r'\dot{z}) + (\rho_f v r^2)' = 0. \tag{4}$$

In the barotropic case only one additional equation of state is required. An additional equation for energy is required for the complete gas dynamics model.

There are branches of longitudinal and transversal elastic waves and gas dynamic waves in this model. Transversal branches do not intersect point of origin as in the case of controlled pressure.

The viscosity of the fluid or gas may be taken into account if the additional term $c_{v2}v''$ is included in (3). The viscosity of the material of the walls is enforced wthin the Kelvin-Voight model [6]

$$\sigma_i \to \sigma_i + \sigma_{vi}, \quad \sigma_{vi} = \nu_{vi}\dot{\lambda}_i, \quad i = 1, 2, 3.$$

It can be shown that for some simplified version of the equations this model leads to the inclusion of the term $c_{v2}v''$ in the equation (3) also [8].

The compressibility of the material of the wall may be taken into account if a stress function (potential) without p is used. A two dimensional potential (potential with hat) can be derived from the equation $\sigma_3 = p_e$, where p_e is the external pressure.

In order to take into account the meridional bending resistance of the thin walls additional "bending resistance pressure" must be formally included in (1) but not in (3) [7, 8]

$$P \to P + P_b,\ P_b = -\cos\alpha\left[\frac{\partial^2}{\partial x^2}\left(D\frac{\partial^2}{\partial x^2}r\right) - \tan\alpha\frac{\partial^2}{\partial x^2}\left(D\frac{\partial^2}{\partial x^2}z\right)\right],\ \frac{\partial}{\partial x} = \frac{\cos\alpha}{z'}\frac{\partial}{\partial Z},\ \tan\alpha = \frac{r'}{z'}.$$

Here α is the angle of inclination of the wall of the tube. This approach is based on the Germain-Lagrange formula. The bending resistance coefficient D must be calculated for the anisotropic linearised state [8]

$$D = \{W_1 + \lambda_1(W_{11} - W_{13}) + [W_3 + \lambda_3(W_{33} - W_{31})]\lambda_3/\lambda_1\}h^3/12.$$

If the geometric nonlinearity is not taken into account then $P_b = P_{bs} = -Dr''''$ and the bending pressure is included only in the equation for r. Only the term $-br''''$ is added to the left part in this case. A problem is not well-posed if $\sigma_1 < 0$ and initial equations are used. The inclusion of the bending resistance enforces well-posedness.

A non-differential term related with the latitudinal (circular) bending resistance [9] may be included in pressure formula. But if the wall of the tube is weakly stretched (this is a typical case when inclusion of meridional bending resistance is required) or if h/R is close to zero this term vanishes.

The term simplified equations is used in the theory of shocks to denote hyperbolic equations describing waves with slow variations. The simplified form of the controlled pressure equations is

$$\dot{u} = q';\quad \rho R\dot{q} = \left(R\sigma_1 u/\lambda_1^2\right)' - P^* r',\quad \lambda_1 = u = z',\quad P^* = \sigma_2/(\lambda_2 r u).$$

No simplification is required for the equations (2), (3), (4). These equations describe non-wave zones for Riemann problem solutions.

Equations for waves in tubes in non-dissipative cases are reversible equations with dispersion caused by terms with derivatives of various order. Dispersion vanishes if wavenumber $\kappa \to 0$. Exception is the case when bending resistance is taken into account in simplified form. Hence we may assume that in some cases only generalized solutions with dissipative shock structures or other singularities may exist.

2. Numerical methods, general approach for non-dissipative and weakly-dissipative models

Standard non-dissipative three-layer time and space centered scheme and space centered two-layer schemes with approximation of time derivatives by Runge-Kutta method were used for solition of equatios of waves in tubes. These schemes are used for all models analysed in the theory of reversible and weakly dissipative shocks [4, 5]. The schemes are explicit usually. Similar implicit numerical schemes are used for the approximation of complex time-space derivatives. If three-layer scheme is used then dissipative terms are approximated on lower layer, by DuFort-Frankel type scheme or implicitly. According to the theory in non-dissipative case homogeneous states and waves zones are observed in solutions of Riemann problem. They are separated by reversible shock structures of solitary wave type and kink type. Length of wave zones is increased with time. In weakly dissipative case increase of wave zones is stopped after some time and their length depends on values of dissipative coefficients. That is why it is essential to use non-dissipative or weakly dissipative numerical schemes.

For the case of controlled pressure the three-layer scheme was used [7]. But for cases of fluid-filled and gas-filled tubes boundary numerical instability is observed. For fluid-filled tubes this instability may be corrected by inclusion of dissipative zones near boundaries. An additional term with second-order derivative must be included in the equation for v. If the space step is decreased then the coefficient before this term must be increased. Good results are achieved if this term is approximated by the DuFort-Frankel method. This is not convenient for calculations. That is why two-layer schemes were used for the cases of fluid-filled and gas-filled tube.

Results of analysis of methods based on Runge-Kutta approximation for time derivatives are given below in order to understand their properties. Generalised transport equation is considered

$$\frac{\partial u}{\partial t} = \mathcal{L}u, \quad \mathcal{L} = \sum_{j=1}^{M} a_j \mathcal{L}_j, \quad \mathcal{L}_j = \frac{\partial^j u}{\partial x^j}$$

Terms with odd order derivatives are non-dissipative terms. Terms with even order derivatives are dissipative terms. Dissipation is correct if $a_{4m-2} > 0$ and $a_{4m} < 0$, $m = 1, 2, \ldots.$ This equation is similar to equation for v of the transformed system of equation for fluid-filled tubes. Centered approximations for space derivatives are used

$$u_x \to \frac{u^n_{k+1} - u^n_{k-1}}{2\tilde{h}}, u_{xx} \to \frac{u^n_{k+1} + u^n_{k-1} - 2u^n_k}{\tilde{h}^2}, u_{xxx} \to \frac{u^n_{k+2} - 3u^n_{k+1} + 3u^n_{k-1} - u^n_{k-2}}{2\tilde{h}^3}, \ldots$$

There are many variants of Runge-Kutta method. It may be shown that in scalar case with fixed coefficients one step representation of this methods is

$$u^{n+1} = \left(1 + L + \frac{L^2}{2!} + \frac{L^3}{3!} + \frac{L^4}{4!} + \ldots\right) u^n$$

Here L/τ is operator of the finite-difference approximation of $\mathcal{L}$, τ is time step. Order of expansion of $\exp L$ corresponds to order of method. Let function $F(A)$ is a result of substitution $u^n_k = q^n \exp ik\phi$ to some operator A, $F(AB) = F(A)F(B)$, $F(A+B) = F(A) + F(B)$. For non-dissipative case we derive

$$|q|^2 = F\left(1 + \frac{L^2}{2!} + \frac{L^4}{4!} + \ldots\right)^2 - F\left(\left(L + \frac{L^3}{3!} + \frac{L^5}{5!} + \ldots\right)^2\right) = 1 - \frac{2G(L^2)}{2!} + \frac{G(L^2)^2}{2!^2} + \frac{2G(L^2)^2}{4!} - \frac{G(L^2)^3}{4!} +$$
$$\frac{G(L^2)^4}{4!^2} \ldots + G(L^2) - \frac{2G(L^2)^2}{3!} + \frac{G(L^2)^3}{3!^2} + \frac{2G(L^2)^3}{5!} - \frac{2G(L^2)^4}{3!5!} + \frac{G(L^2)^5}{5!^2} + \ldots, G(L^2) = -F(L^2) > 0$$

Results of calculation for 1-5 order methods are given below

$$|q|^2 = 1 + G(L^2); \ \tau c^2_{2m-1} 2^{2(2m-2)-1} (-1)^m \frac{\partial^{2m} u}{\partial^{2m} x}$$

$$|q|^2 = 1 + \frac{G(L^2)^2}{4}; \ \tau^3 c^4_{2m-1} 2^{4(2m-2)-1} \frac{-1}{4} \frac{\partial^{4m} u}{\partial^{4m} x}$$

$$|q|^2 = 1 - G(L^2)^2 \left(\frac{1}{12} - \frac{G(L^2)}{3!^2}\right); \ \tau^3 c^4_{2m-1} 2^{4(2m-2)-1} \frac{1}{12} \frac{\partial^{4m} u}{\partial^{4m} x}; \ \tau \le \frac{\sqrt{3}\tilde{h}^{2m-1}}{2^{2m-2} c_{2m-1}}$$

$$|q|^2 = 1 - G(L^2)^3 \left(\frac{1}{72} - \frac{G(L^2)}{4!^2}\right); \ \tau^5 c^6_{2m-1} 2^{6(2m-2)-1} \frac{(-1)^{m+1}}{72} \frac{\partial^{6m} u}{\partial^{6m} x}; \ \tau \le \frac{\sqrt{8}\tilde{h}^{2m-1}}{2^{2m-2} c_{2m-1}}$$

$$|q|^2 = 1 + G(L^2)^3\left[\frac{1}{360} - G(L^2)\left(\frac{1}{960} - \frac{G(L^2)}{5!^2}\right)\right];\ \tau^5 c^6_{2m-1} 2^{6(2m-2)-1}\frac{(-1)^m}{360}\frac{\partial^{6m} u}{\partial^{6m} x}$$

In addition to $|q|^2$ here dissipative term with the lowest order derivative in the differential approximation of numerical scheme for $\mathcal{L} = \mathcal{L}_{2m-1}$ is given. For 1, 2, 5 order it is non-correct (for 6, 7 order it is non-correct also). For 3 and 4 order methods conditions of non-growing of disturbances are given also. Order of growing or decrease of disturbances is $1+O(\tau^{2s}/\tilde{h}^{2s(2m-1)})$. Here $s = 1$ for first-order method, $s = 2$ for 2 and 3 order methods, $s = 3$ for 4 and 5 order methods and so on. According to spectral method scheme is stable if $|q| < 1 + c\tau$, $c > 0$. In the case non-correct dissipation stability is achieved if $\tau \sim \tilde{h}^{2s(2m-1)/(2s-1)}$, but for fixed values of τ long-time calculations are stopped due to growth of disturbances. Calculations must carried out for the interval $0 < t < T$, $T \sim \tau^{-(2s-1)}$. Growing may be eliminated by adding of dissipative term $-(-1)^j a_{2j}\partial^{2j}u/\partial x^{2j}$, $a_{2j} > 0$, $a_{2j} \sim \tau^{2s-1}/\tilde{h}^{2s(2m-1)-2j}$, $\tau < C\tilde{h}^{2m-1}$, C is some value. Correction is made only for shot waves if $2j > 2s(2m-1)$. Optimal value for j is $j = s(2m-1)$, $a_{2j} \sim \tau^{2s-1}$. It was shown for 1, 2 order methods that the given estimates are applicable also for generalized wave equation

$$\frac{\partial^2 u}{\partial t^2} = c_{2m}\frac{\partial^2 u}{\partial x^{2m}},\quad \omega^2 = (-1)^m c_{2m}\kappa^{2m}$$

Dispersion relation is given here also. Equations for walls of tube are similar to this equation. Non-scalar form of equation is used for calculations

$$\frac{\partial u}{\partial t} = b_l\frac{\partial^l v}{\partial x^l} - \left[(-1)^j a_{2j}\frac{\partial^{2j} u}{\partial x^{2j}}\right],\ \frac{\partial v}{\partial t} = b^*_{2m-l}\frac{\partial^{2m-l} u}{\partial x^2} - \left[(-1)^j a_{2j}\frac{\partial^{2j} v}{\partial x^{2j}}\right],\ c_{2m} = b_l b^*_{2-l}\ l = 0,\ 1,\ \ldots 2m$$

For $m = 1$, $l = 1$ these equations are similar to gas dynamics equations. Estimates are applicable for well-posed cases ($\omega \in R$) if $2m - 1$ is replaced by m.

Note that the second-order Runge-Kutta method may be treated as method of predictor-corrector type. Half of time step predictor is used. Growth of disturbances may be excluded if the predictor is applied for the entire time step. But such numerical scheme possesses first-order of τ numerical dissipation and its application for non-dissipative and weakly dissipative problems is not convenient.

For calculation of generalized solution additional dissipative or non dissipative terms may be added. For term with derivative of order m the order of coefficient must be $\tilde{h}^{m-1}$. In this case number of cells for shock structure will be not changed when $\tilde{h} \to 0$. Hence there is essential difference between order of these terms and order of terms for correction of numerical schemes that are proportional to some order of τ. Inclusion of high-order derivatives affects mainly on short waves but do not affects on long waves. That is why it is appropriate to use dissipative terms with high-order derivatives if non-dissipative and dissipative shock structures are calculated at the same time.

3. Riemann problem

Numerical solutions of the Riemann problem are found. Results of calculations correspond to the theory of reversible and weakly dissipative shocks. Typical types of shock structures [5] are found for all cases. Examples of graphs for the case of controlled pressure are given in [7] and for fluid-filled tube they are given in [8]. One more example for fluid-filled tube is presented in fig. 1*a*. Second-order Runge-Kutta method with correction is used. Graphs for great density, $\rho_f = 10$, curve *1*, and low density, $\rho_f = 0.25$, curve *2*, are given, $t = 1200$, $v|_{t=0} = 0.4$, $c_{v4} = c_{z4} = 10^{-5}$, $\tilde{\mu} = 2$, $J_m = 30$, $R = 1$, $\rho = 1$, $\rho^*_f = 1$. Left longitudinal elastic waves left transversal elastic waves, kink structure, right transversal elastic and right longitudinal elastic waves are observed. In the case of low density no homogeneous regions between shock structures are observed. The simplified equations badly describe such solution. These equations may be derived from the transformed equations of fluid-filled tube and it is obvious that in equation for v we cannot withdraw dispersive terms if $\rho_f \to 0$. Model of fluid-filled tube do not transforms to model with controlled pressure. Note also that for the case of great density right shock structure of longitudinal elastic waves is calculated as dissipative shock due to correction of scheme.

Results obtained by fourth-order Runge-Kutta method for gas-filled tube are demonstrated in fig. 1b, $t = 5500$, isothermal case, $p = c\rho_f$, $c/H = 200$, $p_e = 0$, $v|_{t=0} = -0.25$. Left longitudinal elastic waves left gas dynamic waves, kink structure, right gas dynamic waves and right longitudinal elastic waves are observed. Growth of disturbances is not found.

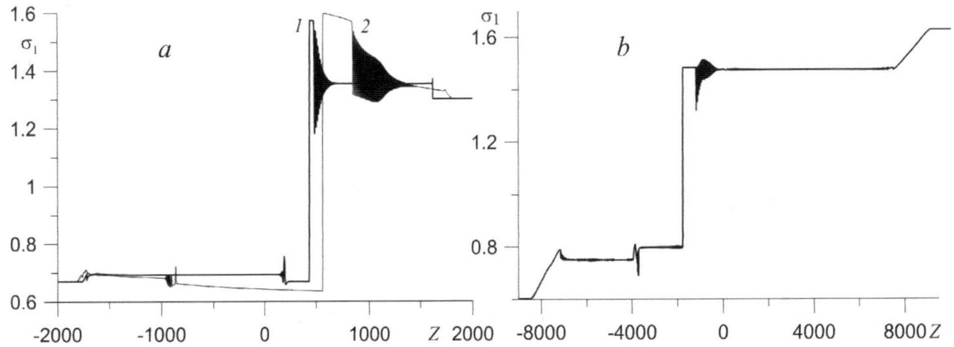

Figure 1. Solutions of Riemann problem

4. Conclusions and discussion

Results of calculations show that the theory of reversible and weakly dissipative shock structures may be applied to all considered models for waves in tubes. Three-layer centered scheme is effective for calculation of such equations with complex dispersion. But number of branches of numerical dispersion relation for this scheme is two times greater than number of branches of calculated equations. It may lead to numerical instability. Two-layer schemes with Runge-Kutta approximation of time derivatives are effective in this cases. These schemes are stable but dissipative properties of these schemes may lead to some dissipative changes in solutions or to growth of disturbances. This effects are readily decreased if order of method is increased.

References

[1] Epstein M., Johnston J. W., Proc. Roy. Soc. London A **457**, 2009, 1195–1213 (2001)
[2] Fu Y. B., Pearce S. P., IMA J. Appl. Math., **75**, 4, 581–602 (2010)
[3] Fu Y. B., Il'ichev A., IMA J. Appl. Math. **75**, 2, 257–268 (2010)
[4] Bakholdin I. B. *Bezdissipativnye razryvy v mekhanike sploshnoy sredy* [Non-dissipative shocks in mechanics of continuum media] (Moscow: Fizmatlit. 2004)
[5] Bakholdin I. B., J. Appl. Math. Mech., **78**, 6, 599–612 (2014)
[6] Kulikovskii A.G., Sveshnikova E. I. *Nonlinear Waves in Elasic Media* (CRC Press, Boca Raton, New York, London, Tokio, 1995)
[7] Bakholdin I.B., J. Comp. Math. and Math. Phys. **55**, 11, 1884–1898 (2015)
[8] Bakholdin I. B., J. Appl. Math. Mech., **81**, 5, 409–419 (2017)
[9] Fu Y.B. , Liu J.L., Francisco G.S., J. of the Mechanics and Physics of Solids **90**, 45-60 (2016)

Simulation of process of small satellites separation from deployer installed on cargo spacecraft

N V Bogomolov[1], A S Anfalov[1], S V Borzykh[1] and V N Bakulin[2,3]

[1] S.P. Korolev Rocket and Space Corporation Energia, Korolev, 141070, Russia
[2] Moscow Aviation Institute,Volokolamsk Highway 4, 125993, Moscow, Russia
[3] Institute of Applied Mechanics of the Russian Academy of Sciences, Leningradsky prt 7, Moscow, 125040, Russia

E-mail: nbogomolov1992@gmail.com

Abstract. A possible way to launch CubeSats is to place them in deployers installed on Progress MS cargo spacecraft. Launching satellites into orbit is potentially feasible both during Progress MS autonomous flight and during its stay in the International Space Station. In order for the CubeSats to achieve their mission targets, the separation process requires that they obtain the necessary linear velocities, the angular disturbances obtained be minimized and that the satellites continue to move without collisions. Addressing these issues involves the development of a dynamic model based on the numerical integration of motion equations. Once the model has been verified, taking into account, in addition to the permissible mass-centric and inertial characteristics of the objects, the contact interactions of satellites with the deployer guide rails and the door, a series of statistical calculations are carried out for different launch scenarios, based on the results of which a conclusion about the possibility of implementing the schemes considered is drawn and recommendations are made (in terms of dynamics) on the choice of the most appropriate layout.

1. Introduction

When planning and carrying out each space mission, special attention should be paid to the dynamics of transient processes (e.g., landing of spacecraft [1, 2, 3, 4], disclosure of elements of their construction [5, 6], separation of space rocket systems and complexes objects [7, 8, 9]).

Currently, many organizations, private companies and educational institutions are engaged in the development of small satellites (SS) such as CubeSat to address various target tasks. For many years there has been a steady demand for small satellite launch services. Frequently, SS are derived as a concomitant payload for the launch of medium and heavy class launch vehicles (LV). Every year, the number of light class LV manufacturers for launching micro- and nanosatellites grows. Nevertheless, despite the popularity of small satellites launching with the help of a LV, there are other popular ways of launching small satellites - for example, from the International Space Station (ISS) with the help of the JEM RMS manipulator and the NanoRacks launch platform. The satellites themselves are delivered to the station inside cargo spacecraft (CS). The SS can also be launched directly from the CS, which makes it possible to launch satellites into a higher or lower orbit than the ISS. Such a scheme is implemented on Cygnus cargo spacecraft.

Progress spacecraft offers an excellent opportunity to launch SS [10, 11]. One of the promising proposals is to place deployers for CubeSats on the outer surface of Progress MS spacecraft (figure 1).

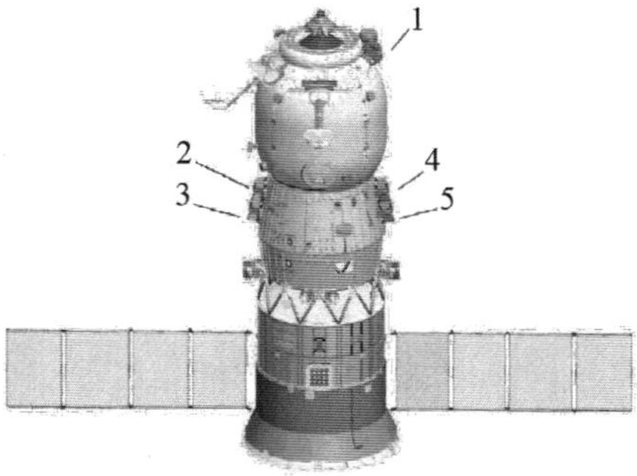

Figure 1. Possible layout scheme of deployers on external surface of Progress MS spacecraft: 1, 2, 3, 4, 5 - deployer locations.

In order to control the fulfillment of the requirements on angular (less than 5,0 deg/s) and linear relative (from 0,5 to 2,0 m/s) velocities of CubeSats after separation, it was necessary to develop a dynamic model of the SS separation process from each possible variant of the deployer and conduct appropriate studies.

2. Study aims and objectives

The aim of the study was to develop an approach, dynamic model and computational algorithms to determine the values of parameters that ensure the separation of CubeSats from the deployers installed on Progress MS spacecraft with a given linear and angular velocity. In case of launch from a spacecraft docked to the ISS and not being in autonomous flight, the mandatory condition is also the absence of collisions between the separated objects and the station.

Due to the variety of possible launch scenarios, deployer versions, their location on the spacecraft and filling in different CubeSat form-factors, the main tasks were to develop a methodology for calculations and analysis of the results, to build a dynamic model of the SS separation process that correctly takes into account their interaction with the elements of the deployer, to give recommendations on the values of the parameters of the objects under study and layout.

3. Description of deployer design

The deployer is a rectangular construction, the main elements of which are a body with mechanisms, electronic switches, electric heaters, guillotines, end probes, thermometer, screen vacuum insulation and cables. Separation of the main payload from the deployer is carried out due to the mechanism of ejection consisting of a pusher plate and one or two (in case of 6U CubeSat) springs. The SS starts from the moment the command to the compartment is given. When the guillotine is triggered, the door is opened (this is the responsibility of the two springs in the hinges attached to the deployer body by means of the pivot pins). When the door is at the desired opening angle, it is fixed in place to prevent it from rebounding back. By triggering the ejection mechanism, the SS exit(s) the deployer.

4. Distinctive features of study

Although it is prohibited for the main payload to collide with CS or station (for the corresponding launch option) after leaving the deployer, satellites may come into contact with both the guide rails and the door when moving inside the deployer. This has become a key feature in comparison with the authors' previous studies in this area.

Compared to the known works carried out earlier in this field by other researchers, a characteristic feature of the analysis described, in addition to the lack of additional means of retaining the main payload in the deployer, is the variety of design cases. CubeSats can be launched from Progress MS spacecraft, which is either in autonomous flight or docked to the ISS. There are several possible options for placing deployers onboard, types of deployers, versions of their layout. Since these deployers are not designed for a specific group of CubeSats with well-known mass-centric and inertial characteristics, it is necessary to take into account possible variations of masses, center-of-mass coordinates, and satellite inertia tensor components in the ranges set by the CubeSat standard for each size. In addition, in order to separate the satellites with a given linear velocities and the lowest possible angular velocities, it is necessary to select the most appropriate characteristics of the pusher springs (stiffness factor, free length, working stroke), torsion springs for the deployer doors, to determine the possible values of gaps between the deployer guide rails and satellites, taking into account the possible transverse displacements of the latter at the beginning of separation. These parameters are also within a certain tolerance.

5. Dynamic process modeling of CubeSat separation from deployer

To carry out the research, a model was used in which the deployer, the pusher plate of the ejection mechanism, the satellite (or a group of 1U + 2U or 1U + 1U + 1U CubeSats) and the door are represented by rigid bodies. Their movement is described by well-known motion equations

$$\frac{\mathrm{d}\boldsymbol{r_i}}{\mathrm{d}t} = \boldsymbol{V_i} \tag{1}$$

$$\frac{\mathrm{d}\boldsymbol{V_i}}{\mathrm{d}t} = \frac{\sum \boldsymbol{F_{k_i}^{(e)}}}{m_i} \tag{2}$$

$$\frac{\tilde{\mathrm{d}}\boldsymbol{\omega_i}}{\mathrm{d}t} = I_i^{-1}\left[\sum \boldsymbol{M_i^{(e)}} + \sum \boldsymbol{M_i}\left(\boldsymbol{F_{k_i}^{(e)}}\right) - \boldsymbol{\omega_i} \times (I_i \boldsymbol{\omega_i})\right], \tag{3}$$

where $\boldsymbol{r_i}$ is the i-th body radius-vector of the mass center in the inertial reference frame, $\boldsymbol{V_i}$ is linear speed of mass center, $\boldsymbol{\omega_i}$ is angular speed, m_i is mass , I_i is inertia tensor, $\boldsymbol{F_{k_i}^{(e)}}$ - external forces, $\boldsymbol{M_i^{(e)}}$ - external moments, $\boldsymbol{M_i}\left(\boldsymbol{F_{k_i}^{(e)}}\right)$ - moments from external forces acting on the body.

All the bodies of the system in question are affected by gravity force. CS (or CS + ISS bundle) may be affected by external controlled forces and moments. The door has one degree of freedom and from the beginning of the separation process until a certain angle of rotation is reached, it is opened by torsional moments and then fixed. The spring force (each of the two springs in the case of 6U CubeSat) of the ejection mechanism is calculated by the formula

$$F_{spr} = k_{spr}\left(s - s_0\right) + f_{spr}\mathrm{d}s, \tag{4}$$

where k_{spr} is stiffness coefficient, s is current spring length, s_0 is free spring length, f_{spr} is damping coefficient, $\mathrm{d}s$ is spring length change rate.

If damping is not taken into account, in the first approximation, the following ratio can be used to select suitable springs that provide the main payload separation with the required relative speed $\boldsymbol{V_r}$:

$$c \cdot k_{spr}\frac{\Delta s^2}{2} = (m + m_{plate})\frac{V_r^2}{2},$$

where m_{plate} is pusher plate mass, m is the main payload mass, Δs is full stroke of the spring, c is number of springs in the ejection mechanism.

In case of combined CubeSats separation (schemes 1U + 2U or 1U + 1U + 1U), in accordance with the standard, additional spring pushers are installed between the satellites, the spring force of which is also according to the formula (4).

There are a large number of models for the contact interaction of objects with each other. In this paper, the contact between the pusher plate of the ejection mechanism and the satellite, as well as the satellite and the door, is determined by the insertion of points on one body into the surface of the other and is considered to be a collision of objects. The contact force is represented by the sum of three components:

$$\boldsymbol{F_{contact}} = \boldsymbol{F_n} + \boldsymbol{F_d} + \boldsymbol{F_{fr}}$$

where $\boldsymbol{F_n}$ is the force of elastic normal reaction, $\boldsymbol{F_d}$ is the force of normal damping, $\boldsymbol{F_{fr}}$ is the friction force.

Force of elastic normal reaction

$$F_n = k_c l_c$$

where k_c is is the normal contact stiffness of the interaction, l_c is the depth of contact.

Body impact consists of two phases: compression (characterized by increased contact depth) and recovery. During the compression phase, the kinetic energy of the interacting bodies passes into the potential energy of the elastic normal reaction force. During the recovery phase, the kinetic energy is partially recovered. The loss of energy is due to friction and normal damping:

$$F_d = \chi_d\left(V_n\right) F_n$$

where V_n is projection of the rate of penetration of the point relative to the penetration surface on the normal to this surface, $\chi_d\left(V_n\right)$ is function of the relative strength of normal damping

$$\chi_d\left(V_n\right) = \begin{cases} \chi_1 & \text{if } k_d V_n > \chi_1 \\ \chi_2 & \text{if } k_d V_n < \chi_2 \\ k_d V_n & \text{otherwise} \end{cases}$$

where

$$\chi_1 = \frac{\left(1 - c_r^2\right) k_\varepsilon}{1 - k_\varepsilon + c_r^2 k_\varepsilon}, \quad \chi_2 = \frac{\left(c_r^2 - 1\right)\left(1 - k_\varepsilon\right)}{1 - k_\varepsilon + c_r^2 k_\varepsilon}$$

are values of the relative limiting force of normal damping in the compression and recovery phases, c_r is rate recovery coefficient $(0 \leq c_r \leq 1)$, k_ε is coefficient of relative energy absorption in the compression phase $(0 \leq k_\varepsilon \leq 1)$, k_d is coefficient of linear transition between damping modes.

The vector of friction force $\boldsymbol{F_{fr}}$ is oppositely directed $\boldsymbol{V_t}$ - projection of the velocity of the embedded point relative to the penetration surface on a plane perpendicular to the normal to this surface.

$$F_{fr} = k_{fr} N \chi_{fr}\left(V_t\right),$$

where N is full force of normal reaction $(N = |\boldsymbol{F_n} + \boldsymbol{F_d}|)$, k_{fr} is friction coefficient, $\chi_{fr}\left(V_t\right)$ is friction force function:

$$\chi_{fr} = \begin{cases} 1 & \text{if } k_t V_t > 1 \\ k_t V_t & \text{otherwise} \end{cases}$$

where k_t is is the coefficient of linear transition between rest (friction force is equal to zero) and full inclusion of force.

As a result of taking into account all force factors in (1, 2, 3) and setting initial conditions, a system of homogeneous differential equation is formed, which can be solved by numerical integration, thus determining the position, orientation, velocities and accelerations of the bodies at any time.

6. Conducting research and analysis of obtained results

The study consisted of two stages. Initially, for various separation scenarios, a series of calculations by Monte Carlo method were carried out using the dynamic model described earlier (numerical integration was carried out by Runge-Kutta method of 4 order with a step of 0,001 s). Inside each series for 10000 calculation cases the masses of the spacecraft (10,0...12,0 kg for 6U; 3,0...4,0 kg for 3U; 2,0...2,66 kg for 2U; 1,0...1,33 kg for 1U), displacements of satellite mass centers coordinates from their geometrical centers (congruent to CubeSat standards) varied according to the uniform distribution law, and, in the range from -10% to +10% of the nominal values, the spring stiffness coefficients of the ejection mechanisms and additional pushers (for the cases with separation of 1U + 2U and 1U + 1U + 1U). Within the range 0...25% of the nominal value of the gap Δ between the satellites and deployer guide rails, the transverse displacement of CubeSats inside the deployer also changed.

Determination of inertia tensor components for each calculation case in the series was carried out as follows. First, the inertia moments were calculated assuming that SS are homogeneous parallelepipeds of corresponding masses and dimensions

$$I_{xx} = \frac{1}{12} k_I m \left(l_y l_y + l_z l_z \right), \quad I_{yy} = \frac{1}{12} k_I m \left(l_x l_x + l_z l_z \right), \quad I_{zz} = \frac{1}{12} k_I m \left(l_x l_x + l_y l_y \right),$$

where m is SS mass at the current run, l_z, l_x, l_y are length, width and height of the satellite, k_I is multiplier of the inertia moments, varying by a uniform law in the range from 1,0 to 1,2. Then the inertia tensor was recalculated from the geometric center of the SS to its center of mass by the formula

$$I_C = I - m \left(r_C^2 E - \boldsymbol{r_C} \boldsymbol{r_C}^T \right),$$

where I is initial inertia tensor, I_C is recalculated inertia tensor, $\boldsymbol{r_C}$ is radius-vector from the center of mass of the satellite to its geometric center for the current calculated case, E is unit tensor.

The uniform law was chosen in order to get more "extreme" values, as the variants with the most unfavorable possible conditions of separation were of the greatest interest. For the same reason, damping of the springs of the ejection mechanisms and additional pushers was introduced into the model.

The result of a series of such calculations was a set of files with initial values of variable parameters for each calculation case and implemented trajectories of the CubeSats relative motion, their angular and relative linear velocities at the time of exit from the deployer. These sets were used at the second stage of the study, which consisted in statistical processing of the obtained data and construction of tubes of trajectories of the separated objects.

As an example of the results obtained in the course of the second stage of the study, figure 2 shows graphs of 6U CubeSat angular and relative linear velocity distribution functions for one of the possible cases of the deployer on Progress MS spacecraft, which is in autonomous flight, after leaving the deployer.

7. Conclusion

The research made it possible to determine the suitable characteristics of the deployer design elements and layout of the CubeSats in it. In particular, to avoid collisions between satellites when separating in combinations of 1U + 2U, 1U + 1U + 1U, it is necessary to place the CubeSats in the deployer in ascending order of mass so that the most massive satellite was the first to leave the deployer. The value of gaps between the deployer guide rails and satellites with the working stroke of the springs of the ejection mechanisms have strong influence on the realized angular velocities of the CubeSats.

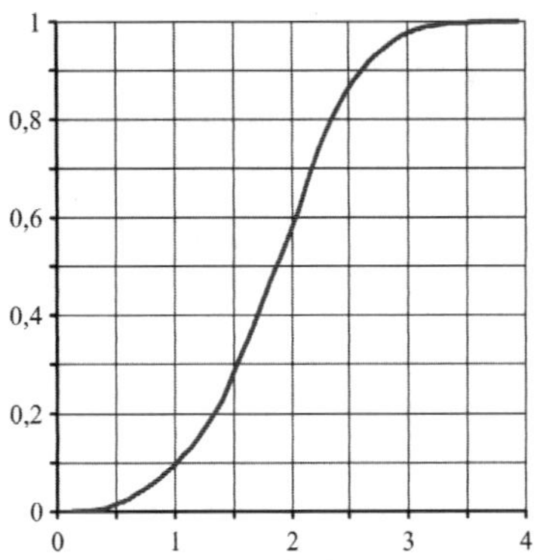

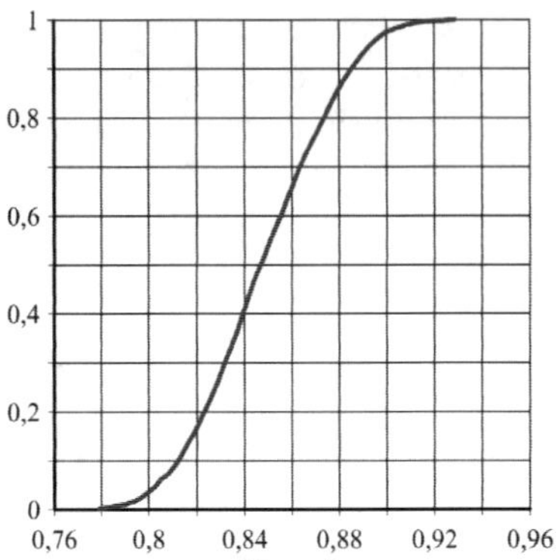

Figure 2. Graphs of 6U CubeSat angular and relative linear velocity distribution functions for one of possible cases.

The analysis of the results also showed that in the whole range of initial conditions the SS exit from the zone of CS and ISS structural elements (under the corresponding scenario) occurs without impact with a relative linear velocity, located within the required range.

For the initially designed gaps between the deployer guide rails and satellites, 3U and 6U angular velocities do not exceed the maximum permissible values at the moment of exit, but for smaller standard sizes this condition is not met in all cases, and therefore it may be necessary to reduce the gaps, introduce more stringent restrictions on the SS data alignment or soften the requirements for angular velocity.

The model developed will be useful for the CubeSat developers. For example, it could be supplemented by aerodynamic forces and moments and the inclusion of the SS control systems, which would allow the movement of specific satellites to be investigated after separation to meet orientation, stabilization and communication objectives.

Acknowledgments

Work is made with financial support from the Russian Foundation of Basic Researches (grant No. 17-08-01410).

References

[1] Bakulin V N, Borzykh S V, Voronin V V and Kokushkin V V 2012 *Aerospace MAI J. Rev.* 19 **5** 45–50
[2] Bakulin V N, Borzykh S V and Voronin V V 2016 *Russian Aeronautics* **1** 22–6
[3] Bakulin V N, Borzykh S V and Voronin V V 2011 *Aerospace MAI J. Rev.* 18 **4** 38–46
[4] Bakulin V N, Borzykh S V, Voronin V V and Kokushkin V V 2014 *Aerospace MAI J. Rev.* 21 **1** 25–32
[5] Bakulin V N, Borzykh S V and Iliasova I R 2011 *Aerospace MAI J. Rev.* 18 **3** 295–302
[6] Borzykh S V, Levitskaya Yu Yu and Shchiblev Yu N 2019 *Space Engr. and Technol.* **1**(24) 95–106
[7] Bakulin V N, Borzykh S V and Reshetnikov M N 2011 *Aerospace MAI J. Rev.* 18 **3** 287–95
[8] Anfalov A S, Bogomolov N V and Borzykh S V 2018 *Proc. of Higher Edu. Inst. Machine Building* **5**(698) 77–84
[9] Anfalov A S, Bogomolov N V and Borzykh S V 2017 *Cosmonautics and Rocket Engr.* **1**(94) 24–9
[10] Markov A V, Matveeva T V, Murtazin R F, Smirnov A V, Soloviev V A, Sorokin I V, Churilo I V and Khamits I I 2015 *Space Engr. and Technol.* **1**(8) 42–52
[11] Belyaev M Yu, Karasev D V, Matveeva T V and Rulev D N 2018 *Space Engr. and Technol.* **1**(20) 23–39

Trajectory portraits for the two perturbed centrally symmetric systems of point vortices

V. D. Boyarintsev, T.E. Boyarintseva, N.P. Gvozdev, V.M. Kobzeva
Bauman Moscow State Technical University, Moscow, Russia
E-mail: t.bojare@mail.ru

Abstract. We describe the types of motion for two centrally symmetric systems of point vortices (in one vortices are located in the corners of a regular hexagon, in the other – in the corners of a regular octagon), caused by spatially periodic perturbation: in which the vortices are in the corners of two regular triangles or two squares with a common center.

1. Introduction

The relevance of the study of vortex structures is due to the practical benefits: vorticity present in various natural phenomena (example: tornadoes, tornadoes); manifested not only on our planet, but also throughout the solar system; in the human heart there are vortices of liquid; occur when moving aircraft and ships. Point vortices, in themselves, do not occur in nature, but the structures of point vortices can be simulated real vortices. Phenomena that are associated with vortex rings are destructive and at the moment there are no effective methods of counteraction and prediction of these phenomena, respectively, a deeper understanding of the processes of formation and growth of vortices will contribute to the creation of methods for solving problems. Interest point vortices lately resumed this subject are devoted [1]-[6].

The scientific novelty of the work is that detailed trajectory portraits even for similar vortex structures have not yet been obtained.

The result of this work was the construction of detailed trajectory portraits for both declared point vortex systems. These results can be used at the Institute of Oceanology, the Institute of mechanics and other institutions that deal with the problem of vortex motion.

2. Solution

Let the vortices be at the vertices of a regular hexagon or octagon with the distance between the opposite vertices 2R.

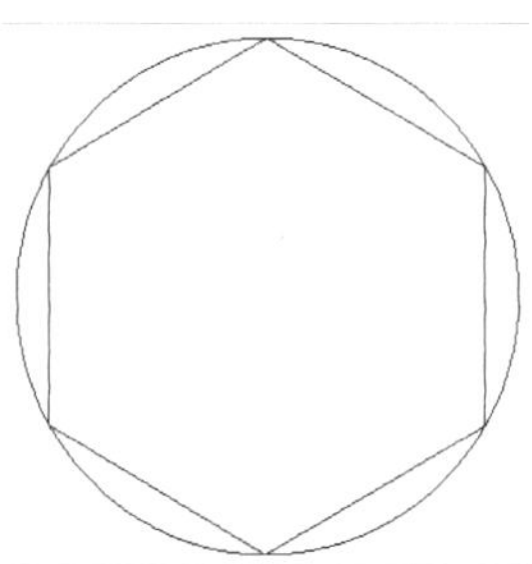

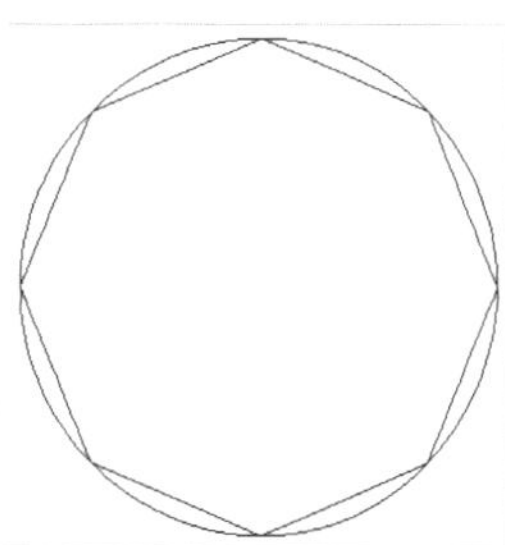

Pic 1a Pic 1b

The self-induced motion of such a system is a motion along a circle of radius R with a constant velocity. Note that the speed of rotation the greater the smaller R.

Suppose the next outrage: vortices using a single shift value ($\delta r, \theta$). Then three (or four) vortices are at the vertices of an equilateral triangle (or square) inscribed in a circle of radius R, we call them the main vortices, and the other three (or four) vortices will be at the vertices of a regular triangle (or square), inscribed in a circle of radius r with the same center as the first circle, and rotated relative to the first triangle (square) at an angle θ.

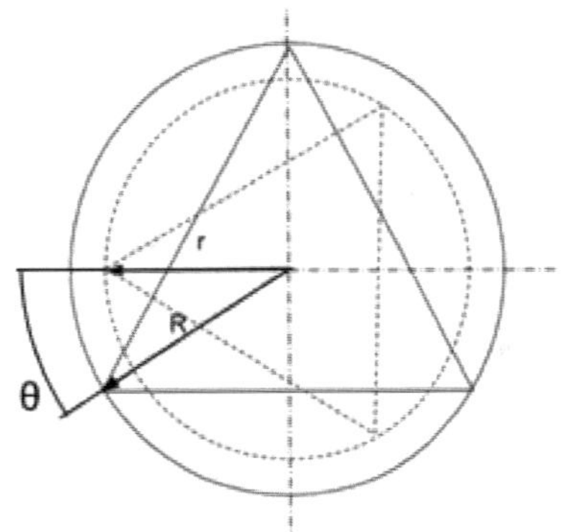

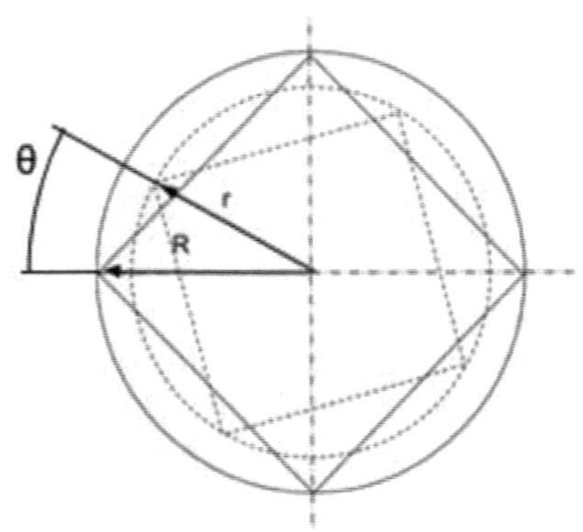

Pic 2a Pic 2b

The system as a whole rotates around a common center, and the mutual arrangement of vortices in the system changes. For a better understanding of the mutual motion, we assume that three (or four) vortices are located on a circle of radius R = 1, and the position of three (or four) other vortices will be calculated. The complete picture of the lines of motion of some vortices relative to others is called a trajectory portrait.

It is required to construct trajectory portraits for these two systems of point vortices.

The formula for calculating trajectory portraits is taken from the article by Boyarintseva T. E. and Savin A. S. [7]

$$C(r,\theta) = (r\sqrt{1-r^2})^{n-1}(r^{2n} + (1-r^2)^n - 2r^n(\sqrt{1-r^2})^n * \cos(3\theta))$$

Trajectory portrait is obtained as a set of a large number of points for each value of C. For the practical calculation of trajectory points was written in C++ , implemented on a computer DELL Inspiron 7577 - 5990.

The program for solving the problem implements the following algorithm:

Set the initial value C = 0.

In nested loops we iterate over all variants of r values from 0 to 1.1 with a given step and all variants of θ values from 0 to 6.4 (up to 2π) with a given step. The range of radius values is set by selection, based on the fact that there are no points of interest for values of radius greater than the specified maximum.

For all pairs of values r and θ calculate the result according to the given formula and compare the result with the selected value C. If it fits into the permissible error (±0.01%), then this pair r and θ is the point of the graph.

Increase the value of C by a given step. Go to 2.

Repeat step 4 until C reaches the pre-selected value 2π.

140 billion points have been tested.

Results

Trajectory portrait for a system of three pairs of point vortices: this system has three types of mutual vortex motion:

1) Movement in the area close to the common center of the system.

2) the offset Rotation of the vortices around unbiased. This type of motion occurs if the displaced vortices are close to the unbiased ones.

3) the Rotation of the displaced vortices near the points of their original position. This type of movement occurs not only with a relatively small initial displacement, its zone is a buffer between the zones of the first and second types of movement.

The points of stable equilibrium are on the same circle as the main vortices. For pic. 3 these points are clearly visible. The locations of the unbiased vortices in pic. 3 not visible, since only the motion of displaced vortices was studied, but it is seen that these points are also points of stable equilibrium. Points of unstable equilibrium are located at the joints of two different zones of movement of the third type, these points are also adjacent to the zone of the first and second types.

Pic. 3

Consider a trajectory portrait for four pairs of point vortices. In this system, there are five types of reciprocal movements of the vortices.

Pic. 4.

1) Movement in the area close to the common center of the system.

2) the offset Rotation of the vortices around unbiased. Also similar to the second type for the system of their three pairs of vortices.

3) No change of the relative position of vortices, that is, the rotation of the system as a single rigid whole point of stable equilibrium are displaced, they are not on the same circle that the main vortices. In those places where there were vortices before the beginning of the perturbation of the system, there are points of unstable equilibrium. And for each such point, two points of stable equilibrium are formed, located on the same radial direction, but a little closer to the center of the system and a little further from it. In addition to them, there are also points of unstable equilibrium on the segments between the center of the system and the points of location of the main, that is, unbiased vortices.

4) The rotation of the displaced vortices around the points of stable equilibrium.

5) The Rotation of the displaced vortices near the points of their initial position, but skirting both zones of oscillation around the corresponding points of stable equilibrium.

3.Conclucsion

It is worth noting a paradoxical fact - in both cases near the center of the system (zone of the first type of movement) there are no trajectories going around this center. There is a large number of different trajectories, which in General can be considered as an area of stochastic motion.

References

1) A.V. Borisov, I. S. Mamaev Interaction of Kirchhoff vortices and point vortices in an ideal fluid *Nonlinear dynamics*, 2006, **2**, 199-213.

2) V. V. Vaskin, N. N. Erdakova Dynamics of two point vortices in an annular region *Nonlinear dynamics*, 2010, vol. 6, **3**, 531-547.

3) V. N. Nosov, A. A. Savin, A. S. Savin The impact of a point source on the free surface of a two-layer liquid *Vestnik MGTU. N. E. Bauman. Ser. «Natural science»,* 2012, 100-110 (in Russian).

4) V. V. Pukhnachev Point vortex in viscous incompressible fluid *Applied mechanics and technical physics,* 2014, **55**, 180-187 (in Russian).

5) I. Y. Vladimirov, N. N. Korchagin, A. S. Savin Modeling of wave action on horizontal structural elements in the upper layer of stratified flow *Mathematical Modeling and Numerical Methods*, 2014, No. 4, 74-87 (in Russian).

6) I. Y. Vladimirov, N. N. Korchagin, A. S. Savin Wave reaction of a dipole at circulation flow by a two-layer flow of finite depth *Izvestiya RAS. Fluid and Gas Mechanics*, 2016, No. 2, 90-99 (in Russian).

7) T. E. Boyarintsev, A. S. Savin The perturbed motion of a centrally symmetric system of point vortices *Izvestiya RAS. Fluid and Gas Mechanics*, 2001, No. 1, 102-107 (in Russian).

Far surface gravity waves fields under unstable generation regimes

V V Bulatov[1] **and Yu V Vladimirov**[2]

[1] Ishlinsky Institute for Problems in Mechanics RAS, 119526 Moscow, Russia
[2] Ishlinsky Institute for Problems in Mechanics RAS, 119526 Moscow, Russia

E-mail: internalwave@mail.ru, vladimyura@yandex.ru

Abstract. The problem of constructing uniform asymptotics of surface perturbations far fields from a localized harmonic source in the flow of a heavy homogeneous fluid of infinite depth is considered. It is shown that the wave pattern of excited long-range fields is, for certain generation parameters, a system of hybrid wave perturbations which simultaneously exhibit the properties of the following two types: annular-type (transverse) and wedge-type (longitudinal) waves. Specific features of the phase structure and wave fronts are studied. Uniform asymptotics of the solutions which describe the hybrid surface wave perturbations at a far distance from the harmonic source are constructed.

1. Introduction

The surface wave motions in the marine environment can either originate due to natural causes (wind waves, flow past underwater obstacles, bottom relief variations, density and flow fields) or be generated by the flow past natural obstacles (platforms, underwater pipelines, complex hydraulic facilities). The general system of hydrodynamic equations describing the surface perturbations is a rather complicated mathematical problem from the standpoint of proving the existence and uniqueness theorems for solutions in the corresponding function classes and from the computational standpoint. In the framework of the linear theory, the surface wave perturbations are analytically studied by integral representation methods and various asymptotic methods. The main results of solving the problems of generation of surface wave perturbations are represented in most general integral form, and to obtain the integral solutions, it is thus necessary to develop asymptotic methods for their investigation which admit a qualitative analysis and rapid estimations of the obtained solutions. Moreover, to analyze the data of the sea surface remote sensing, it is required to know the causes of various surface phenomena [1-7]. To obtain a detailed description of a wide class of physical phenomena related to the dynamics of surface perturbations in inhomogeneous and unsteady natural environments, it is necessary to have sufficiently developed mathematical models. The fact that the structure of the heavy sea surface is three-dimensional is also significant, and there are currently no possibilities for large-scale computational experimental modeling of three-dimensional ocean flows at large times with a sufficient accuracy . But in several cases, the initial qualitative concept of the considered class of wave phenomena can be obtained by using simpler asymptotic models and analytic methods for studying them [8-12]. In this connection, it is necessary to mention the classical hydrodynamic problems of constructing asymptotic solutions which describe the evolution of surface perturbations excited by sources of various nature in heavy homogeneous liquids. The model solutions permit further obtaining representations of surface wave fields with regard to variability and unsteadiness of real natural environments. So several results of asymptotic analysis of linear problems describing different regions of generation and propagation of surface perturbations also underlie the currently actively developing nonlinear theory of generation of ocean waves of extremely large amplitude, the so-called rogue waves (killer waves) [3,6] . The contemporary state of the art in the study of linear and nonlinear surface perturbations can be found in [7]. The problems of constructing uniform far field asymptotics of internal and surface perturbations from a moving source were considered in [2,5,13,14]. Therefore, it is interesting to consider more complicated regimes of surface wave generation due to the unsteady nature of the source of perturbations. The goal in this paper is to construct uniform far field asymptotics of surface perturbations generated in the flow of a heavy homogeneous liquid of infinite depth around a localized harmonic source of perturbations.

2. Main results and discussion

We consider the problem of homogeneous flow of an infinitely deep heavy liquid past a harmonic perturbation source of intensity $q = Q\,exp(\,i\omega t\,)$; the liquid has velocity V at a far distance from the source. The source is located at depth h with respect to the unperturbed free surface, i.e., at the point $(\,0{,}0{,}-h\,)$. To determine a physically realizable solution of the problem, it is necessary to replace the frequency ω by $\omega - i\varepsilon$ and then let ε tend to zero in the obtained solution for the free surface elevation. Further, let $\Omega(\,x,y,z,t\,)$ be the potential of velocity perturbations with respect to the velocity of the homogeneous flow. In the framework of the linear theory, to determine Ω, we have the problem [1,14]

$$\Delta\Omega(x,y,z,t) = Q\exp(i(\omega - i\varepsilon)t)\delta(x)\delta(y)\delta(z+h),\ z<0;\left(\frac{\partial}{\partial t}+V\frac{\partial}{\partial x}\right)^2\Omega + g\frac{\partial\Omega}{\partial z} = 0,\ z=0 \qquad (1)$$

where Δ is the three-dimensional Laplace operator and $\delta(\,x\,)$ is the Dirac delta function. The free surface elevation $H(\,x,y,t\,)$ is determined from the Cauchy-Lagrange integral

$$H(\,x,y,t\,) = -\frac{1}{g}\left(\frac{\partial}{\partial t}+V\frac{\partial}{\partial x}\right)\Omega(\,x,y,z,t\,),\ z=0\,. \qquad (2)$$

We seek the solution of equations (1)-(2) in the form

$$\Omega(\,x,y,z,t\,) = exp(\,i(\,\omega - i\varepsilon\,)t\,)\varphi(\,x,y,z\,),\ H(\,x,y,t\,) = exp(\,i(\,\omega - i\varepsilon\,)t\,)\eta(\,x,y\,).$$

In the dimensionless variables $x_* = gxV^{-2}$, $y_* = gyV^{-2}$, $z_* = gzV^{-2}$, $h_* = ghV^{-2}$, $\omega_* = V\omega/g$, $t_* = gt/V$, $\varepsilon_* = V\varepsilon/g$, $\varphi_* = V^2\varphi/Qg$, $\eta_* = V^3\eta/Qg$, we have the following problem for determining the functions $\varphi_*(\,x_*,y_*,z_*\,)$, $\eta_*(\,x_*,y_*\,)$ (the superscript «*» is further omitted):

$$\Delta\varphi(\,x,y,z\,) = \delta(\,x\,)\delta(\,y\,)\delta(\,z+h\,),\ z<0 \qquad \left(i\omega+\varepsilon+\frac{\partial}{\partial x}\right)^2\varphi + \frac{\partial\varphi}{\partial z} = 0,\ z=0 \qquad (3)$$

$$\eta(\,x,y\,) = -\left(i\omega+\varepsilon+\frac{\partial}{\partial x}\right)\varphi,\quad z=0 \qquad (4)$$

We represent the functions $\varphi(\,x,y,z\,)$, $\eta(\,x,y\,)$ as double Fourier integrals

$$\varphi(\,x,y,z\,) = \frac{1}{4\pi^2}\int_{-\infty}^{\infty} exp(\,-i\mu x\,)d\mu \int_{-\infty}^{\infty} exp(\,-i\nu y\,)f(\,\mu,\nu,z\,)\,d\nu \qquad (5)$$

$$\eta(x,y)=\frac{1}{4\pi^2}\int_{-\infty}^{\infty} exp(-i\mu x)d\mu \int_{-\infty}^{\infty} exp(-i\nu y)\Lambda(\mu,\nu)\,d\nu \tag{6}$$

Substituting expressions (5) in equation (3), we obtain the boundary-value problem

$$\frac{\partial^2 f(\mu,\nu,z)}{\partial z^2}-(\mu^2+\nu^2)f(\mu,\nu,z)=\delta(z+h),\quad z<0 \tag{7}$$

$$(i\omega+\varepsilon - i\mu)^2 f(\mu,\nu,z)+\frac{\partial f(\mu,\nu,z)}{\partial z}=0,\quad z=0$$

The solution of problem (7) in the domain $-h<z<0$ has the form

$$f(\mu,\nu,z)=-\frac{(\omega-\mu)^2\mathrm{sh}(kz)+k\mathrm{ch}(kz)}{k\,exp(kh)((\varepsilon+i(\omega-\mu))^2+k)},\quad k^2=\mu^2+\nu^2$$

The function $\Lambda(\mu,\nu)$ is determined from equation (4) as

$$\Lambda(\mu,\nu)=\frac{i(\omega-\mu)exp(-kh)}{(\varepsilon+i(\omega-\mu))^2+k} \tag{8}$$

The zeros of the denominator in (8) determine the dispersion relation $(\omega-\mu)^2=(\mu^2+\nu^2)^{1/2}$ which can be explicitly written as

$$\nu(\mu)=\pm((\omega-\mu)^4-\mu^2)^{1/2} \tag{9}$$

The set of frequencies $\omega>0$ is divided by two characteristic values $\omega_1=0.25$ and $\omega_2=\sqrt{6}/9$ into three intervals. For $\omega<\omega_1$, the dispersion curve determined from equation (9) consists of three branches: one closed and two unclosed. Then the wave pattern is the sum of two ship (longitudinal) waves with half-opening angle of the wave wedge less than $\pi/2$ and annular (transverse) waves around the source. For $\omega>\omega_2$, the dispersion curve consists of two unclosed branches without extrema. In this case, the wave pattern is the sum of two ship waves with half-opening angle of the wave wedge less than $\pi/2$. If $\omega_1<\omega<\omega_2$, then the dispersion curve consists of two unclosed curves one of which has two local extremes. One branch of the dispersion curve corresponds to usual ship waves with half-opening angle of the wave wedge less than $\pi/2$, and the second branch, to ship waves with half-opening angle of the wave wedge greater than $\pi/2$ (the wave front is directed upstream away from the source). This system of hybrid waves simultaneously has characteristic features of both annular (transverse) and ship (longitudinal) waves. We further consider this case ($\omega=0.255$). Figure 1 shows the dispersion curve branch (further denoted by $\nu_1(\mu)$) which describes the hybrid waves. Figure 2 shows the branch of dispersion curve (9) (further denoted by $\nu_2(\mu)$) which describes the ship waves. For $\varepsilon>0$ and $\mu<\omega$, we have $Im\,\nu_1(\mu)<0$, and for $\mu>\omega$, we have $Im\,\nu_2(\mu)>0$. Then, calculating the inner integral in equation (6) by closing the contour of integration in the variable ν for $y>0$ in the lower half-plane (the poles $\nu_1(\mu)$ and $-\nu_2(\mu)$), and for $y<0$, in the upper half-plane (the poles $-\nu_1(\mu)$ and $\nu_2(\mu)$), we obtain

$$\eta(x,y)=I_1(x,y)+I_2(x,y)$$

$$I_1(x,y)=\frac{1}{2\pi}\int_{-\infty}^{C}\frac{E}{\nu_1(\mu)}exp(-i(\mu x+\nu_1(\mu)|y|))d\mu,\ I_2(x,y)=\frac{1}{2\pi}\int_{D}^{\infty}\frac{E}{\nu_2(\mu)}exp(-i(\mu x-\nu_2(\mu)|y|))d\mu$$

where $E=(\omega-\mu)^3 exp(-(\omega-\mu)^2 h)$, C and D are the abscissas of the rightmost and leftmost points of the dispersion curves $\nu_1(\mu)$ and $\nu_2(\mu)$ in figures 1 and 2, respectively. Points A and G are points of inflection of the dispersion curves $\nu_1(\mu)$ and $\nu_2(\mu)$ in figures 1 and 2, respectively. The integrals $I_2(x,y)$ with the corresponding dispersion dependence $\nu_2(\mu)$, which describe the usual ship waves, were studied in detail, for example, in [2,5,13,14]. A more complicated and yet unstudied wave pattern of amplitude-phase characteristics of hybrid surface wave perturbations is described by the integrals $I_1(x,y)$. We denote the phase by $\Phi=\nu_1(\mu)|y|+\mu x$. Then, using the phase stationary condition in the form

$$\frac{d\nu_1(\mu)}{d\nu}=-\frac{x}{|y|} \tag{10}$$

we obtain the family of lines of constant phase with a parameter μ (the subscript «1» is omitted)

$$x=-\frac{\Phi\nu'(\mu)}{\nu(\mu)-\mu\nu'(\mu)},\quad |y|=\frac{\Phi}{\nu(\mu)-\mu\nu'(\mu)} \tag{11}$$

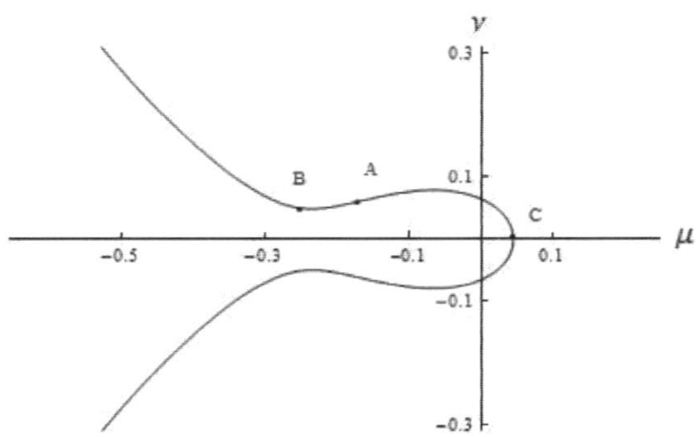

Figure 1. Dispersion curve $\nu_1(\mu)$

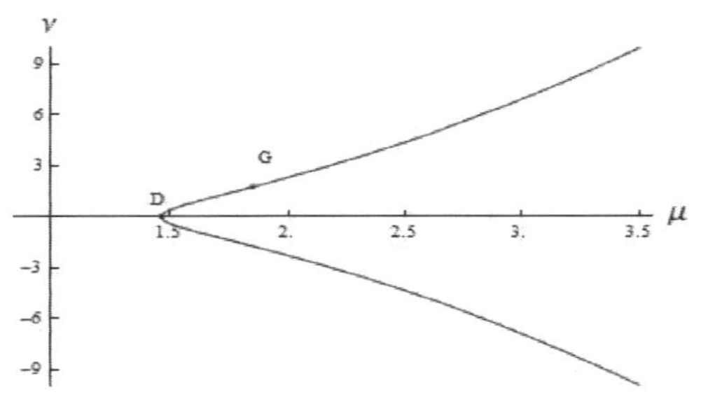

Figure 2. Dispersion curve $\nu_2(\mu)$

Figure 3 presents the lines of equal phase for different values of Φ with step 2π. The part of the dispersion curve from point C to the inflection point A corresponds to annular (transverse) waves, and the set of corner cusps forms the wave front moving upstream, which is depicted in figure 3 by a dotted line. To determine the wave front, it is necessary to supplement equation (11) with the condition $\frac{\partial x}{\partial\mu}\frac{\partial|y|}{\partial\Phi}-\frac{\partial x}{\partial\Phi}\frac{\partial|y|}{\partial\mu}=0$ or, which is the same, $\frac{d^2\nu(\mu)}{d\mu^2}=0$, where the abscissa μ_A of point A is a solution of the equation under study. Then the equation for determining the wave front becomes $x=-\nu'(\mu_A)|y|$ and the corresponding half-opening angle of the wave wedge is equal to $105^0 7'$. The part of the dispersion curve from point A to point B corresponds to the longitudinal crests of waves propagating from the front to infinity (depicted on the left in figure 3 by a dashed line). The dashed line in figure 3 corresponds to the crest of the wave with phase $\Phi=0$ and is described by the equation $x=-\nu'(\mu_B)|y|$ or $x=\sqrt{16\omega^2-1}|y|$, where μ_B is a root of the equation $\nu(\mu)=\mu\nu'(\mu)$

whose solution is $\mu_B = -\omega$, $\nu'(\mu_B) = -\sqrt{16\omega^2 - 1}$. The part of the dispersion curve to the left of point B corresponds to longitudinal crests of waves propagating from infinity to the origin (to the right of the dashed line in figure 3). In this figure, the phases Φ corresponding to the part of the dispersion curve to the right of point B are equal to $2\pi n, n = 1,2,3,4$, and to the left of point B, to $2\pi n, n = -11,-10,...,-1$. At infinity (for large values of x, $|y|$), the equations of crests of longitudinal waves have the form $x = \sqrt{16\omega^2 - 1}|y| - 2\pi k/\omega$, where k is integer, i.e., these crests are crests of a plane wave of length $\lambda = \pi\omega^{-2} \approx 24$. The wave length of annular (transverse waves) in the direction of the axis x is equal to $\lambda = 2\pi/C \approx 142$, i.e., is approximately six times greater than the length of wedge-shaped (longitudinal) waves.

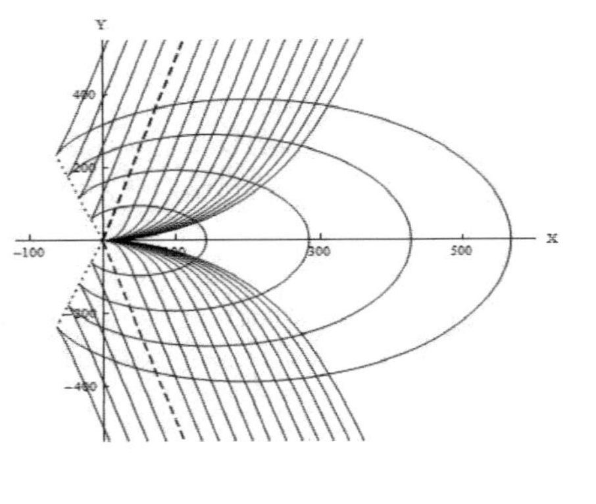

Figure 3. Phase structure of surface gravity waves

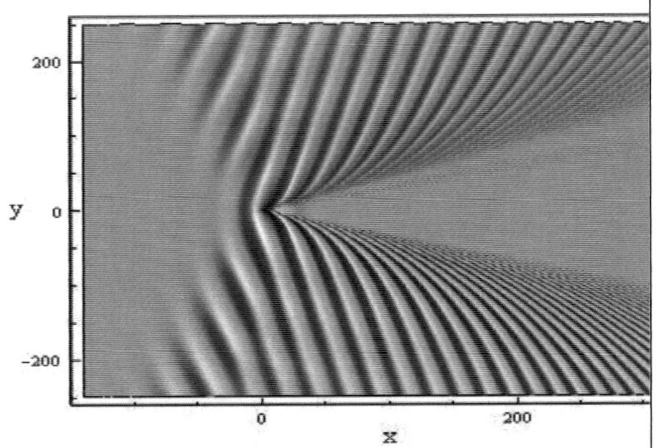

Figure 4. Surface gravity waves from moving unstable source

The integral $I_1(x,y)$ belongs to the class of integrals with two stationary points [15]. On the wave front (the dotted line in figure 3), the stationary points determined by equation. (10) merge. The uniform asymptotics $I_1(x,y)$ for large value of $|y|$ is constructed similarly [13-14] and has the form

$$I_1(x,y) = \frac{T^+(\rho)}{|y|^{1/3}} Ai(|y|^{2/3}\sigma(\rho)) exp(-i|y|a(\rho)) - i\frac{T^-(\rho)}{|y|^{2/3}\sqrt{\sigma(\rho)}} Ai'(|y|^{2/3}\sigma(\rho)) exp(-i|y|a(\rho))$$

$$T^{\pm}(\rho) = \frac{1}{2}\left(F(\mu_2)\sqrt{\frac{-2\sqrt{\sigma(\rho)}}{\theta(\mu_2,\rho)}} \pm F(\mu_1)\sqrt{\frac{2\sqrt{\sigma(\rho)}}{\theta(\mu_1,\rho)}} \right) \quad F(\mu) = (\omega-\mu)^3 exp(-h(\omega-\mu)^2)/\nu(\mu)$$

$$\sigma(\rho) = \left(\frac{3}{4}\left(S(\mu_2,\rho) - S(\mu_1,\rho)\right)\right)^{2/3}, \quad a(\rho) = \frac{1}{2}\left(S(\mu_2,\rho) + S(\mu_1,\rho)\right)$$

$$S(\nu,\rho) = \nu(\mu) - \rho\mu, \quad \rho = -x/|y|, \quad \theta(\nu,\rho) = \frac{\partial^2 S(\mu,\rho)}{\partial\mu^2}, \quad Ai(\tau) = \frac{1}{2\pi}\int_{-\infty}^{\infty} cos\left(\tau u - u^3/3\right) du$$

where $\mu_1 = \mu_1(\rho)$ and $\mu_2 = \mu_2(\rho)$ are roots of the equation $\partial S(\mu,\rho)/\partial\mu = 0$, $|\mu_1| < |\mu_2|$, $Ai(\tau)$ is the Airy function, and $Ai'(\tau)$ is the derivative of the Airy function [15]. To determine the free surface elevation, it is necessary to multiply expression for $I_1(x,y)$ by $exp(i\omega t)$ and take the real part of the obtained result. Figure 4 presents the wave pattern of the free surface elevation for $t = 10$ and $h = 3$ which is calculated in dimensionless coordinates (near the origin, the integral $I_1(x, y)$ was calculated numerically). For the parameter values which are typical of the ocean conditions $Q = 10^3 m^3 / sec, V = 3m / sec$, the elevation amplitudes are of the order of 0.3 meter. Using the asymptotics of the Airy function and its derivative at a far distance from the front, we can obtain a non-uniform asymptotics for $I_1(x, y)$ consisting of two terms. The first of these terms corresponding to the root μ_1 describes wedge-shaped (longitudinal) waves, and the second term corresponding to the root μ_2 describes annular (transverse) waves. Thus, we have shown that, in certain generation regimes, the far fields of surface perturbations from a non-stationary source localized in the flow of a heavy liquid of infinite depth form a hybrid system of waves of the following two types: annular (transverse) and wedge-shaped (longitudinal). The qualitative picture of wave fields at a far distance from a nonstationary source is significantly more complicated compared to the case of wave generation by a moving stationary source when the wave fronts come to a fixed observation point. The unsteadiness of the perturbation source amplitude results not only in the appearance of annular waves diverging on the liquid surface directly from the source but also in generation of hybrid surface perturbations propagating upstream from the source. The obtained asymptotics of surface wave perturbations far field allow one efficiently to calculate the basic amplitude-phase characteristics of wave fields and, in addition, qualitatively to analyze the obtained solutions, which is important in developing of well-posed mathematical models of wave dynamics of surface perturbations of the real natural environments. This opens wide opportunities for investigating the wave fields in general, which is also important for formulating correct statements of mathematical models of wave dynamics and for obtaining express evaluations in the surface field measurements in ocean. The research was carried out in the framework of the Federal target program No. AAAA-A17-117021310375-7.

References

[1] Lighthill M J 1978 *Waves in fluids* (Cambridge: Cambridge University Press).
[2] Bulatov V V and Vladimirov Yu V 2007 *Internal gravity waves: theory and applications* (Moscow: Nauka Publishers).
[3] Kharif C, Pelinovsky E and Slynyaev A 2009 *Rogue waves in the ocean* (Berlin–Heidelberg: Springer).
[4] Pedlosky J 2010 *Waves in the ocean and atmosphere: introduction to wave dynamics* (Berlin–Heidelberg: Springer).
[5] Bulatov V V and Vladimirov Yu V 2012 *Wave dynamics of stratified mediums* (Moscow: Nauka Publishers).
[6] Shurgalina E G and Pelinovsky E N 2015 *Dynamics of irregular wave ensembles in the coastal zone* (Nizhny Novgorod: Nizhny Novgorod State Technical University n.a. R.E.Alekseev Publishing).
[7] Mei C C, Stiassnie M and Yue D K-P 2017 *Theory and applications of ocean surface waves* (London: World Scientific Publishing).
[8] Lannes D 2005 *J.American Math. Society* **18** 605.
[9] Brown M G 2000 *Wave Motion* **32** 247.
[10] Zakharov V E , Dyachenko A I and Vasilyev O A 2002 *European J. Mech.. B Fluids* **21** 283.
[11] Ruban V P 2005 Phys.Lett.A **340** 194.
[12] Dobrokhotov S Yu, Grushin V V , Sergeev S A and Tirozzi 2016 *Russian J. Math. Physics* **23** 455.
[13] Bulatov V V and Vladimirov Yu V 2015 *Fluid Dynamics* **50** 741.
[14] Bulatov V V, Vladimirov Yu V and Vladimirov I Yu 2017 *Fluid Dynamics* **52** 617.
[15] Haberman R 2013 *Applied partila differential aequtions* (NJ: Pearson Educations).

Journal of Physics: Conference Series **1392** (2019) 012006 doi:10.1088/1742-6596/1392/1/012006

On linear stability of shear flows of an ideal stratified fluid: research methods and new results

A A Gavrilieva[1], **Yu G Gubarev**[2,3]

[1] V.P. Larionov Institute of Physical-Technical Problems of the North of Siberian Branch of Russian Academy of Sciences, Yakutsk, 677891 Russia
[2] Lavrentyev Institute of Hydrodynamics, Siberian Branch, Russian Academy of Sciences, Novosibirsk, 630090 Russia
[3] Novosibirsk National Research State University, Novosibirsk, 630090 Russia

E-mail: gav-ann@yandex.ru

Abstract. The results, that obtained by the spectral method with use of integral relations for the problem of linear stability of steady-state shear plane-parallel flows of an inviscid stratified incompressible fluid in the gravity field with respect to plane perturbations in the Boussinesq approximation and without it, are specified, complemented and developed by the most powerful analytical method of the modern mathematical theory of hydrodynamic stability – the second (or direct) Lyapunov method. In both case, the new analytical method made it possible to prove that given steady-state flows of stratified fluid are absolutely unstable in theoretical sense with respect to small plane perturbations and to obtain the sufficient conditions for practical linear instability of considered flows. The illustrative analytical examples of given steady-state flows and small plane perturbations as normal waves imposed on them are constructed. Using the asymptotic method, it is proved that constructed perturbations grow in time irrespective of the fact whether the Miles-Howard and the Miles-type theorems are valid or not.

1. Introduction

The buoyancy effect on the inertial stability/instability of fluid flows is considered: the problem of linear stability of steady-state shear plane-parallel flows of an inviscid stratified incompressible fluid in the gravity field with and without the Boussinesq approximation is investigated. The statement and numerous studies of this problem are included in many monographs on hydrodynamics [1, 2] and geophysical hydrodynamics [3, 4, 5]. Resulting conditions of linear stability or instability of the considered fluid flows are widely used as conditions for the appearance of a laminar-turbulent transition in the processes observed on the ocean surface and in the atmosphere, also in aviation and hydraulic engineering.

The major conditions for the linear stability of considered fluid flows were obtained by the spectral method with use of integral relations [1, 6, 7, 8]. This method includes the use of integral relations and relations derived from the boundary value problems for differential equations with variable coefficients. In this method, the equivalence between differential and integral statements has not been proved, causing that not all possible perturbations are actually investigated. Moreover, except for states of rest given in work [9], it is impossible to find the conditions of theoretical stability for any steady-state flows of studied fluid with respect to small plane perturbations by using energy reasons. All this implies that there is a linear absolute instability for studied fluid flows. The main purpose of this paper is to verify this assumption of linear absolute instability with new analytical method [10] and by comparing with the results obtained early by the spectral method with use of integral relations.

2. The direct Lyapunov method (in the Boussinesq approximation)

We consider the nonsteady-state plane flows of an inviscid stratified incompressible fluid in the gap between two steady-state impermeable solid parallel unbounded surfaces in the gravity field in the Boussinesq approximation [11]. This approximation allows us to disregard the changes in density (which affect inertia), rather than in weight (or bouyancy) of the fluid [3]. In the Cartesian coordinate system $(x,\ y)$ these flows are characterized by evolutionary solutions to an initial-boundary value problem of the form [3, 7, 6, 8]

$$\widehat{\rho}_0 Du = -p_x,\ \widehat{\rho}_0 Dv = -p_y - \rho g,\ D\rho = 0,\ u_x + v_y = 0 \text{ in } \tau;\ v = 0 \text{ on } \partial\tau;$$

$$u(x,y,0) = u_0(x,y), v(x,y,0) = v_0(x,y),\ D \equiv \partial/\partial t + u\partial/\partial x + v\partial/\partial y, \tag{1}$$

$$\tau \equiv \{(x,y) : -\infty < x < +\infty, 0 < y < H\},\ \partial\tau \equiv \{(x,y) : -\infty < x < +\infty; y = 0, H\},$$

where t is time, $\widehat{\rho}_0 \equiv \text{const} > 0$ is the average fluid density; p and ρ are the pressure and density perturbations; $(0,\ g \equiv \text{const} > 0)$ is the acceleration due to gravity, $(u,\ v)$ is the fluid velocity field. Initial values $(u_0,\ v_0)$ of fluid velocity make the fourth and fifth relation of problem (1) into identity.

Initial-boundary value problem (1) has exact steady-state solutions

$$\rho = \rho_0(y),\ u = U(y),\ v = 0,\ p = P(y) \equiv p_0 - g \int_0^y \rho_0(y_1)\, dy_1, \tag{2}$$

where ρ_0, U are arbitrary functions of coordinate y, p_0 is an additive constant.

We linearize the mixed problem (1) in the neighbourhood of exact steady-state solutions (2)

$$\widehat{\rho}_0(u'_t + Uu'_x + v'\frac{dU}{dy}) = -p'_x,\ \widehat{\rho}_0(v'_t + Uv'_x) = -p'_y - \rho' g,$$

$$\rho'_t + U\rho'_x + v'\frac{d\rho_0}{dy} = 0,\ u'_x + v'_y = 0 \quad \text{in} \quad \tau; \tag{3}$$

$$v' = 0 \quad \text{on} \quad \partial\tau;\ u'(x,y,0) = u'_0(x,y), v'(x,y,0) = v'_0(x,y),$$

where u', v', ρ', p' are small plane perturbations of velocity u, v, density ρ and pressure p.

In order to demonstrate a instability of stationary solution (2) of mixed problem (1) with respect to small plane perturbations (3) we need at least one of these perturbations, but with the exponential time growth. The search for such perturbations is carried out below in subclass of plane flows of form Lagrangian displacements field $(\xi_1,\ \xi_2)$ [11, 12]:

$$\xi_{1t} = u' - U\xi_{1x} + \xi_2\frac{dU}{dy},\ \xi_{2t} = v' - U\xi_{2x}. \tag{4}$$

It can be shown that the differential inequality holds for small plane perturbations (3), (4) of stationary flows (2)

$$M \equiv \int_{-\infty}^{+\infty} \int_0^H \widehat{\rho}_0(\xi_1^2 + \xi_2^2)\, dydx : \ \frac{d^2M}{dt^2} - 2\lambda\frac{dM}{dt} + 2(\lambda^2 + \alpha)M \geq 0 \tag{5}$$

with a parameter λ and the constant $\alpha \equiv (g/\widehat{\rho}_0) \max\limits_{0\leq y\leq H} \mid d\rho_0/dy \mid > 0$.

Proposition 1. If $\lambda > 0$ and the following additional conditions

$$M\left(\frac{\pi n}{2\sqrt{\lambda^2+2\alpha}}\right) > 0; \quad n = 0, 1, 2, ...; \frac{dM}{dt}\left(\frac{\pi n}{2\sqrt{\lambda^2+2\alpha}}\right) \geq 2\left(\lambda + \frac{\alpha}{\lambda}\right) M\left(\frac{\pi n}{2\sqrt{\lambda^2+2\alpha}}\right);$$
$$M\left(\frac{\pi n}{2\sqrt{\lambda^2+2\alpha}}\right) \equiv M(0)\exp\left(\frac{\pi n\lambda}{2\sqrt{\lambda^2+2\alpha}}\right), \frac{dM}{dt}\left(\frac{\pi n}{2\sqrt{\lambda^2+2\alpha}}\right) \equiv \frac{dM}{dt}(0)\exp\left(\frac{\pi n\lambda}{2\sqrt{\lambda^2+2\alpha}}\right), \quad (6)$$

are satisfied along with (5) the following a priori exponential lower estimate is obtained

$$M(t) \geq C\exp(\lambda t), \ C \equiv \text{const} > 0.$$

The proof of **Proposition 1** are constructed according to allowability standards for procedure of integrating differential inequality and it can be found in the papers of the authors [10, 13].

Thus, according to the Lyapunov definition of instability [14], stationary flows (2) are absolutely theoretical unstable with respect to small plane perturbations (3), (4). Moreover, inequalities of relations systems (6) can be interpreted as sufficient conditions for practical linear instability [15] of steady-state flows (2), but for small plane perturbations (3), (4) in the form of normal waves – as necessary and sufficient ones.

3. Spectral methods using integral relations (in the Boussinesq approximation)

The finding result of absolute instability is compared with the well-known result of spectral theory on the instability of considered shear flows (2), obtained earlier by the method of integral relations for small plane perturbations (3) as normal waves, – with the Miles-Howard theorem [6, 7].

The small plane perturbations (3) of stationary flows (2) as normal waves with a complex phase velocity $c \equiv c_r + ic_i$ and a wave number $k > 0$ are considered. According to the Miles-Howard theorem [6, 7], the exponentially growing small plane perturbations arise if and only if the local Richardson number

$$\text{Ri} \equiv -\frac{g}{\widehat{\rho}_0}\frac{d\rho_0}{dy}\left(\frac{dU}{dy}\right)^{-2} < 1/4 \quad (7)$$

at least at one point in the fluid flow domain τ.

However, the authors in the work [11] proved that small plane perturbations (3) in the form of normal waves with an amplitude containing a countable set of branches are not covered by the Miles-Howard theorem [6, 7] and the following proposition holds.

Proposition 2. In the Boussinesq approximation, the reverse inequality (7) for local Richardson number in domain τ of the fluid flow is the necessary and sufficient condition for the stability of exact steady-state solutions (2) to mixed problem (1) with respect to one incomplete nonclosed subclass of small plane perturbations (3), (4) as normal waves.

The validity of **Proposition 2** indicates that no contradictions between the Miles-Howard theorem [6, 7] and linear instability of steady-state flows (2) are exist.

Below, we construct an example of exact solutions (2) to mixed problem (1) and the imposed on them small perturbations (3) as normal waves. Asymptotic behaviour of these perturbations indicates that the subclass small plane perturbations (3) of considered steady-state flows (2) which are not covered by the Miles-Howard theorem [6, 7] is not empty set.

We consider the steady-state flows form of

$$U \equiv by + b_1; \ b, b_1 > 0; \ \rho_0 \equiv a - \frac{\widehat{\rho}_0}{g}y. \quad (8)$$

Here, b, b_1 and a are constants.

For the small plane perturbations (3) in the form of normal waves of steady-state fluid flows (2), (8), we obtain the dispersion function $DJ(c,k)$ and the dispersion relation $DJ(c,k)=0$ of forms

$$DJ(c,k) \equiv J_\nu\Big(i\frac{k}{b}[c-b_1]\Big)Y_\nu\Big(i\frac{k}{b}[c-b_1-bH]\Big) - Y_\nu\Big(i\frac{k}{b}[c-b_1]\Big)J_\nu\Big(i\frac{k}{b}[c-b_1-bH]\Big) = 0, \quad (9)$$

where $J_\nu(ik[c-by-b_1]/b)$, $Y_\nu(ik[c-by-b_1]/b)$ are transcendental Bessel functions of order $\nu \equiv \sqrt{1/4-1/b^2}$, which have a countable set of branches [16]. Therefore, this functions will produce counterexamples to the Miles-Howard theorem [6, 7].

Dispersion relation (9) is too complicated, and its roots $c(k)$ for any order ν cannot be found explicitly using analytical methods. However, using the rules of analytical operations with asymptotic expansions of the Bessel functions $J_\nu(z)$, $Y_\nu(z)$ that are valid for a fixed order ν and for large modulus of argument z [17, 18],

- it can be proved
  Proposition 3. Every cylindrical function is uniquely determined by its asymptotic for a fixed order and large modulus of its argument;
- it can be obtained that the zero approximation of the dispersion function $DJ(c,k)$ (9) has no roots, the first approximation of the dispersion function $DJ(c,k)$ (9) has roots, but they are real, and the second approximation of the dispersion function $DJ(c,k)$ (9) has complex roots in the form

$$c_{1,2,3,4} = b_1 + \frac{bH}{2} \pm \frac{1}{2k\sqrt{2}}\Big(2[b^2H^2k^2 + kH\coth(kH) - 1] \pm$$

$$\pm i\coth(kH)\Big[(1/b^4 + 4/b^2 + 16b^2k^2H^2 + 8k^2H^2)\tanh^2(kH) + \quad (10)$$

$$+4(1/b^2+4)kH\tanh(kH) - 4k^2H^2\Big]^{1/2}\Big)^{1/2}.$$

Among complex roots $c_{1,2,3,4}$ (10), one root always has a positive imaginary part, regardless of which value of the local Richardson Ri (7) number is. Therefore, the second approximation of the dispersion function $DJ(c,k)$ (9) has the desired complex root with a positive imaginary part. Then, according to **Proposition 3**, it follows that the exact dispersion relation (9) also has the complex solution with a positive imaginary part, regardless of which value of the local Richardson Ri (7) number is.

4. The direct Lyapunov method (without the Boussinesq approximation)

The same fluid flows are considered, only without the Boussinesq approximation. In this case, the density field of the fluid is not divided into the average density and its perturbations. Then these flows are characterized by a problem in the form of (1) with exact solutions in the form (2), the small plane perturbations imposed on this exact solutions are now solutions to the linearized problem (3) in which the average density $\widehat{\rho}_0$ and its perturbations ρ have to be replaced by the density field ρ, and the perturbations of the pressure field p now simply mean the pressure field.

Following Section 2, growing small plane perturbations (3) are sought in the subclass (4). In this subclass, the differential inequality holds for the auxiliary functional (5), but now with $\widehat{\rho}_0 \equiv \rho$, and $\alpha \equiv g \max\limits_{0\le y\le H} |(d\rho_0/dy)\rho_0^{-1}| > 0$. Therefore, an a priori exponential lower estimate can be constructed (according to **Proposition 1**), which indicates the absolute linear instability of the considered steady-state flows (2) without the Boussinesq approximation.

5. Spectral methods (without the Boussinesq approximation)

According to the Miles-type theorem [2], the exponentially growing small plane perturbations arise if and only if the local Richardson number

$$\mathrm{Ri} \equiv -\frac{g}{\rho_0}\frac{d\rho_0}{dy}\left(\frac{dU}{dy}\right)^{-2} < 1/4 \tag{11}$$

at least at one point in the fluid flow domain τ.

Along the lines of reasoning of Section 3, the authors proved that small plane perturbations (3) imposed on steady-state flows (2) without the Boussinesq approximation in the form of normal waves with an amplitude containing a countable set of branches are not covered by the Miles-type theorem [2], and **Proposition 2** holds for local Richardson number as (11) [11].

It remains to show that the subclass of small plane perturbations as normal waves which are not covered by the Miles-type theorem [2] is not empty set.

We consider steady-state fluid flows in the form

$$U \equiv by + b_1; b, b_1 > 0; \ \rho_0 \equiv b\exp(-my), \ m \equiv \mathrm{const} > 0. \tag{12}$$

The dispersion function and the dispersion relation for small plane perturbations as normal waves of steady-state flows (2), (12) take the following form

$$\begin{aligned} DW(c,k) \equiv\ & M_{\frac{m^*}{k^*},\nu}(k^*[-c+b_1])W_{\frac{m^*}{k^*},\nu}(k^*[-c+bH+b_1]) - \\ & -W_{\frac{m^*}{k^*},\nu}(k^*[-c+b_1])M_{\frac{m^*}{k^*},\nu}(k^*[-c+bH+b_1]) = 0, \end{aligned} \tag{13}$$

where $M_{\frac{m^*}{k^*}\nu}(y)$ and $W_{\frac{m^*}{k^*},\nu}(y)$ are transcendental Whittacker functions [19] of order $(\frac{m^*}{k^*},\ \nu)$, in which $m^* \equiv m/b$, $k^* \equiv \sqrt{m^{*2} + 4k^2/b^2}$, $\nu = \sqrt{1/4 - mg/b^2}$.

The transcendental Whittaker functions have a countable set of branches [18], similarly to the transcendental Bessel functions. Therefore, the Whittaker functions will generate counterexamples to the Miles-type theorem [2].

In analogy with dispersion relation (9), the dispersion relation (13) is also not susceptible to the fact that its roots can be found explicitly by any exact analytical methods. However, approximate solutions to dispersion relation (9) can be found, using the well-known asymptotic expansions for the Whittaker functions $M_{\frac{m^*}{k^*},\nu}(z)$, $W_{\frac{m^*}{k^*},\nu}(z)$ that are admitted for fixed orders $\frac{m^*}{k^*}$, ν and for large modulus of argument z [17]. So, using the rules of analytical operations with this asymptotic expansions for the Whittaker functions $M_{\frac{m^*}{k^*},\nu}(z)$, $W_{\frac{m^*}{k^*},\nu}(z)$ [18],

- it can be proved
  Proposition 4. Every solution $w(z)$ of the Whittaker equation
  $$\frac{d^2w}{dz^2} + (\frac{l}{z} - \frac{1}{4} + \frac{1/4-\mu^2}{z^2})w = 0$$
  is uniquely determined by its asymptotic behaviour as $|z| \to \infty$;
- it can be showed that the zero approximation of the dispersion function $DW(c,k)$ (12) has complex roots in the form

$$\begin{aligned} c_r &= b_1 - bH\frac{\exp(\frac{k^{*2}bH}{2m^*})\cos(\frac{\pi n k^*}{m^*}) - 1}{1 - 2\exp(\frac{k^{*2}bH}{2m^*})\cos(\frac{\pi n k^*}{m^*}) + \exp(\frac{k^{*2}bH}{m^*})}, \\ c_i &= bH\frac{\exp(\frac{k^{*2}bH}{2m^*})\sin(\frac{\pi n k^*}{m^*})}{1 - 2\exp(\frac{k^{*2}bH}{2m^*})\cos(\frac{\pi n k^*}{m^*}) + \exp(\frac{k^{*2}bH}{m^*})}, \quad n \in Z. \end{aligned} \tag{14}$$

The expression (14) indicates that among the complex solutions to the zero approximate dispersion relation (13) there necessarily exists a countable set of solutions with a positive imaginary part (regardless of the truth/falsity of the necessary and sufficient condition of linear instability for the local Richardson number (11), the Miles-type theorem [2]). Then, according to **Proposition 4**, it follows that among solutions to the exact dispersion relation (13) there also necessarily exists the desired complex solution with a positive imaginary part.

6. Conclusion

In this paper, using the direct Lyapunov method, we have proved that steady-state plane-parallel shear flows of an inviscid stratified incompressible fluid in channel in the gravity field are absolutely unstable in theoretical sense with respect to small plane perturbations with and without the Boussinesq approximation. The sufficient conditions for practical linear instability of these flows have been obtained, and for small plane perturbations in the form of normal waves these conditions are the necessary and sufficient ones.

For both cases, the boundaries of applicability of the well-known necessary conditions for linear instability (the Miles-Howard and the Miles-type theorems), previously obtained by the spectral method using integral relations, have been clearly specified. It has been found that the Miles-Howard and the Miles-type theorems are inherently sufficient and necessary statements with respect to certain incomplete nonclosed subclasses of considered small perturbations. We have constructed illustrative analytical examples of given steady-state flows and small plane perturbations as normal waves imposed on them. Using the asymptotic method, we have proved that constructed perturbations grow in time irrespective of the fact whether the Miles-Howard and the Miles-type theorems are valid or not. Therefore, the results obtained earlier by other authors using the method of integral relations for problems of linear stability of steady-state plane-parallel shear flows of an ideal stratified incompressible fluid require a strict description of the specified partial classes of small plane perturbations because otherwise these results can be wrong.

References

[1] Drazin F 2005 *Introduction to the Theory of Hydrodynamic Stability* (Moscow: Fizmatlit)
[2] Yih C H 1980 *Stratified Flows* 2nd ed (New York: Academic Press)
[3] Phillips O M 1966 *Dynamics of the Upper Ocean* (New York: Cambridge Univ. Press)
[4] Dikii L A 1976 *Hydrodynamic Stability and Atmosphere Dynamics* (Leningrad: Gidrometeoizdat)
[5] Kundu P K and Cohen I M 2008 *Fluid Mechanics* (Oxford: Elsevier)
[6] Howard L N 1961 *J. Fluid Mech.* **10**(4) 509–512
[7] Miles J W 1961 *J. Fluid Mech.* **10**(4) 496–508
[8] Kozyrev O R and Stepanyants Y A 1991 *Itogi Nauki Tekh., Ser.: Mekh. Zhidk. Gaza* **25** 3–89
[9] Vladimirov V A 1987 *Fluid Dyn* **22**
[10] Gubarev Y G 2009 *Progress in nonlinear analysis research* (New York: Nova science publishers, inc.) chap The development of Lyapunov's direct method in the application to new types of problems of hydrodynamic stability theory, pp 137–181 ISBN 978-1-60456-359-7
[11] Gavrilieva A A, Gubarev Y G and Lebedev M P 2017 *Lobachevskii Journal of Mathematics* **38**(3) 560–570
[12] Gavrilieva A A, Gubarev Y G and Lebedev M P 2013 *Int. J. Theor. Math. Phys.* **3**(4) 123–129
[13] Gavrilieva A A and Gubarev Y G 2012 *Vestn. Sev.-Vost. Fed. Univ. im. M.K. Ammosova* **9**(3) 15–21
[14] Lyapunov A M 1992 *The general problem of the stability of motion* (London: Taylor&Francis)
[15] La-Salle J P and Lefshets S 1961 *Stability by the Direct Liapunov Method with Applications* (New York: Academic)
[16] Lavrentiev M A and Shabbat B V 1973 *Methods in the theory of functions in a complex variable* (Moscow: Nauka)
[17] Abramovitz M and Stegan I 1979 *Handbook of Mathematical Functions with Formulas, Graphs and Mathematical Tables* (Moscow: Nauka)
[18] Olver F W J 1974 *Asymptotics and special functions* (New York-London: Academic press)
[19] Bateman H and Erdélyi A 1973 *Higher Transcendental Functions, Vol. 1: The Hypergeometric Function, Legendre Functions* 2nd ed (Moscow: Nauka)

The search algorithm for discrete mode of electron plasma

N M Gordeeva[a]* and A A Yushkanov[b]

[a] Bauman Moscow State Technical University, Baumanskaya street, 5/1, Moscow, 105005, Russia
[b] Moscow Region State University, Radio street,10a, Moscow, 105005, Russia
*e-mail: nmgordeeva@bmstu.ru, yushkanov@inbox.ru

Abstract. The question about the shielding of external electric field electron plasma is examined. The search algorithm for discrete modes of electron plasma have been described. The case of arbitrary degree of degeneracy of the electron gas have been considered. Special attention is paid to the case when the degree of degeneracy of the electron gas is large. Have been demonstrated an effective way of finding the discrete mode in this case.

1. Introduction

The question about the response of the plasma on the electric field has a long history [1-7]. But a number of fundamental issues still remain unresolved. It refers to the nature of the shielding of electric field by plasma, with varying degrees of degeneration of the electron gas.

In earlier works [8, 9], the authors concerned the question of the existence of a plasma mode. This mode is also known as a Debye mode. This mode is a discrete root of the dispersion function. In the present work are described in detail the search algorithm of this discrete root. Our results are correct for the general case of plasma with an arbitrary degree of degeneracy of the electron gas.

We proceed from the analytical solution of the problem of the behavior of plasma in an alternating electric field. From this solution it is possible to find the dispersion function. Properties of dispersion functions determine the behavior of discrete mode. Our goal is to find the algorithm that gives the possibility to define a discrete mode according to the properties of the dispersion functions.

2. Statement of the problem and the main equations

In electronic plasma with arbitrary degree of degeneracy the dispersion function is as follows:

$$\Lambda(z) = \Lambda(z, \Omega, \varepsilon) = 1 + \frac{1}{w_0 \eta_1^2} \int_{-\infty}^{+\infty} \frac{\eta_1^2 - tz}{t - z} k(t, \alpha)\, dt. \tag{1}$$

Here Ω is the frequency of the external field divided by the plasma frequency, ε is the collision frequency divided by the plasma frequency, w_0, η_1^2 – some complex constants that depend on plasma parameters, $k(t,\alpha)$ is a real core, depending on the plasma parameters, z is a complex variable, integration is carried out on the real axis.

To establish the existence of finite complex zeros of the dispersion function, we use the argument principle. As on the real y axis the dispersion functions have a peculiarity, then to eliminate this peculiarity it is necessary to conduct a cut along the specified axis. Because of the evenness of the dispersion function we can examine only the upper half-plane as the zeros, if they exist, will differ only by a sign. Take in the upper half plane contour Γ_{ε}. It goes in a positive direction and consists of a semicircle of radius $R=1/\varepsilon$ and of the line γ_ε, separated from the real axis $(-\infty; +\infty)$ at a distance ε.

Contour arbitrarily closely approaches the real axis when ε→0. Because of the argument principle the number of zeros in the region bounded by the contour is equal to

$$N = \frac{1}{2\pi i} \oint_{\Gamma_\varepsilon} d \ln \Lambda(z).$$

Let us turn to the limit in this equality when ε→0. Take into account that the dispersion function is analytical in a neighbourhood of an infinitely distant point, and also the property $\Lambda^+(-z) = \Lambda^-(z)$. Then we get

$$N = \frac{1}{2\pi i} \oint_{\Gamma_\varepsilon} d \frac{\Lambda^+(z)}{\Lambda^-(z)}.$$

Consider depending on a parameter μ, the curve $G(\mu) = \Lambda^+(\mu)/\Lambda^-(\mu)$. The number of zeros N is equal to the number of revolutions of the curve around the origin. Separate out of function G(μ) the real and imaginary parts. Represent it in the form:

$$G(\mu) = \frac{(w_0 - 1)\eta_1^2 + (\eta_1^2 - \mu^2)\lambda(\mu) + is(\mu)(\eta_1^2 - \mu^2)}{(w_0 - 1)\eta_1^2 + (\eta_1^2 - \mu^2)\lambda(\mu) - is(\mu)(\eta_1^2 - \mu^2)}.$$

Here $s(\mu) = \pi\mu/2$.
Given that

$$w_0 = 1 - i\frac{\omega}{\nu}, \quad \eta_1^2 = \varepsilon(\varepsilon - i\Omega)r(\alpha),$$

we obtain

$$G(\mu) = \frac{P^-(\mu) + iQ^-(\mu)}{P^+(\mu) + iQ^+(\mu)}$$

$$P^{\pm}(\mu) = \Omega^2 r(\alpha) + \big(\mu^2 - \varepsilon^2 r(\alpha)\big)\lambda(\mu) \pm s(\mu)\varepsilon\Omega r(\alpha),$$

$$Q^{\pm}(\mu) = \varepsilon\Omega r(\alpha)(\lambda(\mu) + 1) \pm s(\mu)(\varepsilon^2 r(\alpha) - \mu^2).$$

Select the real and imaginary parts of the function G(μ):

$$G(\mu) = \frac{G_1(\mu)}{G_0(\mu)} + i\frac{G_2(\mu)}{G_0(\mu)}.$$

Here

$$G_0(\mu) = \big(P^+(\mu)\big)^2 + \big(Q^+(\mu)\big)^2,$$

$$G_1(\mu) = P^-(\mu)P^+(\mu) + Q^-(\mu)Q^+(\mu),$$

$$G_2(\mu) = P^+(\mu)Q^-(\mu) - P^-(\mu)Q^+(\mu).$$

Consider the curve L, defined implicitly by parametric equations:

$$L\colon G_1(\mu, \Omega, \varepsilon) = 0, \quad G_2(\mu, \Omega, \varepsilon) = 0.$$

Expressing these equations in the explicit form of ε and Ω, we obtain parametric equations for the curve L(α, ε, Ω).

On the borders of change of parameter μ must mention, because there is a critical point μ* from which the parametric equations for the curve have solutions. In other words, to implement the curve L, the parameter μ should vary from this value, μ* to infinity.

After some transformations, we obtain the resulting expression for $\Omega^2 = L_1$, $\varepsilon^2 = L_2$.

$$L(\alpha)\colon \ \Omega = \sqrt{L_1(\mu, \alpha)}, \quad \varepsilon = \sqrt{L_2(\mu, \alpha)}, \qquad \mu^* \le \mu \le +\infty,$$

where

$$L_1(\mu, \alpha) = \frac{s_0(\alpha)}{s_2(\alpha)} \cdot \frac{\mu^2\big[\lambda(\mu, \alpha)\big(1 + \lambda(\mu, \alpha)\big) + s^2(\mu, \alpha)\big]^2}{\big(-\lambda(\mu, \alpha)\big)\Big[\big(1 + \lambda(\mu, \alpha)\big)^2 + s^2(\mu, \alpha)\Big]},$$

$$L_2(\mu,\alpha) = \frac{s_0(\alpha)}{s_2(\alpha)} \cdot \frac{\mu^2 s^2(\mu,\alpha)}{(-\lambda(\mu,\alpha))\left[(1+\lambda(\mu,\alpha))^2 + s^2(\mu,\alpha)\right]}$$

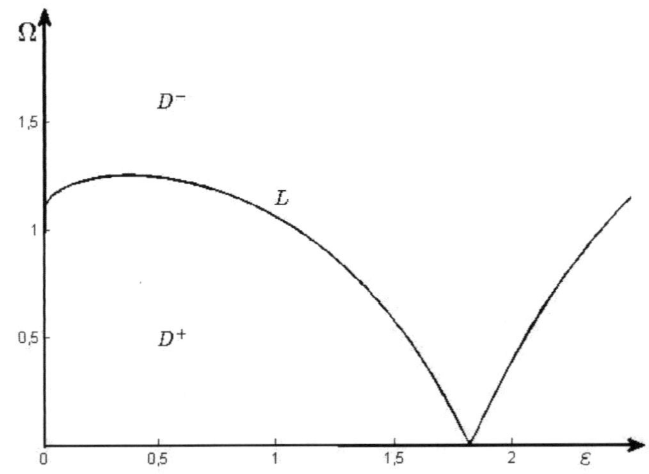

Fig.1 Region of existence and the lack of finite complex root.

Functions L_1 (μ,α) and L_2 (μ,α) determine in an explicit parametric form a curve L that is the boundary of the region D+ (the area where a discrete root of the dispersion function exists) (fig.1). With the increase of the parameter μ movement along the curve happens from right to left. If $\mu \to \mu^*$, this corresponds to infinitely large values of ε and Ω. Further, the curve approaches the origin, at some point (around ε=1.8). The numerator of L_1 (μ,α) becomes zero, the curve touches the axis ε. Then the curve behaves as a smooth line, and for $\varepsilon \to 0$ the corresponding value for $\Omega \to 1$, $\mu \to +\infty$. It is possible to plot the curve G(μ), which illustrate conclusions about the presence of the root. If (Ω, ε) $\in D^+$, the curve G(μ) rotates around the origin, and if (Ω, ε) $\in D^-$ the closed curve does not passes the origin. In [3] and [8], a study was conducted at the border, when (Ω, ε) ∈L. Below we will analyze analogous cases to this problem.

Since the physical meaning have non-negative frequency of the electric field and the frequency of collisions in the plasma, then the image of the curve L can be limited to the first quarter of the plane (Ω, ε). In Fig. 1 one can see how the curve L divides it into two regions, D^+ and D^-. Figure 2 shows the behavior of the curve G(μ) for the two cases of the parameters (Ω, ε). In the first case the curve is constructed for the parameters from the region D^-. It is seen that it does not capture the origin of coordinates. in the second case, the parameters derived from the field D^+. Then the curve passes once the origin of coordinates. This means that in the upper half-plane we have one root z_0. The reverse root $-z_0$ is locates in the lower half-plane.

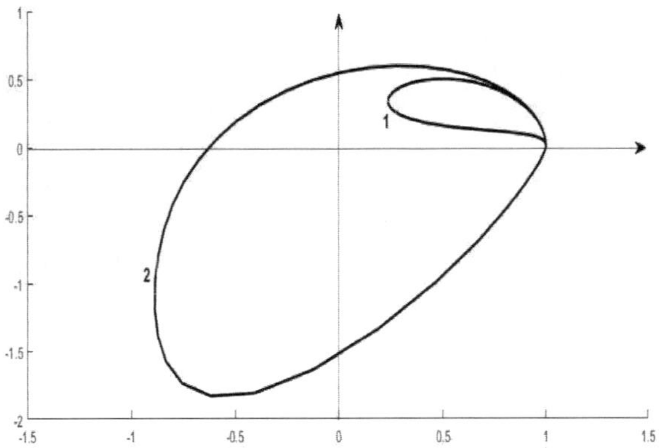

Fig.2 Curve G(μ), built for parameters from different areas. Curve 1 is the case of the (Ω, ε) ∈D, ε=0,002, Ω=1.4, the curve 2 – the case (Ω, ε) ∈D^+, ε=0,002, Ω=1,1.

Because of the evenness of the dispersion functions of its zeros differ only in sign. For z_0 accept the one whose real part is positive. This zero meets the following solution:

$$\varphi_{z_0}(x,\mu) = \exp\left(-\frac{w_0 x}{z_0}\right)\frac{E_0}{w_0}\frac{z_0\mu - \eta_1^2}{z_0 - \mu}, \qquad e_{z_0}(x) = \exp\left(-\frac{w_0 x}{z_0}\right)E_0.$$

3. The dependence of the Debye mode from the problem parameters

The root of the dispersion equation corresponding to the Debye mode z0 can be computed as follows:

$$z_0^2 = z_1^2 + \frac{\Lambda(z_1)}{\Lambda_\infty X(z_1) X(-z_1)}, \tag{2}$$

where z_1 is any point selected from the region $D^+ \backslash \gamma_\varepsilon$, $\Lambda_\infty = 1 - 1/w_0 + \omega_p^2/(\nu\, w_0)^2$.

The function X(z) has the following form:

$$X(z) = \frac{1}{z}\exp\left(\frac{1}{2\pi i}\int_0^{+\infty}\frac{\ln G(t) - 2\pi i}{t - z}dt\right)$$

It is interesting to see what happens for high chemical potentials when increasing the frequency of the external fields interspersed region D+ and D-.

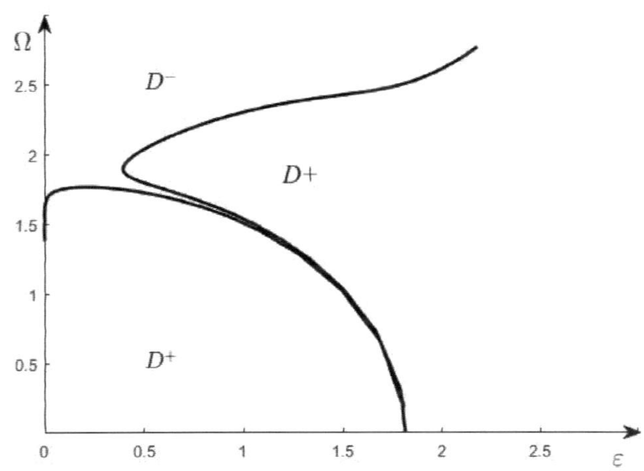

Fig. 3 Curve L with α=50.

Figure 3 represents a curve L when $\alpha=50$. It is possible see how narrow the interval of values of the frequency of the external field, where the discrete root will disappear, and then appear with increasing Ω. According to the formula (2) it is possible to find the value of roots for each part of the region D^+. The results of the calculations are depicted in figure 4.

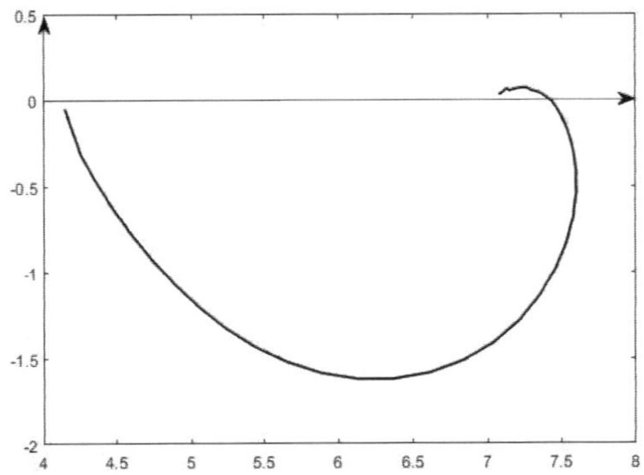

Fig. 4. The locus of point, which is the root of the dispersion function with $\alpha=50$ and the collision frequency of electrons in the plasma, equal to ω_p.

The graph for this case is depicted in figure 4. The left end of the loop corresponds to movement from zero in the direction of increasing the frequency of the external field. When approaching the boundary of a domain D+ graph of root approaches the real axis, the root becomes part of a continuous spectrum. Moving further in the field of D^- the root is not observed. If it enters again in the area D^+ loop continues to move in the first quarter of the complex plane. When approaching the border of the D^+ graph of root again "sits" on the axis. Moreover, the curve looks smooth, as if it was not part of the D^- region. This is due to the narrowness of the transition region D^- between two areas of existence of Debye mode.

4. Conclusion

The algorithm of finding of discrete mode (Debye mode) at screening the external alternative electric field presents. The efficiency of this algorithm is demonstrated by finding Debye mode for the case of the electron gas with a high degree of degeneration. It is shown that the Debye mode shows in this case the unusual behavior. This fact may be important for describing the interaction of electromagnetic radiation with plasma.

References

[1] Vlasov A A 1938 "On vibration properties of electron gas" *J. Phys. USSR* **8** pp 291–318
[2] Landau L D 1946 "On the vibrations of the electronic plasma [in Russian]" in: *Collected Works*, Vol. **2** Nauka, Moscow (1969) pp 7–25; "On the vibrations of the electronic plasma [in Russian]" *J. Phys. USSR* **10** 25–34
[3] Latyshev A V and Yushkanov A A 1992 "Analytic solution of boundary-value problems for nonstationary model kinetic equations" *Theor. Math. Phys.* **92** pp 782–790
[4] Latyshev A V and Yushkanov A A 2006 "Plasma in a high-frequency electric field with a reflective condition on the boundary" *Fluid Dynamics* **41** pp 161–172
[5] Latyshev A V and Yushkanov A A 1993 "Analytic solution of the problem of the behavior of an electronic plasma in the half-space of a metal in a variable electric field [in Russian]" *Surface: Phys. Chem. Mech.* No **2** 25–32
[6] Latyshev A V and Yushkanov A A 2006 "Degenerate plasma in a half-space under an external electric field" *Theor. Math. Phys.* **147** pp 854–867
[7] Latyshev A V and Yushkanov A A 2008 "Analytic solution of the problem of the behavior of a

degenerate electronic plasma [in Russian]" in: *Encyclopedia of Low-Temperture Plasma: Ser. B* (V E Fortov, ed.) Vol. **7-1**, *Mathematical Modeling in a Low-Temperature Plasma* Yanus-K Moscow pp 159–177

[8] Gordeeva N M Yushkanov A A 2018 "Features of the Debye Mode in an Electron Plasma at Various Degrees of the Electron Gas Degeneracy" Technical Physics Letters **44** No 12 pp 1184–1187

[9] Gordeeva N M Yushkanov A A 2018 " On behavior peculiarity of electron plasma" Journal of Physics: Conf. Series **996** 012009

[10] Latyshev A V and Yushkanov A A 1998 "Nonstationary boundary problem for model kinetic equations at critical parameters" *Theor. Math. Phys.* **116** pp 978–989

Journal of Physics: Conference Series **1392** (2019) 012008 doi:10.1088/1742-6596/1392/1/012008

Development of combined search methods for efficiency indicator extreme in variation statement of forecasting tasks for determine characteristics of composite materials

E L Gusev[1,2], V N Bakulin[3,4], V D Chernykh[5]

[1]Institute of Oil and Gas Problems of Siberian Branch of Russian Academy of Sciences, Oktyabrskaya str. 1, Yakutsk, 677980, Russia
[2]Institute of Mathematics and Informatics of the North-East Federal University, Kulakovskogo str.48, Yakutsk, 677013, Russia
[3]Moscow Aviation Institute (National Research University), Volokolamskoye highway 4, Moscow, 125993,Russia
[4]Institute of Applied Mechanics of the Russian Academy of Sciences, Leningradsky avenue 7, Moscow, 125040, Russia
[5]Institute of Physics and Technology of the North-Eastern Federal University, Kulakovskogo str.48, Yakutsk, 677013, Russia

E-mail: [1,2]elgusev@mail.ru

E-mail: [3,4]vbak@yandex.ru

Abstract. Modified search methods for global extreme of multiparameter functions are studied, considering the structural features of efficiency criteria used in problems of forecasting residual life and durability of structures made of composite polymeric materials. Comparative and qualitative studies have been carried out and methods have been developed to select the most promising areas of search for the absolute minimum of the studied performance indicators in refined variation formulation of inverse problems of forecasting residual life of composites.

1. Introduction

In recent decades, one of the most important tasks in development of various structures, machines and mechanisms has been creation of reliable methods for quantifying performance of products from polymeric and composite materials [1-13]. Specificity of studied problems of forecasting strength change of polymer composites affected by extreme climatic factors leads to the fact that extrapolation approach in traditional formulation is not effective for these problems. Conceptual improvement of extrapolation approach can be achieved based on choice of a mathematical model with a focus on physical concepts.

Residual life of composite structure can be considered as its defining characteristic.

2. Problem statement

Models that describe dependence of change in one of defining characteristics (for example, strength R of polymer composites) affected by climatic factors generally form can be represented as functional dependencies of following form:

$$R(u_1, u_2, \ldots, u_n; t) = R_0 + \Phi(u_1, u_2, \ldots, u_n; t) \quad (1)$$

We denote $\Delta\overline{R}_t^3$ as experimental mean values of function $\Delta R = R - R_0$ on experimental samples after t-year ($t = 1, 2, ..., m$). Parameters of optimal forecasting model $u^* = (u_1^*, u_2^*, \ldots, u_m^*)$ deliver extreme value to special criterion of efficiency J, which is a complex function of forecasting problem parameters:

$$J(u) = J\left(\{\Delta R(u,t)\}_{t=1}^m ; \{\Delta\overline{R}_t^s\}_{t=1}^m\right) \quad (2)$$

In the case when experimental data adequately reflect change dependence structure in determining characteristics of composite materials (CM) on impact of extreme environmental factors and operational loads, and experimental data were obtained with minor errors that do not distort behavior patterns of real dependencies, task of restoring parameters of CM models from effects of extremely environmental factors can be reduced to finding absolute minimum of quality function (2) in extreme problem:

$$J(u^*) = \min_u J(u) \quad (3)$$

Task of restoring parameters of models of changes in residual life of a composite structure affected by extreme environmental factors can be reduced to solving an extreme problem (3). Vector of parameters $u^* = (u_1^*, u_2^*, \ldots, u_n^*)$ that delivers minimum value to performance indicator J(u) (3), determines optimal forecast dependence of change in defining characteristics of CM on impact of extreme environmental factors and operational loads.

3. Features of inverse problems of predicting the residual life of composites in the variation formulation

Characteristic feature of inverse problems of forecasting residual resource of composites in a variation formulation with complex efficiency indicators of form (2) is that structure of efficiency indicators has a substantially multi-extremum nature, which does not reliably allow finding absolute minimum of such performance indicators using well-known methods finding a minimum. Simultaneously, conducted studies and corresponding computational experiments (CE) showed that usage of various known methods of finding a minimum can lead to substantially different results. Simultaneously, it is impossible to assess how significantly resulting solutions differ from required global optimal solutions that deliver global minimum to efficiency indicator. Consequence of ineffectiveness of existing approaches to solving inverse problems of forecasting composites in a variation formulation is that predicted solutions of residual resource may differ significantly from actual values of residual resource of composites, even when forecasting for a short period and even when developed predictive mathematical model of durability has a high adequacy of real situation. Nevertheless, conducted studies, CE showed that performance indicators in inverse problems of forecasting residual resources in variation formulation have several specific features. Accounting for these features allow developing solving inverse problems of forecasting residual life of composites in variation formulation more effective methods for finding absolute minimum compared to known approaches, as well as modifying known minimization methods taking into account specific features of considered performance indicators.

4. Combined methods of search of extreme, of performance indicators in the refined variation formulation of the task of remaining life prediction of composites

In [7], method for searching extreme of multiparameter functions with an optimal choice of parameters for solving reverse problems of forecasting strength change of polymer composites affected by extreme climatic factors and operational loads was proposed. Technique has been developed for optimal choice of parameters, which makes it possible to increase efficiency of using gradient methods for solving inverse problem of forecasting determining characteristics of polymer composites and to give them a non-local character. For tasks in which absolute minimums of efficiency indicators are located on directions along which their most significant decrease is observed, proposed approach will be sufficiently effective. However, in the most common situations related to solving forecasting problems, dependence of performance indicators J (u) on determining parameters has rather complicated structure, in which absolute minimum may not be along directions calculated in this way. In [9-10] to solve a more complex range of inverse problems of forecasting changes in strength of polymer composite structures affected by extreme climatic factors of Far North, Arctic zone, described by models in which dependence of efficiency indicators J (u) on determining parameters - parameters u1, u2, ..., on have a more complicated structure, modifications of methods of possible directions with definition of optimal search directions were developed. Structure of modified methods of possible directions includes such additional features as additional procedures for generating a significant number of possible search directions, given by unit vectors $\vec{l} \in \Omega = \overrightarrow{\{l\}}$, whose angles $\varphi_1, \varphi_2, \ldots, \varphi_n$ with coordinate axes represent set of random variables with a uniform distribution law; additional procedures for comparative evaluation of prospects of selected possible directions according to specially constructed criteria $W(\vec{l})$ and finding optimal search direction. Structure of modification of methods of possible directions includes such additional elements that increase their efficiency, such as global optimization of efficiency parameter along the chosen conditionally optimal search direction $l_p^{*'}$; constructing next $(p+1)$ -th approximation to global optimal solution associated with decomposition of efficiency criteria $J(u)$ in Taylor series in vicinity of pth approximation along conditionally optimal search direction l_p^*:

$$J(u) = J\left(u^p\right) + \frac{\partial J(\vec{u}^p)}{\partial l_p^*}\left(u - u^p\right) + \frac{1}{2}\left(u - u^p\right)^T \frac{\partial^2 \vec{J}(\vec{u}^p)}{\partial l_p^* \partial l_p^*}\left(u - u^p\right) + \ldots$$

$$\frac{\partial J(\vec{u}^p)}{\partial l} = \left(\nabla J\left(u^p\right), \vec{\tau}\right) = \sum_{i=1}^{n} \frac{\partial J(u^p)}{\partial u_i}\alpha_i = \sum_{i=1}^{n} \frac{\partial J(u^p)}{\partial u_i}\cos\varphi_i$$

$$\frac{\partial^2 J(\vec{u}^p)}{\partial l^2} = \sum_{i=1}^{n}\sum_{k=1}^{n} \frac{\partial^2 J(u^p)}{\partial u_i \partial u_k}\alpha_i\alpha_k = \vec{\tau}^T \nabla^2 J\left(u^p\right)\vec{\tau}, \quad p = 0, 1, 2, \ldots$$

5. Promising ways to increase the potential in improving the efficiency of methods for predicting the determining characteristics of composites

Question about ways of further effective expansion of potential possibilities in increasing forecasting methods efficiency for solving problems of estimating residual life, reliability, durability of CM and structures for extreme climatic conditions of Far North and Arctic zone is investigated. One of the promising ways to increase potential opportunities is associated with improvement of methods for constructing optimal search directions for absolute minimum of performance indicators (2). And methods outlined in [9-10], which form the basis of combined approach, comparative qualitative analysis of methods of choosing the most promising search directions for absolute minimum of studies performance indicators and mutually conjugate search directions was conducted. Advantages of search methods for absolute extreme of performance indicators, when mutually conjugate directions are selected as promising areas of search, are analyzed. Analysis showed that effectiveness of conjugate directions' method application for solving problems of forecasting residual life of composites is caused by following design features, which make it possible to adequately consider specific features of considered problems of forecasting composites affected by extreme environmental factors.

In methods of combined directions, sufficiently successful selection of next promising search directions is carried out, which quite effectively takes into account specificity of performance indicators in tasks of forecasting residual life of composites. In methods of conjugate directions, next search directions selected as promising, are characterized by property that if vector u^* is a minimum point of performance indicator $J(u)$ (2) along direction ϵ , and vector u^{**} is minimum point of performance indicator $J(u)$, obtained from a different starting point along direction ξ parallel to vector $(u^{**} - u^*)$ will be G -conjugate to direction ξ [14]. Established important properties of G-conjugate directions are applied in studies inverse problems of forecasting residual resource of composites to construct the most promising search directions for absolute minimum of complexly constructed efficiency indicators of form (2).

6. Summary

Conducted research and development of known methods for finding a minimum of multiparameter functions, a comparative assessment of their effectiveness, development of recommendations for their optimal layout and effective design modification in accordance with specific features of considered performance indicators in inverse problems of forecasting residual resource of variation formulation can significantly expand potential for improving effectiveness of forecasting residual life methods, durability and CM structures affected by extreme environmental factors specific to Far North and Arctic areas.

References

[1] Urzhumtsev Y S, Chersky I N 1985 Scientific bases of engineering climatology of polymeric and composite materials *Mech. of comp. mat.* **4** 708-714

[2] Bolotin V V 1984 *Forecasting resource machines and structures.* (Moscow: Mechanical Engineering) 312

[3] Vasiliev V V, Protasov V D, Bolotin V V and other 1990 *Composite materials.Handbook.* (Moscow: Mechanical Engineering) 510

[4] Bakulin V N, Gusev E L, Markov V G 2008 *Methods of optimal design and calculation of composite structures.* vol.1 (Moscow: Fizmatlit) 256

[5] Bakulin V N, Gribanov V M, Ostrik A V and others 2008 *Methods of optimal design and calculation of composite structures.* vol.2 (Moscow: Fizmatlit) 256

[6] Bakulin V N, Ostrik A V 2015 *Complex action of radiation and particles on thin-walled structures with heterogeneous coatings* (Moscow: Fizmatlit) 280

[7] Gusev E L, Babenko F I 2015 Methods of finding the extremum of the optimal choice of parameters for solving the task of recovering the parameters of the prediction models defining characteristics of polymer composites *Proc. papers "Materials for technical devices and structures applied in the Arctic"* (Moscow: VIAM) 132-141

[8] Gusev E L 2016 Using parallel procedures for the searching of the extremum for the decision of the inverse problems prediction of the defining characteristics of the composite materials *Mat. Phys. and Mech.* **26** 70-72

[9] Gusev E. L. 2015 Relaxation methods in predicting the determining characteristics under the influence of extreme environmental factors *Int. J. of the Int. Union of Scien. "Scien. Techn. Prod."* **8** 4-7

[10] Gusev E L, Bakulin V N, Chernykh V D 2017 Promising ways of expanding the potential of predicting the durability of composite materials under the influence of extreme climatic factors *Proc Int. Conf. on Comp. Mech. and Mod. Appl. Soft Sys.* (Moscow: MAI-PRINT) 224-226

[11] Gusev E L, Bakulin V N 2017 Methods for solving inverse problems of predicting the residual life of structures made of composite materials under the influence of extreme climatic factors of the environment in the refined statements *Proc. of the 7th All-Russ. Sc. Conf. with Int.l Part."Mech. of Comp. Mat. and Str., Com. and Het. Med."* (Moscow: IPRIM RAS) 68-70

[12] Gusev E L, Bakulin V N 2018 Refined formulation of problems of predicting the residual life of composites under the influence of extreme environmental factors and methods of their solution *Proc. of the XII Int. Conf. on App. Math. and Mech. in the Aer.Ind.* (Moscow: MAI) 363-365

[13] Gusev E L, Bakulin V N 2017 Mathematical methods of optimum designing of the structure of the composite materials with the required complex of the properties in variations statement*Proc. of the XXVII Int. Conf. "Math. and Comp. Sim. in Mech. of Sol. and Str.* (St. Petersburg) 72-73

[14] Batishchev DI 1984 *Methods of optimal designing* (Moscow: Radio and communication) 248

About orientational instability in nematic liquid crystal films

A G Kalugin, D V Pavlov

Mechanics and Mathematics Faculty, Lomonosov Moscow State University, Leninskiye Gory 1, Moscow, 119991, Russia

E-mail: kalugin@mech.math.msu.su

Abstract. In the presented work the role of the surfacelike (splay-bend) elastic constant in the Frank's energy of nematic liquid crystals for orientational instability in the thin nematic films is considered. It is shown that nontrivial periodical perturbations exist in the nematic film with undisturbed planar director orientation in case of sufficiently large value of splay-bend constant. For these solutions the relations between splay-bend constant, another Frank's constants and layer thickness are received and studied. These correlations allow us to make new explanation for certain experimental effects and to present method for surfacelike constant value measurement.

1. Introduction

The anisotropic properties of nematic liquid crystals (NLC) are described by unit vector $\vec{n}$ (director). In case of Frank-Oseen-Leslie model the Frank's volume free energy has the form

$$2F_V = K_{11}(\mathrm{div}\vec{n})^2 + K_{22}(\vec{n}\cdot\mathrm{rot}\vec{n})^2 + K_{33}|[\vec{n}\times\mathrm{rot}\vec{n}]|^2 + K_{24}\left(\nabla_i n_j \nabla^j n^i - (\nabla_k n^k)^2\right) \quad (1)$$

where K_{ii} are the Frank's constants [1]. On account of nematic symmetry only four independent constants exist, but discussion about Frank's energy expression has a long history [2,3]. The last term in (1) has the divergent form and does not bring the contribution in the equations of continuum motion and director evolution or equilibrium, but it is necessary to take it into account together with the surface energy [4] in the boundary conditions in case of director weak anchoring on the nematic-isotropic interface. The Frank's energy is the positive definite function if K_{ii} satisfy the Ericksen inequalities $2K_1 \geq K_{24} \geq 0$, $2K_2 \geq K_{24} \geq 0$, $K_3 \geq 0$ [1], but some nematics does not satisfy these conditions [5].

Scientific interest to the role of surfacelike part of Frank's energy arose after Oldano-Barbero paradox (continuous solution for director field in the nematic layer does not exist for some boundary conditions) was discovered [6,7]. Instability induced by surfacelike constant K_{24} in the cylindrical pore is studied in [8,9], in the surface waves – in [10], works [11,12] deal with the periodical solutions in the hybrid aligned nematic sell. In the presented work we discuss the periodical solutions as perturbation of the homogeneous director field in the NLC film.

2. Formulation of the problem on equilibrium of NLC film

For the problem of nematic equilibrium the equations governing the director field in the absence of external body forces and the weak anchoring take the form [13]

$$\left(\delta^i_q - n^i n_q\right)\left(\frac{\partial F_V}{\partial n^q} - \nabla_k\left(\frac{\partial F_V}{\partial \nabla_k n^q}\right)\right) = 0 \tag{2}$$

$$\left(\delta^j_k - n^j n_k\right)\left(\frac{\partial F_V}{\partial \nabla_i n^j} m_i + \frac{dF_S}{dn_m} m_j\right) = 0 \tag{3}$$

For surface energy Rapini–Papoular model is used [4]

$$F_S = \gamma + \frac{W}{2}(1 - (\vec{n}\cdot\vec{m})^2) = \gamma + \frac{W}{2}(1 - (\sin\Omega\sqrt{1-n_\nu^2} + \cos\Omega|n_\nu|)^2) \tag{4}$$

where $n_m = \vec{n}\cdot\vec{m}$, $\vec{m}$ is the unit outward normal to the boundary, γ, W, and Ω are constants, and Ω is the angle between the light orientation axis and the normal to the surface. In case of planar orientation for director on the boundary $\Omega = \pi/2$.

We consider the NLC film with undisturbed homogeneous director field $\vec{n} = (0, 1, 0)$, because this configuration is the most unstable for perturbation in the xz plane [14]. In Cartesian coordinates this film is bounded by planes $z = \pm h$. In the disturbed state the director is given as follows: $\vec{n} = (-\sin\varphi\cos\theta, \cos\varphi\cos\theta, -\sin\theta)$, where φ and θ are the angles of deviation from the initial direction of the director. Then the linearized equilibrium equations (2) for perturbations of the angles φ and θ, which were obtained analogously in [15] for another initial state of the director, can be written in the form ($f_z = \partial f/\partial z$)

$$K_{22}\theta_{xx} + K_{11}\theta_{zz} = (K_{22} - K_{11})\varphi_{xz} \tag{5}$$

$$K_{11}\varphi_{xx} + K_{22}\varphi_{zz} = (K_{22} - K_{11})\theta_{xz} \tag{6}$$

We will not take into account the boundary perturbations, thus the boundary conditions (3) can be reduced to the relations at $z = \pm h$ [16]

$$K_{22}\,\varphi_z = \theta_x(K_{22} - K_{24}) \tag{7}$$

$$K_{11}\,\theta_z \pm W\theta = \varphi_x(K_{24} - K_{11}) \tag{8}$$

Condition (7) can be applied to both boundaries. In condition (8) the upper and lower signs relate to upper and lower boundaries, respectively.

3. Nontrivial periodic solutions for perturbations

Without loss of generality we will seek the solutions of equations (5), (6) in the form: $\theta = f(z)\sin kx$, $\varphi = g(z)\cos kx$. As a result of calculations these one can be written as follows

$$\theta = [(A_1 + A_2 z)\exp(kz) + (A_3 + A_4 z)\exp(-kz)]\sin kx$$

$$\varphi = [(B_1 + B_2 z)\exp(kz) + (B_3 + B_4 z)\exp(-kz)]\cos kx$$

where A_i, B_j are constant. By virtue of (5), (6), they are connected by the relations

$$B_2 = A_2\,, \qquad kB_1 - kA_1 = \lambda B_2\,, \qquad B_4 + A_4 = 0 \tag{9}$$

Journal of Physics: Conference Series **1392** (2019) 012009 doi:10.1088/1742-6596/1392/1/012009

$$kB_3 - kA_3 = \lambda A_4\,, \qquad \lambda = \frac{K_{11} + K_{22}}{K_{11} - K_{22}} \tag{10}$$

The boundary conditions (7), (8) together with the relations (9) and (10) make it possible to write a linear homogeneous system for the coefficients A_i and B_i. The nontrivial solutions for perturbations exist if its determinant is equal to zero. As a result we receive the equations

$$\pm 2\left(\frac{1}{K_{11}} - \frac{1}{K_{22}}\right) hK_{24}^2 k^2 + 2W(ch(2kh) \pm 1) = kK_{24}\left(K_{24}\left(\frac{1}{K_{11}} + \frac{1}{K_{22}}\right) - 4\right) sh(2kh) \tag{11}$$

which permit us to find the wave number k as function of K_{ii}, W and h.

We will determine that NLC film is unstable if solutions of (11) exist for $k > 0$. If $K_{24} = 0$, positive k as root of (11) does not exist and film is stable. As a result of analysis of equation (11) then $K_{24} \neq 0$ we can distinguish three domain of splay-bend constant value (all calculations below are made for MBBA parameters, then $K_{11} = 6$ pN, $K_{22} = 4$ pN and $W = 4 \cdot 10^{-5}$ N/m if it is not specified separately), which are corresponding to the Ericksen inequalities.

1) When $0 \leq K_{24} \leq 2K_{11}$ and $0 \leq K_{24} \leq 2K_{22}$ or Ericksen's inequalities for K_{24} [1] are satisfied and $F_V \geq 0$. In this case positive solutions $k = k(h)$ in (11) are absent and NLC film is stable.

2) When $2K_{22} < K_{24} < 2K_{11}$. In this case we can find positive solutions $k(h)$, orientational instability appears. But for some value of K_{24} in this zone function $k(h)$ is defined only for h, which is less than certain critical value h_0. Also h_0 is small and orientational instability exists in very thin films. This result is well compatible with experiments in work [17]. On the Fig. 1 the graph for period $L = 2\pi/k$ of received solutions as function of film thickness $2h$ is shown. Calculations are made for $K_{24} = 8.8$ pN. Graph on the Fig. 2 ($K_{24} = 10$ pN) shows the transition state from the instability in the thin film only to one in the film of arbitrary thickness. Also for this K_{24} value solutions of (11) coincide if $2h > 0.2\mu$m.

3) If $K_{24} > 2K_{22}$ and $K_{24} \geq 2K_{11}$ function $k(h)$ is defined for the films with arbitrary thickness. On the Fig. 3 the graph for $L(2h)$ is shown. Calculations are made for $K_{24} = 14$ pN. We can see, that two solutions of equation (11) coincide for $2h > 3\mu$m.

Also we need to emphasize that only domain 1 for K_{24} can be found precisely, because it is determined by condition, that F_V is positive definite function, only. Properties of solutions in domains 2 and 3 also depend from the anchoring coefficient W. On the Fig. 4 we demonstrate this fact. There are period L as function of K_{24} for some values of W is shown for typical experimental size of NLC films and layers. One can see on this graph, for $2h = 10 \div 500$ μm period L practically does not depend from film thickness. Both solutions of (11) in this case coincide in the investigated range of $2h$ similarly previous values of surfacelike constant.

Also if $W = 0$ period L is linear function of $2h$ and orientational instability exists for $K_{24} = 8 \div 12$ pN. Thus, if anchoring is absent, director field instability appears in case 2 only. Measurements of the splay-bend elastic constant in [5] give us that K_{24} can fall outside the specified limits and this type of the director instability is rarely observed for shear flows, when we can assume the zero anchoring on the NLC layer boundary [18–20].

4. Conclusions

In the presented work we investigate orientational instability in the NLC films with undisturbed homogeneous planar orientation of director. If we take into account divergent part in the Frank energy and the K_{24} constant is rather great, orientational instability in the nematic film is possible. This effect appears in the nontrivial periodical solutions, which exist under the action of uniform external forces and boundary conditions. Also this type of instability is occur if Ericksen inequalities for splay-bend constant are not fulfilled.

In case of orientational instability two possibilities exist. In the first one very thin films are instable only, in the second orientational instability does not depend on the film thickness. It is shown relation between two types of instabilities and Ericksen inequalities.

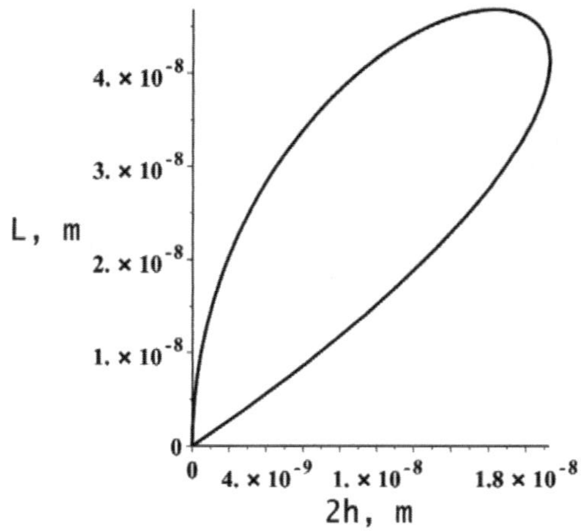

Figure 1. Period L as function of film thickness, case 2, $K_{24} = 8.8$ pN.

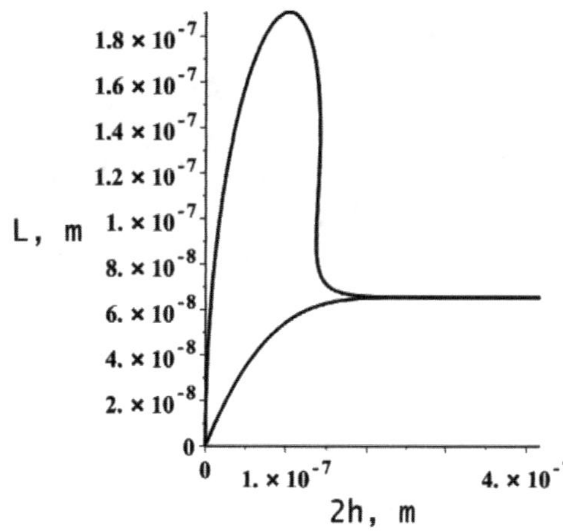

Figure 2. Period L as function of film thickness $2h$, transition from case 2 to 3, $K_{24} = 10$ pN.

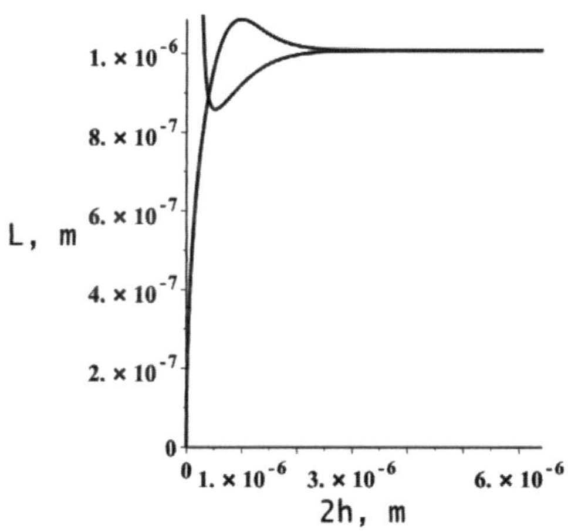

Figure 3. Period L as function of film thickness $2h$, case 3, $K_{24} = 14$ pN.

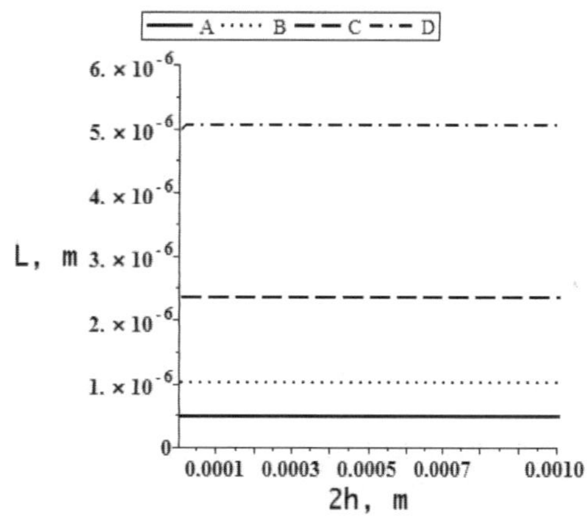

Figure 4. Period L as function of film thickness $2h$, case 3. Lines A and C are constructed for $K_{24} = 12$ pN, B and D – for $K_{24} = 14$ pN; lines A and B are constructed for $W = 40$ μN/m, C and D – for $W = 8$ μN/m.

In addition we analyze how the period of the received nontrivial periodical solutions depends on the anchoring coefficient. It is shown that for zero anchoring instability exists if splay-bend constant is bounded above and below.

This model of director field instability permit us to explain some experimental facts. Also we can calculate relation between period of received periodical solutions and film thickness. This function can be used for splay-bend constant K_{24} measurement.

References

[1] Ericksen J L 1966 *The Physics of Fluids* **9** 1205
[2] Barbero G and Strigazzi A 1984 *J. Phys. Lett.* **45** L-857
[3] Hinov H P 1987 *Mol. Cryst. Liq. Cryst.* **148**(1) 197
[4] Rapini A and Papoular M 1969 *J. Phys. (Paris) Colloq.* **30**(C4) 54
[5] Polak R D, Crawford G P, Kostival B C, Doane J W and Zumer S 1994 *Phys. Rev. E* **49** R978
[6] Barbero G and Oldano C 1985 *Il Nuovo Cimento D* **6** 479
[7] Oldano C and Barbero G 1985 *J. Phys. Lett.* **46**(10) L-451
[8] Kiselev A and Reshetnyak V 1995 *Proceedings of SPIE* **2408** 167
[9] Kiselev A and Reshetnyak V 1995 *Functional Materials* **2** 240
[10] Golubyatnikov A N and Kalugin A G 2001 *Vestnik Moskovskogo Universiteta, Matematika Mekhanika* **1** 58
[11] Strigazzi A 1987 *Molec. Cryst. and Liquid Cryst. Incorp. Nonlin. Optics* **152**(1) 435
[12] Sparavigna A and Strigazzi A 1992 *Molec. Cryst. and Liquid Cryst. Science and Techn. Sect. A. Molec. Cryst. and Liquid Cryst.* **221**(1) 109
[13] Kalugin A G and Golubyatnikov A N 1998 *Proc. of the Steklov Institute of Mathematics* **223** 168
[14] Kalugin A G 2013 *Moscow Univ. Mech. Bull.* **68** 32
[15] Alexe-Ionescu A L, Barbero G and Lelidis I 2002 *Phys. Rev. E* **66** 061705-1
[16] Kalugin A G 2016 *Fluid Dynamics* **51**(1) 81
[17] Delabre U, Richard C and Cazabat A M 2009 *Phys. Chem. B* **113**(12) 3647
[18] Golovanov A V, Kaznacheev A V and Sonin A S 1998 *Izv. RAN, Ser. Fiz.* **62**(8) 1658
[19] Pasechnik S V, Shmeleva D V, Tsvetkov V A, Krekhov A P and Nasibullaev I Sh 2005 *J. Exp. Theor. Phys.* **100** 804
[20] Kalugin A G 2016 *Moscow Univ. Mech. Bull.* **71** 142

Aeroelastic stability of a shell supported by a cylinder with a hole of linearly varying radius

M A Konopelchev[1], V N Bakulin[1,2] and A Ya Nedbai[3]

[1] Moscow Aviation Institute, Volokolamskoe shosse 4, Moscow, 125993 Russia
[2] Institute of Applied Mechanics of the Russian Academy of Sciences, Leningradsky prt 7, Moscow, 125040 Russia
[3] Moscow Institute of Thermal Technology, Berezovaya al 10, Moscow, 127237 Russia

E-mail: konopelchevm@gmail.com

Abstract. The aeroelastic stability of a shell supported by an elastic cylinder of varying thickness with the outer surface exposed to supersonic gas flow is studied. The supporting cylinder is considered as an elastic foundation, whose bedding value is determined from the equations of the three-dimensional theory of elasticity. The solution is obtained by the Bubnov-Galerkin method in the form of a trigonometric series in the axial coordinate. The characteristic equation is approximated by the Lagrange polynomial, whose stability is investigated with the use of the Routh-Hurwitz criterion. The dependences of the critical flow velocity on the length of the cylinder section with the tapered hole are shown for various values of the axial force, the damping decrement, and the shell thickness.

1. Introduction

The problem of flutter and dynamic stability of composite shells supported by an elastic cylinder was studied in a number of papers [1–10]. However, the flutter of shells supported by a cylinder with a hole of varying radius remains unexplored.

2. Formulation of the problem

In this paper, we investigate the stability of a cylindrical shell made of composite material supported by an elastic cylinder. The cylinder has a coaxial hole consisting of a tapered and a cylindrical section. The shell is loaded with axial forces and external supersonic gas flow. The ends of the shell are simply supported. The problem geometry is shown in figure 1.

The motion of the shell is described by the equation of the orthotropic shell theory. The elastic cylinder is considered as a Winkler foundation, whose bedding value is determined from the solution of the equations of the three-dimensional elasticity theory. The inertial properties of the cylinder are taken into account by the use of the Rayleigh method [11].

We introduce a cylindrical coordinate system nondimensionalized by the radius of the shell.

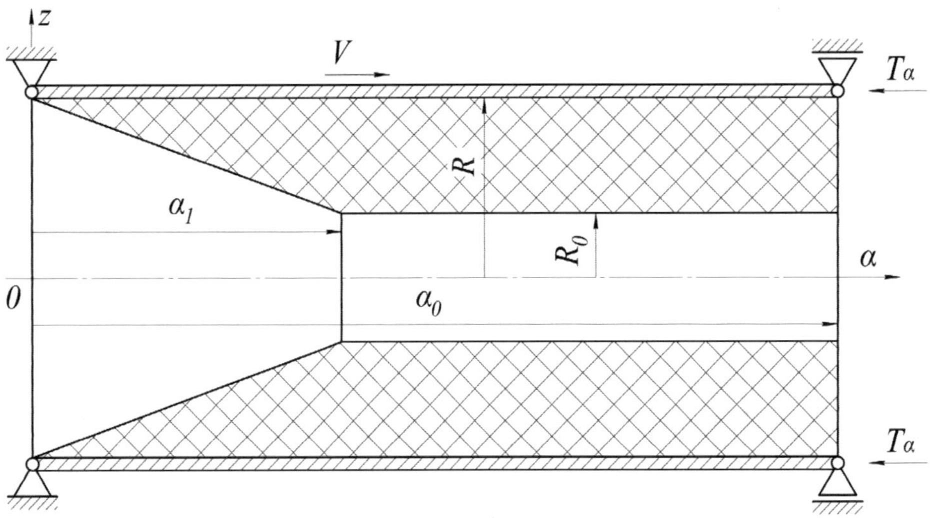

Figure 1. Problem geometry.

Then, the equation of motion of the shell can be written as [12]

$$\left\{a_3\nabla^8 + a_2\frac{\partial^4}{\partial\alpha^4} + \nabla^4\left(a_5\frac{\partial^2}{\partial\alpha^2} + a_8\frac{\partial}{\partial\alpha} + a_9\frac{\partial}{\partial t}\right) + \right.$$
$$\left. + \nabla^4 B\sum_{j=1}^{2}\left(K_j + a_{7j}\frac{\partial^2}{\partial t^2}\right)\left[\sigma_0\left(\alpha-\alpha_{j-1}\right) - \sigma_0\left(\alpha-\alpha_j\right)\right]\right\}w = 0, \quad (1)$$

where ∇^8, ∇^4 are differential operators of the form

$$\nabla^8 = a_1\frac{\partial^8}{\partial\alpha^8} + \left[a_4 + a_6\left(2a_1-\nu_\beta\right)\right]\frac{\partial^8}{\partial\alpha^6\partial\beta^2} +$$
$$+\left\{a_2\left(\frac{\partial^2}{\partial\beta^2}+1\right)^2 + \left[2a_6\left(a_4-a_6\nu_\beta\right)+a_2\right]\frac{\partial^4}{\partial\beta^4}\right\}\frac{\partial^4}{\partial\alpha^4} +$$
$$+ a_4\left[\left(a_4-a_6\nu_\beta\right)\left(\frac{\partial^2}{\partial\beta^2}+1\right)^2 + 2a_1a_6\frac{\partial^4}{\partial\beta^4}\right]\frac{\partial^4}{\partial\alpha^2\partial\beta^2} + a_2a_4\left(\frac{\partial^2}{\partial\beta^2}+1\right)^2\frac{\partial^4}{\partial\beta^4};$$

$$\nabla^4 = a_1\frac{\partial^4}{\partial\alpha^4} + \left(a_4-a_6\nu_\beta\right)\frac{\partial^4}{\partial\alpha^2\partial\beta^2} + a_2\frac{\partial^4}{\partial\beta^4};$$

$$a_1 = \frac{G_{\alpha\beta}\left(1-\nu_\alpha\nu_\beta\right)}{E_\alpha}; \quad a_2 = a_1a_4; \quad a_3 = \frac{h^2}{12R^2}; \quad a_4 = \frac{E_\beta}{E_\alpha}; \quad a_5 = \frac{BT_\alpha}{2\pi R^3}; \quad a_6 = 2a_1+\nu_\beta;$$
$$a_{71} = h\left(\frac{\alpha}{\alpha_1}\mu+\rho\right); \quad a_{72} = h\left(\mu+\rho\right); \quad a_8 = \frac{B\varkappa_0 p_0 M}{R}; \quad a_9 = B\left(\frac{\varkappa_0 p_0}{C_0}+\varepsilon\right);$$
$$B = \frac{R^2\left(1-\nu_\alpha\nu_\beta\right)}{E_\alpha h}; \quad K_1 = \frac{\alpha}{\alpha_1}K_0; \quad K_2 = K_0; \quad \mu = \frac{R^2-R_0^2}{6Rh}\rho_c;$$
$$C_0 = \sqrt{\frac{\varkappa_0 p_0}{\rho_0}}; \qquad M = \frac{V}{C_0}; \qquad \varepsilon = \frac{\delta\rho h}{\pi R}\sqrt{\frac{E_\alpha}{\rho}};$$

α and β denote the axial and circumferential nondimensional coordinates; w the radial shell displacement; R and h the radius and the thickness of the shell; R_0 the radius of the cylindrical part of the hole; E_α and E_β the axial and circumferential moduli of elasticity; $G_{\alpha\beta}$ the transverse shear modulus; ν_α and ν_β Poisson's ratios; K_0 the bedding value of the elastic cylinder section with the cylindrical hole; ρ and ρ_c the shell and the cylinder density; δ the damping decrement; $\varkappa_0$, p_0 and ρ_0 the polytropic index, pressure, and speed of sound in the freestream; M the Mach number; V the air velocity; α_1 the length of the tapered part of the hole.

3. Solution of the equation of motion

A solution of equation (1) is assumed of the form

$$w = \cos n\beta \sum_{m=1}^{\infty} A_m \sin \gamma_m \alpha \, e^{\omega t}, \tag{2}$$

where $\gamma_m = m\pi/\alpha_0$; $\alpha_0 = L/R$; L is the length of the shell; m the number of axial half-waves; n the number of circumferential full waves; ω the complex frequency.

Substituting (2) into (1) and using Galerkin's procedure, we obtain an infinite system of algebraic equations

$$\left(a_{10}\omega^2 + a_9\omega + \psi_k\right) A_k + \sum_{m=1}^{\infty} f_{mk} A_m = 0, \qquad (k = 1, 2, 3, \dots), \tag{3}$$

where $a_{10} = Bh\rho$; $\quad a_{11}^{(m)} = B\left(K_{0m} + h\mu\omega^2\right)$; $\quad \psi_k = \left(a_3\nabla_k^8 + a_2\gamma_k^4\right)/\nabla_k^4 - a_5\gamma_k^2$;

$$f_{mk} = a_{11}^{(m)} \left[\frac{1}{\alpha_1} H_1^{(mk)} + H_2^{(mk)}\right] + a_8 H_3^{(mk)};$$

$$H_1^{(mk)} = \begin{cases} \frac{\alpha_1^2}{2\alpha_0}, & \text{when } m = k_0; \\ \frac{1}{\alpha_0\theta_{mk}^2}\left(\alpha_1\theta_{mk}\sin\theta_{mk}\alpha_1 + \cos\theta_{mk}\alpha_1 - 1\right) - & \\ -\frac{1}{\alpha_0\varphi_{mk}^2}\left(\alpha_1\varphi_{mk}\sin\varphi_{mk}\alpha_1 + \cos\varphi_{mk}\alpha_1 - 1\right), & \text{when } m \neq k; \end{cases}$$

$$H_2^{(mk)} = \begin{cases} \frac{\alpha_0 - \alpha_1}{\alpha_0}, & \text{when } m = k_0; \\ \frac{1}{\alpha_0}\left(\frac{1}{\varphi_{mk}}\sin\varphi_{mk}\alpha_1 - \frac{1}{\theta_{mk}}\sin\theta_{mk}\alpha_1\right), & \text{when } m \neq k; \end{cases}$$

$$H_3^{(mk)} = \begin{cases} \frac{4}{\alpha_0}\frac{mk}{k^2 - m^2}, & \text{if } m \pm k_0 \text{ is odd;} \\ 0, & \text{if } m \pm k \text{ is even;} \end{cases}$$

$$\varphi_{mk} = \pi\left(m + k\right)/\alpha_0; \qquad \theta_{mk} = \pi\left(m - k\right)/\alpha_0; \qquad \nabla_k^4 = a_1\gamma_k^4 + \left(a_4 - a_6\nu_\beta\right)\gamma_k^2 n^2 + a_2 n^4;$$

$$\nabla_k^8 = a_1\gamma_k^8 + \left[a_4 + a_6\left(2a_1 - \nu_\beta\right)\right]\gamma_k^6 n^2 + \left\{a_2\left(1 - n^2\right)^2 + \left[2a_6\left(a_4 - a_6\nu_\beta\right) + a_2\right]n^4\right\}\gamma_k^4 + \\ + a_4\left[\left(a_4 - a_6\nu_\beta\right)\left(1 - n^2\right)^2 + 2a_1 a_6 n^4\right]\gamma_k^2 n^2 + a_2 a_4\left(1 - n^2\right)^2 n^4.$$

For the chosen model of the elastic foundation, the bedding value can be expressed as [12]

$$K_{0m} = \frac{2\mu_0\Delta}{RQ}; \; Q = \sum_{r=1}^{6} F_r D_{6r}; \; F_1 = -\frac{n^2}{\gamma_m} I_n(\gamma_m); \; F_3 = \frac{q_1}{2}\gamma_m\left(\frac{n^2}{\gamma_m^2} + 1\right) I_n(\gamma_m); \; F_5 = -I_n'(\gamma_m),$$

where Δ and D_{6r} are the determinant and a cofactor, respectively, of the matrix a_{ij}:

$$a_{11} = \frac{n^2}{x_m} I_n(x_m); \quad a_{13} = -I_n'(x_m) - q_1\left(\frac{n^2}{x_m^2}+1\right)x_m I_n(x_m); \quad a_{15} = 2I_n'(x_m);$$

$$a_{21} = \left(\frac{n^2}{x_m^2}+\frac{1}{2}\right)I_n(x_m) - \frac{1}{x_m}I_n'(x_m); \quad a_{23} = \frac{q_1}{2}\left[\frac{1}{x_m}I_n'(x_m) - \left(\frac{n^2}{x_m^2}+1\right)I_n(x_m)\right];$$

$$a_{25} = \frac{1}{x_m}\left[I_n'(x_m) - \frac{1}{x_m}I_n(x_m)\right]; \quad a_{31} = \frac{n^2}{x_m}\left[\frac{1}{x_m}I_n(x_m) - I_n'(x_m)\right];$$

$$a_{33} = \frac{q_1}{2}\left[\left(\frac{n^2}{x_m^2}+1\right)x_m I_n'(x_m) - \left(\frac{n^2}{x_m^2}-q_2\right)I_n(x_m)\right]; \quad a_{35} = \frac{1}{x_m}I_n'(x_m) - \left(\frac{n^2}{x_m^2}+1\right)I_n(x_m);$$

$$a_{41} = \frac{n^2}{\gamma_m} I_n(\gamma_m); \quad a_{43} = -I_n'(\gamma_m) - q_1\left(\frac{n^2}{\gamma_m^2}+1\right)\gamma_m I_n(\gamma_m); \quad a_{45} = 2I_n'(\gamma_m);$$

$$a_{51} = \left(\frac{n^2}{\gamma_m^2}+\frac{1}{2}\right)I_n(\gamma_m) - \frac{1}{\gamma_m}I_n'(\gamma_m); \quad a_{53} = \frac{q_1}{2}\left[\frac{1}{\gamma_m}I_n'(\gamma_m) - \left(\frac{n^2}{\gamma_m^2}+1\right)I_n(\gamma_m)\right];$$

$$a_{55} = \frac{1}{\gamma_m}\left[I_n'(\gamma_m) - \frac{1}{\gamma_m}I_n(\gamma_m)\right]; \quad a_{61} = \frac{n^2}{\gamma_m}\left[\frac{1}{\gamma_m}I_n(\gamma_m) - I_n'(\gamma_m)\right];$$

$$a_{63} = \frac{q_1}{2}\left[\left(\frac{n^2}{\gamma_m^2}+1\right)\gamma_m I_n'(\gamma_m) - \left(\frac{n^2}{\gamma_m^2}-q_2\right)I_n(\gamma_m)\right]; \quad a_{65} = \frac{1}{\gamma_m}I_n'(\gamma_m) - \left(\frac{n^2}{\gamma_m^2}+1\right)I_n(\gamma_m);$$

$$q_1 = \frac{\lambda+\mu_0}{\lambda+2\mu_0}; \quad q_2 = \frac{\mu_0}{\lambda+\mu_0}; \quad \mu_0 = \frac{E_0}{2(1+\nu_0)}; \quad \lambda = \frac{E_0\nu_0}{(1+\nu_0)(1-2\nu_0)}; \quad x_m = z_0\gamma_m; \quad z_0 = \frac{R_0}{R},$$

where E_0 and ν_0 are the elastic modulus and Poisson's ratio of the cylinder material; $I_n(x)$ is the modified Bessel function of the first kind of order n; here the prime denotes the derivative with respect to the respective argument. To obtain even columns of the matrix a_{ij} and F_i with even indices, I_n in the previous odd elements should be replaced with the modified Bessel function of the second kind K_n of the same argument.

4. Stability analysis

Reducing the system (3) and setting the determinant of its matrix equal to zero, we obtain the characteristic equation. However, it is quite difficult to investigate this equation directly. It is more convenient to analyze its stability using the method described in reference [13].

Using the Lagrange polynomial approximation, we express the left-hand side of the characteristic equation $P(\omega) = 0$ as a polynomial in ω

$$P(\omega) = \sum_{j=0}^{2r} b_j \omega^{2r-j} = 0, \tag{4}$$

where b_j are unknown constant coefficients.

Calculating $P(\omega_i)$ for a number of ω_i, we find the values of b_j from the system of equations

$$\sum_{j=0}^{2r} b_j \omega_i^{2r-j} = P(\omega_i), \qquad (i = 0, 1, \ldots, 2r).$$

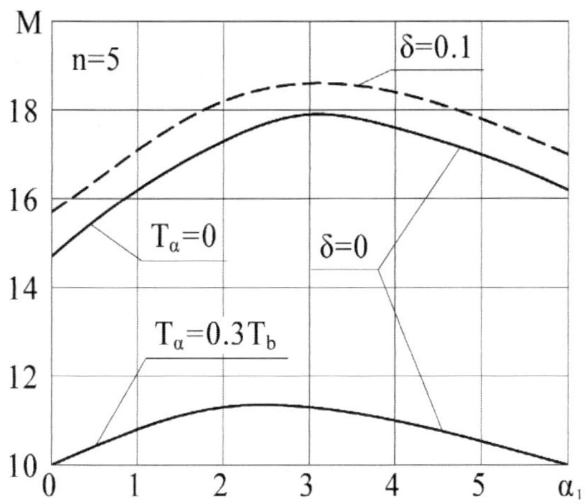

Figure 2. Critical Mach number vs. length of tapered hole ($h/R = 0.008$).

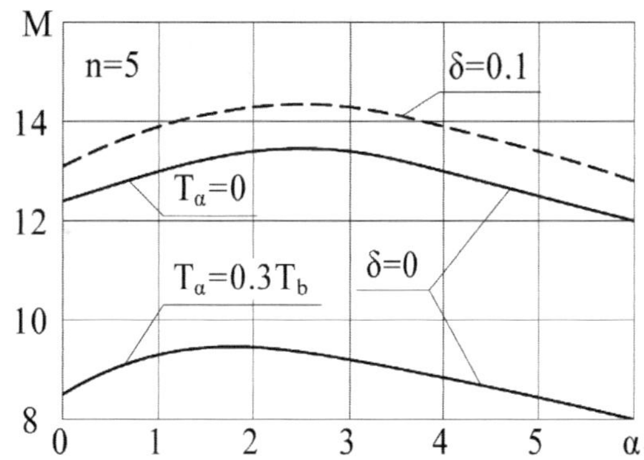

Figure 3. Critical Mach number vs. length of tapered hole ($h/R = 0.006$).

The unperturbed state of the shell is stable if the real parts of all the complex frequencies are negative $\mathrm{Re}\,\omega_i < 0$. A necessary and sufficient condition for that is the positivity of the Hurwitz determinants of the polynomial (4) [14]

$$D_j(M) = \begin{vmatrix} b_1 & b_0 & 0 & 0 & \dots\dots\dots & 0 & 0 \\ b_3 & b_2 & b_1 & b_0 & 0 & \dots\dots\dots & 0 \\ \dots & \dots & \dots & \dots & \dots & \dots & \dots \\ 0 & \dots\dots\dots & & 0 & b_j & b_{j-1} & b_{j-2} \\ 0 & 0 & \dots\dots\dots & & 0 & 0 & b_j \end{vmatrix} > 0, \qquad (j = 0, 1, \dots, 2r).$$

As the Mach number M increases, this inequality will cease to hold when either b_{2r} or $D_{2r-1}(M)$ becomes zero. The former case corresponds to divergence and the latter to flutter of the shell.

5. Numerical example

As an example, two shells of different thicknesses were considered. The parameters of the shells are as follows:

$$L/R = 6; \quad E_\alpha/E_0 = 1.5 \times 10^4; \quad E_\beta/E_0 = 2.3 \times 10^4; \quad G_{\alpha\beta}/E_0 = 2.5 \times 10^4;$$

$$\nu_\alpha/E_0 = 0.15; \quad \nu_\beta/E_0 = 0.23; \quad \nu_0 = 0.49; \quad \varkappa_0 = 1.4; \quad p_0/E_0 = 0.03; \quad (\rho;\rho_c)C_0^2/E = 109.$$

In figure 2 the critical Mach number is plotted vs. the nondimensional length of the tapered part of the hole for various values of the axial force and the damping decrement. Here T_b is the axial buckling force. Figure 3 shows these dependences for the shell with a different thickness-to-radius ratio.

6. Conclusions

We have proposed a mathematical model of the behavior of an orthotropic cylindrical shell supported by an elastic isotropic cylinder of linearly varying thickness in a supersonic gas flow. A characteristic equation is obtained, which, with the use of the Routh-Hurwitz asymptotic stability criterion, allows us to determine the dependence of the critical flow velocity on a number of design parameters. The effect of the cylinder thickness gradient on the occurrence of flutter has been shown.

Acknowledgments
The work has been performed under the state order. Official registration number: AAAA-A 19-119012290177-0.

References
[1] Solomonov Yu S et al 2012 *Proc. of Moscow Inst. of Therm. Tech.* **12**(1) 5–13

[2] Bakulin V N, Volkov E N and Nedbaj A Ya 2015 Flutter of a sandwich cylindrical shell supported with annular ribs and loaded with axial forces *Doklady Phys.* **60**(8) 360–363

[3] Volkov E N, Danilkin E V and Nedbai A Ya 2015 *Proc. of Moscow Inst. of Therm. Tech.* **15**(1) 54–64

[4] Bakulin V N, Volkov E N and Nedbai A Ya 2016 Dynamic stability of a cylindrical shell reinforced by longitudinal ribs and a hollow cylinder under the action of axial forces *J. Engineering Phys. and Thermophysics* **89**(3) 747–753

[5] Bakulin V N, Danilkin E V and Nedbai A Ya 2018 Dynamic stability of a cylindrical shell stiffened with a cylinder and longitudinal diaphragms at external pressure *J. Engineering Phys. and Thermophysics* **91**(2) 537–543

[6] Korbut B A and Nagornyi Yu I 1972 *Dynamics and Strength of Machines* vol 15 (Khar'kov: Khar'kov State Univ.) pp 70–77

[7] Bakulin V N, Volkov E N and Simonov A I 2017 Dynamic stability of a cylindrical shell under alternating axial external pressure *Russian Aeronautics* **60**(4) 508–513

[8] Bakulin V N, Bokov M A and Nedbai A Ya 2018 Aeroelastic stability of a cylindrical composite shell at a bilateral flow *Mech. Comp. Materials* **53**(6) 801–808

[9] Bakulin V N, Konopelchev M A and Nedbai A Ya 2018 Flutter of a laminated cantilever cylindrical shell with a ring-stiffened edge *Russian Aeronautics* **61**(4) 517–523

[10] Bakulin V N, Nedbai A Ya and Shepeleva I O 2019 Dynamic stability of orthotropic cylindrical shell of piecewise constant thickness under the action of external pulsating pressure *Izv. VUZ. Aviatsionnaya Tekhnika* **2** 19–25

[11] Timoshenko S 1937 *Vibration problems in engineering* (New York: D. Van Nostrand Company, Inc.)

[12] Solomonov Yu S et al 2014 *Applied Problems of the Mechanics of Composite Cylindrical Shells* (Moscow: Fizmatlit)

[13] Moskvin V G 1973 *Theory of Shells and Plates: Proc. of the 8th All-Union Conf. (Rostov-on-Don, September 16-22, 1971)* (Moscow: Nauka) pp 527–531

[14] Chetaev N G 1961 *The Stability of Motion* (New York: Pergamon Press)

Model Estimates of Circulating Flow around Objects in the Marine Medium and Atmosphere

[1,2]**Korchagin, N.N., and** [1,2]**Vladimirov, I.Yu.**

[1] Shirshov Institute of Oceanology of the Russian Academy of Sciences, Moscow
[2] Bauman Moscow State Technical University

e-niknik@mail.ru

Abstract. Model studies of force action on streamlined fragments of technical structures by a stratified sea current have been carried out. Analytical expressions are obtained for the hydrodynamic load on a point dipole during its circulation flow around a two-layer fluid flow of finite depth. The dependences of the wave resistance and the lifting force on the velocity of the current and circulation are investigated. It is shown that taking into account the velocity circulation can significantly affect the force effect on the dipole. The effect of a sharp (reversible) change in the direction of lift in a relatively narrow range of the flow velocity of a pipeline fragment simulated by a dipole has been found. The effect can also appear when moving self-propelled underwater objects and aircraft in the atmosphere.

1. Statement of the problem and its solution

Viscous drag acts on a streamlined obstacle in a stratified flow. In addition, it is additionally experiencing power effects due to the development of internal waves. An example of such effects is suspended suspension flows occurring in the bottom layers of the sea when the flow is mixed with micro- and fine-grained soil particles, which leads to an excess flow density (relative to the surrounding water) and the formation of a sharper pycnocline at its boundary [1]. The compaction of pycnocline leads to an increase in internal waves and a corresponding increase in the dynamic load on the streamlined technical object.

Of particular interest are the force effects arising from the flow around an obstacle localized near the density jump layer. At the same time, the difference in its localization relative to the density jump boundary: above or below the boundary of such a layer is important. In addition, in the general case, the flow around an obstacle can occur with a certain circulation, which additionally introduces unaccountable force effects of the medium on the object, and which can be significant.

In general, the problem statement by the authors was considered in [3]. Below (Fig. 1) there is a flow diagram of a dipole that simulates the elements of a pipeline localized under a density jump (task 1) and above a jump (task 2).

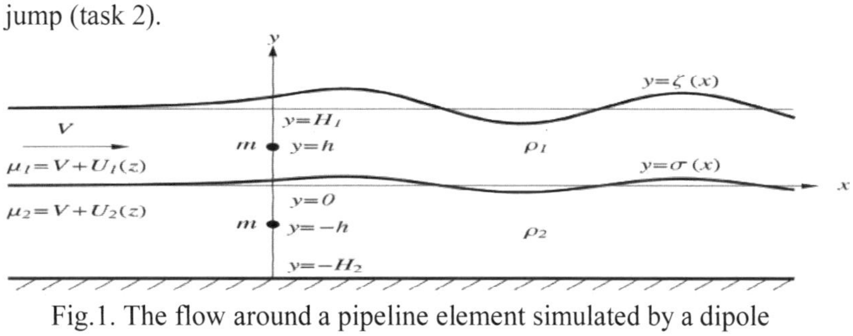

Fig.1. The flow around a pipeline element simulated by a dipole

The thickness of the upper layer H_1, lower - H_2 density, respectively, ρ_1 and ρ_2, $\rho_1 < \rho_2$. The origin of coordinates is located on the unperturbed boundary between the fluid layers, the x axis is directed along this boundary, and the y axis is vertically upwards. Dipole and vortex are combined in points $(0, \mp h)$. Note that density jumps are widespread in sea waters almost from the surface to the bottom. In general, they determine the structure of stratified waters, in the region of which internal waves arise. Therefore, the parameter $\varepsilon = (\rho_2 - \rho_1)/\rho_2$ can be defined as the energy characteristic of the marine environment.

We look for the solution of the model problem in the format of the complex-conjugate velocity of a two-layer flow disturbed by a dipole in the form $\mu_k(z) = V + U_k(z)$, where $z = x + iy, U_k(z)$ are the velocity disturbances, $k = 1, 2$. In order to universalize the problem and generalize the model parameters of the dipole, a transition is made to dimensionless variables

$$X = \frac{x}{R},\ Y = \frac{y}{R},\ Z = \frac{z}{R} = X + iY,\ \delta = \frac{\rho_1}{\rho_2},\ h_1 = \frac{H_1}{R},\ h_2 = \frac{H_2}{R}$$

$$h = \frac{H}{R},\ u_k = \frac{U_k}{V},\ \gamma = \frac{\Gamma}{2\pi VR},\ E = \frac{gR}{V^2} = \frac{1}{\mathrm{Fr}^2}$$

Here g is the acceleration due to gravity, Fr is the Froude number along the cylinder radius. Then, in the problem of perturbations of the complex-conjugate velocity u_k introduced by a dipole and a vortex under a density jump, it is required to find $u_1(Z)$ and $u_2(Z)$ with boundary conditions:

$$\operatorname{Im}\left[i\frac{du_1}{dZ} - Eu_1\right] = 0 \quad \text{at } Y = h_1 \tag{1.1}$$

$$\delta \operatorname{Im}\left[i\frac{du_1}{dZ} - Eu_1\right] = \operatorname{Im}\left[i\frac{du_2}{dZ} - Eu_2\right] = 0 \quad \text{at } Y = 0 \tag{1.2}$$

$$\operatorname{Im} u_1 = \operatorname{Im} u_2 \quad \text{at } Y = 0 \tag{1.3}$$

$$\operatorname{Im} u_2 = 0 \quad \text{at } Y = -h_2, \tag{1.4}$$

so that $u_1(Z)$ is regular function in the band $-\infty < X < +\infty$, $0 < Y < h_1$ and $u_2(Z)$ in the band $-\infty < X < +\infty$, $-h_2 < Y < 0$ everywhere except for point $Z = -ih$, at which it has a second-order pole:

$$u_2(Z) = -\frac{1}{(Z + ih)^2} + \frac{\gamma}{i}\frac{1}{Z + ih} + f(Z),\ f(Z) \text{ is a regular function.}$$

The boundary condition (1.1) describes the constancy of pressure on the free surface, (1.2) – the continuity of pressure at the interface of the layers, (1.3) determines the absence of fluid flow through this boundary, and (1.4) is the condition of no-flow at the bottom (derivation of (1.1)–(1.4) see in [3]).

After some transformations using the Chaplygin formula [3] we find the coefficient of the complex-conjugate hydrodynamic load on the cylinder $\Delta C^* = C_x - i\Delta C_Y = \frac{F_X - i\Delta F_Y}{\rho_2 V^2 R}$ added to the generalized force of Zhukovsky, where F_X is the wave drag, ΔF_Y is the additional lifting force of the dipole (dimensional)

$$\begin{aligned} C_X &= 2\pi^2 \sum_{j=1}^{s} \operatorname*{res}_{\xi=\xi_j} \left\{-(\gamma+\xi)\left[\xi C_m(\xi) + \gamma C_\gamma(\xi)\right]e^{\xi h} + (\gamma-\xi)\left[\xi D_m(\xi) + \gamma D_\gamma(\xi)\right]e^{-\xi h}\right\} \\ \Delta C_Y &= -2\pi i \int_0^{+\infty} \left\{(\gamma+\xi)\left[\xi C_m(\xi) + \gamma C_\gamma(\xi)\right]e^{\xi h} + (\gamma-\xi)\left[\xi D_m(\xi) + \gamma D_\gamma(\xi)\right]e^{-\xi h}\right\} d\xi \end{aligned} \tag{1.5}$$

$$C_m(\xi)=\frac{-(1-\delta)(\xi^2-E^2)\operatorname{sh}\xi(h_1-h)+[(1+\delta)\xi^2+(1-\delta)E^2]\operatorname{sh}\xi(h_1+h)-2\xi E\operatorname{ch}\xi(h_1+h)}{2\operatorname{ch}\xi h_1\operatorname{ch}\xi h_2 e^{\xi h_2}\left\{\xi^2+[\delta\xi^2+(1-\delta)E^2]\operatorname{th}\xi h_1\operatorname{th}\xi h_2-\xi E(\operatorname{th}\xi h_1+\operatorname{th}\xi h_2)\right\}},$$

$$D_m(\xi)=\frac{\operatorname{ch}\xi(h_2-h)\left\{-[(1+\delta)\xi^2+2\xi E+(1-\delta)E^2]e^{-\xi h_1}-(1-\delta)(\xi^2-E^2)e^{\xi h_1}\right\}}{2\operatorname{ch}\xi h_1\operatorname{ch}\xi h_2\left\{\xi^2+[\delta\xi^2+(1-\delta)E^2]\operatorname{th}\xi h_1\operatorname{th}\xi h_2-\xi E(\operatorname{th}\xi h_1+\operatorname{th}\xi h_2)\right\}},$$

$$C_\Gamma(\xi)=\frac{(1-\delta)(\xi^2-E^2)\operatorname{ch}\xi(h_1-h)+[(1+\delta)\xi^2+(1-\delta)E^2]\operatorname{ch}\xi(h_1+h)-2\xi E\operatorname{sh}\xi(h_1+h)}{2\operatorname{ch}\xi h_1\operatorname{ch}\xi h_2 e^{\xi h_2}\left\{\xi^2+[\delta\xi^2+(1-\delta)E^2]\operatorname{th}\xi h_1\operatorname{th}\xi h_2-\xi E(\operatorname{th}\xi h_1+\operatorname{th}\xi h_2)\right\}},\tag{1.6}$$

$$D_\Gamma(\xi)=\frac{\operatorname{sh}\xi(h_2-h)\left\{[(1+\delta)\xi^2+2\xi E+(1-\delta)E^2]e^{-\xi h_1}+(1-\delta)(\xi^2-E^2)e^{\xi h_1}\right\}}{2\operatorname{ch}\xi h_1\operatorname{ch}\xi h_2\left\{\xi^2+[\delta\xi^2+(1-\delta)E^2]\operatorname{th}\xi h_1\operatorname{th}\xi h_2-\xi E(\operatorname{th}\xi h_1+\operatorname{th}\xi h_2)\right\}}.$$

The integral in (1.5) is meant in the sense of the Cauchy principal value, and the residues are taken over all s poles ξ_j of the corresponding function located on the positive real axis. From (1.6) it is clear that these poles are the positive roots of the equation

$$\xi^2+[\delta\xi^2+(1-\delta)E^2]\operatorname{th}\xi h_1\operatorname{th}\xi h_2-\xi E(\operatorname{th}\xi h_1+\operatorname{th}\xi h_2)=0.$$

Note that the expression for the lift coefficient does not include the force of Archimedes.

In a similar formulation, problem 2 on the hydrodynamic reaction in a circulating flow around a cylinder simulated by a point dipole located in the upper layer of the flow (above the jump). Then, omitting the cumbersome calculations (see [5]), we numerically calculate the final results of the model with their presentation in graphical form.

2. Calculations of the force action

Calculations of the force action on a streamlined cylinder were carried out for real values of the characteristics of the marine medium using the formulas $\Delta F_Y=\rho_2 V^2 R\Delta C_Y$ (lower layer), $\Delta F_Y=\rho_1 V^2 R\Delta C_Y$ (upper layer), where the coefficients ΔC_Y are expressed through (1.5), and also (2.2) from [5]. The density in the upper layer is $\rho_1=1024$ kg/m^3, and the relative density difference between the lower and upper layers varies in the range $\rho_2/\rho_1=1.001-1.005$. In the bottom layer, the differential ρ_2/ρ_1 may vary within $1.01-1.05$, which corresponds to the suspended flow in the lower layer. In the first case, this is a relatively weak weighing stream, in the second, a turbulent flow, which manifests itself near the bottom on the slopes of submarine elevations [2]. The total thickness of the flow is 60m, the thickness of the upper layer $H_1=50$m, the radius of the cylinder (pipeline) $R=0.71$m. The dipole is localized in the lower or upper layer at a distance $h=2$m from the interface between the layers. The dimensionless circulation was defined as $\gamma=\Gamma/(2\pi VR)$, where V is the flow velocity, Γ is the dimensional circulation. Note that by virtue of the equivalence of positive and negative vorticities in the flow, when analyzing the variability of the lifting force, one may choose any of $\gamma^\pm$: ΔF_Y^+ for positive, and ΔF_Y^- for negative circulation.

Analysis of model solutions in [4, 5] showed that the amplification of the effect of flow on an obstacle is due to an increase in the power of the density jump ε, and together with the distance h they are the main parameters of the problem under consideration. Figure 2 shows the dependences of the force effects on the dipole on the relative density drop ε.

In Fig. 2, curves with unidirectional positive circulations ΔF_Y^+ depending on localization with respect to a density jump differ significantly from each other both in form and in values. In this case, the curves on different sides of the jump in the nature of variability are opposite to each other. According to [5], the differences in the graphs of ΔF_Y^+ shown in Fig. 2 and constructed close to, but on different sides of the interface between the layers are caused by different directions of circulation, which in some cases leads to an increase in the flow velocity (between the jump and the cylinder), in others, to its decrease.

Thus, the power loads, additionally arising in the presence of circulation, act in different directions along the section of the pipeline. In this case, despite their insignificance, but acting for a long time, they can lead to the deformation of the pipeline, and in the future – to its possible damage. Therefore, these effects must be considered when laying underwater communications and pipelines. It should be noted that changes in the lift force identified in the model, depending on the circulation γ_i, were not previously taken into account in such studies.

In continuation of the analysis of the results of the model problem, it is of interest to investigate a characteristic feature in the variability of lift and its corresponding interpretation. In this connection, we will move from the above-considered motionlessly streamlined point model to a kinematically similar physical model of a steadily moving object. Without disturbing the generality of the model problem under consideration, a limited fragment of the pipeline can be represented as a cigar-shaped (or ovoid) body, in the form of a simulating a certain type of self-propelled underwater object. At the same time, in order to adequately compare the design (geometry) of a model ovoid with the original point model, it is necessary to separate the lateral component of the force vector perpendicularly distributed along the body length of the physical model as an impact on the underwater object.

On the basis of such a transformation of the original model, it is hypothetically possible to use the calculations of the dependence of the lifting force of the dipole located below the density jump on the flow velocity – Fig.3. Here, the dipole models the flow around an ovoid-shaped underwater object and clearly represents reversing changes in the direction of lifting force $\Delta F_Y^+(V)$ at the corresponding intervals of velocity of flow around the object.

In order to substantiate the possibility of a hypothetical transformation of a point model into the corresponding form of an ovoid object, one can refer to a typical example from maritime practice. Thus, when an object moves in a marine environment, situations arise regarding the rapid and dangerous submergence of an underwater object ten or more meters deep [5]. This effect is usually explained by the presence of internal waves generated at jumps in the density of the medium. However, the parameters of the internal waves, depending on the power of the density jump ε (with a minimum period of about ten minutes and a length of hundreds of meters), reflect the slower dynamics of the internal wave compared to the relatively sharp lowering of the object. In this case, a similar effect may occur either when the direction of movement of the object changes with the appearance of a lateral component of the flow, or by variations in the direction of the flow itself (also with a selected side component to the object body). And as the calculations showed, the selection of the lateral component of the flow with non-zero circulation can lead to the appearance of an additional force directed vertically downwards – Fig. 3

A similar effect during the circulation flow around an object can also manifest itself in the atmosphere, for example, during the flight of an airplane. In this case, the configuration (geometry) of such an object as a whole is also represented in the form of an ovoid. Then the airflow around it can lead to a reverse change in lift, which may be the cause of an unexpectedly sharp and dangerous decrease in the height of the aircraft. Therefore, the use here of model results (albeit hypothetical) in the study of the flight of an aircraft appears to be legitimate.

In addition, the equations of hydro- and aerodynamics (without taking into account the compressibility of air in the latter) are identical. And in the case of a flow around at a speed of no more than 1/5 of the speed of sound (or a flight speed of up to 250-300 km/h), the presence of air compressibility can be completely neglected [6]. However, here, as for the underwater object, the calculated model parameters are also taken hypothetically and therefore should be considered only as qualitative assessments. And such estimates are significant. Therefore, in reality, we can expect the manifestation of such effects.

Figure 4 clearly shows the relatively powerful force action on the aircraft, which occurs when it is landing with a decrease in flight speed within 300-400 km/h. At the same time, a sharp increase in the module of vertical force $\Delta F_Y^{\pm}$ for all γ_i in the graphs of Fig. 4 indicates a sharp decrease (dip) in the airplane's trajectory.

As a result, the total impact of forces of the same nature, but applied to different parts of the aircraft – wings, fuselage elements, etc. can, depending on their direction, lead to sharp vertical fluctuations in the flight path, which is not uncommon in flight practice.

The authors wish to thank A.T. Il'ichev for fruitful discussions of the model and calculation results. The work was carried out within the framework of the State Program of IORAS (theme № 0149-2019-0004).

References

[1] *Zhmur, V.V., and Sapov, D.A.* Catastrophic gravitational suspension-carrying currents in the near-bottom layer of the ocean, in: *World Ocean. Vol. 1. Ocean Geology and Tectonics. Catastrophic Phenomena in the Ocean*, Moscow: Nauchnyi Mir, 2013, pp. 499–524.

[2] *Loytsianskii L.G.* Mechanics of Liquids and Gases, Oxford: Pergamon Press, 1966.

[3] *Vladimirov, I.Yu., Korchagin, N.N., and Savin, A.S.* Surface disturbances in stratified finite-depth flow past obstacles, Okeanologiya, 2012, vol. 52, no. 6, pp. 825–835.

[4] *Vladimirov, I.Yu., Korchagin, N.N., and Savin, A.S.* Wave impact of a suspension-carrying stream on an obstacle in flow, Dokl. Ross. Akad. Nauk, 2015, vol. 461, no. 2. pp. 223–227.

[5] *Vladimirov, I.Yu., Korchagin, N.N., and Savin, A.S.* Wave reaction of a dipole in a circulatory double-layer flow of finite depth, Fluid Dynamics, 2016, vol. 51, no. 2, pp. 214–223.

[6] *Sedov, L.I.* Mekhanika sploshnoi sredy (Continuum Mechanics), vol. 2, Moscow: Nauka, 1973.

Figure captions

Fig.1. The flow around a pipeline element simulated by a dipole.

Fig.2. Lift force of a dipole as a function of the relative density drop between the layers; model of a pipeline. The dipole is located above (a) and beneath (b) the density jump at $V = 0.7$ m/s: curves *1–4* correspond to $\gamma_i = 0, 0.1, 0.2$, and 0.3, respectively.

Fig.3. Lift force of a dipole located above the density jump as a function of the flow velocity for $\rho_2/\rho_1 = 1.005$, $H_1 = 100$ m, $H_2 = 200$ m, $h = 6$ m, and $R = 3.5$ m; model of the underwater object: curves *1–4* correspond to $\gamma = 0.0, 0.1, 0.2$, and 0.3, respectively. Black arrows indicate the intervals with the reversible variation in curves $\Delta F_Y^{+}(V)$.

Fig.4. Lift force of a dipole located beneath the density jump layer as a function of the flow velocity for $\rho_2 = 1.26$ kg/m^3, $\rho_2/\rho_1 = 1.3$, $H_1 = 1000$ m, $H_2 = 2000$ m, $h = 6$ m, and $R = 3.5$ m: curves *1–4* correspond to $\gamma = 0.0, 0.1, 0.2$, and 0.3, respectively.

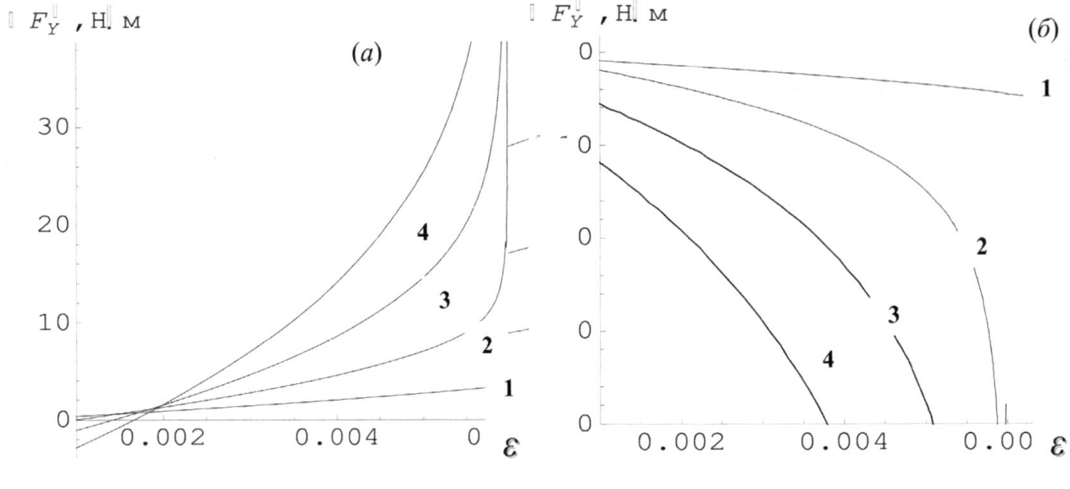

Fig. 2.

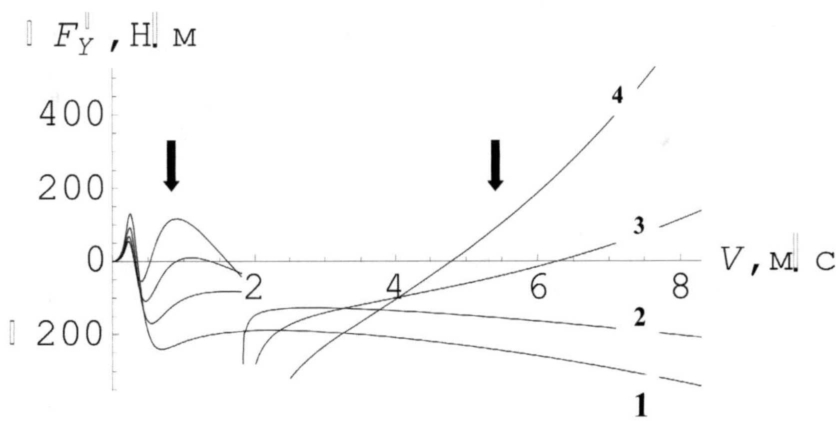

Fig.3.

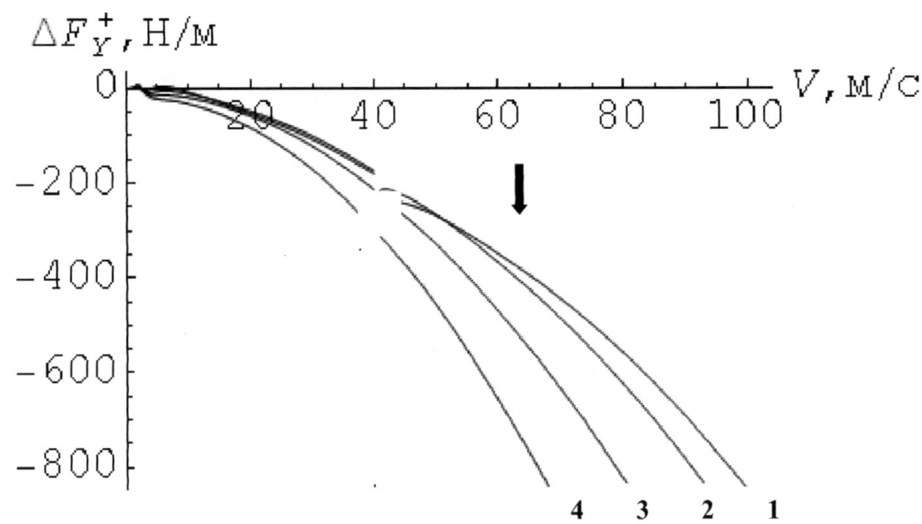

Fig.4.

Non-uniqueness of transonic flow in an intake-type channel

A. Kuzmin

St. Petersburg State University, 28 University Ave., St. Petersburg 198504, Russia

E-mail: a.kuzmin@spbu.ru

Abstract. The two-dimensional turbulent airflow in a 9-degrees-bent channel is studied numerically. Inner surfaces of the top and bottom walls are parallel to each other upstream and downstream of the bend. The free stream is supersonic, whereas the flow is subsonic at the channel exit. Solutions of the Reynolds-averaged Navier-Stokes equations are obtained with a finite-volume solver using the Spalart-Allmaras and Shear Stress Transport $k - \omega$ turbulence models. The solutions reveal a flow hysteresis and non-uniqueness in considerable bands of the free-stream Mach number, angle of attack, and exit pressure. At the endpoints of the bands, there are abrupt changes of the shock wave system. The non-uniqueness admits different losses of the total pressure, which may cause different trusts of an air breathing engine.

1. Introduction

Both inviscid and turbulent supersonic airflows in convergent channels involve shock waves which are known to be unstable. The instability can be explained using quasi-one-dimensional equations which govern mass flow rate and stagnation temperature across the normal shock [1]. Recently, instability of shocks was also revealed in bent channels whose cross section is either constant or slightly increasing [2, 3]. In the latter case, the instability is caused by an interaction of a shock wave formed in front of the concave wall with an expansion flow developed over the opposite (convex) wall. The interaction causes a hysteresis phenomenon under gradual variations of the outlet pressure, free-stream Mach number M_∞ or angle of attack α.

In practice, transonic flow in curved/bent channels occurs, e.g., in Y-shaped intakes of battle-plane and fighter aircraft engines. Kotteda and Mittal [4] studied the effect of sideslip angle on the performance of a slightly divergent Y-intake and identified non-unique flow regimes as a result of these variations; also an effect of the initial condition was investigated. Feng *et al.* [5] performed a numerical and experimental study of supersonic flow in a bent channel modeling a variable geometry dual mode combustor. The study indicated that the pressure distribution on the wall had an obvious hysteresis under continuous variations of the outlet cross section area.

In [6] we considered an intake-type channel with a short top wall and parallel inner surfaces of walls before and after the bend section. A dependence of 2D shock positions on M_∞ at zero and negative angles of attack was studied, and considerable attention was paid to a subsonic exit condition in contrast to the supersonic condition used in a previous paper [7].

In the present paper, we examine a channel with a displacement of the bottom wall bend downstream with respect to the top wall bend. This results in a smaller exit section of the channel and larger Mach numbers that produce the instability of shocks. In Sections 2 and 3 we formulate the problem and outline a numerical method. Section 4 addresses locations of a shock wave versus M_∞ at a given angle of attack; the flow non-uniqueness and losses of the total pressure are discussed. In Section 5 we study the shock position as a function of the angle of attack at a given M_∞. Effects of the exit pressure and a replacement of the bottom bend by a local circular arc are analyzed in Section 6.

2. Problem formulation

We consider a channel whose top wall is a compression corner

$$y = 30 \quad \text{at} \quad -28 \le x \le 0, \qquad y = 30 - x\,\tan 9^\circ \quad \text{at} \quad 0 \le x \le 50,$$

whereas the bottom wall is an expansion corner

$$y = 0 \quad \text{at} \quad -68 \le x \le 8, \qquad y = (8 - x)\,\tan 9^\circ \quad \text{at} \quad 8 \le x \le 50, \tag{1}$$

see Fig. 1. Here and further in the paper, the Cartesian coordinates (x, y) are dimensional and given in millimeters. The outer surfaces of the top and bottom walls are a corner and segment, respectively. The thickness of leading edges of the walls is 0.05.

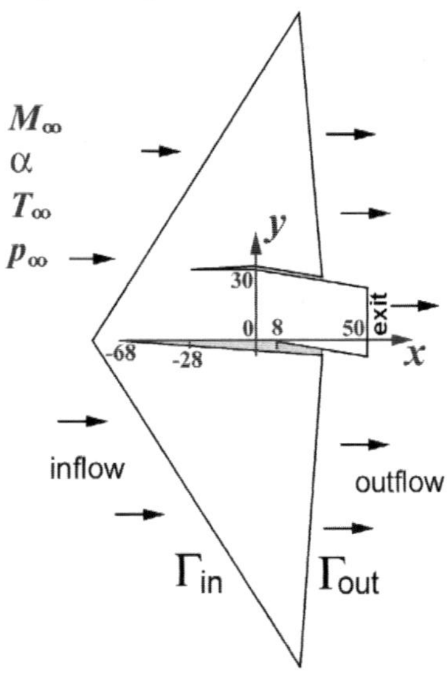

Figure 1. Sketch of the channel and computational domain.

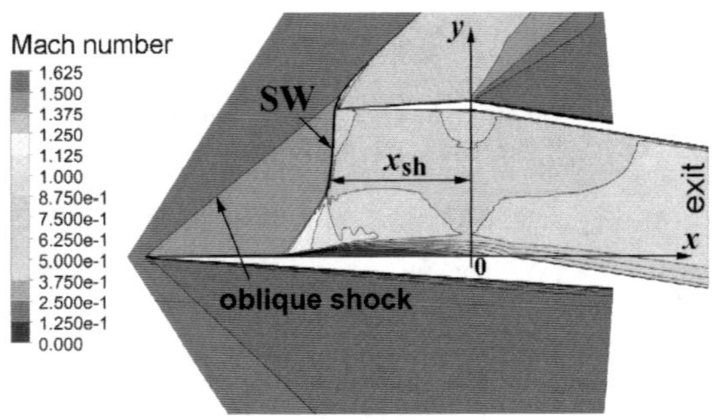

Figure 2. Mach number contours and expelled shock wave SW at $\alpha = -5^\circ$, $M_\infty = 1.62$, $p_{\text{exit}} = 2 \times 10^5$ N/m^2.

The inflow boundary Γ_{in} of the computational domain is formed by two segments with the beginnings at the point $y = 0$, $x = -72$ and ends at $x = 20$, $y = \pm 140$. We prescribe the $x-$ and $y-$components of the velocity on Γ_{in} as follows: $U_\infty = M_\infty a_\infty \cos\alpha$, $V_\infty = M_\infty a_\infty \sin\alpha$, where $M_\infty > 1$. In addition, we impose on Γ_{in} the static pressure p_∞, the turbulence level of 1%, and static temperature $T_\infty = 200$ K which determines the sound speed $a_\infty = 283.58$ m/s.

The outflow boundary Γ_{out} of the computational domain is constituted by two segments with the beginnings on the outer surfaces of walls at $x = 30$ and ends at $x = 20$, $y = \pm 140$. We prescribe the condition $M > 1$ on Γ_{out}. At the channel exit $x = 50$ we impose the subsonic condition $M_{\text{exit}} < 1$ along with the static pressure p_{exit}. The vanishing heat flux and no-slip condition are imposed on the walls. The air is treated as a perfect gas, whose specific heat at constant pressure is 1004.4 J/(kg K) and the ratio of specific heats is 1.4. We use the Sutherland formula for the molecular dynamic viscosity and adopt the value of 28.96 kg/kmol for the molar mass. The pressure p_∞ is set to 8×10^4 N/m^2 throughout the paper, except for Fig. 9 in which we compare results obtained for this and tripled free-stream pressure. The Reynolds number based on $p_\infty = 8 \times 10^4$ N/m^2, $M_\infty = 1.6$, and height of the channel is 1.4×10^6. Initial data are either parameters of the free stream or a flow field calculated for other α, M_∞ or p_{exit}.

3. Numerical method

Solutions of the unsteady Reynolds-averaged Navier-Stokes equations were obtained with an ANSYS-18.2 CFX finite-volume solver of second-order accuracy. An implicit backward Euler scheme was employed for the time-accurate computations. We used the Shear Stress Transport $k-\omega$ and Spalart-Allmaras turbulence models, which are known to reasonably predict aerodynamic flows with boundary layer separations [8, 9].

Computations were performed on hybrid unstructured meshes constituted by quadrangles in 38 layers on the walls and by triangles in the remaining region. The non-dimensional thickness y^+ of the first mesh layer on the walls was less than 1. The sizes of triangles essentially decreased in the channel to resolve shock waves accurately. Test solutions obtained on uniformly refined meshes of approximately 2×10^5, 4×10^5, and 8×10^5 cells showed that a discrepancy between shock wave coordinates obtained on the second and third meshes did not exceed 1%. Global time steps of 5×10^{-7} s and 10^{-6} s yielded indistinguishable solutions. For this reason, the time step of 10^{-6} s and mesh of 4×10^5 cells were employed for the study of 2D flow hysteresis at various α, M_∞, and p_{exit}. The root-mean-square CFL number (over mesh cells) was about 2.

For the solver validation, we recomputed several benchmark transonic flow problems and obtained good agreement with results available in the literature, see [6, 7]. Computations were run on Huawei RH2288H computational nodes using a splitting of a task among 96 cores.

4. Shock wave location as a function of M_∞ at $\alpha = -5°$.

First, we solved the problem at $M_\infty = 1.55$, $\alpha = -5°$ using the free-stream parameters for initialization of the solution. Computations showed a convergence of time-dependent mean parameters of turbulent flow to a steady state in less than 0.1 s of physical time. The obtained flow field exhibits an oblique shock attached to the leading edge of bottom wall, and a shock SW located closer to the entrance ($x_{\text{entrance}} = -28$, $0 < y < 30$). When M_∞ increases step-by-step from 1.55 to 1.62, the shock SW shifts downstream and becomes nearly normal, see Fig. 2; at each step for initial conditions we used the flow field obtained at the previous M_∞. Such a flow pattern is called hereafter the flow regime with an expelled SW.

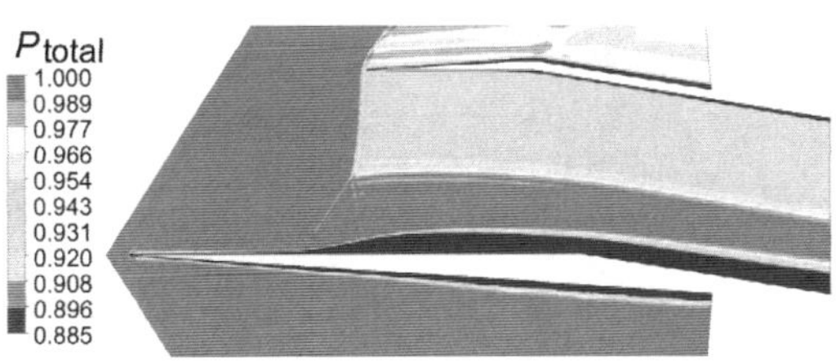

Figure 3. Contours of the non-dimensional total pressure P_{total} in the flow depicted in Fig. 2.

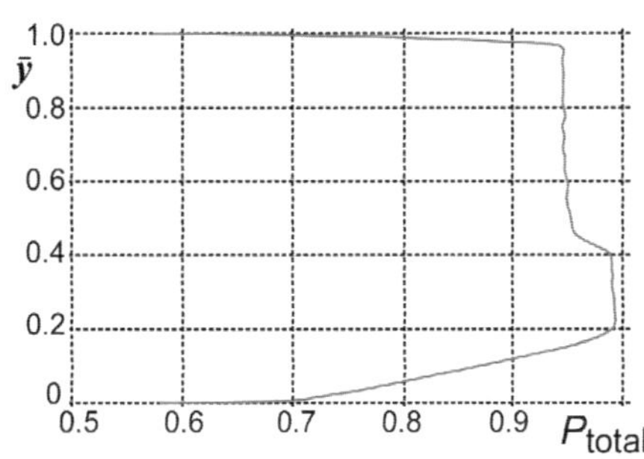

Figure 4. P_{total} in the exit section of the flow depicted in Fig. 2 versus the non-dimensional distance $\bar{y}$ from the bottom wall; $\bar{y} = 1$ corresponds to the top wall.

Figures 3 and 4 illustrate distributions of the non-dimensional total pressure $P_{\text{total}} = p_{\text{total}}/p_{\text{total},\infty}$ in the flow displayed in Fig. 2, where $p_{\text{total},\infty} = 350,280$ N/m^2 is the stagnation pressure.

To analyze streamwise positions of SW, we will trace its coordinate $x_{\rm sh}$ at the height $y = 15$. The lower branch of solid curve 1 in Fig. 5 shows that $x_{\rm sh}$ gradually increases when M_∞ increases from 1.55 to 1.625. Further increase of M_∞ triggers the swallowing of SW and transition from the lower branch of curve 1 to the upper one. After that a slight decrease of M_∞ to 1.62 yields a flow with the swallowed SW, see Fig. 6, which essentially differs from that displayed in Fig. 2 for the same M_∞.

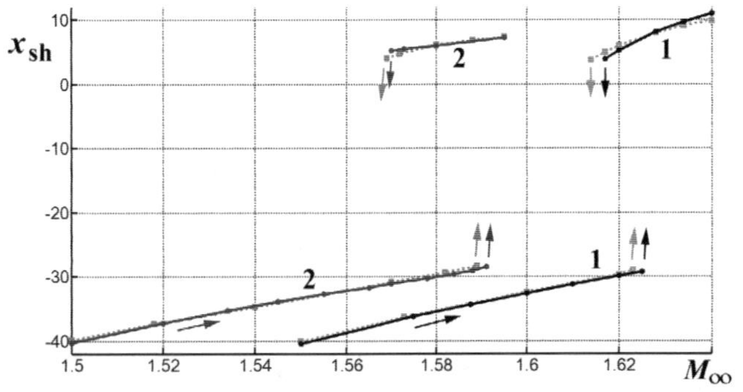

Figure 5. Coordinate $x_{\rm sh}$ of the shock wave SW versus M_∞ at $\alpha = -5°$: curve 1 – $p_{\rm exit} = 2 \times 10^5$ N/m^2, curve 2 – $p_{\rm exit} = 1.8 \times 10^5$ N/m^2. Solid curves – SST $k - \omega$ turbulence model [8]; dashed curves – Spalart-Allmaras turbulence model [9]. The upper (lower) branches of the curves correspond to flow regimes with the swallowed (expelled) SW.

Further decrease of M_∞ from 1.62 to values smaller than 1.616 triggers an expulsion of SW from the channel. This means a transition to the lower branch of curve 1 in Fig. 5. Due to the hysteresis, the flow is non-unique in the band $1.617 \le M_\infty \le 1.625$.

Computations using the Spalart-Allmaras turbulence model instead of SST $k - \omega$ produced results which are very close to those discussed above (cf. dashed and solid curves in Fig. 5).

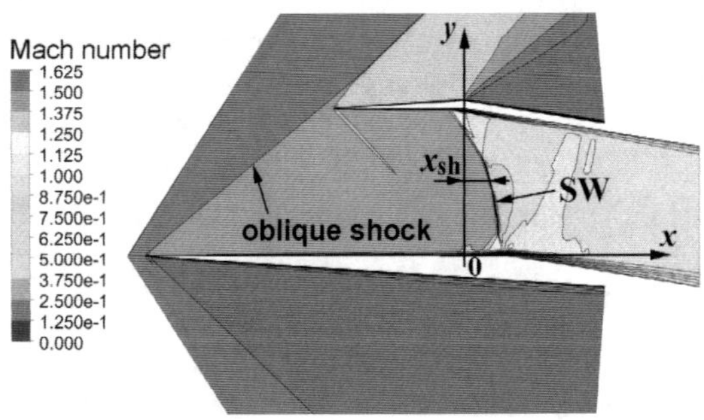

Figure 6. Mach number contours and swallowed shock wave SW at $\alpha = -5°$, $M_\infty = 1.62$, $p_{\rm exit} = 2 \times 10^5$ N/m^2.

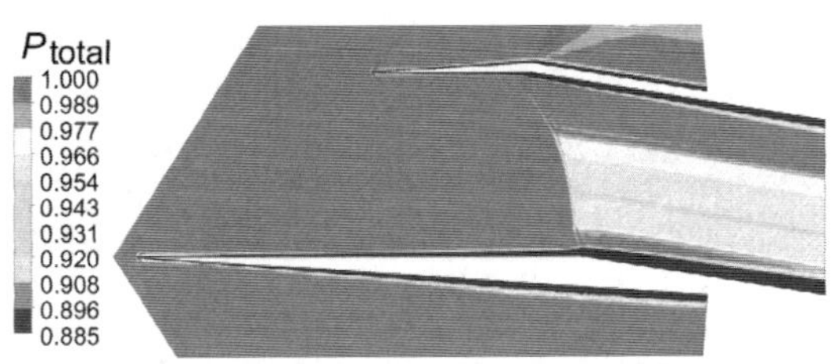

Figure 7. Contours of the non-dimensional total pressure in the flow depicted in Fig. 6.

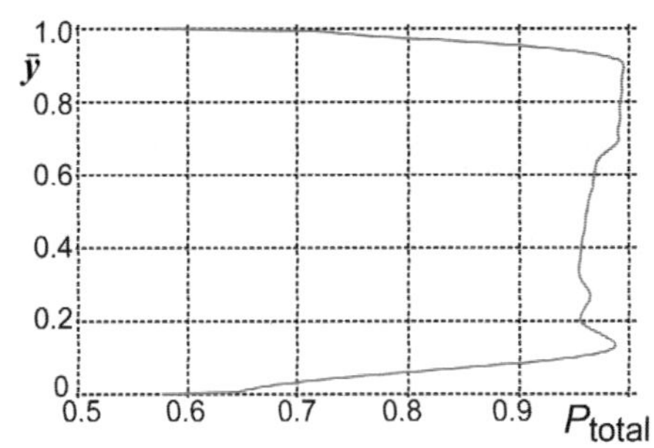

Figure 8. $P_{\rm total}$ in the exit section of the flow depicted in Fig. 6 versus the distance $\bar{y}$ from the bottom wall.

In the same way we calculated a dependence $x_{\rm sh}$ on M_∞ at a smaller pressure $p_{\rm exit} = 1.8\times10^5$ N/m^2, see curve 2 in Fig. 5. The reduction of $p_{\rm exit}$ yields a displacement of SW downstream to the exit. Therefore one needs smaller M_∞ to return SW to a position near the wall bend which admits its expulsion from the channel due to a small perturbation. That is why curve 2 in Fig. 5 is shifted to smaller M_∞ with respect to curve 1.

Figures 7, 8 illustrate the total pressure distribution in the flow depicted in Fig. 6. As seen, the total pressure losses in this regime noticeably differ from the ones displayed in Figs. 3, 4.

5. Shock wave location as a function of of the angle of attack α.

Curve 1 in Fig. 9 presents a dependence of $x_{\rm sh}$ on α for $M_\infty = 1.62$. Both the numerical solution and oblique shock relations show that at $\alpha = -7°$ (see the left endpoint of curve 1) the Mach number behind the bow oblique shock is $M_{\rm bow} = 1.378$. This value coincides with $M_{\rm bow}$ in the conditions $M_\infty = 1.55$, $\alpha = -5°$ which correspond to the left endpoint of curve 1 in Fig. 5. However, the SW locations are different in these two cases, as they depend not only on $M_{\rm bow}$ but also on a triple shock configuration formed in front of the top wall. A detachment of the triple shock from the wall is much smaller at $M_\infty = 1.62$, $\alpha = -7°$ than at $M_\infty = 1.55$, $\alpha = -5°$. This explains a displacement of the whole shock system downstream and, as a consequence, the larger ordinate of the left endpoint of curve 1 in Fig. 9 than in Fig. 5.

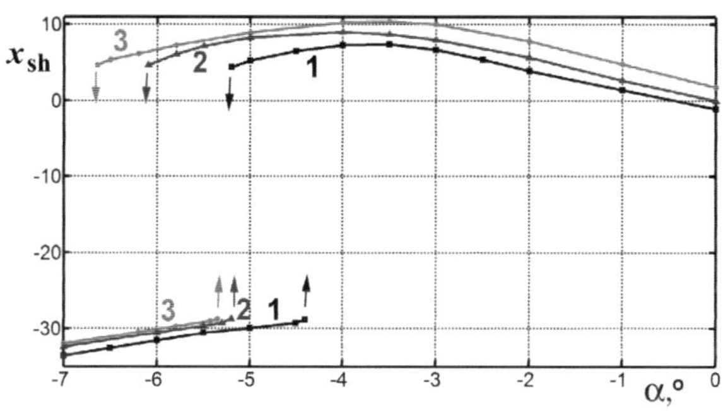

Figure 9. Coordinate $x_{\rm sh}$ of SW versus the angle of attack α: 1 – $M_\infty = 1.62$, $p_\infty = 8\times10^4$, $p_{\rm exit} = 2\times10^5$; 2 – $M_\infty = 1.63$, $p_\infty = 8\times10^4$, $p_{\rm exit} = 2\times10^5$; 3 – $M_\infty = 1.63$, $p_\infty = 24\times10^4$, $p_{\rm exit} = 6\times10^5$.

With increasing α from $-7°$ to $-5°$ at $M_\infty = 1.62$, the Mach number $M_{\rm bow}$ behind the bow shock increases, and the detachment of triple shock from the top wall becomes small, see Fig. 2. Further increase of α to values larger than $-4.4°$ triggers the swallowing of SW and transition from the lower branch of curve 1 in Fig. 9 to the upper one.

At a larger free-stream Mach number, $M_\infty = 1.63$, the Mach number $M_{\rm bow}$ decreases. Therefore, one needs smaller values of α to trigger the swallowing/expulsion of SW. That is why curve 2 in Fig. 9 is shifted to smaller M_∞ with respect to curve 1. Curve 3 presents results obtained for the tripled pressures p_∞ and $p_{\rm exit}$, which imply a tripled Reynolds number. As seen, the larger pressures and Reynolds number produce a noticeable expansion of the hysteresis.

6. Shock wave location as a function of the exit pressure $p_{\rm exit}$.

Computations of $x_{\rm sh}$ versus $p_{\rm exit}$ were performed in a similar way. As seen from Fig. 10, the shock position exhibits a hysteresis, which is insignificantly larger at $M_\infty = 1.60$ than at $M_\infty = 1.62$. We notice that pressures $p_{\rm exit} < 1.8\times10^5$ N/m^2 lead to formation of local supersonic velocities at the exit and, consequently, require a revision of the exit condition.

To study an effect of smoothing the bottom wall corner, we performed flow computations for a modified channel obtained by a local replacement of corner (1) with a circular arc of radius R in a short interval $x_1 \le x \le x_2$; the arc is tangential to corner (1) at $x = x_1 < 8$ and $x = x_2 > 8$. We examined four values of R : (a) 38.12, (b) 63.52, (c) 101.65, (d) 190.60. In the cases (a) and (b), computations showed only insignificant distinctions of the plots $x_{\rm sh}(p_{\rm exit})$

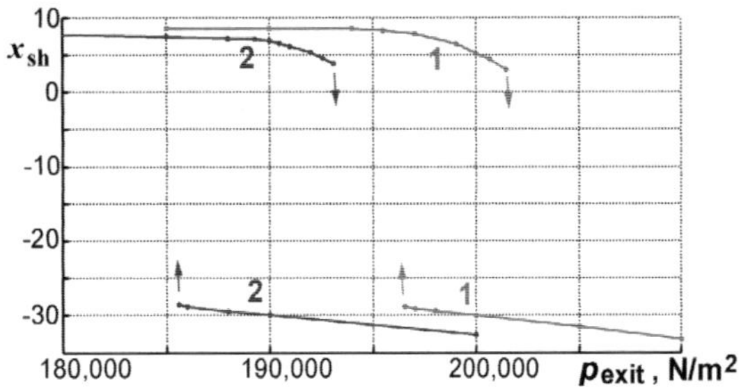

Figure 10. Coordinate x_{sh} of SW versus p_{exit} at $\alpha = -5°$: $1 - M_\infty = 1.62$, $2 - M_\infty = 1.60$.

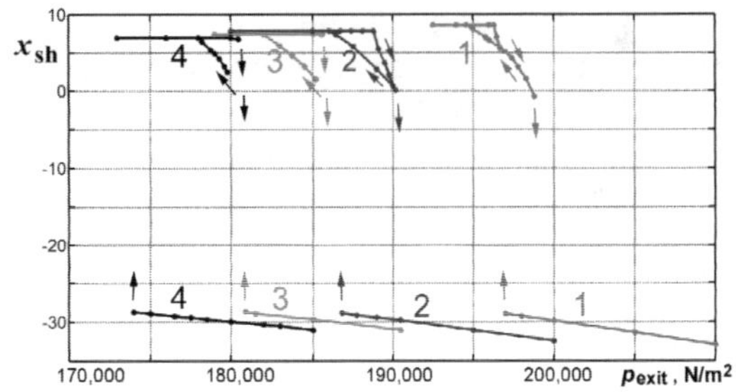

Figure 11. Coordinate x_{sh} of SW versus p_{exit} at $\alpha = -5°$ and radius $R = 190.6$ of the arc smoothing the bottom corner (1): $1 - M_\infty = 1.62$, $2 - M_\infty = 1.60$, $3 - M_\infty = 1.59$, $4 - M_\infty = 1.58$.

from those displayed in Fig. 10. Meanwhile, in the cases (c) and (d), computations revealed an extra hysteresis of SW location. In particular, Fig. 11 demonstrates a dependence of x_{sh} on p_{exit} obtained at $R = 190.60$. The extra hysteresis arises in the regime with a swallowed SW, see the upper branches of curves 1 - 4. It is attributed to the separated boundary layer like a flow hysteresis in straight nozzles [10]. The upper branches of curves 1 and 2 create loops, which illustrate steady positions of SW under step-by-step changes of p_{exit}. Transitions from the most top branches of curves 3 and 4 to slightly lower branches (whose descent is about 45°) were performed through unsteady states as indicated by arrows.

7. Conclusions

Transonic flow in channels with a shift of the bottom wall corner downstream with respect to the top corner is studied numerically. The obtained solutions demonstrate non-unique flow regimes and significant hysteresis under gradual variations of the free-stream Mach number, angle of attack, or exit pressure. The Mach numbers under consideration are larger than those in [6] due to the smaller area of the exit. This enhances boundary layer separation from the lower wall and admits a double hysteresis when the bottom wall corner (1) is smoothed with a circular arc of sufficiently large radius.

Acknowledgments

This research was performed using computational resources provided by the Computational Center of St. Petersburg State University (http://cc.spbu.ru). The work was partially supported by the Russian Foundation for Basic Research under grant no. 19-01-00242.

References

[1] Torenbeek E and Wittenberg H 2009 *Flight Physics* (Dordrecht: Springer)
[2] Guo S, Wang Z and Zhao Y 2014 *J. National University of Defense Technology* **36** 10
[3] Kuzmin A 2016 *Archive Appl. Mech.* **86** 787
[4] Kotteda V M K and Mittal S 2016 *J. Propulsion and Power* **32** 171
[5] Feng S, Chang J, Zhang Ch, Wang Y, Ma J and Bao W 2017 *Aerospace Sci. and Technology* **67** 96
[6] Kuzmin A 2019 *Advances in Aircraft and Spacecraft Science* **6** 19
[7] Kuzmin A 2017 *Archive Appl. Mech.* **87** 1269
[8] Menter F R 2009 *Int. J. Comput. Fluid Dynamics* **23** 305
[9] Spalart P R and Allmaras S R 1994 *La Recherche Aérospatiale* no. 1, 5
[10] Setoguchi T, Matsuo S, Alam M M A, Nagao J and Kim H D 2010 *J Thermal Science* **19** 526

The problem of suspension injection into a porous medium within a two-velocity model of deep bed filtration

N E Leontiev
Lomonosov Moscow State University, Moscow, Russia
E-mail: leontiev_n@mail.ru

Abstract. We consider flows of a suspension through a porous medium with account of the deposition of the suspended particles onto the internal surface of a porous matrix (deep bed filtration). The velocities of the suspended particles and the carrier fluid are plausibly supposed to differ by a known factor that depends on the suspension concentration. We present analytical and numerical solutions to the problem of suspension injection into a porous sample. The main qualitative result of the investigation is that — under certain circumstances — the smooth solutions (i.e., the ones that do not, in contrast to usual assumptions, involve concentration jumps) to the problem may exist. This feature may be of use for experimental data interpretation.

1. Introduction
Flows of suspensions through porous media with gradual contamination of porous matrix [1, 2] arise in water treatment problems, oil industry, etc. The standard way of investigating the parameters of deep bed filtration models is the performance of the laboratory experiments with cylindric porous samples under controlled conditions. In the injection tests, aside from direct measuring the concentration at different positions within the porous sample, the experimenters typically obtain the dependence of the effluent particle concentration on the time (so called breakthrough curve). In such experiments, the data treatment rests upon the exact solutions with jumps in suspended particles concentration that propagate along the porous sample and separate its contaminated and clear parts. In this paper we call special attention to the fact that under certain (practicable) circumstances the solution to the injection problem may be free of jumps, so the analysis of the experimental data should be thoroughly performed with great attention to details.

2. Two-velocity model of deep bed filtration
In this section we briefly present the model of suspension flow through a porous medium that accounts for the retardation of suspended particles relative to the carrier fluid [3].

For one-dimensional flows with plane waves, the mathematical model comprises the continuity equations for suspended particles and carrier fluid

$$\frac{\partial}{\partial t}(m\alpha) + \frac{\partial}{\partial x}(m\alpha v) = \frac{\partial m}{\partial t}, \quad \frac{\partial}{\partial t}((1-\alpha)m) + \frac{\partial}{\partial x}(m(1-\alpha)w) = 0,$$

the equation for the kinetics of particle deposition (or entrainment) with known right-hand side term

$$\frac{\partial m}{\partial t} = f(m, \alpha, v, w, \ldots),$$

and the relation between the velocities of the suspended particles and carrier fluid [4]

$$v = g(w, \alpha, m, \ldots),$$

where v and w are the velocities of the particles and the carrier fluid, m is the porosity, α is the concentration of the suspended particles, t and x are the time and the spatial coordinate along the direction of the flow.

At the discontinuities, the mass flux conditions give the jump conditions

$$[(\alpha - 1)m]D - [\alpha m v] = 0, \quad [(1 - \alpha)m]D - [(1 - \alpha)mw] = 0,$$

where D is the velocity of the discontinuity, and brackets denote the jump in the quantity under consideration across the discontinuity. Some additional empirical conditions, e.g., the continuity of the porosity $[m] = 0$, should also be added.

In this paper we consider the simplest kinetic equation with local rate of change in porosity being proportional to the particle flux through a given section of the porous sample

$$\frac{\partial m}{\partial t} = -\gamma \alpha m |v|,$$

where γ is a known constant, and a linear dependence of the particles velocity on the carrier fluid velocity

$$v = \varkappa(\alpha) \cdot w.$$

The system of equations is then reduced to the hyperbolic system for $\alpha(x,t)$ and $m(x,t)$

$$\frac{\partial \alpha}{\partial t} + \frac{qF'(\alpha)}{m}\frac{\partial \alpha}{\partial x} = -\frac{\gamma q(1-\alpha)F(\alpha)}{m}, \quad \frac{\partial m}{\partial t} = -\gamma q F(\alpha)$$

where $q = m\alpha v + m(1 - \alpha)w = \text{const}$, $q > 0$, is the total volumetric flow rate of the suspension (for the sake of simplicity being assumed constant), and

$$F(\alpha) = \frac{\varkappa(\alpha)\alpha}{\varkappa(\alpha)\alpha + 1 - \alpha}.$$

We will consider the poroblem of the injection of the suspension into an uncontaminated ($\alpha(x, 0) = 0$, $m(x, 0) = m_0 = \text{const}$) porous sample through the inlet section $x = 0$ with the boundary condition $\alpha(0,t) = \alpha_0 = \text{const}$, $t > 0$.

The formal solution to the problem [3] involves the jump in concentration propagating with the velocity

$$D = \frac{q}{m} \cdot \frac{[F(\alpha)]}{[\alpha]},$$

the time-independent concentration distribution behind the jump being determined by the equation

$$\gamma x = G(\alpha(x)), \quad G(\alpha) = \int_{\alpha}^{\alpha_0} \frac{F'(\alpha_1)\, d\alpha_1}{(1-\alpha_1)F(\alpha_1)}.$$

The jump conditions allow determining the implicit dependence of the concentration behind the jump on the time $\alpha_{\mathrm{f}}(t)$

$$\gamma q\, t = -\int_{\alpha_0}^{\alpha_{\mathrm{f}}} \frac{m_0 \alpha F'(\alpha)\, d\alpha}{(1-\alpha)F^2(\alpha)},$$

and the porosity distribution behind the jump is determined by a simple integration.

Journal of Physics: Conference Series **1392** (2019) 012013 doi:10.1088/1742-6596/1392/1/012013

3. Jumpless solutions. Numerical solutions to the problem of the injection into a porous sample

The presented formal solution may fail to exist if the concentraition jump is not evolutionary [3], so the actual solution turns out to be smooth. In order to find the form of the solution, we use the second-order accuracy finite difference scheme. The discontinuous solutions are evaluated through the use of artificial viscosity [5] and quite simple smoothing procedure.

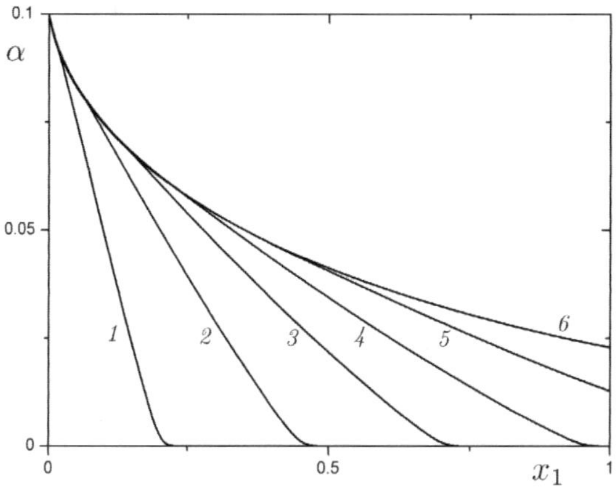

Figure 1. Smooth dependence of concentration α on dimensionless coordinate $x_1 = \gamma x$ for $\beta = 5$, $m_0 = 0.3$, $\alpha_0 = 0.1$. Curves *1* to *6* correspond to $t = 0.06$, 0.135, 0.21, 0.285, 0.39, 0.72.

In numerical calculations, we consider the velocities ratio $\varkappa$ to be of the form $\varkappa(\alpha) = 1 - \beta\alpha$, $\beta = \text{const}$, which has a simple and clear physical meaning: the more the concentration of suspended particles is, the more the retardation of particles reveals itself, and, on the other hand, for small concentrations, $\alpha \ll 1$, the velocities tend to coincide.

For positive β considered, the distribution of α consists of an uncontaminated zone $\alpha \equiv 0$ ahead of the wave, an expanding wave with gradual change in concentration, and a time-independet zone immediately adjacent to the inlet section. An example of temporal evolution of α for $\beta = 5$ is given in figure 1.

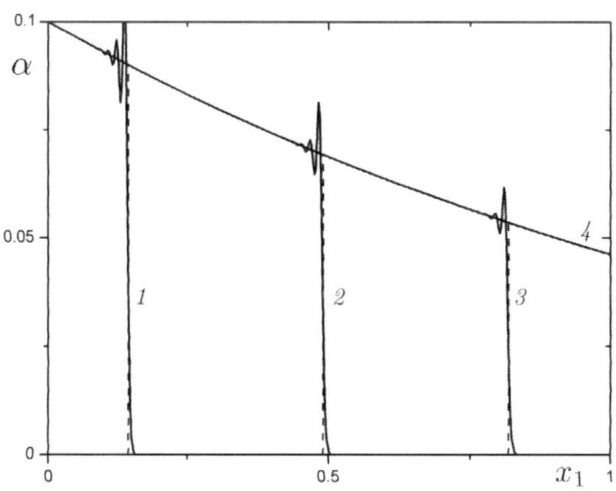

Figure 2. Discontinuous solution for $\beta = -5$, $m_0 = 0.3$, $\alpha_0 = 0.1$. Curves *1* to *4* correspond to the dimensionless time $t_1 \equiv \gamma q t = 0.0307$, 0.107, 0.184, 0.261. Solid lines represent the numerical calculations, dash lines, exact analytical solution.

The leftmost zone for $\beta \neq 1$ can be analyticaly described as

$$(\beta - 1)\gamma x = \beta \ln \frac{(\beta\alpha_0 - 1)(\beta\alpha^2 - 1)}{(\beta\alpha - 1)(\beta\alpha_0^2 - 1)} + 2\sqrt{\beta}(\operatorname{arctanh}(\sqrt{\beta}\alpha_0) - \operatorname{arctanh}(\sqrt{\beta}\alpha))$$
$$+ (\beta - 1)\ln\frac{\alpha_0}{\alpha} + \ln\frac{\alpha_0 - 1}{\alpha - 1}.$$

It is notable that the intermediate zone, as opposed to, for example, analogous Rieman problem in two-phase Backley–Leverett model, has not the self-similar form $\alpha(x/t)$, as it follows from the consideration of the sole second-order equation for α

$$-\gamma q F(\alpha)\left(\frac{\partial \alpha}{\partial t}\right)^2 + H(\alpha)\frac{\partial^2 \alpha}{\partial t^2} = \frac{\partial \alpha}{\partial t}\frac{\partial H(\alpha)}{\partial t}, \quad H(\alpha) = -\gamma q(1-\alpha)F(\alpha) - qF'(\alpha)\frac{\partial \alpha}{\partial x}.$$

For negative β the function $F(\alpha)$ may be, for certain combinations of governing parameters, convex downwards, so the jump in analytical solution appears to be evolutionary, and the discontinuous solution is realizable. It is worth noting that in this case $\varkappa > 1$, and this solution corresponds to the velocity of the particles being greater than the carrier fluid velocity.

An example of temporal evolution of α for $\beta = -5$ is presented in figure 2. The minor oscillations at the jump arise, as is well known, due to the numerical nature of calculations, and are not physical. For discontinuous solutions, the concentration jump decelerates, as follows from graphs in figure 3.

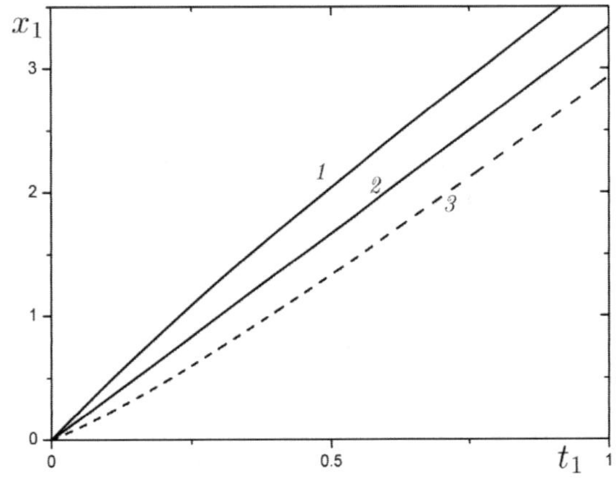

Figure 3. The dependence of the dimensionless jump position $x_1 = \gamma x$ on the dimensionless time $t_1 = \gamma q t$ for $m_0 = 0.3$, $\alpha_0 = 0.1$. Curve *1* corresponds to $\beta = -5$; curve *2*, to $\beta = 0$ (standard one-velocity model); dash curve *3*, to formal (unreliazable) discontinuous solution for $\beta = 5$.

4. Conclusion

In this paper we numerically investigate the problem of suspension injection of into a porous sample. Although the vast majority of investigations of the problem is based on the solutions with strong discontinuities in concentration of suspended particles, we show that even for simple realistic relations between the velocities of the particles and the carrier fluid it is possible that the solutions without concentration jump exist. This fact may affect the interpretation of typical laboratory experiments and should therefore be borne in mind.

Acknowledgments

The work was supported by the RFBR (project 17-01-00037).

References

[1] Herzig J P, Leclerc D M, Le Goff P 1970 Flow of suspensions through porous media. Application to deep filtration *Ind. and Eng. Chem.* **62**(5) 8–35

[2] Civan F 2000 *Reservoir Formation Damage: Fundamentals, Modeling, Assessment, and Mitigation* (Houston, Texas: Gulf Publ.)

[3] Leont'ev N E 2017 Exact solutions to the problem of deep-bed filtration with retardation of a jump in concentration within the framework of the nonlinear two-velocity model *Fluid Dynamics* **52**(1) 165–170

[4] Kuznetsov A Yu, Poslavskij S A 2009 Investigation of a mathematical model of the mechanical suffusion *Visn. Khark. Univ., Ser. Mat. Prykl. Mat. Mekh.* **875** 57–68 (in Russian)

[5] Fletcher C A J 1988 *Computational Techniques for Fluid Dynamics 2* (Berlin, Heidelberg: Springer-Verlag)

Solution of model boundary value problems on oscillations of mechanical systems with moving boundaries by the duhamel method

Vladislav L. Litvinov

Samara State Technical University, 45, Sovetskaya str., Syzran', 446001, Russian Federation

vladlitvinov@rambler.ru

Abstract. The Kantorovich – Galerkin method in conjunction with the Duhamel method is considered as applied to solving problems describing the oscillations of visco–elastic objects with conditions on moving boundaries. The mathematical formulation of the problem includes an inhomogeneous partial differential equation with respect to the desired displacement function, inhomogeneous boundary and initial conditions. By introducing a new function into the problem, the boundary and initial conditions are reduced to homogeneous. The solution is made in dimensionless variables with an accuracy of the second order of smallness with respect to small parameters characterizing the velocity of the boundary and viscoelasticity. Using the Kantorovich – Galerkin method and the Duhamel method, an approximate solution of the problem of forced longitudinal oscillations of a viscoelastic rod of variable length is found.

1. Introduction

Systems whose boundaries are moving are widely distributed in the technique (cables of cargo–lifting installations [1, 2, 9, 14–19, 21, 22], flexible links of transmission [3, 7, 9], etc.). The presence of moving boundaries causes considerable difficulties in describing such systems. Exact methods of solution are limited to the wave equation and relatively simple boundary conditions [2, 3, 8, 9, 11, 12, 15]. Of the approximate methods, the Kantorovich – Galerkin method, described in [2, 10, 14–19, 21, 22], is most effective. However, the Kantorovich – Galerkin method in conjunction with the Duhamel method can also be applied in more complex cases when solving a completely inhomogeneous initial – boundary value problem, taking into account the effect on the system of viscoelastic properties of an oscillating object [14, 16, 22].

2. Problem statement

Suppose you want to get a solution to a hyperbolic equation

$$U_{\tau\tau}(\xi,\tau) + L[U(\xi,\tau)] + \varepsilon_1 L_1[U(\xi,\tau)] = \varphi(\xi,\tau); \tag{1}$$

under initial conditions

$$U(\xi,0) = \Phi_0(\xi);\ \ U_\tau(\xi,0) = \Phi_1(\xi) \tag{2}$$

and boundary conditions

$$Y_{ji}\left[U\left(\ell_j(\varepsilon\tau),\tau\right)\right] = F_{ji}(\tau); \tag{3}$$

$$i = \overline{1,m};\ j = \overline{1,2},$$

where L – is a linear homogeneous differential operator ξ in order of $2m$; L_1 – linear differential operator (its order is τ equal to one, but ξ does not exceed $2m$); Y_{ji} – linear homogeneous differential operators on ξ; $\varphi(\xi,\tau)$ – given class function C^1; $\Phi_0(\xi)$, $\Phi_1(\xi)$, $F_{ji}(\tau)$ – specified class functions C^2; ε, ε_1 – small parameters ($\varepsilon = V/a$, V – border speed, a – the speed of propagation of oscillations).

Record of the laws of movement of boundaries in the form $\ell_j(\varepsilon\tau)$ corresponds to the slow motion mode.

3. The solution of the problem

To reduce the problem (1) – (3) to the problem with homogeneous boundary conditions, a new smooth function $u(\xi,\tau)$ is introduced that satisfies the boundary conditions (3).

By virtue of the linearity L, L_1, Y_{ji} of the operators for the difference $\tilde{U}(\xi,\tau) = U(\xi,\tau) - u(\xi,\tau)$ we get

$$\tilde{U}_{\tau\tau}(\xi,\tau) + L\left[\tilde{U}(\xi,\tau)\right] + \varepsilon_1 L_1\left[\tilde{U}(\xi,\tau)\right] = \tilde{\varphi}(\xi,\tau); \tag{4}$$

$$\tilde{U}(\xi,0) = \tilde{\Phi}_0(\xi);\ \tilde{U}_\tau(\xi,0) = \tilde{\Phi}_1(\xi); \tag{5}$$

$$Y_{ji}\left[\tilde{U}\left(\ell_j(\varepsilon\tau),\tau\right)\right] = 0; \tag{6}$$

$$i = \overline{1,m};\ j = \overline{1,2},$$

where

$$\tilde{\varphi}(\xi,\tau) = \varphi(\xi,\tau) - u_{\tau\tau}(\xi,\tau) - L\left[u(\xi,\tau)\right] - \varepsilon_1 L_1\left[u(\xi,\tau)\right];$$

$$\tilde{\Phi}_0(\xi) = \Phi_0(\xi) - u(\xi,0);\ \tilde{\Phi}_1(\xi) = \Phi_1(\xi) - u_\tau(\xi,0).$$

To obtain problems with a single in homogeneity, either in the equation or in the initial condition, we introduce into the problem (4) – (6) a new function

$$\tilde{U}(\xi,\tau) = V(\xi,\tau) + H(\xi,\tau), \tag{7}$$

where

$$V(\xi,\tau)_{\tau\tau} + L\left[V(\xi,\tau)\right] + \varepsilon_1 L_1\left[V(\xi,\tau)\right] = 0; \tag{8}$$

$$V(\xi,0) = \tilde{\Phi}_0(\xi);\ V_\tau(\xi,0) = \tilde{\Phi}_1(\xi); \tag{9}$$

$$Y_{ji}\left[V\left(\ell_j(\varepsilon\tau),\tau\right)\right] = 0; \tag{10}$$

$$i = \overline{1,m};\ j = \overline{1,2},$$

and the function $H(\xi,\tau)$ is found as a solution to the following problem:

$$H(\xi,\tau)_{\tau\tau} + L\left[H(\xi,\tau)\right] + \varepsilon_1 L_1\left[H(\xi,\tau)\right] = \tilde{\varphi}(\xi,\tau); \tag{11}$$

$$H(\xi,0) = 0;\ H_\tau(\xi,0) = 0; \tag{12}$$

$$Y_{ji}\left[H\left(\ell_j(\varepsilon\tau),\tau\right)\right] = 0; \tag{13}$$

$$i = \overline{1,m};\ j = \overline{1,2}.$$

According to the Duhamel method [26], the solution to problem (11) – (13) is

$$H(\xi,\tau) = \int_0^\tau z(\xi,\tau-\tau_0,\tau_0)\,d\tau_0,$$

where $z(\xi,\tau,\tau_0)$ – is the solution of the impulse problem

$$z(\xi,\tau)_{\tau\tau}+L\left[z(\xi,\tau)\right]+\varepsilon_1 L_1\left[z(\xi,\tau)\right]=0; \tag{14}$$

$$z(\xi,0)=0;\ z_\tau(\xi,0)=\tilde{\varphi}(\xi,\tau_0); \tag{15}$$

$$Y_{ji}\left[z(\ell_j(\varepsilon\tau),\tau)\right]=0. \tag{16}$$

4. The problem of longitudinal vibrations of a viscoelastic rod

As an example, consider the forced longitudinal oscillations of a viscoelastic rod, the change in the length of which occurs at the free end (for example, the rod extends at a speed v_0). If the attenuation of transverse oscillations is mainly due to the action of external damping forces, then in the case of longitudinal oscillations, the main influence on the damping is provided by the elastic imperfections of the material of the oscillating object.

The differential equation (viscoelasticity is taken into account based on the Voig–t hypothesis), has the form [1]

$$Z_{tt}(x,t)-a^2[Z_{xx}(x,t)+\mu Z_{xxt}(x,t)]=f(x,t). \tag{17}$$

Initial conditions

$$Z(x,0)=\psi_0(x);\ Z_t(x,0)=\psi_1(x). \tag{18}$$

System boundary conditions

$$Z(0,t)=f_1(t);\ Z_x(l_0(v_0t),t)=f_2(t). \tag{19}$$

In problem (17) – (19), the following is indicated: $Z(x,t)$ – the longitudinal displacement of a point with a coordinate x at time t; $a=\sqrt{E/\rho}$ – speed of propagation of longitudinal waves in the rod; E – long–term modulus of elasticity of the material of the rod; ρ – linear mass density; μ – small parameter, taking into account the viscoelasticity; $l_0(v_0t)=L_0-v_0t$ – the law of movement of the border; $f(x,t)$ – given class function C^1; $\psi_0(x)$, $\psi_1(x)$, $f_1(t)$, $f_2(t)$ – specified class functions C^2.

Let us introduce dimensionless variables into the problem (17) – (19):

$$\xi=\frac{\omega_0}{a}x;\ \tau=\omega_0 t+\varphi_0;\ \varphi_0=-\frac{\omega_0 L_0-a}{v_0};\ Z(x,t)=U(\xi,\tau),$$

where ω_0 – is a constant value.

After the transformations we get:

$$U_{\tau\tau}(\xi,\tau)-U_{\xi\xi}(\xi,\tau)-\varepsilon_1 U_{\xi\xi\tau}(\xi,\tau)=\varphi(\xi,\tau); \tag{20}$$

$$U(\xi,0)=\Phi_0(\xi);\ U_\tau(\xi,0)=\Phi_1(\xi); \tag{21}$$

$$U(0,\tau)=F_1(\tau);\ U_\xi(l(\varepsilon_0\tau),\tau)=F_2(\tau), \tag{22}$$

where

$$\varepsilon_1=\mu\omega_0;\ l(\varepsilon_0\tau)=1+\varepsilon_0\tau;\ \varepsilon_0=-v_0/a;\ \varphi(\xi,\tau)=\frac{1}{\omega_0^2}f\left(\frac{a}{\omega_0}\xi;\ \frac{\tau-\varphi_0}{\omega_0}\right);$$

$$\Phi_0(\xi)=\psi_0\left(\frac{a}{\omega_0}\xi\right);\ \Phi_1(\xi)=\psi_1\left(\frac{a}{\omega_0}\xi\right);\ F_1(\tau)=f_1\left(\frac{\tau-\varphi_0}{\omega_0}\right);\ F_2(\tau)=f_2\left(\frac{\tau-\varphi_0}{\omega_0}\right).$$

Following the method described above, we get

$$\tilde{U}_{\tau\tau}(\xi,\tau)-\tilde{U}_{\xi\xi}(\xi,\tau)-\varepsilon_1\tilde{U}_{\xi\xi\tau}(\xi,\tau)=\tilde{\varphi}(\xi,\tau); \tag{23}$$

$$\tilde{U}(\xi,0)=\tilde{\Phi}_0(\xi);\ \tilde{U}_\tau(\xi,0)=\tilde{\Phi}_1(\xi); \tag{24}$$

$$\tilde{U}(0,\tau)=0;\ \tilde{U}_\xi(l(\varepsilon_0\tau),\tau)=0, \tag{25}$$

where the function $\tilde{U}(\xi,\tau)$ has the form (7).

The function $V(\xi,\tau)$ is found as a solution to the following problem:

$$V_{\tau\tau}(\xi,\tau)-V_{\xi\xi}(\xi,\tau)-\varepsilon_1 V_{\xi\xi\tau}(\xi,\tau)=0; \quad (26)$$

$$V(\xi,0)=\tilde{\Phi}_0(\xi); V_\tau(\xi,0)=\tilde{\Phi}_1(\xi); \quad (27)$$

$$V(0,\tau)=0; V_\xi(l(\varepsilon_0\tau),\tau)=0. \quad (28)$$

To solve problem (26) – (28), we use the Kantorovich – Galerkin method [9, 10, 15, 19, 23]. The solution will be sought in the form

$$V(\xi,\tau)=\sum_{n=1}^{\infty} f_n(\tau)X_n(\xi,\varepsilon_0\tau). \quad (29)$$

Here are $X_n(\xi,\varepsilon_0\tau)=\sin(\omega_{0n}(\varepsilon_0\tau)\xi)$, $\omega_{0n}(\varepsilon_0\tau)=\dfrac{\pi n-\pi/2}{l(\varepsilon_0\tau)}$ – the eigenfunctions and eigenfrequencies of the problem:

$$X_{n\xi\xi}(\xi,\varepsilon_0\tau)+\omega_{0n}^2(\varepsilon_0\tau)X_n(\xi,\varepsilon_0\tau)=0; \quad (30)$$

$$X_n(0,\varepsilon_0\tau)=0; X_{n\xi}(l(\varepsilon_0\tau),\varepsilon_0\tau)=0. \quad (31)$$

Substituting the n – th term of series (29) into equation (26) we get:

$$[f_n(\tau)X_n(\xi,\varepsilon_0\tau)]_{\tau\tau}+\omega_{0n}^2(\varepsilon_0\tau)f_n(\tau)X_n(\xi,\varepsilon_0\tau)+\varepsilon_1\omega_{0n}^2(\varepsilon_0\tau)f_n'(\tau)X_n(\xi,\varepsilon_0\tau)=0. \quad (32)$$

As in [23], the function $f_n(\tau)$ will be determined from the condition of orthogonality of the left–hand side of equation (32) with a function $X_n(\xi,\varepsilon_0\tau)$ on the interval $[0,l(\varepsilon_0\tau)]$. In this case we will have:

$$\int_0^{l(\varepsilon_0\tau)} [f_n(\tau)X_n(\xi,\varepsilon_0\tau)]_{\tau\tau}X_n(\xi,\varepsilon_0\tau)d\xi+ \\ +A_{1n}(\varepsilon_0\tau)\omega_{0n}^2(\varepsilon_0\tau)f_n(\tau)+\varepsilon_1 A_{1n}(\varepsilon_0\tau)\omega_{0n}^2(\varepsilon_0\tau)f_n'(\tau)=0, \quad (33)$$

where $A_{1n}(\varepsilon_0\tau)=\displaystyle\int_0^{l(\varepsilon_0\tau)} X_n^2(\xi,\varepsilon_0\tau)d\xi=\frac{l(\varepsilon_0\tau)}{2}$.

Equation (33) with an accuracy of values of the order of smallness ε_0^2 will have the form

$$f_n''(\tau)+2A_n(\varepsilon_0\tau)f_n'(\tau)+\omega_{0n}^2(\varepsilon_0\tau)f_n(\tau)=0, \quad (34)$$

where

$$A_n(\varepsilon_0\tau)=\frac{\varepsilon_0 A_{2n}(\varepsilon_0\tau)}{A_{1n}(\varepsilon_0\tau)}+\frac{\varepsilon_1}{2}\frac{(\pi n-\pi/2)^2}{l^2(\varepsilon_0\tau)}, \quad (35)$$

$$\varepsilon_0 A_{2n}(\varepsilon_0\tau)=\int_0^{l(\varepsilon_0\tau)} X_{n_\tau}(\xi,\varepsilon_0\tau)X_n(\xi,\varepsilon_0\tau)d\xi=-\frac{\varepsilon_0 l'(\varepsilon_0\tau)}{4}. \quad (36)$$

Note that in [2], equation (34) has a similar form:

$$l^2(\varepsilon\tau)f_k''(\tau)-\varepsilon l(\varepsilon\tau)f_k'(\tau)+(R-1/2)^2\pi^2 f_k(\tau)=0.$$

Substituting (36) into (35) we get

$$A_n(\varepsilon_0\tau)=-\frac{\varepsilon_0 l'(\varepsilon_0\tau)}{2l(\varepsilon_0\tau)}+\frac{\varepsilon_1}{2}\frac{(\pi n-\pi/2)^2}{l^2(\varepsilon_0\tau)}.$$

If we introduce in equation (34) a new function

$$f_n(\tau)=A_{0n}(\varepsilon_0\tau)y_n(\tau),$$

where

$$A_{0n}(\varepsilon_0\tau)=\exp\left[-\int_0^\tau A_n(\varepsilon_0\zeta)d\zeta\right]=\sqrt{l(\varepsilon_0\tau)}\exp\left[-\int_0^\tau \frac{\varepsilon_1}{2}\frac{(\pi n-\pi/2)^2}{l^2(\varepsilon_0\zeta)}d\zeta\right],$$

then the equation can be transformed so that it will not contain a member with $y_n'(\tau)$:

$$y_n''(\tau)+\omega_{0n}^2(\varepsilon_0\tau)y_n(\tau)=0.$$

Performing transformations similar to transformations [23], for the amplitude of oscillations corresponding to the n–th dynamic mode, we obtain the following expression:

$$A_n(\tau)=\frac{l(\varepsilon_0\tau)}{\sqrt{\pi n-\pi/2}}\cdot e^{-\int_0^\tau \frac{\varepsilon_1}{2}\frac{(\pi n-\pi/2)^2}{l^2(\varepsilon_0\zeta)}d\zeta} *$$

$$*\left(D_n\cos\left(\frac{\pi n-\pi/2}{\varepsilon_0}\ln(1+\varepsilon_0\tau)\right)+E_n\sin\left(\frac{\pi n-\pi/2}{\varepsilon_0}\ln(1+\varepsilon_0\tau)\right)\right) \quad (37)$$

The solution to problem (26) – (28) is

$$V(\xi,\tau)=\sum_{n=1}^{\infty}\sin\left(\omega_{0n}(\varepsilon_0\tau)\xi\right)\cdot\frac{l(\varepsilon_0\tau)}{\sqrt{\pi n-\pi/2}}\cdot e^{-\int_0^\tau \frac{\varepsilon_1}{2}\frac{(\pi n-\pi/2)^2}{l^2(\varepsilon_0\zeta)}d\zeta} *$$

$$*\left(D_n\cos\left(\frac{\pi n-\pi/2}{\varepsilon_0}\ln(1+\varepsilon_0\tau)\right)+E_n\sin\left(\frac{\pi n-\pi/2}{\varepsilon_0}\ln(1+\varepsilon_0\tau)\right)\right),$$

where the constants D_n, E_n are determined from the initial conditions (27).

The exact solution of problem (26) – (28) without regard to viscoelasticity, obtained in [11], has the form:

$$V_\xi(\xi,\tau)=\sum_{n=1}^{\infty}\cos\left\{\frac{k_n}{2}\left[\varphi(\tau+\xi)-\psi(\tau-\xi)\right]\right\}*$$

$$*\left(A_n^*\cos\left\{\frac{k_n}{2}\left[\varphi(\tau+\xi)+\psi(\tau-\xi)\right]\right\}+B_n^*\sin\left\{\frac{k_n}{2}\left[\varphi(\tau+\xi)+\psi(\tau-\xi)\right]\right\}\right);$$

$$A_n^*=4\int_{-1}^{0}r(\zeta)\cos(k_n\zeta)d\zeta;\ B_n^*=4\int_{-1}^{0}r(\zeta)\sin(k_n\zeta)d\zeta.$$

The function $H(\xi,\tau)$ is found as a solution to the following problem:

$$H_{\tau\tau}(\xi,\tau)-H_{\xi\xi}(\xi,\tau)-\varepsilon_1 H_{\xi\xi\tau}(\xi,\tau)=\tilde{\varphi}(\xi,\tau); \quad (38)$$

$$H(\xi,0)=0;\ H_\tau(\xi,0)=0; \quad (39)$$

$$H(0,\tau)=0;\ H_\xi(l(\varepsilon_0\tau),\tau)=0, \quad (40)$$

According to the Duhamel method [26], the solution to problem (38) – (40) is

$$H(\xi,\tau)=\int_0^\tau\left(\sum_{n=1}^{\infty}\sin\left(\omega_{0n}(\varepsilon_0(\tau-\tau_0))\xi\right)\cdot\frac{1+\varepsilon_0(\tau-\tau_0)}{\sqrt{\pi n-\pi/2}}\cdot e^{-\int_0^{\tau-\tau_0}\frac{\varepsilon_1}{2}\frac{(\pi n-\pi/2)^2}{(1+\varepsilon_0\zeta)^2}d\zeta} *\right.$$

$$\left.*\left(F_n\cos\left(\frac{\pi n-\pi/2}{\varepsilon_0}\ln(1+\varepsilon_0(\tau-\tau_0))\right)+G_n\sin\left(\frac{\pi n-\pi/2}{\varepsilon_0}\ln(1+\varepsilon_0(\tau-\tau_0))\right)\right)\right)d\tau_0,$$

where $\omega_{0n}(\varepsilon_0(\tau-\tau_0)) = \dfrac{\pi n - \pi/2}{1+\varepsilon_0(\tau-\tau_0)}$, and the constants F_n, G_n are determined from the initial conditions (15).

The general solution of problem (23) – (25) is

$$\tilde{U}(\xi,\tau) = \sum_{n=1}^{\infty} \sin\left(\omega_{0n}(\varepsilon_0\tau)\xi\right)\cdot\frac{l(\varepsilon_0\tau)}{\sqrt{\pi n-\pi/2}}\cdot e^{-\int_0^{\tau}\frac{\varepsilon_1}{2}\frac{(\pi n-\pi/2)^2}{l^2(\varepsilon_0\zeta)}d\zeta} \, *$$

$$*\left(D_n\cos\left(\frac{\pi n-\pi/2}{\varepsilon_0}\ln(1+\varepsilon_0\tau)\right)+E_n\sin\left(\frac{\pi n-\pi/2}{\varepsilon_0}\ln(1+\varepsilon_0\tau)\right)\right)+$$

$$+\int_0^{\tau}\left(\sum_{n=1}^{\infty}\sin\left(\omega_{0n}(\varepsilon_0(\tau-\tau_0))\xi\right)\cdot\frac{1+\varepsilon_0(\tau-\tau_0)}{\sqrt{\pi n-\pi/2}}\cdot e^{-\int_0^{\tau-\tau_0}\frac{\varepsilon_1}{2}\frac{(\pi n-\pi/2)^2}{(1+\varepsilon_0\zeta)^2}d\zeta} \, *\right.$$

$$\left.*\left(F_n\cos\left(\frac{\pi n-\pi/2}{\varepsilon_0}\ln(1+\varepsilon_0(\tau-\tau_0))\right)+G_n\sin\left(\frac{\pi n-\pi/2}{\varepsilon_0}\ln(1+\varepsilon_0(\tau-\tau_0))\right)\right)\right)d\tau_0.$$

In conclusion, we note that the results presented here make it possible to study oscillations for systems that are described by problem (1) – (3).

References

[1] Goroshko OA, Savin G. N. Introduction to the mechanics of deformable one–dimensional bodies of variable length // Kiev: Naukova Dumka, 1971. 270 p.
[2] Lezhnev A.A. Longitudinal oscillations of a beam of variable length: Uchenye zapiski / A.A. Lezhnev. – Perm: Perm. Un – t, 1966. – №156. – C. 133–142.
[3] Vesnitsky A.I. Waves in systems with moving boundaries and loads. – M .: Fizmatlit, 2001. – 320 p.
[4] Erofeev, VI, Kolesov, DA, Lisenkova, E.E. Investigation of wave processes in a one–dimensional system, lying on an elastic–inertial base, with a moving load // Bulletin of scientific and technical development. № 6 (70), 2013. – p. 18–29.
[5] Erofeev V.I., Kazhaev V.V., Semerikova N.P. Waves in the rods. Dispersion. Dissipation. Nonlinearity – M .: Science. Fizmatlit, 2002. – 208c.
[6] Astashev V.K., Krupenin V.L. Waves in distributed and discrete vibro–impact systems and highly nonlinear media // Problems of mechanical engineering and machine reliability. – 1998. – №5. – p.13–30.
[7] Krupenin V.L. Oscillations of a string colliding with a limiter consisting of a line and a point // Bulletin of Scientific and Technological Development. – 2017. – №9 (121). – P.34–45.
[8. Samarin Yu.P. On a nonlinear problem for a wave equation in a one–dimensional space // Applied Mathematics and Mechanics. – 1964. – V. 26. – V. 3. – P. 77–80.
[9] Anisimov V.N., Litvinov V.L. Resonant properties of mechanical objects with moving boundaries: monograph. – Samara: Samar. state tech. Un – t, 2009. – 131 p.
[10] Anisimov V.N., Litvinov V.L. Investigation of resonant properties of mechanical objects with moving boundaries using the Kantorovich – Galerkin method // Herald of the Samara State Technical University. Ser. "Physical – Mathematical Sciences", 2009, 1 (18), 149–158.
[11] Anisimov V.N., Litvinov V.L., Korpen I.V. On a method of obtaining an analytical solution of the wave equation describing oscillations of systems with moving boundaries // Herald of the Samara State Technical University. Ser. "Physics and Mathematics", 2012, 3 (28), 145–151.
[12] Litvinov V.L. Solution of boundary value problems with moving boundaries using the method of changing variables in a functional equation // Journal of the Middle Volga Mathematical Society. V. 15, No. 3. 2013. – P. 112–119.

[13] Anisimov V.N., Litvinov V.L. Analysis of the influence of the movement of boundaries in the study of the resonant properties of systems with damping // Bulletin of the Samara State Technical University. Ser. "Physics and Mathematics", 2009, 2 (19), 147–152.
[14] Anisimov V.N., Litvinov V.L. Calculation of the natural frequencies of transverse vibrations of a viscoelastic rope moving in the longitudinal direction and having flexural rigidity // Mathematical modeling and boundary value problems: Proceedings of the Fifth All–Russian Scientific Conference with international participation. Part 1: Mathematical models of mechanics, strength and reliability of structural elements. – Samara: Samara State Technical University, 2008. – 358 p.
[15] Litvinov V.L., Anisimov V.N. Mathematical modeling and study of vibrations of one–dimensional mechanical systems with moving boundaries: monograph. // Samar. state tech. Un – t, 2017. – 149 p.
[16] Litvinov V.L. Transverse vibrations of a viscoelastic rope lying on an elastic foundation, taking into account the influence of environmental resistance forces. Vestnik nauchno–tekhnicheskogo razvitiya. № 4 (92), 2015. – P. 29–33.
[17] Anisimov V.N., Litvinov V.L. Calculation of the natural frequencies of the rope moving in the longitudinal direction // Journal of the Middle Volga Mathematical Society. – 2017. – Vol. 19, No. 1. – P. 130–139.
[18] Litvinov V.L. Longitudinal oscillations of a rope of variable length with a load at the end // Bulletin of Scientific and Technical Development. № 1 (101), 2016. – p. 19–24.
[19] Litvinov V.L. Application of the Kantorovich–Galerkin method in the study of the resonant properties of systems with damping. Herald of Scientific and Technological Development. № 2 (114), 2017. – P. 37–46.
[20] Anisimov V.N., Litvinov V.L. An Analytical Method for Solving the Wave Equation with a Wide Class of Conditions on Moving Boundaries // Bulletin of Scientific and Technological Development. № 2 (102), 2016. – pp. 28–35.
[21] Anisimov V.N., Litvinov V.L. Mathematical models of longitudinal–transverse vibrations of objects with moving boundaries // Vestn. Sam.gos tech. un–that. Ser. Phys–mat. Science, 2015. Vol. 19, №2. Pp. 382–397.
[22] Anisimov V.N. Longitudinal resonant oscillations of a viscoelastic rope of a load–lifting installation // Proceedings of the Samara Scientific Center of the Russian Academy of Sciences. 2016. V. 18. № 4–1. Pp. 128–133.
[23] Anisimov V.N., Litvinov V.L., Korpen I.V. Use of the Kantorovich – Galerkin method for solving boundary value problems with conditions on moving boundaries // Izvestia of the Russian Academy of Sciences. Solid mechanics. 2018. №2. Pp. 70–77.
[24] Smirnov V.I. The course of higher mathematics, 4. Fizmatgiz, M., 1958.
[25] Kamke E. Handbook of ordinary differential equations. M .: Nau–ka, 1976. 576 p.
[26] Kotlyar Ya.M. Methods of mathematical physics and hydroaerodynamic problems: [Proc. manual for technical colleges] / Ya. M. Kotlyar. – M.: Higher. school, 1991. – 207 p.

About the algorithm for calculating the final probabilities

Mastikhin A.V., Mastikhina A. A.

Department of Higher Mathematics, Bauman Moscow State Technical University, 2nd Bauman St., 5, 105005 Moscow, Russia

E-mail: mastikhin@bmstu.ru

Abstract. We consider a Markov process with continuous time and a countable number of states, known as the (general) Bartlett–Mac-Kendrick epidemic. Kolmogorov's equation for the exponential (double) generating function of the final probabilities is a hyperbolic partial differential equation. Its solution requires the calculation of the final probabilities for trajectories specifying one of the boundary conditions. It is shown that the desired trajectory is described as the Dick paths, enumerated by Catalan numbers. For arbitrary trajectories of the final probabilities of the general epidemic process, a calculation algorithm based on the contex-free grammar is proposed. The enumerative problem is solved.

1. Introduction

There are two basic models in the mathematic theory of epidemics: the simple epidemics (also named after G.Weiss) and the general epidemics. We consider some problems of the last one. Definition of the process is as follows.

On the set of states $N^2 = \{\alpha = (\alpha_1, \alpha_2),\ \alpha_1, \alpha_2 = 0, 1, 2, \dots\}$ we consider a homogeneous (in time) Markov process $\xi(t) = (\xi_1(t), \xi_2(t))$, $t \in [0, \infty)$, with transition probabilities $P_{(\beta_1,\beta_2)}^{(\alpha_1,\alpha_2)}(t) = \mathbf{P}\{\xi(t) = (\beta_1, \beta_2) \,|\, \xi(0) = (\alpha_1, \alpha_2)\}$. Suppose the transition probabilities have the following form as $\Delta t \to 0+$:

$$\begin{aligned} P_{(\alpha_1,\alpha_2-1)}^{(\alpha_1,\alpha_2)}(\Delta t) &= \mu\alpha_1\alpha_2\Delta t + o(\Delta t), \\ P_{(\alpha_1-1,\alpha_2)}^{(\alpha_1,\alpha_2)}(\Delta t) &= \rho\alpha_1\Delta t + o(\Delta t), \\ P_{(\alpha_1,\alpha_2)}^{(\alpha_1,\alpha_2)}(\Delta t) &= 1 - (\mu\alpha_1\alpha_2 + \rho\alpha_1)\Delta t + o(\Delta t), \end{aligned}$$

where $\mu > 0, \rho > 0$.

The Markov process $\xi(t)$ can be interpreted as a model for the spread of epidemic through a population. The state of the process $(\alpha_1, \alpha_2, \alpha_3)$ means the presence of α_1 particles of type T_1 (the carriers of infection) and α_2 particles of type T_2 (the susceptible individuals).

In the stochastic theory of epidemics the general stochastic epidemic (i.e. Bartlett–Mac-Kendric epidemic process) and the simple stochastic epidemic (i.e. Weiss epidemic process) are basic models; both of them are considered on the set of states N^2.

In Bartlett–Mac-Kendric epidemic process an ineracting pair of infected and susceptible particles tranform into a pair of two infected particles (or two carriers)

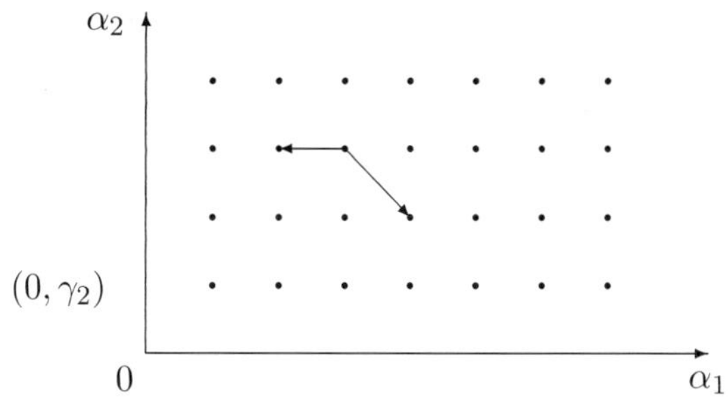

Fig.1. Jumps of the process.

For the process $\xi(t)$ define *final probabilities* for absorbing states $(0, \gamma)$, $\gamma = 0, 1, 2, \dots$ (i.e. all carriers were removed),

$$q_{(0,\gamma)}^{(\alpha_1,\alpha_2)} = \lim_{t\to\infty} P_{(0,\gamma)}^{(\alpha_1,\alpha_2)}(t), \ (\alpha_1, \alpha_2) \in N^2; \quad \sum_{\gamma=0}^{\infty} q_{(0,\gamma)}^{(\alpha_1,\alpha_2)} = 1. \tag{1}$$

For final probabilities (1) we have generating functions ($|s| \le 1$)

$$\begin{aligned} \Phi_{(\alpha_1,\alpha_2)}(s) &= \sum_{\gamma=0}^{\infty} q_{(0,\gamma)}^{(\alpha_1,\alpha_2)} s^{\gamma}; \\ \Phi(z_1, z_2; s) &= \sum_{\alpha_1,\alpha_2=0}^{\infty} \frac{z_1^{\alpha_1} z_2^{\alpha_2}}{\alpha_1!\alpha_2!} \Phi_{(\alpha_1,\alpha_2)}(s), \end{aligned} \tag{2}$$

and Kolmogorov differential equation

$$\mu z_2\left(\frac{\partial\Phi}{\partial z_1} - \frac{\partial^2\Phi}{\partial z_1 \partial z_2}\right) + \rho\left(\Phi - \frac{\partial\Phi}{\partial z_1}\right) = 0, \tag{3}$$

with initial condition $\Phi(0, z_2; s) = e^{z_2 s}, \Phi(z_1, 0; s_) = e^{z_1}$. The solution of equation (5) is based on Riemann method function and we need some extra initial condition

$$\frac{\partial\Phi(0, z_2; s)}{\partial z_1} = \sum_{\alpha_2,\gamma=0}^{\infty} \frac{z_2^{\alpha_2}}{\alpha_1!} q_{(0,\gamma)}^{(1,\alpha_2)} s^{\gamma}.$$

2. Grammar for final probabilities

Consider the probabilities of diagonal and horizontal jumps

$$q_{(\alpha_1+1,\alpha_2-1)}^{(\alpha_1,\alpha_2)} = \frac{\mu}{\alpha_2 + \mu}, \quad q_{(\alpha_1-1,\alpha_2)}^{(\alpha_1,\alpha_2)} = \frac{\alpha_2}{\alpha_2 + \mu}.$$

For fixed initial $(\alpha_1, \alpha_2)$and the ultimate $(0, \gamma)$ points one trajectory is a polyline consisting of diagonal and horizontal jumps, which correspond to the multipliers in the total sum of all trajectories for the final probabilities $q_{(0,\gamma)}^{(\alpha_1,\alpha_2)}$. Each summand contains the same set of multipliers,corresponding (vertical) to diagonal jumps. The probability of horizontal jumps depend only on the vertical level. The following algorithm for calculating the final probabilities $q_{(0,\gamma)}^{(1,\alpha_2)}$ is proposed.

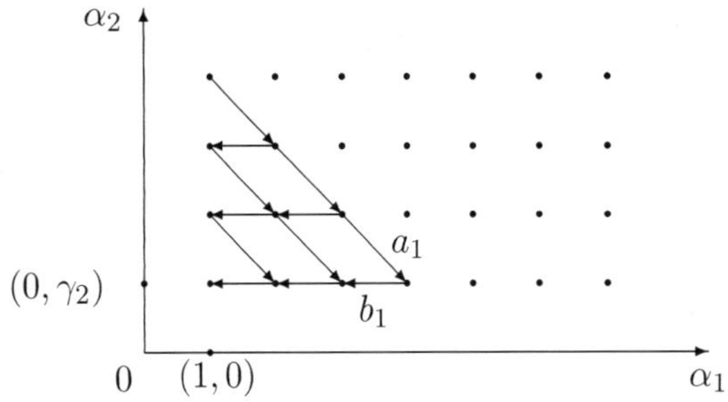

Fig.2. The set of trajectories

Denote the number of vertical transitions $n = \alpha_2 - \gamma$, the probabilities of the diagonal transitions $a_k = q^{(\alpha_1,\alpha_2)}_{(\alpha_1+1,\alpha_2-1)} = \frac{\mu}{\alpha_2+\mu}$, $k = 1,\dots,n$, horizontal transition probabilities $b_k = q^{(\alpha_1,\alpha_2)}_{(\alpha_1-1,\alpha_2)} = \frac{\alpha_2}{\alpha_2+\mu}$, $k = 1,\dots,n$. For example, for $n = 3$ the required probability is calculated as the sum of five terms corresponding to five possible trajectories

$$a_1a_2a_3b_1(b_1^3 + b_1^2b_2 + b_1^2b_3 + b_1^2b_2^2 + b_1b_2b_3).$$

The terms in brackets and their products of horizontal probabilities are of interest. Consider them in layers for convenience.

THEOREM. 1 Trajectories determined by final probabilities $q^{(1,\alpha_2)}_{(0,\gamma)}$ of general epidemics are generated by the grammar

$$\begin{aligned}
r_0 &\to b_1r_1a_1;\\
r_1 &\to b_1r_2a_2|b_2r_3a_2;\\
r_2 &\to b_1r_4a_3|b_2r_5a_3|b_3r_6a_3;\\
r_3 &\to b_2r_5a_3|b_3r_6a_3;\\
&\dots\\
r_{k-n+1} &\to b_1r_{k+1}a_n|b_2r_{k+1}a_n|\dots|b_{n-1}r_{k+1}a_n|b_nr_{k+1}a_n;\\
&\dots\\
r_k &\to b_{n-1}r_{k+1}a_n|b_nr_{k+1}a_n;\\
r_{k+1} &\to \lambda;\\
\lambda &\to 0.
\end{aligned}$$

Number of trajectories determined by final probabilities $q^{(1,\alpha_2)}_{(0,\gamma)}$ of general epidemics is equal to Catalan number $c_n = \frac{1}{n+1}C^n_{2n}$, $k = (n(n+1))/2$.

3. Calculation of number of final probabilities

Let's denote the number of all trajectories c^m_n, determined by final probabilities $q^{(\alpha_1,\alpha_2)}_{(0,\gamma)}$. Then $c^1_n = c_n$. It follows from the definition that $c^{m-1}_n = c^m_n + c^{m-2}_{n-1}$ or $c^m_n = c^{m-1}_n - c^{m-2}_{n-1}$. For example, $c^2_n = c_n - c_{n-1}$. Repeatedly applying this formula, we obtain

$$\begin{aligned}
c^2_n &= c_n - c_{n-1},\\
c^3_n &= c_n - 2c_{n-1},\\
c^4_n &= c_n - 3c_{n-1} + c_{n-2},\\
c^5_n &= c_n - 4c_{n-1} + 3c_{n-2},
\end{aligned}$$

$$\begin{aligned}
c_n^6 &= c_n - 5c_{n-1} + 6c_{n-2} - c_{n-3},\\
c_n^7 &= c_n - 6c_{n-1} + 10c_{n-2} - 4c_{n-3},\\
c_n^8 &= c_n - 7c_{n-1} + 15c_{n-2} - 10c_{n-3} + c_{n-4},\\
c_n^9 &= c_n - 7c_{n-1} + 21c_{n-2} - 20c_{n-3} + 5c_{n-4},\\
\ldots
\end{aligned}$$

Let A_k^m is the coefficient of the number of Catalan c_{n-k} in the sum

$$c_n^m = c_n - A_1^m c_{n-1} + A_2^m c_{n-2} - \ldots + (-1)^k A_k^m c_{n-k} + \ldots + (-1)^{[\frac{m}{2}]} A_{[\frac{m}{2}]}^m c_{n-[\frac{m}{2}]}. \tag{4}$$

It's easy to find the first coefficients

$$A_0^m = 1, \quad A_1^m = m-1, \quad A_2^m = \frac{(m-2)(m-3)}{2}.$$

THEOREM 2. Number of trajectories c_n^m, determined by final probabilities $q_{(0,\gamma)}^{(\alpha_1,\alpha_2)}$ of general epidemics is equal to

$$c_n^m = \sum_{k=1}^{[\frac{m}{2}]} (-1)^k \frac{(k+1)\ldots(m-k)}{(m-2k)!} c_{n-k}.$$

The proof is based on the following four lemmas

LEMMA 1. Coefficients A_k^{m-1} in the formula (6) are subject to the ratio

$$A_k^m = A_k^{m-1} + A_{k-1}^{m-2}.$$

LEMMA 2. Exponential generating function $f_1(t)$ for the coefficients A_1^{m-1} is

$$f_1(t) = \sum_{m=2}^{\infty} A_1^{m-2} \frac{t^{m-2}}{(m-2)!} = 1 + (t-1)e^t.$$

LEMMA 3. Exponential generating function $f_k(t)$ for the coefficients A_k^{m-1}satisfies the recurrence relation

$$f_k(t) = -\int_0^t f_{k-1}(\tau)d\tau + e^t \int_0^t e^{-\tau} f_{k-1}(\tau)d\tau.$$

LEMMA 4. For images obtained by the Laplace transform $F_k(p)$ of exponential generating functions $f_k(t)$ we have

$$F_k(p) = \frac{F_1(p)}{p^{k-1}(p-1)^{k-1}} = \frac{1}{p^k(p-1)^{k+1}}.$$

4. Conclusion.

The resulting Catalan number decomposition is strikingly different from a similar result for a simple epidemic. The problem of the number of trajectories of the General epidemic process can be generalized to the multidimensional case. However, the results show its non-triviality.

References

[1] Kalinkin A V 1998 *Theory Prob Appl* **43** 633
[2] Kalinkin A V 2002 *Russian Math Surveys* **57** 241
[3] Mastikhin A V 2007*Math. Notes.* **82** 787
[4] Kalinkin A V, Mastikhin A V 2019 *JOTP* **32** 1 163

Journal of Physics: Conference Series **1392** (2019) 012017 doi:10.1088/1742-6596/1392/1/012017

Construction of the equations of state for polycrystalline solids for the purpose of the numerical solution of problems of continuous medium mechanics

A V Ostrik and D N Nikolaev

Institute of Problems of Chemical Physics of the Russian Academy of Sciences, Academician Semenov Avenue 1, Chernogolovka, Moscow Region 142432, Russia

E-mail: ostrik@ficp.ac.ru, nik@ficp.ac.ru

Abstract. The equations of state for polycrystalline solids which are in region of compression are developed on experimentally defined shock Hugoniot adiabat and dependence of specific heat on density and temperature. These equations are convenient for the numerical solution of problems of continuous medium mechanics. The numerical method for definition of specific cold energy and the Gruneizen function at $T = 0$ K depending on compression ratio is proposed. Results of comparison for temperatures behind shock wave in sapphire by means of the developed equations of state and other methods are given.

1. Introduction

The equations of state (EOS) [1, 2] are required for the solution of problems of continuous medium mechanics (CMM) by numerical methods [3]. In general case it is convenient to represent EOS in the form of dependences of specific internal energy and pressure on density and temperature [3, 4]. The construction of EOS in such a form requires to meet a conditions of thermodynamic compatibility.

Development of EOS becomes more simple for polycrystalline bodies due to their isotropy and a possibility of application of the Debye theory of specific heat. Moreover the class of polycrystalline bodies includes constructional materials which are important for practice (in particular, metals and their alloys). The representative data sets required for construction of EOS is accumulated for such materials.

The CMM equations (conservation laws) are complemented with tabular wide-range and semi-empirical EOS [2, 5] that leads to a set of difficulties [4, 6] in the numerical solution of these equations. Large times of calculation and deviation from a condition of thermodynamic compatibility at approximation on a two-dimensional grid of temperatures and density are the main of them. The requirements for tabular EOS and their approximations for the solution of CMM problems are formulated in [3, 4].

Thus the wide-range EOS, where the approximations on two-dimensional grids are not required, are necessary for development of effective numerical CMM codes. A significant amount of works (in relation to area of shock compression see, for example, [7, 8, 9]) is devoted to development of the simplified variants of analytical EOS for the solution of problems of CMM. Unlike these works the Gruneizen function at absolute zero and cold pressure depending on density is offered to be represented in a tabular form (a grid is one-dimensional in this case). Methods of creation of these functions and EOS which takes into account a contribution of electronic components in specific heat are considered in the present article.

2. Basic relations

Data from shock wave experiments in combination with data on its thermophysical properties are traditionally used for construction of the equations of state for substances. In particular the dependence of specific heat on density and temperature can be the basis for creation of EOS. In this case the equations of state can be constructed in the form of ratios like Mi-Gruneizen

$$E = E(\rho, T) = E_c(\rho) + E_T(\rho, T) == E_c(\rho) + \int_0^T C_v(\rho, T)dT, \tag{1}$$

$$P = P(\rho, T) = P_c(\rho) + \Gamma(\rho, t)\rho E_T = \rho^2 \frac{dE_c}{d\rho} + \Gamma(\rho, t) \int_0^T C_v(\rho, T)dT, \tag{2}$$

where ρ is substance density; E, E_c, E_T are full cold and thermal specific energy; P, P_c, P_T are full cold and thermal pressure; $\Gamma = \Gamma(\rho, T)$ is Gruneizen function; C_v is specific heat.

Two unknown functions $E_c(\rho)$ and $\Gamma(\rho, T)$ (the dependence of specific heat on density and temperature is considered known) are included into EOS (1), (2). They can be found from data of shock – wave compression experiments using a condition of thermodynamic compatibility

$$\left(\frac{\partial E}{\partial(1/\rho)}\right)_T = T\left(\frac{\partial P}{\partial T}\right)_\rho - P \tag{3}$$

We receive the equation for definition of the Gruneizen function by substituting (1), (2) in (3)

$$\left(\frac{\partial T E_T \Gamma}{\partial T}\right)_\rho - 2\frac{T E_T \Gamma}{T} = \left(\frac{\partial E}{\rho \partial(1/\rho)}\right)_T, \tag{4}$$

Then we find

$$\Gamma(\rho, T) = \frac{T}{E_T(\rho, T)} \int_0^T \left(\frac{\partial E_T(\rho, T)}{\rho \partial(1/\rho)}\right)_T \frac{dT}{T^2} \tag{5}$$

integrating (4) with an initial condition (it's considered that at $T \to 0$ and $C_v \to 0$)

$$\left.\frac{T E_T \Gamma}{T^2}\right|_{T=0} = \lim_{T\to 0} \frac{C_v T \Gamma}{T} = \lim_{T\to 0} C_v \Gamma = 0.$$

By substituting the expression for thermal energy through thermal capacity in (5) and integrating by parts ($v = 1/\rho$) we receive

$$\Gamma(\rho, T) = \frac{E_{T0}}{E_T}\frac{T}{T_0}\left(\Gamma_0(\rho) + \left(1 - \frac{T_0}{T}\right)\frac{dE_{T0}}{E_{T0}\rho dv}\right) + \frac{T}{E_T}\int_{T_0}^T \left(\frac{\partial C_v}{\rho \partial v}\right)_T \frac{dT}{T} - \frac{1}{E_T}\int_{T_0}^T \left(\frac{\partial C_v}{\rho \partial v}\right)_T dT, \tag{6}$$

where $\Gamma_0(\rho)$ is the Gruneizen function at a normal temperature.

Let's note that the constant value of thermal capacity at compression when temperature exceeds normal temperature of T_0 is accepted in the simple models of EOS [7, 8, 9] ($E_T(T) = E_{T0}(\rho) + C_v(T - T_0)$). Then the Gruneizen function (6) takes a form

$$\Gamma(\rho, T) = \frac{E_{T0}}{E_T}\frac{T}{T_0}\left(\Gamma_0(\rho) + \left(1 - \frac{T_0}{T}\right)\frac{dE_{T0}}{E_{T0}\rho dv}\right) \tag{7}$$

Two opportunities follow from (7). If function $\Gamma_0(\rho)$ is taken from experimental data then the Gruneizen function depends also on temperature. Besides we can demand independence from T that gives according to (7)

$$\Gamma(\rho) = \Gamma_0(\rho) = \frac{1}{C_vT_0 - E_{T0}}\frac{dE_{T0}}{\rho d(1/\rho)} \tag{8}$$

Integrating by parts (5) for Debye approach when $E_T(\rho, T) = Te_T(\theta_D(\rho)/T)$ and $C_v(\rho, T) = C_v(\theta_D(\rho)/T)$, we receive

$$\Gamma(\rho) = \Gamma_{00}(\rho) = -\frac{dln\theta_D}{dln(1/\rho)}, \tag{9}$$

where $\Gamma_{00}(\rho)$ is the Gruneizen function at absolute zero. According to (9) the Gruneizen function does not depend on temperature in Debye approach. Let's note that the dependence (9) provides implementation of a condition of thermodynamic compatibility.

In more common case the specific internal energy is defined considering a contribution of electronic components in thermal capacity

$$E_T(T) = Te_T\left(\frac{\theta_D(\rho)}{T}\right) + \frac{C_eT^2}{2}\left(\frac{\rho_0}{\rho}\right)^{\gamma_e}, \tag{10}$$

where ρ_0 is substance density under normal conditions; C_e is coefficient of electronic heat conductivity; γ_e is an electronic analog of Gruneizen coefficient.

We find from (5), (9) and (10)

$$\Gamma(\rho, T) = \frac{T}{E_T(\rho, T)}\left[\Gamma_{00}(\rho)e_T\left(\frac{\theta_D(\rho)}{T}\right) + \gamma_e\frac{C_eT}{2}\left(\frac{\rho_0}{\rho}\right)^{\gamma_e}\right], \tag{11}$$

According to (11) Gruneizen function depends also on temperature when the contribution of electronic heat capacity is considered. But this function is still defined by functions of one variable $\theta_D(\rho)$, $\Gamma_{00}(\rho)$.

Thus in all considered special cases the Gruneizen function can be calculated from the known functions of one variable ρ. These functions can be presented in a tabular form.

3. Definition of cold energy and the Gruneizen function in the compression region

The shock compressibility is used for finding the specific cold energy $E_c = E_c(\rho)$. Hugoniot relation at the front of a shock wave has the form (the index zero notes the values at normal conditions):

$$E - E_0 = \frac{1}{2}(P_H(\rho) + P_0)\left(\frac{1}{\rho_0} - \frac{1}{\rho}\right)$$

Using (1), (2) and Hugoniot relation for Debye approach ($\Gamma(\rho) = \Gamma_{00}(\rho)$) we find

$$E_T(\rho, T_H(\rho)) = \psi(\rho) - E_c^* \tag{12}$$

$$\psi(\rho) = \frac{1}{2}(P_H(\rho) + P_0)\left(\frac{1}{\rho_0} - \frac{1}{\rho}\right) + E_{T0}, \quad E_{T0} = E_T(\rho_0, T_0), \quad E_c^* = E_c(\rho) - E_c(\rho_0).$$

Here $T_H(\rho)$ is temperature on a shock Hugoniot curve. Let's note that if the experimental Hugoniot can be approximated by linear dependence $D = \alpha + \beta U$ between wave and particle speeds (D is wave speed; U is particle speed, α, β are experimentally defined coefficients) then the shock compressibility is determined by the known relation

$$P_H(\rho) = \rho_0 DU = \rho_0 \frac{\alpha^2(1-\rho_0/\rho)}{[1-\beta(1-\rho_0/\rho)]^2}.$$

The equations of state are applicable on a shock Hugoniot curve too and we have for $\Gamma(\rho) = \Gamma_{00}(\rho)$

$$-\frac{dE_c^*}{d(1/\rho)} + \Gamma_{00}(\rho)\rho E_T(\rho, T_H(\rho)) = P_H(\rho). \tag{13}$$

Function $\Gamma_{00}(\rho)$ is calculated for a polycrystalline solid from the known and common relation ($\gamma = 2t/3$)

$$\Gamma_{00}(\rho) = -\frac{2-t}{3} - \frac{1}{2\rho}\frac{\frac{d^2}{d(1/\rho)^2}\left(\rho^{-\gamma}P_c(\rho)\right)}{\frac{d}{d(1/\rho)}\left(\rho^{-\gamma}P_c(\rho)\right)}, \tag{14}$$

where $t = 0$ corresponds to Landau and Slater theory, $t = 1$ corresponds to Dugleyla and MacDonald hypothesis and $t = 2$ corresponds to theory of free volume. Parameter t entering in (14) can be determined from a condition for existence of tangent of the third order between a Hugoniot curve and a zero isentrope

$$\Gamma_{00}(\rho_{00}) = 2\beta - \frac{2+t}{3}.$$

The ordinary differential equation of the third order for definition of specific cold energy depending on density follows from (12)-(14)

$$\frac{d^3E_c^*}{d\rho^3} = 2\left(\frac{2+t}{3} + \frac{P_H(\rho) - \rho^2 dE_c^*/d\rho}{\rho(\psi(\rho) - E_c^*)} - 3\right)\left(\frac{d^2E_c^*}{\rho d\rho^2} + (2-\gamma)\frac{dE_c^*}{\rho^2 d\rho}\right) + (3-\gamma)(2-\gamma)\frac{dE_c^*}{\rho^2 d\rho}. \tag{15}$$

Initial conditions for the equation (15) are set at the normal density $\rho = \rho_0$

$$E_c^*|_{\rho=\rho_0} = 0, \quad \left.\frac{dE_c^*}{d\rho}\right|_{\rho=\rho_0} = \frac{P_0}{\rho_0^2} - \frac{\Gamma_0 E_{T0}}{\rho_0}, \quad \left.\frac{d^2E_c^*}{d\rho^2}\right|_{\rho=\rho_0} = -\left.\frac{dP_H}{\rho^2 d\rho}\right|_{\rho=\rho_0} - \frac{2P_0}{\rho_0^3} - \frac{\Gamma_0 E_{T0}}{\rho_0^2}(\Gamma_0 - 2).$$

The first condition is obvious from determination of E_c^*. The second condition follows from (13). The third condition for the second derivative can be received from a condition that a shock Hugoniot and isentrope line have common tangent ($(\partial P/\partial\rho)_{H0} = (\partial P/\partial\rho)_{S0}$).

The differential equation (15) is wrote in a form of a system of three ordinary differential equations of first order and integrated numerically by Runge – Kutta method of the fourth order of accuracy with a variable step on density.

Debye temperature is defined after calculation of function $E_x^* = E_x^*(\rho)$ and its derivatives

$$\theta_D(\rho) = \theta_{D0}\left(\frac{\rho}{\rho_0}\right)\sqrt{\frac{d(\rho^{-\gamma}P_c)/d\rho}{d(\rho^{-\gamma}P_c)/d\rho|_{\rho=\rho_0}}}, \qquad P_c = -\frac{dE_c^*}{d(1/\rho)}.$$

Then temperature on a shock Hugoniot curve is calculated as the solution of the nonlinear equation (12)

$$T_H(\rho)e_T(\theta_D(\rho)/T_H(\rho)) = \psi(\rho) - E_c^*(\rho)$$

The system of five differential equations is integrated in more common case when contribution of electronic components in the specific heat (see (10), (11)) taking into account. The differential equations for temperature and Debye temperature having the corresponding initial conditions are added

4. Equations of state for Sapphire

Sapphire is widely used for measurements of temperature of shock compression of metals and their alloys [10, 11, 12, 13]. Up to now the shock compressibility of sapphire is experimentally defined in pressure range of 80-340 GPa by means of the two stage gas gun [14]. Action of powerful laser radiation was used for measurement of shock compressibility at pressure of 0.7-2 TPa [15, 16, 17].

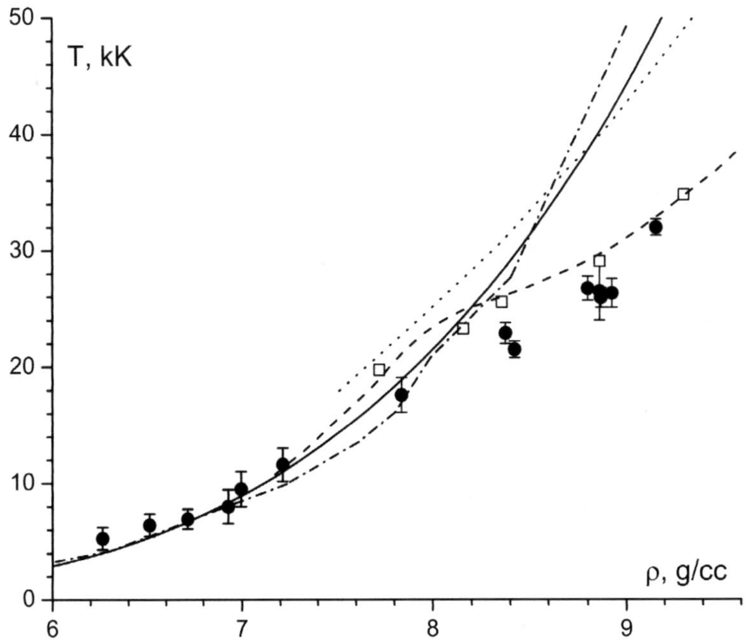

Figure 1. Hugoniot temperature vs density. Experiment: ● this work, □[19, 20]. Calculations: ——this work, - - - -semiempirical EOS [19], ······SESAME table EOS [20], — · —QMD simulation [21]

Pressure range of of 340-700 GPa experimentally was not studied. We used plane shock compression generator [18] for creation of pressure of this range. Temperature behind the front of the shock wave propagating in sapphire was measured by optic pyrometry methods.

Data on a shock Hugoniot curve [19] were used for construction of sapphire EOS. Specific heat was calculated from Debye theory. The contribution of electronic components in thermal capacity was not considered as sapphire is dielectric under normal conditions.

Numerical and experimental results for sapphire temperature on a shock Hugoniot are shown on a figure 1. Our calculations were made with use of the developed EOS. It is visible that our data of pyrometric measurements are slightly below than results of these calculations.

5. Conclusions

(i) New method for construction of the wide-range EOS for polycrystalline materials is proposed. In this method the experimental data on dynamic compressibility of materials are reproduced with any required accuracy (5% as a rule).

(ii) The proposed EOS are convenient for the solution of problems of CMM as the interpolation of data is required only on one-dimensional tables.

(iii) Comparison of numerical and experimental data for temperatures behind the front of a shock wave demonstrates satisfactory agreement.

Acknowledgments

Work is made with financial support from the Russian Foundation of Basic Researches (grant No. 18-08-00094-a, 18-08-00964-a).

References

[1] Bushman A V, Lomonosov I V, Fortov V E 1992 *The equations of state of metals at the high density of energy* (Chernogolovka: IPCP RAS)
[2] Fortov V E 2016 *Thermodynamics and equations of state for Matter: from ideal gas to quark-gluon plasma* (USA: World Scientific Publishing)
[3] Kuropatenko V F 2007 *Models of mechanics of continuous media* (Chelyabinsk: Chelyabinsk state University)
[4] Kim V V, Lomonosov I V, Ostrik A V 2015 *Constructions from composite materials* **2**(138) 39–45
[5] Sapozhnikov A T, Gershchuk P D, Malyshkina E L, et al 1991 *VANT: Math. mod. of phys. proc.* **1** 9–16
[6] Prokopov V G 2005 *VANT: Math. mod. of phys. proc.* **4** 98–101
[7] Belkheeva R K 2009 *The Bulletin of the Nizhny Novgorod State University: Math., Mech. and Inf.* **9:3** 23–32
[8] Kinelovsky S A, Mayevsky K K 2014 *Thermophysics of High Temp.* **52, 6** 843–852
[9] Ostrik A V, Utkin A V 2017 *J. Mat. Phys. and Mech.* **31, 1-2** 48–51
[10] Tan H, Ahrens T J 1990 *High Pressure Research* **2, 3** 159–182 10
[11] McQueen R G, Isaak D G 1990 *J. of Geophys. Res.* **95, 21** 753–765 11
[12] Anderson W A, Ahrens T J 1996 *J. of Geophys. Res.* **101(B3)** 5627–5642 12
[13] Hare D E, Webb D J, Lee S H, et al 2002 *AIP Conf. Proc.* **2** 1231–1234 13
[14] Erskine D 1994 *High-Press. Sci. Techn.-1993, ed. S. C. Schmidt et al* 141–143 14
[15] Hicks D G, Celliers P M, Collins G W, et al 2003 *PRL* **91, 3** 035502
[16] Root S, Magyar R, Lemke R, et al 2013 *Top. Conf. on the Sh. Comp. of Mat.*
[17] Ozaki N, Nellis W J, Mashimo T, et al 2016 *Scientific Reports 6:26000*
[18] Nikolaev D, Ternovoi V, Kim V, et al 2014 *J. of Phys: Conf. Ser.* **14** 1–5
[19] Kerley G I 2009 *Scientific Reports 1306.6877*
[20] Miller J, Boehly T R, Meyerhofe D D, et al 2006 *Presetnation at http://www.lle.rochester.edu/media/publications/presentations/documents/APS06/Miller_ APS06.pdf*
[21] Liu H, Tse J S, Nellis W J 2015 *Scientific Reports 5:12823*

Mathematical modeling of creep and viscoplastic flow of a cylindrical layer material

L V Kovtanyuk and G L Panchenko

Institute of Automation and Control Processes, Far Eastern Branch, Russian Academy of Sciences, Radio st. 5, Vladivostok, 690041, Russia

E-mail: panchenko.21@yandex.ru

Abstract. Within the framework of a mathematical model of large elastoplastic deformations, taking into account viscous properties of materials, a boundary value problem of creep and viscoplastic flow of a material in the gap between two rigid coaxial cylindrical surfaces under rectilinear motion of the inner surface is considered. Irreversible deformations accumulate from the beginning of the deformation process and are initially associated with the creep process of the material, which is described using the Norton creep power law. With a subsequent increase in the speed of movement of the inner cylinder at some time point, the stress state will reach the yield surface and a viscoplastic flow will begin in the material. As a plastic potential, we use the Tresca yield criterion, generalized in the case of taking into account the viscous properties of the material. The viscoelastic deformation, the emergence and development of the viscoplastic flow under uniformly accelerated motion of the inner surface, the flow under constant speed, and the flow inhibition under equal slow motion of the surface to a full stop are considered. Stresses, reversible and irreversible deformations, displacements are calculated, stress relaxation after the complete stop of the cylinder is investigated.

1. Introduction

The most preferred method of processing some structural materials at the present time is a technology of cold forming, when large irreversible deformations are accumulated at the expense of the slow creep process. The difficulty in applying this technology lies in the fact that at impact a rigging to the moldable material in the places of their contact plastic areas are formed. Their presence substantially redistributes the stress field and, therefore, directly affects the creep process as a whole. Since in such industrial technologies of processing materials by pressure the materials significantly change the shape, therefore, they acquire large deformations. Estimated forecasting of processes in such technologies is based mainly on rigid-plastic analysis. The neglect of reversible deformations present in technological operations is absolutely not satisfied for technological practice, since exceptionally important effects are not taken into account: spring back and the formation of residual stresses, which significantly reduce the performance characteristics of metal structures. Thus, the development of mathematical modeling tools is an urgent task for calculating the optimal parameters of intensive deformation processes.

In [1], in the framework of the model of large elastoplastic deformations [2], it was proposed to divide the irreversible deformations acquired by the body on the deformations of creep and plastic flow by the mechanism of their production. Elastoplastic boundaries are surfaces on which the mechanism of accumulation of irreversible deformations changes from viscous to plastic and vice versa. The laws of creep and plastic flow must be harmonized so that on such surfaces a continuous growth of irreversible deformations takes place, which is achieved by an appropriate choice of plasticity conditions and creep laws. In [3], using the example of solving the boundary value problem of the theory of small deformations, an approach to such an agreed choice is given in the Norton creep power law [4] and in the von Mises yield criterion [5]. In the present paper, we consider the problem of deformation of a cylindrical layer and point out at this example the matching in the Norton creep law and Tresca yield criterion [6].

2. Basic model relations

Consider the basic model relations of the theory of large elastoplastic deformations. In a rectangular system of spatial Cartesian coordinates of Euler x_i the kinematics of a medium is given by the dependencies

$$\begin{gathered}
d_{ij} = 0.5\left(u_{i,j} + u_{j,i} - u_{k,i}u_{k,j}\right) = e_{ij} + p_{ij} - 0.5e_{ik}e_{kj} - e_{ik}p_{kj} - p_{ik}e_{kj} + e_{ik}p_{km}e_{mj},\\
\frac{De_{ij}}{Dt} = \varepsilon_{ij} - \gamma_{ij} - 0.5\left((\varepsilon_{ik} - \gamma_{ik} + z_{ik})e_{kj} + e_{ik}(\varepsilon_{kj} - \gamma_{kj} - z_{kj})\right),\\
\frac{Dp_{ij}}{Dt} = \gamma_{ij} - p_{ik}\gamma_{kj} - \gamma_{ik}p_{kj}, \quad \frac{Dy_{ij}}{Dt} = \frac{dy_{ij}}{dt} - r_{ik}y_{kj} + y_{ik}r_{kj},\\
\varepsilon_{ij} = 0.5\left(v_{i,j} + v_{j,i}\right), \quad v_i = \frac{du_i}{dt} = \frac{\partial u_i}{\partial t} + u_{i,j}v_j, \quad u_{i,j} = \frac{\partial u_i}{\partial x_j},\\
r_{ij} = w_{ij} + z_{ij}\left(\varepsilon_{sk}, e_{sk}\right), \quad w_{ij} = 0.5\left(v_{i,j} - v_{j,i}\right).
\end{gathered} \tag{1}$$

Here u_i, v_i are components of displacements and velocities of medium points; d_{ij} are components of the Almansi strain tensor; e_{ij} and p_{ij} are their reversible and irreversible components; D/Dt is an operator of the used objective derivative of tensors with respect to time, which is written for an arbitrary tensor y_{ij}; γ_{ij} and $\varepsilon^e_{ij} = \varepsilon_{ij} - \gamma_{ij}$ are rates of accumulation of irreversible and reversible deformations. The rotational tensor r_{ij} differs from the classical velocity vortex tensor w_{ij} by the presence of a nonlinear part z_{ij}, which is completely given in [2].

Stresses in the medium are completely determined by reversible deformations and for an incompressible medium are calculated from the relation

$$\sigma_{ij} = -p\delta_{ij} + \frac{\partial W}{\partial e_{ik}}\left(\delta_{kj} - e_{kj}\right). \tag{2}$$

In equation (2) σ_{ij} are components of the Euler–Cauchy stress tensor, p is unknown function of additional hydrostatic pressure, W is an elastic potential having a form

$$\begin{gathered}
W = -2\mu I_1 - \mu I_2 + bI_1^2 + (b-\mu)I_1I_2 - \chi I_1^3 + ...,\\
I_1 = e_{kk} - 0.5e_{ks}e_{sk}, \quad I_2 = e_{ks}e_{sk} - e_{ks}e_{st}e_{tk} + 0.25e_{ks}e_{st}e_{tn}e_{nk}.
\end{gathered} \tag{3}$$

Here μ is the shear modulus, b, χ are material constants.

The dissipative mechanism of irreversible deformation is associated with the rheological and plastic properties of the material. Farther we believe that irreversible deformations accumulate from the beginning of the deformation process and are initially associated with the creep process of the material. In areas where the stress state has not yet reached the yield surface, or where the plastic flow has been occuring but stopped, the corresponding dissipative mechanism of deformation we set in the form of Norton creep power law

$$V\left(\sigma_{ij}\right) = B\Sigma^n\left(\sigma_1, \sigma_2, \sigma_3\right), \quad \Sigma = \max\left|\sigma_i - \sigma_j\right|, \quad \gamma_{ij} = \varepsilon^v_{ij} = \frac{\partial V\left(\Sigma\right)}{\partial \sigma_{ij}}, \tag{4}$$

wherein $\sigma_1, \sigma_2, \sigma_3$ are principal values of stress tensor, B, n are material creep parameters, ε_{ij}^v are components of tensor of creep strain rates.

Over time the stress state reaches the yield surface, the dissipative mechanism of deformation changes and plastic flow begins in the material. Without separation of irreversible deformations into components, we assume that irreversible deformations of creep accumulated by the time of the onset of plastic flow are initial values for plastic deformations that accumulate further in the flow region. In the case of taking into account the viscous properties of the medium during plastic flow, the coincidence of the rates of irreversible deformations is also required when the deformation mechanism changes from viscous to plastic.

According to the Mises maximum principle the relationship between plastic strain rates ε_{ij}^p and stresses is established by the associated law of plastic flow

$$\alpha_{ij} = \lambda \frac{\partial F}{\partial \sigma_{ij}}, \quad F(\sigma_{ij}, \alpha_{ij}) = k, \quad \lambda > 0, \quad \alpha_{ij} = \varepsilon_{ij}^p - \varepsilon_{ij}^{v0}, \tag{5}$$

where F is a plastic potential, k is the yield strength, ε_{ij}^{v0} are components of tensor of creep strain rates at the start time of plastic flow.

As a plastic potential we will use the Tresca yield criterion, generalized in the case of taking into account the viscous properties of the material

$$\max |\sigma_i - \sigma_j| = 2k + 2\eta \max |\alpha_n|. \tag{6}$$

Here η is a viscosity coefficient.

3. Boundary value problem statement and viscoelastic deformation

Consider the boundary value problem about creep and viscoplastic flow in an incompressible material located in the gap between two rigid cylindrical surfaces. The inner cylindrical surface of radius $r = r_0$ moves straightforward, and the outer surface of radius $r = R$ is rigidly fixed. We consider that sticking conditions are satisfied on cylindrical walls. Then in the cylindrical coordinate system r, φ, z boundary conditions of the problem take the form

$$v|_{r=R} = u|_{r=R} = 0, \quad v|_{r=r_0} = v_0, \quad u|_{r=r_0} = \int_0^t v_0 dt, \quad \sigma_{rr}|_{r=r_0} = a_0. \tag{7}$$

Here $v = v_z(r,t)$, $u = u_z(r,t)$ are the only non-zero components of the velocity and displacement vectors, respectively; $v_0 = v_0(t)$, $a_0 = a_0(t)$ are given functions.

From relations (1) in this case we establish that the kinematics of the medium is described by the dependencies

$$d_{rr} = -0.5u_{,r}^2, \quad d_{rz} = 0.5u_{,r}, \quad \varepsilon_{rz} = 0.5v_{,r}. \tag{8}$$

In (8) index after the comma denotes partial derivative with respect to the spatial coordinate.

To simplify the solution of the problem, we confine ourselves to the first-order terms along the diagonal components and the second-order terms along non-diagonal components. From relations (2) and (3) in this case we find the stresses in the medium

$$\sigma_{rr} = -P + 2\mu e_{rr}, \quad \sigma_{\varphi\varphi} = -P - 3\mu e_{rz}^2, \quad \sigma_{zz} = -P + 2\mu e_{zz}, \quad \sigma_{rz} = 2\mu e_{rz},$$
$$P = -(p + 2\mu) + 2b(e_{rr} + e_{zz}) + \mu e_{rz}^2.$$

In the framework of the quasistatic approximation we write the equilibrium equations for our problem

$$\sigma_{rr,r} + \sigma_{rz,z} + (\sigma_{rr} - \sigma_{zz}) r^{-1} = 0, \quad \sigma_{rz,r} + \sigma_{zz,z} + \sigma_{rz} r^{-1} = 0. \tag{9}$$

We integrate equations (9) under the assumption that the stresses are finite, i.e. $P_{,z} = 0$:

$$\sigma_{rz} = c(t)r^{-1}, \quad e_{rz} = 0.5\mu^{-1}c(t)r^{-1}, \quad P = f(r,t). \tag{10}$$

We rewrite relations (4) for our problem

$$V(\sigma_{ij}) = B\left(4\sigma_{rz}^2 + (\sigma_{rr} - \sigma_{zz})^2\right)^{\frac{n}{2}}, \quad \varepsilon_{rz}^v = (-1)^n 2^{n-1} Bn\sigma_{rz}^{n-1},$$
$$\varepsilon_{rr}^v = -\varepsilon_{zz}^v = 0.5\varepsilon_{rz}^v (e_{rr} - e_{zz}) e_{rz}^{-1}. \tag{11}$$

Using the dependencies $\varepsilon_{rz} = \varepsilon_{rz}^e + \varepsilon_{rz}^v$, (8), (10), (11) and sticking conditions (7) on the surface $r = R$ we get the ratios for the velocity and displacement

$$v = F_1(\dot{c}, r, R) + F_2\left(c^{n-1}, r, R\right), \quad u = F_1(c, r, R) + F_2(c_1, r, R)$$
$$F_1(c, r, R) = \frac{c}{\mu}\ln\frac{r}{R}, \quad F_2(c_1, r, R) = \frac{(-1)^n 2^n Bnc_1}{2-n}\left(\frac{1}{r^2} - \frac{1}{R^2}\right), \quad c_1(t) = \int_0^t c^{n-1}dt. \tag{12}$$

Overdot denotes the time derivative. Taking into account the boundary conditions (7) on the inner wall $r = r_0$, from (12) we obtain an ordinary differential equation for the unknown function $c = c(t)$

$$F_1(\dot{c}, r_0, R) + F_2\left(c^{n-1}, r_0, R\right) = v_0, \quad c(0) = 0.$$

Using the equation for changing the components of irreversible deformations from (1), we get that $\dot{p}_{rz} = \varepsilon_{rz}^v$. Integrating this equation we find

$$p_{rz} = (-1)^n 2^{n-1} Bnc_1 r^{n-1}. \tag{13}$$

With increasing speed of movement of the inner cylinder v_0, the obtained solution of the problem will be valid up to a time point $t = t_0$ at which on the inner wall $r = r_0$ the plasticity condition (6) will be fulfilled, in our case taking the form $\sigma_{rz}|_{r=r_0} = -k$. From the first relation (10), we obtain an equation for determining the time point $t = t_0$ at which the viscoplastic flow begins: $c(t_0) = -kr_0$.

4. Viscoplastic flow

The region of the viscoplastic flow $r_0 \le r \le m(t)$ developing from the time point $t = t_0$ is bounded by surfaces $r = r_0$ and $r = m(t)$. Viscoelastic deformation still occurs in the area $m(t) \le r \le R$. Surface $r = m(t)$ is a moving boundary of the viscoplastic flow region.

The dependencies of the previous section with an unknown integration function $c = c(t)$ remain true in the region $m(t) \le r \le R$.

Integrating the equilibrium equations (9) in the region of the viscoplastic flow $r_0 \le r \le m(t)$ and using the condition of continuity of the stress components on the elastoplastic boundary $r = m(t)$, we obtain that dependencies (10) also hold in this region.

From relations (5), (6) and (10) we find

$$\sigma_{rz} = -k + \eta\left(\varepsilon_{rz}^p - \varepsilon_{rz}^{v_0}\right), \quad \varepsilon_{rz}^p = \eta^{-1}(cr^{-1} + k) + \varepsilon_{rz}^{v_0}. \tag{14}$$

From the condition of coincidence of the rates of irreversible deformations on the elastoplastic boundary $r = m(t)$ we find from (14) that $c(t) = -km(t)$.

In the flow region $r_0 \le r \le m(t)$ using relations $\varepsilon_{rz} = \varepsilon_{rz}^e + \varepsilon_{rz}^v$, (8), (10), (14) and boundary condition for the velocity on the surface $r = r_0$ (7) we find

$$\begin{aligned} v &= F_1(\dot{c}, r, r_0) + F_3(c, r, r_0) + v_0, \\ F_3(c, r, r_0) &= 2\eta^{-1}\left(c\ln(rr_0^{-1}) + k(r - r_0)\right) - 2^n k^{n-1} Bn(r - r_0). \end{aligned} \tag{15}$$

The position of the elastoplastic boundary $r = m(t)$ at each moment of time is given by an ordinary differential equation that follows from (12), (15) and the condition of continuity of the velocity of material points on this boundary

$$F_1(-k\dot{m}, r_0, R) + F_2\left((-km)^{n-1}, m, R\right) - F_3(-km, m, r_0) = v_0.$$

Integrating the rate of plastic deformations ε_{rz}^p (14) on time and taking into account (10) and (11), we find the component of irreversible deformations p_{rz} in the flow region

$$p_{rz} = \frac{1}{\eta}\left(\frac{c_2}{r} + kt\right) - 2^{n-1}k^{n-1}Bnt + g, \quad c_2 = \int_{t_0}^{t} c dt. \tag{16}$$

Here $g = g(r)$ is an unknown integration function, which we find from the condition of continuity of irreversible deformations (13) and (16) on the moving elastoplastic boundary

$$g = (-1)^n 2^{n-1} Bnc_1(\zeta) r^{1-n} - \eta^{-1}\left(c_2(\zeta)r^{-1} + k\zeta\right) + 2^{n-1}k^{n-1}Bn\zeta. \tag{17}$$

In dependency (17) the function $\zeta = \zeta(r)$ is determined from an ordinary differential equation

$$\zeta_{,r}\left(F_2\left((-kr)^{n-1}, r, R\right) - F_3(-kr, r, r_0) - v_0(\zeta)\right) = F_1(k, r_0, R), \quad \zeta(r_0) = t_0.$$

Taking into account the relations (10), (16), equality $u_{,r} = 2(e_{rz} + p_{rz})$ and the condition of continuity of displacements on the elastoplastic boundary $r = m(t)$, we find an expression for the displacements in the region of viscoplastic flow

$$u = F_1(c, r, R) + F_2(c_1, m, R) + tF_3\left(t^{-1}c_2, r, m\right) + 2\int_m^r g dr.$$

When performing calculations, the function v_0 was assumed linear: $v_0 = \xi_1 t$ in the interval $0 \le t \le t_1$, $v_0 = \xi_1 t_1$ in the interval $t_1 \le t \le t_2$, $v_0 = \xi_1 t_1 - \xi_2(t - t_2)$ in the interval $t_2 \le t \le t_4$ and $v_0 = 0$ for $t \ge t_4$. The calculations were carried out in dimensionless variables rR^{-1}, $\tau = \xi_1 t^2 r_0^{-1}$ at constant values $k\mu^{-1} = 0.003$, $r_0 R^{-1} = 0.2$, $n = 3$, $B\mu^2\sqrt{r_0\xi_1^{-1}} = 3$, $\xi_1\xi_2^{-1} = 0.5$. Fig. 1 shows the graphs of the elastoplastic boundary as a function of time in the deformation process (at time point τ_3 the elastoplastic boundary has a maximum value; at time point τ_5 the elastoplastic boundary has a minimum value $m = r_0$). The change of irreversible deformations in the interval $0 \le \tau \le \tau_5$ on the inner surface $r = r_0$ is shown in Fig. 2. Fig. 3 illustrates the distribution of displacements in the cylindrical layer at different time points and relaxation of the stress component σ_{rz} from time point τ_5 to $\tau = 1000$.

5. Conclusion

Here is given the solution to the problem of viscoelastic deformation and viscoplastic flow of the material at the increasing speed of movement of the internal cylinder. The following stages of deformation are considered, but not shown here: the viscoplastic flow at a constant speed of the inner cylinder, flow deceleration at a decreasing speed and medium unloading. We note that the ratios for these stages in general have the same form as the above dependencies.

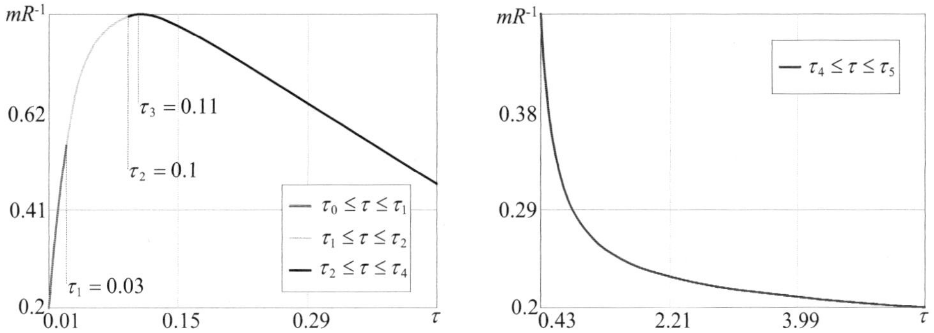

Figure 1. Dependencies of the elastoplastic boundary on time.

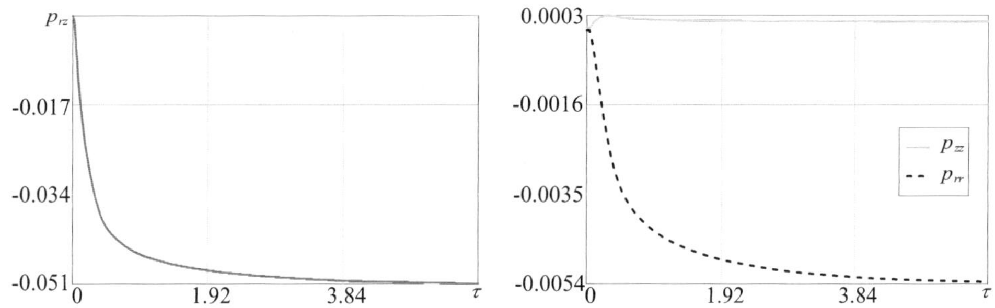

Figure 2. Irreversible deformations depending on time at the point $r = r_0$.

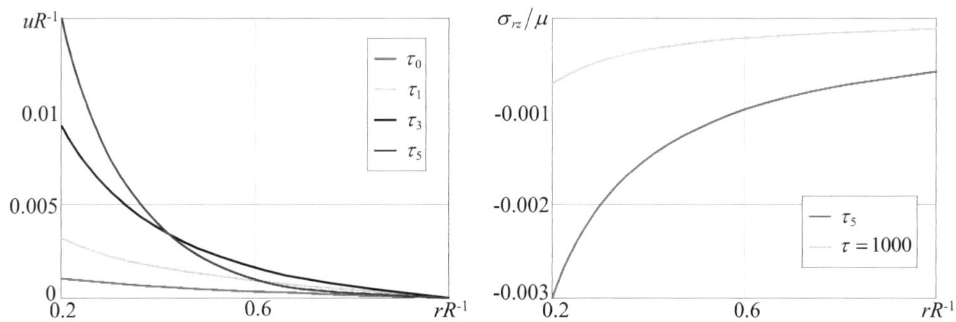

Figure 3. Displacements and stresses at different time points.

Acknowledgments

The present study was financially supported by RFBR (project No. 18-01-00038).

References

[1] Begun A S, Burenin A A and Kovtanyuk L V 2016 *Dokl. Phys.* **61** 463–6
[2] Burenin A A, Bykovtsev G I and Kovtanyuk L V 1996 *Dokl. Phys.* **41** 127–9
[3] Burenin A A, Galimzyanova K N, Kovtanyuk L V and Panchenko G L 2018 *Dokl. Phys.* **63** 407–10
[4] Norton F H 1929 *The creep of steel at high temperatures* (New York: McGraw Hill)
[5] von Mises R 1913 Mechanik der festen Körper im plastisch deformablen Zustand *Göttin. Nachr. Math. Phys.* **1** 582–92
[6] Tresca H 1864 Mémoire sur l'écoulement des corps solides soumis à de fortes pressions *C. R. Acad. Sci.* **59** 754

The effect of an additional air inlet in the hot outlet area on the overall effectiveness of a vortex tube

L Y Privalov[1], **C I Mikhaylenko**[2]
[1]Ufa State Aviation Technical University, Ufa, Russia
[2]Mavlyutov Institute of Mechanics, UFRC RAS, Ufa, Russia
E-mail: HungryDoodles@yandex.ru

Abstract. In this paper the modification of a vortex tube is shown. The feature of this vortex tube is that it has an additional air inlet in the hot outlet area along the tube axis designed to redistribute air consumption between cold and hot outlets. Computational experiments are done in openFOAM toolbox using sonicFoam solver. In order to increase computational speed the mesh is being decomposed into equal parts, which allows for the parallel run using MPI on a single computer. The results of the experiment show temperature values and air consumption on the corresponding outlets in relation to the change of the vortex tube length. A conclusion about the effective vortex tube length is made.

1. Introduction

The vortex tube is a device that separates a stream of compressed gas from vortex generators into two streams so that one of the streams is colder and the other one is hotter. This phenomenon is called Ranque–Hilsch effect and named after French engineer Georges–Joseph Ranque who discovered it in 1931 [1, 2].

This device is notably simple, has no moving parts and easy to manufacture. Due to this it is being used in numerous fields where colling or fractionation of gases or fluids is required. In this regard, research in the field of vortex generating devices is one of the focus areas, especially because the mchanism of temperature separation is not yet clear [3]. There are theoretical approaches to explain this phenomenon related to acoustics [4], and purely experimental [5]. The effects of changing the number of inlet nozzles (from 1 to 5), tube length and cold outlet diameter have been shown [6]. It is also demostrated, that the inlet pressure is important and there is a small optimal conical angle of primary nozzle [7].

In general, described effect is of theoretical and pratical interest and hence there are a lot of works dedicated to various aspects of study and use of vortex tubes [8–10]. This fact shows the persistent interest to this effect and the continuing relevance, that is noted in reviews [11–13].

Currently, studies of non-standard vortex tube configurations and their effect on it's characteristics are of particular interest. There are mentions of a double-circuit vortex tube [14], a tube with alternative shapes of the hot end [15] and even modified main cylinder [16].

In this paper the effect of an additional inlet along tube axis in the hot outlet area on overall vortex tube effectiveness is shown.

2. The model and the methods

2.1. The simulation domain

In this paper, a numerical modeling of a counterflow vortex tube with an additional inlet is made. For this case a vortex tube mesh was made according to schematics, that are presented in figure 1. Parameter L of the vortex tube, which is the length of the tube, is the only one set to be changed during computational stage in order to examine the effect of the additional inlet. Other parameters (length ℓ; tube diameter D; cold outlet nozzle diameter d and widening angle α; hot outlet ring width h; inlet dimenstions p, q and m; additional inlet nozzle diameter r) stay fixed.

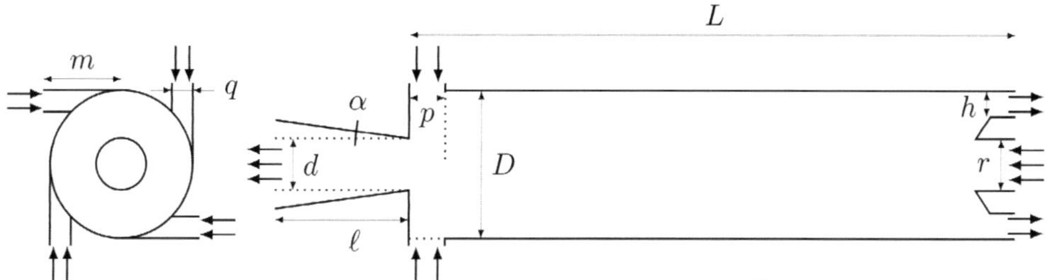

Figure 1. The vortex tube schematics and dimensions (a — cold outlet side; b — sideview section)

2.2. The mathematical model

Gas dynamics in the vortex tube canal is described on the basis of the standard system of equations (continuity, momentum and energy) closed by the equation of state of an ideal gas. In the momentum and energy equations, viscous terms are introduced for calculations using the $k - \varepsilon$ model of turbulence:

- mass conservation (continuty) equation

$$\frac{\partial \rho}{\partial t} + \nabla(\rho \mathbf{U}) = 0\,;$$

- momentum conservation equation

$$\frac{\partial \rho \mathbf{U}}{\partial t} + \nabla(\rho \mathbf{U} \times \mathbf{U}) = -\nabla p + \nabla \tau\,;$$

- energy conservation equation

$$\frac{\partial \rho E}{\partial t} + \nabla(\rho \mathbf{U} E) = -\nabla p \mathbf{U} + \nabla(\tau \times \mathbf{U})\,;$$

- state equation for an ideal gase

$$p = (\gamma - 1)\rho\varepsilon\,.$$

Here the standard notation is used: U is velocity vector; ρ is density; p is pressure; E is specific total energy; ε is specific internal energy; γ is adiabatic exponent; τ is viscous stress tensor, which elements are calculated as

$$\tau_{ij} = (\mu + \mu_t)\left[\left(\frac{\partial v_i}{\partial x_j} + \frac{\partial v_j}{\partial x_i}\right) - \frac{2}{3}\delta_{ij}\frac{\partial v_k}{\partial x_k}\right],$$

where δ_{ij} is Kronecker symbol; v_i is velocity vector components; μ is substance viscosity; $\mu_t = \rho C_\mu \frac{k^2}{\varepsilon}$, is turbulent viscosity calculated using $k - \varepsilon$ turbulence model [17]:

- turbulent kinetic energy equation

$$\frac{\partial \rho k}{\partial t} = \nabla \left(\frac{\mu_t}{\sigma_k} \nabla k \right) - \frac{2}{3} \rho \left(\nabla \cdot u \right) k - \rho \varepsilon \,,$$

- turbulence dissipation rate equation

$$\frac{\partial \rho \varepsilon}{\partial t} = \nabla \left(\frac{\mu_t}{\sigma_\varepsilon} \nabla \varepsilon \right) - \frac{2}{3} C_1 \rho \left(\nabla \cdot u \right) \varepsilon - C_2 \rho \frac{\varepsilon^2}{k} \,,$$

where $C_\mu = 0.09$, $C_1 = 1.44$, $C_2 = 1.92$, $\sigma_k = 1$ and $\sigma_\varepsilon = 1.3$.

2.3. Algorithms and software

The calculations were performed in OpenFOAM software using sonicFoam solver. Special attention was paid to uniformity of orthogonalized finite-difference mesh, since this approach allows to increase timestep without a significant rise in error [9]. In addition, with the increase of length of the vortex tube, cell count along the main axis of the tube is also increased to prevent cell stretching that could cause error.

The important thing before the computational stage is mesh decomposition, that separates mesh into several approximately equal parts that allows to make a parrallel run and reduce computational time by the factor of 3.5 on 6 proccesses. By the means of openFoam decomposition utility the mesh has been sliced into 6 submeshes using planes perpendicular to the main tube axis, minimizing the number of interacting cells between submeshes.

3. Results and discussion

A series of computational experiments was performed on multiple configurations of the vortex tube with following parameters: $D = 4.7$ cm; $r = d = 0.9$ cm; $h = 0.5$ cm; $p = 0.45$ cm; $q = 1$ cm; $m = 4$ cm; $l = 2.5$ cm. The parameter L is variable and ranges between 20 and 80 cm.

As for initial state and boundary conditions for the vortex tube: the inside volume of the vortex tube there is air resting still at the temperatre of 300 K and the pressure of 10^5 Pa; on the inlets air with the temperature of 300K is being supplied at a constant volume flow rate (0.2 m^3/s on the vortex generators and 0.02 m^3/s on the additional inlet); It should be noted that on the outlets wave transmissive boundary condition is set. The tube walls are thermally insulated and have no friction.

Graphs of temperatures, velocity and mass rate are presented in figures 2 and 3 for hot and cold outlet respectively. Starting with minimal length behaviour of the tube is typical to a vortex tube: growth of the temperature on the hot outlet and drop on the cold outlet. This proccess goes on until it reaches the length of 50 cm. All simulations are ought to reach at least 150 ms of simulation time where last 50 ms are used to calculate average of the physical values. It's assumed that up to 100 ms and further the flow is stable what can be observed as a set of similar values of averages on slices on the cold and hot side at discrete moments of time. For lengths over 50 cm simulation time has been risen up to 210 ms because the longer the tube, the longer it takes for the flow to stabilize. After simulation of all tubes with lengths between 20 and 80 cm with the step of 10 cm, the addittional lengths 55, 65 and 75 cm were also taken due to their interesting behaviour.

It is evident that in all presented graphs the characteristics of the vortex tube are improving and are going to their supposed optimum at the length of 50 cm. Hot oulet shows notably smooth picture, nearly straight line in the temperature graph in figure 2 (a) and slightly jittery

Journal of Physics: Conference Series **1392** (2019) 012019 doi:10.1088/1742-6596/1392/1/012019

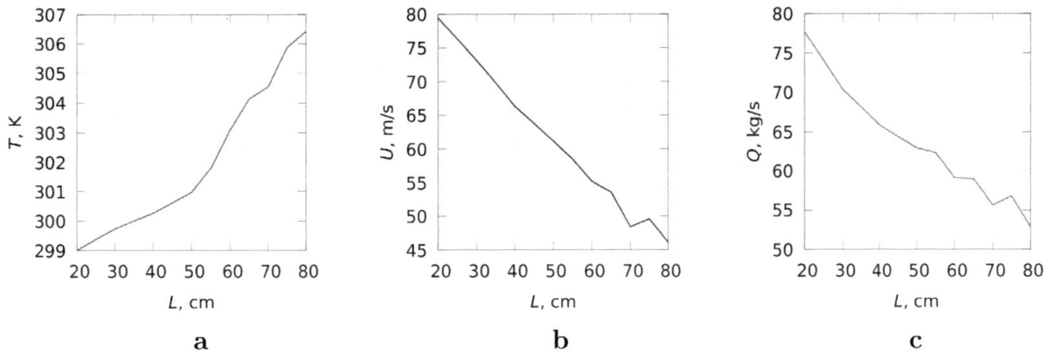

Figure 2. Dependencies of the average temperature T (a), the velocity U (b) and the mass flow rate Q (c) on the vortex tube channel length L on the hot outlet cross section

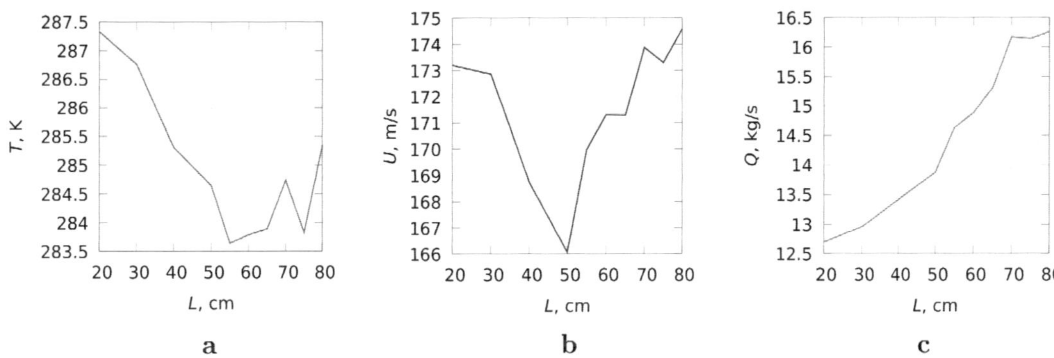

Figure 3. Dependencies of the average temperature T (a), the velocity U (b) and the mass flow rate Q (c) on the vortex tube channel length L on the cold outlet cross section

in the end of graphs of mass rate and velocity in figures 2 (b) and 2 (c). However, for the cold outlet the graph line fluctuates a lot more. It reaches it's local minimum at 55 cm (see figure 3 (a)) where the temperature first slowly climbs up and then leaps up and down. The same happens for the velocity in figure 3 (b) and for the mass rate in figure 3 (c). Supposedly, the core physics of the flow for the lengths over 50 cm changes dramatically: while being faced against fast incident flow an additional inlet makes significant turbulent disturbance that wether breaks smoothness of transition between outer and inner vortex flows or causes an oscillation. In any case, configurations with the lengths over 50 cm are yet to be carefully revised with possible recomputation with longer average interval.

In general, found dependencies demonstrate that such configuration of a vortex tube is able to reproduce Ranque–Hilsch effect which tells about a potential ability of using an additional inlet. Nethertheless, the lengths of the tube of 50 cm and more require special attention because the results on the cold outlet indicate particularly uneven transitions when in this exact range while being hidded behind instability the optimum could exist. Mainly, the points of the graph at 65 cm and over clearly appear as sporadic movement. Worth noting a hard to notice alignment around some average at the lengths between 70 and 80 cm that has occured multiple times in different experiments.

4. Conclusion

Thus, the presented configuration of a vortex tube with an additional inlet in the hot outlet area demonstates the Ranque–Hilsch effect as well as a classical configuration by gradually improving characteristics with the increase in length. However, after a ceratin point the core physics of the flow is undergoing a serious change that needs to be looked into more carefully at least because it's hard to say anything unequivocal about what happens at the lengths over 50 cm. The assumption was made that the nature of such behaviour might be the turbulence itself, mostly due to collision of incident streams.

Acknowledgments

This work was supported by the state budget funds for the state assignment 0246-2019-0052 of RAS.

References

[1] Hilsch R 1947 *Review of Sientific Instruments* **18** 108–113.
[2] Ranque G J 1933 *J. Phys. Radium* **4** 112–114.
[3] Borisoglebskiy I K, Metusova M V, Mikhaylenko C I 2018 *Multiphase Systems* **13 (3)** 52–58.
[4] Xue Y, Arjomandi M, Kelso R 2010 *Experimental Thermal and Fluid Science* **34 (8)** 1367–1374.
[5] Majidi D, Arjomandi M, Kelso R 2017 *International Journal of Refrigeration.* **85** 282–291.
[6] Moraveji A, Toghraie D 2017 *Int.J. of Heat and Mass Transfer* **1137** 432–443.
[7] Thakare H R, Monde A, Parekh A D 2015 *Renewable and Sustainable Energy Reviews.* **52** 1043–1071.
[8] Mikhaylenko C I 2017 *Proc. of Mavlyutov Institute of Mechanics* **12(2)** 174–179.
[9] Mikhaylenko C I 2016 *Proc. of Mavlyutov Institute of Mechanics* **11 (1)** 112–118.
[10] Mikhaylenko C I 2019 *Journal of Physics: Conference Series.* **1158(3)** 032032.
[11] Gutsol A F 1997 *Physics-Uspekhi* **167(6)** 665–687.
[12] Eiamsa-Ard S, Promvonge P 2008 *Renewable and Sustainable Energy Reviews* **1** 1822–1842.
[13] Subudhi S, Sen M 2015 *Renewable and Sustainable Energy Reviews* **52** 172–178.
[14] Khait A, Noskov A, Alekhin V, Bianco V 2018 *Applied Thermal Engineering.* **130** 1391–1403.
[15] Rafiee S, Sadeghiazad M 2017 *Applied Thermal Engineering.* **110** 648–664.
[16] Liu J, Chen S, Gan M, Chen Q 2018 *Applied Thermal Engineering.* **144** 702–710.
[17] Launder B E, Spalding D B 1974 *Computer Methods in Applied Mechanics and Engineering.* **3(2)** 269–289.

Simulation of the fracture of materials and structures under dynamic loads using parallel computing

Pavel Radchenko, Stanislav Batuev, Andrey Radchenko
Tomsk State University of Architecture and Building, Tomsk, Russia
E-mail: radchenko@live.ru

Abstract. The present work suggests algorithm enabling to describe adequately fracture of materials and structures under impulse loadings. The methodology of reconfiguration finite element mesh with re-computation of concentrated masses in nodes is considered. The algorithm uses all the capabilities of modern parallel computing to the full extent and can be easily scaled for all possible objectives. Numerical methodology demonstrates good qualitative and quantitative comparison with the experimental data.

1. Introduction

Description of materials fracture in the area of contact interaction is a challenging issue in numerical simulation of high-velocity interaction of solid bodies. High-velocity loading induces occurrence of the areas with intensive fracture of interacting bodies. Local distortion of computational mesh can lead to both fall of step size over time as well as to the structural failure of computational mesh. Generally it is suggested to remove the fractured finite elements out of numerical calculation.

Current research related to this issue offer various computational mesh adaptive algorithms [1] [2] [3]. Contact problems are solved using different numerical approaches [4] [5] [6]: Lagrange multiplier method, penalty and regularization method, quadratic programming, modifications of Newton method, Schwarz decomposition method, pseudo-medium, and others. Multiple algorithms modify the mesh only in two-dimensional case [7]. Parallel computations have recently become of great concern while developing and modification of finite-element meshes [8].

When solving the objective of solid bodies penetration an extensive area of fractured material is being formed inside the barrier crater; this area influences significantly the step size over time for the evident calculation models. In such case removing of elements and nodes with mass influences negatively the validity of solution; previously authors suggested the recalculation algorithm for local mass in the nodes of computational mesh providing high-quality description of objectives of solid bodies dynamic loading [9].

2. Algorithm description

The erosion algorithm is being implemented when the criterion of fracture in finite element is followed [9] the finite element being on the contact surface. Let us consider the element E_1, which is formed by the nodes $N_1N_2N_3N_4$ (Fig. 1a). When following the criterion of fracture in finite element on the surface of a solid body, this element is removed from the computational

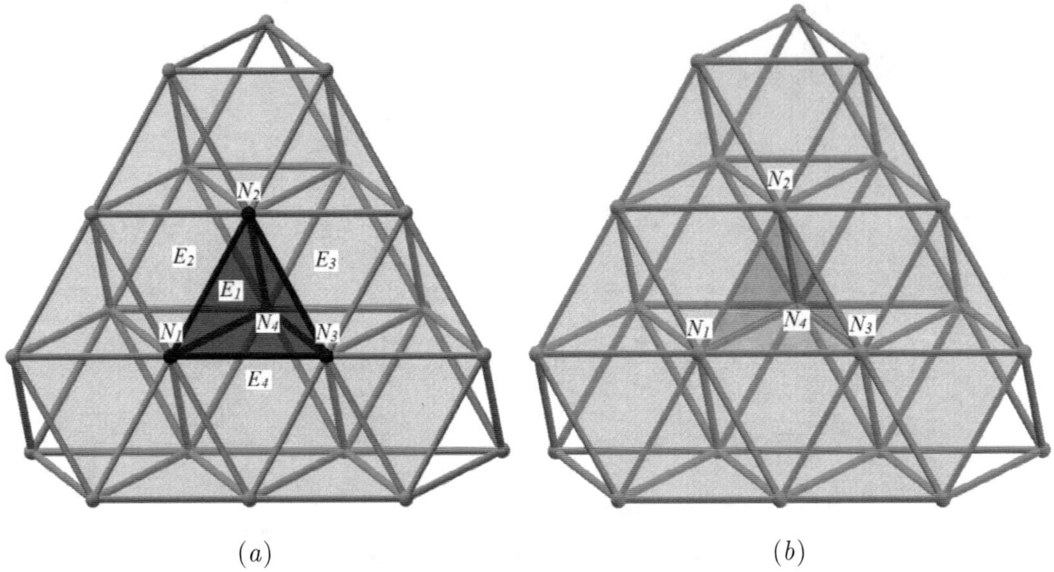

Figure 1. 3D finite element mesh: (a) before (b) after element removal

grid (Fig. 1a). The element E_1 had mutual planes with the elements E_5, E_9, E_{13}, after removal of the element E_1 from the grid, the new surface triangles are being built: $N_1N_2N_4$, $N_1N_4N_3$, $N_2N_3N_4$ (Fig. 1b).

One of the crucial issues occurring while numerical implementation of the erosion fracture algorithm is application of the law of conservation of mass. There are various approaches to implementation of surface erosion of interacting bodies under high-velocity loading. Authors used the following approach: in case when one or several nodes of the fractured element no longer belong to the other elements, than their mass is equally distributed between the remaining nodes of this element.

In spite of all the advantages of erosion fracture algorithm data preparation for its implementation can take as much time as the calculation itself. The data required for its implementation is:

(i) Four data sets indicating which surface each of the four finite elements planes belongs to.
(ii) Data set containing information on all of the adjacent planes, namely element number, number of its plane, the number of adjacent element and of its plane.

In order to accelerate data preparation for erosion fracture algorithm searching algorithm of adjacent planes of elements was optimized.

Optimization was performed using the method of sorting of all planes of all elements and their correlation with each other. The exact optimization of the algorithm includes the following steps: initially all finite element mesh is divided into N areas and searching for the adjacent plane for the element i is performed first in the area it belongs to and then it proceeds to adjacent areas, the maximum number of areas where the search of adjacent element can be performed is 27. The areas for search were defined according to the maximum capability of an adjacent plane to get into it, and the process of sorting was taking place according to the order mentioned below:

- the area where the element belongs to (the main, variant a);
- the group of areas which have mutual plane with the main area (6 areas, variant b);
- the group of areas which have mutual edge with the main area (12 areas, variant c);
- the group of areas which have mutual node with the main area (8 areas, variant d).

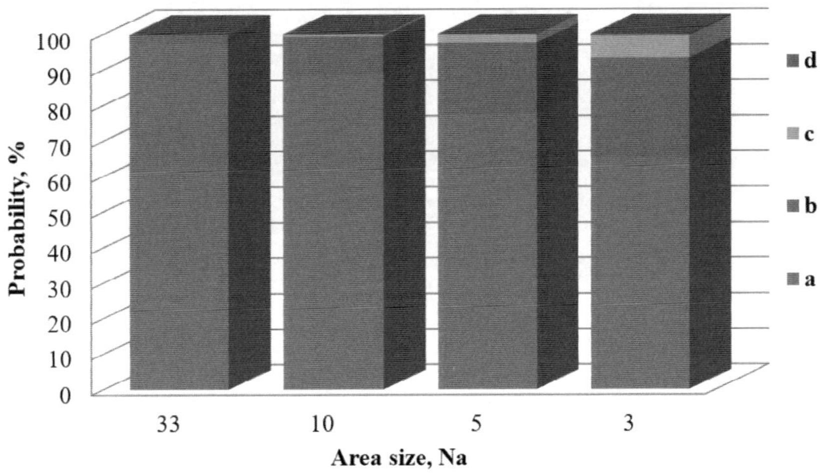

Figure 2. The dependency of probability of getting of element into the definite areas of searching on the size of area

3. Test calculations

Here are the results of investigations of the body in the shape of cube with the edge 1 m, which was divided into 5.3 million finite elements. The diagram demonstrating the number of elements getting into each of groups of areas is given in Fig. 2.

As can be seen from Fig. 2, when the size of area is decreasing it results in increasing probability that the elements will be able to find the adjacent plane in the other area. As the searching algorithm is occurring consequently in all areas, finding the adjacent plane in other area leads to increasing the time of calculation due to additional sorting of areas. However, this variant of algorithm ensures finding adjacent planes for each element.

Performance improvement relies on substantial decrease in the number of sorting variants. Before optimization the number of sorting was $8 \times N_e^2$, where N_e is the number of finite elements (4 planes of each element were compared with 4 planes of all other elements). After algorithm optimization this number cannot exceed $27 \times 4 \times N_a \times N_{ea}^2$, where N_a is the number of areas, N_{ea} is the number of elements in the current area, i.e. only planes of the elements within the area are compared between each other and then in case the adjacent plane is not found the searching proceeds to the adjacent areas.

For the computation mesh which is divided into 1000 areas (N_a=1000), the number of elements in one area will be approximately 1000 times less than the number of elements in the whole body ($N_e/N_{ea} = N_a$). The number of iterations before optimization is $8 \times (5 \times 10^6)^2 = 2 \times 10^{14}$. The maximum number of iterations after optimization is $27 \times 4 \times 1000 \times (5 \times 10^3)^2 = 2,7 \times 10^{12}$. After optimization the algorithm will be completed at least 74 times faster.

It is necessary to consider that the optimized algorithm requires extra processing time for areas division. And this time grows along with the increase in the number of areas (Fig. 3). The number of areas along each of the dimensions is defined as N_x, N_y, N_z. Yet in this example the number of areas along each dimension is the same and is denoted as follows: N_{xyz}.

Dividing the mesh of 5.3 mln. elements into 125000 areas ($50 \times 50 \times 50$) requires preparation time of about 90 seconds.

Figure 4 shows the time required for the searching algorithm of adjacent planes after optimization on a workstation with four 16-core CPU AMD Opteron 6376.

The largest performance improvement is reached when the area size is 3 mesh spaces. When the size of area increases, time for algorithm data preparation decreases. However, it is less

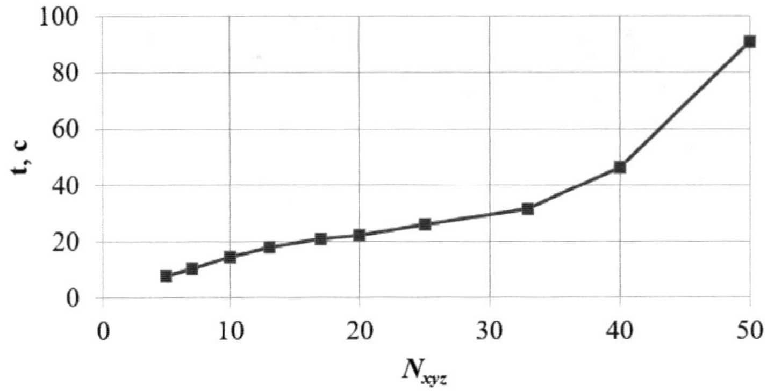

Figure 3. Processing time for data preparation in each area

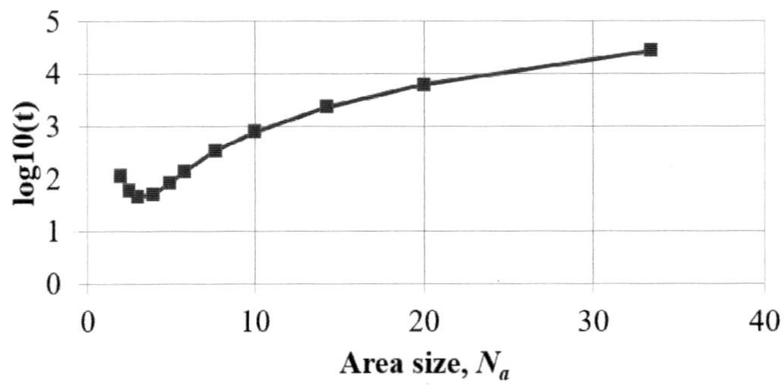

Figure 4. Time for calculation of searching algorithm of adjacent planes after optimization

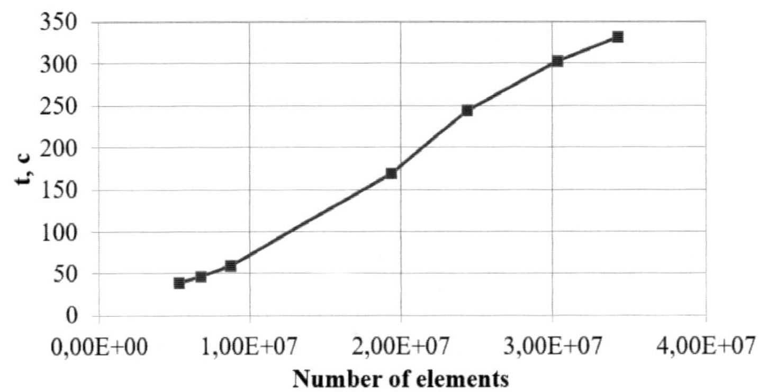

Figure 5. Time for calculation of searching algorithm of adjacent planes for different number of finite elements

than increase in searching time of adjacent plane, therefore, total time for algorithm completion increases.

Figure 5 indicates time comparison required for searching algorithm of adjacent planes for different meshes with different number of finite elements, while the number of areas along each dimension equals to 33.

It can be seen from the graph that increase in the calculation time and increase in the number of elements in the mesh occur almost linearly. Meanwhile, it should be highlighted

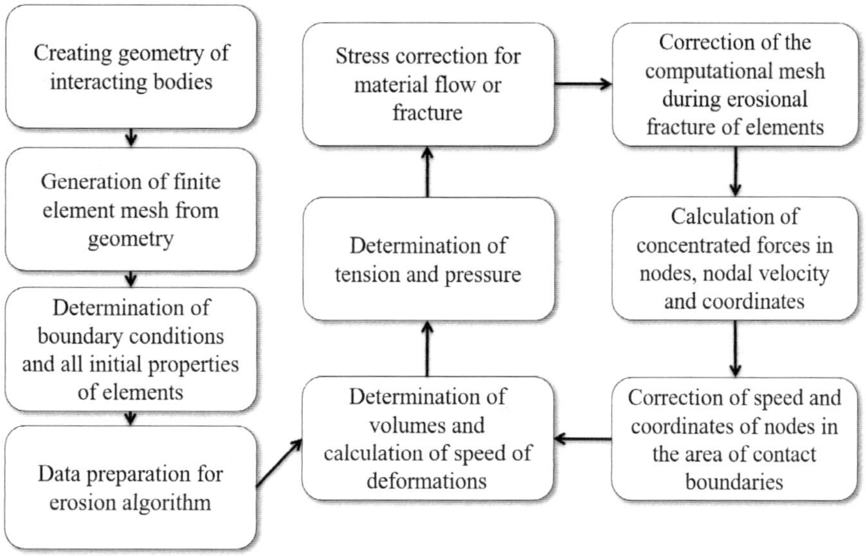

Figure 6. Integration block

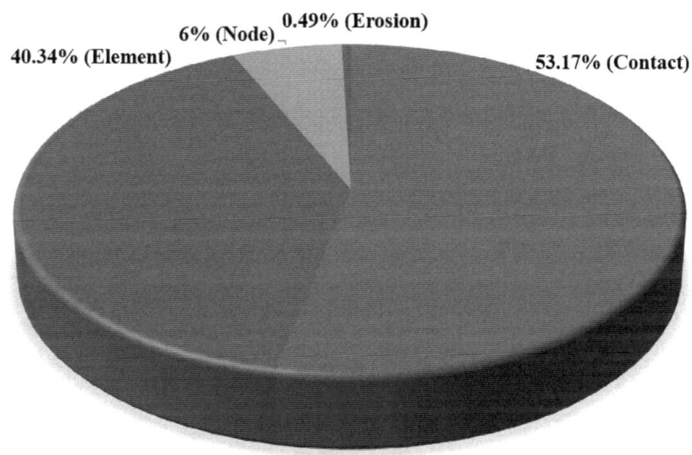

Figure 7. Diagram of labour input for computation completion

that the optimal calculation time by increasing number of elements in computation mesh can be reached when number of divisions along each dimension is different from 33. This issue demands further investigation.

After the algorithm of finite element mesh preparation integration cycle is launching (Fig. 6).

Figure 7 shows percentage-based labour input for calculation of corresponding variables in nodes and mesh elements, as well as for implementation of algorithms of contact interaction and erosion fracture. As can be seen from the diagram the algorithm of erosion fracture is less labour-consuming and takes less than 0.5% of processing time.

The efficiency of parallelization of software package was also analyzed (Fig. 8). For the test calculations workstation with four 16-core CPU AMD Opteron 6376 was used.

Unfortunately, not full program code can be realized simultaneously, there are some areas

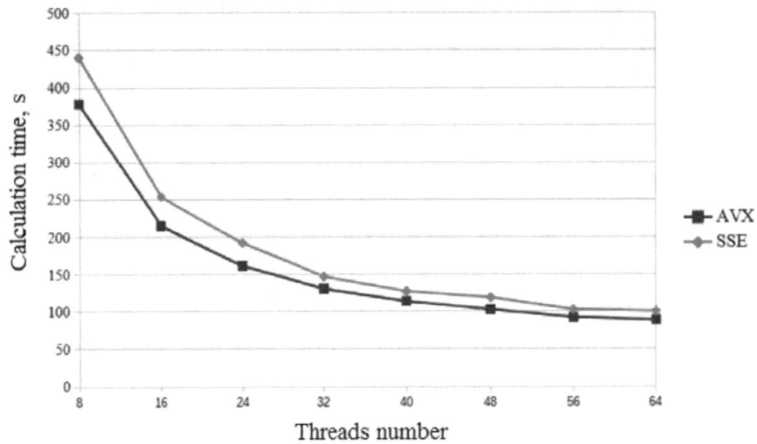

Figure 8. The graph of parallelization efficiency

which need to be calculated consequently; therefore performance improvement with increasing the number of cores is non-linear. However, along with increase in the size of computation mesh the efficiency of using larger number of cores grows significantly.

4. Conclusions

(i) Application of erosion fracture algorithm is required for appropriate description of the processes of fracture propagation under high-velocity interaction of solid bodies.

(ii) Dividing of computation mesh into areas provides significant performance improvement while data preparation for erosion fracture algorithm.

(iii) Decrease in the size of area results in reduced relative number of elements getting into the main area, which influences negatively the processing time.

(iv) Time which the processor spends on data preparation should be taken into account, still it contributes to the total time to the less extent as compared to the iterations on searching the adjacent planes of finite elements.

Acknowledgments

This research was partly supported by RFBR (grant No. 18-48-700035, No. 18-41-703003).

References

[1] Renev S A and Shelofast V V 2017 *Marine Intellectual Technologies* vol. 3 4 (38). pp. 72-78.
[2] Benin A V, Semenov A S, Semenov S G and Melnikov B E 2012 *Magazine of Civil Engineering* 7. pp. 32-42.
[3] Yadrov V I 2014 *Omsk Scientific Bulletin* 3 (133). pp. 152-156
[4] Bourago N G and Kukudzhanov V N 2005 *Mechanics of Solids* vol. 40 1, pp. 3571.
[5] Laursen T A 2003 *New York: Springer* xv+454 p. DOI: 10.1007/978-3-662-04864-1
[6] Karavaev A S and Kopytov S P 2017 *The Bulletin of Udmurt Univ. Mechanics.* Vol. 27. pp. 396-413
[7] Lukashevich A A 2007 *The Bulletin of Pacific National University* 1 (4). pp. 69-82.
[8] Andre K, Gluschenko O N, Ivanov E G and Kudryavtsev A N 2008 *Comput. Math. and Math. Phys.* Vol. 48 pp. 1448-1457.
[9] Radchenko P A, Batuev S P and Radchenko A V 2018 *AIP Conference Proceedings* doi: 10.1063/1.5065273

Method of Finite Bodies for Mathematical Modeling of the Stress-strain State of Cylindrical Orthotropic Shell with the Reinforced Rectangular Hole

Victor P. Revenko[1], **Vladimir N. Bakulin**[2]

[1] Pidstrigach Institute for Applied Problems of Mechanics and Mathematics National Academy of Sciences of Ukraine, 3-b, Naukova str., L'viv, 79060 Ukraine
[2] Institute of Applied Mechanics of Russian Academy of Sciences, Leningradsky prt. 7, Moscow, 125040 Russia

E-mail: victor.rev.12@gmail.com; vbak@yandex.ru

Abstract. An analytical-numerical method (finite bodies) for calculating the stress state of the cylindrical orthotropic shell with a large reinforced rectangular hole is proposed. The shell and reinforcement consist of various materials. The systems of equations that exactly satisfies the equilibrium equations of the orthotropic cylindrical shell and the orthotropic reinforced are used. The presentation of the solution is divided into basic and self-balanced states. The finite-body method uses the conditional partition of the three-connected surface of the shell and the reinforcement into simpler doubly-connected and simply connected parts. An algorithm is developed for analytic-numerical solution of boundary-value problems, based on the proposed universal method of reducing the satisfaction of all contact conditions of shell parts and boundary conditions to minimization of the generalized quadratic form. We establish criteria under which the construction of approximate solution coincides with the exact one.

1. Introduction

Cylindrical orthotropic shell stress strain state investigation is an important scientific and practical problem [1]–[10]. In [3, 6] the authors introduced the methods that allowed to describe the stress strain state for a cylindrical shell with a small hole, [7] helps us to consider the stress state of the shell weakened by an elliptic hole. In papers [8, 9] the finite element method was used to find the stress state for a shell with a rectangular hole, and [9, 10] experimental studies of the shells reinforced in the hole zone were carried out.

2. Statement of the problem.

Consider the stress-strain state of the orthotropic cylindrical shell: here R is the radius of the central surface, h is the cylinder thickness and $2H$ is its length. We also assume that the shell is weakened by the sizeable rectangular hole Π with sides $2a$ and $2b$ directed along the forming and directing lines of the shell surface. The edges of the hole are reinforced by an orthotropic elastic strip (reinforcement): here h_1 is the reinforcement strip thickness, c is its width (see Figure 1). We denote the curve coordinate along the shell directrix by γ, $\gamma = R\varphi$ and that along the shell generatrix by z. The angle φ is taken from the middle of the hole, and the coordinate z is taken from the center of the cylinder.

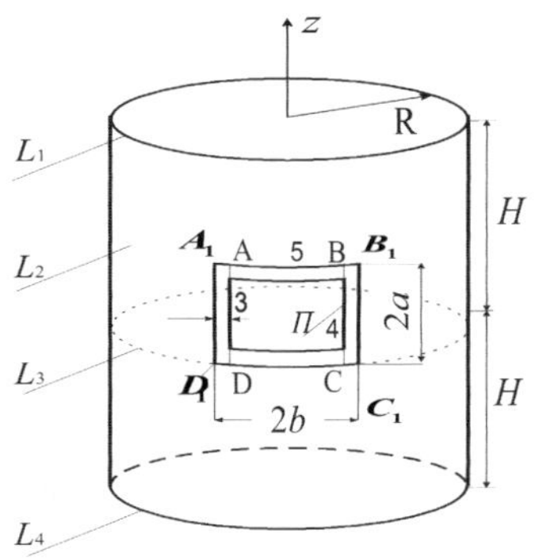

Figure 1. Scheme of the shell mental decomposition into parts.

In [11] the authors presented the equation system that identically satisfies all the equilibrium equations for the orthotropic cylindrical shell and can be written down in the cylindrical coordinates as follows:

$$V_1\Phi = \varepsilon E_1 \frac{\partial^2 w}{\partial z^2}, \quad V_2 w = -(1-\nu_1\nu_2)\frac{12\varepsilon}{h^4 E_2}\frac{\partial^2 \Phi}{\partial z^2} - \frac{\varepsilon^2}{h^2 R^2}\frac{\partial^2 w}{\partial \varphi^2}, \tag{1}$$

here $\Phi(z,\varphi)$ is the stress function, $w(z,\varphi)$ is the middle surface deflection function (these two functions are often called basic), $\varepsilon = h/R$; and V_1, V_2 are fourth order differential operators [11]; E_1, E_2 are Young's modules in the directions of the axial and circumferential variable, respectively; ν_1, ν_2 are Poisson's coefficients; $\nu_1 E_2 = \nu_2 E_1$ [1]. The stress state of the reinforced strip (index 1 at the top) is described by functions $\Phi^1(z,\varphi)$, $w^1(z,\varphi)$, which satisfy the equation (1) with its orthotropy characteristics: E_1^1, E_2^1, ν_1^1, ν_2^1; $\varepsilon_1 = h_1/R$, where h_1 is thickness.

We can find the forces T_1, T_2, S_{12}, S_{21}, moments $M_1, M_2, H_{12} = H_{21}$ and shear forces N_1, N_2, which defined on the shell middle surface by formulas [11].

Let us find the stress state of the shell loaded at its ends by the balanced symmetric forces and moments:

$$T_1(\pm H,\varphi) = P_1(\varphi), \quad M_1(\pm H,\varphi) = P_2(\varphi), \quad N_1(\pm H,\varphi) = P_3(\varphi),$$

$$S_{12}(\pm H,\varphi) = P_4(\varphi), \quad H_{12}(\pm H,y) = P_5(\varphi), \tag{2}$$

here $P_j(\varphi)$, $j = \overline{1,3}$ are even functions, $P_4(\varphi)$ is an odd function and $P_5(\varphi) = 0$. The contour of the reinforced hole $\{z,\varphi\} \in \Pi$ is not loaded:

$$T_n(z,\varphi) = 0, \quad S_{12}(z,\varphi) = 0, \quad M_n(z,\varphi) = 0,$$

$$N_n(z,\varphi) = 0, \quad H_{12}(z,\varphi) = 0. \tag{3}$$

Paper [11] shows us that in order to satisfy simultaneously the boundary conditions both at the ends and at the hole contour one should mentally divide the three connected shell surface into simpler simply connected and doubly connected parts. Let us mentally divide the shell along the circles $z = \pm a$ relative to the lines L_2, L_3 into three parts. Cut out the reinforcement along the intercepts A_1D_1, B_1C_1 and cut it along the intercepts AD, BC.

We get seven separate parts. The ideal contact conditions should be met along these cuts. Due to the symmetry of the loads, it is sufficient to consider only the points of the shell $H \geq z \geq 0$. The part of the shell is $H \geq z \geq a$, denoted D_1, and the segment of the shell is $a \geq z \geq -a$, $\varphi \in [\theta_1,\, 2\pi - \theta_1]$, where $\theta_1 = b/R$, we denote by D_2 (figure 1).

The line L_2 is divided into four arcs: I is $L_{2,1}$, where $\varphi \in [\theta_1,\, 2\pi - \theta_1]$ it corresponds to the contact line of the shell segments D_1 and D_2, the rest corresponds to the contact line of the shell segment D_1 and the reinforcement: II is $L_{2,2}$, where $\varphi \in [-\theta_1,\, -\theta_2]$, III is $L_{2,3}$, where $\varphi \in [-\theta_2,\, \theta_2]$, IY is $L_{2,4}$, where $\varphi \in [\theta_2,\, \theta_1]$, $\theta_2 = (b-c)/R$. The parts of the reinforcement that contain the arc $L_{2,4}$ we denote D_3 and accordingly: $L_{2,3}$ we denote D_5; $L_{2,2}$ we denote D_4. We denote the free straight edges of the reinforcement, respectively: Π_1, where $\varphi = \theta_1 - \theta_2$; Π_2, where $\varphi = -\theta_1 + \theta_2$, here $z \in [-a+c, a-c]$; Π_3, where $\varphi \in [-\theta_2,\, \theta_2]$, $z = a - c$.

To solve the problem, it is necessary to satisfy the conditions of ideal contact along the circle L_2, intercepts A_1D_1, B_1C_1, and parts of intercepts AD where ($\varphi = \theta_1 - \theta_2$), BC where ($\varphi = -\theta_1 + \theta_2$), here, $z \in [a-c, a]$, which we denote S_1, S_2. And also satisfy boundary conditions along the contour of the hole Π and the circle L_1.

The stress state of separate parts of the shell will be divided into two components: the basic stress state, which is determined by the main vectors of forces and moments; and a self-balanced state (as a series), which decrease with distance from the edges of the shell segments. The method of calculating the stress state by identifying and finding only the basic state was proposed by de Saint-Venant and has received the wide practical application. Note that the basic and self-balanced stress states have different physical nature and accordingly must be described by different functions. The main stress state will assume known [11] and mark with the index 0 at the top.

3. Representation of functions describing the self-balanced shell parts state.

We write down the basic functions of shell parts D_1, D_2 are found in [11]

$$\Phi_j(\alpha,\varphi) = h^3 E_1 \sum_{k=1}^{M_j} b_k^j \varphi^j(k,\alpha,\varphi), \quad w_j(\alpha,\varphi) = h \sum_{k=1}^{M_j} b_k^j w^j(k,\alpha,\varphi), \tag{4}$$

where $j = \overline{1,2}$, $\alpha = z/R$, $M_1 = 8N$, $M_2 = 12N$, N is the number of harmonics decomposition.

For parts of the reinforcement D_j, $j = \overline{3,5}$, we define the boundaries of the change in dimensionless coordinates $\alpha_j = z/R$, $\varphi_j = \varphi$:

$$\alpha_j \in [J_1^j,\, J_2^j],\ j = \overline{3,5},\ J_j^3 = J_j^4 = (-1)^j \frac{a}{R},\ j = \overline{1,2},\ J_1^5 = \frac{a-c}{R},\ J_2^5 = \frac{a}{R},$$

$\varphi_3 \in [\theta_2,\, \theta_1]$, $\varphi_4 \in [-\theta_1,\, -\theta_2]$, $\varphi_5 \in [-\theta_2,\, \theta_2]$. Similarly to [11], for each part of the lining basic functions are built that allow to satisfy the conditions specified on separate arcs $L_{2,j}$, $j = \overline{2,4}$.

Construction of the shell functions that satisfy the boundary conditions on intercepts parallel to the axis Oz. For example, consider the part of the reinforcement D_3, where we will find for them in this form:

$$\Phi_3(\alpha_3,\varphi_3) = h_1^3 E_1^1 \mathrm{Re}[a_3 \exp((-1)^n \omega\eta(\varphi - \theta_n))] \cos(k\omega\alpha),$$

$$w_3(\alpha_3,\varphi_3) = h_1 \mathrm{Re}[(\eta) b_3 \exp((-1)^n \omega\eta(\varphi - \theta_n))] \cos(k\omega\alpha),\ n = \overline{1,2}, \tag{5}$$

here $\omega = k_1 R/a$, $4 > k_1 > \pi$, the choice of the parameter k_1 allows to improve the convergence, a_3 is the complex coefficient, η is the complex spectral parameter, which is determined from the following characteristic equation

$$[\eta^4 + (1 - k^2 q_2^1)\eta^2 + k^4 \frac{E_1^1}{E_2^1}][\eta^4 - q_1^1 k^2 \eta^2 + \frac{E_1^1}{E_2^1} k^4] - (1 - \nu_1^1 \nu_2^1) \frac{12a^4}{h_1^2 R^2 k_1^4} \frac{E_1^1}{E_2^1} k^4 = 0, \tag{6}$$

Using the first equation (1) and the representation (5), we find the connection between

$$b_3 = \delta_3(\eta)a_3,\ \delta_3(\eta) = -\frac{h\omega^2}{Rk^2}[\eta^4 - q_1^1k^2\eta^2 + \frac{E_1^1}{E_2^1}k^4].$$

An analysis of the roots of the equation (6) is given in [11].

Considering the above and formulas (4), (6) we write the functions of the parts of the reinforcement D_j, which are determined by the roots $\mu_{k,m}^1$, $\eta_{k,m}^1$, $m = \overline{1,2}$, in the following compact form:

$$\Phi_j(\alpha,\varphi) = h^3E_1^1\sum_{k=1}^{M_j} b_k^j\varphi^j(k,\alpha,\varphi), \quad w_j(\alpha,\varphi) = h_1\sum_{k=1}^{M_j} b_k^j w^j(k,\alpha,\varphi), \tag{7}$$

here $j = \overline{3,5}$, $M_3 = M_4 = 12N$, $M_5 = 16N$, b_k^j are real unknowns.

4. Representation of boundary conditions, conditions of ideal contact and the method of their satisfaction.

On the circle L_1 and the arc $L_{2.1}$, the boundary conditions are considered in [11]. On arcs $L_{2,j}$, $j = \overline{3,5}$, which are the connection line of the shell D_1 and the parts of the reinforcement D_j, $j = \overline{3.5}$, ideal contact conditions are met, which can be submitted to as follows:

$$T_1 + T_1^0 = T_{1,j}^1,\quad S_{12} + S_{12}^0 = S_{12,j}^1,\quad N_1 + N_1^0 = N_{1,j}^1,\quad M_1 + M_1^0 = M_{1,j}^1,$$

$$w_1 + w_1^0 = w_j^1,\quad \frac{(\partial w_1 + w_1^0)}{\partial z} = \frac{\partial w_j^1}{\partial z},\ u_1 + u_1^0 = u_j^1,\ v_1 + v_1^0 = v_j^1,\ j = \overline{3,5}. \tag{8}$$

Express the conditions of ideal contact on the intercepts A_1D_1, B_1C_1

$$T_2 + T_2^0 = T_{2,j}^1,\quad S_{21} + S_{21}^0 = S_{21,j}^1,\quad N_2 + N_2^0 = N_{2,j}^1,\quad M_2 + M_2^0 = M_{2,j}^1,$$

$$w_2 + w_2^0 = w_j^1,\quad \frac{(\partial w_2 + w_2^0)}{\partial\varphi} = \frac{\partial w^j}{\partial\varphi},\ u_1 + u_1^0 = u_j^1,\ v_1 + v_1^0 = v_j^1,\ j = \overline{3,4} \tag{9}$$

and intercepts S_1 , S_2

$$T_{2,5}^1 = T_{2,j}^1,\quad S_{21,5}^1 = S_{21,j}^1,\quad N_{2,5}^1 = N_{2,j}^1,\quad M_{2,5}^1 = M_{2,j}^1,$$

$$w_5^1 = w_j^1,\quad \frac{(\partial w_5^1}{\partial\varphi} = \frac{\partial w^j}{\partial\varphi},\ u_5^1 = u_j^1,\ v_5^1 = v_j^1,\ j = \overline{3,4}. \tag{10}$$

Let us write together the boundary conditions on the arcs L_1, $L_{2,1}$, systems of equations (8)–(10), and also the relations (3), after transformations of which we get

$$\sum_{k=1}^{M} c_k A_{m,k}(\gamma_m) = P_m(\gamma_m),\ \ m = \overline{1,K},\ \ \gamma_m \in [\beta_m, \alpha_m], \tag{11}$$

here
$K = 84$; $\beta_m = 0$, $\alpha_m = 2\pi$, $m = \overline{1,5}$; $\beta_m = \theta_1$, $\alpha_m = 2\pi - \theta_1$, $m = \overline{6,13}$; $\beta_m = \theta_2$, $\alpha_m = \theta_1$, $m = \overline{14,21}$; $\beta_m = -\theta_1$, $\alpha_m = -\theta_2$, $m = \overline{22,29}$; $\beta_m = -\theta_2$, $\alpha_m = \theta_2$, $m = \overline{30,37}$; $\beta_m = 0$, $\alpha_m = J_2^3$, $m = \overline{38,53}$; $\beta_m = J_1^5$, $\alpha_m = J_2^5$, $m = \overline{54,69}$; $\beta_m = 0$, $\alpha_m = J_1^5$, $m = \overline{70,79}$; $\beta_m = J_1^5$, $\alpha_m = J_2^5$, $m = \overline{80,84}$; functions $P_m(\gamma_m)$ are determined by the the basic stress state of shell.

In order to meet conditions (11) we apply analytical and numerical method that was proposed in [12] for plane problem solving for the rectangular domain, in [13] for boundary problems in the homogeneous cylinders, and in [14] in a multilayer cylinder using homogeneous solutions. The given in [11]–[14] method allows us reduce to numerical solution of the system eighty-four equations (11) to finding the minimum $F(N)$ of such a generalized quadratic form:

$$\sum_{m=1}^{K} \|f_{m,N}(\gamma) - P_m(\gamma)\|_m^2 = \sum_{k,j=1}^{M} c_k c_j W_{kj} - 2\sum_{k=1}^{M} c_k V_k + P^2, \tag{12}$$

here

$$W_{kj} = \sum_{m=1}^{K} (A_{m,k}, A_{m,j})_m, \;\; W_{jk} = W_{kj}, \;\; V_k = \sum_{m=1}^{K} (A_{m,k}, P_m)_m, \;\; k, j = \overline{1, M},$$

$$P^2 = \sum_{m=1}^{K} \|P_m\|_m^2, \;\; (f, g)_m = \int_{\beta_m}^{\alpha_m} f(\gamma) g(\gamma) d\gamma - \text{scalar product.}$$

The found variables c_k, on which the minimum $F(N)$ is reached, are denoted c_k^N, and the basic functions of the parts of the shell which defined by the coefficients c_k^N , we denote $\Phi_j^N(z,\varphi)$, $w_j^N(z,\varphi)$. According to the known formulas we find the forces and moments in the shell and the reinforcement.

4.1. Criteria for convergence of the constructed solution.

Lemma ([13]). *The function $F(N)$ is nonnegative and nonincreasing.*

Theorem ([14]). *If there exists N such that $F(N) < \frac{\varepsilon^2}{4}$ for any $\varepsilon > 0$, then displacements, forces and moments expressed through functions $\Phi_j = \lim_{N\to\infty} \Phi_j^N$, $w_j = \lim_{N\to\infty} w_j^N$, will exactly satisfy meet conditions (11) in the metrics of $L_2[\beta_m, \alpha_m]$.*

Note that for minimum $F(N)$ the equality is fulfilled

$$F(N) = \sum_{m=1}^{K} \|f_{m,N}(\gamma) - P_m(\gamma)\|_m^2, \tag{13}$$

which serves as an estimate of the accuracy of the found solution.

The results review. Numerical calculations show us that the process of simultaneous boundary conditions satisfaction by the Fourier series for the triply connected cylindrical shell with a large rectangular hole is unstable. This is explained by the fact that the representation of the solution includes exponential functions attached to individual sides of the hole or the ends of the shell, whose performance increases linearly with increasing N. They cannot (without conditional cutting the shell into pieces) rationed so that they do not grow on some sides. In the works [12, 13] it is shown that the computational process of construction using the Fourier series method of solving a boundary value problem in a rectangular region is stable. Since the exponential functions can be normalized so that they will be equal unity the given sides of the rectangle and decrease with distance from them. On this basis, it has been proposed to divide the shell with a reinforced hole into simpler, rectangular segments in the plan.

5. Conclusions.

The finite-body method is substantiated, based on a special separation of the three-connected shell surface with the reinforced rectangular hole into simpler doubly connected or simply connected parts, at the edges of which the exponential functions entering the solution are normalized by one. This allowed us to construct a stable algorithm for solving the boundary problem. An approach has been developed that allowed simultaneously to take into account a large number of boundary or contact conditions, both in the forces and in the displacements specified on different edges of the shell. The principle of de Saint-Venant is confirmed. The mathematical justification for constructing a representation of the self-balanced state (in the form of series) is given, which rapidly decreases when away from the edges of the shell segments. The technique has been developed to reduce satisfaction of all boundary conditions and conditions of contact of shell segments to finding the minimum of the generalized quadratic form.

The algorithm has been developed for the calculation of the shell with the hole and the reinforcement of various materials. Its advantage is that there are no restrictions on the basic systems of functions, in particular, non-orthogonal functions can be used. The additive generalized quadratic form was introduced, which is not linear, but its minimum tends to zero with increasing N, which determines the convergence and accuracy of the solution. The developed technique allowed generalizing the well-known Fourier decomposition method for a single equation to the case of the system of equations with non-orthogonal coefficients. The criteria for convergence of the proposed analytical-numerical method are established, which are based on the analysis of the value of one number - the minimum of the generalized quadratic form.

References

[1] Ambarzumyan S A 1974 *General Anisotropic Shell Theory.* (Moscow: Nauka).
[2] Grigorenko Ya M, Vlaikov G G and Grigorenko A Ya 2006 *Numerical-Analytical Solution of the Shell Mechanics Problems Based onDifferent Models.* (Kiev: Akademperiodica).
[3] Guz A N, Chernyshenko I S, Chekhov Val N, Chekhov Vik N and Shnerenko K I 1980 *Cylindrical Shell Weakened by Holes. Shell Calculation Methods.* Vol. 1. (Kiev: Naukova Dumka).
[4] Vasiliev V V 1988 *Mechanics of structures made of composite materials.* (Moscow: Mashinostroenie).
[5] Bakulin V N 2018 *Doklady Physics. vol.* 63 No. 1, pp. 23-27. DOI: 10.1134/S1028335817110076
[6] Ciarlet P G 2000 *Mathematical elasticity.* V. 3 *Theory of shells.* (Amsterdam: North Holland).
[7] Revenko V P 1988 *Sov. Appl.Mech.* **24** No. 4, pp 368373.
[8] Bakulin V N and Repinskii V V 1988 *Numerical Method Studies of Aircraft Strength.* (Moscow: Publisher MAI) pp 5–10.
[9] Bakulin V N and Repinskii V V 1989 *Thematic collection of scientific papers MAI.* (Moscow: Publisher MAI) pp 4–6.
[10] Vorobey V V 1979 *Sov. Appl.Mech.* **24** No. 4, pp 82–85.
[11] Bakulin V N and Revenko V P 2016 *Russian Mathematics.* No. 6, pp 1-11.
[12] Revenko V P 2008 *Int. Appl.Mech.* **44** No. 1, pp 73-80.
[13] Revenko V P 2009 *Int. Appl. Mech.* **45** No. 7, pp 730-741.
[14] Bakulin V N and Revenko V P 2017 *Materials physics and mechanics.* **31** No. 1/2, pp 93-96.

Neural network approach to solving the inverse problem of surface-waves generation

E.A.Voronin,V.N.Nosov, A.S.Savin

Federal Research Center "Informatics and Management" RAS, Moscow
Vernadsky Institute RAS, Moscow
Bauman Moscow State Technical University, Moscow

E-mail: e.voronin1@gmai.com

E-mail: viktor_nosov@mail.ru

E-mail: assavin@list.ru

Abstract. The paper proposes an approach to the inverse tasks of the surface-waves generation, based on information technologies of machine learning. On the basis of this approach, a method has been developed for determining the parameters of a cylinder and a wing profile moving in fluid according to a laboratory experiment about a free surface liquid disturbed by them.

1. Introduction

Radar and optical means of remote sensing of the marine environment do not allow one to directly observe the processes occurring at a considerable depth. But, with their help, parameters of the sea surface can be measured. The need to study the processes occurring in the marine deepness according to parameters of its surface led to the creation of computer radio tomography of the marine environment [1].

The theoretical basis of this section of science and technology consists of various approaches to solving the task of restoring the internal structure of a heavy fluid flow according to its free surface. In particular, the inverse task of the generation of surface waves, related to determining the parameters of sources of disturbances from the features of the waves generated by them on the liquid surface, is of considerable interest.

This task of continuing the fields in the direction of their sources belongs to the class of incorrectly stated tasks [2]. Its solution is unstable in relation to small changes in the original data, which makes it difficult to obtain information about the sources of disturbances.

For the first time, some approaches to solving the inverse task of generating surface waves when a plane flow goes around a point-like hydrodynamic feature were proposed in [3, 5]. However, during test application of the proposed algorithms to processing the data of laboratory experiments [6, 7], in which measurements were made of the angles of inclination of the water surface perturbed by the moving cylinder and wing profile aroused some difficulties appropriate to incorrect tasks of determining the parameters of sources of disturbances.

In this paper, a fundamentally new approach to solving the inverse task of generating surface waves is proposed, based on the methods of machine learning, namely, the theory of neural networks, which is now widely used in data processing and analysis [8, 10].

This is explained by the fact that modern learning algorithms for neural networks such as a stochastic gradient descent, an impulse or a packet method are based on the principles of overcoming mathematical incorrectness.

2. Solution

To use machine learning technology, you must have a training set. It should be noted here that such a set can be obtained in various ways. For example, in the current task, it can be obtained by multiple numerical determinations of waves on the surface of a stream flowing around point-like hydrodynamic features that simulate an obstacle in the stream, using the formulas [?].

To simulate real conditions, some noise can be applied to the solutions obtained in this way. In any case, this will be a certain model solution, since neither the imposed noise, nor the system of hydrodynamic features, which simulates the obstacle streamlined by the flow, can not exactly correspond to reality.

In the present work, the training samples were obtained by measuring the slopes of the surface of the water, perturbed by the wing profiles and cylinders of various diameters moving in its thickness. The experiments were carried out on an experimental facility [3, 5], which is a 300 cm $x = 50$, $x = 50$ cm channel filled with water. In the , coordinate system moving with the model, $y = S(x)$ is the deviation of the free surface of the liquid from the equilibrium position of $y = 0$. At various points x_n, the tangent of the angle $\alpha_n(n = 1, , N)$ of the inclination of the free water surface to the horizon, that is, the derivative $S'(x_n$, was measured. In [6], these results are presented in tabular form. One series of experiments was carried out with a symmetrical wing profile moving at a depth of $H = 175mm$ with different velocities $V = 482 - 938mm/s$. The other series - with a circular cylinder with a diameter of $D = 6mm$, moving with velocities $V = 280 - 1056mm/s$ at depths of $H = 66 - 126mm$. The third one - with a circular cylinder with a diameter of $D = 20mm$, moving with velocities $V = 324 - 565mm/s$ at depths $H = 103 - 133mm$.

Formally, the task of machine learning is reduced to the construction of a neural network and is formulated as a mathematical problem of approximation. Let a region $\Omega : \vec{r} = (x_1, x_2, x_3) \in \Omega \subset R^3$ be given and two sets of points be given: the input data set $Q_N = \{r_n\} \in \Omega(n = 1, N)$ and the target output data set $Q_M = \{r_m\} \in \Omega(m = 1, M)$. Let $f(Q_N)$ be the function values at the points of the training set, and then the approximating function $F(r$ is realized with the help of a neural network as a continuous mapping $F : Q_n \to Q_m$. The coordinates of the point $r \in \Omega$ arrive at the input. After the signal is converted in the neural network, the output is a signal $F(r)$.

In such tasks, training with a teacher is always used, that is, for each training input vector there is a training output vector. For this, a network of direct signal propagation of the multilayer perceptron type was used, which was taught by algorithms based on the method of back propagation of error [8].

An important problem, along with the acquisition of data for training and their processing, is a search of the characteristic features of the training set. In our case, the training set was a sequence of tangent values of the angles of inclination of the free surface of the liquid, measured at different points xn. To select the main features of such a set that characterize the properties and behavior of the process or object under study, recently, deep learning networks [12, 15] are used, but for this, besides a large amount of initial information, considerable computational resources or a long time of the learning process are required, which unattainable in real conditions and does not meet the requirements of operational information processing when determining the parameters of the source of the perturbation of the liquid. Therefore, to identify the characteristic features of the observed process and reduce the training set to the data obtained during the experiment, namely, to a sequence of tangents of the angles of inclination of the liquid surface, measured at different points, spectral analysis was applied. It is known that the spectrum of the observed signal is invariant with respect to the displacement along the spatial coordinate.

This feature made it possible to move from the whole sequence of observed coordinates of the training vector to several lines of the spectrum, which significantly reduced the size of the neural network and the duration of its learning process. As a result of spectral analysis of experimentally obtained sequences of values of tangents of angles of inclination of the water surface, perturbed by different models towed in the fluid, three main frequencies were identified and found to have similar values for all series of laboratory experiments.

Statistical analysis of a sample of the amplitudes of the spectral lines obtained as a result of processing the experimental data revealed a strong correlation between them and the characteristics of the sources of disturbances. Therefore, for training the neural network, the calculated parameters of the training vector with three coordinates that have the values of the main amplitudes of the spectrum were used.

After conducting a sufficient number of numerical experiments, the neural network architecture was chosen. From the initial data, a larger sample was formed by 5% noise with a uniform density. The choice of the noise interval was identified by computational experiment.

For learning on the variability of factors, data were normalized, that is, reduced to relative units in the interval (0, 1 by dividing each parameter by the maximum value. Neural network training is reduced to minimizing the error function, by adjusting the weights of synaptic connections between neurons. The expression for the root-mean-square error is represented as follows.

$$E = \frac{1}{2p} \sum_{j=1}^{p} (d_j - y_j)^2,$$

where y_j, is the actual result, d_j is the desired network response.

The classical gradient descent method is based on choosing the value of the weights of connections between neurons in order to achieve the minimum approximation error of the objective function in the minimum number of learning steps. The disadvantage of this method is that in some cases the minimization process does not converge to a global value. This task can be solved by the method of stochastic gradient descent (SGD) [9], [12, 13, 14, 15, 16, 17].

Features of this approach are that the method of stochastic gradient descent helps to avoid "sticking" the model at a local minimum point, is not sensitive to small disturbances, is simpler to calculate, is not sensitive to small disturbances, and is not picky about computational resources. Numerical experiments with a neural network confirmed their coincidence with the results of laboratory experiments with an accuracy of 5.

3. Conclusion

Methods of machine learning and preparation of training samples require significant computational resources and can be successfully implemented by means of parallel programming on computing clusters or graphic processors like Tesla.

Since they have shown their effectiveness in solving the inverse task of determining the parameters of a moving body according to the disturbances of the free surface of a fluid created by this moving body, it is necessary to extend the described approaches to other classes of incorrectly posed tasks of fluid mechanics and geophysics.

References

[1] Methods, procedures and means of aerospace computer radio tomography of the sur-face areas of the Earth / Ed. Nesterov S.V., Shamaev, A.S., Shamaev S.I.. - M .: Sci-entific world, 1996. - 272 p.
[2] Tikhonov A.N., Arsenin V.Ya. Methods for solving ill-posed problems. - M .: Nauka, 1979. - 286 p.
[3] Konovalov A.V., Levchenko E.S., Savin A.S. Restoration of the plane flow of a heavy ideal fluid by the shape of its free surface // Report of Academy of Sciences of the USSR. - 1989. 305, 2. - P. 294 - 296.

[4] Isichenko I.V., Konovalov A.V., Levchenko E.S., Savin A.S. The inverse problem of streamlining features with a plane flow of an ideal fluid with a free boundary // PMTF. - 1989. - No. 6 - P.86 - 91.

[5] Savin A.S. Determination of parameters of hydrodynamic features in a flat flow from the parameters of its free surface. // Izv. RAS.MZHG.2001. 2. Pp. 139146.

[6] Boyarintsev V.I., Lednev A.K., Prudnikov A.S., Savin A.S., Savina E.O. Modeling and experimental study of disturbances of the free boundary of a plane flow by im-mersed sources. Preprint / Institute for Problems in Mechanics of the Russian Acade-my of Sciences. - M., 2002. No. 720.- 37 p.

[7] Boyarintsev V.I., Lednev A.K., Prudnikov A.S., Savin A.S., Savina E.O. Disturbance of the free surface of a liquid by an airfoil, Izv. RAS. Fluid and gas mechanics .- 2004.- No. 6.- p. 145-152.

[8] Callan R. Basic concepts of neural networks: Trans. from English M.: Williams Pub-lishing House, 2001. 288 c.

[9] Gudeflou J., Bengio I., Courville A. Deep learning / Trans. from English. A.A.Slinkin. M.: DMK Press, 2017. 652 p.: Ill.

[10] Khaikin, S. Neural networks: a full course, 2nd ed., Rev. / S. Khaikin. - M.: Publish-ing house "Williams", 2006. 1104.

[11] Sretensky L.N. Theory of wave motion of a fluid. M.: Science, 1977. 815s.

[12] C. C. Manning, P. Raghavan, H. Schutze. - Cambridge University Press, 2008. - 544 p.

[13] Max Kuhn, Kjell Johnson. Applied Predictive Modeling // Springer, 2013. 458p.

[14] Nathan M, James W. Big Data: Principles and best practices of scalable realtime data systems: Manning Publications. 2015. 325 c.

[15] Sebastian Raska. Python and machine learning. DMK-Press, 2017. p. 11 408.

[16] Ryshard T. Elementary introduction to the technology of neural networks with exam-ple programs. M.: Fizmatlit, 2011. 408 c.

[17] Galushkin A.I. Neural networks. Fundamentals of the theory. M .: Hotline. Telecom, 2010. 450 p.

Journal of Physics: Conference Series **1392** (2019) 012023 doi:10.1088/1742-6596/1392/1/012023

Mathematical Simulation of local transfer for non-Newtonian fluid in porous fabrics

Yu.I. Dimitrienko, Shuguang Li

Computational Mathematics and Mathematical Physics Department, Bauman Moscow State Technical University, 2nd Baumanskaya Street, 5, Moscow, 105005, Russia.

E-mail: dimit.bmtstu@gmail.com, shuguangli2008@gmail.com

Abstract. Many processes in the polymer composite, textile, food, and pharmaceutical industries are associated with the flow of non-Newtonian fluids in porous media. This paper considers the mathematical model of a multi-scale process for the filtration of non-Newtonian fluid in periodic porous fabrics. The model should be based on a three-dimensional Navier-Stokes equation with non-Newtonian Carreau-Yasuda viscosity using the asymptotic averaging method. A numerical algorithm was developed for solving local problems of non-Newtonian fluids in periodic cells, and the distribution of velocity, pressure and non-Newtonian viscosity in a single pore was obtained. The algorithm for calculating the permeability tensor is developed, and the effects of fluid rheology are also highlighted.

1. Introduction

The study and simulation of the flow of non-Newtonian fluids in porous media is very important in many biological systems or processes of the petroleum, pharmaceutical, food, cosmetic, textile, paper and polymer composite industries [1]. In the above-mentioned fields, the flowing fluids involved generally exhibit complex behaviors such as shear thinning/thickening effects, elasticity, anisotropy, and yield stress [1, 4]. Therefore, it is extremely important to adequately simulate the flow of a liquid binder in a porous composite structure having a complex space geometry. Modeling various process problems in porous systems, especially with reference to the mechanics of composite materials, is quite interesting. Various aspects of modeling the technological processes of composite materials production are considered in [2 − 4, 6].

In most existing literature, the permeability coefficients of a porous medium are determined experimentally or by using various empirical and approximate relationships to determine the local filtration processes. Thus, a rough estimates of the actual processes occurring within the pores with complex geometries are obtained, which results in large deviations in the permeability measurement. Therefore, an important part of the filtering study is to use the solution of the non-Newtonian Navier-Stokes equation, and derive the average filtration equation from the "first principle" to analyze the local process of fluid space flow in a single pore, rather than based on phenomenological theory. This approach to the modeling of filtration in porous structures is called a microscale simulation, which is a joint study of the liquid flow in individual pores (a local problem).

However, the non-Newtonian characteristics of the fluid pose a significant challenge to the numerical simulation of the local transmission process. Although there has been a small amount of research on local transport of non-Newtonian fluids in regular geometries, this is clearly insufficient. In order to better understand the above complex fluid flows, we performed a multi-scale simulation for the flow of non-Newtonian fluids in a porous fabrics. The effect of non-Newtonian viscosity on the local transport process is highlighted in this study.

2. Mathematical model

2.1. Main assumptions of the model

It is assumed that the composite structure is characterized by periodicity. And individual microscopic periodic cells also have geometric mirror symmetry.

Take the following assumptions regarding the properties of the phases:
(1) fluids are isotropic non-Newtonian viscous incompressible media;
(2) the porous skeleton is assumed to be non-deformable, i.e. his movement is not considered;
(3) fluid movement is isothermal;
(4) mass forces are absent.

2.2. Principal equations

The motion of non-Newtonian fluid in the composite occupying the pore, within the framework of the assumptions made, is described by the system of the incompressible Navier-Stokes equations, which, together with the adhesion conditions on a solid surface, have the following form [6]:

$$\nabla \cdot \mathbf{v} = 0 \tag{1}$$

$$\rho_0(\frac{\partial \mathbf{v}}{\partial t} + \mathbf{v} \cdot \nabla \mathbf{v}) = -\nabla p + \nabla \cdot 2\mu \mathbf{D}; \tag{2}$$

$$\mathbf{v}|_{\Sigma} = 0; \tag{3}$$

$$p|_{t=t_0} = p_0. \tag{4}$$

where ρ_0 -density; $\mathbf{v}$ -velocity of fluid; p - pressure; and μ - non-newtonian viscosity.

The strain rate tensor:

$$\mathbf{D} = \frac{1}{2}(\nabla \mathbf{v} + \nabla \mathbf{v}^T). \tag{5}$$

The A_I and A_V models of Voigt isotropic viscous media [6] are considered, and non-Newtonian viscosity satisfies the following defining relation [1, 9]:

$$\frac{\mu - \mu_\infty}{\mu_0 - \mu_\infty} = (1 + (\lambda I_2)^2)^{\frac{1-n}{2}}. \tag{6}$$

which depends on the second invariant of the strain rate tensor [5]:

$$I_2(\mathbf{D}) = \sqrt{2\mathbf{D} \cdot \cdot \mathbf{D}}. \tag{7}$$

Then we used a dimensionless form, here:

$$\bar{\mathbf{v}} = \frac{\mathbf{v}}{v_0}, \quad \bar{p} = \frac{p}{p_0}, \quad \bar{t} = \frac{t}{t_0}, \quad \bar{\mathbf{x}} = \frac{\mathbf{x}}{x_0}, \quad t_0 = \frac{x_0}{v_0}. \tag{8}$$

Here $\bar{\mathbf{v}}$, $\bar{p}$ are the dimensionless velocity and pressure of fluid, respectively; v_0, p_0 are their typical magnitudes. t_0 is the dimensionless time. The equations for the flow of an incompressible fluid can be written in a dimensionless form as follows (the symbol is omitted below):

$$\nabla \cdot \mathbf{v} = 0 \tag{9}$$

$$\frac{\partial \mathbf{v}}{\partial t} + \mathbf{v} \cdot \nabla \mathbf{v} = -Eu \nabla p + \frac{1}{Re} \nabla \cdot 2\mu \mathbf{D}; \tag{10}$$

$$\mu = s + (1-s)(1+(CuI_2)^2)^{\frac{1-n}{2}}, \quad s = \frac{\mu_\infty}{\mu_0}; \tag{11}$$

$$\mathbf{D} = \frac{1}{2}(\nabla\mathbf{v} + \nabla\mathbf{v}^T), \quad I_2(\mathbf{D}) = \sqrt{2\mathbf{D}\cdot\cdot\mathbf{D}} \tag{12}$$

$$\mathbf{v}|_\Sigma = 0; \tag{13}$$

$$p|_{t=t_0} = p_0. \tag{14}$$

where $Eu = \frac{p_0}{\rho_0 v_0^2}$ - Euler number, $Re = \frac{\rho_0 v_0 x_0}{\mu_0}$ - Reynolds number, $Cu = \frac{\lambda v_0}{x_0}$ - Carreau number.

3. Statement of the local problem

In accordance with the method of asymptotical averaging [3, 4, 7, 8], composite structure can be presented by a periodical structure consisting of repeating elements, i.e., periodicity cells V_ξ. Local coordinates ξ introduced as

$$\xi = \frac{\bar{\mathbf{x}}}{\kappa}. \tag{15}$$

can change within limits of periodicity cells V_ξ, where $\bar{\mathbf{x}} = \frac{\mathbf{x}}{x_0}$ are the nondimensional global coordinates changing inside the whole region V occupied by the material; x_0 is the characteristic global size of the whole body V; l_0 is the characteristic size of the periodicity cell V_ξ; κ is the small parameter:

$$\kappa = \frac{l_0}{x_0} << 1. \tag{16}$$

Then all main functions f describing a fluid flow in pores can be considered to be quasi-periodic (i.e. depending on ξ and $\bar{\mathbf{x}}$) and periodic in ξ. Differentiation of the functions is realized by the following rule:

$$\nabla f = \nabla_x f + \frac{1}{\kappa}\nabla_\xi f. \tag{17}$$

where $\nabla_{x,\xi}$ are nabla-operators over coordinates $\mathbf{x}, \xi$.

We introduce the operator of the mean value $< \cdot >$ over the periodicity cells V_ξ for quasi-periodic in local $\xi \in V_\xi$ coordinates of the functions f:

$$< f >= \frac{1}{\varphi_p}\int_{V_\xi} f dV, \quad \varphi_p = \int_{V_\xi} dV. \tag{18}$$

However, due to periodicity of the structure, the solution may be represented in the form of asymptotic expansion in terms of parameter κ [7, 8]:

$$\mathbf{v} = \mathbf{v}^{(0)}(\mathbf{x},\xi) + \kappa\mathbf{v}^{(1)}(\mathbf{x},\xi) + \kappa^2\mathbf{v}^{(2)}(\mathbf{x},\xi) + \cdots; \tag{19}$$

$$p = p^{(0)}(\mathbf{x},\xi) + \kappa p^{(1)}(\mathbf{x},\xi) + \kappa^2 p^{(2)}(\mathbf{x},\xi) + \cdots; \tag{20}$$

$$\mu = \mu^{(0)}(\mathbf{x},\xi) + \kappa\mu^{(1)}(\mathbf{x},\xi) + \kappa^2\mu^{(2)}(\mathbf{x},\xi) + \cdots. \tag{21}$$

Having substituted expansions (19) – (21) into Equations (9) – (14), then having collected in them terms at the same powers of κ and putting terms at the lowest powers of κ equal to zero, we obtain the local problems of the zero level 'over the periodicity cell':

$$\nabla_\xi \cdot \mathbf{v}^{(0)} = 0 \tag{22}$$

$$-\nabla_\xi p^{(1)} + \eta_0 \nabla_\xi \cdot 2\mu^{(0)}\mathbf{D}^{(0)} = -\nabla_x p^{(0)}; \tag{23}$$

$$\mathbf{D}^{(0)} = \frac{1}{2}(\nabla_\xi \mathbf{v}^{(0)} + \nabla_\xi {\mathbf{v}^{(0)}}^T); \tag{24}$$

$$\mu^{(0)} = s + (1-s)(1+(Cu^0)^2 Y^{(0)})^{\frac{n-1}{2}}, \quad Y^{(0)} = 2\mathbf{D}^{(0)}\cdot\cdot\mathbf{D}^{(0)}; \tag{25}$$

$$\mathbf{v}^{(0)}|_{\Sigma_\xi} = 0, \quad < p^{(1)} >= 0. \quad [[\mathbf{v}^{(0)}]] = 0, \quad [[p^{(1)}]] = 0. \tag{26}$$

where unknown $\mathbf{v}^{(0)}$ and $p^{(1)}$, the symbol $[[\cdot]]$ denotes the periodicity conditions. $\nabla_x p^{(0)}$ is considered as "input data". And $Re\kappa^0 = Re^0 = O(1)$, $Eu\kappa^2 = Eu^0 = O(1)$ and $Cu = \kappa^0 Cu^0$. Then $\eta_0 = \frac{1}{Re^0 Eu^0}$.

4. Numerical Simulation

Next, the finite element method will be used to solve the local problems (22) – (26). We only consider finite elements that meet LBB conditions [10]. Here, the chemical industry raw material benzene is taken as an example for numerical study, so $\eta_0 = 0.0652$. Applying the physical symmetry of the periodic cells in the hypothesis, the distribution of the results of 1/8 periodic cells can be solved.

4.1. Distribution of results on periodic cells

Here we consider reinforced composites and the composites are isotropic. Here we consider the power-law index $n = 0.25$.

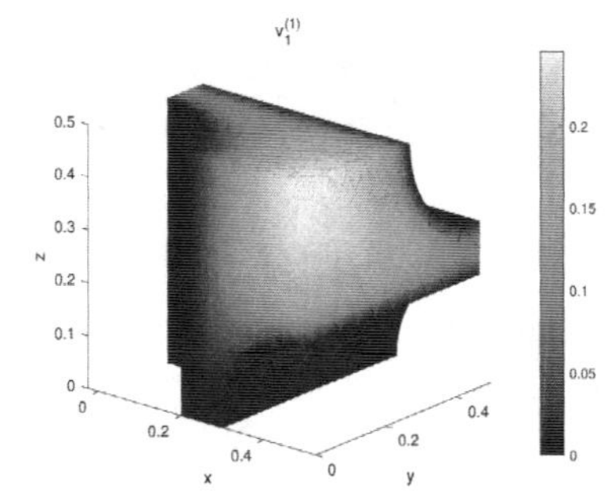

Figure 1. Distribution of the velocity component $v_1^{(0)}$ in 1/8 of the tissue structure periodicity cell.

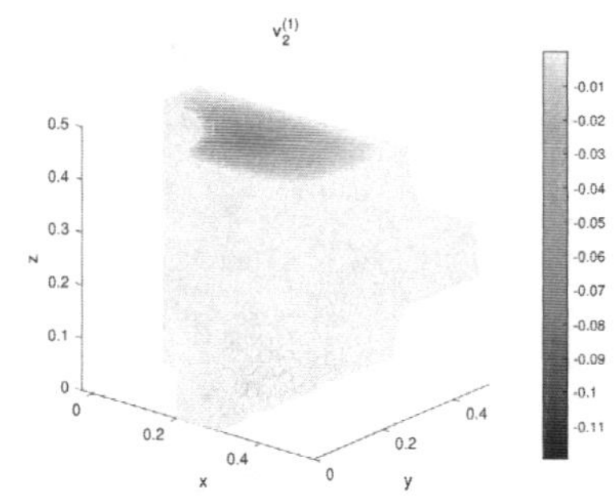

Figure 2. Distribution of the velocity component $v_2^{(0)}$ in 1/8 of the tissue structure periodicity cell.

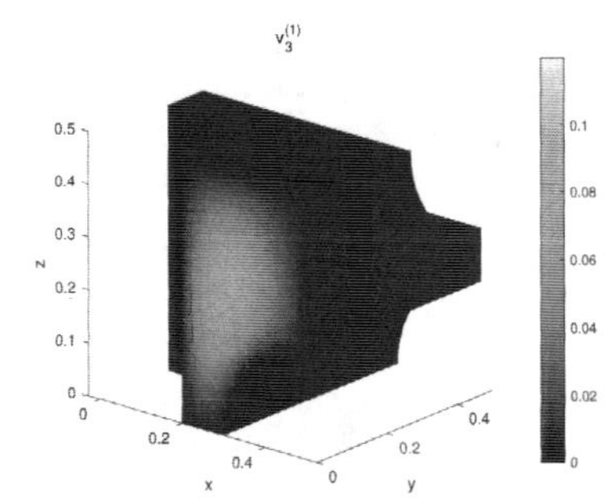

Figure 3. Distribution of the velocity component $v_3^{(0)}$ in 1/8 of the tissue structure periodicity cell.

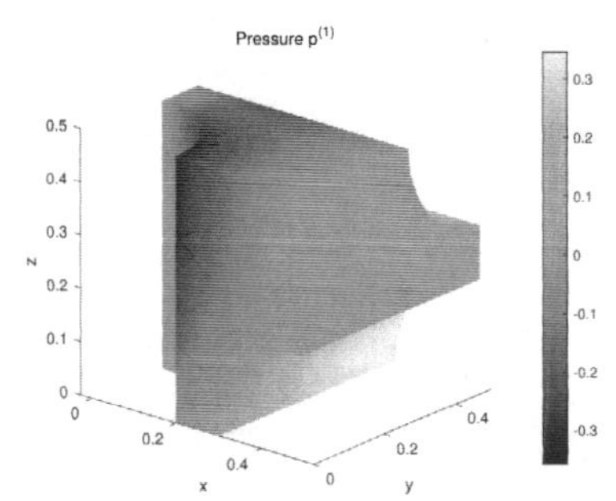

Figure 4. Distribution of pressure $p^{(1)}$ in 1/8 of the tissue structure periodicity cell.

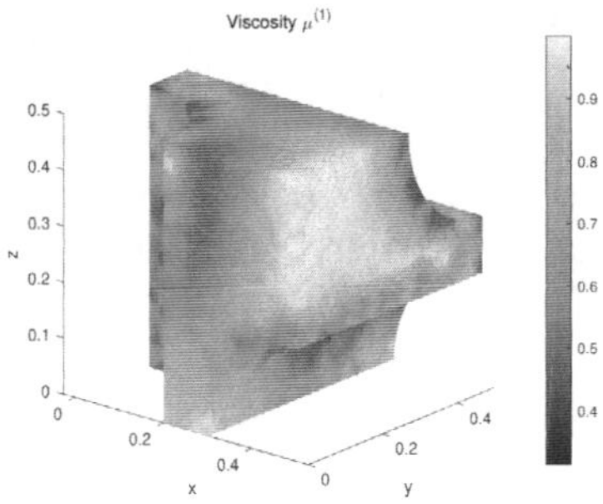

Figure 5. Distribution of viscosity $\mu^{(0)}$ in 1/8 of the tissue structure periodicity cell.

4.2. Compare the permeability of Newtonian fluid

The averaging of the local equations of Newtonian fluids allows one to obtain the classical Darcy's law of filtration:

$$< \mathbf{v} >= -\mathbf{K}\nabla_x p. \tag{27}$$

where $< \mathbf{v} >$ - the filtration velocity, $\mathbf{K}$ - the permeability tensor of the porous medium.

For Newtonian fluids, the components of the seepage rate tensor is constant. For non-Newtonian fluids, however, the components of the permeability tensor are a function related to the constitutive equation of viscosity. To describe this difference, we compared the seepage rate and viscosity changes of non-Newtonian fluids and Newtonian fluids in porous media.

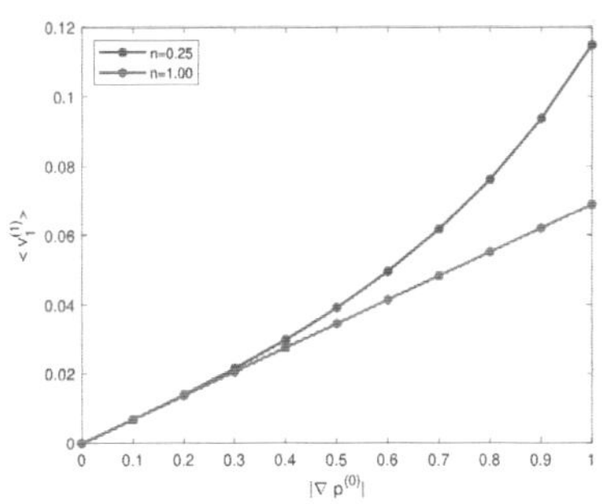

Figure 6. Distribution of the relationship between permeability velocity $< \mathbf{v}^{(0)} >$ and pressure gradient $|\nabla_x p^{(0)}|$.

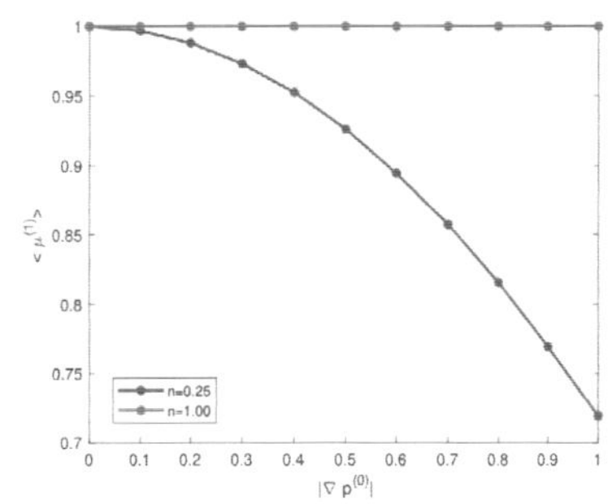

Figure 7. Distribution of the relationship between average viscosity $< \mu^{(0)} >$ and pressure gradient $|\nabla_x p^{(0)}|$.

5. Conclusions

The proposed method allows calculation of the distribution of pressure and non-Newtonian viscosity micro-regions, as well as the filtration velocity component within a single pore, and can also numerically calculate the permeability coefficient of the porous medium without any additional experimental research. For this purpose, modeling of non-Newtonian viscosity resin flow in the periodicity cells of composite structure using an asymptotic homogenization method. And the finite element method is used to calculate the local problem. The results of numerical modeling of non-Newtonian fluid local spatial flows on the periodicity cell of composite structures are obtained, showing the effectiveness of the proposed algorithm for solving local problems. Comparing the difference between the filtration speed of Newtonian fluid and non-Newtonian fluid, the particularity of the local transmission process of non-Newtonian fluid is fully verified.

6. References

[1] Bird R.B, Armstrong R.C. and Hassager O. 1987 *Dynamics of polymeric liquids. Vol. 1: Fluid mechanics* (New York: John-Wiley) P 649.

[2] Golovatov D., Mikhaylov M., Bosov A. 2016 Optimization of Technological Parameters of Impregnation of Load-Bearing Rod Elements of Reflector made of Polymer Composite Materials by Transfer Molding Method *Indian J. of Science and Technology.* **9(46)**.

[3] Idris Z., Orgeas L., Geindreau C., Bloch J.F. and Auriault J.L. 2004 Microstructural effects on the flow law of power-law fluids through fibrous media *Model. Simul. Mater. Sc.* **12(5)** 995.

[4] Orgeas L., Geindreau C., Auriault J.L. and Bloch J.F. 2007 Upscaling the flow of generalised Newtonian fluids through anisotropic porous media *J. Non-Newtonian Fluid Mech.* **145** 15-29.

[5] Dimitrienko Yu.I. 2002 *Tensor analysis and nonlinear tensor functions* (Berlin: Springer) P 662.

[6] Dimitrienko Yu.I. 2010 *Nonlinear continuum mechanics and large inelastic deformations* (Berlin: Springer) P 747.

[7] Dimitrienko Yu.I. 2015 *Thermomechanics of composites under high temperatures* (Berlin: Springer) P 434.

[8] Dimitrienko Yu.I. and Dimitrienko I.D. 2013 Simulation of local transfer in periodic porous media *Eur. J. Mech. B-Fluid.* **37** 174-179.

[9] Dimitrienko Yu.I. and Li S.G. 2018 Mathematical simulation of non-isothermal steady flow of non-Newtonian fluid by finite element method *Mathematical Modeling and Computational Methods.* **2** 70-95.

[10] Zienkiewicz O.C., Taylor R.L. and Zhu J.Z. 2005 *The Finite Element Method: Its Basis and Fundamentals: Its Basis and Fundamentals* (Amsterdam: Elsevier) P 733.

Journal of Physics: Conference Series **1392** (2019) 012024 doi:10.1088/1742-6596/1392/1/012024

On a theorem of impulse and energy to determine parameters of the electric jet engine KA

Nikolay Sidnyaev

Moscow State Technical University Bauman, Moscow, ul. Baumanskaya 2-ya, 5, Russia

E-mail: sidnyaev@yandex.ru

Abstract. The report presents the modern physical principles of acceleration of charged particles and plasma in electric and magnetic fields, which are the basis for the classification of electric rocket engines. It is shown that for reactive means of correction of orbit parameters the lifetime depends to a greater extent on the reserves of the working fluid and the corresponding thrust and energy for the propulsion system. It is postulated that the special value of the determination of pulses and energy for electric propulsion (ERD) is that their application makes it possible to obtain an idea of these phenomena on the boundary surface of a certain area, without knowledge of the individual phenomena occurring within the ERD. That is, often in cases where the differential equations of plasma flow can not be composed or at least can not be integrated, the theory of pulses or energy makes it possible to know the nature of the movement in its General features. It is convincingly proved that the theorems of impulses and energy are a convenient means for checking the correctness of the results, including experimental ones.

1. Introduction

A new, rapidly developing branch of electric propulsion is the space engine industry, which combines the design, development and manufacture of space propulsion systems. Electric propulsion systems (EDU) are commonly referred to as propulsion systems of spacecraft (SC) designed to perform ballistic maneuver in space flight [1,2]. During spacecraft flights on orbits of satellites of celestial bodies or on interplanetary trajectories creation of necessary control forces and moments for the purpose of preservation of parameters of orbits (trajectories) and the set position of the satellite in space. A characteristic element of the EDU is the engine, in which the input energy is converted into kinetic energy of the expiring working substance, and the reaction force arising at the expiration is directly the driving force (thrust force) [3]. The possible values of the effective flow rate in chemical engines are limited both by the relatively low energy of chemical bonds and by the high molecular weight of combustion products. The achievable effective flow rates in chemical engines do not exceed 4.5 - 5.0 km/s. To date, electric rocket engines have been used in orientation and correction systems of various spacecraft [4-6]. Both in Russia and abroad have repeatedly conducted flight and space tests of different classes of ERD, on a number of space objects ERD performed operational functions. It provides for the use of EDR in advanced propulsion systems. For a number of practically important tasks ERD essentially have no competitors [7-10]. Their main advantages, confirmed in the process of research and development, are as follows: high specific impulse, large resource, the possibility of a large number of inclusions up to 100 - 120 and above, extremely small single impulse up to 10 DIN-s, etc. The level of ERD characteristics achieved so far is not the maximum possible, work on further more in-depth study of accelerators and engines is continuing [8]. At the new stage of research, we can expect further success: the expansion of the range of stable operation of accelerators with high efficiency and low vibration, the emergence of new types of accelerators with fundamentally important advantages [11,12].

2. Problem statement. Method of research

The test program of a pilot electric rocket engine usually includes an experimental determination of the grid of its orbital characteristics over the entire range of flight conditions [7].The results of such an experiment are necessary for the analysis of satellite flight data, since the calculated orbital characteristics of the engine used in the early stages of satellite development do not always fully reflect the actual impact of flight conditions. Such, often very significant, flight factors as the change in density in all elements of the flow part, the violation of geometric similarity due to thermal and elastic deformations of parts, the change in the actual characteristics of the control system and the limitation of modes and a number of others, are not fully taken into account in the calculations, even if the mathematical model used is identified by the results of the earth bench tests of the engine [2]. The article deals with the problem of choosing the design characteristics of the spacecraft (SC), which has in its composition an electric propulsion system for correcting the orbit parameters. Mathematical models are considered to describe the relationship of environmental factors, design parameters of on-Board systems (including erdu) and design characteristics of the low-orbit SPACECRAFT as a whole. In modern electric propulsion engines (EDU), the jet principle of creating a driving force is used. A characteristic element of the EDU is the engine, in which the input energy is converted into kinetic energy of the expiring working substance, and the reaction force arising at the expiration is directly the driving force (thrust force). Propulsion systems with chemical and gas jet engines are widely used in modern SPACECRAFT. In chemical engines, the fuel stored on Board is both a source of energy and a source of working substance, and there are no special systems for converting and dissipating energy. The working process of chemical engines consists of two main stages: first, in the combustion chamber, the chemical energy of the fuel is converted into thermal energy of gaseous combustion products, and then in the nozzle, the thermal energy passes into the kinetic energy of the jet of gases. The possible values of the effective flow rate in chemical engines are limited both by the relatively low energy of chemical bonds and by the high molecular weight of combustion products. The carried out development and flight and space tests revealed a number of advantages of ERD in comparison with engines of other classes used for solving similar problems (see Fig.1), for example, in comparison with gas reactive systems and micro-LRE. For a number of practically important tasks ERD essentially have no competitors [1].

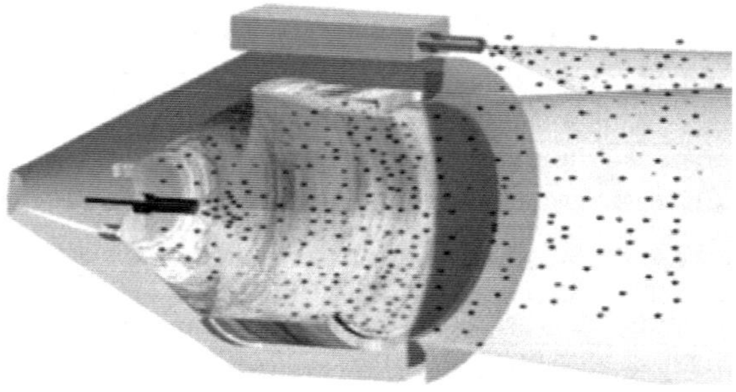

Figure 1. The principle of creating thrust ERD

3. The impetus for the steady motion of ERD

The special value of the definition of impulses and energy for ERD is that their application makes it possible to get an idea about these phenomena only their knowledge of the state on the boundary surface of a certain area, without knowing separately the phenomena occurring within the ERD, without understanding the "mechanism" of the phenomenon. That is, often in cases where the differential equations of plasma flow can not be composed or at least can not be integrated, the theory of pulses or energy makes it possible to know the nature of the movement in its General features, without going into the details of the phenomenon. On the other hand, the momentum and energy theorems are a convenient means to verify the correctness of the results, including experimental ones. The theorem of impulses is of practical importance only for the steady-state phenomena of motion or for the "average" steady-state movements, i.e. for the such vortex and seemingly irregular movements, which allow us to notice a steady main motion of the plasma. While the momentum theorem can be applied to phenomena in which energy loss occurs due to the flow of the anode grid - for the energy theorem, this is impossible, since here the thermal energy formed as a result of the influence of the magnetic field would remain as unknown, so that the applied theorem would no longer give the opportunity to draw conclusions about But with unsteady movements, the energy theorem in some cases makes it possible to draw conclusions about the nature of the movement; its application to steady movements always leads to trivial results. Let us first give the first conclusion, we proceed, therefore, from the theorem of General mechanics on the amount of motion of the system: referred to the unit of time of pulses or quantities of motion of a ($\sum m\omega$) limited system of material points is the sum of forces acting on this system from the outside, i.e.

$$\frac{d}{dt}\left(\sum m\omega\right) = \sum P,$$

where $\sum P$ means the sum of all external forces, i.e. such forces with which the masses of the system are not belonging to the system (internal forces, i.e. those forces with which the masses belonging to the ERD act on each other, are mutually destroyed according to the principle of equality of action and counteraction. In the transition from the system of material points to the plasma, considered as a continuum, the sum $\sum m\omega$ passes into the integral $\int \omega dm = J$, and, therefore, in this case we will have

$$\frac{dJ}{dt} = \frac{d}{dt}\int \omega dm = \sum P.$$

Consider a certain amount of plasma limited by the plasma surface. The time change t_1 in the amount of motion of this limited plasma occurs because, on the one hand, the velocities within the region under consideration can change and, on the other hand, the limiting plasma surface moves. F If the movement is steady, hence each particle is replaced within the element of time in a fixed location particle plasma with the same velocity, the time variation of momentum of the mass of the plasma bounded by a surface F, will be only in the change of momentum caused by the displacement of the plasma surface.

Take a stationary surface in space. Let it coincide with the plasma surface at any time t_1. Then when you move the plasma surface F of the shift pulses through fixed in the space of the surface F. Let it AA mean a part of a stationary surface BB in space, and – a part of a plasma surface at a time $t_1 + dt$. If the external normals of closed surfaces are considered positive, then the volume passing into the element of time dt through the elementary platform dF of a fixed surface AA in space is equal $dF \circ \omega dt$, and, consequently, the amount of motion transferred per unit time through the elementary platform dF is equal $\rho dF \circ \omega\omega$. Thus, the change in the unit time of the total pulse caused by the movement of the liquid bounding surface is equal to the resulting pulses passing per unit time through the fixed in space bounding surface, ie.

$$\frac{dJ}{dt} = \iint \rho dF \circ \omega\omega,$$

$$\frac{dJ_x}{dt} = \iint \rho dF \,|\omega| \cos(n,\omega)u, \frac{dJ_y}{dt} = \iint \rho dF \,|\omega| \cos(n,\omega)v, \frac{dJ_z}{dt} = \iint \rho dF \,|\omega| \cos(n,\omega)w,$$

where it n means outward positive normal. If, in particular, dF perpendicular to the direction x, the displaced mass ρdFu will be, and if dF perpendicular to the direction, y the displaced mass will be ρdFv, and

$$\frac{dJ_x}{dt} = \iint \rho dFuv.$$

A closed stationary surface in space will be used as a reference surface. Thus, a theorem was obtained showing that with the steady-state motion phenomenon, the second change in the total impulse of the plasma quantity under consideration is equal to the flow of impulses through the control surface.

The right side of equality, i.e. the sum of all external forces is formed under the assumption of an ideal plasma - from:

1. pressure forces along the control surface $\iint \rho dF$ (vector integration; the component x is equal to: - $\iint \rho dF \cos(n,x)$, etc.). The negative sign is set because the pressure refers to the inward force acting on the reference surface, i.e. in the direction opposite to the outer normal, considered positive.

2. mass forces, in particular gravity $\iiint \rho g dV$.

3. extraneous forces, i.e. forces acting from the outside on the catchers and the anode grid, which are in the plasma: $\sum P_f$. The fact is that in a limited part of the plasma there may be bodies, for example, bodies that, with their fixing devices, protrude from a limited part electroract engineof, and these bodies act outside the electroract engine, through the devices for holding them, forces act. The surface pressure integral ρ can also be extended to the surface of foreign bodies, and in this case, especially foreign bodies should not be taken into account. However, it is often more convenient to deal directly with outsiders. So, for the pulse method we get the following expression:

$$\iint \rho dF \circ \omega\omega = \iiint \rho g dV - \iint p dF + \sum P_f, \iint \rho dF \circ \omega\omega + \iint p dF = \sum P_f.$$

If we proceed from the theorem of areas of the mechanics of a system of material points, according to which the change in angular momentum relative to some point or axis relative to a unit of time is equal to the resulting moment of forces, i.e. the corresponding momentum theorem is obtained:

$$\frac{d}{dt}\left(\sum mr \times \omega\right) = \sum r \times P, \iint \rho dF \circ \omega r \times \omega = \iiint \rho r \times g dV - \iint p r \times dF + \sum r \times P_f.$$

4. Findings

The design of several types of engines is based on the interaction between the electric and magnetic fields, which allows you to create a thrust at a high level. In the West, the main research went in the direction of the magnetnoplan-dynamic (MTD) engine. In the former USSR engines based on the Hall effect were developed. The technology of their creation was handed over to the United States - in the framework of a joint program. Relatively cheap installations were used in many satellites. An external radial magnetic field is necessary for the operation of such an engine. Inside it there is a constant axial electrical discharge in xenon, which is used as

Journal of Physics: Conference Series **1392** (2019) 012024 doi:10.1088/1742-6596/1392/1/012024

fuel and is under low pressure. As a result of the interaction between axial and radial fields, a Hall current arises, perpendicular to the electric field and the azimuthal field inside the engine. Hall current interacts with the magnetic field responsible for the axial acceleration of the fuel. The performance depends on the size of the engine. The stated results indicate that the use of mathematical planning methods for an experiment really allows to increase the information content of one of the important types of testing of electric rocket engines on high-altitude stands and should be introduced into practice.

5. Conclusion

The analysis shows that, depending on the conditions, the electroract engine can operate in different modes, each of which has its own specific features. Acceleration of plasma in the electroract engine in typical conditions occurs with the simultaneous action of various mechanisms. Such a nature of the processes in a turbofan engine, on the one hand, complicates the regulation of its characteristics, and also complicates its calculation. To date, it has been possible to develop satisfactory calculation methods only for individual special cases. On the other hand, the possibility of using various engine operating modes undoubtedly expands the practical possibilities of its use. For example, the contribution of the gas-dynamic and dissipative acceleration mechanisms makes it possible to obtain flow rates that exceed the value determined by the applied potential difference. Conducting a comparison of end and linear Hall engines based on materials, it is easy to notice between them a lot in common. In both cases, the processes of anomalous conductivity and the development of oscillations are very similar.

6. References

[1] Gusev Yu G and Pilnikov A V 2012 The role and place of electric rocket engines in the Russian space program *Electronic journal "Trudy MAI"* **60**

[2] Makridenko L A, Volkov S N, Khodnenko V P and Golden S A 2010 Conceptual issues of the creation and use of small spacecraft *Electromechanical Issues* **114** 15-26.

[3] Grishin S D, Leskov L V and Kozlov N P 1975 *Electric rocket engines* (Moscow: Mashinostroenie) p 272

[4] Yarygin V I, Ruzhnikov V A and Sinyavsky V V 2012 *Space nuclear power plants: past, present, future. Part I. Space nuclear power plants of the first generation* (Obninsk: Ed. IATE NRNU MEPhI) p 50

[5] Grodzovsky G L, Ivanov Yu N and Tokarev V V 1975 *Space Flight Mechanics: Optimization Problems* (Moscow: Science) p 702

[6] Garrigues L 2011 Study of a Hall effect thruster working with ambient atmospheric gas as propellant for low earth orbit missions *Presented at the 32nd Int. Electric Propulsion Conf. (Wiesbaden, Germany)* IEPC-2011-142

[7] Diamant K D 2010 A 2-Stage Cylindrical Hall Thruster for Air Breathing Electrical Propulsion *46th AIAA/ASME/SAE/ASEE Joint Propulsion Conf. and Exhibit (United States)* AIAA Paper 2010-6522

[8] Hruby V, Pote B, Brogan T, Hohman K, Szabo J and Rostler P 2003 *Air Breathing Electrically Powered Hall Effect Thruster Patent* WO 03/098041 A2

[9] Malyshev G V 2006 The use of electric propulsion engines for the removal, correction of the orbit and the maintenance of satellite system groups *Flying* 34-40

[10] Sidnyaev N I, Onufriev V V and Govor S A 2017 Determination of the height characteristics of electric rocket engines of a spacecraft using experiment planning methods *Control problems* **1** 75-85

[11] Sidnyaev N I, Onufriev V V and Gecha V Ya 2017 Parameters of the critical orbit in the conditions of the existence of the spacecraft *Abstracts of the International Scientific Conf. "Fundamental and applied problems of mechanics" (Moscow)* (Moscow: MGTU im. N.E. Bauman) 163-164

Estimation of maximum water levels during spring flood on Lena river parts using artificial neural networks

G. P. Struchkova, T. A. Kapitonova, V. V. Timofeeva, D.D. Nogovitsyn

V.P. Larionov's Institute of Physical-Technical Problems of the North, Siberian Branch of the Russian Academy of Sciences, Yakutsk, Russian Federation

E-mail: pandoramy8@list.ru

Abstract. The Republic of Sakha (Yakutia) possessing a vast territory located in various climatic zones and a developed network of water bodies is exposed to a wide range of natural emergency situations. Among them, spring-summer floods are the most frequent and bringing enormous damage, causing inundations of vast areas and national economy objects, which determines relevance of development and improvement of flood forecast methods for implementation of timely measures to prevent and reduce an inundation risk.

Artificial neural networks have proven their effectiveness in solving various forecast problems, especially when using statistical data. In particular, usage of a neural network approach based on the forecast of a time series from previous values gives good results. The artificial neural networks, unlike statistical methods of analysis, are based on parallel data processing, have an ability of self-learning and recognition of nonlinear relationships between input and output data sets. A choice of parts of the Lena river to predict maximum water levels during the floods was determined based on locations of potentially hazardous objects, the inundation of which can cause the considerable material damage. On the basis of the hydrological data obtained during 70 years, the neural network models are obtained, which make it possible to predict flood hazards from the spring floods, two variants of the transformed initial data are considered and different network structures are compared. Relative errors of the forecasts obtained during the work vary considerably (7–20%), which indicates necessity for the processing of the initial data and careful selection of the structure.

1. Introduction

The Republic of Sakha (Yakutia) has the vast territory located in the different climatic zones. A large number of rivers flowing from south to north contribute, especially during a flood period, to an occurrence of natural emergency situations (ES). The most characteristic of them are the spring-summer floods, causing the inundations of the vast areas, the objects, infrastructure and bringing the enormous damage to the national economy, destabilization processes in a technosphere, which determines the relevance of the development and improvement of the flood forecast methods to reduce the hazard levels and damage. During 13 years, from 2001 to 2013, the spring flood of the rivers in the Republic caused the damage amounting to a total of about 16 billion rubles [1]. The total damage from the flood in Yakutia in the spring and summer of 2018, as a result of which more than 5.3 thousand residents of Yakutia were harmed, exceeded 1.4 billion rubles. In the Republic during the spring and summer flood, 63 settlements in 15 districts and in the territory of Yakutsk were affected; 1482 residential houses were flooded [2]. On the basis of the statistical data obtained during 75 years and regression modeling, the usage of neural networks to develop the model allowing one to predict the flood hazards from the spring floods is presented. The choice of the forecast part was determined based on the locations of the potentially hazardous objects, the inundation of which can cause the considerable material damage, natural factors affecting the level of the spring floods have been determined.

The time series modeling for the forecast of hydrological variables is an important step in planning and operational analysis of water resources. At present, a sufficiently large number of the methods of the forecast of the inundation hazard from the spring floods have been developed [3,4,5,6]. Traditionally, the autoregressive moving average models (ARIMA) were used to model the time series of the water resources, since such models are accepted as a standard representation of the stochastic time series. However, these models do not attempt to represent nonlinear dynamics inherent in the hydrological process and may not always work well. The time series analysis requires mapping of the complex relationships between the input (inputs) and output data (outputs), since the predicted values are displayed as a function of observed patterns in the past.

The intellectual computational methods, such as the artificial neural networks (ANN), have proven their effectiveness in solving the various problems, especially for working with the statistical data. Specifically, in the forecast the usage of the neural network approach, based on construction of the time series from the previous values, gives the good results [7, 8, 9]. The artificial neural networks, unlike the statistical methods of analysis, are based on the data parallel processing, have the self-learning ability and are able to recognize the nonlinear relationships between the input and output data sets. The theory of artificial neural networks is currently undergoing a stage of formation, which causes a variety of problem statements and basic definitions. The simplest way of application of the artificial neural networks in the forecast problems is to use a conventional perceptron with one, two or (in extreme cases) three hidden layers.

The time series analysis requires the mapping of the complex relationships between the input and output data, since the predicted values are displayed as the function of the observed patterns in the past. Due to difficulties associated with identification of the structure of the nonlinear model and estimation of parameters, the very few truly theoretical hydrological models of the nonlinear system were presented [9,10,11]. Particularly, in a prediction the use of the neural network approach, based on the construction of the time series from the previous values, gives the good results.

2. Materials and research methods

Using the data on the maximum water levels on the Lena river in the area of Tabaga during the period from 1938 to 2013, we will make the forecast till 2017 using the neural network. To increase prognostic capabilities of the neural networks, it is necessary to preliminary make a transformation of the data for the neural network [12,13]. We will consider the two variants of the transformed initial data - the series smoothed by the moving average and the time series with a subsequent seasonal adjustment. After the data transformation, we proceed to the choice of a configuration of the neural network itself. As practice shows, the multi-layer perceptron is best suited for the modeling and forecast of the time series. Using the sliding window method for the neural network, we form a block of training samples. A window size, i.e. a time series segment representing the input vector is difficult to exactly indicate due to a complex behavior of the time series. When choosing the number of the inputs of the neural network to predict the time series, quality of training should be taken into account, since an increase of the number of the inputs of the neural network accordingly reduces the size of the training sample. We adjust the model for the training: we set the value of hidden neurons, select the activation functions for the hidden and output neurons. We carry out the process of the training of the networks, then from the trained neural networks we choose the best one.

3. Results and discussions

We will analyze the data obtained to select the best model, comparing each model in terms of performance and errors. We will consider the two variants of the transformed initial data - the series smoothed by the moving average and the time series with the subsequent seasonal adjustment.

In the first case, the following models are considered with the higher performance and the smallest error:

1. MLP 12-4-1 is with an architecture of the network of a type of the multi-layer perceptron with 12 inputs, with 7 hidden neurons and with 1 output.
2. MLP 20-6-1 with the architecture of the network of the type of the multi-layer perceptron with 20 inputs, with 2 hidden neurons and with 1 output.
3. MLP 40-8-1 with the architecture of the network of the type of the multi-layer perceptron with 40 inputs, with 5 hidden neurons and with 1 output.

In the second case, the following models are selected:

1. MLP 12-7-1 with the architecture of the network of the type of the multi-layer perceptron with 12 inputs, with 7 hidden neurons and with 1 output.
2. MLP 20-2-1 with the architecture of the network of the type of the multi-layer perceptron with 20 inputs, with 2 hidden neurons and with 1 output.
3. MLP 40-5-1 with the architecture of the network of the type of the multi-layer perceptron with 40 inputs, with 5 hidden neurons and with 1 output.

Table 1. Comparative characteristic of results of retro-forecasts of water level in Lena river in area of Tabaga.

	2014		Error	2015		Error	2016		Error	2017		Error
	Observed	Predicted	%	Observed	Predicted	%	Observed	Predicted	%	Observed	Predicted	%
The series smoothed by the moving average												
MLP 12-4-1	744	846	13	1037	948	8	825	933	13	745	829	11
MLP 20-6-1	744	836	12	1037	959	7	825	890	8	745	822	10
MLP 40-8-1	744	898	20	1037	899	13	825	925	12	745	874	17
The series with the subsequent seasonal adjustment												
MLP 12-7-1	744	803	7	1037	971	6	825	756	8	745	684	12
MLP 20-2-1	744	789	6	1037	1001	3	825	770	7	745	797	7
MLP 40-5-1	744	814	9	1037	923	10	825	924	12	745	909	22

The best results were shown by the MLP 20-2-1 model with the subsequent seasonal adjustment with the architecture of the network of the type of the multi-layer perceptron with 20 inputs, with 2 hidden neurons and with 1 output. From the data in Table 1 it follows that the quality of the forecast of the time series of the maximum water levels in the river during the spring flood depends on the choice of the optimal window size wherein the quality of the prognostic capabilities of the neural networks deteriorates with the excessive increase of the window size (a phenomenon of overtraining).

References

1. Burtseva E.I., Parfenova O.T. Economic damage from floods on the rivers of the Republic of Sakha (Yakutia) // Problems of modern economy, 2015, no 1 (53), pp. 256-259
2. http://ulus.media/2019/02/26/zhiteli-sela-kyuptsy-do-sih-por-ne-poluchili-kompensatsii-za-proshlogodnij-pavodok/

3. Kireeva M. B., Frolova N. L. Pozhidaeva D. S. Analysis of occurrence of extreme hydrological events on the rivers of the Russian Arctic // Changing climate and socio-economic potential of the Russian Arctic. M: League-Vent, 2015, pp. 21-35
4. Koronkevich N.I., Barabanova E.A., Zaytseva I.S. Extreme hydrological situations. M .: Media-PRESS, 2010, 464 p.
5. Struchkova G.P., Timofeeva V.V., Kapitonova T.A. The statistical model of floods during tides and ice drift on the Lena River near Yakutsk // Russian meteorology and hydrology, 2015, v 40, no 5, pp. 343-346. DOI: 10.3103/S1068373915050088
6. Krasnogorskaya N.N., Ferapontov Yu.I., Nafikova E.V. Development of methods for forecasting hydrological processes for water management tasks // Scientific notes of the Russian State Hydrometeorological University, 2013, no 28, pp. 43-50
7. Grebnev Y. V., Yarovoy A. V. Monitoring and forecasting of floods on the territory of the Krasnoyarsk region, using neural network algorithms // The Siberian fire and rescue Bulletin, 2018, v 10, no 3, pp. 13-16
8. Chen Y.H., Chang F.J. Evolutionary artificial neural networks for hydrological systems forecasting J. Hydrol., 367 (2009), pp. 125-137
9. Fazel S.A.A., Mirfenderesk H., Tomlinson R., Blumenstein M.Towards robust flood forecasts using neural networks, Neural Networks (IJCNN), 2015 International Joint Conference on, IEEE(2015), pp. 1-6
10. Volchak A., Kostyuk D., From Lavrentyev. neural forecasts for integrated flood monitoring system // proceedings of the 6th International scientific conference on the global energy and water cycle and the 2nd scientific conference on the earth's ecosystem and atmosphere (iLEAPS), 24-28 August 2009, Melbourne, Australia. Thom. II-Pp. 713-714.
11. Holger R. Maier, Ashu Jain, Graeme C. Dandy, Sudheer K.P. Methods used for the development of neural networks for the prediction of water resource variables in river systems: Current status and future directions, Environmental Modelling & Software, Volume 25, Issue 8, August 2010, Pages 891-909
12. Osovsky S.A. Neural networks of information processing / S.A. Osovsky. - M .: Finance and Statistics, 2002, 344 p.
13. Borovikov V.P. Neural networks. Statistica Neural Networks. Methodology and technology of modern data analysis, 2nd ed., Revised and add. - M .: Hotline - Telecom, 2008, 392 p.

Journal of Physics: Conference Series **1392** (2019) 012026 doi:10.1088/1742-6596/1392/1/012026

Estimation of Near-Bottom Suspension-Carrying Flow around a Pipeline

[1,2]Vladimirov, I.Yu., and [1,2]Korchagin, N.N.

[1] Shirshov Institute of Oceanology of the Russian Academy of Sciences, Moscow
[2] Bauman Moscow State Technical University

iyuvladimirov@rambler.ru

Abstract. In a model format, the force effect of a stratified ideal fluid flow on streamlined elements of engineering structures is investigated. Integral representations of wave drag and lift force are obtained. Numerical calculations are performed for the real marine medium. The conditions under which significant changes in hydrodynamic reactions to streamlined elements of technical structures occur are revealed.

1. Formulation and solution of the model problem

The underwater obstacles streamlined by a stratified sea current are subject to viscous drag. In addition, hydrodynamic reactions arise due to the generation of internal waves at jumps in the density of the medium. Near the bottom, this effect can be amplified by bottom-weighted (turbid) streams arising when micro- and fine-grained particles of bottom soil are mixed in them. The velocities of such flows can exceed 1 m/s, and the thicknesses – several tens of meters. At the same time, the effective density of water in the streams may be more than ten percent more than the density of the surrounding water, which causes a more sharp drop in density at the flow boundaries, leading, in turn, to increased stratification of the medium [1] and a corresponding intensification of internal waves.

Thus, the underwater elements of engineering structures (in particular, transport pipelines) streamlined by turbid flows can additionally experience quite significant power effects associated with the generation of internal waves. Therefore, the corresponding problem is considered below in the model version: a dipole is considered as an obstacle in the flow; it simulates a transverse flow past a cylinder (as an element of a pipeline).

To estimate the force effects on the obstacle being flown, we consider a two-layer flow of an ideal fluid, bounded below by a horizontal bottom, that flows stationary around a cylinder with a cross-section of radius R The cross section of the pipeline is modeled by a point dipole with a moment $m = 2\pi VR^2$, where V is the incoming flow velocity. The thickness of the upper layer is H_1, the lower – H_2 and the densities are ρ_1 and ρ_2 respectively ($\rho_1 < \rho_2$). The origin of coordinates is placed on the undisturbed boundary between the fluid layers, the x axis is directed along this boundary, the y axis – vertically upwards.

The complex velocity of a flow disturbed by a dipole is sought in accordance with [2-4]. Further, within the framework of the linear theory, two model problems are considered: the dipole is located below the density jump, i.e. at the point $(0,-h)$, and above the jump (at the point $(0,h)$). To calculate the hydrodynamic forces applied to the cylinder under consideration, we represent the complex-conjugate velocity in each of the layers in the form $\mu_j = V + U_j$, $j = \{1, 2\}$. Then the mathematical problem of finding the velocity perturbations U_j introduced into the flow by a dipole localized in the lower layer is formulated as follows: it is required to find analytical functions $U_1(z)$ and $U_2(z)$ satisfying the boundary conditions

$$\operatorname{Im}\left[i\frac{dU_1}{dz} - \nu U_1\right] = 0 \text{ at } y = H_1, \tag{1}$$

$$\delta \operatorname{Im}\left[i\frac{dU_1}{dz} - \nu U_1\right] = \operatorname{Im}\left[i\frac{dU_2}{dz} - \nu U_2\right] \text{ at } y = 0, \tag{2}$$

$$\operatorname{Im} U_1 = \operatorname{Im} U_2 \text{ at } y = 0, \tag{3}$$

$$\operatorname{Im} U_2 = 0 \text{ at } y = -H_2. \tag{4}$$

Here, $\delta = \rho_1/\rho_2$, $z = x + iy$, $\nu = g/V^2$, g is acceleration of gravity; the function $U_1(z)$ is regular in the band $-\infty < x < +\infty$, $0 < y < H_1$, and $U_2(z)$ – in the band $-\infty < x < +\infty$, $-H_2 < y < 0$ everywhere except for the point $z = -ih$, at which it has a pole of the second order: $U_2(z) \sim -m/2\pi(z+ih)^2$, $z \to -ih$.

The boundary condition (1) describes the constancy of pressure on the free surface, (2) – the continuity of pressure when passing through the interface of the layers, (3) – the absence of fluid flow through this surface, (4) is the no-flow condition on the bottom.

The complex-conjugate hydrodynamic reaction of a cylinder $F^* = X - iY$ is calculated using the Chaplygin formula [5]. Here, X is the wave drag, Y is the lift force of the dipole.

When a dipole is localized under a density jump, we have

$$F^* = \frac{i\rho_2 m^2}{2\pi}\int_0^{+\infty} k(C(k)e^{kh} - D(k)e^{-kh})dk - \frac{\rho_2 m^2}{2}\sum_{j=1}^{s} \operatorname*{res}_{k=k_j}\left\{k(C(k)e^{kh} + D(k)e^{-kh})\right\}, \tag{5}$$

$$C(k) = \frac{k\{-(1-\delta)(k^2-\nu^2)\operatorname{sh} k(H_1-h) + [(1+\delta)k^2 + (1-\delta)\nu^2]\operatorname{sh} k(H_1+h) - 2k\nu \operatorname{ch} k(H_1+h)\}}{2\operatorname{ch} kH_1 \operatorname{ch} kH_2 e^{kH_2}\left\{k^2 + [\delta k^2 + (1-\delta)\nu^2]\operatorname{th} kH_1 \operatorname{th} kH_2 - k\nu(\operatorname{th} kH_1 + \operatorname{th} kH_2)\right\}}$$

$$D(k) = \frac{k \operatorname{ch} k(H_2-h)\left\{-[(1+\delta)k^2 + 2k\nu + (1-\delta)\nu^2]e^{-kH_1} - (1-\delta)(k^2-\nu^2)e^{kH_1}\right\}}{2\operatorname{ch} kH_1 \operatorname{ch} kH_2\left\{k^2 + [\delta k^2 + (1-\delta)\nu^2]\operatorname{th} kH_1 \operatorname{th} kH_2 - k\nu(\operatorname{th} kH_1 + \operatorname{th} kH_2)\right\}}. \tag{6}$$

The integral in (5) is meant in the sense of the Cauchy principal value, and the residues are taken over all s poles k_j of the function $k(C(k)e^{kh} + D(k)e^{-kh})$ located on the positive real axis. From (6) it is clear that these poles are the positive roots of the equation

$$k^2 + [\delta k^2 + (1-\delta)\nu^2]\operatorname{th} kH_1 \operatorname{th} kH_2 - k\nu(\operatorname{th} kH_1 + \operatorname{th} kH_2) = 0. \tag{7}$$

In addition, it is obvious that the points $k = k_j$ (and only they) are singular for the integrand in the first term of (5) (that is, the poles located on the integration contour).

The analysis carried out in [2,3] showed that equation (7) has two positive roots when the following condition is satiafied

$$V < V_{cr}^{\text{int}} = \sqrt{\frac{gH_1(\beta + 1 - \sqrt{(\beta+1)^2 - 4\varepsilon\beta})}{2}}$$

and a single positive root if

$$V_{cr}^{\text{int}} < V < V_{cr}^{\text{sur}} = \sqrt{\frac{gH_1(\beta+1+\sqrt{(\beta+1)^2-4\varepsilon\beta})}{2}};$$

When $V > V_{cr}^{\text{sur}}$ there are no positive solutions. Here, $\varepsilon = 1-\delta = (\rho_2-\rho_1)/\rho_2$ – relative density difference between the layers, $\beta = H_2/H_1$. From the physical point of view, the critical velocities V_{cr}^{int} and V_{cr}^{sur} mean the maximum flow velocity at which waves are formed in the flow behind the streamlined cylinder, due to the presence of a layer of density jump and a free surface respectively (i.e. internal and surface waves).

Separating in (5) the real and imaginary parts, we finally get the following expressions for the wave drag and lift force:

$$X = -2\pi^2\rho_2 V^2 R^4 \sum_{j=1}^{s} \operatorname*{res}_{k=k_j} \left\{k\left(C(k)e^{kh} + D(k)e^{-kh}\right)\right\},$$

$$Y = -2\pi\rho_2 V^2 R^4 \int_0^{+\infty} k\left(C(k)e^{kh} - D(k)e^{-kh}\right)dk.$$

Here, the calculation of residues is carried out according to the formula [6]

$$\operatorname*{res}_{k=k_j} \left\{k\left(C(k)e^{kh} + D(k)e^{-kh}\right)\right\} = \left.\frac{k(g_1(k)e^{kh} + g_2(k)e^{-kh})}{dg_3(k)/dk}\right|_{k=k_j},$$

$$g_1(k) = \frac{k\{(\delta-1)(k^2-\nu^2)\operatorname{sh} k(H_1-h) + [(1+\delta)k^2+(1-\delta)\nu^2]\operatorname{sh} k(H_1+h) - 2k\nu \operatorname{ch} k(H_1+h)\}}{2e^{kH_1}\operatorname{ch} kH_1 \operatorname{ch} kH_2}$$

$$g_2(k) = \frac{k \operatorname{ch} k(H_2-h)\left\{-[(1+\delta)k^2+2k\nu+(1-\delta)\nu^2]e^{-kH_1} - (1-\delta)(k^2-\nu^2)e^{kH_1}\right\}}{2\operatorname{ch} kH_1 \operatorname{ch} kH_2},$$

$$g_3(k) = k^2 + [\delta k^2 + (1-\delta)\nu^2]\operatorname{th} kH_1 \operatorname{th} kH_2 - k\nu(\operatorname{th} kH_1 + \operatorname{th} kH_2).$$

Further, in the same formulation, we study the problem of determining the hydrodynamic load on a dipole located in the upper layer of the flow (above a density jump). Solving this problem in the way indicated above, we obtain:

$$F^* = \frac{i\rho_1 m^2}{2\pi}\int_0^{+\infty} k\left(A(k)e^{-kh} - B(k)e^{kh}\right)dk - \frac{\rho_1 m^2}{2}\sum_{j=1}^{s}\operatorname*{res}_{k=k_j}\left\{k\left(A(k)e^{-kh} + B(k)e^{kh}\right)\right\}, \quad (8)$$

$$A(k) = \frac{k\left[(\delta k + (1-\delta)\nu)\operatorname{th} kH_2 - k\right]\left[\nu \operatorname{ch} k(H_1-h) - k \operatorname{sh} k(H_1-h)\right]}{\operatorname{ch} kH_1\left\{k^2 + [\delta k^2 + (1-\delta)\nu^2]\operatorname{th} kH_1 \operatorname{th} kH_2 - k\nu(\operatorname{th} kH_1 + \operatorname{th} kH_2)\right\}},$$

$$B(k) = \frac{k(k+\nu)e^{-kH_1}\left\{[(1-\delta)\nu \operatorname{ch} kh - \delta k \operatorname{sh} kh]\operatorname{th} kH_2 - k \operatorname{ch} kh\right\}}{\operatorname{ch} kH_1\left\{k^2 + [\delta k^2 + (1-\delta)\nu^2]\operatorname{th} kH_1 \operatorname{th} kH_2 - k\nu(\operatorname{th} kH_1 + \operatorname{th} kH_2)\right\}}.$$

Hence, we have the following expressions for the wave drag and lift force:

$$X = -2\pi^2\rho_1 V^2 R^4 \sum_{j=1}^{s} \operatorname*{res}_{k=k_j} \left\{k\left(A(k)e^{-kh} + B(k)e^{kh}\right)\right\},$$

$$Y = -2\pi\rho_1 V^2 R^4 \int_0^{+\infty} k\left(A(k)e^{-kh} - B(k)e^{kh}\right)dk.$$

The integral in (8), as in (5), is meant in the sense of the Cauchy principal value, k_j are the poles of the function $k(A(k)e^{-kh}+B(k)e^{kh})$ (solutions of equation (7)), and the residues in (8) are calculated similarly to the first problem.

2. Calculation results

Calculations of the hydrodynamic effects on the model cylinder were carried out for the values of the characteristics of the medium corresponding to the actual sea conditions. Thus, the density in the upper layer is $\rho_1 = 1024\,\mathrm{kg/m^3}$, and the density drops between the upper and lower layers are $\rho_2/\rho_1 = \{1.01;\, 1.10\}$, the second value of ρ_2/ρ_1 corresponds to the presence of turbid flow in the lower layer. The total depth of the stream is $H_0 = H_1 + H_2 = 50\,\mathrm{m}$, the thickness of the upper layer is $H_1 = 40$ m, the radius of the cylinder is $R = 0.71\mathrm{m}$. The dipole was successively localized at distances $h = \{1.5\,\mathrm{m}, 2\,\mathrm{m}, 3\,\mathrm{m}, 4.5\,\mathrm{m}\}$ from the interface between the layers.

The results of the calculation of the wave drag X and lift force Y (per linear meter of cylinder length), as functions of the speed of the incident flow V, are shown in Fig. 1–2. Each graph shows 4 curves corresponding to different distances h, and, as can be seen in the figures, the increase of h leads to a decrease in the maximum of drag and lift force.

Attention is drawn to the significant effect of the density jump power ε on changes in the magnitude of the maximum of wave drag and lift force. So, the comparison of graphs in fig. 1-2 and constructed for two different values of the jump power ε, shows that such changes are almost directly proportional to the magnitude of ε. The values of the maxima themselves shift towards an increase in the flow velocity when the cylinder is shifted from the jump layer. At the same time, calculations showed that the values of X and Y practically do not depend on the thicknesses of the flow layers H_1 and H_2. Thus, the power of the density jump ε and the distance h from the dipole modeling the cylinder to the density jump are the main parameters of the problem; their variations can lead to significant changes in the hydrodynamic reactions to the obstacle streamlined.

It should be noted that at flow rates greater than the critical speed V_{cr}^{int}, the wave drag $X(V)$ experiences a discontinuity of the first kind, sharply decreasing to almost zero. Then, it again reaches significant values only at significantly higher values of V ($V > 4\,\mathrm{m/s}$), i.e. at stream speeds not characteristic to real sea conditions. At the same time, the value of the critical velocity itself increases with increasing power of the density jump layer ε, which corresponds to the expansion of the range of speeds inside which the cylinder experiences significant wave drag. In this case, the graphs (Fig. 1) are generally shifted toward larger values of V.

The graphs of function $Y(V)$ presented in Fig. 2 reflect an important feature in the variability of lift force. Thus, in a relatively narrow range of flow velocity, the lift force abruptly changes its direction to the opposite. The question arises about the possible consequences of such a drastic and reversible change of $Y(V)$. To answer it, one may use the Bernoulli integral, which relates the change in pressure with the change in the velocity of the stream. Since in real sea conditions the flow velocity is non-uniform in space, the pressure field is also non-uniform. And the change in pressure directly affects the change in force effects on the streamlined underwater obstacles. As a result, alternating force effects can lead to the deformation of individual horizontally extended elements of streamlined structures (for example, transport pipelines). This result is new and must be taken into account when designing and constructing underwater structures on the seabed.

The authors wish to thank A.T. Il'ichev for fruitful discussions of the model results. The work was carried out within the framework of the State Program of IORAS (theme № 0149-2019-0004).

References

[1] *Zhmur, V.V., and Sapov, D.A.* Catastrophic gravitational suspension-carrying currents in the near-bottom layer of the ocean, in: *World Ocean. Vol. 1. Ocean Geology and Tectonics. Catastrophic Phenomena in the Ocean*, Moscow: Nauchnyi Mir, 2013, pp. 499–524.

[2] *Vladimirov, I.Yu., Korchagin, N.N., and Savin, A.S.* Surface effects related to the flow of a nonuniformly stratified fluid around obstacles, Doklady Earth Sciences, 2011, vol. 440, №2, pp. 1474-1477.

[3] *Vladimirov, I.Yu., Korchagin, N.N., and Savin, A.S.* Surface perturbations generated by a stratified flow of finite depth over obstacles, Oceanology, 2012, vol. 52, № 6, pp. 760-770.

[4] *Gorlov, S.I.* Hydrodynamic characteristics of a vortex source performing translational motion in a multilayer heavy fluid, Journal of Applied Mechanics and Technical Physics, 2000, vol. 41, № 5, pp. 887-894.

[5] *Loytsianskii L.G.* Mechanics of Liquids and Gases, Oxford: Pergamon Press, 1966.

[6] *Sidorov, Yu.V., Fedoryuk, M.V., and Shabunin, M.I.* Lektsii po teorii funktsii kompleksnogo peremennogo (Lectures on the Theory of Functions of a Complex Variable), Moscow, Nauka, 1976.

Figure captions

Fig. 1. Wave drag of a dipole as a function of the flow velocity. The dipole is located beneath the density jump (the left pair of graphs), dipole is located above the jump (the right pair); $\rho_2/\rho_1 = 1.01$ for the top pair of graphs; $\rho_2/\rho_1 = 1.1$ for the lower pair; curves 1–4 correspond to $h = 1.5\,\mathrm{m}$, $2\,\mathrm{m}$, $3\,\mathrm{m}$, and $4.5\,\mathrm{m}$, respectively.

Fig. 2. Lift force of a dipole as a function of the flow velocity. The dipole is located beneath the density jump (the left pair of graphs), dipole is located above the jump (the right pair); $\rho_2/\rho_1 = 1.01$ for the top pair of graphs; $\rho_2/\rho_1 = 1.1$ for the lower pair; curves 1–4 correspond to $h = 1.5\,\mathrm{m}$, $2\,\mathrm{m}$, $3\,\mathrm{m}$, and $4.5\,\mathrm{m}$, respectively.

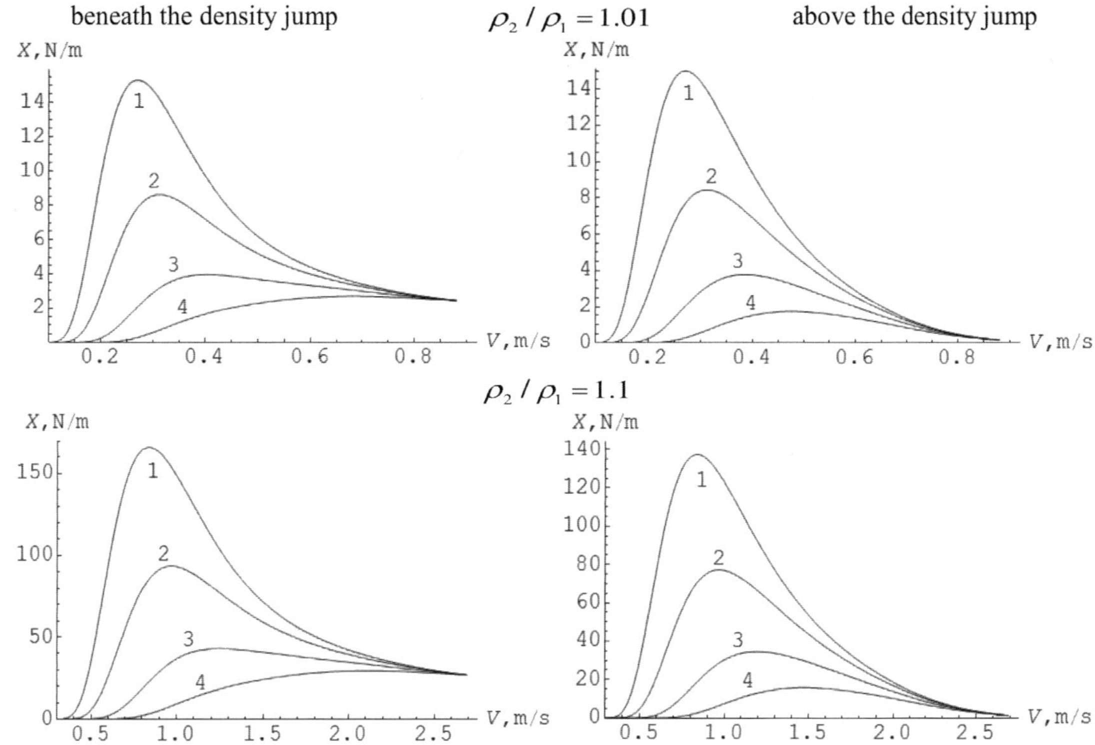
beneath the density jump
$\rho_2 / \rho_1 = 1.01$
above the density jump
X, N/m
V, m/s
$\rho_2 / \rho_1 = 1.1$

Fig. 1

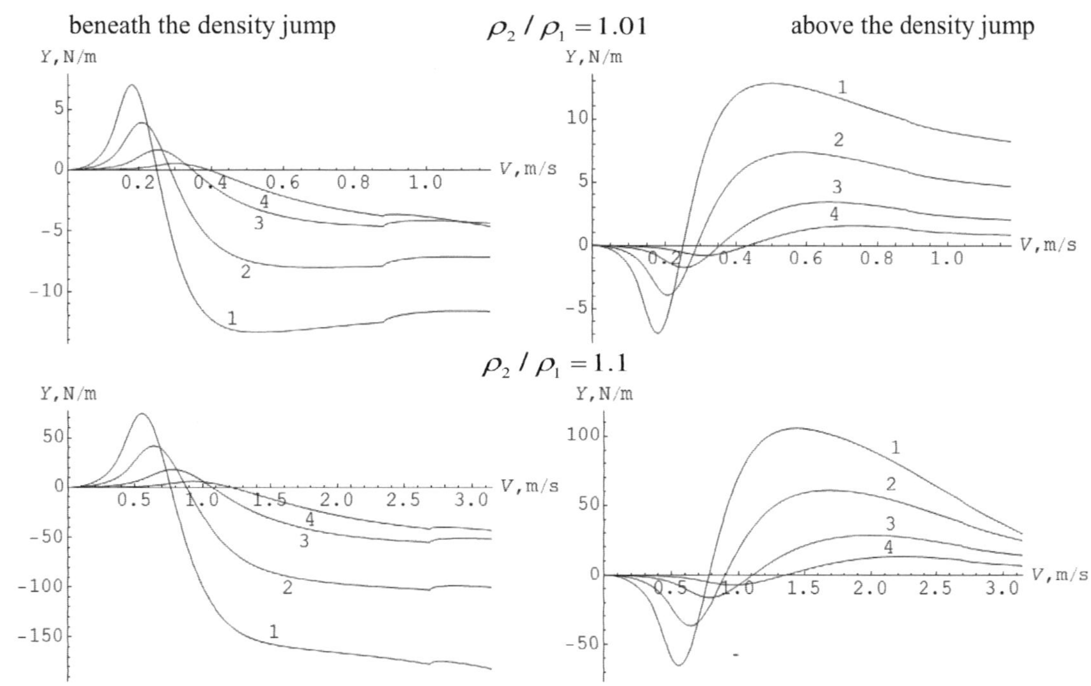
beneath the density jump
$\rho_2 / \rho_1 = 1.01$
above the density jump
Y, N/m
V, m/s
$\rho_2 / \rho_1 = 1.1$

Fig. 2

On hyperbolized nonlinear Schredinger type equations

A.D.Yunakovsky, Ya.L.Bogomolov, N.V. Sapogova

Institute of Applied Physics, Russian Academy of Sciences, Nizhny Novgorod, Russia

E-mail: yun@appl.sci-nnov.ru, bogomol@appl.sci-nnov.ru, ns@ufp.appl.sci-nnov.ru

Abstract. A nonlinear Schredinger type equation in the case of an unbounded in space operator having, a non-divergent form, is considered. A new approach based on the hyperbolization procedure is suggested. The possibilities of the method proposed are tested.

1. Introduction

A nonlinear Schredinger equation (NLS) is actively applied to describe a lot of problems in physics, e.g., in the theory of surface waves, in plasma oscillations, in nonlinear optics, in biophysics etc. Hyperbolization of NLS equation is provided by an adding of the second derivative of an unknown function in time with a small parameter. At the same time the hyperbolization method [1, 2] is intensively developed. This idea was first proposed for electrodynamics by A. Melani [3]. For the problems of hydrodynamics, the fundamental work in this direction is the work of B.N. Chetverushkin [1]. Due to its suitable adaptation for a high-performance parallel computing this method is widely used nowadays (in hydro and gas dynamics especially) [4]. This procedure makes it possible to construct explicit schemes possessing the better stability condition in comparison with classic schemes (both for parabolic equations and for non-stationary Schredinger type ones [5]). This fact is very important in the case of small mesh size. Moreover, hyperbolization permits to avoid the phenomenon of random phases taking place for NLS [5].

2. Relaxation method for NLS

Dynamics of a plasma density jump in a given field is governed by the nonlinear Schredinger type equation [6],

$$-2i\frac{\partial E}{\partial t} + \frac{\partial^2 E}{\partial x^2} + (-x + |E|^2)E - B = 0, \tag{1}$$

where $\frac{\partial^2 E}{\partial x^2} - xE$ is the unbounded operator, the continuous spectrum of which covers the entire real axis. Its eigenfunctions are expressed in terms of the Airy functions. Consequently, Fourier components of E in (17) are varied according to the $exp(ik^3t)$ rule in time. This fact yields a severe restriction for the time step ($\triangle t \sim \triangle x^3$).

The heuristic substitution

$$E = Ue^{i(t+T)(x/2+(t+T)^2/24)} \tag{2}$$

leads to a semi-bounded operator in the spatial coordinate for U:

$$-2i\frac{\partial U}{\partial t} + \frac{\partial^2 U}{\partial x^2} + i(t+T)\frac{\partial U}{\partial x} + NU = Be^{-i(t+T)(x/2+(t+T)^2/24)}. \tag{3}$$

Here $N = |E|^2$. The meaning of the newly introduced parameter T will be clear later on.

Calculations for U are possible in a small time interval due to random phases appearing in high frequency components. To avoid this effect, it is necessary return to E function after few time steps considering it as the new initial data for $t = 0$. Then the previous procedure is repeated again and again.

Further regularization for (3) is a hyperbolization procedure: adding of the second derivative in time with a small parameter:

$$\mu\frac{\partial^2 U}{\partial t^2} = -2i\frac{\partial U}{\partial t} + \frac{\partial^2 U}{\partial x^2} + i(t+T)\frac{\partial U}{\partial x} + NU - Be^{-i(t+T)(x/2+(t+T)^2/24)}. \tag{4}$$

We introduce two additional functions H and G and represent the hyperbolized equation as a system of three equations with three unknown functions U, H and G:

$$\frac{\partial U}{\partial t} = i\frac{\partial H}{\partial x} + iG,\ \ i\mu\frac{\partial H}{\partial t} = 2H + \frac{\partial U}{\partial x} + i(t+T)U. \tag{5}$$

For the function G, we obtain the equation, reducing this system to a hyperbolized equation (4). To do this, we differentiate the first of the equations (5) with t, multiply the result by μ and make a substitution under the sign of the derivative with respect to x using the second equation of the system (5).

$$\mu\frac{\partial^2 U}{\partial t^2} = i\mu\frac{\partial}{\partial x}\frac{\partial H}{\partial t} + i\mu\frac{\partial G}{\partial t} = 2\frac{\partial H}{\partial x} + \frac{\partial^2 U}{\partial x^2} + i(t+T)\frac{\partial U}{\partial x} + \mu i\frac{\partial G}{\partial t}. \tag{6}$$

Replacing $\frac{\partial H}{\partial x}$ by virtue of the first equation of the system (5), we get

$$\mu\frac{\partial^2 U}{\partial t^2} = -2i\frac{\partial U}{\partial t} + \frac{\partial^2 U}{\partial x^2} + i(t+T)\frac{\partial U}{\partial x} - 2G + i\mu\frac{\partial G}{\partial t}. \tag{7}$$

This equation coincides with the hyperbolized equation (4) if the function satisfies the equation

$$i\mu\frac{\partial G}{\partial t} - 2G = NU - Be^{-i(t+T)(x/2+(t+T)^2/24)}. \tag{8}$$

The equation for the function H, and the function U itself, have the meaning of stream relaxation with the relaxation parameter μ. The equation for the function G sets the relaxation of the part of the original operator, which is responsible for stabilizing the spreading of the wave packet due to the nonlinear dispersion law in accordance with the expansion of the initial dispersion equation in a series in powers of the amplitude E.

We introduce in the space (x,t) uniform in each direction calculation grids with a chess arrangement of knots: with integer indices for conservative variables U and G

$$\omega_c = \{x_j = jh_x, -N_x \le j \le N_x,\ t_k = kh_t,\ 0 \le k \le M\}$$

and with fractional indices of coordinates of nodes

$$\omega_f = \left\{x_{j+1/2} = (j+1/2)h_x, -N_x \le j \le N_x,\ t_{k+1/2} = (k+1/2)h_t,\ 0 \le k \le M\right\}$$

for stream variable.

After each step in t we will return to the original values of E, i.e. we will recalculate the initial data for U, G and (somewhat more complicated) for H at the point with fractional coordinates.

In order not to clutter up the formulas, we will use superscripts k to denote points by t, and lower j by x only where it is difficult or impossible to understand the results.

The equation for the function G is a linear equation of the first order. We can integrate it at the step of t, replacing the function NU with a linear combination of the extreme points.

$$G(t) = G(0)e^{-i2t/\mu} - \frac{i}{\mu}\int_0^t \left(\frac{\tau}{t}NU(t) - \frac{\tau - t}{t}NU(0)\right) e^{i2(\tau-t)/\mu} d\tau - \hat{B} = \tag{9}$$

$$= G(0)D - NU(t)P + NU(0)Q - \hat{B}.$$

Here $D = e^{-i2t/\mu}$, $P = \frac{1}{2}\left(1 - \frac{\mu}{2}(1-D)\right)$, $Q = \frac{1}{2}\left(D + i\frac{\mu}{2}(1-D)\right)$,

$$\hat{B} = \frac{B}{\mu}\int_0^t e^{i(2(\tau-t)/\mu - (\tau+T)(x/2+(\tau+T)^2/24))} d\tau. \tag{10}$$

$$H(t/2) = H(-t/2)D - \frac{i}{\mu}\int_{-t/2}^{t/2} \left(\frac{\partial}{\partial x}U(0) + iTU(0)\right) e^{i2(\tau-t/2)/\mu} d\tau \tag{11}$$

$$= H(-t/2)D + \frac{1}{2}\left(\frac{\partial}{\partial x}U(0) + iTU(0)\right)(1-D)$$

Here it becomes clear the meaning of the introduction of the parameter T - we more accurately consider the relaxation of the flows.

$$U(t)_j = U(0)_j + \frac{i}{h_x}\int_0^t \left(H^{k+\frac{1}{2}}_{j+\frac{1}{2}} - H^{k+\frac{1}{2}}_{j-\frac{1}{2}}\right) d\tau - \frac{i}{2}\int_0^t (G(t)_j + G(0)_j) d\tau = \tag{12}$$

$$= U(0)_j + \frac{it}{h_x}\left(H^{k+\frac{1}{2}}_{j+\frac{1}{2}} - H^{k+\frac{1}{2}}_{j-\frac{1}{2}}\right) - \frac{it}{2}G(0)_j(1-D) - \frac{itP}{2}NU(t)_j + \frac{itQ}{2}NU(0)_j - \frac{it}{2}\hat{B}_j$$

We rewrite the last relation, dropping the duplicate index j on U and putting all the terms with $U(t)$ to the left side

$$U(t)\left(1 + \frac{itP}{2}N(U(t))\right) = R = \tag{13}$$

$$= U(0) + \frac{it}{h_x}\left(H^{k+\frac{1}{2}}_{j+\frac{1}{2}} - H^{k-\frac{1}{2}}_{j+\frac{1}{2}}\right) - \frac{it}{2}G(0)(1-D) + \frac{itQ}{2}NU(0) - \frac{it}{2}\hat{B}$$

Hence we obtain the equation for determination of $|U(t)|^2$

$$|U(t)|^2\left(\left(1 + \frac{tImP}{2}N(U(t))\right)^2 + \frac{t^2(ReP)^2}{4}N(U(t))^2\right) = |R|^2 \tag{14}$$

Since hyperbolization is a method of finding an approximate solution of the original Schredinger equation for the function E, after each step of finding the function $U(t)$ we return to the function $E(t)$. We consider that $E_{new}(0) = E(t)$. It remains for us to go to U_{new}, and set the initial data for auxiliary functions, based on the condition $\frac{\partial U_{new}}{\partial t}$ at the point $t = 0$ is determined from the original equation . Just ask

$$U_{new}(-t) = E(0)e^{i(t-T)(x/2-(t-T^3/24))} = U(0)e^{itx/2+i(t^3+3tT^2-3t^2Y)/24},$$

$$U_{new}(0) = E(t)e^{-iT(x/2+T^3/24)} = U(t)e^{itx/2+i(t^3+3tT^2+3t^2T)/24}.$$

The new value of $H(-t/2)$ is obtained as the half of sum of the corresponding expressions on the upper and lower layers

$$H_{new}(\tfrac{-t}{2}) = \tfrac{1}{4h_x}(U_{new}(0)_{j+1} - U_{new}(0)_j + U_{new}(-t)_{j+1} - U_{new}(-t)_j) +$$

$$+\tfrac{i}{2}(T(U_{new}(0)_{j+1} + U_{new}(0)_j) + (-t+T)(U_{new}(-t)_{j+1} + U_{new}(-t)_j)). \quad (15)$$

Respectively

$$G_{new}(0) = (N_{new}U_{new}(0) - Be^{-iT(x/2+T^2/24)})/2. \quad (16)$$

In fact, we assume that at each step for the auxiliary functions H and G we set new initial data based on the condition that their time derivatives are equal to zero. This is equivalent to specifying the second initial given $\frac{\partial U}{\partial t}$ for equation (4) from the original equation (3).

The described technique is simple, easily parallelized and gives hope for an increase in the time step due to the explicit solution of the equation for nonlinear dispersion. However, it makes it impossible to determine the maximum allowable values of the relaxation parameter μ and the time step. To find them, we construct an approximate solution of the original equation.

3. Slip-step Fourier method for hyperbolized Schredinger equation

In [5] the method of operator exponents constructed a scheme of the SSFM type for the NLS. Application of this method to the hyperbolized Schredinger equation also allows us to find the optimal ratio between the time step and the parameter μ.

Consider a hyperbolized equation in the form more suited to the operator exponent method, adding an additional attenuation with the b parameter

$$a^2\mu^2\left(\frac{\partial^2 U}{\partial t^2} + b\frac{\partial U}{\partial t}\right) + 2i\frac{\partial U}{\partial t} - \frac{\partial^2 U}{\partial x^2} - it\frac{\partial U}{\partial x} + (\frac{t^2}{4} - N - 2d)U + Be^{-it(x/2+d)} = 0. \quad (17)$$

Previously we introduce a new unknown function, taking into account a small parameter, so that the resulting hyperbolic operator has an explicitly computed Green function. Introduce a new time

$$t = t_{old}/(a\mu). \quad (18)$$

Let's make a replacement

$$U = Ve^{-(ba\mu+2i/a\mu)t/2} \quad (19)$$

$$\frac{\partial^2 V}{\partial t^2} - \frac{\partial^2 V}{\partial x^2} - ia\mu t\frac{\partial V}{\partial x} + \frac{(a\mu t)^2}{4}V + \left(\frac{1}{a^2\mu^2} - \frac{b^2a^2\mu^2}{4} - 2d - N\right)V - ibV + Be^{it(a\mu x/2+\eta)} = 0. \quad (20)$$

To simplify further calculations, we fix the parameters

$$b = a\mu/2,\ d = \frac{1}{2a^2\mu^2} - \frac{a^4\mu^4}{32}.\alpha = \frac{a\mu}{2},\ \breve{D} = De^{-it\eta}\delta(k - a\mu t/2),\ \eta = -\frac{1}{2a\mu} - \frac{a^5\mu^5}{32} + i\frac{a^2\mu^2}{4}. \quad (21)$$

Therefore, the application of operator methods for solving the Cauchy problem is simpler and at the same time uniformly carried out by making the Fourier transform with respect to the spatial variable(s).

$$\frac{\partial^2 FV}{\partial t^2} + \left(\frac{a\mu}{2}\right)^2\left(t + \frac{2k}{a\mu}\right)^2 FV - i\frac{a\mu}{2}FV - F(NV) + Be^{-it\eta}\delta(k - a\mu t/2). \quad (22)$$

This equation can be written in the form of a system

$$\frac{\partial}{\partial t}\begin{pmatrix} Fv \\ Fw \end{pmatrix} = \begin{pmatrix} 0 & 1 \\ -\alpha^2(t+2k/\alpha\mu)^2 & 0 \end{pmatrix}\begin{pmatrix} Fv \\ Fw \end{pmatrix} + i\alpha\begin{pmatrix} 0 & 0 \\ 1 & 0 \end{pmatrix}\begin{pmatrix} Fv \\ Fw \end{pmatrix} + \tag{23}$$

$$+F\begin{pmatrix} 0 & 0 \\ N & 0 \end{pmatrix}F^{-1}\begin{pmatrix} Fv \\ Fw \end{pmatrix} + \breve{B}\begin{pmatrix} 0 \\ 1 \end{pmatrix}.$$

The fundamental solution looks rather complicated even in the one-dimensional space case

$$G(t) = e^{i\alpha(t^2+4kt/\alpha\mu)/2}\int\limits_0^t e^{-i\alpha(s^2+4ks/\alpha\mu)}ds,\ G_1(t) = e^{i\alpha(t^2+4kt/\alpha\mu)/2}.$$

We reduce our system to diagonal form by means of the transformation

$$\begin{pmatrix} Fv \\ Fw \end{pmatrix} = \begin{pmatrix} G_1 & G \\ i(\alpha t+\frac{2k}{\mu})G_1 & i(\alpha t+\frac{2k}{\mu})G+\overline{G}_1 \end{pmatrix}\begin{pmatrix} A \\ B \end{pmatrix} = Q(t)\begin{pmatrix} A \\ B \end{pmatrix}, \tag{24}$$

We introduce a new vector function

$$V = \begin{pmatrix} A \\ B \end{pmatrix},\ \tilde{B}(t) = Q^{-1}(t)\begin{pmatrix} 0 \\ \breve{B} \end{pmatrix}. \tag{25}$$

We obtain the equation for the function V with a bounded operator $C(t)$:

$$\frac{dV}{dt} = C(t)V + \tilde{D}(t) = Q^{-1}(t)F\left(\begin{pmatrix} 0 & 0 \\ N(y) & 0 \end{pmatrix}F^{-1}Q(t)V\right) + \tilde{B}(t). \tag{26}$$

Further replacements are completely analogous to those described in [5] and give us the approximate solution on the interval $0 \le \tau \le t$: At the point $\tau = t$ we obtain:

$$\begin{pmatrix} v \\ w \end{pmatrix} = \begin{pmatrix} 1 & 0 \\ -tN(t)/2 & 1 \end{pmatrix}F^{-1}Q(t)Q^{-1}(0)F\left(\begin{pmatrix} 1 & 0 \\ -tN(0)/2 & 1 \end{pmatrix}F^{-1}Q(0)F\begin{pmatrix} v_0 \\ w_0 \end{pmatrix}\right) + \begin{pmatrix} te^{i(\eta+\alpha\mu x/2)t}/2 \\ e^{i(\eta+\alpha\mu x/2)t}/2 \end{pmatrix}. \tag{27}$$

In what follows we need only the first row of the matrix relation (27)

$$v(t) \approx F^{-1}G_1Fv_0 + F^{-1}GF(-\frac{t}{2}N(t)v_0) + F^{-1}GFw_0 + te^{i(\eta+\alpha\mu x/2)t}/2 \tag{28}$$

since the initial data for $Fw(t) = \frac{\partial v(t)}{\partial t}$ are found from the condition that at the point t the original SchrEdinger equation is satisfied.

In order to carry out stable calculations, excluding education random phase by repeated multiplication by $G_1 = e^{-i\alpha(t^2+4kt/\alpha\mu)}$, it is necessary to satisfy that condition

$$\alpha(t^2+4kt/\alpha\mu) < \pi/2. \tag{29}$$

The calculations take place in a bounded spectral region $|K| \le K_{max}$, therefore, returning to the original time, we get

$$\frac{t_{old}^2}{\mu} + \frac{8K_{max}t_{old}}{\mu^2} \le \pi/2.$$

From the equality we find

$$t_{old} \approx \pi\mu^2/8K_{max}.$$

Taking, for example, $\mu^2 = \frac{1}{2K_{max}^{1/2}}$ we get for the hyperbolized nonlinear Schredinger equation (HNSE) restriction $t_{old} \le \frac{\pi}{16K_{max}^{3/2}}$. Recall that for for the NSE this restriction has the form $t_{NSE} \le \frac{\pi}{2K_{max}^2}$

At $K_{max} = 2^7 = 128$ the step can be increased by 4 times. At $K_{max} = 2^9 = 10248$ in 8 times.

Note that operator replacements have automatic parallelism and are simply and practically error-free programmable.

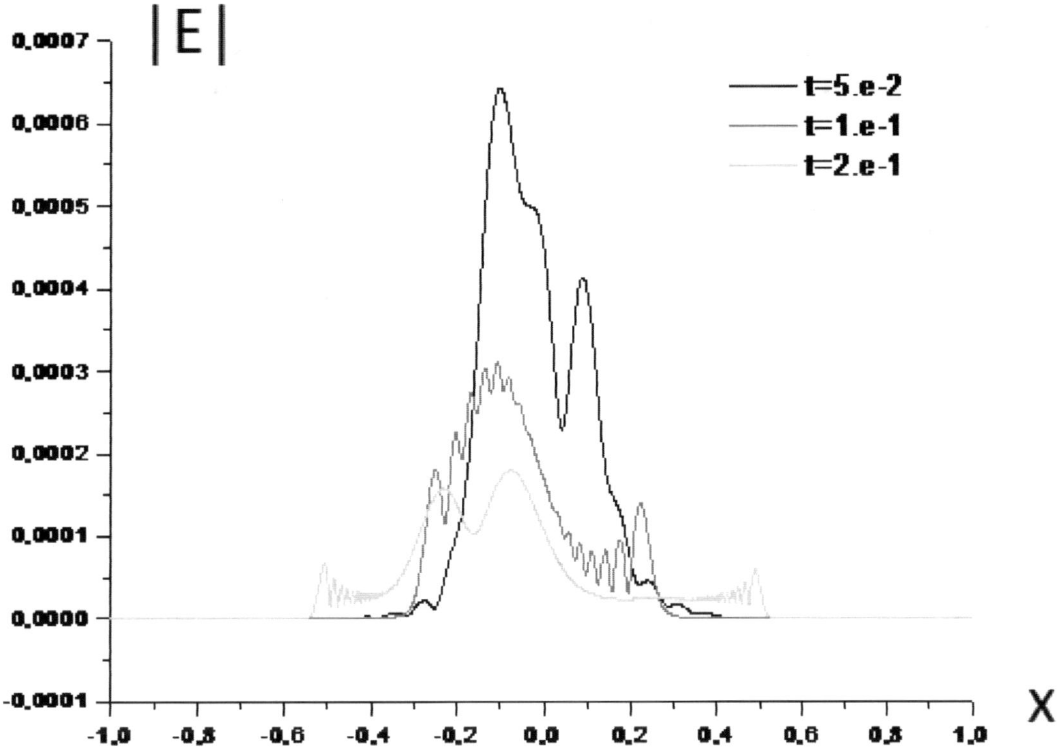

Figure 1. Amplitude decay of the field $|E|$ in time. Both formation and evolution of travelling waves of the Airy type is observed.

4. Conclusion

Typical dynamics of the numerical solution of Eq. (17) obtained by the relaxation method for HNLS (4) is presented in Fig. 1. The initial data are zero, the small external parameter $B = 0.1$. The principal result is the essential difference of wave flow U relaxation for the HNLS, achieved by adjusting the phase of the oscillations in comparison with a heat flow obtained by amplitude constancy.

Initially, a field growing due to external influences quickly falls apart into scattering waves, such as Airy functions, but with a growing frequency. It is characteristic that smooth leading wave fronts are directed in different directions. This is due to the fact. that the waves generated at $x > 0$ and $x < 0$ are solutions of the equation $\frac{\partial^2 \varphi}{\partial x^2} \pm x\varphi$, i. e. the Airy functions $Ai(\pm x)$. Thus we get the wave $Ai(-x)$ in the region of negative x. The figure shows that it is the solution of the Schr?dinger equation that is found: there is no characteristic sign of the solution of the hyperbolic equation two identical waves $\varphi(x-at)$ and $\varphi(x+at)$, traveling in opposite directions.

References

[1] Chetverushkin B.N. 2004 *Kinetic schemes and a quasi-gas-dynamic system of equations.* (Moscow: MAKS Press,) p 332
[2] Chetverushkin B.N. 2012 *Mat. modelling* **24** number 11, 33-52
[3] Albert Milani. 1985 *Pacific Journal of Mathematics.***116** nom. 1 – pp. 111-129.
[4] A.E.Lutsky, B.N.Chetverushkin. 2019 *Differential equation (Russia)* **55** nom. 4 pp. 588-596
[5] A.D. Yunakovsky. 1996 *Modeling of the nonlinear Schredinger equation.* (N. Novgorod) p 160
[6] Landau L.D., Lifshits E.M. 2005 *Teoreticheskaya fizika, Vol. 8* (M.: Fizmatlit) p. 656

The time reversibility property in analysis of sound points in balance-characteristic difference methods

N A Afanasiev[1], **V M Goloviznin**[1]

[1] Lomonosov Moscow State University, Faculty of Computational Mathematics and Cybernetics; Leninskie Gory, Moscow, 119991, Russia;

E-mail: nafanas.cmc@gmail.com

Abstract. A new computational technology of flux-type variables calculation on new time layer is introduced for the balance-difference methods for solving quasilinear hyperbolic systems of PDEs. The proposed method allows to build a scheme, which is locally implicit in the sound points and explicit in all other nodes of the mesh. It also has the second order of approximation in time and space and the property of time reversibility if the flux correction procedure is disabled. Results of test computations for the problem of transonic dam-break are presented.

1. Introduction

High-accuracy numerical solution of quasilinear hyperbolic systems of PDEs has always been an important problem of mathematical modelling. Particular concern is raised by the problem of solving such equations in case of transonic flows. A wide variety of transonic problems includes, for example, the modelling of dam-break using the shallow-water equations or the modelling of gas flow in de Laval nozzle using the Euler's equations of gas dynamics.

In recent years, the balance-characteristic schemes [1] have become very popular for solving the quasilinear hyperbolic systems of PDEs. There are two types of variables in these methods: conservative-type and flux-type variables. Conservative-type variables are used to formulate the mesh analogues of conservation laws, and flux-type variables are computed with the method of inverse characteristics. This allows to utilize all the properties of hyperbolic systems of PDEs and build such second order schemes as the CABARET scheme [2].

The strengths of this scheme include its flexibility, easy parallelization and high modernization potential. One of its weaknesses is the complications with robust computations in case of transonic flows. There a lot of types of sonic points in these flows, so most of the previously proposed sonic point processing algorithms [2, 3, 4, 5] were either ineffective or too complicated trying to process every type of the point individually. This article presents a description of a new approach to processing of the sonic points, which is based purely on such concepts as approximation and time reversibility.

2. Quasilinear hyperbolic systems of PDEs

Let's consider a system of m one-dimensional quasilinear laws of conservation without any sources or sinks:

$$\frac{\partial \mathbf{u}}{\partial t} + \frac{\partial \mathbf{F}(\mathbf{u})}{\partial x} = 0, \tag{1}$$

here $\mathbf{u}$ - m-dimensional vector of unknowns, $\mathbf{F}(\mathbf{u})$ - m-dimensional vector-function of fluxes. Transforming the spatial derivative in (1), another form of the system can be obtained:

$$\frac{\partial \mathbf{u}}{\partial t} + A(\mathbf{u})\frac{\partial \mathbf{u}}{\partial x} = 0, \tag{2}$$

here $A(\mathbf{u})$ is Jakobi matrix of $\mathbf{F}(\mathbf{u})$. Finally, assuming that the system (2) is of hyperbolic type (i.e. all eigenvalues of matrix $A(\mathbf{u})$ are real and different under all possible values of $\mathbf{u}$) and that the analytical expressions for Riemann invariants can be found explicitly, a characteristic form of the system can be obtained:

$$\frac{\partial \mathbf{I}}{\partial t} + \Lambda(\mathbf{I})\frac{\partial \mathbf{I}}{\partial x} = 0, \tag{3}$$

here $\mathbf{I} = \mathbf{I}(\mathbf{u})$ is the m-dimensional vector of Riemann invariants, $\Lambda(\mathbf{I})$ - diagonal matrix with the eigenvalues of $A(\mathbf{u})$ on the diagonal. Therefore, the initial system (1) can be reduced to m independent non-linear transport equations. Hence, the considered system must possess the property of time reversibility if the characteristics of the same family do not intersect. It means that if the system turns $\mathbf{u_0} = \mathbf{u}(t_0)$ into $\mathbf{u_1} = \mathbf{u}(t_0 + \tau)$ over time τ, then it also turns $\overline{\mathbf{u}}_\mathbf{1}$ into $\mathbf{u_0}$ over time τ, where $\overline{\mathbf{u}}_\mathbf{1}$ is equal to $\mathbf{u_1}$ in which all velocities are changed to the opposite ones.

Summarizing these facts about the system of laws of conservation (1), we can state that a good numerical scheme for its solving should capture all or at least some of the properties of the mathematical model, including: conservation of mesh analogues of $\mathbf{u}$ (connection with (1)), transfer of Riemann invariants over the characteristics (connection with (3)), time reversibility (corollary of (3)).

3. Balance-characteristic schemes

Many of numerical schemes for solving quasilinear hyperbolic systems of PDEs were derived using different forms of the stated problem. For example, the conservative finite volume schemes are based on the conservation laws (1), characteristic methods - on the characteristic form (3) and do not rely on the laws of conservation. The balance-characteristic methods (like the CABARET scheme) are based both on forms (1) and (3) and use all the information provided by the mathematical model of the problem.

In this work we will only consider the 3-phase balance-characteristic schemes. The family of such method includes the well-known CABARET scheme [2], Sharp scheme [6] and others.

To describe the main principles of this type of methods, we will introduce a space-time mesh and mesh functions on it. Assume that the system (1) is considered in the interval $x \in [a, b]$ during time $t \in [0, T]$. Then let $\omega = \omega_h \times \omega_\tau$ be the space-time mesh, where $\omega_h = \{x_i \,|\, a = x_0 < x_1 < \ldots < x_{N-1} < x_N = b;\ x_{i+1} - x_i = h_{i+1/2},\ i = \overline{0, N-1}\}$ - irregular in general case mesh in space, $\omega_\tau = \{t_n \,|\, 0 = t_0 < t_1 < \ldots < t_{K-1} < t_K = T;\ t_{n+1} - t_n = \tau_n,\ n = \overline{0, K-1}\}$ - irregular in general case mesh in time. We also introduce two sets of mesh functions: conservative-type and flux-type functions (see figure 1). The values of conservative-type functions are situated in the centres of space cells on integer and half-integer time layers: $(x_{i+1/2}, t_n) \mapsto \phi^n_{i+1/2}$, $(x_{i+1/2}, t_{n+1/2}) \mapsto \phi^{n+1/2}_{i+1/2}$. The values of flux-type functions are situated in the nodes of the mesh ω_h: $(x_i, t_n) \mapsto \psi^n_i$.

One time step of a 3-phase balance-characteristic scheme consists of three phases. Phases 1 and 3 represent the calculation of conservative-type variables according to the mesh analogues of laws of conservation (1). For example, in second order schemes these phases appear as follows:

$$\textbf{Phase 1:}\ \frac{\mathbf{u}^{n+1/2}_{i+1/2} - \mathbf{u}^n_{i+1/2}}{\tau_n/2} + \frac{\mathbf{F}(\mathbf{u}^n_{i+1}) - \mathbf{F}(\mathbf{u}^n_i)}{h_{i+1/2}} = 0. \tag{4}$$

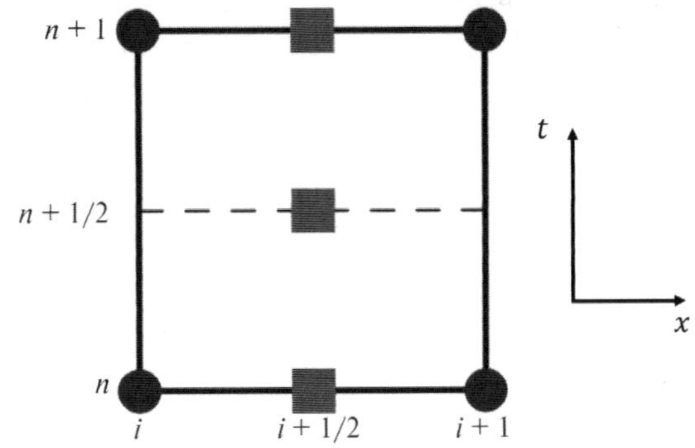

Figure 1: Types of mesh functions: conservative (squares) and flux (circles).

$$\textbf{Phase 3:}\quad \frac{\mathbf{u}_{i+1/2}^{n+1}-\mathbf{u}_{i+1/2}^{n+1/2}}{\tau_n/2}+\frac{\mathbf{F}(\mathbf{u}_{i+1}^{n+1})-\mathbf{F}(\mathbf{u}_{i}^{n+1})}{h_{i+1/2}}=0. \tag{5}$$

The second phase represents the calculation of flux variables $\mathbf{u}_i^{n+1}$ on the new time layer. For that the values of Riemann invariants $\mathbf{I}$ are extrapolated or interpolated and then transferred in the direction of the respective characteristics which slopes are approximated using the values in the centres of space-time cells $\mathbf{u}_{i+1/2}^{n+1/2}$. In cases of strictly subsonic or supersonic flows, when all eigenvalues of matrix A in (2) don't change signs, every point (x_i, t_{n+1}) receives exactly one value for each of m Riemann invariants, so the flux variables $\mathbf{u}_i^{n+1}$ can be uniquely restored. But if at least one of the eigenvalues $\mathbf{I}_k$ changes its sign in two adjacent space-time cells, then the node (x_i, t_{n+1}) between these cells will either receive two different values of a corresponding Riemann invariant on the next time layer or receive none of them. Hence, it won't be possible to uniquely restore the values of the flux variables. In this case we will say that the space-time node (x_i, t_{n+1}) is a sonic point along the invariant $\mathbf{I}_k$.

4. Locally implicit sonic point processing algorithm

Several different algorithms were proposed for the processing of sonic points in quasilinear hyperbolic systems of PDEs [2, 4, 5], most of them were tested with the CABARET scheme. Algorithm from [2] based on choosing the right side of transfer according to the sum of eigenvalues beside the sonic point, although being time reversible, can lead to non-physical discontinuities on the rarefaction waves, which is shown in [4]. Explicit [4] and implicit [5] methods also possess the property of time reversibility, but only work for a restricted class of problems.

We will propose a new second order time reversible method of sound point processing based on inverse characteristic method. Assume that the space-time node (x_i, t_{n+1}) is a sonic point along the invariant $\mathbf{I}_k$, which means that $(\lambda_k)_{i-1/2}^{n+1/2} * (\lambda_k)_{i+1/2}^{n+1/2} \le 0$. Consider the area of two space-time cells around the sonic point (see figure 2). Let's trace a straight characteristic with a slope determined by $(\lambda_k)_i^n = \lambda_k(\mathbf{u}_i^n)$ from the node (x_i, t_n) and transfer the invariant $(I_k)_i^n$ along the line to the half-integer time layer. Then let's trace a straight characteristic with a yet unknown slope determined by $(\widehat{\lambda}_k)_i^{n+1} = \lambda_k(\widehat{\mathbf{u}}_i^{n+1})$ from the node (x_i, t_{n+1}). Next we will introduce a local coordinate system:

$$\xi = x - x_{i-1/2}, \quad x_{i-1/2} = 0.5(x_i + x_{i-1})$$

and construct a second order interpolation $P_2(\xi)$ of I_k on the half-integer time layer based on constraints:

$$P_2(0) = (I_k)_{i-1/2}^{n+1/2}, \;\; P_2(\overline{\xi}_k) = (I_k)_i^n, \;\; P_2(h_i) = (I_k)_{i+1/2}^{n+1/2},$$

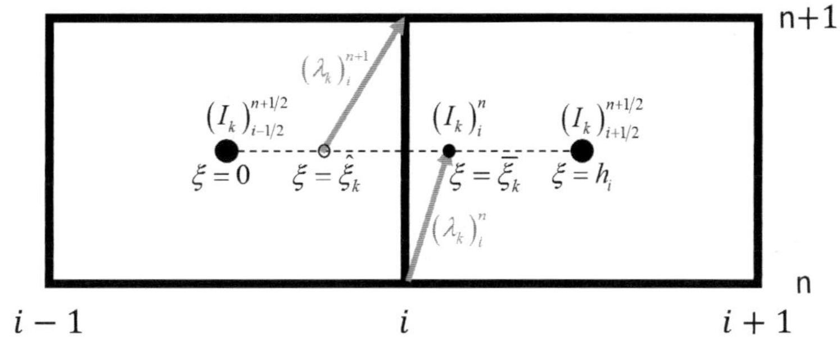

Figure 2: The template of the locally implicit sonic point processing method.

$$\bar{\xi}_k = 0.5h_{i-1/2} + 0.5\tau_n(\lambda_k)_i^n, \quad h_i = x_{i+1/2} - x_{i-1/2}.$$

After that we can find the value of the designated invariant in the point of intersection of $t = t_{n+1/2}$ and the straight characteristic from (x_i, t_{n+1}) and transfer this value to the upper time layer:

$$(\widehat{I}_k)_i^{n+1} = P_2(\widehat{\xi}_k), \quad \widehat{\xi}_k = 0.5h_{i-1/2} - 0.5\tau_n(\widehat{\lambda}_k)_i^{n+1}.$$

Hence, returning to the initial coordinate system we get the equation for the invariant I_k:

$$\begin{aligned} I_k(\widehat{\mathbf{u}}_i^{n+1}) = {} & \tau_n(I_k)_{i-1/2}^{n+1/2}\frac{(h_{i+1/2} + \tau_n\lambda_k(\widehat{\mathbf{u}}_i^{n+1}))((\lambda_k)_i^n + \lambda_k(\widehat{\mathbf{u}}_i^{n+1}))}{4h_i\bar{\xi}_k} - \\ & - \tau_n(I_k)_{i+1/2}^{n+1/2}\frac{(h_{i-1/2} - \tau_n\lambda_k(\widehat{\mathbf{u}}_i^{n+1}))((\lambda_k)_i^n + \lambda_k(\widehat{\mathbf{u}}_i^{n+1}))}{4h_i(h_i - \bar{\xi}_k)} + \\ & + (I_k)_i^n\frac{(h_{i-1/2} - \tau_n\lambda_k(\widehat{\mathbf{u}}_i^{n+1}))(h_{i+1/2} + \tau_n\lambda_k(\widehat{\mathbf{u}}_i^{n+1}))}{4\bar{\xi}_k(h_i - \bar{\xi}_k)}. \end{aligned} \tag{6}$$

Equation (6) is used for invariant I_k in point (x_i, t_{n+1}) only if it is a sonic point along the invariant I_k. In other case the value of $(I_k)_i^{n+1}$ can be found via classic phase 2 algorithms (for example, extrapolation algorithm of the CABARET scheme) and can be considered as a known value. Then a simple functional relationship between the invariants and the initial variables can be used:

$$I_k(\widehat{\mathbf{u}}_i^{n+1}) = (I_k)_i^{n+1}. \tag{7}$$

Finally, we formulate the whole algorithm for sound points. If a space-time point (x_i, t_{n+1}) is a sound point along at least one of the invariants of the system, a system of m non-linear equations is built. For each invariant with a sign change in its eigenvalue, equation (6) is written. Equation (7) is written for all other invariants. Then the system of m non-linear equations with m unknowns $\widehat{\mathbf{u}}_i^{n+1}$ can be solved using, for example, Newton's method, and a set of new flux variables can be found.

The proposed algorithm has a second order of approximation in time and space and does not violate the property of conservation when used in balance-characteristic methods. Moreover, due to the uniqueness of parabolic function built using values in 3 points and the symmetry of the template (see figure 2), the algorithm possesses a time reversibility property. It is also important to notice, that the method is only locally implicit when used with the CABARET scheme: the system of non-linear equations must only be solved in the sound points, in all other points the values are computed explicitly.

In some problems the monotonicity of the solution is more important than the property of time reversibility. In this case the algorithm (6 - 7) can be extended with a flux correction procedure similar to the one in the CABARET scheme: the value of invariants $(\widehat{I}_k)_i^{n+1} = I_k(\widehat{\mathbf{u}}_i^{n+1})$ are changed to stay between the values of $(I_k)_{i-1/2}^{n+1/2}$, $(I_k)_{i+1/2}^{n+1/2}$ and $(I_k)_i^n$.

5. Test results

The proposed algorithm, coupled with the CABARET scheme [2], was tested on one-dimensional shallow-water equations for flows above a flat bottom:

$$\begin{cases} \dfrac{\partial H}{\partial t} + \dfrac{\partial Hu}{\partial x} = 0, \\ \dfrac{\partial Hu}{\partial t} + \dfrac{\partial Hu^2}{\partial x} + \dfrac{g}{2}\dfrac{\partial H^2}{\partial x} = 0, \end{cases} \tag{8}$$

here H is height of fluid, u - velocity of column of fluid, g - acceleration of free fall.

Although many test have been carried out, we will provide the results for only one of them: the comparison of the proposed algorithm with the explicit method [4] on the problem of transonic dam-break. The initial conditions of the problem have a discontinuity in the middle of the considered interval $x \in [0, 1]$:

$$H_{left} = 100.0, \quad H_{right} = 1.0, \quad u_{left} = 0.0, \quad u_{right} = 0.0. \tag{9}$$

Figures 3 and 4 show the results for the above-described problem for the moment of time

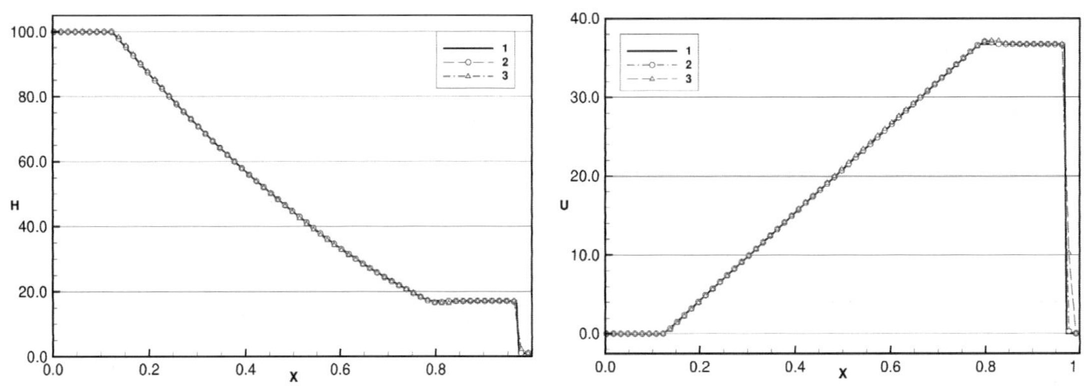

(a) (b)

Figure 3: Graphs of height (a) and velocity (b) on $x \in [0, 1]$ in the problem with initial conditions (9). 1 - analytical solution, 2 - proposed algorithm, 3 - explicit algotithm [4].

$t = 0.012$ computed on a uniform mesh of 200 cells. Figure 3 contains the graphs of results on interval $x \in [0, 1]$ (the values are shown in every third cell of the mesh), figure 4 - on interval $x \in [0.7, 0.9]$ (the values are shown in every cell inside the designated interval).

Both algorithms lead to smearing of the shock wave over 3 cells, and they also capture the rarefaction wave very well. However, the locally implicit algorithm deals better with the nonmonotonicity which occurs right after the rarefaction wave (see figure 4). Such nonmonotonicity presumably appears due to the imbalance between the conservative-type and flux-type variables during the first time steps. It is possible to smooth this effect by synchronizing

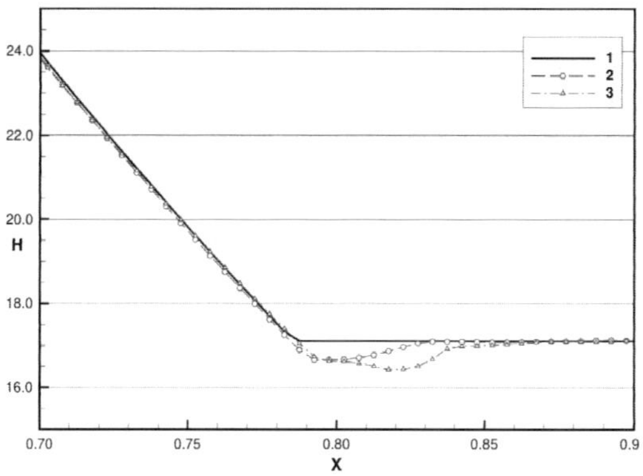

(a)

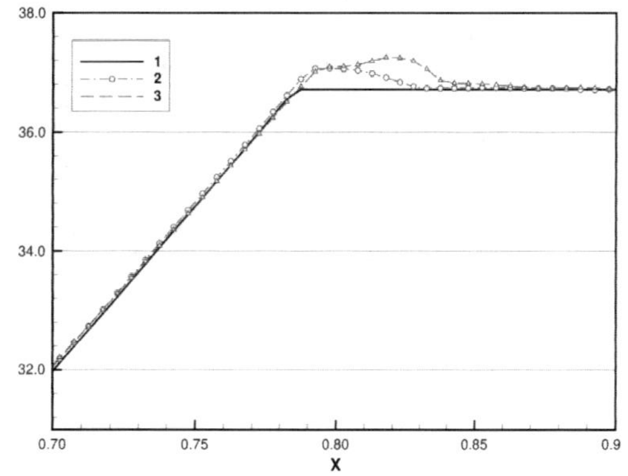

(b)

Figure 4: Graphs of height (a) and velocity (b) on $x \in [0.7, 0.9]$ in the problem with initial conditions (9). 1 - analytical solution, 2 - proposed algorithm, 3 - explicit algotithm [4].

the variables in the beginning of the computation, but a general rule of synchronization is yet to be revealed. The proposed scheme allows to minimize width and amplitude of this nonmonotonicity, which decreases on finer meshes.

The proposed scheme has also showed satisfying results on the Toro's tests [7] for transonic flows in shallow-water equations. The results of these tests will be published in future papers.

6. Conclusion

A new locally implicit algorithm of sonic point processing in balance-characteristic schemes was introduced. The resulting conservative method has a second order approximation and property of time reversibility when the flux correction procedure is disabled. Results of test computations for the problem of transonic dam-break were presented. It was shown that the proposed method allowed to get more accurate results as compared with some previously proposed algorithms. The author hopes to generalize the scheme for the case of non-zero right-hand side in the initial equations and the multidimensional case.

Acknowledgments

This work was supported by the Russian Science Foundation, project no. 19-11-00104.

References

[1] Goloviznin V M, Karabasov S A and Kobrinskii I M, Balance-characteristic schemes with separated conservative and flux variables, 2003, *Mat. Model.* **15** 29-48

[2] Goloviznin V M, Zaitsev M A, Karabasov S A and Korotkin I A 2013 *New CFD Algorithms for Multiprocessor Computer Systems* (Moscow: Moscow State University Press)

[3] Goloviznin V M, Karabasov S A and Kondakov V G, Generalization of CABARET scheme for two-dimensional orthogonal computational grids, 2013, *Mat. Model.* **25** 10336

[4] Goloviznin V M, Solovjev A V and Isakov V A, An approximation algorithm for the treatment of sound points in the CABARET scheme, 2016, *Vychisl. Metody Programm.* **17** 16676

[5] Goloviznin V M and Isakov V A, Balance-characteristic scheme as applied to the shallow water equations over a rough bottom, 2017, *Zh. Vychisl. Mat. Mat. Fiz.* **57** 1140 - 57

[6] Solovjev A V and Danilin A V, Using the sharp scheme of higher-order accuracy for solving some nonlinear hyperbolic systems of equations, 2019, *Vychisl. Metody Programm.* 45-53

[7] Toro E F 2001 *Shock-Capturing Methods for Free-Surface Shallow Flows* (New York: Wiliey)

The algorithm for generation of structured grids in deformed volumes of revolution

A I Anuchina[1], N A Artyomova[2], V A Gordeychuck[1], O V Ushakova[2,3]

[1] Federal State Unitary Enterprise "Russian Federal Nuclear Center – Academician E.I. Zababakhin All-Russian Research Institute of Technical Physics" (FSUE "RFNC-VNIITF"), Vasiliev st. 13, PO box 245, 456770, Snezhinsk, Chelyabinsk region, Russia
[2] Institute of Mathematics and Mechanics named after academician N.N. Krasovskii, S.Kovalevskaya st. 16, 620990 Ekaterinburg, Russia
[3] Ural Federal University named after first President of Russia B.N. Yeltsin, Mira st. 19, 620002, Ekaterinburg, Russia

E-mail: ana@imm.uran.ru

Abstract. For the volume of revolution deformed by another volume of revolution, a grid generation algorithm is suggested. The algorithm is designed for multimaterial hydrodynamic simulation and for solving other physical and engineering problems. The algorithm represents the non-stationary procedure generating three-dimensional structured grids in domains with moving boundaries. The algorithm is developed within the variational approach for constructing optimal curvilinear grids. The volume of revolution is obtained by the rotation about the axis through 180° of a plane generatrix curve consisting of straight line segments and arcs of circles. The non-stationary algorithm is the iterative process at each stage of which the deformation of a grid and then its optimization are carried out. The deformation of a grid is implemented within the geometrical approach, and the optimization of a grid within the variational approach. Iterations are continued until the necessary deformation is reached. The algorithm is realized in the computer code written in C++.

1. Introduction

The algorithm for generation of structured grids in volumes of revolution which are subject to deformation is suggested. Considered volumes are obtained by the rotation around the axis through 180° of a plane generatrix curve consisting of straight line segments and arcs of circles. We consider the pressure caused by volumes of revolution (deforming body) on the another volume of revolution (main body). As the result of such application the main body is deformed: the boundary of the main body is moved inside the body and takes the shape of the additional (deforming) body in the region of deformation while the shape of the other part of the main body (which is not subject to deformation) does not change. Since in the general case, the body of revolution is formed by conical, cylindrical and spherical surfaces, the consideration of a pressure by means of a conus, a cylinder and a sphere onto the main body is the necessary stage for elaborating the algorithm for the cases of pressure by means of a general body of revolution onto the general body of revolution. Such cases are considered first. In the paper, further experience of the development of the algorithm for deformation by the more general bodies of revolution is described. Some examples are given.

Grids in deformed volumes are used for mathematical modeling the processes of multi-material hydrodynamics [1]. The general approach used for grid generation is the mapping approach [2]. According to this approach constructing a structured grid in the domain of a complicated geometrical form called the physical domain is carried out by means of a continuous mapping of the domain of a simpler form (in our case a rectangular parallelepiped) called the computational domain [3]. Thus, the physical domain in this way of grid construction is represented as a curvilinear hexahedron and the computational grid is the image of a uniform rectangular grid in the rectangular parallelepiped under the considered mapping and consists of images of unit cubes which are hexahedral cells. The way of representation of a domain as a curvilinear hexahedron defines the configuration of the domain and the structured grid.

Within the mapping approach a method for generation of structured grids in the volumes of revolution [4, 5] has been developed using the variational technique [6] which allows to construct grids of desired quality, nondegenerate [3, 7, 8] closed to uniform and orthogonal ones and, if necessary, adaptive to the solution of the physical problem [6]. It would be attractive to apply the method developed for the volume of revolution for the case of the deformed volume of revolution.

In this connection, it is obvious the idea to elaborate the non-stationary algorithm for generation of moving grids changing the boundary of the volume of revolution according to the deformation process. The algorithm represents the iterative procedure at each step of which the boundary of the main body is deformed under the pressure of the additional body (additional body moves inside the main body), boundary grid nodes of the main body remove on the deforming surface and then optimization of the deformed grid is carried out by the variational technique [4, 5, 6].

The problems of constructing moving grids are described by many authors (see, for example, a general review and underlying principles in [9] and a review, some examples and a comparison of different mesh morphing methods applied for the case of 3D shape optimization in [10]). The main difficulty of non-stationary algorithms (see [11]) is to provide the generation of a grid satisfying the given quality criteria (nondegeneracy, closeness to uniformity, orthogonality and etc. [5, 6]) automatically and without any interference of the user at each step of the process. So, the described algorithm is a solution of a complicated problem and it represents a new computational technique for multimaterial hydrodynamic simulation and for solving other significant scientific and engineering problems.

The considered problem of grid generation in the volume of revolution deformed by the another volume of revolution is formulated in section 2. In section 3, the non-stationary grid generation algorithm is described. Examples of calculations of grids are given in section 4.

2. Formulation of the problem

In the Cartesian coordinate systems, X with axis $\{x^1, x^2, x^3\}$ (main system) and Ξ with the axis $\{\xi^1, \xi^2, \xi^3\}$ (additional system), the volumes (bodies) of revolution, the main one G_x ((Fig. 1(b))) and the additional one Q_ξ (Fig. 1(a)), are given correspondingly. Each of the volumes is obtained by the the rotation of the plane generatrix curve consisting of straight line segments and arcs of circles (elements) through the angle $\varphi = \pi$ around the axis. The generatrix of the main volume is given in the plane $\{x^1, x^3\}$, and the generatrix of the additional volume is given in the plane $\{\xi^1, \xi^3\}$. The rotation is carried out the around axis x^3 for the main body and around the axis ξ^3 for the additional body.

The origin $O_{\Xi X} = (x_o^1, x_o^2, x_o^3)$ of the coordinate system Ξ in the coordinate system X and the matrix C of the transformation from the coordinate system Ξ to the coordinate system X are given. The vector of deformation $\vec{V}_X$ in the coordinate system X which shows the direction of the deforming body pressure onto the main body is also given. The origin $O_{\Xi X}$ and the matrix C are given in such a way that after the transformation of coordinates (to the coordinate system

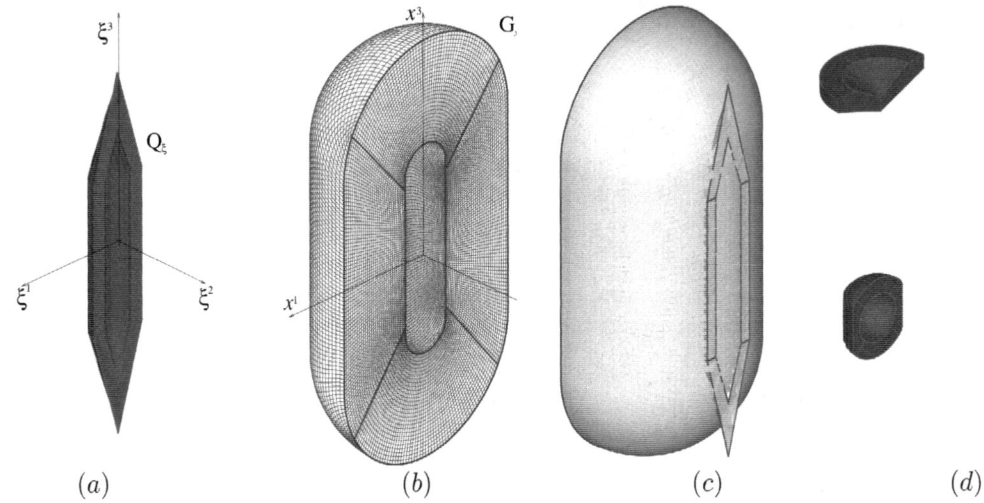

Figure 1. The generation of the deformed grid: (a) domain Q_ξ and (b) grid in the domain G_x, (c) domains Q_x and G_x in the coordinate system X, (d) additional domains Q_ξ.

X) the deforming body Q_x defines its final position inside the main body (Fig. 1(c)).

The structured three-dimensional grid $\mathbf{x}_{ijk} = (x^1_{ijk}, x^2_{ijk}, x^3_{ijk})$, $i = 0, .., N-1$, $j = 0, .., M-1$, $k = 0, .., L-1$ where integers N, M, L give the number of grid nodes is constructed in the domain G_x by the algorithms [4, 5]. The main body corresponding to the domain G_x is deformed by pressure of the additional body. The problem consists in constructing the structured grid in the deformed domain $D = G_x \setminus (G_x \cap Q_x)$.

3. Non-stationary grid generation algorithm

The given shape of the main body is deformed little by little in the iterative process until the deformation reaches the necessary degree (the deforming body turns out in its final position, see Fig. 1(c)). The algorithm consists of three stages. At the first (auxiliary) stage, the parameters (the number and the form of the elements of generatrix) of the deforming body are defined and the movement of the deforming body out of the main body is carried out. The second stage is the deformation of a grid in the main body. At the third stage, the grid is optimized in the correspondence with given quality criteria by the optimization algorithm 3.3. The second and the third stages are continued up to that moment when the additional body reaches its final position (see Fig. 1(c)).

3.1. Definition of parameters

By the text files with the description of the geometries of the main and additional bodies, we define the parameters of the considered case. Each element of the generatrix (a straight line segment or an arc of a circle) defines the form of the corresponding surface of revolution (conical, cylindrical or spherical) which further defines the way of projection of nodes on this surface. Then, we prepare for the process of deformation. During the process of deformation the deforming body has to move inside the main body in the direction given by the vector of deformation. For simplicity of computations, we move not the additional body but the main body and perform all computations in the auxiliary coordinate system Ξ. First, we move the main body out the additional body. The touch is allowed. Then we begin to move the main body. Thus, the coordinates of deforming surface of the auxiliary body do not change in the process of deformation, instead of this the main body moves.

3.2. Deformation of a grid

At this stage, such step of deformation (movement of the main body) is chosen that deformation does not affect (the additional body after movement of the main body does not touch) inner nodes of a grid of the main body. The boundary nodes of the main body subjected to deformation (got inside the additional body) remove on the deforming surface of the additional body: these nodes are projected ([12, 13]) on this surface. Here, it is defined the element of the generatrix forming the corresponding surface of revolution at which the nodes are projected. Then we perform the back transformation to the main coordinate system X and carry out the optimization stage.

3.3. Optimization of a grid

Initial [4] and deformed grids often contain cells unsatisfactory in their quality and differ very much in shape and size and close to degenerate. To construct the grid satisfying the users by its quality from the deformed grid, the optimization [5] or global conservative reconstruction (remapping) of a grid is applied. The conservative property is understood as the property to conserve the volume both locally and globally in the whole domain. The reconstruction is carried out by the variational method [5] via minimization of the discrete functional

$$D = D_{\text{u}} + A_{\text{o}} D_{\text{o}} \tag{1}$$

where $A_{\text{o}} > 0$ is a weight coefficient regulating the closeness of a grid to orthogonal one, D_{u} and D_{o} are measures of closeness of a grid to a uniform one and an orthogonal one, correspondingly:

$$D_{\text{u}} = \sum_{ijk} \left\{ [r_{i+1,j,k} - r_{i-1,j,k}]^2 \left(\frac{1}{r^2_{i+1,j,k}} + \frac{1}{r^2_{i-1,j,k}} \right) \right.$$

$$\left. + [r_{i,j+1,k} - r_{i,j-1,k}]^2 \left(\frac{1}{r^2_{i,j+1,k}} + \frac{1}{r^2_{i,j-1,k}} \right) + [r_{i,j,k+1} - r_{i,j,k-1}]^2 \left(\frac{1}{r^2_{i,j,k+1}} + \frac{1}{r^2_{i,j,k-1}} \right) \right\},$$

$$D_{\text{o}} = \sum_{ijk} \sum_{p=1}^{4} \left(\frac{1}{\sin^2 \varphi^p_{ij}} + \frac{1}{\sin^2 \varphi^p_{ik}} + \frac{1}{\sin^2 \varphi^p_{jk}} \right).$$

Here $r_{i\pm1,j,k} = |\mathbf{h}_{i\pm1}| = |\mathbf{x}_{i\pm1,j,k} - \mathbf{x}_{ijk}|$. Analogously, $r_{i,j\pm1,k} = |\mathbf{h}_{j\pm1}|, r_{i,j,k\pm1} = |\mathbf{h}_{k\pm1}|$. Values φ^p_{ij} ($p = 1, 2, 3, 4$) denote the angles between vectors $\mathbf{h}_{i\pm1}$ and $\mathbf{h}_{j\pm1}$ (see. Fig. 2). Summation is carried out over all inner nodes of the grid.

In [5] the continuous analogue of this functional is supplied and different variational problems are considered. The distinctive feature of the functional of optimality (1) is a special way of a uniformity criterion formalization. It defines the type of Euler equations for constructing grid (hyperbolic in the wide sense), permits to consider different types of boundary conditions in the variational problems for grid generation (fixed and free nodes, orthogonality of coordinate lines to the boundary) and provides good computational properties of grids (see [5, 6]). The construction of the functional of orthogonality D_{o} allows to refer the method to barrier methods [14], which possess a "barrier" for degenerate elements (for details, see also [5]).

In [5], during optimization when boundary grid nodes are considered to be free the movement of nodes is carried out along the surface composed of ruled faces of grid cells. For this movement grid nodes are close to the surface of revolution but do not belong to it. This can lead to the

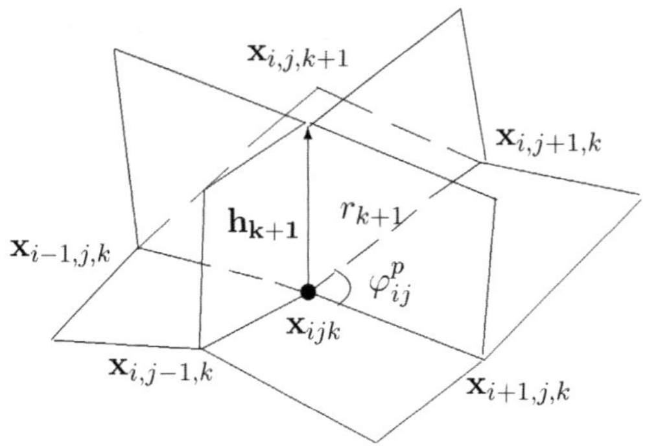

Figure 2. The elements of cells of a three-dimensional grid.

lost of the volume of the domain and the accuracy in the approximation of boundary conditions, so the algorithm for correction of grid nodes to the surfaces of revolution [12, 13, 15, 16] has been developed. Its main idea is in the following. The correcting node is projected on the corresponding surface of revolution. Deformed nodes are projected on the surfaces of revolution of the additional body, others (arranged on the ruled surfaces of cell faces of the main body) on the surfaces of revolution of the main body. In the last case what a surface of revolution has to be chosen is defined by the third coordinate of the node. Each element of the generatrix curve has its own interval of the change of this coordinate. For each face the numbers of elements generating the face are specified. What interval includes the third coordinate of the node defines the number of the element of the generatrix curve generating the corresponding surface for which projection has to be carried out. Thanks to this algorithm the movement of grid nodes can be carried out along surfaces of revolution and the reconstruction process can be conservative.

As a rule only a slight part of the domain is subject to deformation therefore it is not worthwhile to optimize the grid in the whole domain at each step of deformation. It is sufficient to optimize the fragment of a grid which is subjected to deformation.

4. Examples of calculations

The calculations are carried out on the personal computer (quad-core processor Intel Core i5-4570, 3.2 GHz, 4Gb of RAM, 64-bit OS). In Fig. 3 examples of grids for the main volume shown in Fig. 1(b) deformed by the additional volumes (Fig. 1(a), (d)) are given. The grid in these volumes contains 81x81x30=196830 nodes. In dependence of the given parameters (the weight of orthogonality, the numbers of iteration of deformation and optimization of a grid, and the degree of deformation) the calculations can be lasted from some minutes up to some hours. For optimization of a grid in the deformed fragment, the time of calculation reduces several times. Testing of grids is carried out according to criteria [8]. According to [8] cells degenerating into prisms (admissible in [1]) arise along edges of joining faces lying in one plane, other cells are nondegenerate hexahedral cells [7]. Among the last ones there are no twisted hexahedral cells exotic in shape. The closeness of grids to uniform and orthogonal one is estimated by the values of functionals $D_{\mathrm{U}}, D_{\mathrm{O}}$ [6].

Conclusion

Described algorithms are realized in computer codes written in *C++*. Further development of the technology will be performed for the cases of deformation of the volume of revolution by the more complicated volumes of revolution.

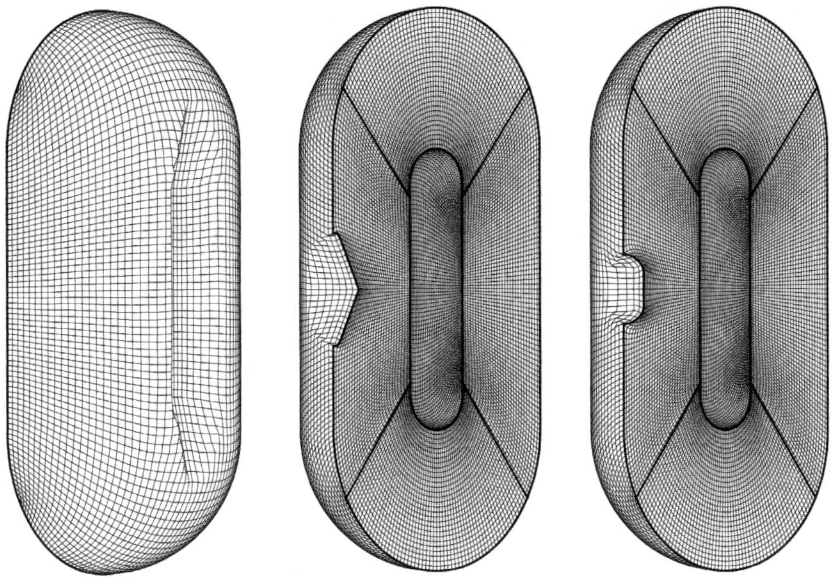

Figure 3. Examples of grids in deformed volumes.

References

[1] Anuchina N, Volkov V, Gordeychuk V, Es'kov N, Ilyutina, and Kozyrev O 2007 Numerical simulation of 3D multi-component vortex flows by MAH-3 code *Advances in Grid Generation* ed Ushakova O (New York: Nova Science Publishers Inc) Chapter 8 pp 215–240

[2] Godunov S, Zabrodin A, Ivanov M, et al 1976 *Numerical Solution of Multidimensional Problems of Gas Dynamics* (Moscow: Nauka)

[3] Ushakova O 2011 Nondegeneracy tests for hexahedral cells *Comput. Method Appl. M.* **200** 1649–58.

[4] Bronina T 2008 Algorithm for constructing initial three-dimensional structured grids for the domains of revolution *Proc. Steklov Inst. Math.* **261, Suppl 1,** S36–S43

[5] Ushakova O 2008 Optimization algorithms for three-dimensional grids in domains of rotations *Proc. Steklov Inst. Math.* **261, Suppl** 1 S228–S59

[6] Khairullina O, Sidorov A and Ushakova O 1999 Variational methods of construction of optimal grids *Handbook of Grid Generation*, ed J Thompson, B Soni et al (Boca Raton: CRC Press) pp 36-1–36-25

[7] Ushakova O 2001 Conditions of nondegeneracy of three-dimensional cells A formula of a volume of cells *SIAM J. Sci. Comp.* **23(4)** 1273–89.

[8] Ushakova O 2018 Criteria for hexahedral cell classification *Appl. Numer. Math.* **127** 18–39

[9] Zegeling P Moving grid techniques 1999 *Handbook of Grid Generation* ed J Thompson, B Soni et (Boca Raton: CRC Press) pp 37-1–37-22

[10] Staten M, Owen S, Shontz S, Salinger A and Coffey T 2011 A comparison of mesh morphing methods for 3D shape optimization *Proceedings of the 20th International Meshing Roundtable* ed W Quadros (Berlin, Heidelberg: Springer) pp 293-311

[11] Prokopov G 2007 Moving mesh calculation in unsteady two-dimensional problems *Advances in Grid Generation* ed Ushakova O (New York: Nova Science Publishers, Inc) pp. 127–160

[12] Ushakova O 2016 An algorithm of correcting a grid with respect to the surface of revolution *Vopr. At. Nauki Tekh. Mat. Model. Fiz. Protsessov* **1** 16–27

[13] Ushakova O 2016 Application of the algorithm of correcting a grid with respect to the surface of revolution *Vopr. At. Nauki Tekh. Mat. Model. Fiz. Protsessov* **2** 31–37

[14] Ivanenko S, Charakhch'yan A 1988 Curvilinear grids of convex quadrilaterals *USSR Comput. Math. Math. Phys.Zh.* **28(2)** 126–33

[15] Ushakova O 2017 An algorithm of correcting a grid with respect to a deformed domain of revolution *Vopr. At. Nauki Tekh. Mat. Model. Fiz. Protsessov* **2** 53–65

[16] Ushakova O 2018 An algorithm of correcting a grid for a region formed by surfaces of revolution with parallel axis of revolution it Vopr. At. Nauki Tekh. Mat. Model. Fiz. Protsessov **1** 30–41

Numerical modelling of wall-bounded turbulent flows based on the CABARET scheme

Danil Asfandiyarov

Nuclear safety institute of Russian academy of sciences, 52, Bolshaya Tulskaya Street, Moscow, 115191, Russia

E-mail: dasfandiyarov@ibrae.ac.ru

Abstract. In this study, the classical problem of turbulent plane channel flow at $Re_m = 13750$ and $Re_m = 21900$ is investigated using a numerical algorithm based on the CABARET scheme. Simulations are performed on the fine grids, resolving all the spatial scales of the turbulence according to a direct numerical simulation approach, as well as on coarse grids standard for large eddy simulation of wall-bounded turbulent flows. In the latter case, in order to obtain a more accurate representation of the momentum transfer towards the wall, artificial boundary conditions are introduced. It allows modelling with high accuracy the mean flow characteristics. Numerical results obtained by the CABARET scheme are compared with numerical results obtained by pseudospectral method.

1. Introduction

At the present time, modelling of turbulent flows is a fundamental problem, which, apart from its scientific importance, has been at the core of numerous technical applications. The most complete description of turbulence can be obtained within direct numerical simulation (DNS) where all spatial scales of turbulence are resolved. At the present stage, DNS approach is constrained by computational performance of modern supercomputers and numerical algorithms. Therefore, considerable efforts have been made to develop and verify various scale-resolving methods that would provide a reasonable balance between the depth of physical description and computational resources needed to solve the Navier-Stokes equations.

A classical large eddy simulation (LES) approach is based on separation of the large-scale motions from the small-scale ones, which are closer to isotropic. The large-scale components of flow variables are explicitly integrated in time, and interactions with the unresolved small-scale components are modelled. LES demonstrated for a wide range of problems that it calculates with high accuracy the mean characteristics of flow as well as the moments of a higher order. Near-wall turbulence poses the most difficult problem to the LES approach. The flow ceases being isotropic and homogeneous near the wall. In the vicinity of the solid boundary the scales of large and small eddies are overlapped and generation of the turbulent energy occurs at the scales comparable with the distance from the wall. As well as in a DNS approach LES requires very detailed structural information on the fluctuations to yield accurate results.

Development of novel efficient LES-algorithms that would allow for a correct representation of the complex dynamical behaviour in the near-wall layer is one of the main problems in LES modelling. Development of implicit LES methods (ILES) based on high-resolution numerical schemes is at the cutting edge of LES modelling. In such schemes, the dissipative mechanism (smoothing filter) is provided by the operator approximating convective terms. A required amount of dissipation is introduced by a non-linear flux correction procedure.

The CABARET scheme [1], which belongs to this group of schemes, is distinguished by its simplicity, computational efficiency, improved dissipative and dispersive properties. The scheme is defined on a compact numerical template. It has a second order of approximation both in time and in space and allows the introduction of non-linear flux correction based on the maximum principle.

This paper presents the results of numerical simulation of incompressible turbulent channel flow based on the CABARET scheme for Reynolds numbers $Re_m = 13750$ and $Re_m = 21900$. Simulations are performed on both fine and coarse grids. In the latter case artificial boundary conditions are added. They can be viewed as an improvement of flow variables calculated in the grid cells in the vicinity to the wall and are introduced by using explicit subgrid-scale model accounting for shear effects in wall-bounded turbulent flows.

2. Governing equations

The Navier-Stokes equations for incompressible fluid were considered:

$$\frac{\partial u_i}{\partial x_i} = 0, \tag{1}$$

$$\frac{\partial u_i}{\partial t} + \frac{\partial (u_i u_j)}{\partial x_j} = -\frac{1}{\rho}\frac{\partial p}{\partial x_i} + \nu \nabla^2 u_i. \tag{2}$$

Time integration of the set of equations (1)-(2) was provided by a computational method based on the CABARET scheme for orthogonal grids under the assumption of incompressible fluid [2]. Simulations on coarse grids were provided in the LES formulation with the closure in the form of an eddy viscosity model:

$$\frac{\partial \bar{u}_i}{\partial x_i} = 0, \tag{3}$$

$$\frac{\partial \bar{u}_i}{\partial t} + \frac{\partial (\bar{u}_i \bar{u}_j)}{\partial x_j} = -\frac{1}{\rho}\frac{\partial \bar{p}}{\partial x_i} + \nu \nabla^2 \bar{u}_i + \frac{\partial (2\nu_T S_{ij})}{\partial x_j}, \tag{4}$$

where S_{ij} is the filtered rate-of-strain tensor. In near-wall boundary cells the shear-improved Smagoronsky model [3] for the eddy viscosity ν_T was used:

$$\nu_T(\mathbf{x}, t) = (C_s \Delta)^2 (|S(\mathbf{x}, t)| - |\langle S(\mathbf{x}, t)\rangle|), \tag{5}$$

where $|S| = \sqrt{2 S_{ij} S_{ij}}$. Here, the angle brackets $\langle \ldots \rangle$ denotes averaging over characteristic time intervals (in present work averaging is performed over 10000 time steps) [4]. The characteristic subgrid-scale $\triangle$ is determined from the volume of the cell $\triangle = (\triangle x \triangle y \triangle z)^{\frac{1}{3}}$. The Smagorinsky constant C_s is assumed to be 0.16.

Shear-improved Smagorinsky model (5) allows for modelling of the energy scatter on the subgrid scale caused by both turbulent fluctuations and the mean gradient of velocity [3].

The components of the rate-of-strain tensor S_{ij} and the eddy viscosity ν_T are calculated only in near-wall boundary cells (at the faces center). At the wall the value of the eddy viscosity ν_T is assumed to be zero. In other cells $\nu_T = 0$ and numerical dissipation works as a subgrid-scale model, due to the non-linear flux correction based on the maximum principle (ILES approach)[4].

3. Numerical simulations

The computational domain is a rectangular channel with two solid walls. Periodic boundary conditions were imposed in the streamwise and spanwise directions and no-slip conditions at the wall. At initial time, velocity distributions were designed to satisfy a Poiseuille profile plus a small random perturbation. The flow rate was kept constant. As soon as fully developed turbulent flow had been achieved, statistics were accumulated.

Non-uniform meshes, like in [4] and [5], with gradual concentration towards the walls and symmetrical relative to the central horizontal plane of the channel were used. In the periodic directions the meshes were uniform.

Fine meshes were used for the DNS simulations. The size of the boundary cells in the normal direction are to be $\Delta y_W^+ \approx 0.1 - 0.15$ in wall units. A LES approach mitigates the requirements to some extent with $\Delta y_W^+ \approx 1 - 1.5$. Here $y^+ = \frac{y u_\tau}{\nu}$, where u_τ is the friction velocity, ν is the kinematic viscosity.

The parameters of the calculation domain and the meshes are shown in the table 1. Here, the Reynolds number is defined as $Re_m = \frac{2\delta u_m}{\nu}$, where u_m is the averaged velocity, δ is the channel half width and Δy_C^+ is the size of the cells in the middle of the channel.

Table 1. Computational domain parameters.

Re_m	Dimensions	Mesh size	Δx_+	δz_+	δy_W^+	Δy_C^+
13750	$2\pi\delta \times 2\delta \times \pi\delta$	$512 \times 256 \times 256$	4.82	4.82	0.11	9.03
13750	$4\pi\delta \times 2\delta \times 2\pi\delta$	$160 \times 64 \times 160$	30.8	15.4	1.18	20.9
21900	$2\pi\delta \times 2\delta \times \pi\delta$	$512 \times 256 \times 512$	7.2	3.6	0.14	13.8
21900	$4\pi\delta \times 2\delta \times 2\pi\delta$	$256 \times 128 \times 256$	28.8	14.4	1.18	14.1

4. Numerical results

In figures 1-2 the mean velocity $u^+(y^+)$ is displayed as a function of the wall-normal distance y^+ for $Re_m = 13750$ and $Re_m = 21900$ respectively. The DNS and ILES results are represented at the same plot. The numerical results obtained by the CABARET scheme are compared with the DNS results obtained by the pseudospectral method [6].

The three normal components of the Reynolds stress tensor (normalized by the squared friction velocity) for both DNS and ILES are shown in figures 3-6.

The DNS results obtained by the CABARET scheme demonstrate a high degree of consistency with the results obtained by the pseudospectral method [6]. The average velocity profiles were calculated quite accurately on coarse meshes, while a slight deviation manifests in the profiles of normal stresses. The calculated values of the skin-friction coefficient are presented in the table 2.

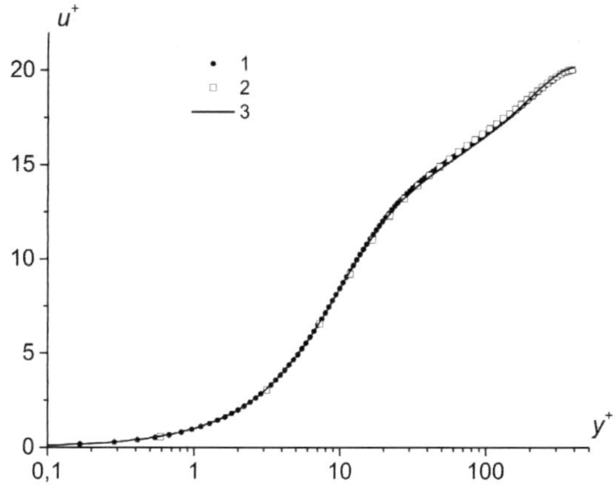

Figure 1. The mean velocity profile normalized by the friction velosity in turbulent channel flow at $Re_m = 13750$. 1 - DNS (present), 2 - ILES (present), 3 - DNS of channel flow by pseudospectral method [6].

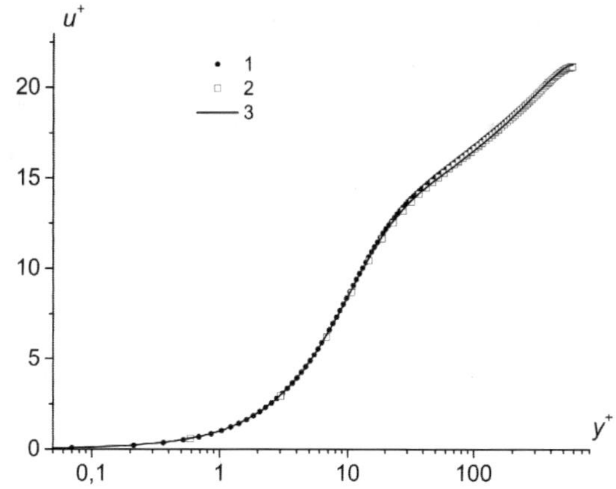

Figure 2. The mean velocity profile normalized by the friction velosity in turbulent channel flow at $Re_m = 21900$. 1 - DNS (present), 2 - ILES (present), 3 - DNS of channel flow by pseudospectral method [6].

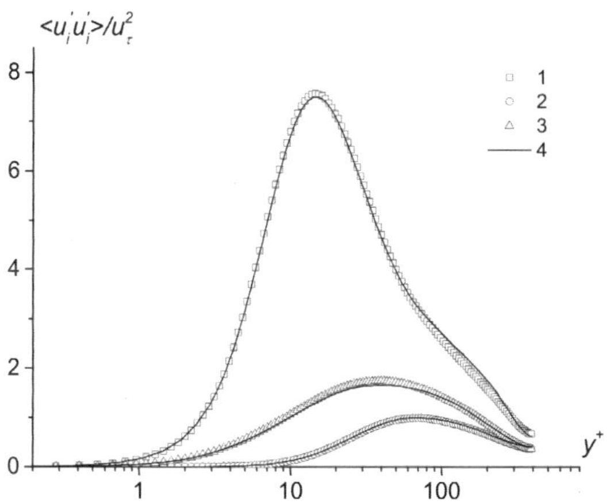

Figure 3. Profiles of Reynolds stresses normalized by the friction velosity in turbulent channel flow at $Re_m = 13750$. DNS results. 1 - $\overline{u'^2}^+$, 2 - $\overline{v'^2}^+$, 3 - $\overline{w'^2}^+$ 4 - DNS of channel flow by pseudospectral method [6].

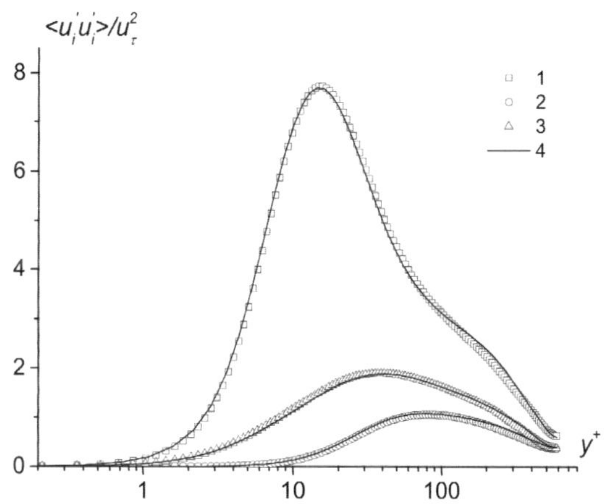

Figure 4. Profiles of Reynolds stresses normalized by the friction velosity in turbulent channel flow at $Re_m = 21900$. DNS results. 1 - $\overline{u'^2}^+$, 2 - $\overline{v'^2}^+$, 3 - $\overline{w'^2}^+$ 4 - DNS of channel flow by pseudospectral method [6].

5. Conclusions

In this study, the classical problem of viscous turbulent channel flow for various Reynolds numbers was investigated. The simulation results obtained by the CABARET scheme following the DNS approach demonstrate a good agreement with those calculated by the pseudospectral method, which is considered to be the most accurate for this type of flow. Moreover, ILES simulation results on relatively coarse meshes are given. The proposed computational algorithm with the use of artificial boundary conditions allows for accurate calculation of the first and second moments of the flow characteristics on such meshes.

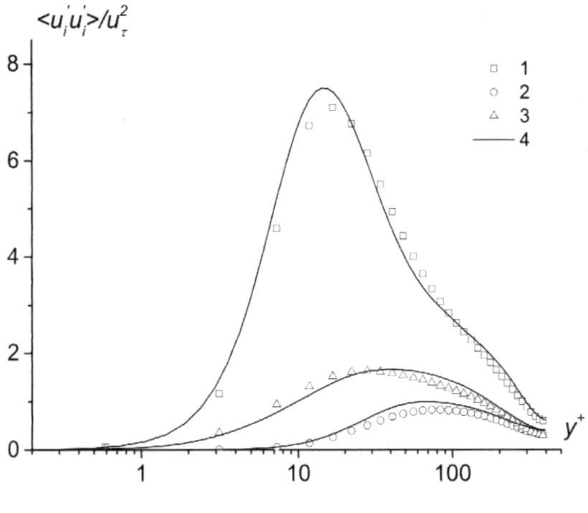

Figure 5. Profiles of Reynolds stresses normalized by the friction velocity in turbulent channel flow at $Re_m = 13750$. ILES results. 1 - $\overline{u'^2}^+$, 2 - $\overline{v'^2}^+$, 3 - $\overline{w'^2}^+$ 4 - DNS of channel flow by pseudospectral method [6].

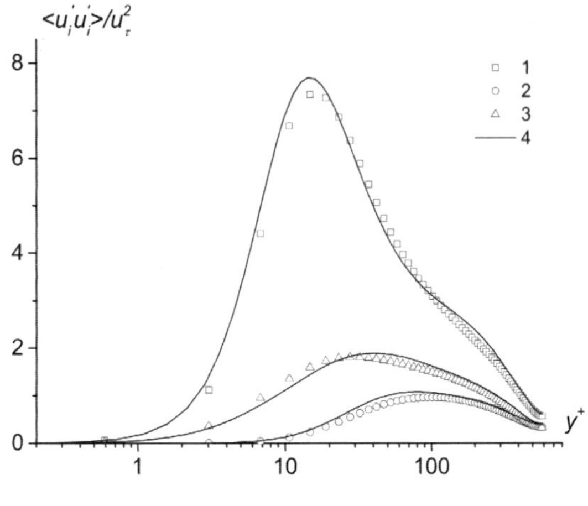

Figure 6. Profiles of Reynolds stresses normalized by the friction velocity in turbulent channel flow at $Re_m = 21900$. ILES results. 1 - $\overline{u'^2}^+$, 2 - $\overline{v'^2}^+$, 3 - $\overline{w'^2}^+$ 4 - DNS of channel flow by pseudospectral method [6].

Table 2. The skin-friction coefficient.

Re_m	C_f	C_f, *DNS Moser et al* [6]
13750	6.51×10^{-3} (*DNS*)	6.50×10^{-3}
13750	6.50×10^{-3} (*ILES*)	6.50×10^{-3}
21900	5.75×10^{-3} (*DNS*)	5.75×10^{-3}
21900	5.78×10^{-3} (*ILES*)	5.75×10^{-3}

Acknowledgments

The research was carried out using the equipment of the shared research facilities of HPC computing resources at Lomonosov Moscow State University.

References

[1] Goloviznin V M, Zaitsev M A, Karabasov S A and Korotkin I A 2013 *New algorithms of computational fluid dynamics for multiprocessor computing systems* (Moscow: Moscow University Press)

[2] Asfandiyarov D G, Goloviznin V M and Finogenov S A 2015 Parameter-free method for computing the turbulent flow in a plane channel in a wide range of Reynolds numbers *J. Comp. Math. Math. Phys.*. **55** 1515 26

[3] Leveque E, Toschi F, Shao L, Bertoglio J P 2007 Shear-improved Smagorinsky model for large-eddy simulation of wall-bounded turbulent flows *J. Fluid Mech.* **570** 491 502

[4] Asfandiyarov D G 2019 *J. Num. Meth. Prog.* **20** 12 20

[5] Asfandiyarov D G, Goloviznin V M and Finogenov S A 2015 *J. Prob. Atom. Sci. Tech. Series: Math. Model. Phys. Proc.* **2** 48 58

[6] Moser R D, Kim J, Mansour N N 1999 Direct numerical simulation of turbulent channel flow up to $Re_\tau = 590$ *Phys. Fluids.* B **11** 943 45

Journal of Physics: Conference Series **1392** (2019) 012031 doi:10.1088/1742-6596/1392/1/012031

Models of dynamics of contamination propagation in surface waters using simulation of hydrodynamical processes of different level of complexity.

Anatoly Cherepanov, Olga Sorokovikova
Nuclear Safety Institute of The Russian Academy of Sciences, 52, Bolshaya Tulskaya Street, Moscow, 115191, Russia
E-mail: lasados@mail.ru

Abstract. This paper uses the ideology of solving hyperbolic equations using the Euler balance-characteristic approach, which allows economically and correctly resolving the dynamics of pollution propagation processes with large concentration gradients characteristic of an accident. The purpose of the study is to develop a three-dimensional model of advection-diffusion in a non-uniform velocity field and non-isotropic field of turbulent diffusion, taking into account some physical processes: gravity sedimentation, sediment accumulation and their dynamics. As well as the implementation of the interaction model with the hydrodynamic module, based on solving the equations of the theory of shallow water in the approximation of Saint-Venant.

1. The equations of advection diffusion and deformation of the bottom

Consider the advection-diffusion equation in differential form for the impurity:

$$\frac{\partial C}{\partial t} + \frac{\partial UC}{\partial x} + \frac{\partial VC}{\partial y} + \frac{\partial (W - w_g)C}{\partial z} = \nabla \left(\nu \cdot \nabla C\right) - q_{sed} \tag{1}$$

where C - is the concentration of radionuclides; U, V, W- components of the flow velocity, wg gravitational speed of sedimentation of suspended particles;; ν- is the coefficient of turbulent diffusion, qsed is the flow of particles to the bottom. Equation (1) is also valid for the dissolved form of radionuclides with w_g=0 q_{sed}=0.

Together with (1), we also consider the one-dimensional bottom deformation equation:

$$\rho_s(1-\varepsilon)\frac{\partial Z}{\partial t} + \frac{\partial p_{tr}}{\partial x} = Q_{in} \tag{2}$$

where Z - is the height of sedimentation of the impurity,ρ_s - is the density of particles, ε - is the porosity of sediments, p_{tr} is the elementary consumption of traction sediment $\frac{kg}{m \cdot s}$, Q_{in} - is the source term.

In turn, the flow of sediment load is proposed to be approximated by the formula proposed by I.I. Levi [2]:

$$p_{tr} = k_0 \left(\frac{u(x)}{\sqrt{gd}}\right)^3 \left(u(x) - v_{beg}\right) \left(\frac{d}{L_z - Z(x)}\right)^{0,25} \tag{3}$$

where k_0 - is a parameter, $u(x)$- is the flow velocity above the bottom, d- is the average particle diameter, g - is the acceleration of gravity, v_{beg} - is the average flow velocity at which particles are captured by the flow and start moving, $L_z - Z(x)$ is the depth flow. In [2], expressions for the calculation at the beginning velocity are also given, for example, the formula of V.N. Goncharov. In this paper, the specification of formulas is not so important, in all the calculations performed, the beginning velocity is constant. Let us consider the problem of sedimentation of suspended particles in a uniform flow under boundary conditions of free flow on the walls of the region, total reflection at the upper boundary and the condition of complete absorption at the lower boundary. Deposition occurs from the bottom settlement cells,q_{sed} is zero everywhere except $z = 0$. Based on this, we arrive at the equation:

$$\rho_s(1-\varepsilon)\frac{\partial Z}{\partial t} + \frac{\partial p_{tr}}{\partial x} = \frac{1}{h_z}\left(\nu\frac{\partial C_{bot}}{\partial z} + w_g C_{bot}\right) \tag{4}$$

where C_{bot} - is the impurity concentration at the bottom, h_z - is the height of the computational cell. The first term on the right (4) is the diffusion flow of particles due to the condition of complete absorption, and the second is the advective flow caused by the vertical movement of particles under the action of gravity.

The vertical derivative of the concentration of suspension in the bottom cell, taking into account the condition of complete absorption will be approximately equal to:

$$\left.\frac{\partial C_{bot}}{\partial z}\right|_{z=0} = \frac{C_{bot} - 0}{h_z/2} = \frac{2C_{bot}}{h_z} \tag{5}$$

Because of the law of conservation of mass:

$$q_{sed}|_{z=0} = \frac{\partial C_{bot}}{\partial t} = -\frac{1}{h_z}\left(\nu\frac{\partial C_{bot}}{\partial z} + w_g C_{bot}\right) = -\frac{1}{h_z}\left(\nu\frac{2C_{bot}}{h_z} + w_g C_{bot}\right) \tag{6}$$

For the time step , the average concentration in the bottom cell due to precipitation processes will change as follows:

$$C_{bot}\left(t+\tau\right) = C_{bot}\left(t\right)\exp\left\{-\tau\left(\frac{2\nu_z}{h_z^2} + \frac{w_g}{h_z}\right)\right\} \tag{7}$$

If we divide the deformation equation (4)

$$\rho_s(1-\varepsilon)\frac{\partial Z}{\partial t} = \frac{1}{h_z}\left(\nu\frac{\partial C_{bot}}{\partial z} + w_g C_{bot}\right) \tag{8}$$

$$\rho_s(1-\varepsilon)\frac{\partial Z}{\partial t} + \frac{\partial p_{tr}}{\partial x} = 0 \tag{9}$$

That level of bottom sediments according to the equation (8) will change as follows:

$$Z\left(t+\tau\right) = Z\left(t\right) + \frac{C_{bot}\left(t\right)h_z}{\rho_s\left(1-\varepsilon\right)}\left[1 - \exp\left\{\tau\left(\frac{2\nu_z}{h_z^2} + \frac{w_g}{h_z}\right)\right\}\right] \tag{10}$$

And the equation (9) is calculated according to the CABARE method in the next paragraph of the paper.

2. CABARE scheme for the nonlinear bottom deformation equation

Simplify the representation of the equation (9) to the form:

$$\frac{\partial Z}{\partial t} + \frac{\partial p}{\partial x} = 0 \tag{11}$$

Before proceeding with the calculation, the initial agreement of the conservative and flow variables is performed in the following procedure:

$$Z_i^0 = \frac{1}{2}\left(Z_{i-1/2}^0 + Z_{i+1/2}^0\right) \tag{12}$$

Fractional indices correspond to the centers of the calculated cells. Integer indices correspond to boundaries.

In the first phase, the calculation of conservative variables on the half-integral layer takes place:

$$\frac{Z_{i+1/2}^{n+1/2} - Z_{i+1/2}^{n}}{\tau/2} + \frac{p_{i+1}^n - p_i^n}{x_{i+1} - x_i} = Q \tag{13}$$

In this equation, $Q = 0$ is the term approximating the source and diffusion. Thus, the values of heights are calculated through time $\tau/2$ in the centers of the calculated cells. In the next step, the initial values of heights are calculated through time at the boundaries of the calculated cells, that is, the flow variables:

$$\tilde{Z}_i^{n+1} = \{ \quad 2Z_{i-1/2}^{n+1/2} - Z_{i-1}^n, U > 0; Z_i^n, U = 0; 2Z_{i+1/2}^{n+1/2} - Z_{i+1}^n, U < 0. \tag{14}$$

Where $U = \frac{\partial p}{\partial Z}$- nominal transfer rate in the equation.

Then differentiating the expression (11) by the x coordinate and substituting (3) :

$$\frac{\partial p}{\partial x} = U\frac{\partial z}{\partial x} = \frac{1}{4}\frac{u^3\left(u - v_{beg}\right)}{\rho_s(1-\varepsilon)h_x g^{3/2} d^{5/4} \left(L_z - Z(x)\right)^{5/4}}\frac{\partial z}{\partial x} \tag{15}$$

Here we consider u to be a constant independent of coordinates. The limit on the time step is determined by the Courant expression:

$$\tau < CFL \cdot \frac{h_x}{U}; \tau < \frac{1}{2} \cdot \frac{h_x}{\max(U)} \tag{16}$$

Where $CFL = \frac{1}{2}$ is Courant number. Next, a nonlinear correction of flow variables is performed.

$$Z_i^{n+1} = MIN\left(MAX\left(\tilde{Z}_i^{n+1}, Z_{min}\right), Z_{max}\right) \tag{17}$$

In these equations, $Zmin$ and $Zmax$ is the estimate of the maximum and minimum height of the impurity at the time layer $n+1$ in the cell from which the extrapolation was made, which depends on the direction of the flow velocity. At the next stage, the concentrations in the centers of the cells on the $n+1$ layer are calculated:

$$\frac{Z_{i+1/2}^{n+1} - Z_{i+1/2}^{n+1/2}}{\tau/2} + \frac{p_{i+1}^{n+1} - p_i^{n+1}}{x_{i+1} - x_i} = Q_{i+1/2}^{n+1/2} \tag{18}$$

The right side is zero in this case. This completes the method step.

3. Model problems of rupture

3.1. Discharge Wave

The initial gap cannot be infinite. It is defined as a linear function with a large slope. To find a solution to equation (15), an integral method was used to find the exact solution. Each point of the profile $X(z)$ moves with speed:

$$U = \frac{1}{4}\frac{u^3\left(u - v_{beg}\right)}{\rho_s(1-\varepsilon)h_x g^{3/2} d^{5/4} (L_z - Z(x))^{5/4}} = \frac{ac}{(L_z - Z(x))^{a+1}} \tag{19}$$

In this way

$$X^{n+1}(z) = X^n(z) + U \cdot \tau \tag{20}$$

The solution obtained by this method will be indicated on the graphs by a red line, and the solution obtained by the CABARE method will be indicated by blue triangular points. Below are the results of the calculation at nominal time points of 200 and 500 s for a problem with the following initial conditions in nominal values.

$$Z_0 = \left\{100, if x > \delta; \frac{50}{\delta}x + 50, if x \subset [0,\delta]; 50, if x < 0\right. \tag{21}$$

where $\delta = 1$.

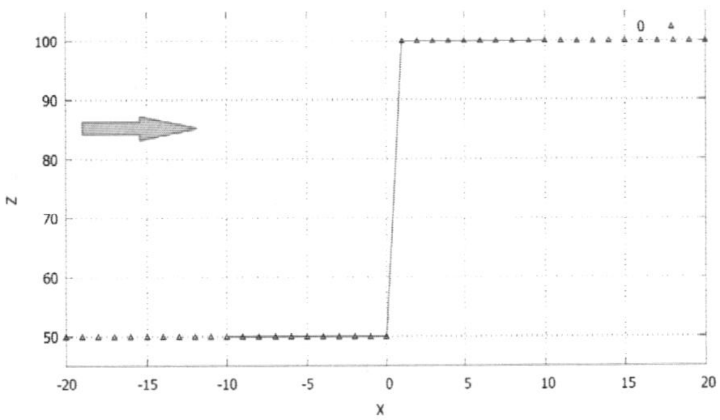

Figure 1. Initial distribution

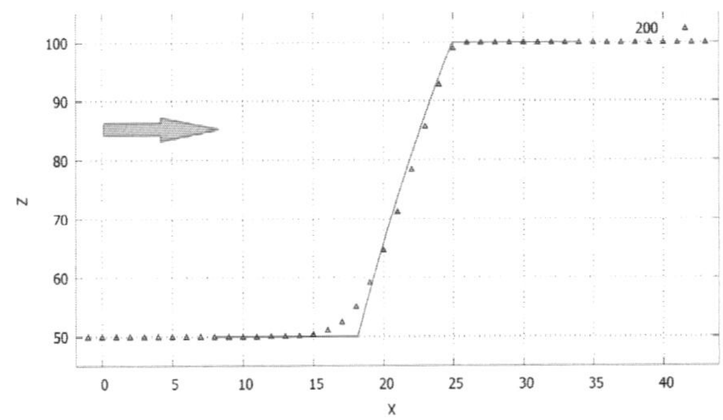

Figure 2. Result of the calculation at 200 s

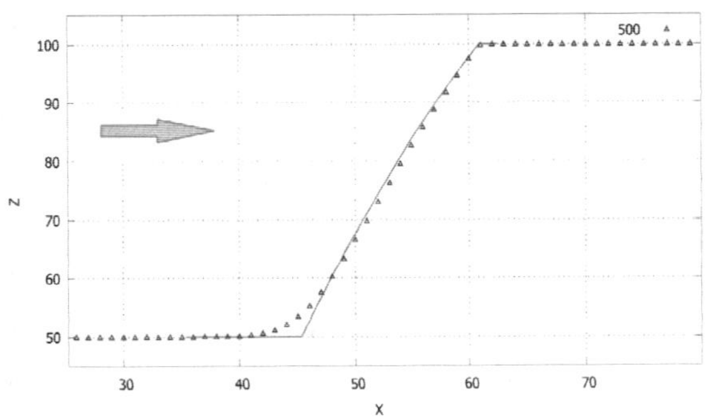

Figure 3. Result of the calculation at 500 s

3.2. Shock wave

We represent equation (11) in a divergent form:

$$\int_G \left(\frac{\partial p}{\partial x} + \frac{\partial Z}{\partial t} \right) dx dt = \oint_{\partial G} Z dx - p dt \quad (22)$$

$$(Z_R - Z_L)\,\Delta x - (p_R - p_L)\,\Delta t = 0 \quad (23)$$

$$V_{gap} = \frac{\Delta x}{\Delta t} = \frac{p_R - p_L}{Z_R - Z_L} = \frac{c}{Z_R - Z_L}\left(\frac{1}{(L_z - Z_R)^a} - \frac{1}{(L_z - Z_L)^a} \right) \quad (24)$$

Expression (24) represents the velocity of the gap.

Below are the results of the calculation at nominal time points of 150 and 450 s for a problem with the following initial conditions in nominal values.

$$Z_0 = \left\{ 50, if x > \delta; 100 - \frac{50}{\delta} x, if x \subset [0, \delta]; 100, if x < 0 \right. \quad (25)$$

where $\delta = 1$.

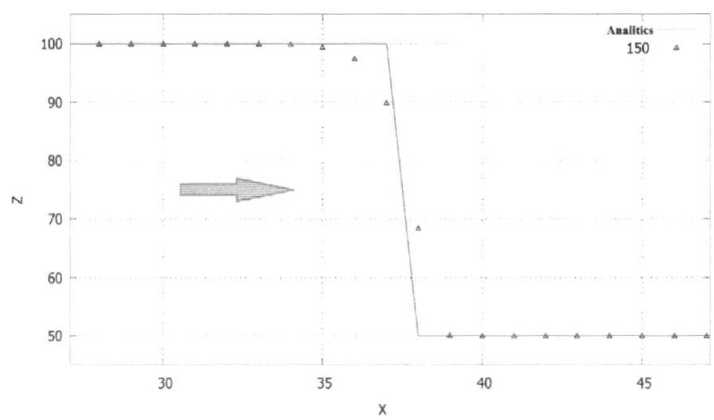

Figure 4. Result of the calculation at 150 s

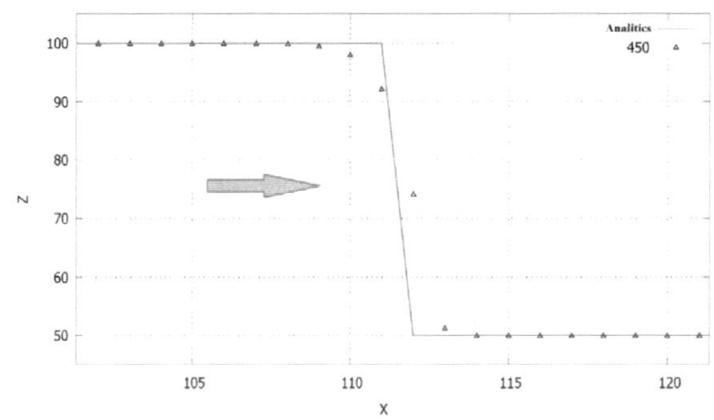

Figure 5. Result of the calculation at 450 s

3.3. The task of precipitation of impurities on the bottom, taking into account the dynamics

Consider the complex problem of advection-diffusion of impurities in the aquatic environment and the deformation of bottom sediments. We take the initial distribution of concentrations in the form of three homogeneous strips of different shapes at different distances from each other. Let be

$$C_0 = \left\{ 1, if x \subset [x_1, x_2] \bigcup [x_3, x_4] \bigcup [x_5, x_6]; 0, else \right. \quad (26)$$

$$Z_0(x) = 0 \quad (27)$$

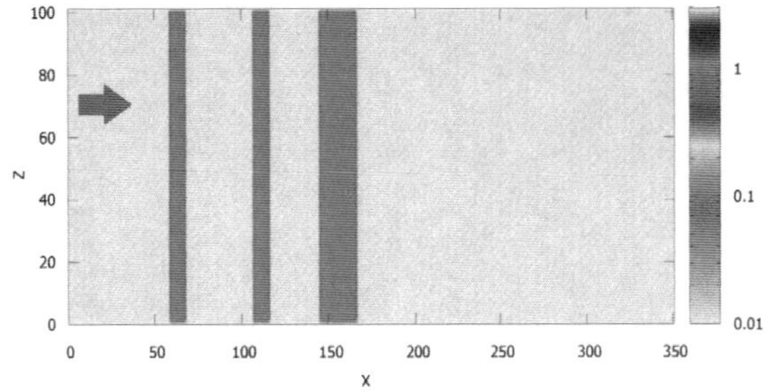

Figure 6. Initial concentration profile

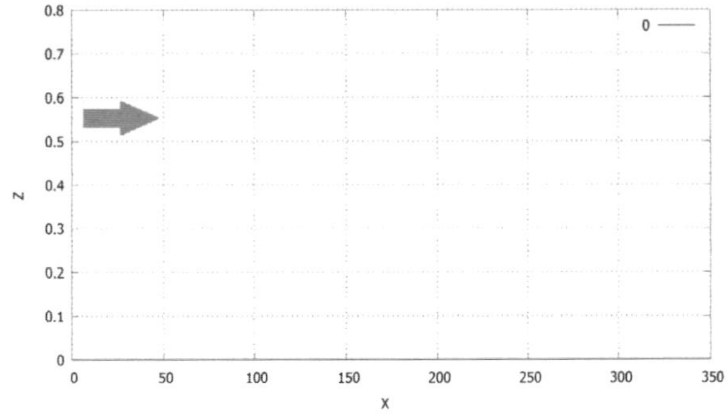

Figure 7. Initial profile of deposition heights

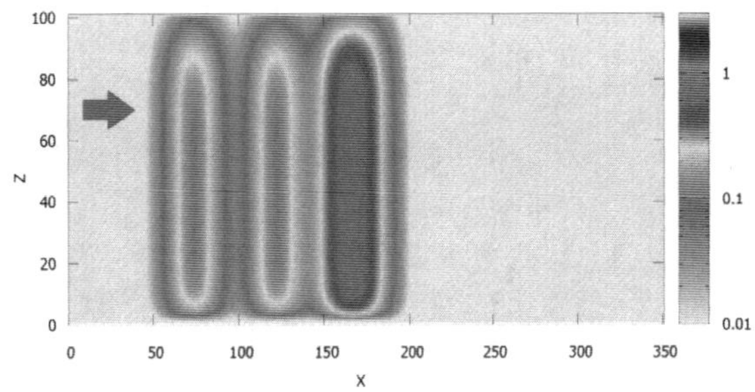

Figure 8. Result of the calculation at 10 s

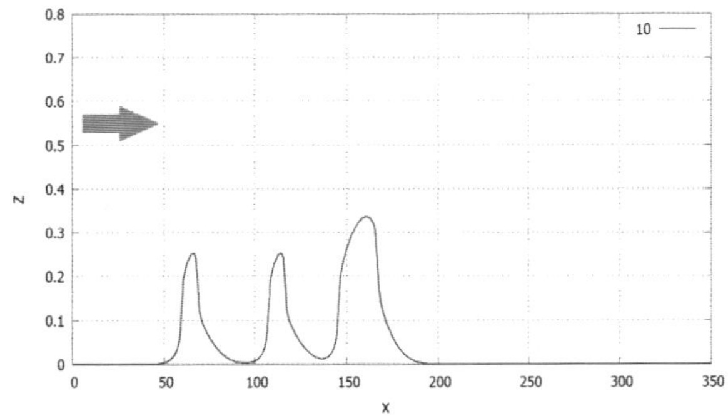

Figure 9. Result of the calculation at 10 s

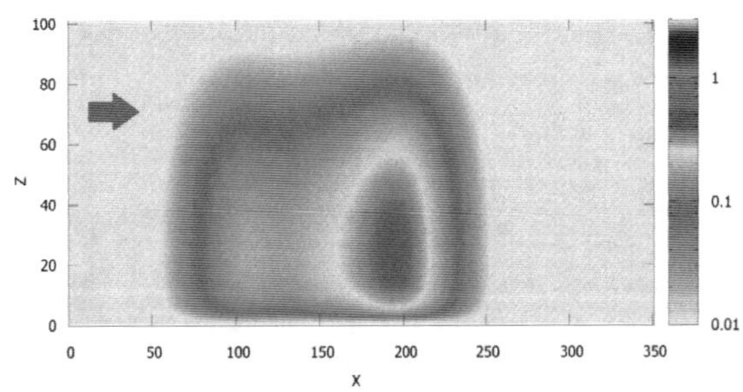

Figure 10. Result of the calculation at 40 s

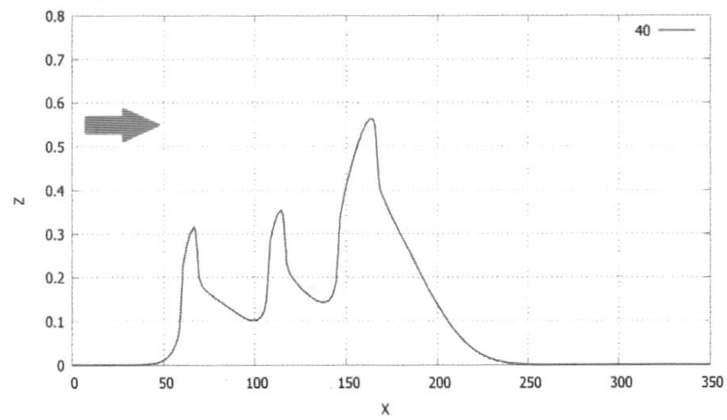

Figure 11. Result of the calculation at 40 s

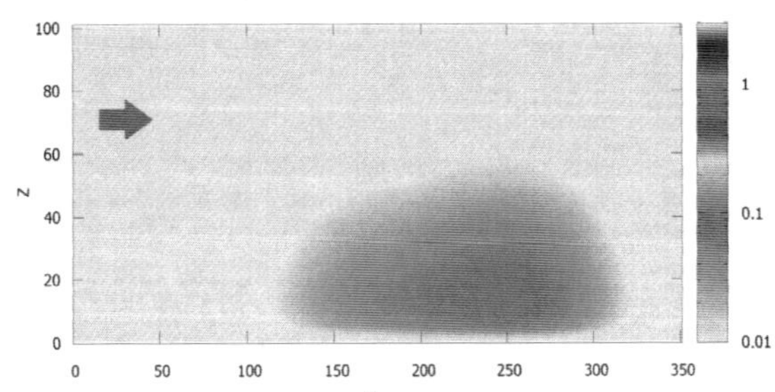

Figure 12. Result of the calculation at 100 s

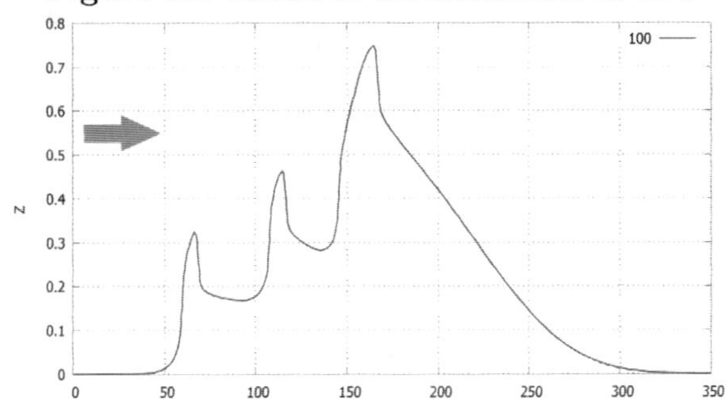

Figure 13. Result of the calculation at 100 s

References

[1] Goloviznin V.M., Zaitsev M.A., Karabasov S.A., Korotkin I.A. 2013 *New Algorithms of Computational Fluid Dynamics* (Moscow:Moscow State University P

[2] A.V. Karaushev 1977 *Theory and Methods for* *ıtion of River Sediments* (Publishing House - Gidrometeoizdat)

Journal of Physics: Conference Series **1392** (2019) 012032 doi:10.1088/1742-6596/1392/1/012032

Probabilistic models of pollution of the surface waters of large marine areas

Olga Sorokovikova, Dmitry Dzama, Danil Asfandiyarov
Nuclear Safety Institute Russian Academy of Science, Moscow, Russia
E-mail: dmitry.dzama@yandex.ru

Abstract. Version of Monte-Carlo model for calculation of propagation of contamination in large marine areas with taking into account a coastline is presented. Based on long standing database of sea currents, bathimetry, and mixed layer depth, the probabilistic model of zoning marine areas by concentration of pollutant is offered. In accordance with modern IAEA recommendations general principles of estimation of radiation situation on radiation hazardous object in normal expluataion mode, conclusion of normal expluatation mode, or in accident mode exists. Practical methodology of safety case and tactics of monitoring of state of radiation hazardous object should rely not only on measuring, but also on model calculations and forecasts. Optimal tactics of monitoring includes definition of probabilistic picture of possible pollution different marine areas assuming hypothetical leak of pollutant with violation of defence barriers. One way of building such probabilistic pictures is execution of multi-variant calculations using long standing databases. Different calculations in multi-variant mode executes assuming different accident start time, in advance models multi-variant dynamic of marine area polluton. Not only maximum values of pollution to be interested, but and statistical estimations with different confidence levels (for filtration of rarery big values) for hypothetical leak of pollutant. In the base of the numerical algorithm of proposed technique lies Monte-Carlo method using lagrangian particles with non-zero sizes.

1. Introduction

In the areas of Arctic (Arctic Ocean) and Far East (Pacific Ocean) radiation hazardous objects, containing spent nuclear fuel, liquid or solid radioactive waste, flooded in the time period from 1946 to 1993 years, is localaized. So, Arctic and Far East should be considered as potentially dangerous regions with radiation risk factor. Accident situations has very low probability, but for analys of radiation state and radiation safety, model calculations and forecasts of possible consequences of hypothetical accidents for terrestrial and marine environment, should be executed.

Last few years in NSI RAS develops the specialized computer code for calculation of propagation of radioactive pollution spot in marine area of Arctic or Pacific Ocean. Two version of model and computer code is developing. First version – is quasi three-dimensional model, intended for calculation consequences of accidents, occurring in upper layer of the ocean, higher than mixed layer depth. Second version – is fully three-dimensional model, intended for accident, occuring in arbitrary depth. In this paper both models is presented. Both computer codes allows define probabilistic characteristics of levels of possible radioactive contamination and location of most important marine areas for choose of places of monitoring and sampling points. Detection of crucial marine areas, where expected the highest levels of contamination and measuring acts aimed at definition of content of radionuclide in this areas, may serves as a signal (in case of increasing levels of content) for carriage out further events for controlling radiation situ, and about possible violation of defense barriers. Probabilistic approach to modeling consequences of leak of radioactive pollution in coastal marine areas of Arctic and Pacific Oceans is needed owing to seasoning variability of the ocean currents and mixed layer depth, uncertainty in accident beginning time. Developed probabilistic approach lays in carriage out big amount of calculations, each with different moment of the beginning of accident and using long standing, variable in time, and heterogeneous in space, characteristics of the ocean flow (and mixed layer depth). Ocean flow characteristics cannot be typify, in contrast of atmospheric conditions (wind velocity and direction, stability, precipitation, ground type), which, in case of short-term atmospheric releases, can be considered homogeneous in space and constant in time.

2. Available databases to provide input parameters for developed models

For big marine areas in Arctic and Pacific Oceans long standing databases with ocean flow characteristics with acceptable spatial and time resolution come in sight relatively recently. Four-dimensional databases of ocean flow may be received by reanalysis of common circulation of ocean and atmosphere, with assimilation of measurement dataset. The problem is that the measurements in ocean area carriage out much rarely, than in the atmosphere. However, situation changed for the better last years.

Brief description of databases, used in the computer code, provides in this chapter. Source database for building coastline in quasi three-dimensional version of model is *ASTER GDEM* [1]. This database contains high resolution data of height of the Earth relief (topography). Spatial resolution of the data is $1/3600°$. Missing values restores by the database with lower spatial resolution ($1/60°$) – *Smith and Sandwell surface relief* [2]. *Smith and Sandwell surface relief* database contains topography and bathimetry data, while *ASTER GDEM* – only topography. Source database for mixed layer depth in quasi three-dimensional version of model is *ORAS4* [3]. Mixed layer depth corrects by the depth of the ocean: it must be less or equal to deep in current point. Spatial resolution of the data is $1°$. Horizontal components of the velocity for Arctic Ocean are taken form the database *ORAS4*, for Far East (Pacific ocean) - *FRA-JCOPEC* [4]. Spatial resolution of *FRA-JCOPEC* is $1/12°$.

Detailed description of these databases may be founded by specified above references.

3. Quasi three-dimensional model for accidents in upper mixed layer

In this section the quasi three-dimensional version of model is presented. This model is intended for modeling propagation of pollutant in upper ocean layer. In this version of model is assumed, that pollutant uniformly mixing in the upper ocean layer, until mixed layer depth. Mixed layer depth is limited by seasonal thermocline, turbulent mixing through this boundary is negligible small, but upper water layer is subject to intensive turbulent mixing. But there are exists the vertical advection through this boundary due to horizontal macroscale divergent space-heterogeneous ocean flows. In this version of the model it is assumed, that pollutant propagate in soluble form or as fine aerosol. The model takes into account the following processes:

- horizontal advection with ocean flow in the upper layer;
- non-uniform horizontal turbulent diffusion (overgrid and subgrid);
- non-uniform and seasonaly variable 2D field of mixed layer depth field;
- horizontal macroscale ocean currents divergence, causing macroscale downvellings in the lower layers.

For modeling of horizontal subgrid turbulent diffusion the algebraic Ozmidov model is used [5]. In this model turbulent diffusion depends only from the linear scale of the pollutant spot. For modeling of horizontal overgrid turbulent diffusion the algebraic Smagorinsky model is used. In this model turbulent diffusion calculates using shear flow deformations. The main differential equation in the model – is two-dimensional advection-diffusion equation in spherical coordinate system:

$$\frac{\partial HC}{\partial t}+\frac{\partial(uHC)}{R\cos\theta\partial\phi}+\frac{\partial(vHC)}{R\partial\theta}=\frac{\partial}{R^2\cos^2\theta\partial\phi}\left(K\frac{\partial HC}{\partial\phi}\right)+\frac{\partial}{R^2\partial\theta}\left(K\frac{\partial HC}{\partial\theta}\right)+Q(\phi,\theta,t) \quad (1)$$

In the equation (1):

- ϕ, θ, t – longitude, lattitide, and time (sec);
- R – Earth radius (m);
- H – mixed layer depth, depends from time, longitude and lattitude;
- C – volume concentration of the pollutant, depends from time, longitude, and lattitude (Bq/m^3);
- u, v – zonal and meridional velocities, depends from time, longitude, and lattitude (m/s);
- K – turbulent diffusion coefficient (m^2/s);
- Q – power of the pollutant source ($Bq/s/m^2$).

Reducing of H for time integration step do not lead to increasing the concentration of pollutant.

Equation (1) solves in lagrangian variables, using Monte-Carlo method. Pollution spot represents as a big amount of points. Each point moves in accordance with ocean flow, and subject to random (stochastic) displacement, which models turbulent diffusion. Volume concentration of points (count of points in a unit volume) associates with volume concentration of pollutant. Each point has a certain amount of pollutant (A). If pollutant is radioactive contamination, then this amount of pollutant is radioactivity of a point. Regular and stochastic displacement of a point is described by a system of Ito differential equations (in Cartesian coordinate system): $\dot{x}=u+u'$, $\dot{y}=v+v'$. Omitting intermediate mathematical explanations, give the final (finite difference) formulas, which describes displacement of a point:

$$\phi^{n+1}=\phi^n+\frac{u\tau}{R\cos\theta^n}+\xi\frac{\sqrt{2K\tau\beta}}{R\cos\theta^n} \quad (2)$$

$$\theta^{n+1}=\theta^n+\frac{v\tau}{R}+\eta\frac{\sqrt{2K\tau\beta}}{R} \quad (3)$$

In equations (2) and (3) τ – time integration step, ξ and η – random variables with normal distribution, ϕ^n and θ^n – longitude and lattitude of a point in n-th time step, β – is the model parameter form 0 to 1. Explain meaning of β below.

The lack of this way of definition volume concentration of pollutant is the need of using a large number of calculation points for a smooth result (without strong fluctuations). Thus, has been developed hybrid method of *wondering clouds*. In this method, a point is represented as a cloud, which has horizontal non-zero size σ. Volume concentration near the center of a cloud obeys Gauss distribution:

$$G(x,\mu,\sigma)=\frac{1}{\sqrt{2\pi}\sigma}\exp\left(-\frac{(x-\mu)^2}{2\sigma^2}\right) \quad (4)$$

$$C(\phi,\theta)=\frac{A}{H}G\left((\phi-\phi^n)R\cos\theta^n,0,\sigma\right)G\left((\theta-\theta^n)R,0,\sigma\right) \quad (5)$$

In equation (5) $C(\phi,\theta)$ – is concentration in point (ϕ,θ). In this version of model the size of a point enlarge in each time integration step: $(\sigma^2)^{n+1}=(\sigma^2)^n+2(1-\beta)K\tau$. Thus, β is the tuning model parameter. If β equal 1, then realized model with points. If 0 – model with clouds with non-zero size, turbulent diffusion only enlarge size of a cloud. In the presented model $\beta = 0,9$. If center of a cell is located farther then 3σ, $C(\phi, \theta) = 0$. Concentrations is calculated on uniform longitude-lattitude computational grid. Each computational cell is marine area, limited two values of longitude and two values of lattitude.

Consider downvelling phenomena. This phenomena leads to decreasing of the pollutant in a cloud A (radioactivity) in accordance with equation:

$$\dot{A} = -\frac{A \cdot W}{H} \tag{6}$$

In equation (6) W – is vertical component of the flow, induced by its non-zero macro-scale horizontal divergent. Then, for the time integration step τ radioactivity of the cloud decrease in accordance with equation:

$$A^{n+1} = A^n \exp\left(-\frac{W\tau}{H}\right) \tag{7}$$

Further consider the turbulence models. The Ozmidov turbulence model is used to parametrize sub-grid turbulence:

$$K_O = 0,1 \cdot \epsilon^{\frac{1}{3}} \cdot L^{\frac{4}{3}} \tag{8}$$

In equation (8) ϵ – turbulent energy dissipation rate (m^2/s^3), L – linear scale of pollutant spot. If $L < 10\ km$, then $\epsilon = 10^{-8}\ m^2/s^3$, else – $\epsilon = 10^{-9}\ m^2/s^3$. The linear scale of pollutant spot calculates as standard deviation from the center of mass of all clouds, where mass of a cloud is its radioactivity (A).

The Smagorinsky model [6] is used to parametrize over-grid turbulence. Consider the flow deformation rate tensor:

$$S_{ij} = \frac{1}{2}\left(\frac{\partial v_i}{\partial x_j} + \frac{\partial v_j}{\partial x_i}\right) \tag{9}$$

In equation (9) $v_1 = u$, $v_2 = v$. Turbulent diffusion coefficient is defined as below:

$$K_S = C_S^2 \Delta^2 \sqrt{2 S_{ij} S_{ij}} = C_S^2 \Delta^2 \sqrt{2\left(\frac{\partial u}{\partial x}\right)^2 + 2\left(\frac{\partial v}{\partial y}\right)^2 + \left(\frac{\partial u}{\partial y} + \frac{\partial v}{\partial x}\right)^2} \tag{10}$$

In relation (10) C_S – is the model dimensionless parameter from 0,6 to 0,28. In described model $C_S = 0,28$.

In described model for a cloud choosing maximum from K_O and K_S.

Distinctive feature of the model is modeling of interaction of clouds with coastline. This interaction subdivided on two processes.

In first process a cloud considers as a point with zero size. Trajectory of a cloud on each time integration step – is a line segment, connecting initial and final position of a cloud. Final position calculates as summa of advection and turbulent diffusion. Coastline in the model considers as aggregate of segments – common sides of marine and terrestrial cells. Thus, model coastline – is broken line, whose segments oriented along longitude or latitude. If trajectory segment of a cloud intersects coastline, then segment specular reflects.

In second process clouds considers as a circle with radius of 3σ. It is necessary to provide zero concentration in terrestrial cells, which located in a circle. For this it is necessary to correct the radius of a circle thus, that in a circle do not appears terrestrial cells. Let the distance from circle center to nearest terrestrial cell be equal to r_t. If $3\sigma \geq r_t$, then σ becomes equal to $r_t/3$.

4. Three-dimensional model for accidents in arbitrary deep

In three-dimensional model there is no assumption about instantaneous mixing in vertical up to mixed layer depth. Vertical turbulent diffusion models explicitly, using Munk-Anderson algebraic model. Besides, in this version of model ocean flow depends not only from longitude and latitude, but and from deep. Thus advection of the pollutant spot defines by three-dimensional field of horizontal flow velocities, vertical advection do not taking into account.

Consider Munk-Andersen vertical turbulence model [7]. Turbulent diffusion coefficient in this model depends on vertical gradient of horizontal flow and sea water density:

$$Ri = \frac{g}{\rho_{00}} \frac{\partial \rho}{\partial z} \frac{1}{\left(\frac{\partial u}{\partial z}\right)^2 + \left(\frac{\partial v}{\partial z}\right)^2} \tag{11}$$

If $Ri < 0$, then $K_{MA} = 0,1$. Else

$$K_{MA} = \alpha + \frac{\beta}{(1+\gamma Ri)^n} \tag{12}$$

In formulas (11) and (12): Ri – Richardson number, g - acceleration of gravity (m^2/s), ρ – density of marine salinity water (kg/m^3), ρ_{00} – density of pure water at 0°C, z - deep (m), $\alpha = 5 \cdot 10^{-6}$m^2/s, $\beta = 5 \cdot 10^{-3}$m^2/s, $\gamma = 5$, $n = 3$. Z axis directs down. Density $\rho = \rho(T,S)$ – is the function of temterature T (°C) and salinity S ($^\circ/_{\circ\circ}$):

$$\rho(T,S) = \rho_p(T)\left(1 + \frac{S}{1000}\right) \tag{13}$$

In relation (13) $\rho_p(T)$ – is density of pure water at temperature T. $T = T(z)$, $S = S(z)$. For calculation of $K_M A$ three-dimensional fields of temperature and salinity should be available.

In this version of model clouds have not only horizontal, but vertical size too. As in the quasi three-dimensional model, here arises the problem of interaction of clouds and ocean floor. Floor of the ocean is the three-dimensional analog of coastline in quasi three-dimensional model. In general algorithm of interaction with ocean floor repeats algorithm of interaction with coastline in quasi three-dimensional version of model. First sub-problem is calculation of trajectory of a cloud under the assumption of specular reflection from faces of cells. Trajectory – is three-dimensional broken line. Second sub-problem is correction of sizes of a cloud thus, that no one cell under the floor of the ocean do not lay in cuboid with sides $3\sigma_{lon}$, $3\sigma_{lat}$, $3\sigma_z$, where $3\sigma_{lon}$ – is size of a cloud along longitude, $3\sigma_{lat}$ – along latitude, $3\sigma_z$ – along vertical. Unlike quasi three-dimensional model, here instead circle considering cuboid, and sizes of cuboid correcting independently. Both models have been parallelized and may run using arbitrary processor count.

5. Examples of application of the model

Give an example of a probabilistic picture of pollutant spot in coastal marine area of Japan Sea – in the bay Razboinik (Far East, Russia). Location of the source – 132,36° of east longitude, 42,89° of north lattitude. Radioactivity of the source – $1,87 \cdot 10^{15}$ Bq of ^{90}Sr, $6,22 \cdot 10^{16}$ Bq of ^{137}Cs. Release of radioactivity is instantaneous. Forecast time – 3 month. Multi-variant calculation has been carride out. Variable parameter – is accident start moment. It varyed from 1-jan-2008 to 31-jan-2008 using step of 1 day (total number of calculation is 31). On the Fig. 1 presented time integrated concentration (TIC) field. TIC value in each point is equal maximum value over all calculations at the level of trust 95%.

Consider the example of application of three-dimensional version of model. Location of the source – 132,69° of east longitude, 41,54° of north lattitude. Radioactivity of the source – 10^{11} Bq of ^{137}Cs. Accident start moment – 12:00 1-jun-2007, source duration – 5 days, source deep – 19 m, forecast time – two month. Vertical size of the computational area is 300 m. TIC field calculated on the same deep as source – 19 m. This field is presented at Fig. 2.

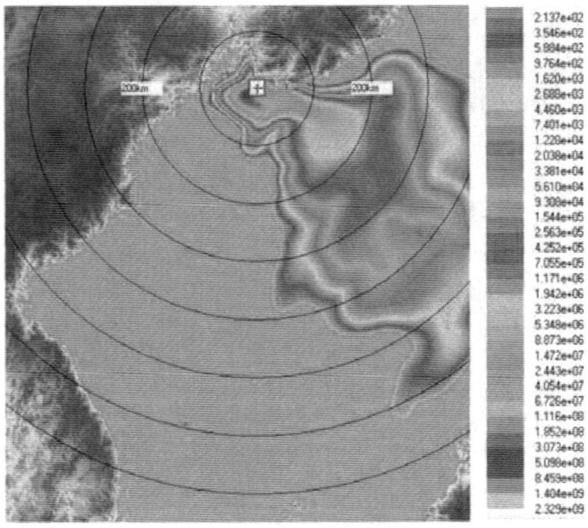

Figure 1. *TIC* – result of calculation using quasi three-dimensional model.

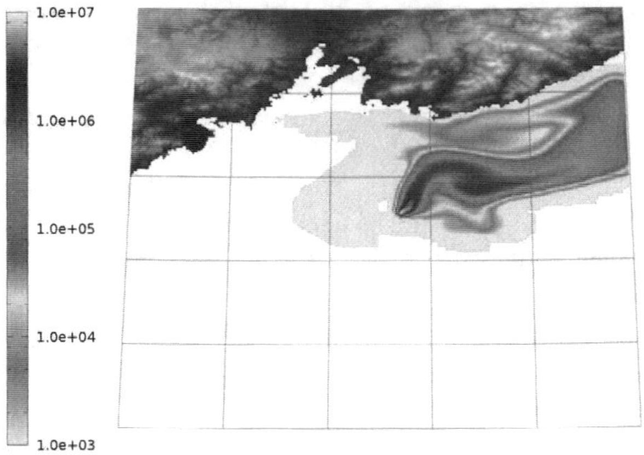

Figure 2. *TIC* – result of calculation using three-dimensional model.

References

[1] ASTER global digital elevation map announcement. `https://asterweb.jpl.nasa.gov/gdem.asp`
[2] Walter H. F. Smith and David T. Sandwell. Global Sea Floor Topography from Satellite Altimetry and Ship Depth Soundings. Science 26 Sep 1997. - Vol. 277, - Issue 5334, - P. 1956-1962. - DOI: 10.1126/science.277.5334.1956
[3] ECMWF Ocean Reanalysis System 4 (ORAS4). `http://apdrc.soest.hawaii.edu/datadoc/ecmwf_oras4.php`
[4] The FRA-JCOPE2 17 year NW Pacific Ocean reanalysis data. `http://apdrc.soest.hawaii.edu/datadoc/fra-jcope2.php`
[5] Diffusion of Contaminants in the Ocean. R.V. Ozmidov, 1990, ISBN 978-0-7923-0611-5
[6] Smagorinsky, J., General circulation experiments with the Primitive Equations. Monthly Weather Review, 1963. 91(3): p. 99-164
[7] Munk, W. H. and Anderson, E. R. Notes on a theory of the thermocline / W.H. Munk, E. R. Anderson // J. Mar. Res., 7, –1948. – p. 276

Data Mining using the Theory Cooperative Games in C-core to form clusters

Anastasia A. Egorova[1]
[1]Mirny Polytechnic Institute (branch) of Ammosov North-Eastern Federal University (NEFU)
E-mail: nastyaegorova@mail.ru

Abstract. For data analysis, objects can be clustered in a variety of ways and methods. The tree of possible options appears. To make decisions, we use the methods of game theory. The game-theoretic model can be considered as a coalition game. The characteristic function is considered as a function that depends on the degree of similarity of objects. Thus, we will form clusters. The game is considered as a differential control game. To make a decision, we will form a highly dynamically stable solution in the C-core.

Keywords: data mining, clustering, cooperative games

Because solving the classification problems, the features that characterize the groups of objects of the studied data set - classes was founding; according to these features, the new object can be attributed to a particular class. To solve the problem of classification can be used methods are nearest neighbor, k-nearest neighbor; Bayesian Networks [1]; induction of decision trees; neural networks [2]. Clustering is a logical extension of the idea of classification. The peculiarity of clustering is that the object class was not initially presenting. The result of clustering is the division of objects into groups.

Recently, active development of new clustering algorithms capable of processing very large databases is underway. They focus on scalability. These algorithms include a generalized representation of clusters [3], as well as the selection and use of data structures supported by the DBMS. Algorithms in which hierarchical clustering methods integrated with other methods. When clustering objects of different nature must was dividing into groups so that within each group were similar objects. The result strongly depends on many factors [5]. When analyzing the data, there are situations of uncertainty when making a decision on classification to a particular group [7], [8].

A new approach is considered in [6] Bure V. M., Staroverova K. V., where clustering is considered as a cooperative game with a coalition structure. In [22] Zenkevich N. A., Petrosyan L. A. the Classical Theory of Cooperative Games simulates a single cooperative interaction of the parties To take into account the time factor, it is necessary, to use yhe theory of cooperative dynamic and differential games. The principles of optimality of the classical cooperative game theory are not dynamically stable and, therefore, inapplicable in strategic management. The problem of time consistency of cooperative solutions is considered. Which is considered including the procedure (PRD) , the regularization of the game in order to form a highly dynamic stable game. The distribution procedure of the game division Petrosyan L. A is considered by in [22].

Authors F. Kidland and E. Prescott [15] solved a special problem. Dynamic stability (temporal consistency) of the optimality principles in differential games was studying in detail in the works of game theory specialists. A. Haurie noticed a temporary failure of the arbitration scheme Nash when using it as the principle of optimality in differential games. Petrosyan L. A. in 1977 mathematically formalized the concept of dynamic stability (temporal consistency), introduced the concept of "division distribution procedures" for cooperative solutions [17]. The paper [Tolwinski, Haurie, Leitmann, 1986] investigated the cooperative equilibrium in differential games, when the threat system ensures the development of the game on the cooperative path. Further, in [18] the detailed analysis of dynamic stability in cooperative differential games was carrying out and the regularization method for construction of dynamically stable (time-dependent) solutions was offering.

Suppose there is a certain amount of data. On a certain basis, we combine data. For example, there are data, which relatively speaking, was calling into question where they were considering as one group or another. There is a conflict situation with the environment. Consider from the point of view of the game-theoretic approach. We consider each Association as a player and see whether it is profitable for them to unite in a coalition. Here, the win function can represent the degree of similarity or the degree of data proximity. Thus, we determine the characteristic function of the coalition game. Find a solution to this game, in which the players benefit is in the coalition (group). The decision of the game must be sustainable. Find the C-core so that it is strongly dynamically stable. Thus, we will approach the Pareto optimal solution of the game. The most common concept of optimal play of many individuals is Nash equilibrium.

The Pareto optimal solution (PO - solution) is a cooperative solution in a broad sense, since the adoption of such a decision requires a coordinated choice of strategies by all players and has the property of group rationality in the strategic behavior of players. The concept of Pareto optimality is equally true for games with transferable and games with transferable winnings (winnings players cannot pass each other). A typical representative of the Pareto optimal solution is the Nash arbitration award [16]. The Nash award (NB - decision)is a cooperative decision in the broad sense of the Pareto case of an optimal solution. Another representative of the Pareto optimal solution is the arbitration decision of Kalai-Smorodinsky[14].

Unfortunately, the condition of strong dynamic stability ceases to be fulfilled when considering more complex optimization problems. This is the case for multi-criteria optimization.

There are several clusters. Assign a number to each cluster. There may be a dependency between clusters or a fuzzy defined set of one cluster that may overlap with another set. Clusters consider as players. Consider a cooperative version of the game in which players enter into an agreement (functions of dependence on each other) on joint actions. The game starts at t_0 from state x_0 and ends at T. The phase variable changes according to the dynamics.

$$x^{'} = \sum_{i=1}^{n} u_i = f(x(t), u(t)), x \in R^m, u \in U \subset R^l, t \in [t_0, T], x(t_0) = x_0, \tag{1}$$

where u_i is the control of player i.

Let be a set of players $\|N\| = n$. The win function is:

$$K_i(x_0, T - t_0, u) = \int_{t_0}^{T} (D_i(u(\tau)) - R_i(x(\tau))d\tau, \tag{2}$$

D_i-positive effect, R_i-negative impact on the game:

$$D_i(u_i(\tau)) = (b_i - 1/2u_i(t))u_i(t), R_i(x(\tau)) = d_i x(t), d_i b_i > 0. \tag{3}$$

It is necessary to find the optimal control when maximizing (minimizing) K_i. Let us find the optimal control $u^* = (u_1^*, u_2^*, ..., u_n^*)$ delivering the maximum to the total gain $\sum_{i=1}^{n} K_i \to max$.

We have the following optimization task:

$$\sum_{i=1}^{n} K_i(x_0, T-t_0; u) = \sum_{i=1}^{n} \int_{t_0}^{T} ((b_i - 1/2u_i)u_i - d_i x)dt \to max. \tag{4}$$

To solve the problem in the class of program strategies, we use the Pontryagin maximum principle. Get $u^*(t)$ in order to prevent a violation of the agreement, a payment scheme was proposing, called the procedure for distributing the division, which ensures the implementation of the decision a C-core, which fiending was using often as a solution. The C-core in the subgame $G_V = (x^*(t), T-t)$ is defined as a subset of the set of divisions

$$C(x^*(t), T-t) \subset L(x^*(t), T-t), C(x^*(t), T-t) = \{\{a_i^t\}_{i=1}^{n}\}, t \in [t_0, T],$$

such that:

$$\sum_{i \in N \setminus S} a_i^t \le V(x^*(t), T-t; N) - V(x^*(t), T-t; N \setminus S), \forall S \subseteq N. \tag{5}$$

If for games with such a characteristic function the condition

$$V(x_0, T-t_0; S) + V(x_0, T-t_0; P) \le V(x_0, T-t_0; S \cup P) + V(x_0, T-t_0; S \cap P), \tag{6}$$

then, the C core is not empty. The characteristic function will be as follows:

$$V(x^*(t), T-t; i, j) = -d_i(T-t)x(t) + 1/2(T - t(b_{ij} + d_{ij}d_s(T-t)^2 - b_s d_{ij}(T-t)).$$

Value of the game

$$V(x_0, T-t_0, N) = \alpha_1(t, T)x(t) + \alpha_0(t, T).$$

Bayesian regularization methods, neural network model regularization (MLP, RBF), regularization method Lavrent'eva MM.

A second step for certainty the creation of groups it is proposed to introduce regularization by using the DWP (distribution division) the Procedure for the allocation of the division, first proposed in [17]. Petrosyan L. A. proved that the procedure of division distribution leading to a dynamically stable cooperative solution exists and is realizable [18].

We are interested in a dynamically stable sub-core based on the distribution procedure of the division. (Petrosyan L.A.)

Lemma 1: The set $B(t) \neq \emptyset, \forall t \in [t_0, T]$ if and only if the characteristic function $V(x^*(t), T-t; S), S \subseteq N, t \in [t_0, T]$ is a continuously differentiable function with respect to t, for which the following condition was fulfilling along the cooperative trajectory:

$$d/dt V(x^*(t), T-t; N) \le d/dt[V(x^*(t), T-t; N \setminus S) + V(x^*(t), T-t; S)]. \tag{7}$$

Lemma 2: The set $C(x^*(t), T-t) \neq \emptyset t \in [t_0, T]$ if and only if the characteristic function $V(x^*(t), T-t; S), S \subseteq N$ is a continuously differentiable function with respect to $t, t \in [t_0, T]$, for which inequality (7) holds.

Theorem: Let $C(x^*(t), T-t) \neq \bar{C}(x^*(t), T-t), t \in [t_0, T]$. Then the sub core C $(x^*(t), T-t) \subseteq C(x^*(t), T-t)$ is strongly dynamically stable in the game $\Gamma_V(x_0, T-t_0)$.

Algorithm for constructing a highly dynamically stable subcore:

1. Optimal controls $u^* = (u_1^*, ..., u_n^*)$ maximizing the total gain of all players, as well as the corresponding cooperative trajectory $x^*(t)$ find.
2. The characteristic function $V(x^*(t), T-t; S), S \subseteq N$ calculated.
3. The superadditivity condition for the fulfillment characteristic function is check.
4. The condition for non-emptiness of the C-core is check.

5. The fulfillment of a sufficient the condition for a non-empty subcore is check.
6. The set B (t) is constructing.
7. The subcore $C(x^*(t), T-t)$ is constructing.

That is, it is necessary to give some "point of growth" for the data to form a cluster. The paper provides an example.

Let us consider an example with the introduction of points-elements of "growth" (we make regularization). Let $n = 2$. We will look for the solution in the game in the control class $u_i(t)$. It is necessary to find the optimal control while maximizing the total gain. The initial data was dividing into 3 classes: in class 1 -1, in the second - 0.9, in the third class 0.

If, for example, 0.5 was defending, then we will introduce the element of "fluctuations", "provocation", points-elements of "growth", matrix A. Add the value $\delta = 0.2.$, matrix B. We obtain the data

$$A = \begin{pmatrix} 1.0 & 0.0 & 0.0 \\ 0.9 & 1.0 & 0.9 \\ 0.5 & 0.9 & 1.0 \end{pmatrix}, B = \begin{pmatrix} 1.2 & 0.2 & 1.1 \\ 1.1 & 1.2 & 1.1 \\ 0.7 & 1.1 & 1.2 \end{pmatrix}$$

From which it is clear that the data was clustering into two clusters. Thus, it can was arguing that the introduction of points of "growth" allows you to determine which cluster belongs to the number. In this paper, a constructive approach was considered, which allows the selection of functions that can be used as PRDs to build a highly dynamically stableprinciple of optimality (subcore) belonging to the C - core.

Thus, we make some "payments" and the game acquires some certainty and dynamic stability over time. It is important for multi-criteria problems.

It was proposing to use a two-stage decision-making model. At the first stage, a cooperative division is selecting in accordance with the chosen principle of optimality. At the second stage, it is necessary to regularize this decision, using a special procedure for the distribution of the division. As a result, a system of "payments" was forming, which should be implementing in the course of the project. The final cooperative solution will have the property of dynamic stability.

References

[1] Afsar M 2015 Energy-E cient Coalition Formation in Sensor Networks: a Game-Theoretic Approach. *Computing Research Repository.* **abs/1512.08019**

[2] Avrachenkov K E, Kondratev A Yu, Mazalov V V 2017 Cooperative Game Theory Approaches for Network Partitioning *International Computing and Combinatorics Conference* LNCS **10392** 591-602.

[3] Badami M, Hamzeh A, Hashemi S 2013 An enriched game-theoretic framework for multi-objective clustering. *Applied Soft Computing* **13** 1853-1868

[4] Bandyopadhyay S, Saha S 2007: GAPS A Clustering Method Using a New Point Symmetry-Based Distance Measure. *Pattern Recognition* **40**, 3430-3451

[5] Bulo S R, Pelillo M 2009 A game-theoretic approach to hypergraph clustering. Advances in Neural Information Processing Systems

[6] Bure V, Staroverova K 2018 Applying Cooperative Game Theory With Coalitional Structure For Data Clustering **10 v 1** 23-29.

[7] Dempster A, Laird N, Rubin D 1977 Maximum Liklihood From Incomplete Data via the EM Algorithm. *Journal of The Royal Statistical Society* **B 39** 1-38

[8] Dhamal S , Bhat S , Anoop K.R., Embar V.R. 2012 Pattern Clustering using Cooperative Game Theory // *Computing Research Repository* **abs/1201.0461** 653-658

[9] Garg V K, Narahari Y 2013 Novel biobjective clustering (BiGC) based on cooperative game theory // *IEEE Transactions on Knowledge and Data Engineering* **25 Is. 5.** 1070-1082.

[10] Gupta U, Ranganathan N 2010 A game theoretic approach for simultaneous compaction and equipartitioning of spatial data sets. *IEEE Transcation on Knowledge and Data Engineering* **22** 465-478

[11] Handl J, Knowles J 2007 An Evolutionary Approach to Multiobjective Clustering. *IEEE Transaction on Evolutionary Computation* **11** 56-76

[12] Handl J, Knowles J D 2004 Evolutionary Multiobjective Clustering. *PPSN LNCS,* **3242** 1081-1091 (Springer, Heidelberg Yao X.)

[13] Handl J, Knowles J 2005 MultiObjective Clustering Around Medoids *Proceedings of IEEE Congress on Evolutionary Computation* 632-639
[14] Kalai E., Smorodinskiy M. 1975. Other solutions to Nash's bargaining problem. *Econometrica* **43** 513-518
[15] Kydland F E, Prescott E C 1977. Rules rather than discretion: The inconsistency of optimal plans. *Journal of Political Economy* **85** 473-490
[16] Nash J F 1950 The bargaining problem. *Econometrica* **18 (2)** 155-162
[17] Petrosyan L A, Danilov N N 1979 *Vestnik Leningrad University* **1** 52-59
[18] Petrosjan L A, Zenkevich N A 1996 *Game Theory* (World Scientific Publishing Co. Pte. Ltd., Singapore)
[19] Sedakov A, Parilina E, Volobuev Yu, Klimuk D 2013 Existence of Stable Coalition Structures in Three-person Games // *Contributions to Game Theory and Management* **6** 407-422
5Shapley L S 1953 A value for n-person games. Contributions to the Theory of Games II (Princeton University Press, Princeton) 307-317.
[20] Strehl A, Ghosh J 2003 Cluster Ensembles: A Knowledge Reuse Framework for Combining Multiple Partitions.*Machine Learning Research* **3** 93-98
[21] Yeung D W K, Petrosyan L A 2006 *Cooperative Stochastic Differential Games* (Springer)
[22] Zenkevich N A, Petrosjan L A 2006 *Time-consistency of cooperative solutions. Discussion Paper* **8(R)-2006** (SPb.Institute of Management, St. Petersburg State University)

An effective algorithm of the numerical solution to the Stefan problem

Galina I. Kurbatova[1] **and Nadezhda N. Ermolaeva**[2]

[1],[2] Faculty of Applied Mathematics and Control Processes, Saint-Petersburg State University, Universitetskii pr. 35, Petergof, Saint Petersburg, Russia

E-mail: g.kurbatova@spbu.ru[1], n.ermolaeva@spbu.ru[2]

Abstract. The method of the frontal interval of the numerical solution to Stefan problems is proposed. The method allows the calculation of the position of the phase transition boundary with high accuracy at any of time. Using the example of a plane two-phase Stefan problem, which has an exact analytical solution, the advantages of the proposed method in comparison with a variable time stepping method are demonstrated with respect to the calculation of the position of the phase transition boundary. An example of the solution to the one-phase Stefan problem using the frontal interval method is also given.

1. Introduction

Practically, it is important to know the phase transition boundary position with high accuracy in the Stefan ploblems. It is known numerical methods that explicitly track the motion of the phase transition boundary, for example, a variable time stepping method [1, 2], a front-fixing method [3], a phase-field method [4], a combined finite difference method [5] and numerous other. Reviews of papers utilizing the front-tracking techniques can be found, for example, in [5] and [7].

In present paper we propose a *frontal interval method* which allows for the calculation of the position of the phase transition boundary with high accuracy at any time. This method relates to the class of front-tracking methods and has similarities with the front-fixing method and the combined finite difference method. However, unlike the variable time stepping method and the front-fixing method, the frontal interval method uses a constant spatial grid and time step. The proposed method, in contrast with the combined finite difference method [5], is implicit and utilizes different algorithms for the temperature calculation near the phase transition boundary.

2. The frontal interval method

In using this method, assume that there are two stationary phases (for example, the frozen and unfrozen soil) separated by a phase interface (the phase transition boundary). The movement of the phase transition boundary occurs due to changes in the temperature of the primary phase on the outer boundary. For numerical solutions similar Stefan problems a uniform spatial grid is introduced. The *frontal interval* is defined as the only interval between two adjacent grid nodes, that either contains both phases or through one of the nodes of this interval the phase front passes at the considered point in time. The frontal interval method is based on the computation of both the frontal interval position and the position of the phase interface in it at the every moment of time.

Out of the frontal interval the heat equations in each phase are solved using an implicit difference scheme. At every time step, a starting point of the frontal interval is determined and the movement of the phase interface is computed using difference analog of the Stefan condition, in which the heat fluxes are determined using temperature values of the primary and secondary phases at the previous temporal level. The computational scheme of the front interval method is *an implicit non-iterative scheme.*

3. Model problem

Let us demonstrate the algorithm of the numerical solution to the Stefan problems using the proposed frontal interval method with the example the well-known 1D two-phase Stefan problem [8] which has an exact analytical solution depending on the values x and t through the self-simulated variable $\xi = x/(a\sqrt{t})$. We have used the following designations: T_∞ is the temperature of the secondary phase at infinity (K), T_* is the phase transition temperature (K), T_0 is the temperature (K) of the primary phase at $x = 0$, $k = 1, 2$ is the phase number, c_k is the coefficient of specific volumetric heat capacity of the k-th phase (J m^{-3} K^{-1}), λ_k is the thermal conductivity coefficient of the k-th phase (W m^{-1}K^{-1}), and Q is the specific (per unit volume of ice) heat of phase transition (latent heat of melting), (J m^{-3}).

The statement of this Stefan problem is written as follows:

$$c_1 \frac{\partial T1}{\partial t} = \lambda_1 \frac{\partial^2 T1}{\partial x^2}, \quad t > 0, \quad x \in (0,\ y(t)); \tag{1}$$

$$t > 0, \quad x = 0 : \ T1 = T_0; \tag{2}$$

$$c_2 \frac{\partial T2}{\partial t} = \lambda_2 \frac{\partial^2 T2}{\partial x^2}, \quad t > 0, \quad x \in (y(t),\ \infty); \tag{3}$$

$$t = 0,\ x \in (0,\ \infty) : \quad T2(x) = T_\infty; \tag{4}$$

$$t > 0, \quad x \to \infty : \ T2 = T_\infty; \tag{5}$$

$$t > 0, \quad x = y(t) : \ T1 = T2 = T_*; \tag{6}$$

$$\lambda_1 \frac{\partial T1}{\partial x}\bigg|_{y-0} - \lambda_2 \frac{\partial T2}{\partial x}\bigg|_{y+0} = Q \frac{dy}{dt}; \tag{7}$$

$$t = 0 : \ y = 0, \tag{8}$$

where $Tk(x,t)$ is the temperature of the k-th phase ($k = 1, 2$) and $y(t)$ is the coordinate of the phase change boundary.

We assume all thermal and physical characteristics of the phases and the given temperatures are constant. The problem (1)–(8) has a known analytical solution [8]:

$$T1(x,t) = A_1 + B_1 \operatorname{erf}(\frac{x}{2a_1\sqrt{t}}), \quad T2(x,t) = A_2 + B_2 \operatorname{erf}(\frac{x}{2a_2\sqrt{t}}), \quad y(t) = \alpha\sqrt{t}. \tag{9}$$

The constants a_k are equal to: $a_k = (\lambda_k/c_k)^{1/2}$, $k = 1,\ 2$; constants A_1, A_2, B_1, and B_2 in a known manner [8] are expressed through the parameters of the problem (1)–(8) and the value of α, which is calculated by the transcendental equation given in [16], depends on all parameters of the problem, namely T_0, T_*, T_∞, c_1, c_2, λ_1, λ_2 and Q.

4. Numerical solution to the Stefan problem by the frontal interval method

Let us consider the numerical solution to the Stefan problem (1)–(8) for non-zero initial conditions by the proposed frontal interval method. We assume that the primary phase thickness is $y_0 \neq 0$ in the initial instant of time $t = t_0$. Both, the initial moment t_0 and the initial temperatures distributions $T1^{(0)}(x, t_0)$ and $T2^{(0)}(x, t_0)$ are calculated using the given y_0 from the analytical solution (9). The constant step h was chosen so that y_0 was a multiple of h. The heat equations in the primary and secondary phases are approximated using an implicit scheme. The difference analogs can be represented as:

$$Ak\; Tk_{i-1}^{n+1} - Bk\; Tk_i^{n+1} + Ck\; Tk_{i+1}^{n+1} = Dk_i^n,$$

$$Ak = Ck = m_k, \quad Bk = 1 + 2m_k, \quad Dk_i^n = -Tk_i^n, \quad m_k = \tau\, a_k^2/h^2, \quad k = 1, 2,$$

Tk_i^{n+1} is the temperature of the k-th phase at the i-th nodes on the $(n+1)$-th temporal level. At the (n)-th temporal level, the following values are considered to be known: the position of the front interval, which is between the node $s = k^n$ in the primary phase and the node 1 in the secondary phase; temperature arrays $T1_i^n$, $T2_i^n$ of the primary and secondary phases; the value $\triangle y^n$ is equal to the sum of all previous displacements inside the frontal interval beginning at the k-th node; the position y^n of the phase interface.

5. Transition algorithm from the n-th to the $(n+1)$-th temporal level

The displacement δy^{n+1} of the phase interface for the time from t^n to t^{n+1} is calculated from the difference analog of the Stefan condition (7), and the heat fluxes using temperature values at the n-th temporal level are calculated as:

$$\delta y^{n+1} = (\tau/(h\,Q))(\lambda_1(T1_s^n - T1_{s-1}^n) - \lambda_2(T2_2^n - T2_1^n)). \tag{10}$$

The resulting displacement of the phase interface inside the frontal interval equals

$$\triangle y^{n+1} = \triangle y^n + \delta y^{n+1}. \tag{11}$$

The relative fraction θ^{n+1} of the primary phase into the frontal interval is defined from the value $\triangle y^{n+1}$ as

$$\theta^{n+1} = \triangle y^{n+1}/h. \tag{12}$$

If $\theta^{n+1} < 1$, the position of the frontal interval at the $(n+1)$-th temporal level remains unchanged and the temperatures of the primary and secondary phases on the boundaries of the frontal interval are calculated as follows:

$$T1_s^{n+1} = T_* - \theta^{n+1}(T1_s^n - T1_{s-1}^n), \quad T2_1^{n+1} = T_* + (1 - \theta^{n+1})(T2_2^n - T2_1^n). \tag{13}$$

If $1 \leq \theta^{n+1} < 2$, the frontal interval at the $(n+1)$-th temporal level moves to the adjacent node and the temperatures of the primary and secondary phases are:

$$T1_s^{n+1} = T1_s^n, \quad T2_1^{n+1} = T_*. \tag{14}$$

The summary displacement $\triangle y^{n+1}$ of the phase change boundary into the new frontal interval is assigned the value

$$\triangle y^{n+1} = \triangle y^n + \delta y^{n+1} - h. \tag{15}$$

Temperatures in both phases are found from the boundary conditions (2) and (5). The calculation of all other temperature values of the primary and secondary phases at the $(n+1)$-th temporal level is made by the tridiagonal matrix algorithm. Which scenario of the front interval movement realizes at the $(n+1)$-th temporal level is defined via the found phases temperatures.

I. If $T2_1^{n+1} > T_*$, then the frontal interval remains in the same position. In this case, the node number defining the beginning of the frontal interval and position of the change phase boundary are calculated by the following formulas:

$$k^{n+1} = k^n = s. \tag{16}$$

$$y^{n+1} = k^n\, h + \triangle y^{n+1}, \tag{17}$$

where $\triangle y^{n+1}$ is defined by the equality (11).

II. If $T2_1^{n+1} \le T_*$ and $T2_2^{n+1} > T_*$, then the frontal interval moves to the adjacent node and the node number defining a position of the frontal interval is assigned the value

$$k^{n+1} := k^n + 1 = s, \tag{18}$$

where the node $(k^n + 1)$ was placed at time instant t^n in the secondary phase becomes the node belonging to the primary phase at time instant t^{n+1}. The phase boundary position is calculated using the formula (17).

In the frontal interval method it is necessary that for every step in time the boundary of the phase transition either remains in the same spatial interval or passes into the adjacent interval. With this approach it is possible only the considered above sceneries *I* and *II*, but fulfillment of this requirement imposes restrictions on the permissible value of step τ for specified values of h and other problem parameters.

6. Basis for the steps selection in the frontal interval method

The formulated requirement for the value of the front displacement during time τ is written as follows:

$$y^{n+1} = y^n + \sigma\, h, \quad 0 < \sigma < 1. \tag{19}$$

Condition (19) must be fulfilled for all time moments in the considered problem. If the speed of front moving is a decreasing function of time (as in problem (1)–(8)), then the maximum limit on τ occurs when calculating the first step, i.e. at the time instance $t_0 + \tau$. In this case, an estimation of the admissible value τ is easy to calculate from an analytical solution ($y = \alpha\sqrt{t}$). Indeed, condition (19) implies the equality: $\alpha\sqrt{t_0 + \tau} = \alpha\sqrt{t_0} + \sigma\, h$, which for τ less than t_0 leads to the following estimation:

$$\tau < (2\sqrt{t_0}\,\sigma\, h)/\alpha. \tag{20}$$

At the initial instance of time, the primary phase area must contain at least three spatial nodes, since the x-derivative of the primary phase temperature is used in the calculations for the frontal interval method. Hence the following condition must be satisfied: $y_0 = s_0\, h$, $s_0 \ge 3$, and the estimation (20) can be written in the form:

$$\tau < (2\, s_0\, \sigma\, h^2)/\alpha^2, \quad s_0 \ge 3, \quad 0 < \sigma < 1. \tag{21}$$

7. Comparison of the numerical solution to the Stefan problem by the frontal interval method with the exact analytical solution

For the model problem (1)–(8), the following thermophysical parameters and conditions specific to glaciation problems in saline water were set as:

$$T_\infty = 272.15,\ T_* = 271.236,\ T_0 = 265.236,\ c_1 = 2150500,$$

$$c_2 = 3995450,\ \lambda_1 = 2.0,\quad \lambda_2 = 0.6,\ \ Q = 2.805 \cdot 10^8, \tag{22}$$

where all values in (22) are given in the above units.

At the initial time instance, the thickness of the primary phase is considered equal to $y_0 = 0.0025$ m. The value $t_0 = 76.27$ sec was found from the analytical solution (9). For the dimensional spatial step $h = 5 \cdot 10^{-4}$ m at the initial time instance, the number of spatial steps in the first phase is $s_0 = 5$, and the permissible value of the time step for $\sigma = 0.5$ in accordance with (21) should be less than 15 sec. We have set that $\tau = 15$ sec.

The results of the calculations of position y^n of the phase transition boundary by the proposed method for model problem (1)–(8) under the selected initial conditions and the parameters (22) are shown in Table 1. Table 1 also shows the deviations δ_n ($\delta_n = y^a(t^n) - y^n$) of the numerically calculated values y^n from the exact $y^a(t^n) = \alpha\sqrt{t^n}$ found using the analytical solution (9).

Table 1. The values of y^n and the deviations δ_n.

$(t - t_0)$, days	1	4	10	16	22	30
y^n, cm	8.42	16.83	26.61	33.66	39.46	46.08
δ_n, cm	$-6.6 \cdot 10^{-4}$	$-8.0 \cdot 10^{-4}$	$4.7 \cdot 10^{-4}$	$2.7 \cdot 10^{-3}$	$4.8 \cdot 10^{-3}$	$8.3 \cdot 10^{-3}$

The data in the table demonstrates a high computational accuracy of the phase transition boundary position by the frontal interval method. The calculations show that similar accuracy is maintained under other parameters of the problem leading to larger, as well as smaller, Stefan numbers.

Generalizations of the frontal interval method on the multidimensional Stefan problems and the study of the convergence of this method are beyond the scope of this paper.

Application of the proposed method on the one-dimensional one- and two-phase Stefan problems with other boundary conditions does not pose any difficulties. As an example we presented the solution of the one dimensional one-phase Stefan problem for an infinite cylinder with radius $R = 0.67$ m under a given constant temperature $T1(R, t) = T_0$ on the lateral surface and with a given heat flux $q = 30$ ($\mathrm{J\,m^{-2}} \cdot \mathrm{sec^{-1}}$) into the phase transition boundary from the ambient. The thermophysical parameters of growing ice (the primary phase), the specific heat of phase transition, the values of steps h and τ, as well the initial conditions are equal to the above. When the condition $y_0 \ll R$ is met, it is permissible to set the initial temperature distribution $T1(r, t_0)$ in an cylindrical layer with the thickness y_0 in the form of the distribution $T1^{(0)}(x, t_0)$ found from the analytical solution to the plane problem (1)–(8).

The calculation results of the ice layer thicknesses y^n in the time instances t^n for the one-dimensional problem of the cylinder glaciation by the frontal interval method are presented in Table 2.

Table 2. The values of the ice layer thicknesses y^n.

$(t - t_0)$, days	1	3	6	9	14	20	26	30
y^n, cm	7.78	12.58	16.60	19.28	22.35	24.84	26.59	27.50

8. Comparison of the numerical solution to the Stefan problem by the frontal interval method and the variable time stepping method

In Table 3 we presented the calculation results of the numerical solution to the model problem by the variable time stepping method for the parameters (22), the same step h and under the same initial conditions as in the calculations in Table 1. yl^n is the position of the phase transition boundary calculated by the variable time stepping method at the time instance t^n, the values $\delta l_n = y^a(t^n) - yl^n$ are the deviations of the numerically calculated values y^n from the exact $y^a(t^n) = \alpha\sqrt{t^n}$.

Table 3. The values of the ice layer thicknesses y^n.

$(t - t_0)$, days	0.80	4.34	8.84	12.72	28.58	35.27
yl^n, cm	7.5	17.5	25.0	30.0	45.0	50.0
δl_n, cm	0.023	0.022	0.016	0.011	−0.017	−0.030

Comparison of Tables 1 and 3 shows that for the considered model problem, the calculation accuracy of the phase transition boundary position by the frontal interval method is two orders of magnitude higher than by the variable time stepping method. This same level of accuracy was similar for long-time processes and other parameters of the problem.

9. Conclusion

In the present study the frontal interval method of the numerical solution to Stefan problems is proposed. This method allows for calculation of the position of the phase transition boundary with a high accuracy at any instance of time. The proposed method is an implicit non-iterative scheme and is distinguished for the algorithmic simplicity. The results of the test calculations of the plane one-dimensional two-phase Stefan problem are presented using the frontal interval method over a long period of time. A comparison with an analytical solution to this problem demonstrates a high accuracy of the phase transition boundary calculation. The comparative results of this problem solution by the frontal interval method and the variable time stepping method are presented. These results indicated the advantage of the proposed method in the accuracy of the phase transition boundary calculation. An example of the one-phase Stefan problem solution for a cylinder using the frontal interval method is given.

10. References

[1] Vasiliev F On finite difference methods for the solution of Stefans single- phase problem 1963 *USSR Comput. Math. Math. Phys.* **3** 5 1175–91. [in Russian]
[2] Budak B and Uspenskii A 1963 A difference method for the solution to a two-phase Stefan problems, *USSR Comput. Math. Math. Phys.* **5** 3 874 – 86. [in Russian]
[3] Budak B and Uspenskii A 1969 Difference front-fixing method for the solution to the Stefan problems, *USSR Comput. Math. Math. Phys.* **9** 6 1299 – 1315. [in Russian]
[4] Karma A and Rappel W 1998 Quantitative Phase-Field Modeling of Dendritic Growth in Two and Three Dimensions, *Physical Review E* **57** 4 4323 – 49.
[5] Nikitenko N. I. 1978 *Study of the processes of heat and mass transfer by the method of grids* (Kiev: Naukova Dumka) [in Russian]
[6] Dar'in N and Mazhukin V 1989 Mathematical Modeling of Nonstationary Two-Dimensional Boundary-Value Problems on Grids with Dynamic Adaptation, *Mathematical Modeling* **1** 3 29 – 43. [in Russian]
[7] Samarsky A, Vabishchevich P, Iliev O and Churbanov A 1992 Numerical modeling of convective-diffusion processes during phase transitions, Moscow Center of Mathematics and Mathematics *Preprint* 7. [in Russian]
[8] Tichonov A and Samarskii A 1963 *Equations of mathematical physics* (USA: Pergamon Press Ltd)

Hyperbolic decomposition for hydrostatic approximation of free surface flow problems

V M Goloviznin[1], **P A Mayorov**[2], **P A Mayorov**[2]

[1] Lomonosov Moscow State University, Faculty of Computational Mathematics and Cybernetics; Leninskie Gory, Moscow, 119991, Russia;
[2] Nuclear Safety Institute Russian Academy of Sciences; Bolshaya Tulskaya st., 52, Moscow, 113191, Russia;

E-mail: gol@ibrae.ac.ru, pavel.a.mayorov@gmail.com, maiorov.peter@gmail.com

Abstract. The article is devoted to the description of a new method of numerical solution for hydrostatic approximation of incompressible hydrodynamic problems with free surfaces and variable density. This technique develops a model of multilayer shallow water, taking into account the mass and momentum exchange between the layers.

1. Introduction

Multilayer shallow water equations [1] are not always hyperbolic [2], which leads to the fact that initial - boundary value problem becomes incorrect in the process of solution , and the computational algorithms polished for always hyperbolic equations lose stability. The loss of hyperbolicity is believed to occur in situations when conditions for the development of Kelvin – Helmholtz arise instability at the interfaces between layers. In these cases, an intensive mass and momentum exchange between the layers should occur, which is forbidden in the classical multilayer approximation. This disadvantage can be eliminated by including so-called "turbulent" viscosity in the multilayer equations. This viscosity depends in a complex way on the parameters of calculated flow and contains empirical parameters that can be adjusted to different types of flows. The disadvantage of this approach is the critical dependence of the calculation results on the experience of the researcher.

Another approach to solving this problem is to modify the original model based on the derivation of the classical multilayer shallow water equations. We remove the ban on exchange of mass and momentum between the layers [3]. In fact, this means a transition from the Lagrangian description of the motion of boundaries of layers in the vertical direction to the mixed Eulerian – Lagrangian description. In this case, the regularizing mass and momentum flows arise between the layers.

The implementation of this approach uses the method of splitting into physical processes. First, the values of physical variables on the new time step are calculated using the classical equations of multilayer shallow water with variable density. Then, the prescribed vertical coordinates of the interfaces between layers are determined in some way. Finally, these coordinates determine the values of exchange fluxes of mass and momentum between the layers. When calculating the prescribed vertical coordinates of the boundaries of the layers, following considerations are used: the uppermost layer, which is the free boundary of the fluid, remains

Lagrangian, the lowermost one is fixed. If the coordinates of intermediate boundaries are found using the condition of equality of the thicknesses of the layers on each vertical line, then such computational grids are called "sigma - grids". Another class of grids often used in computational oceanology is the so-called "Z-grids", when the boundaries of all layers, except the uppermost one, remain fixed. Intermediate options are also possible.

Many different computational algorithms with their own advantages and disadvantages have been developed for the numerical solution of the classical equations of multilayer shallow water with variable density. Usually, computational algorithms are based on the finite volume method, and some variant of downstream transferring of the values is used to calculate convective flows.

It should be noted that high-resolution methods based on solving the Riemann problem and well-proven in the one-layer case are not applicable for the multilayer case, since the Riemann problem in the multilayer case does not have a simple solution.

In the new numerical method, the values of horizontal flows are proposed to be calculated by the method of characteristics, as is done in the CABARET scheme [4, 5]. For this, the multilayer model is represented as a sum of one-layer models (hyperbolic decomposition) interacting through reaction forces applied to the interfaces of the layers. The CABARET scheme is used for each of the one-layer models, taking into account the nonlinear correction of the fluxes based on the maximum principle. The condition of incompressibility and the conservation laws of mass and momentum are also satisfied. As a result, conservative and flux variables are found at the new time step without the exchange of mass and momentum between the layers. Since this problem is incorrect, new flux variables are filtered, not violating the conservation laws.

The resulting algorithm has a second order of approximation and maintains a state of rest at a constant density over any bottom topography (well-balanced algorithm). At the second stage of splitting the prescribed vertical coordinates of the layer boundaries are specified according to physical processes. Using new vertical coordinates, the exchange flows between the layers are calculated with second order of approximation.

The properties of the new technique are illustrated on the test problem.

2. The equations of multilayer shallow water with variable density and considering external pressure and bottom relief

The equations of multilayer shallow water with variable density without mass and momentum exchange between the layers can be represented as the sum of one-layer equations related to each other by the forces of pressure and reaction acting on their interfaces:

$$\frac{\partial h_k}{\partial t} + \frac{\partial (hu)_k}{\partial x} = 0$$
$$\frac{\partial \rho_k h_k}{\partial t} + \frac{\partial (\rho h u)_k}{\partial x} = 0$$
$$\frac{\partial (\rho h u)_k}{\partial t} + \frac{\partial (\rho h u^2)_k}{\partial x} + \frac{g\rho_k}{2}\frac{\partial (h^2)_k}{\partial x} + g\rho_k h_k \frac{\partial Z_{k+1}}{\partial x} + P_k \frac{\partial Z_{k+1}}{\partial x} - P_k \frac{\partial Z_k}{\partial x} + \frac{\partial (hP)_k}{\partial x} = 0$$
$$k = 1, ..., N \; Z_1 = H \; Z_{N+1} = B \; h_k = Z_k - Z_{k+1} \; P_{k+1} = P_k + g\rho_k h_k$$

here k is layer number counted from the free surface, Z_k is coordinates of the upper boundary of kth layer, $P_1 = P_T$ is pressure on the free surface Z_1.

The so-called "simple form" of this system appears as follows:

$$\frac{\partial \vec{\psi}}{\partial t} + \mathbf{G}\frac{\partial \vec{\psi}}{\partial x} = \vec{D}, \; \vec{\psi} = (h_1, ...h_N, \rho_1, ..., \rho_N, u_1, ...u_N)^T$$

where $\mathbf{G}$ is matrix of dimension $N \times N$ and $\vec{D}$ - the right side, the specific form of which does not matter to us. It is well known that, even when $N = 2$, the matrix $\mathbf{G}$ can have complex roots and the system is not always hyperbolic. This creates known computational difficulties. For a larger number of layers, the situation is only getting worse. To overcome this difficulty, we can apply the technique we call the "hyperbolic decomposition" of the problem.

If we consider the forces acting on the interface as "external", then this system can be represented as:

$$\frac{\partial \vec{\varphi}_k}{\partial t} + \mathbf{A}_k \frac{\partial \vec{\varphi}_k}{\partial x} = \vec{d}_k,\ \vec{\varphi}_k = (h_k, \rho_k, u_k)^T,\ \vec{d}_k = (0, 0, F_k/\rho_k h_k\ ),\ k = 1, ..., N$$

Where

$$\mathbf{A}_k = \begin{pmatrix} u_k & 0 & h_k \\ 0 & u_k & 0 \\ P_{k+1}\,(\rho h)_k^{-1} & 0 & u_k \end{pmatrix} \quad F_k = P_k \frac{\partial Z_k}{\partial x} - P_{k+1} \frac{\partial Z_{k+1}}{\partial x} - h_k \frac{\partial P_k}{\partial x}$$

This leads to a system of independent characteristic equations:

$$\left(\frac{\partial u_k}{\partial t} + \frac{c_k}{h_k}\frac{\partial h_k}{\partial t}\right) + (u_k + c_k)\left(\frac{\partial u_k}{\partial x} + \frac{c_k}{h_k}\frac{\partial h}{\partial x}\right) = (F_k/\rho_k h_k\ )$$

$$\left(\frac{\partial u_k}{\partial t} - \frac{c_k}{h_k}\frac{\partial h_k}{\partial t}\right) + (u_k - c_k)\left(\frac{\partial u_k}{\partial x} - \frac{c_k}{h_k}\frac{\partial h}{\partial x}\right) = (F_k/\rho_k h_k\ )$$

$$\frac{\partial \rho_k}{\partial t} + u_k \frac{\partial \rho_k}{\partial x} = 0$$

Therefore, it is possible to use for the numerical solution of the complete system of multilayer equations algorithms that are well proven in the one-layer case. The incorrectness of the total system does not disappear and arises in the form of high-frequency distortion of the boundaries. They can be regularized by using special filters that do not violate the approximation and lead to the exchange of mass and momentum between the layers, which does not violate the conservation laws.

3. Features of numerical implementation

The CABARET scheme uses two types of variables: flux-type and conservative-type. The flux-type variables belong to the middle of the vertical edges of the calculated cells, the conservative-type variables belong to their centers. The computational algorithm consists of the following elements:

- the calculation of conservative-type variables at an intermediate time step using conservative difference schemes obtained by the finite volume method (Phase 1)
- calculation of flux-type variables at a new time step by extrapolation of local Riemann invariants (Phase 2)
- calculation of conservative-type variables at a new time step using finite-volume schemes (Phase 3).

Short-wave perturbations are generated in the areas of loss of hyperbolicity. Computed flux-type variables have to be filtered to regularize the problem.

4. The exchange of mass and momentum between the layers

Regularization by filtering flux-type variables that does not affect conservative-type values does not always make a full system correct for long enough periods of time. It is necessary to adjust the conservative-type variables, which leads to the exchange of mass and momentum between the layers.

In the proposed algorithm, the restructuring of grid is carried out twice at one time step. The first time after the calculation of conservative-type variables on the intermediate layer in phase 1, the second time after phase 3.

5. Example of calculations. Kurganov's Test

The initial data:

$$u_1(x,0) = 0.4,\ \rho_1 = 0.98,\ h_1(x,0) = \begin{cases} 1,\ 2 \geq |x| \geq 1 \\ 1-0.25\sin(2\pi x),\ |x|<1 \end{cases}$$

$$u_2(x,0) = -0.4,\ \rho_2 = 1,\ h_2(x,0) = \begin{cases} 1,\ 2 \geq |x| \geq 1 \\ 1+0.25\sin(2\pi x),\ |x|<1 \end{cases}$$

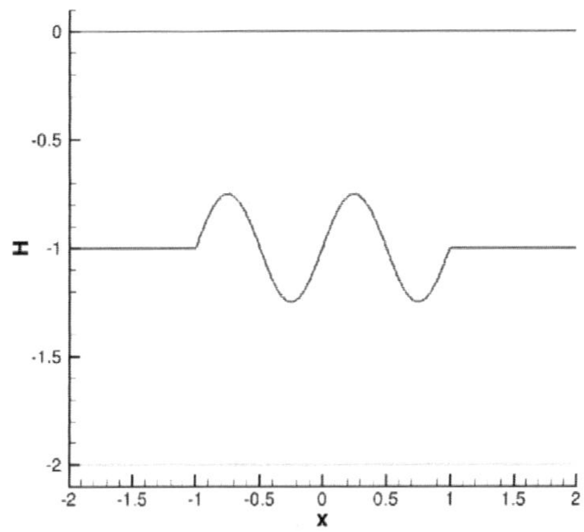

Figure 1: The initial layer thickness data

The bottom topography is $B(x) = -2, x \in [-2, 2]$, $g = 10$.

The computational grid consists of 801 nodes along the x axis. The surfaces (left) and velocities (right) of the layers calculated according to the CABARET scheme at time T = 0.5 are given below.

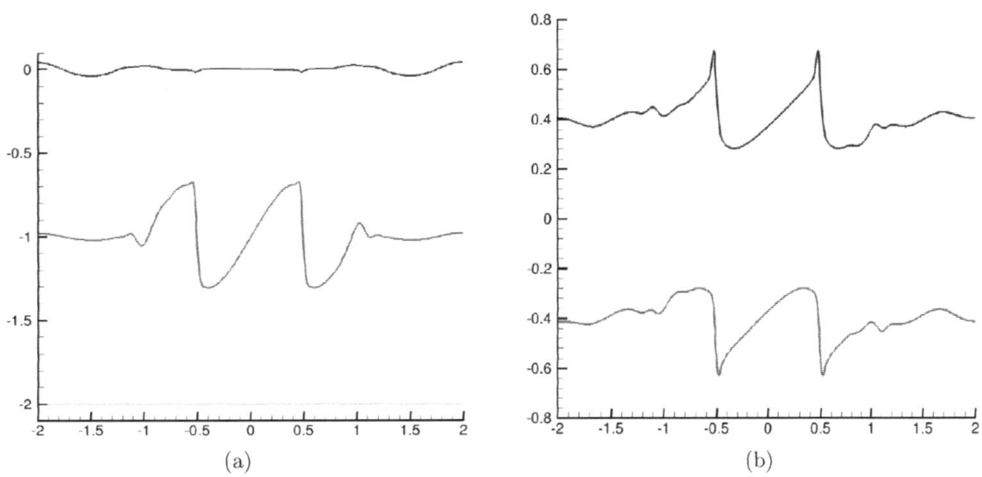

Figure 2: Calculation by CABARET scheme on 800 cells at t = 0.5. The shape of boundaries of the layers (a), velocities (b)

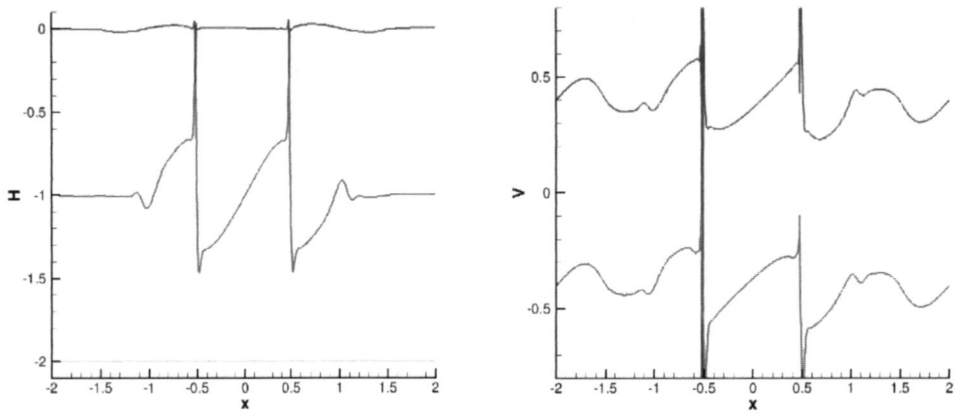

(a) (b)

Figure 3: Calculation by CABARET scheme on 800 cells at t = 0.55. The shape of boundaries of the layers (a), velocities (b)

In the work of A. Kurganov[6], the calculation of this problem was made with the same parameters, and similar results were obtained. Further calculation according to the CABARET scheme showed that the solution loses its stability, becomes irregular and the calculation stops (Fig.3). In the calculation on a denser grid with 1601 nodes loss of stability occurs much earlier. On a coarser grid (201 nodes), the instability develops more slowly (Fig. 4), but poor conditionality of the problem inevitably leads to abnormal termination of the calculation (Fig. 5). Such behavior of the solution will most likely be in calculations by A. Kurganov's scheme[6]. The restructuring of the computational grid and, as a consequence, the exchange of mass and momentum between the layers regularizes the problem.

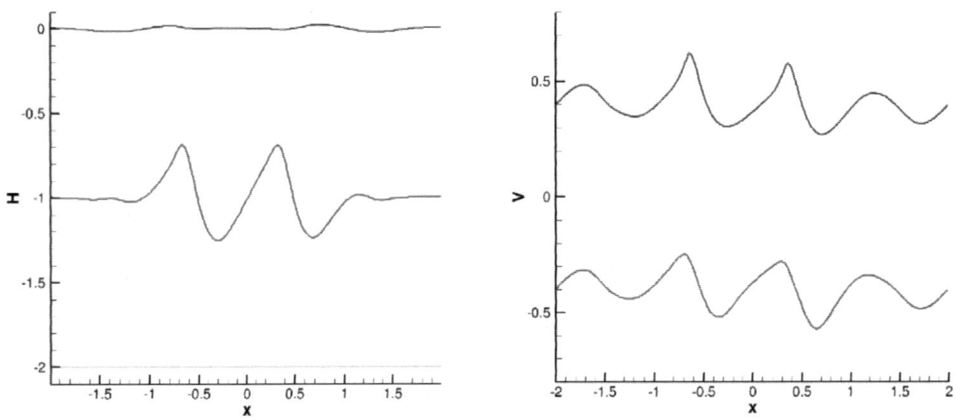

(a) (b)

Figure 4: Calculation by CABARET scheme on 200 cells at t = 0.55. The shape of boundaries of the layers (a), velocities (b)

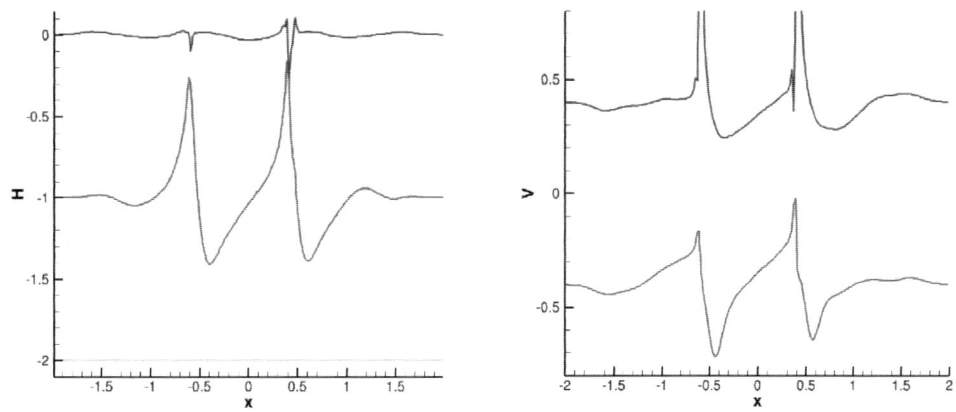

(a) (b)

Figure 5: Calculation by CABARET scheme on 200 cells at t = 1.05. The shape of boundaries of the layers (a), velocities (b)

6. Conclusion

Multilayer hydrostatic model with a free surface, excluding the mass and momentum exchange between the layers, in most cases of interest are incorrect and cannot be used in practical calculations. The restructuring of the computational grid at each time step allows to regularize the problem. Multilayer models should be used to adequately describe the diffusion of mass and momentum associated with the loss of hyperbolicity properties.

Acknowledgments

This work was supported by the Russian Science Foundation, project no. 18-11-00163.

References

[1] E. Audusse, A multilayer Saint-Venant system: Derivation and numerical validation, 2005, *Discrete Contin. Dyn. Syst.* **5** 189–214

[2] Vincent Duchene, A note on the well-posedness of the one-dimensional multilayer shallow water model, 2013, *hal-00922045*

[3] E. Audusse, M.-O. Bristeau, B. Perthame, J. Sainte-Marie, A multilayer Saint-Venant system with mass exchanges for shallow water flows. Derivation and numerical validation, 2011, *ESAIM: M2AN*, **45** 169-200

[4] Goloviznin V M, Zaitsev M A, Karabasov S A and Korotkin I A 2013 *New CFD Algorithms for Multiprocessor Computer Systems* (Moscow: Moscow State University Press)

[5] Goloviznin V M and Isakov V A, Balance-characteristic scheme as applied to the shallow water equations over a rough bottom, 2017, *Zh. Vychisl. Mat. Mat. Fiz.* **57** 1140 - 57

[6] Kurganov A., Petrova G., Central-upwind schemes for two-layer shallow water equations, 2009, *SIAM J. Sci. Comput.* **31** 1742–1773

The Balance-Characteristic Numerical Method on Triangle Grids.

D Y Gorbachev[1], V M Goloviznin[1]

[1] Lomonosov Moscow State University, Faculty of Computational Mathematics and Cybernetics, Laboratory of Industrial Mathematics, Russia

E-mail: `lim@cs.msu.ru`

Abstract. In this paper we introduce new balance-characteristic method on triangle grids. We describe main steps of the algorithm on the two-dimensional shallow water equations. Method verified on basic test for shallow water equations with flat bottom and compared with the results of other authors.

1. Introduction

The balance-characteristic numerical method CABARET [1] shows good results in wide area of CFD problems. For example, CABARET was verified on solving simple transfer equation [2], gas dynamics problems [3] and industrial problems such as: aeroacoustics [4], shallow water over a rough bottom [5], containment modeling and many others. Despite all the advantages of this method it has restriction to use quadrangular computational grids. But in many industrial problems it is not acceptable because of many reasons: computational domain is too complex to be meshed with quadrangle, time step is too small on quadrangular grids in comparison with triangle grids with the same quality, legacy applications that are still in use accepts only triangle grids and so on.

We introduce a new balance-characteristic numerical method that aims to solve this problem without losing the properties inherent in the CABARET method.

2. Problem definition

We consider two-dimensional shallow water equations with flat bottom and subsconic flows:

$$\begin{cases} \dfrac{\partial H}{\partial t} + \dfrac{\partial H u_x}{\partial x} + \dfrac{\partial H u_y}{\partial y} = 0 \\ \dfrac{\partial H u_x}{\partial t} + \dfrac{\partial \left(H u_x^2 + \frac{gH^2}{2} \right)}{\partial x} + \dfrac{\partial H u_x u_y}{\partial y} = 0 \\ \dfrac{\partial H u_y}{\partial t} + \dfrac{\partial H u_x u_y}{\partial x} + \dfrac{\partial \left(H u_y^2 + \frac{gH^2}{2} \right)}{\partial y} = 0 \end{cases} \tag{1}$$

Here H - water depth, g - acceleration due to gravity, u_x and u_y velocity components in the x and y directions respectively.

Journal of Physics: Conference Series **1392** (2019) 012036 doi:10.1088/1742-6596/1392/1/012036

3. Computational grid

Described equations are solved on an unstructured grid with triangular cells. In every triangle cell we have ten nodes: four nodes on every edge and one node at the intersection of medians. Variables related to nodes on edges and to the cell center we're naming flux variables and conservative variables respectively. As shown on Figure 1 nodes with conservative variables will be labeled as blue circles and flux variables as black circles. Such labeling will be used in entire article. In each node we have a full set of variables: H, u_x and u_y.

In the context of the time dimension the template of the numerical scheme have structure as it shown on Figure 2. Like in the CABARET in variables calculation at new time step $n+1$ involved variables from current time step n and half time step $n+1/2$. On half step $n+1/2$ only conservative variables are used.

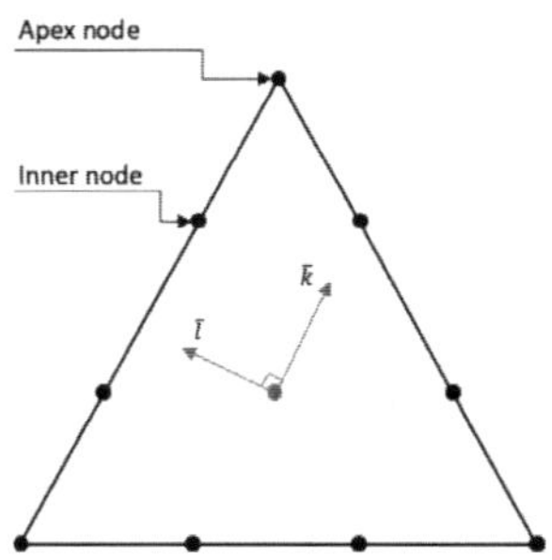

Figure 1. Cell template.

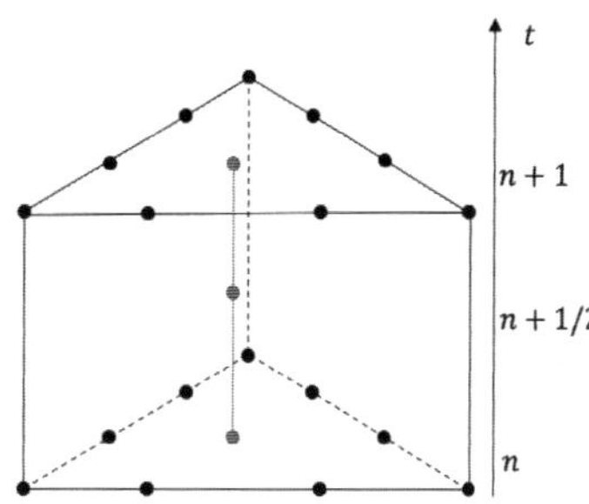

Figure 2. Scheme template.

4. Numerical method

Variables calculation at the new time step is divided into three phases:

1. On the first phase conservative variables at the half time step are calculated.
2. On the second phase flux variables at the new time step are calculated.
3. Finally, On the third phase conservative variables at the new time step are calculated.

Lets describe each of three phases in details.

4.1. First phase

In this phase flux and conservative variables at the current time step are involved in calculation. First phase template to calculate conservative variables at the half time step shown on Figure 3.

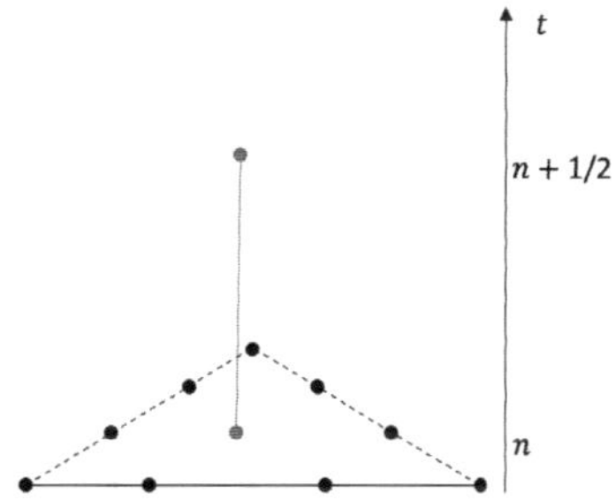

Figure 3. First phase template.

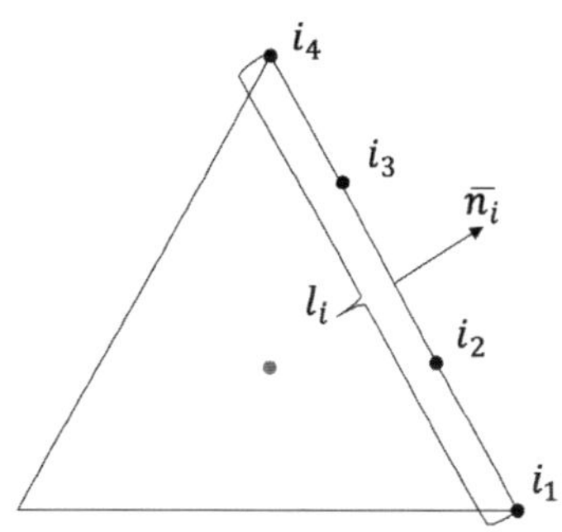

Figure 4. Sketch of variables involved in divergence calculation for ith edge in cell.

Based on this template we can calculate new variables with the following system of equations:

$$\begin{cases} \dfrac{H_c^{n+1/2} - H_c^n}{\tau/2} + \dfrac{1}{S}\sum_{i=1}^{3} integrate\left(H_i\overline{u}_i, \overline{n_i}, l_i\right) = 0 \\ \dfrac{(Hu_x)_c^{n+1/2} - (Hu_x)_c^n}{\tau/2} + \dfrac{1}{S}\sum_{i=1}^{3} integrate\left(\overline{\left(Hu_x^2 + \dfrac{gH^2}{2}, Hu_xu_y\right)_i}, \overline{n_i}, l_i\right) = 0 \\ \dfrac{(Hu_y)_c^{n+1/2} - (Hu_y)_c^n}{\tau/2} + \dfrac{1}{S}\sum_{i=1}^{3} integrate\left(\overline{\left(Hu_xu_y, Hu_y^2 + \dfrac{gH^2}{2}\right)_i}, \overline{n_i}, l_i\right) = 0 \end{cases} \quad (2)$$

Here H_c^n, $(Hu_x)_c^n$ and $(Hu_y)_c^n$ related to cell center node at n time step. S is the square of the triangle. Finally, *integrate* is an operation that is described as follows:

$$integrate(\overline{a}_i, \overline{n}_i, l_i) = \left(\frac{\overline{a_{i_1}} + 3\overline{a_{i_2}} + 3\overline{a_{i_3}} + \overline{a_{i_4}}}{8}, \overline{n_i}\right) l_i$$

Description is done for ith edge in triangle according to the Figure 4. So $\overline{n_i}$, l_i is ith edge normal and length respectively. Concrete values of $\overline{a_i}$ is taking from nodes i_1, i_2, i_3 and i_4. This operation calculate integral with third order of accuracy and it is also known as Simpson's 3/8 rule.

4.2. Second phase

This phase is most important because on this phase we calculate flux variables at the new time step. This phase has a lot of variation and it is sensitive to the method used. In our method we have different procedures for calculation flux variables in edge inner nodes and triangle apex nodes (Figure 1). In this phase we operate with Riemann invariants and not with initial variables. For example, for shallow water equations for the dimension $\overline{k}$ from Figure 1 we have next Riemann invariants with corresponding velocity

$$\begin{cases} R = (\overline{u}, \overline{k}) + 2\sqrt{gH} & \lambda_R = (\overline{u}, \overline{k}) + \sqrt{gH} \\ Q = (\overline{u}, \overline{k}) - 2\sqrt{gH} & \lambda_Q = (\overline{u}, \overline{k}) - \sqrt{gH} \\ S = (\overline{u}, \overline{l}) & \lambda_S = (\overline{u}, \overline{k}) \end{cases}$$

4.2.1. Edge inner nodes We have the following template for calculation new values in edge inner nodes:

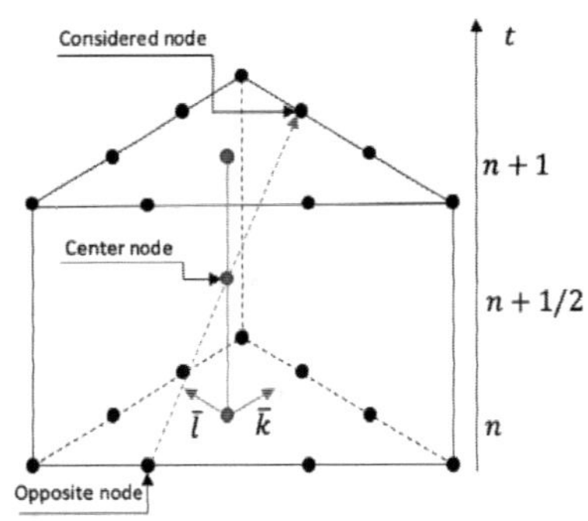

Figure 5. Second phase template for edge inner nodes

For every edge inner node there is an opposite node from the same cell. Also, straight line connecting them passes through cell center node at the half time step. This facts allow us to use the same extrapolation procedure for Riemann invariants as in the CABARET.

Lets consider node from Figure 5. For this node we calculate new values of Riemann invariants as follows:

$$I^{n+1}_{considered} = 2I^{n+1/2}_{center} - I^{n}_{opposite}$$

Then we apply nonlinear correction procedure for this new value:

$$I^{n+1}_{considered} = \begin{cases} max + \tau Q, & I^{n+1}_{considered} \geq max + \tau Q \\ min + \tau Q, & I^{n+1}_{considered} \leq min + \tau Q \\ I^{n+1}_{considered}, & otherwise \end{cases} \quad (3)$$

Here $max = max\{I^{n}_{considered}, I^{n}_{center}, I^{n}_{opposite}\}$, $min = min\{I^{n}_{considered}, I^{n}_{center}, I^{n}_{opposite}\}$ and Q calculates as follows:

$$Q = \frac{I^{n+1/2}_{c} - I^{n}_{c}}{\tau/2} + \lambda_I \left(grad\, I, \overline{k}\right)$$

$grad\, I$ calculates in the same manner as it done for divergence in first phase.

When all new values of Riemann invariants are calculated we can find initial variables. This can be done uniquely in subsonic flows. To choose concrete set of Riemann invariants to perform this operation we compare values of corresponding velocity from adjacent cells.

4.2.2. Triangle apex nodes The calculation of new variables in triangle apex nodes occurs after calculation is performed in edge inner nodes. This allows us to use variables at the new time step $n+1$ from edge inner nodes.

Lets consider some apex node with adjacent cells. We can calculate average value of H and $\overline{u}$ at this node by values of variables from adjacent edge inner nodes at the new time step $n+1$. For every Riemann invariant we can select cell from which it transfers base on these average H and $\overline{u}$. This is done by comparison values of projected corresponding Riemann invariant velocity. Projection of the velocity is done in the directions which are constructed by connecting this apex node and center node of adjacent cells.

For example, some cell was selected for one of the Riemann invariants and we have the following template:

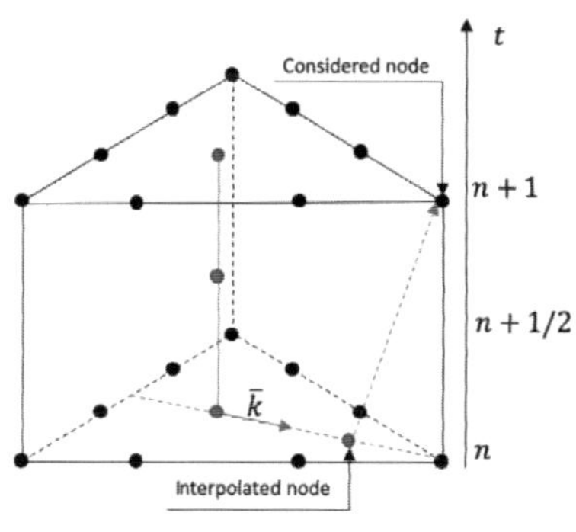

Figure 6. Second phase template for triangle apex nodes

Direction $\overline{k}$ for this step is different from direction that we select for edge inner nodes. In this direction we are searching node at the intersection of line corresponding to $\overline{k}$ and line corresponding to characteristic curve. We reconstruct value of the Riemann invariant in this node by using interpolation polynomial in this cell. This interpolation polynomial is built on ten nodes and has third order of accuracy.

Finally, new value of the Riemann invariant we calculate as follows:

$$I^{n+1}_{considered} = I^{n}_{interpolated} + \tau Q$$

Here Q calculates in the same way as it done for edge inner nodes.

So we can calculate new values for each Riemann invariant in considered node and reconstruct values of initial variables.

4.3. Third phase
In this phase we calculate conservative variables at new time step $n + 1$. In this calculation flux variables at new time step $n + 1$ and conservative variables at half time step $n + 1/2$ are involved. The system of equations for this phase can be constructed similar to the one from the first phase.

5. Water drop problem
This test is taken from the article [7] from section 4.5. For this test we have 7776 number of cells and use CFL equal to 0.3.

Figure 7. t = 0.0

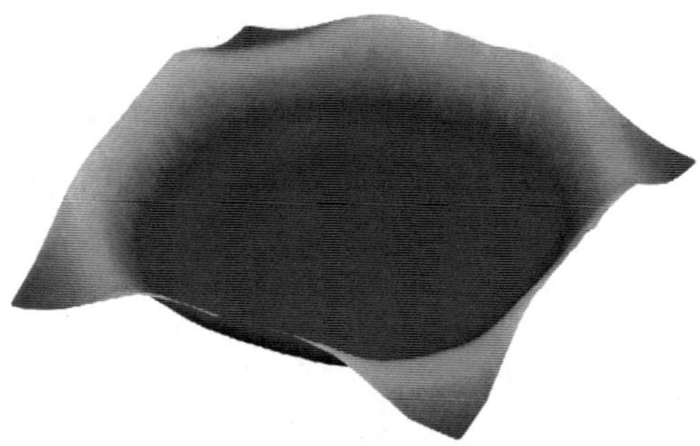

Figure 8. t = 0.15

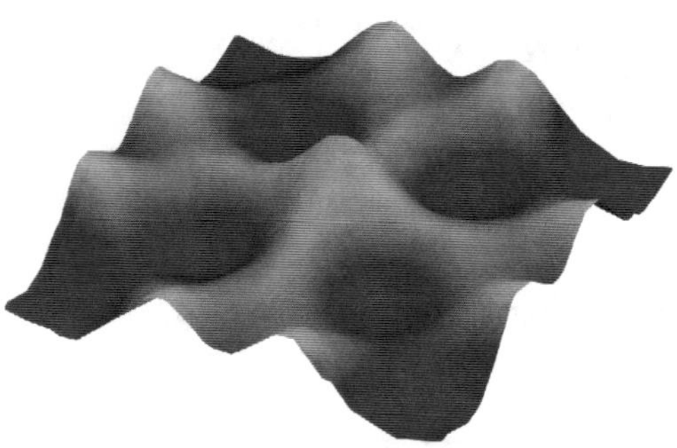

Figure 9. t = 0.3

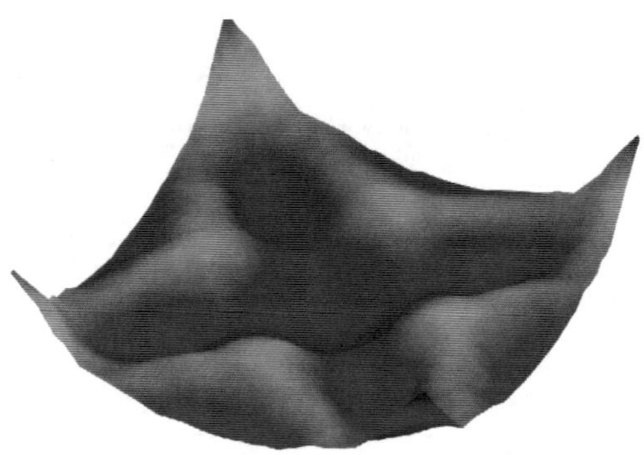

Figure 10. t = 0.5

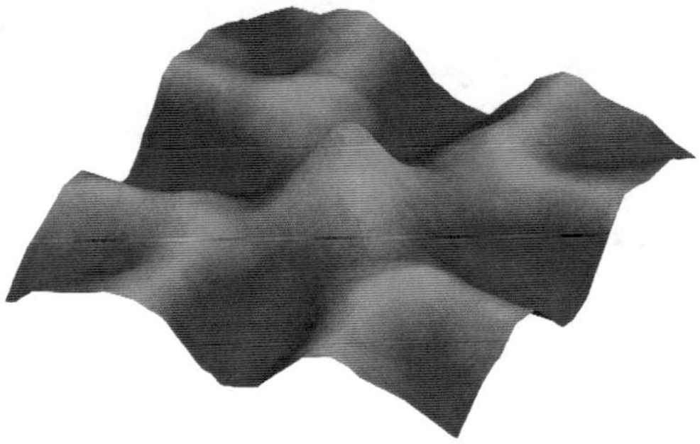

Figure 11. t = 0.7

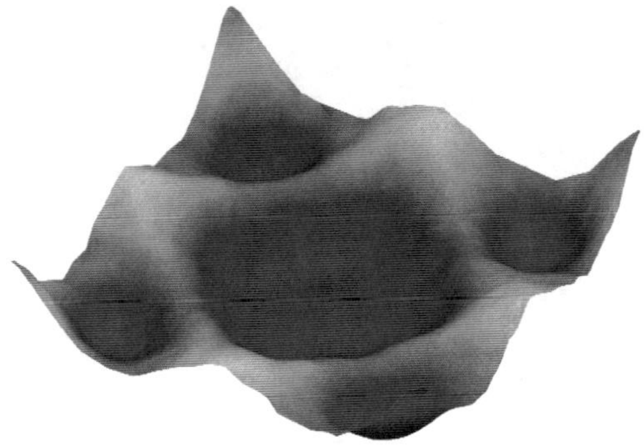

Figure 12. t = 0.8

Our results are consistent with the results from the mentioned article with flat bottom and from the article [8].

6. Conclusions

A new balance-characteristic method has been developed and tested to simulate shallow water flows. Advantages of the present method: the capabilities of handling complex geometry by using unstructured grids, both usage conservative and characteristic representation of initial hyperbolic equations in the natural manner.

Acknowledgments

This work was supported by the Russian Science Foundation, project no. 19-11-00104.

References

[1] Goloviznin V M, Zaitsev M A, Karabasov S A, Korotkin I A 2013 *Novel Algorithms of Computational Hydrodynamics for Multicore Computing* (Moscow: Moscow State University Press)
[2] Goloviznin V M, Karabasov S A, Kobrinski I M 2003 Balanced-characteristic schemes with staggered conservative and transport variables *Math. Model.* **15** 2948
[3] Goloviznin V M 2006 Balanced-characteristic method for numerical solution of one-dimensional Euler equations *Math. Model.* **18**:11(2006) 14-30
[4] Goloviznin V M, Karabasov S A, Kozubskaya T K, Maksimov N V 2009 CABARET scheme for the numerical solution of aeroacoustics problems: Generalization to linearized one-dimensional Euler equations *Comput. Math. Math. Phys.* **49**:12 (2009) 21682182
[5] Goloviznin V M, Isakov V A 2017 Balance-characteristic scheme as applied to the shallow water equations over a rough bottom *Comput. Math. Math. Phys.* **57**:7(2017) 402-414
[6] Goloviznin V M, Chetverushkin B N 2018 New generation algorithms for computational fluid dynamics *Comput. Math. Math. Phys.* **58**:8 (2018) 1271-1225
[7] Xing Y, Zhang X 2013 Positivity-preserving well-balanced discontinuous galerkin methods for the shallow water equations on unstructured triangular meshes *Journal of Scientific Computing* **57(1)** 1941
[8] San O, Kara K 2011 High-order accurate spectral difference method for shallow water equations *Int. J. Res. Rev. Appl. Sci.* **6** 41-45
[9] Yoon T H, Kang S K 2004 Finite volume model for two-dimensional shallow water flows on unstructured grids *Journal of Hydraulic Engineering* **130(7)** 678-688
[10] Anastasiou K, Chan C T 1997 Solution of the 2D shallow water equations using the finite volume method on unstructured triangular meshes *International Journal for Numerical Methods in Fluids* **24(11)** 1225-1245

CUDA UNIFIED MEMORY ACCELERATION OF THE PROCESS OF DEVELOPING MULTI-GPU APPLICATIONS

A Y Gorchakov [1,2,3] and A V Solovjev [2]

[1] Dorodnicyn Computing Centre, Federal Research Center "Computer Science and Control" of Russian Academy of Sciences, Moscow, Russia

[2] Nuclear Safety Institute of the Russian Academy of Sciences, Moscow, Russia

[3]Moscow Institute of Physics and Technology,Dolgoprudnyi, Moscow oblast, Russia

andrgor12@gmail.com

Abstract. The experience of using Cuda Unified Memory technology to develop Multi-GPU applications is described. The paper deals with the method of transferring computations on the MGDS-GPU model two-dimensional problem described by the equations of single-layer shallow water with variable density over an uneven bottom. To solve the problem was applied balance - characteristic scheme "CABARET". The minimum steps necessary to go from computing on the GPU to computing on the MULTI-GPU are given. The computational efficiency is investigated depending on the size of the grid.

Introduction

The solution of modern practical problems of mathematical modeling, problems of assimilation of measurement data requires the use of high-performance computing. In this case, the typical performance of the computing complex should be tens to hundreds of TFlops or more. Such characteristics are achieved either on clusters equipped with thousands of CPUs, or on single servers with several graphic accelerators like Nvidia Tesla V100 (for reference, the performance of this graphic accelerator is 7.8 teraflops during double precision operations). The Message Passing Interface (MPI) standard is used in software development. Using MPI, the researcher instead of solving purely mathematical problems begins to solve technical problems, for example, the problem of choosing the type of messaging – blocking/non-blocking (MPI_send/MPI_isend), or the problem of choosing algorithms for collective operations. If we move on to computing on graphics accelerators, then for calculations on a single graphics accelerator, we forget about these problems, but when moving to Multi-GPU calculations, again the question arises how to exchange data between devices.

In this work, we investigated the possibility of using Cuda Unified Memory technology for calculations on several graphics accelerators. The main idea of the technology is the use of a single address space, which includes both the entire system memory and the memory of all installed graphics accelerators (supporting CUDA). To allocate unified-memory, it is enough to call the cudaMallocManaged function, and then without any effort from the developer, the system using the Page Migration engine (starting with the Pascal architecture and further), independently organizes the exchange of CPU-GPU/GPU-GPU data. A side effect of using unified-memory is to increase the available memory on the GPU.

1. Statement of the model problem

To test the effectiveness of Cuda Unified Memory, a relatively simple model two-dimensional problem is considered [1]. The problem contains the main features of this class of problems, namely, the problem of fluid dynamics with a free surface and a variable density in the approximation of a single-layer shallow water over an uneven bottom. Usually shallow water equations are written for the case of constant density, and in this case we consciously go to the complexity of the problem to bring it closer to reality.

The system of differential equations describing the dynamics of single-layer shallow water with variable density follows from the laws of conservation of mass, momentum, and continuity equations and has the form.

$$\frac{\partial \rho h}{\partial t}+\frac{\partial \rho hu}{\partial x}+\frac{\partial \rho hv}{\partial y}=0;\quad \frac{\partial \rho hu}{\partial t}+\frac{\partial \rho hu^2}{\partial x}+\frac{\partial \rho huv}{\partial y}+\frac{g}{2}\frac{\partial \rho h^2}{\partial x}=-g\rho h\frac{\partial B}{\partial x};$$
$$\frac{\partial \rho hv}{\partial t}+\frac{\partial \rho hvu}{\partial x}+\frac{\partial \rho hv^2}{\partial y}+\frac{g}{2}\frac{\partial \rho h^2}{\partial y}=-g\rho h\frac{\partial B}{\partial y};\quad \frac{\partial h}{\partial t}+\frac{\partial hu}{\partial x}+\frac{\partial hv}{\partial y}=0; \tag{1.1}$$

Here ρ - the density, u, v - the velocity components along the directions x, y, respectively, $B(x, y)$ - the bottom relief function, $h = H - B$ - the thickness of the liquid layer, $H(x, y, t)$ - the level of the free surface. From the first and last equations of this system follows the density transport equation.
Let the area G in which the problem is considered has the form of a rectangle with sides L_x, L_y (Figure 1).

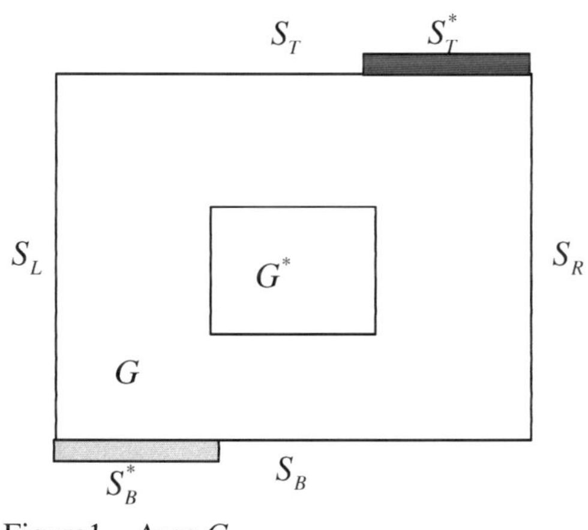

Figure1 – Area G

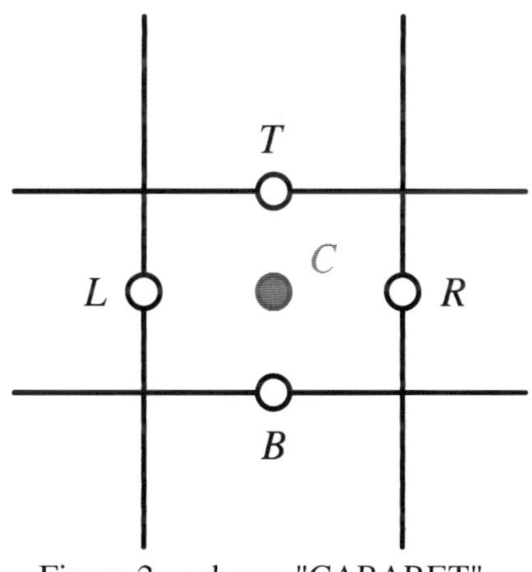

Figure 2 - scheme "CABARET"

The boundaries of this region will be denoted by S_R, S_L, S_T, S_B, which corresponds to the right, left, upper and lower sides of the rectangle. The bottom relief function is chosen as a piecewise constant:

$$B(x,y)=\{\Delta z,\quad if\quad (x,y)\in G^*\ and\ 0,\quad if\quad (x,y)\in G/G^*\} \tag{1.2}$$

As the boundary conditions on the sides S_R, S_L, we set : $u|_{S_R} = u|_{S_L} = 0$.

On the part of the upper boundary S_T^*, we define the "upper reservoir" condition, on the lower boundary part S_B^* - the "lower reservoir" condition, which determine the values of the Riemann invariants coming from outside the region:

$$\begin{aligned} &H_T^* = const > H(x,y,t_0); u_T^* = v_T^* = 0; \rho_T^* = f(t); (x,y) \in S_T^* \\ &H_D^* = const < H(x,y,t_0); u_B^* = v_B^* = 0; \rho_B^* = \rho_0; (x,y) \in S_B^* \end{aligned} \quad (1.3)$$

From the "upper reservoir" liquid flows into the region G, and from the region G it flows into the "lower reservoir". On the remaining sections of the upper and lower boundaries, the non-percolation condition is set: $v|_{S_T/S_T^*} = v|_{S_B/S_B^*} = 0$

We set the initial data as follows. We assume that at the initial moment t_0 the liquid is at rest, has a constant density, and its free surface is horizontal, and

$$H(x,y,t_0) = H_0; \quad H_T^* > H_0 > \Delta z; \quad H_0 > H_B^*; \quad \Delta z < H_B^*$$

The direct problem is solved until the time t_1 at which the quasi-steady vortex flow can be considered to be formed. On this time interval, the density of the inflowing liquid is considered equal to ρ_0. Then, another density $\rho_1 > \rho_0$ is applied to the input and a dynamic substitution of one liquid for another takes place. We assume that the values of the initial parameters are given in such a way that the flow remains subsonic.

In the second time interval, a discrete spatial-temporal monitoring network Θ_m is specified that fixes the density, velocity components and the height of the free surface $\{\rho_k, u_k, v_k, H_k\}, k \in \Theta_m$.

The task of four-dimensional assimilation of monitoring data consists in determining the values of the density, velocity, and height of the free surface from these values at the instant of time t_1.

2. Algorithm for the numerical solution of a direct problem

For the numerical solution of a direct problem, we use the CABARET scheme [2], combining the merits of conservative and characteristic methods. The area G is covered by a regular grid with rectangular cells with sides $\Delta x, \Delta y$ (Figure 2). The calculated physical fields will be referred to both the centers of the calculated cells (ρ_C, u_C, v_C, H_C) and to the centers of their faces (ρ_*, u_*, v_*, H_*), where the asterisk takes on values $\{R, L, T, B\}$. The values relating to the centers of the cells will be called "conservative", and to the midpoints of the faces – "fluxshape".

In the CABARET scheme three consecutive stages are distinguished - three phases. In the **first phase**, conservative variables are calculated on the intermediate time layer:

$$\frac{\vec{\psi}_C^{n+1/2} - \vec{\psi}_C^n}{\tau/2} + \vec{\Lambda}\left(\vec{\varphi}_f^n\right) = 0; \quad (2.1)$$

where $\vec{\psi}_C^n = \left\{\left[(h),(\rho h),(\rho h u),(\rho h v)\right]_C^n\right\}^T$, $\vec{\Lambda}$ is the operator approximating the spatial derivatives in (1.1), the small index f indicates fluxshape variables on the faces of the cell. Conservative variables $\left\{h_C^{n+1/2}, \rho_C^{n+1/2}, u_C^{n+1/2}, v_C^{n+1/2}\right\}$ are evaluated sequentially explicitly.

Phase 2 calculates the values of fluxshape variables on a new time layer t_{n+1}. At this stage, the characteristic forms of equations (1.1) along the axis x are used:

$$\left\{\frac{\partial u}{\partial t}+\frac{g}{c}\cdot\frac{\partial h}{\partial t}+\frac{c}{2\rho}\cdot\frac{\partial \rho}{\partial t}\right\}+(u+c)\cdot\left\{\frac{\partial u}{\partial x}+\frac{g}{c}\cdot\frac{\partial h}{\partial x}+\frac{c}{2\rho}\cdot\frac{\partial \rho}{\partial x}\right\}=D_1^x;$$
$$\left\{\frac{\partial u}{\partial t}-\frac{g}{c}\cdot\frac{\partial h}{\partial t}-\frac{c}{2\rho}\cdot\frac{\partial \rho}{\partial t}\right\}+(u-c)\cdot\left\{\frac{\partial u}{\partial x}-\frac{g}{c}\cdot\frac{\partial h}{\partial x}-\frac{c}{2\rho}\cdot\frac{\partial \rho}{\partial x}\right\}=D_2^x; \tag{2.2}$$
$$\frac{\partial v}{\partial t}+u\cdot\frac{\partial v}{\partial x}=D_3^x; \qquad \frac{\partial \rho}{\partial t}+u\cdot\frac{\partial \rho}{\partial x}=D_3^x;$$

A similar equation can be written along the axis y. Here $c=\sqrt{gh}$ is the speed of sound, the right-hand parts $D_k^x, D_k^y; k=1,2,3,4$ are fairly complex expressions that depend on the derivatives along the orthogonal direction. The form of these expressions in the CABARET scheme is not used.

It should be noted that the characteristic equations (2.2) in contrast to shallow water equations at constant density, can not be represented in total differentials and there are no Riemann invariants for the entire flow. However, when going from a differential problem to a discrete one for each computed space-time cell, you can enter a so-called "Local" Riemann invariants:

$$I_1^x=u+h\cdot\left(\frac{g}{c}\right)_c^{n+1/2}+\rho\cdot\left(\frac{c}{2\rho}\right)_c^{n+1/2}; \quad I_2^x=u-h\cdot\left(\frac{g}{c}\right)_c^{n+1/2}-\rho\cdot\left(\frac{c}{2\rho}\right)_c^{n+1/2}; \quad I_3^x=u; \quad I_4^x=\rho; \tag{2.3}$$

In terms of these expressions, the characteristic equations (1.8) are approximated as follows:

$$\frac{\left(I_k^x\right)_C^{n+1/2}-\left(I_k^x\right)_C^n}{\tau/2}+\left(\lambda_k\right)_C^{n+1/2}\cdot\frac{\left(I_k^x\right)_R^n+\left(I_k^x\right)_L^n}{\Delta x}=\left\{D_k^x\right\}_C^{n+1/2}+O\left(\tau,\Delta x^2\right); \quad k=1,\ldots 4; \tag{2.4}$$

Similar formulas are obtained in the direction y. After the computation of phase 1, the conservative variables on the intermediate time layer are already known to us, therefore the left-hand sides of formulas (2.4) are also known and serve as a definition for the unknown right-hand sides.

Calculation of the values of local Riemann invariants on the new time layer is carried out by linear extrapolation. For the first two invariants in subsonic flows we obtain:

$$\left(I_1^x\right)_R^{n+1}=2\left(I_1^x\right)_C^{n+1/2}-\left(I_1^x\right)_L^n; \quad \left(I_2^x\right)_L^{n+1}=2\left(I_2^x\right)_C^{n+1/2}-\left(I_2^x\right)_R^n; \tag{2.5}$$

The direction of extrapolation for the third and fourth invariants depends on the direction of their transference speed on the intermediate time layer, which on each edge is assumed to be equal to half the sum of the conservative velocities in the adjacent cells:

$$\left.\begin{matrix}\left(I_3^x\right)_R^{n+1}=\left(I_3^x\right)_C^{n+1/2}-\left(I_3^x\right)_L^n\\ \left(I_4^x\right)_R^{n+1}=\left(I_4^x\right)_C^{n+1/2}-\left(I_4^x\right)_L^n\end{matrix}\right\} \text{if } u_R^{n+1/2}>0; \qquad \left.\begin{matrix}\left(I_3^x\right)_L^{n+1}=\left(I_3^x\right)_C^{n+1/2}-\left(I_3^x\right)_R^n\\ \left(I_4^x\right)_L^{n+1}=\left(I_4^x\right)_C^{n+1/2}-\left(I_4^x\right)_R^n\end{matrix}\right\} \text{if } u_L^{n+1/2}\le 0; \tag{2.6}$$

Thus, four local invariants are calculated for the vertical faces of the calculation cells located within the computational domain. Similar formulas are obtained for horizontal faces T,B .

For the faces lying on the boundary of the domain, at best, three invariants will be computed. The missing invariant is replaced by the boundary condition - the vanishing of the normal component of the velocity vector on an impermeable wall, or by a local invariant coming outside the region from the "upper" or "lower" reservoir.

A very important operation in the CABARET scheme, as in all other high resolution schemes, is the non-linear correction of flows, which prevents the appearance of nonphysical oscillations in the solution. Here it is implemented on the basis of the maximum principle, which must obey local invariants - their values should be in the interval between the maximum and minimum, determined by the expressions:

$$\max\left(I_k^x\right)=\max\left\{I_L^x,I_C^x,I_R^x\right\}_k^n+\tau\cdot\left\{D_k^x\right\}_C^{n+1/2}; \min\left(I_k^x\right)=\min\left\{I_L^x,I_C^x,I_R^x\right\}_k^n+\tau\cdot\left\{D_k^x\right\}_C^{n+1/2}; k=1,\ldots,4; \tag{2.7}$$

Nonlinear correction consists in the fact that if the value of the local invariant goes beyond the interval defined in this way, then the value of the nearest boundary of this interval is assigned to it.

After this, according to the corrected invariants, the values of the fluxshape variables on the new time layer are calculated.

The process of calculating one time step is completed by **phase 3**, where new values of conservative variables are found.

$$\frac{\vec{\psi}_C^{n+1}-\vec{\psi}_C^{n+1/2}}{\tau/2}+\vec{\Lambda}\left(\vec{\varphi}_f^{n+1}\right)=0; \quad \vec{\psi}_C^{n}=\left\{\left[(h),(\rho h),(\rho hu),(\rho hv)\right]_C^n\right\}^T \tag{2.8}$$

The CABARET difference scheme belongs to the class of high resolution schemes and has the second order of approximation on smooth solutions, when the flow correction procedure does not "work". It allows to resolve of thin vortex flow patterns on relatively coarse computation grids and is well parallelized.

3. The numerical experiment

To perform the calculations, the computing resources of FIC IU RAS were used [3] - 2xIBM Power9 3.8 Hz, 1024 GB RAM, 4xNvidia Tesla V100, 16GB RAM. The Nvidia CUDA Toolkit 9.2 was used. The area was covered with a regular grid NxN - N cells along the x axis, N cells along the y axis. Where T took the values 500, 700, 1000, 1400, 2000, 2800, 4000. The height of the liquid in the "upper" reservoir $H_T^*=1.1$, the height of the liquid in the "lower" reservoir $H_B^*=0.9$, the density of the inflowing liquid $\rho_0=1$ and $\rho_1=1.1$. Parameters of the bottom relief function $B(x,y)$ - $\Delta z=0.3$, G^* - a rectangle located in the center of the area G with sides $0.4L_x, 0.4L_y$ (see Figure 1).

Memory allocation was carried out by the function cudaMallocManaged. It should be noted that by default, the data exchange between graphics accelerators goes along the path GPU-CPU-GPU, in order to allow the exchange of GPU-GPU, we used the function cudaDeviceEnablePeerAccess. For each graphics accelerator, a stream was created through the cudaStreamCreate function. The regular grid is divided into 4 parts along the x-axis, according to the number of graphic accelerators. The kernels corresponding to each phase of the CABARET scheme were launched for each of the 4 areas of the regular grid in the corresponding stream. Kernel synchronization was carried out through the event mechanism (cudaEventSynchronize). The results of acceleration compared to calculations on one graphics accelerator are presented in Figure 3.

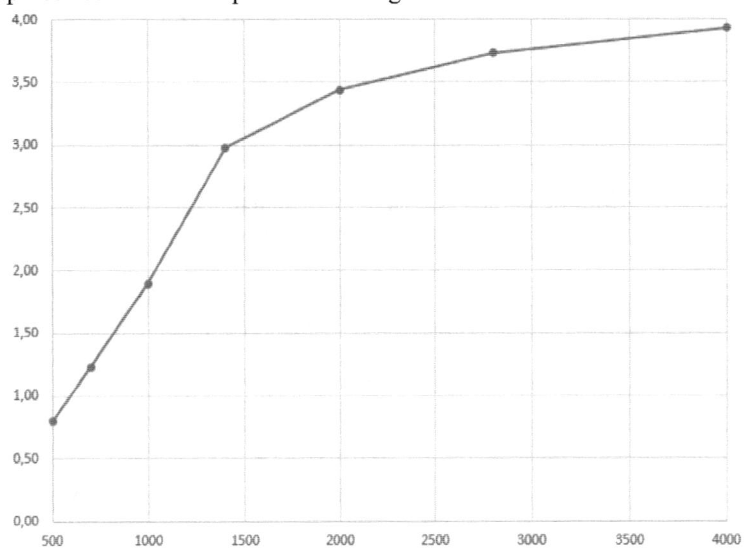

Figure 3 - acceleration of calculations depending on the size of a regular grid.

As can be seen from figure 3 for small grids, we do not get acceleration. This is due to the fact that the exchange is not implemented effectively enough, and is made page by page. The page size, according to information from the Unified profiling memory, is 64 KB result. Increasing the size of the grid we come to an acceleration of four times.

Conclusion

The use of technology Cuda Unified Memory allows without the application of any significant labor costs to move to the calculations on Multi-GPU systems. In some cases, you can get a performance boost. At the moment, this technology is advisable to use for calculations on single servers with large grids. It can be expected that the further development of this technology will expand the scope of its application.

Acknowledgments The authors express their gratitude to the postgraduates Mayorov Pavel and Mayorov Petr for their assistance in the calculations.
The work was supported by the RSCF, project No. 18-11-00163.

References

[1] Evtushenko Y. G., Gorchakov A. Y., Goloviznin V. M. Fast automatic differentiation in problems variations four-dimensional data assimilation (4Dvar) //Journal of Physics: Conference Series. – IOP Publishing, 2018. – T. 1128. – №. 1. – C. 012001.

[2] Goloviznin V M, Zaytsev M A, Karabasov S A and Korotkin I A 2013 New algorithms of computational hydrodynamics for multiprocessor computer systems (Moscow: Moscow University)

[3] Federal Research Center “Computer Science and Control” of Russian Academy of Sciences [Electronic resource]: site. – (Moscow: FRC CSC RAS) - URL: http://frccsc.ru

Journal of Physics: Conference Series **1392** (2019) 012038 doi:10.1088/1742-6596/1392/1/012038

Application of non-stationary gas dynamic functions for mathematical modeling of gas dynamic processes

Y A Grishin[1], **V N Bakulin**[2,3]

[1]Bauman State Technical University, 2nd Bauman str. 5, Moscow, 107005, Russia
[2]Moscow Aviation Institute (National Research University), Volokolamskoye highway 4, Moscow, 125993,Russia
[3]Institute of Applied Mechanics of the Russian Academy of Sciences, Leningradsky avenue 7, Moscow, 125040, Russia

E-mail: [1]grishin.bmstu@mail.ru

E-mail: [2,3]vbak@yandex.ru

Abstract. At numerical calculations of non stationary flow in extended channels with use of basic provisions of disintegration of discontinuities method noticeable acceleration of calculations gives application of specially developed non-stationary gas dynamic functions from Mach number. At the same time the greatest advantages of these functions' application is shown during recording of boundary conditions. Here they are applied in a combination with the known functions of a steady flow as at disintegration of discontinuities, the idea of quasi-steady flow between the waves which passed and reflected from a boundary condition is used. By means of the program complex developed on the basis of non-stationary and stationary functions a number of practical tasks, in particular, of the organization of dynamic pressurization of the piston engine was solved.

1. Introduction

Mathematical model methods enable to assign and solve a wide range of practical gas dynamic problems [1-5]. Among the most efficient numerical calculation methods dedicated to extended channel non-stationary processes is a breakdown of an arbitrary discontinuity (Godunov scheme) [4-8]. This method is successfully used for inertia supercharging wave adjustment of individual inlet pipes of piston engines; description of pneumatic system processes, gas pipes and other gas dynamic problems [7-11].

Non-stationary flow in extended channels is considered in one dimension (length-time coordinates, $x-t$), based on the basic gas dynamic equation system [4-8]:

$$\left.\begin{aligned}\oint(\rho u)dt-\rho dx=0,\ \oint\left(\rho u^2+p\right)dt-\rho u dx=-\tfrac{\lambda}{2D}\iint\rho u|u|dxdt,\\ \oint\rho u\left(e+\tfrac{p}{\rho}+\tfrac{u^2}{2}\right)dt-\rho\left(e+\tfrac{u^2}{2}\right)dx=-\tfrac{4\alpha}{D}\iint\tfrac{(T-T_w)}{\rho}dxdt, p=\rho RT.\end{aligned}\right\}\tag{1}$$

Where: ρ, u, p, T, are density, velocity, pressure, temperature, specific internal energy ($e=c_vT$) respectively; D and T_w are hydraulic diameter and pipe wall temperature; λ and α are coefficients of friction and heat transfer to channel walls [12]. In case of numerical solution, a computational region is divided into intervals Δx, a calculation time step Δt is assumed using a certain maximum possible Courant grid number $Co=a\Delta t/\Delta x$ (where α is a speed of sound), which can ensure the calculation procedure stability. Mass calculations in most cases involve simplified dependencies (linearization) [5,6] to define velocities $u_{i+1/2}$ and pressures $p_{i+1/2}$ of a flow crossing a boundary $i+1/2$ between cells i and $i+1$ of a numerical solution (fig.1):

$$p_{i+1/2} - p_i + a_i\rho_i \left(u_{i+1/2} - u_i\right), \quad p_{i+1} - p_{i+1/2} - a_{i+1}\rho_{i+1} \left(u_{i+1} - u_{i+1/2}\right). \tag{2}$$

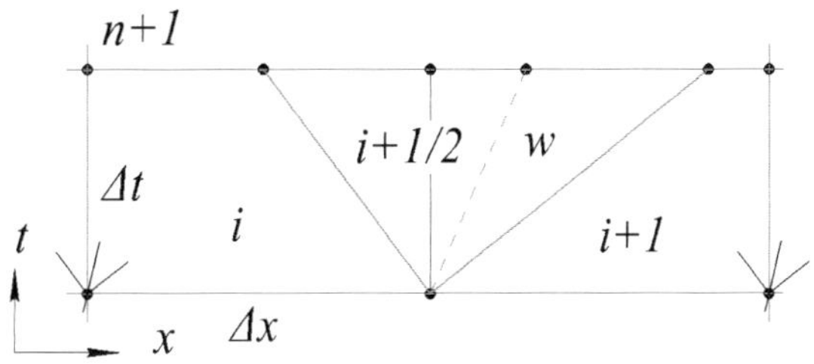

Figure 1. Breakdown of an arbitrary discontinuity on a boundary between numerical solution cells.

However, numerical analysis of high amplitude waves processes, which, for example, are involved in inlet pipes of inertia discharging-adjusted piston engines, cannot ensure a necessary accuracy of results even with significantly reduced calculation step Δt. The best calculation results may be obtained through Rankine-Hugoniot relations as to forward wave front, i.e. the front shall be considered as a shock front to correspond to basic idea of a breakdown of an arbitrary discontinuity [4-7]. Meanwhile, an expansion wave dividing zones i and $i+1/2$, is presented as simple Riemann wave, the parameter variations of which are described by Poisson adiabatic law without a simplifying linearization. By equating velocity expressions $u_{i+1/2} = u_w$ on both sides of the contact area, we shall obtain the pressure equation $p_{i+1/2}$:

$$u_i - \frac{2a_i}{k-1}\left[\left(\frac{p_{i+1/2}}{p_i}\right)^{\frac{k-1}{2k}} - 1\right] = u_{i+1} - \frac{p_{i+1} - p_{i+1/2}}{\sqrt{\left[(k-1)p_{i+1} + (k+1)p_{i+1/2}\right]\rho_{i+1}/2}}, \tag{3}$$

and find $u_{i+1/2}$ using either left or right part. However, the implementation of such calculation pattern leads to some difficulties in using shock-capturing method, since it requires making pre analysis of wave shape happening during a breakdown of an arbitrary discontinuity and is nonuniform in terms of direct and return wave calculation. Besides, the solution of equation (3) requires iterations, thus significantly increasing calculation time.

2. Formulation of problem

The experience has shown that as far as simulation of even high intensity gas dynamic impulses is concerned, a compression wave can also involve non-linearized simple wave relations. Thus, a uniform algorithm can be obtained without additional wave shape analysis and time expenditures for making iterations at each calculation step. Here we can use Mach non-stationary gas dynamic functions $M = u/a$ [13,14].

Thus, Riemann invariant r^+ may be found as follows:

$$r^+ = \frac{2}{k-1}a + u = \frac{2}{k-1}a''^*,$$

where a''^* is a speed of sound stopped by $(u = 0)$ a simple wave, traveling backwards (upwind).

This value is obviously definitely related to Riemann invariant value like corresponding temperature T''^{*}, density ρ''^{*} and pressure p''^{*}. Then the equation of upwind wave front sound speed, temperature, density and pressure variation may be written as follows:

$$a_{i+1/2} = a_i \frac{a''^{*}}{a''^{*}} \frac{\alpha''(M_{i+1/2})}{\alpha''(M_i)} = a_i \frac{\alpha''(M_{i+1/2})}{\alpha''(M_i)}, \quad T_{i+1/2} = T_i \frac{T''^{*}}{T''^{*}} \frac{\tau''(M_{i+1/2})}{\tau''(M_i)} = T_i \frac{\tau''(M_{i+1/2})}{\tau''(M_i)},$$

$$\rho_{i+1/2} = \rho_i \frac{\rho''^{*}}{\rho''^{*}} \frac{\varepsilon''(M_{i+1/2})}{\varepsilon''(M_i)} = \rho_i \frac{\varepsilon''(M_{i+1/2})}{\varepsilon''(M_i)}, p_{i+1/2} = p_i \frac{p''^{*}}{p''^{*}} \frac{\pi''(M_{i+1/2})}{\pi''(M_i)} = p_i \frac{\pi''(M_{i+1/2})}{\pi''(M_i)},$$

where we shall have corresponding non-stationary gas dynamic functions

$$\alpha''(M) = 1/\left(1 + \tfrac{k-1}{2}M\right), \quad \tau''(M) = [\alpha''(M)]^2,$$
$$\varepsilon''(M) = [\alpha''(M)]^{\frac{2}{k-1}}, \pi''(M) = [\alpha''(M)]^{\frac{2k}{k-1}}.$$

The opposite family waves shall involve the equation $r^- = \frac{2}{k-1}a - u = \frac{2}{k-1}a'^{*}$, where is α'^{*} speed of sound stopped by a simple wave, traveling forwards (downwind). Then the equation of downwind wave front sound speed, temperature, density and pressure variation may be written as follows:

$$a_w = a_{i+1} \frac{a'^{*}}{a'^{*}} \frac{\alpha'(M_w)}{\alpha'(M_{i+1})} = a_{i+1} \frac{\alpha'(M_w)}{\alpha'(M_{i+1})}, T_w = T_{i+1} \frac{T'^{*}}{T'^{*}} \frac{\tau'(M_w)}{\tau'(M_{i+1})} = T_{i+1} \frac{\tau'(M_w)}{\tau'(M_{i+1})},$$

$$\rho_w = \rho_{i+1} \frac{\rho'^{*}}{\rho'^{*}} \frac{\varepsilon'(M_w)}{\varepsilon'(M_{i+1})} = \rho_{i+1} \frac{\varepsilon'(M_w)}{\varepsilon'(M_{i+1})}, \quad p_w = p_{i+1} \frac{p'^{*}}{p'^{*}} \frac{\pi'(M_w)}{\pi'(M_{i+1})} = p_{i+1} \frac{\pi'(M_w)}{\pi'(M_{i+1})},$$

where non-stationary gas dynamic functions are:

$$\alpha'(M) = 1/\left(1 - \tfrac{k-1}{2}M\right), \tau'(M) = [\alpha'(M)]^2,$$
$$\varepsilon'(M) = [\alpha'(M)]^{\frac{2}{k-1}}, \quad \pi'(M) = [\alpha'(M)]^{\frac{2k}{k-1}}.$$

A contact area breakdown of an arbitrary discontinuity involves the following relations: $u_{i+1/2} = M_{i+1/2}a_{i+1/2} = u_w = M_w a_w$ and $p_{i+1/2} = p_w$. Pressure variations between wave shape zones can be written as follows:

$$\frac{p_i}{p_{i+1}} = \frac{p_i}{p_{i+1/2}} \cdot \frac{p_{i+1/2}}{p_{i+1}} = \frac{\pi''(M_i)\,\pi'(M_w)}{\pi''(M_{i+1/2})\,\pi'(M_{i+1})} = \left[\frac{\alpha''(M_i)\,\alpha'(M_w)}{\alpha''(M_{i+1/2})\,\alpha'(M_{i+1})}\right]^{\frac{2k}{k-1}}. \tag{4}$$

Equality of velocities upon contact area is represented as:

$$M_{i+1/2}a_i \frac{\alpha''(M_{i+1/2})}{\alpha''(M_i)} = M_w a_{i+1} \frac{\alpha'(M_w)}{\alpha'(M_{i+1})}. \tag{5}$$

Let us designate complexes including initial parameters $p_i, p_{i+1}, a_i, a_{i+1}, M_i$ and M_{i+1} as follows: $F = \alpha'(M_{i+1})/\alpha''(M_i); X = F(p_i/p_{i+1})^{\frac{k-1}{2k}}; Y = Fa_i/a_{i+1}$. Discovering values $\alpha'(M)$ and $\alpha''(M)$ of M and transformations lead to the following simple relations:

$$M_{i+1/2} = \frac{2}{k-1}\frac{X-1}{Y+1}, M_w = \frac{Y}{X}M_{i+1/2}. \tag{6}$$

To continue calculation, we need only $M_{i+1/2}$. After that using corresponding non-stationary gas dynamic function formula we can find velocity $u_{i+1/2}$ and pressure $p_{i+1/2}$. By comparing the test calculation results of shock breakdown of discontinuity in terms of pressure variations

$p_i/p_{i+1} = 2$ and using non-stationary gas dynamic functions we shall obtain insignificant errors $\delta_p = \delta_u = 0.0005$ as against using the equation (3). In case of calculation of breakdown of an arbitrary discontinuity with similar pressure variation using linearized equations (2) we shall have errors two degrees higher: $\delta_p = \delta_u = 0.05$. Besides, the grid number may be reduced $1.5 - 2$ times even as against a linearized pattern. Thus, the calculation time becomes shorter.

The greatest benefits of non-stationary gas dynamic functions are seen in cases of practical computations on one-dimension non-stationary flow calculation regions boundaries, with some local resistances being always present, which are considered as lumped boundary conditions. The most efficient calculation methods for stationary flows over such elements are based on gas dynamic functions [15] of Mach $M = u/a$ or specific speed $\lambda = v/a_{\text{kp.}}$

A construction of calculation models for interaction of a non-stationary flow and such boundary conditions should involve the idea of breakdown of an arbitrary discontinuity. It is convenient to apply non-stationary gas dynamic functions combined with known stationary flow gas dynamic functions, since at each step a breakdown of an arbitrary discontinuity involves the idea of quasi-stationary flow between the waves, which passed and were reflected by boundary conditions [16, 17].

3. Boundary condition like a piston engine valve

Let us review a calculation of interaction between a non-stationary flow and boundary condition like a piston engine valve in an inlet pipe adjusted for inertia discharge. The calculation diagram of a corresponding discontinuity breakdown is shown in fig. 2, where the last calculation cell has indexed m.

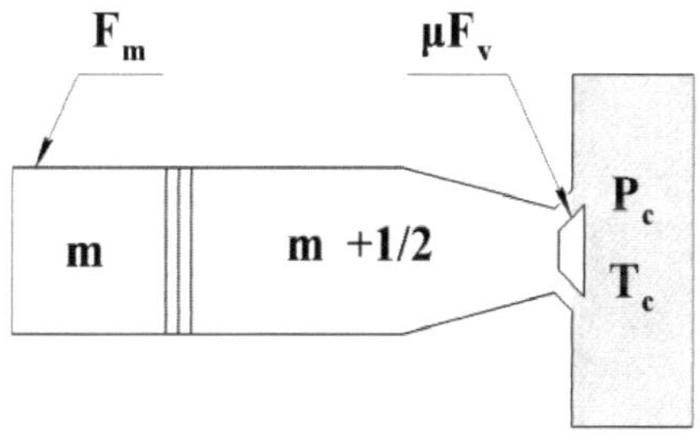

Figure 2. Discontinuity breakdown calculation diagram in case of flowing into a cylinder through a valve.

The through-the-valve flow in a cylinder head is supposed as adiabatic, i.e. stationary slow down temperature remains constant $\left(T^*_{m+1/2} = T^*_v\right)$. Besides, fluid and gas dynamics state that full pressure loss $\left(p^*_{m+1/2} = p^*_v\right)$ in short converging duct like a valve inlet pipe is negligible. Then using Mach stationary and non-stationary functions we can obtain the following equation as to pressure variation between calculation areas (fig. 2):

$$\frac{p_m}{p_c} = \frac{p_m}{p_{m+1/2}} \times \frac{p_{m+1/2}}{p_c} = \frac{\pi''(M_m)\,\pi\left(M_{m+1/2}\right)}{\pi''\left(M_{m+1/2}\right)\pi(M_v)}. \tag{7}$$

For the calculation, the equation (7) is supplemented with the following flow relation [15]:

$$m\frac{F_m p^*_{m+1/2} q\left(M_{m+1/2}\right)}{\sqrt{T^*_{m+1/2}}} = m\frac{\mu F_v p^*_v q(M_v)}{\sqrt{T^*_v}}. \tag{8}$$

Here μ is a flow coefficient reflecting unsteady flow through the valve hole F_v. It is defined using experimental valve blowdown. Parameters p_m, M_m and F_v at calculation step are known. Using iteration methods to solve the equation system (7) and (8) with two unknowns $M_{m+1/2}$ and M_v, we shall define number $M_{m+1/2}$ at this calculation step, required for making further calculation using discontinuity breakdown method with regard to in-pipe wave process and in-cylinder flow. The description of in-cylinder procedures involved methods described in [8]. Meanwhile, it is necessary to specify mass $G\Delta t$ of a charge coming through the valve and the value of total enthalpy $H^*\Delta t = h^*G\Delta t$.

4. Numerical example

The mentioned methods have been used to simulate a wave process in terms of inertia supercharging of diesel engine 1Ch8,5/11 at $n = 1500rpm$. Fig.3 shows corresponding wave pictures on entering into a cylinder head obtained in experiment [11] and calculation, as well as dependency of coefficient of admission v on inlet pipe length L (fig.4). The calculation results are well consistent with the experimental data and prove that having a pipe 1.65 and 2.15m-long shall ensure inertia supercharging with v increased over 1.05. As against the unadjusted system, the coefficient of admission and engine power are increased by 24 %. The results of the calculations mentioned in fig. 3 and 4 prove the validity of the proposed computational simulation methods as to high intense wave processes and can be recommended for solving practical problems of non-stationary gas dynamics, such as design of piston engine intake systems with wave adjustment to ensure inertia supercharging with power much increased.

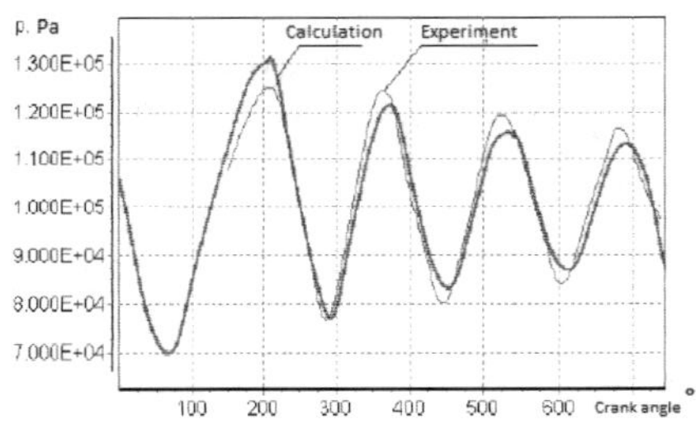

Figure 3. Inertia discharge intake pressure, $L = 1.65m$.

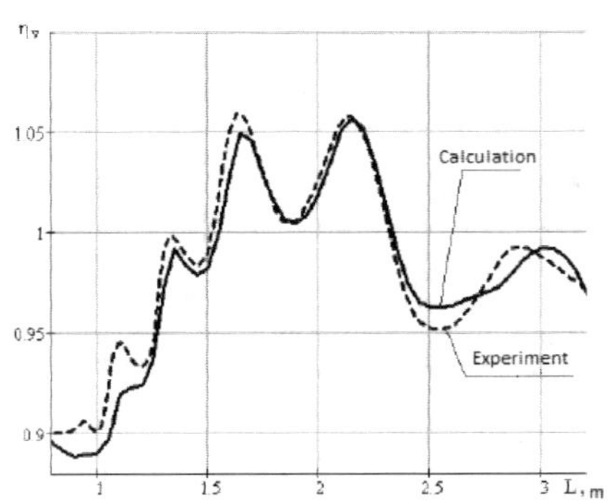

Figure 4. Admission coefficient η_v vs. length L of adjusted inlet pipe.

5. Conclusion

In paper the new modification of a numerical method of disintegration of discontinuities based on specially developed gasdynamic functions of nonstationary flow from Mach number is offered. Application of functions allows to carry out the shock-capturing calculation of wave processes of considerable intensity without use of iterations at the calculation step during setting of increased values mesh Courant number. It increases the speed of calculations and increases the accuracy of results. The greatest advantages of application of these functions are shown at formulation of boundary conditions on local resistance where they during recording of ratios of discontinuities disintegration are combined with the known gasdynamic functions of a stationary flow from Mach number. It simplifies calculations up to obtaining, in certain cases, analytical decisions on boundaries.

References

[1] Grishin Y A, Bakulin V N 2015 New Calculation Schemes Based on the Large-Particle Method for Modeling Gas-Dynamic Problems *Dok. Ph.* **60** 12 555 - 558
[2] Grishin Y A, Bakulin V N 2015 Numerical Investigation of Flow in a Centrifugal Compressor. *J. of Eng. Ph. and Termoph.* **88** 5 1274 - 1279
[3] Grishin Y A, Zenkin V A, Bakulin V N 2014 Implicit Schemes of Method of Large Particles in Problems of Calculating Gas Dynamics *Proc of 17th Int. Conf. on the Meth. of Aeroph. Res.* Novosibirsk:Ins. Theor and Appl. Mech. SB RAS) 76 - 78
[4] Godunov S K et al. 1976 *Numerical Solution of Multi-Dimensional Gas Dynamic Problems* (Moscow: Science) 400
[5] Pirumov U G, Rosliakov G S 1987 *Numerical Methods of Gas Dynamics.* (Moscow: Hight school) 232
[6] Rozhdestvenskii B L, Ianenko N N 1978 *Quasi-Linear Equation Systems* (Moscow: Science) 688
[7] Gogrichiani G V, Shipilin V 1986 *Pneumatic System Transients* (Moscow: VIAM) 160
[8] Grishin Y A, Zenkin V A, Khmelev R N 2017 Boundary Conditions for Numerical Calculation of Gas Exchange in Piston Engines *J. of Eng. Ph. and Termoph.* **90** 4 965 - 970
[9] Grishin Y A, Sonkin V I 2013 *Intke Systems. Internal Combustion engines* (Moscow: Mechanical Engineering) **IV** 14 461 - 474
[10] Stehlig J, Taylor J, Dingelstadt R and Gurney D 2012 Variable-length Air Intake Module for Turbocharged Engines *MTZ* **73** 34 - 41
[11] Semenov B N, Komov A P 1972 *Acoustic Supercharge as a Tool for Reducing Specific Effective Fuel Consumption* (Moscow: NIIINFORMTiAZhMASh-MASh)4 71 13 30 - 35.
[12] Yudaev B N 1988 *Engineering Thermodynamics. Heat Transfer* (Moscow: Hight School) 479
[13] Grishin Y A, Bakulin V N, Zenkin V A 2017 Breakdown of an Arbitrary Discontinuity and Gas Dynamic Functions of Non-Stationary Flow*Proc. of 20th Int. Conf. on Comp. Mech. and Appl. Syst.* (Moscow: MAI publ.) 426 - 428.
[14] Grishin Y A , Kruglov G, Rudoi B P 1977 Gas Dynamic Functions for Calculating Non-Stationary Gas Flows *Proc. of High.Ed. Inst. Mach. Build.* **3** 52 - 58
[15] Abramovich G N 1991 *Applied Gas Dynamics* (Moscow: Science) **1** 600
[16] 1963 Shock Tunnel Experiment Results ed. Ferry A (Nuclear Science and Equipment Publishing Office) 442
[17] 1962 Shock Tunnels: collected articles ed. Rakhmatullina K A and Semenova S S (Moscow: Foreign Literature Publishing) 700

Journal of Physics: Conference Series **1392** (2019) 012039 doi:10.1088/1742-6596/1392/1/012039

Mathematical modelling of hydrogen safety problems with CABARET scheme

A.A. Kanaev, V.Yu. Glotov, V.M. Goloviznin, V.G. Kondakov, A.E. Kiselev

Nuclear Safety Institute of Russian Academy of Science, 52, Bolshaya Tulskaya Street, Moscow, 115191, Russian Federation

E-mail: kanaev@ibrae.ac.ru, glotov-v@yandex.ru, gol@ibrae.ac.ru, kondakov@ibrae.ac.ru, ksv@ibrae.ac.ru

Abstract. Experiments on overall fluid transfer and mixing in the large scale experimental facility (PANDA), where the initial helium-rich layer is eroded by a flow resulting from the interaction of a jet/plume with horizontal flow obstruction, were simulated with the CABARET scheme. Computational simulation results are provided for two mesh resolutions and are compared with the experimental data. Good agreement between calculated and measured gas temperature and the breakup of the helium layer dynamics was achieved by refining the mesh in the jet area and by adding the model of radiation heat transfer.

1. Introduction

During a severe accident with loss of coolant, a large amount of hot steam and hydrogen can enter the atmosphere of the containment of a pressurized water reactor. The non-uniform distribution and stratification of hydrogen in the containment can lead to the formation of an explosive mixture areas that threatens the integrity of the containment. To ensure hydrogen safety of nuclear power plants, systematic experimental studies are carried out on hydrogen stratification issues (international projects: ERCOSAM-SAMARA [1], OECD/NEA HYMERES [2], OECD/NEA HYMERES-2).

HYMERES project HP1 test series addressed helium (simulating hydrogen) layer erosion by the flow, resulting from the interaction of a steam jet with flow obstructions. One of these tests was selected for a benchmark exercise. The comparison between calculated and experimental results, as well as between simulations, raised a few questions about the actual importance of considering radiative heat transfer, the relation between mesh topology and other modelling aspects (turbulence model, numerical methods, boundary conditions, etc.) [3]

At Nuclear Safety Institute (IBRAE RAS), a software tool for numerical modeling of hydrogen safety problems based on innovative approaches to constructing detailed multidimensional and multiphysical transport models for gas mixtures is developed. It is based on the CABARET scheme [4], which is intended to solve systems of hyperbolic equations. The method is second-order accurate on non-uniform grids in space and time, has a very small dispersion error and computational stencil defined within one space–time cell. For shock-capturing, the scheme uses a conservative non-linear correction procedure which is directly based on the maximum principle. Unlike the semi-empirical approaches used in modern engineering CFD-codes, there are no tuning parameters in the CABARET scheme. The use of a parameter-free approach can make it possible to estimate the effect of thermal radiation on helium stratification process.

2. Experimental facility and Initial data

The experimental investigation is performed in the PANDA facility. The computational model of the PANDA facility was constructed on the basis of the data in [5, 6]. The main modeling elements are the pressure vessels Vessel 1 and Vessel 2, the interconnecting pipe IP and the tubes for injecting steam and helium (Figure 1) with total internal volume $184m^3$, height $8m$, diameter of the cylindrical part of vessels $4m$. The vessels were carefully insulated using rock-wool mats, which was simulated by a heat transfer coefficient based on experimentally determined heat loss rate.

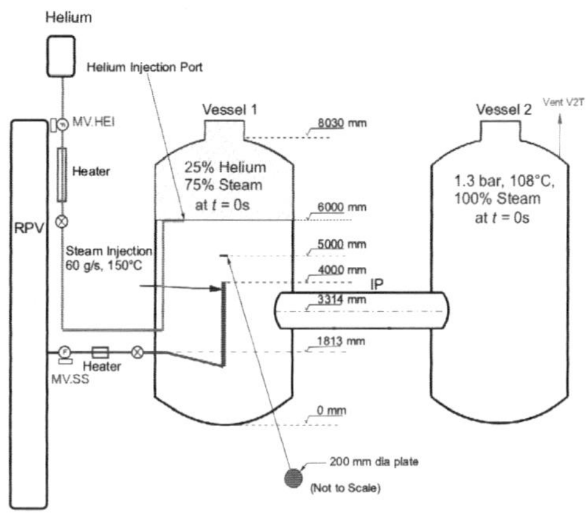

Figure 1. Experimental configuration and nominal initial condition for HP1_6_2.

The test is aimed at assessing the helium-rich layer erosion and overall fluid transport and mixing in Vessel 1 and Vessel 2 for two-gases, for fluid conditions without condensation and with constant pressure, in the presence of a horizontal flow obstruction by a jet characterized by high momentum (Table 1).

Table 1. The main conditions of HP1_6_2 experiment.

Parameters	Nominal Value
Initial vessel pressure, *bar*	1.3
Vessel 1 helium molar fraction	25% (above $6m$)
Vessel 1 steam + air molar fraction	100% steam (below $6m$), no air
Vessel 2 steam + air molar fraction	100% steam, no air
Vessel wall temperature, °C	108
Vessel fluid temperature, °C	108
Steam injection flow rate, g/sec	60
Steam injection temperature, °C	150
Pressurization	-
Flow obstruction	+ (horizontal circular plate)
Steam condensation	-

3. Mathematical model

The low-speed turbulent convective flow of a multi-component (helium-air-vapor) mixture that occurs in the field of gravity is simulated. The system of defining equations includes the equations of balance of mass, momentum and energy for the mixture as a whole, as well as the equations of components transfer:

$$\begin{cases} \frac{\partial \rho}{\partial t} + div\left(\rho \vec{u}\right) = 0, \\ \frac{\partial \rho u_i}{\partial t} + div\left(\rho u_i \vec{u}\right) = -\nabla_i P^* + \nabla_i \tau_{ij} + \left(\rho - \bar{\rho}\right) g_i, i, j = 1, \ldots, 3 \\ \frac{\partial \rho C_V T}{\partial t} + div\left(\rho C_P T \vec{u}\right) = -div\left(\vec{q} + \left(\sum_{k=1,N} C_{p,k} \vec{j}_k\right) T\right), \\ \frac{\partial \rho_k}{\partial t} + div\left(\rho_k \vec{u}\right) = -div\left(\vec{j}_k\right), k = 1, \ldots, N. \end{cases} \quad (1)$$

Here ρ, P, u and T are the density, pressure, speed and temperature of the mixture. When $u/c \ll$(c - speed of sound), the pressure P might be represented as a sum of the thermodynamic pressure, averaged over the volume of the mixture $P_0=P_0(t)$, and the dynamic pressure $P^*=P^*(x,t)$, ensuring the propagation of sound waves in the medium [7]. In this case, the propagation of small perturbations in the medium is a "fast" process as compared to diffusion phenomena and convective mixing of gases. The change in the internal energy during the passage of the sound wave through the medium occurs due to compressibility. Thus, the dynamic pressure can be found from the equation

$$P^* = c^2\left(\rho - \rho_0\right), \rho_0 = \rho\left(P_0/P\right)^{1/\gamma}. \quad (2)$$

To increase the time step ($\tau \sim 1/c$), artificial sound speed $c_a<c$ can be used (artificial compressibility method). Choosing c_a from the condition $u/c_a 10^{-1}$, we obtain a weakly compressible approximation of the medium $\partial\rho/\rho \sim (u/c_a)^2 < 10^{-2}$.

The total diffusion transfer is represented in the system (1) by a stress tensor τ_{ij}, and vectors of heat q and mass j_k flux density for which gradient approximations are used in the form of Newton, Fourier and Fick laws.

3.1. Radiation heat transfer model

For the problems of low-temperature gas dynamics, a radiation has an effect only on the redistribution of energy in the medium. However, this is often quite enough to drastically change the whole picture of the phenomenon. Only polyatomic gases (CO_2, H_2O, NH_3, etc.) have a significant ability to emit and absorb radiant energy. For the problems of containment thermal hydraulics, the most interesting is water vapor, which comes in a large volume from the break in the primary circuit during serve accident. With increasing temperature and vapor concentration in the containment rooms, the medium becomes optically dense. For example, with a partial pressure of steam equal to $1 bar$ and a temperature equal to $500K$, the optical thickness of $1m$ thick layer is $\tau \approx 120 \gg 1$ [8]. In this case, the radiation field becomes isotropic; therefore, the diffusion approximations become valid.

In the Rosseland diffusion model of radiative thermal conductivity, the radiative heat flux is expressed through the temperature gradient, as well as the conductive flux in the Fourier law

$$\vec{q}_{rad} = -k_{rad} \cdot gradT, k_{rad} = \frac{16\sigma T^3}{3a_R}. \quad (3)$$

Here $a_R=a_R(x_v\, T,\, P)$ is an average steam absorption coefficient according to Rosseland, σ is a Stefan-Boltzmann constant. On the walls, the isotropic condition of radiation is violated. In practice, an approximate formula is used

$$q_w = \varepsilon_{gw}\sigma\left(T_g^4 - T_w^4\right), \varepsilon_{gw} = \frac{1}{1/\varepsilon_g + 1/\varepsilon_w - 1}. \tag{4}$$

Here ϵ_g and ϵ_w are gas and wall emissivity coefficients, ϵ_{gw} is reduced emissivity.

4. Results

In the experiment, the time histories of gas temperatures and helium concentrations above the injection describe the progression of the erosion process, which is characterized, at each elevation, by helium concentration drop and temperature increase as a result of the rise of the leading edge of the steam jet and its associated mixing zone.

To study the effect of mesh resolution on stratification erosion dynamics, two hexagonal unstructured meshes of 1 million (coarse) and 3.6 million (finer) cells were used. The mesh cells were modified locally to reduce the cell size from $1.5cm$ to $0.75cm$ in the jet area (Figure 2).

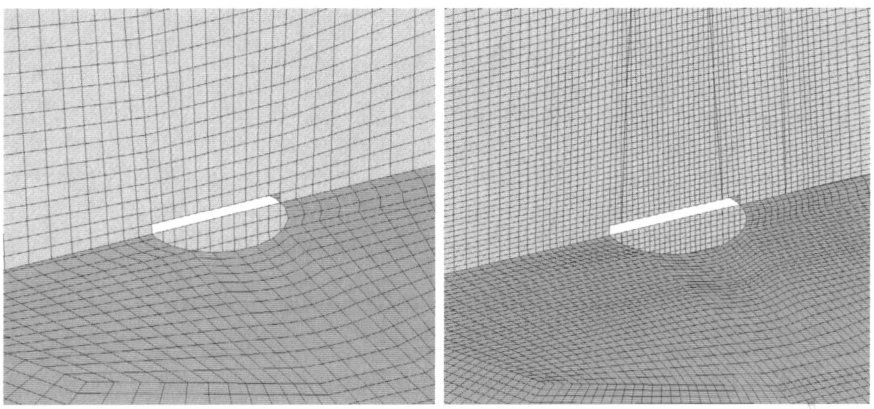

Figure 2. Mesh refinement in near jet area from $1.5cm$ (left) to $0.75cm$ (right).

The calculations without the included thermal radiation model showed that for a coarse mesh the time of helium stratification breakup is substantially underpredicted. This is related to the difficulties in modelling the resulting recirculation flow above the plate due to the underpredicted broadening of the free jet on the coarse mesh (Figure 3, right). On the finer mesh, erosion dynamics at the $6926m$ level is close to the experimental (Figure 3, left), but at the levels above, the mixing is smoother. Also on both meshes an overprediction of temperature of 5-7 °C is observed.

Journal of Physics: Conference Series **1392** (2019) 012039 doi:10.1088/1742-6596/1392/1/012039

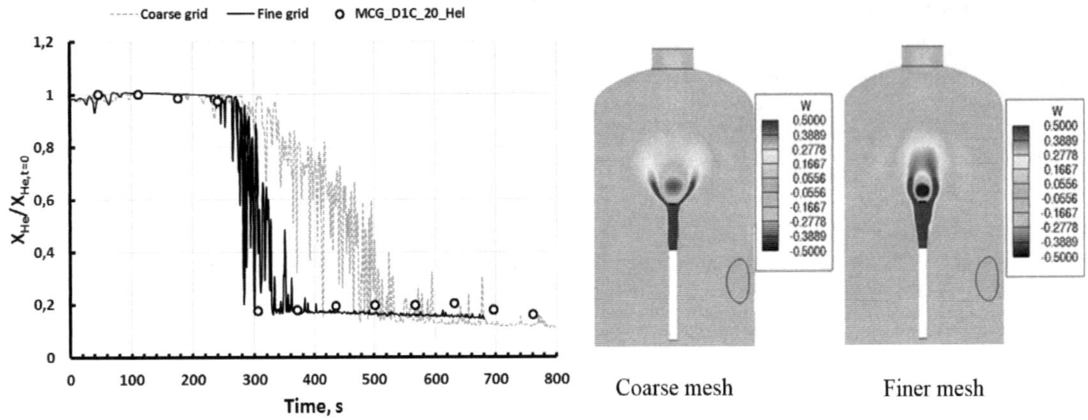

Figure 3. Time history of helium concentration at the level 6926m for two meshes (left) and comparison of averaged vertical speed fields for two meshes (right).

Despite the fact that the vapor temperature is relatively small (below 150 °C), thermal radiation may have a noticeable role. The calculation with the enabled Rosseland thermal heat transfer model showed significantly better agreement with the measured temperature (Figure 4). The dynamics of the helium layer breakup in the upper region of the facility also demonstrates a better convergence (Figure 5, left), which may be due to increased buoyancy of the plume resulting from the interaction of a jet with the disk.

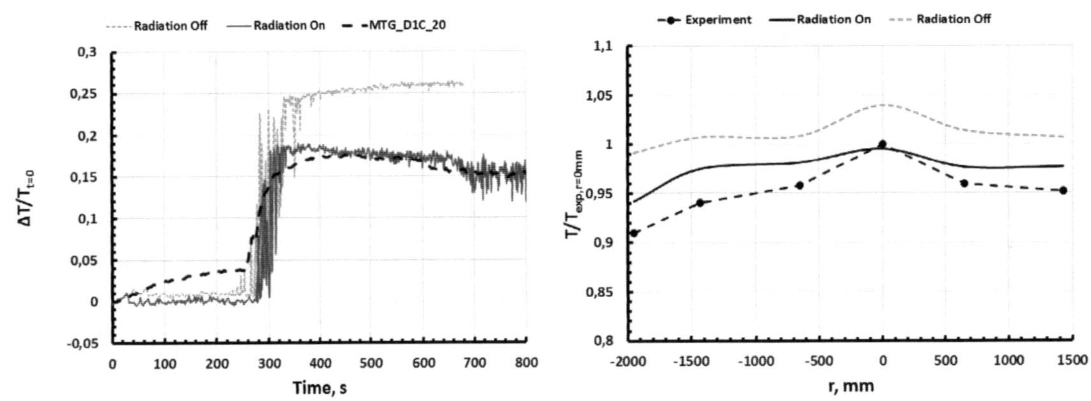

Figure 4. Time history of temperature on the axis line at the level 6926m (left) and gas temperature horizontal profile at level 5626m at 300sec (right).

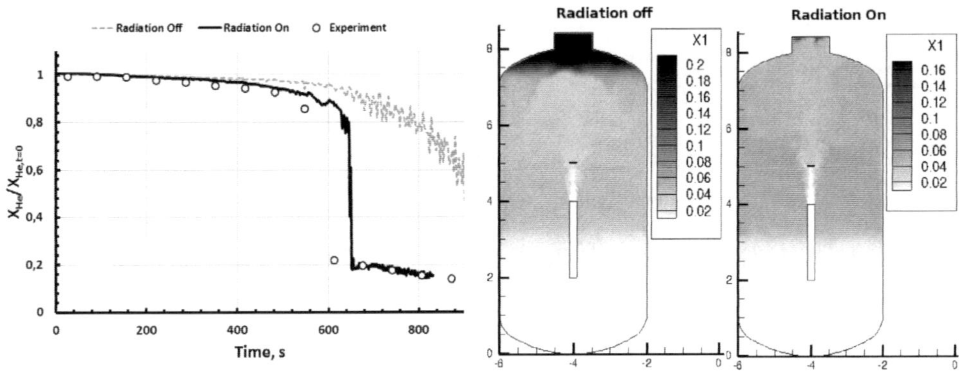

Figure 5. Time history of of helium concentration at the level 8030m (left) and Helium distribution comparison in Vessel 1 at the time of 680sec (right).

5. Conclusions

An important advantage of the CABARET technique for the modeling of a turbulence is the lack of semi-empirical models and tuning parameters. The only source of uncertainty is the grid itself, the selection criterion of which is based on the analysis of grid convergence. Since convergence to experimental data is possible only with modeling of all the important processes, the use of CABARET can significantly simplify the interpretation of the results of experiments in a complex multiphysical formulation. According to the results of the calculations, significant influence of considering of radiation heat transfer to erosion process was observed. At the moment, the question of the effect of thermal radiation at low temperatures of steam and the choice of numerical models of thermal radiation is open and requires further research.

Acknowledgments

This research uses the data from the high quality experiments conducted in PSI in the framework of HYMERES project. We thank our colleagues from HYMERES-1 group who provided insight and expertise that greatly assisted the research.

The research is carried out using the equipment of the shared research facilities of HPC computing resources at Lomonosov Moscow State University.

References

[1] Paladino D et al. 2016 Outcomes from the EURATOM-ROSATOM ERCOSAM SAMARA projects on containment thermal-hydraulics for severe accident management *Nuclear Eng. and Design* **308** 103-114
[2] Paladino D et al. 2014 *Proc. of ICAPP 2014* (Charlotte)
[3] Andreani M et al. 2016 *CFD for Nuclear Reactor Safety Applications* (Cambridge)
[4] Karabasov S and Goloviznin V 2009 Compact Accurately Boundary-Adjusting high-REsolution Technique for fluid dynamics *Journal of Computational Physics* **228** 7426-51
[5] Mignot G, Kapulla R, Paranjape S, Zboray R, Fehlmann M, Bissels W,Suter S,Paladino D 2013 *OECD/NEA HYMERES project: PANDA test facility description and geometrical specifications* (Villigen:Paul Scherrer Institut)
[6] Paranjape S, Fehlmann M, Kapulla R, Mignot G, Suter S, Andreani M and Paladino D 2015 *OECD/NEA HYMERES project: PANDA test HP1_6_2 specifications for the blind benchmark with initial and boundary conditions* (Villigen:Paul Scherrer Institut)
[7] Rieper F 2008 *On the Behaviour of Numerical Schemes in the Low Mach Number Regime* (Cottbus:Brandenburg University of Technology)
[8] Ozicik M 1985 *Heat Transfer, A Basic Approach* (Scranton:McGraw-Hill Book Company)

Algorithms for numerical solution of integral equations of three-dimensional scalar diffraction problem

A A Kashirin, S I Smagin

Computing Center, Far Eastern Branch, Russian Academy of Sciences; 65 Kim Yu Chena St., Khabarovsk, 680000, Russian Federation

E-mail: `elomer@mail.ru`, `smagin@ccfebras.ru`

Abstract. The three-dimensional diffraction problem of stationary acoustic waves on a homogeneous inclusion is considered. It is reduced to the weakly singular boundary Fredholm integral equations of the first kind with one unknown function, each of which is conditionally equivalent to the original problem. By using the original method of averaging the integral operators kernels, these equations are approximated by systems of linear algebraic equations. The resulting systems are solved numerically by the generalized minimal residual method (GMRES). Then the solution of the initial problem is calculated. To find the solution on the spectrum of integral operators, where the condition of equivalence of integral equations to the original problem is violated, the interpolation solution method is proposed. It does not require knowledge of the spectrum and allows us to find the approximated solutions with high accuracy. The proposed algorithms have been implemented in the computing cluster of Computing Center FEB RAS. The results of the calculations that allow us to assess the possibilities of this approach are presented.

1. Introduction

Three-dimensional diffraction problems are found in various fields of science and technology. Their analytical solutions can be found only in exceptional cases. Therefore, the diffraction problems are mainly solved numerically.

In this work, the initial problem is reduced to weakly singular boundary integral Fredholm equations of the first kind with one unknown function. Each of these equations is conditionally equivalent to the diffraction problem. Using the special method of averaging the integral operators kernels with weak singularities, these equations are approximated by systems of linear algebraic equations. The resulting systems are solved numerically by means of the generalized minimal residual method (GMRES). Then approximate solutions of the original problem are calculated [1], [2].

Integral operators of these equations have a spectrum. It is a countable set of positive wave numbers. For each of these numbers the condition of integral equations equivalence to the initial problem is violated. Herein the method of solution interpolation is proposed. It does not require knowledge of the spectrum and allows us to find a solution of the diffraction problem for all positive wave numbers [3].

This research was supported in through computational resources provided by the Shared Facility Center Data Center of FEB RAS (Khabarovsk) [4]. The results of numerical experiments are presented. They allow us to assess the potential of this approach.

2. The initial problem and its equivalent integral equations

Let us formulate the initial problem.

Problem 1. In bounded domain Ω_i of three-dimensional Euclidean space R^3 and in unbounded domain $\Omega_e = R^3 \backslash \bar{\Omega}_i$ separated by a closed surface $\Gamma \in C^{r+\beta}$, $r+\beta > 1$, find complex-valued functions $u_{i(e)} \in H^1(\Omega_{i(e)}, \Delta)$, satisfying integral identities

$$\int\limits_{\Omega_{i(e)}} \nabla u_{i(e)} \nabla v^*_{i(e)} dx - k^2_{i(e)} \int\limits_{\Omega_{i(e)}} u_{i(e)} v^*_{i(e)} dx = 0 \qquad \forall v_{i(e)} \in H^1_0\left(\Omega_{i(e)}\right), \tag{1}$$

matching conditions at the interface of media from Ω_i and Ω_e

$$\left\langle u_i^- - u_e^+, \mu \right\rangle_\Gamma = \langle f_0, \mu \rangle_\Gamma \quad \forall \mu \in H^{-1/2}(\Gamma), \tag{2}$$

$$\langle \eta, p_i N^- u_i - p_e N^+ u_e \rangle_\Gamma = \langle \eta, p_e f_1 \rangle_\Gamma \quad \forall \eta \in H^{1/2}(\Gamma),$$

as well as the radiation condition at infinity for u_e

$$\partial u_e / \partial |x| - i k_e u_e = o\left(|x|^{-1}\right), \quad |x| \to \infty, \tag{3}$$

if functions $f_0 \in H^{1/2}(\Gamma)$ and $f_1 \in H^{-1/2}(\Gamma)$ are set on the boundary Γ.

Here v^* is a complex conjugate function to v, $\langle \cdot, \cdot \rangle_\Gamma$ is duality ratio on $H^{1/2}(\Gamma) \times H^{-1/2}(\Gamma)$, generalizing the scalar product in $H^0(\Gamma)$, $u^\pm \equiv \gamma^\pm u$, $\gamma^- : H^1(\Omega_i) \to H^{1/2}(\Gamma)$, $\gamma^+ : H^1(\Omega_e) \to H^{1/2}(\Gamma)$ are trace operators, $N^- : H^1(\Omega_i, \Delta) \to H^{-1/2}(\Gamma)$, $N^+ : H^1(\Omega_e, \Delta) \to H^{-1/2}(\Gamma)$ are operators of normal derivatives [5], $f_0 = u_0^+$, $f_1 = N^+ u_0$,

$$k^2_{i(e)} = \omega\left(\omega + i\gamma_{i(e)}\right) / c^2_{i(e)}, \quad \mathrm{Im}(k_{i(e)}) \geq 0, \quad p_{i(e)} = c^2_{i(e)} k^{-2}_{i(e)} \rho^{-1}_{i(e)},$$

ω is a circular oscillation frequency, $c_{i(e)} > 0$, $\rho_{i(e)} > 0$, $\gamma_{i(e)} \geq 0$. Definitions of the functional spaces used hereafter are available in [5].

Let us introduce the following notations:

$$\left(A_{i(e)} q\right)(x) \equiv \left\langle G_{i(e)}(x, \cdot), q \right\rangle_\Gamma, \quad \left(B_{i(e)} q\right)(x) \equiv \left\langle N_x G_{i(e)}(x, \cdot), q \right\rangle_\Gamma, \tag{4}$$

$$\left(B^*_{i(e)} q\right)(x) \equiv \left\langle N_{(\cdot)} G_{i(e)}(x, \cdot), q \right\rangle_\Gamma, \quad G_{i(e)}(x, y) = \exp\left(i k_{i(e)} |x - y|\right) / (4\pi |x - y|).$$

We will search for the solution of problem 1 in the form of potentials

$$u_e(x) = (A_e q)(x), \quad x \in \Omega_e, \tag{5}$$

$$u_i(x) = \left(p_{ei} A_i \left(N^+ u_e + f_1\right) - B_i^* \left(u_e^+ + f_0\right)\right)(x), \quad x \in \Omega_i,$$

where $q \in H^{-1/2}(\Gamma)$ is an unknown density, $f_0 \in H^{1/2}(\Gamma)$, $f_1 \in H^{-1/2}(\Gamma)$, $p_{ei} = p_e / p_i$.

The kernels of the integral operators here are the fundamental solutions of the Helmholtz equations and their normal derivatives. Therefore, they satisfy the identities (1) and the radiation condition at infinity (3). Furthermore, the fulfillment of the first matching conditions (2) for them automatically means the fulfillment of the second. By substituting potentials (5) into the first condition, we obtain a weakly singular Fredholm integral equation of the first kind to find unknown density q:

$$\langle Cq, \mu\rangle_\Gamma = \langle f_2, \mu\rangle_\Gamma \ \forall \mu \in H^{-1/2}(\Gamma), \tag{6}$$

where

$$C = (0.5 + B_i^*)\, A_e + p_{ei} A_i\,(0.5 - B_e)\,, \quad f_2 = -\,(0.5 + B_i^*)\, f_0 + p_{ei} A_i f_1.$$

Problem 1 allows for another equivalent formulation in the form of an integral Fredholm equation of the first kind with a weak singularity in the kernel. We will seek its solution in the form of

$$u_i(x) = (A_i q)\,(x), \quad x \in \Omega_i, \tag{7}$$

$$u_e(x) = \left(A_e\left(f_1 - p_{ie} N^- u_i\right) - B_e^*\left(f_0 - u_i^-\right)\right)(x), \quad x \in \Omega_e,$$

where $q \in H^{-1/2}(\Gamma)$ is an unknown density, $f_0 \in H^{1/2}(\Gamma)$, $f_1 \in H^{-1/2}(\Gamma)$, $p_{ie} = p_i/p_e$.

In this case, problem 1 is reduced to equation

$$\langle Dq, \mu\rangle_\Gamma = \langle f_0, \mu\rangle_\Gamma \ \forall \mu \in H^{-1/2}(\Gamma), \tag{8}$$

$$D = (0.5 - B_e^*)\, A_i + p_{ie} A_e\,(0.5 + B_i)\,.$$

Theorem 1. Let $f_0 \in H^{1/2}(\Gamma)$, $f_1 \in H^{-1/2}(\Gamma)$, $\gamma_e > 0$ or ω is not the eigenfrequency of the following problem

$$\Delta u + k_e^2 u = 0, \quad x \in \Omega_i, \qquad u^- = 0. \tag{9}$$

Then equations (6) and (8) are correctly solvable in the class of densities $q \in H^{-1/2}(\Gamma)$ and formulas (5) and (7) give the solution of problem 1.

3. Numerical method

Let us construct surface cover Γ by system $\{\Gamma_m\}_{m=1}^M$ of the neighborhoods of node points $x_m' \in \Gamma$ lying inside the spheres of radii h_m with the centers in x_m', and denote its subordinate unity partition as $\{\varphi_m\}$. We shall use functions

$$\varphi_m(x) = \varphi_m'(x)\left(\sum_{k=1}^M \varphi_k'(x)\right)^{-1}, \qquad \varphi_m'(x) = \begin{cases} \left(1 - r_m^2/h_m^2\right)^3, & r_m < h_m, \\ 0, & r_m \geq h_m, \end{cases}$$

where $x \in \Gamma$, $r_m = |x - x_m'|$, $\varphi_m \in C^1(\Gamma)$ if $\Gamma \in C^{r+\beta}$, $r + \beta > 1$.

The approximate solutions of equations (6) and (8) will be sought on the grid $\{x_m\}$,

$$x_m = \frac{1}{\bar\varphi_m}\int\limits_\Gamma x\varphi_m d\Gamma, \quad \bar\varphi_m = \int\limits_\Gamma \varphi_m d\Gamma,$$

whose nodes are the centers of gravity of functions φ_m. We assume that for all $m = 1, 2, \ldots, M$ inequalities

$$0 < h' \leq |x_m - x_n|\,, \quad m \neq n, \quad n = 1, 2, \ldots, M,$$

$$h' \leq \sigma_m \leq h_m \leq h, \quad h/h' \leq q_0 < \infty,$$

are satisfied where h, h' are positive numbers depending on M, q_0 does not depend on it, $\sigma_m^2 = 0.5\bar\varphi_m$.

Integral operators from (4) on Γ are approximated by expressions [2]

$$\left\langle A_{i(e)}q,\varphi_m\right\rangle_\Gamma \approx \sum_{n=1}^{M} A_{i(e)}^{mn} q_n, \quad m=1,2,...,M, \tag{10}$$

$$A_{i(e)}^{mn} \equiv A_{mn}\left(k_{i(e)}\right), \quad A_{mn}(k) = \frac{\bar{\varphi}_m\bar{\varphi}_n}{8\pi r_{mn}} \exp\left(\mu_{mn}^2-\gamma_{mn}^2\right)\left(w\left(z_{mn}^-\right)-w\left(z_{mn}^+\right)\right), \quad n\neq m,$$

$$A_{mm}(k) = \frac{\bar{\varphi}_m^2}{4\pi}\exp\left(\mu_{mm}^2\right)\left(ikw\left(\mu_{mm}\right)+\frac{\sqrt{2\pi}}{\bar{\varphi}_m}\left(\frac{\bar{\varphi}_m}{\pi\sigma_m}+2\sigma_m-\frac{k^2\sigma_m^3}{3}\right)\right),$$

$$\sigma_{mn}^2=\sigma_m^2+\sigma_n^2, \quad \mu_{mn}=0.5k\sigma_{mn}, \quad z_{mn}^\pm=\mu_{mn}\pm i\gamma_{mn}, \quad \gamma_{mn}=r_{mn}/\sigma_{mn}, \quad i^2=-1,$$

$$w(z) = -\frac{2i}{\sqrt{\pi}}\exp\left(-z^2\right)\int_z^\infty \exp\left(t^2\right)dt,$$

$$\left\langle aq+B_{i(e)}q,\varphi_m\right\rangle_\Gamma \approx \sum_{n=1}^{M} B_{i(e)}^{mn} q_n, \quad m=1,2,...,M, \quad a=\pm 0.5, \tag{11}$$

$$\left\langle aq+B_{i(e)}^{*}q,\varphi_m\right\rangle_\Gamma \approx \sum_{n=1}^{M} B_{i(e)}^{nm} q_n, \quad m=1,2,...,M, \tag{12}$$

$$B_{i(e)}^{mn} = \frac{\eta_{mn}}{4\pi r_{mn}^2}\exp\left(ik_{i(e)}r_{mn}\right)\left(ik_{i(e)}r_{mn}-1\right)\bar{\varphi}_m\bar{\varphi}_n, \quad n\neq m,$$

$$B_{i(e)}^{mm} = \left(-|a|+a+\mathrm{Gs}_m\right)\bar{\varphi}_m, \quad \eta_{mn}=\sum_{l=1}^{3} n_{ml}\frac{x_{ml}-x_{nl}}{r_{mn}}, \quad \mathrm{Gs}_m=-\sum_{n\neq m}^{M}\frac{\eta_{nm}\bar{\varphi}_n}{4\pi r_{mn}^2},$$

n_{ml} are components of the unit vector of the external normal to surface Γ at the point x_m.

Operators on the left-hand sides of equations (6) and (8) are approximated by compositions of operators (10)–(12):

$$\langle Cq,\varphi_m\rangle_\Gamma \approx \sum_{n=1}^{M} C_{ie}^{nm} q_n, \quad \langle Dq,\varphi_m\rangle_\Gamma \approx -\sum_{n=1}^{M} C_{ei}^{nm} q_n, \quad m=1,2,...,M, \tag{13}$$

$$C_{ie}^{nm} = B_i^{nm}A_e^{mn} - p_{ei}A_i^{mn}B_e^{mn},$$

whereas those on the right-hand sides of equations (6) (8) – by formulas

$$\langle f_2,\varphi_m\rangle_\Gamma \approx \sum_{n=1}^{M}\left(p_{ei}A_i^{mn}f_{1n}-B_i^{nm}f_{0n}\right), \quad \langle f_0,\varphi_m\rangle_\Gamma = \bar{\varphi}_m f_{0m}, \tag{14}$$

$$f_{lm} = \langle f_l,\varphi_m/\bar{\varphi}_m\rangle_\Gamma, \quad l=0,1, \quad m=1,2,...,M.$$

By solving the corresponding systems of linear equations, we find the approximate values of the densities of integral equations at the points of discretization. After that, the diffraction problem solution can be easily and accurately calculated with the help of integral representations both in the near and far zones.

Journal of Physics: Conference Series **1392** (2019) 012040 doi:10.1088/1742-6596/1392/1/012040

3.1. Problem solution on the spectrum

For the numerical solution of integral equations on the spectrum, we are going to use the following scheme. Let $k_e > 0$ denotes some eigenvalue of the problem (9), and $q(k_e)$ is a particular solution of the inhomogeneous equation (6) or (8) depending on it. We choose a number $\delta > 0$. Then the interpolation formula

$$q(k_e) = 4q(k_e + i\delta) - q(k_e - \delta + i\delta) - q(k_e + \delta + i\delta) - q(k_e + 2i\delta) + O(\delta^4) \tag{15}$$

is valid for the particular solution of the integral equation. Here all densities in the right part are solutions of correctly solvable integral equations. The substitution of the density in formulas (5) or (7) gives an approximate solution of the original problem.

4. Results of numerical experiments

The program for numerical solution of diffraction problems is written in Fortran 90, which is a console application designed to work on multiprocessor computing systems. Intel Fortran is used as a compiler, complimented by Intel Math Kernel Library (Intel MKL). OpenMP standard is implemented for this compiler.

The correctness and accuracy of the numerical method have been verified by means of numerical solution of test problems with known exact solutions. As such internal and external boundary value Dirichlet and Neumann problems for the Helmholtz equation in areas bounded by a unit sphere and a three-axis ellipsoid, as well as the problem of plane wave diffraction on a unit ball were used. The convergence of approximate solutions to exact solutions of the same problems on thickening grids was also checked. In all the cases considered, there was a good agreement of decisions [2].

Example 1. The diffraction problem of a plane acoustic wave on a unit ball with a center at the origin and three different sets of parameters of the medium and inclusion is considered. The complex amplitude of the initial pressure wave field has the form $u_0(x) = \exp(ik_e x_3)$, media parameters: I) $k_i = 12.5$, $\rho_i = 4$, $k_e = 7.725251836937$, $\rho_e = 3$; II) $k_i = 7$, $\rho_i = 2$, $k_e = 16.9236212852138$, $\rho_e = 5$; III) $k_i = 21$, $\rho_i = 7$, $k_e = 13.6980231532492$, $\rho_e = 4.5$.

The eigenvalues of the problem (9) are chosen as k_e.

The main aim has been to demonstrate the potential of the solution interpolation method. In this regard, the problem from example 1 was solved twice with and without interpolation of the solution (formula (15), $\delta = 0.01$). It was formulated in the form of equation (8). The number of discretization points M varied from 500 to 128,000. The exact solution of the diffraction problem is given in [2].

Figures 1 and 2 show the values of relative solution errors of the diffraction problem from example 1 calculated in the norm of grid functions spaces $H_h^0(\Omega_{i(e)})$ (fig. 1 – without interpolation of the solution, fig. 2 – with interpolation of the solution). The solid line denotes the errors of functions u_i, the dotted line denotes the errors of functions u_e. The results of calculations relating to the first, second and third set of parameters are marked on the charts by triangles, circles and squares, respectively. It is seen that the solution interpolation method can be effectively used to find approximate solutions of diffraction problems on the spectrum of the problem (9). For sufficiently large M the order of solution error is not lower than $h^2 \sim M^{-1}$.

The results of other numerical experiments are given in [2], [3].

5. Conclusions

The work presents the numerical method for the diffraction problem of acoustic waves on three-dimensional homogeneous inclusions. This problem formulated in the form of boundary integral Fredholm equations of the first kind with one unknown function. Based on the results of computational experiments, it can be concluded that the proposed numerical method has a

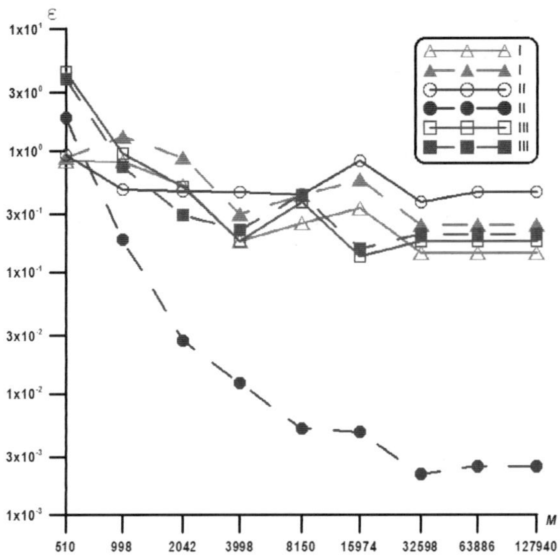

Figure 1. Diffraction problem solution errors from example 1 found without solution interpolation.

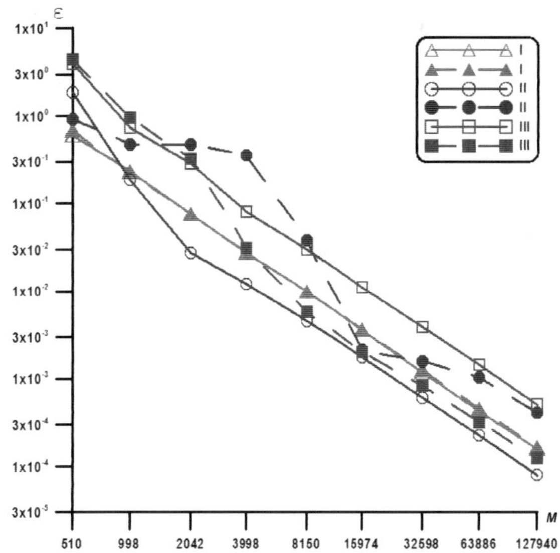

Figure 2. Diffraction problem solution errors from example 1 found with solution interpolation.

sufficiently high accuracy. With a large number of discretization points M it is a second-order method with respect to the "step" h given the surface grid. The method allows for calculations in a wide range of wave numbers and can be used to solve other problems of mathematical physics formulated in the form of weakly singular boundary or volume Fredholm integral equations. Further development of the proposed approach to the numerical solution of diffraction problems may be connected with a decrease in its computational complexity, for example, by using fast methods of matrix-vector multiplication [6].

Acknowledgments
This work was supported by the Russian Foundation for Basic Research (project 17-01-00682) and Basic Research Program of the Far Eastern Branch of the Russian Academy of Sciences (project 18-5-100).

References
[1] Kashirin A A and Smagin S I 2014 Numerical solution of integral equations for a scalar diffraction problem *Doklady Mathematics* **90** 549-552
[2] Kashirin A A and Smagin S I 2018 Numerical solution of integral equations for the three-dimensional scalar diffraction problems *Comput. Technologies* **23** 20-36 (In Russ.)
[3] Kashirin A A and Smagin S I 2018 Numerical solution of integral equations on the spectrum for acoustic waves diffraction *Informatika i Sistemy Upravleniya* **58** 141-149 (In Russ.)
[4] Sorokin A A, Makogonov S V and Korolev S P 2017 The information infrastructure for collective scientific work in the Far East of Russia *Scientific and Technical Information Processing* **44** 302-304
[5] McLean W 2000 *Strongly Elliptic Systems and Boundary Integral Equations* (Cambridge: Cambridge University Press)
[6] Kashirin A A, Smagin S I and Taltykina M Yu 2016 Mosaic-skeleton method as applied to the numerical solution of three-dimensional Dirichlet problems for the Helmholtz equation in integral form *Comput. Math. and Math. Phys.* **56** 612-625

Lagrange-Eulerian method for numerical integration of the gas dynamics equations: parallel implementation on GPUs

Sergey Khrapov[1*], Alexander Khoperskov[1], Sergey Khoperskov[2]

[1]Volgograd State University, Volgograd, 400062, Russia

[2] Institute of Astronomy, Russian Academy of Sciences, Pyatnitskaya st., 48, 119017 Moscow, Russia

E-mail: *khrapov@volsu.ru

Abstract. We describe a new CSPH-TVD method for numerical integration of hydrodynamical equations. The method is based on combined Lagrange-Euler approaches, and it has been devoted to simulations of hydrodynamical flows in various astrophysical systems with non-homogeneous gravitational fields and the non-steady boundary between gas and vacuum. A numerical algorithm was tested on analytical solutions for various problems, and a detailed comparison of our method with the MUSCL scheme is also presented in the paper. It is shown that the CSPH-TVD scheme has a second order of accuracy for smooth solutions (well-balanced approach) and it provides reliable solutions in the vicinity of strong shock waves and at the open gas-vacuum interfaces. We also study the effectiveness of parallel implementations of CSPH-TVD method for various NVIDIA Tesla K20/40/80, P100 graphics processors.

1. Introduction

The existing practice of gas-dynamic numerical experiments develops mainly in two directions. First is an improvement in the quality of numerical schemes, which allows one to describe a tiny three-dimensional non-steady structure of flows. Such refined algorithms (numerical integration of hydrodynamic equations) provide a high order of accuracy in space and time [1, 3, 11, 13, 16]. Another way is to increase the efficiency of parallel computing where the main aim is the massive transition to graphics processors (GPUs) [2, 5, 7].

A promising approach is the use of hybrid numerical schemes based on both Euler and Lagrange methods, combining the advantages of each [6, 9, 12]. Earlier in the works [6, 9] we proposed a new numerical scheme CSPH-TVD (Combined Smoothed Particle Hydrodynamics – Total Variation Diminishing) for integrating the Saint-Venant equations describing the dynamics of surface water in the approximation of shallow water on irregular topography terrain containing kinks and sudden drops in water levels. The method is based on the joint use of Lagrangian (SPH) and Eulerian (TVD) approaches. The CSPH-TVD algorithm for the Saint-Venant equations is well balanced, conservative and allows a stable calculation of the unsteady "water – dry bottom" boundaries on a substantially non-uniform bottom topography.

In this paper, we generalize the CSPH-TVD numerical scheme to the three-dimensional case of a full system of hydrodynamical equations for an ideal non-viscous gas with potential forces and analyze the quality of the scheme by comparing the results of calculations with known analytical solutions and numerical calculations based on the MUSCL (Monotonic Upwind Scheme for Conservation Laws) approach. The main focus is CUDA parallel implementation of our CSPH-TVD algorithm and on the well balance (WB) study of the properties of a numerical scheme for a classical problem of supersonic gas flow through a gravitational potential well.

2. Method CSPH–TVD

2.1. Basic equations

We will proceed from the integral laws of conservation of mass, momentum, and energy for a liquid particle of a volume $\Omega(t)$:

$$\frac{d}{dt}\int\limits_{\Omega(t)} \rho\, dV = 0\,, \tag{1}$$

$$\frac{d}{dt}\int\limits_{\Omega(t)} \rho\mathbf{u}\, dV = -\int\limits_{\Omega(t)} (\nabla p - \rho\mathbf{f})\, dV\,, \tag{2}$$

$$\frac{d}{dt}\int\limits_{\Omega(t)} e\, dV = -\int\limits_{\Omega(t)} (\nabla(\mathbf{u}p) - \rho\mathbf{u}\mathbf{f})\, dV\,, \tag{3}$$

where $dV = dx\,dy\,dz$, ρ is the mass density, $\mathbf{u} = \{u, v, w\}$ is the velocity vector, p is the gas pressure, $e = \rho(|\mathbf{u}|^2/2 + \varepsilon)$ is the total energy per unit volume, ε is the internal energy per unit mass, $\mathbf{f} = -\nabla\psi$ is the potential external force per unit volume, ψ is the gravitational potential. Equations (1)–(3) is completed by the equation of state $p = (\gamma - 1)\rho\varepsilon$ with adiabatic index γ.

2.2. Numerical scheme

To build a numerical model, we use the standard procedure for discretization of a continuous medium on the spacetime grid defined by the grid cells ($x_i = x_{i-1} + h, y_j = y_{j-1} + h, z_k = z_{k-1} + h, t_n = t_{n-1} + \Delta t$), and for each function we have $f(x, y, z, t) \to f(x_i, y_j, z_k, t_n,) = f^n_{i,j,k}$. Since the representation of a continuum described by a system of equations (1) - (3), Lagrangian "liquid particles" (further a particle) can be carried out arbitrarily with continuous coverage of the flow region (there are no gaps between the particles), then at the initial moment of time the particles are combined with the cells of the Eulerian grid.

The CSPH-TVD method consists of two main steps. At the first Lagrangian stage, the modified SPH algorithm [8, 10] is used to approximate the equations (1) - (3) and the pressure and the external forces. The second Euler stage is based on an explicit method of Godunov type[15], including TVD approach. At this stage, the solutions are combined with the Lagrange stage and the mass, momentum, and energy fluxes associated with gas advection through the boundaries of Eulerian cells. The general scheme of the algorithm, presented in reverse order, as follows:

$$\mathbf{U}^{n+1}_{i,j,k} = \widetilde{\mathbf{U}}^{n+1}_{i,j,k} - \frac{\Delta t}{h}\left(\mathbf{F}^{n+1/2}_{i+1/2,j,k} - \mathbf{F}^{n+1/2}_{i-1/2,j,k} + \mathbf{G}^{n+1/2}_{i,j+1/2,k} - \mathbf{G}^{n+1/2}_{i,j-1/2,k} + \mathbf{H}^{n+1/2}_{i,j,k+1/2} - \mathbf{H}^{n+1/2}_{i,j,k-1/2}\right), \tag{4}$$

where

$$\mathbf{U} = \begin{pmatrix} \rho \\ \rho\mathbf{u} \\ E \end{pmatrix}, \quad \mathbf{F} = \begin{pmatrix} \rho u \\ \rho u\mathbf{u} \\ uE \end{pmatrix}, \quad \mathbf{G} = \begin{pmatrix} \rho v \\ \rho u\mathbf{u} \\ vE \end{pmatrix}, \quad \mathbf{H} = \begin{pmatrix} \rho w \\ \rho u\mathbf{u} \\ wE \end{pmatrix}.$$

In the equation (4), the value of $\widetilde{\mathbf{U}}^{n+1}_{i,j,k}$ is calculated at the Lagrangian stage by using a second-order predictor-corrector scheme:

Predictor

$$\widetilde{\mathbf{U}}^{*}_{i,j,k} = \mathbf{U}^{n}_{i,j,k} + \Delta t\,\mathbf{Q}_{i,j,k}\left(\mathbf{U}^{n},\, \mathbf{r}^{n}\right), \qquad \mathbf{r}^{*}_{i,j,k} = \mathbf{r}^{n}_{i,j,k} + \Delta t\frac{\mathbf{u}^{n}_{i,j,k} + \mathbf{u}^{*}_{i,j,k}}{2}, \tag{5}$$

Corrector

$$\widetilde{\mathbf{U}}^{n+1}_{i,j,k} = \frac{\mathbf{U}^{n}_{i,j,k} + \widetilde{\mathbf{U}}^{*}_{i,j,k}}{2} + \frac{\Delta t}{2}\mathbf{Q}_{i,j,k}\left(\mathbf{U}^{*},\, \mathbf{r}^{*}\right), \qquad \mathbf{r}^{n+1}_{i,j,k} = \frac{\mathbf{r}^{n}_{i,j,k} + \mathbf{r}^{*}_{i,j,k}}{2} + \frac{\Delta t}{2}\frac{\mathbf{u}^{n}_{i,j,k} + \mathbf{u}^{n+1}_{i,j,k}}{2}, \tag{6}$$

where $\mathbf{r}$ is the coordinate vector of SPH-particles which represent continuous medium at the Lagrangian stage of integration. At every single moment, t_n, before the start of the predictor-corrector procedure, all the particles are stored in the centers of the Euler cells.

The value of $\mathbf{Q}_{i,j,k}$ in the equations (5) and (6) is defined as follows:

$$\mathbf{Q}_{i,j,k} = -\begin{pmatrix} 0 \\ \varphi_{i,j,k}\displaystyle\sum_{i'=i-1}^{i+1}\sum_{j'=j-1}^{j+1}\sum_{k'=k-1}^{k+1}\varphi_{i',j',k'}\nabla W_{i,j,k} + \rho_{i,j,k}\mathbf{f}_{i,j,k} \\ \varphi_{i,j,k}\displaystyle\sum_{i'=i-1}^{i+1}\sum_{j'=j-1}^{j+1}\sum_{k'=k-1}^{k+1}\varphi_{i',j',k'}\frac{\mathbf{u}_{i,j,k} + \mathbf{u}_{i',j',k'}}{2}\nabla W_{i,j,k} + \rho_{i,j,k}\mathbf{u}_{i,j,k}\mathbf{f}_{i,j,k} \end{pmatrix}, \tag{7}$$

where $\varphi = \sqrt{2p}$, $W_{i,j,k} = W(|\mathbf{r}_{i,j,k} - \mathbf{r}_{i',j',k'}|, h)$ is a smoothing kernel used to approximate spatial derivatives in the equations (1)-(3) in the SPH approach.

Flux values $\mathbf{F}^{n+1/2}_{i\pm1/2,j,k}$, $\mathbf{G}^{n+1/2}_{i,j\pm1/2,k}$ $\mathbf{H}^{n+1/2}_{i,j,k\pm1/2}$ calculated at the cell boundaries using solutions of the Riemann problem at each half-time step layer $t_{n+1/2}$.

To solve numerically the Riemann problem we use the approximate methods of Lax-Friedrichs (LF), Harten-Lax-van Lier (HLL), and the modified Harten-Lax-van Lier method (HLLC) [14] where the flux values at the boundaries depend on the gas parameters on both left (L) and right (R) of a cell boundary.

To construct a second-order accuracy predictor-corrector scheme, we used a piecewise linear approximation of the values of $\mathbf{U}$ inside the cells. In the CSPH-TVD method, the values of the flow parameters on the left (L) and on the right (R), for example, from the border $i + 1/2$ are as follows:

$$\mathbf{U}^{L} = \mathbf{U}^{n+1/2}_{i,j,k} + \frac{1}{2}\left(h - \xi^{n+1}_{i,j,k}\right)\mathbf{\Theta}^{n}_{i,j,k}, \quad \mathbf{U}^{R} = \mathbf{U}^{n+1/2}_{i+1,j,k} - \frac{1}{2}\left(h + \xi^{n+1}_{i+1,j,k}\right)\mathbf{\Theta}^{n}_{i+1,j,k}, \tag{8}$$

where $\xi^{n+1}_{i,j,k}$ is the offset of the SPH particle relative to the center of the cell (i, j, k) at time t_{n+1}. In the piecewise linear reconstruction (8), the slope of function $\mathbf{\Theta}$ should satisfy the TVD-condition.

2.3. Testing numerical model

Below we compare the accuracy of CSPH numerical scheme compare to purely grid-based TVD MUSCL (Monotone Upstream Schemes for Conservation Laws) method [17], based on piecewise-linear reconstruction. For time integration in the CSPH TVD and MUSCL schemes, we apply the second-order Runge Kutta method that satisfies TVD condition [4].

The accuracy, convergence, and computational efficiency of CSPH-TVD and MUSCL numerical schemes were studied for a problem of linear transfer of smooth Riemann wave through the compressible grids with the number of cells $N = 400, 800, 1600, 3200, 6400$ in various runs. For the CSPH-TVD scheme ($N = 6400$) the discrepancy of the numerical solution from analytic

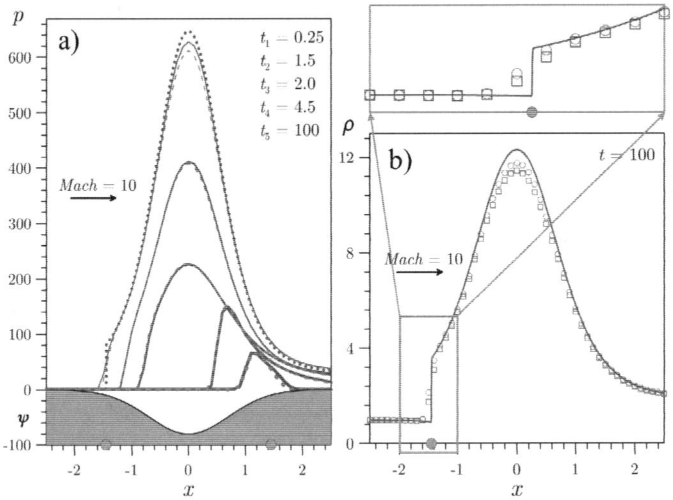

Figure 1. Supersonic flow through a potential well. The left shows the pressure distribution $p(x)$ at different times. CSPH - TVD for $N = 100$ (solid line), MUSCL for $N = 100$ (dashed line). The dotted line shows the steady solution at $t = 100$ for $N = 10^4$. The positions of the stationary shock wave at time point $t = 100$ are shown in the right frame. The exact position of the shock wave front is marked by green point.

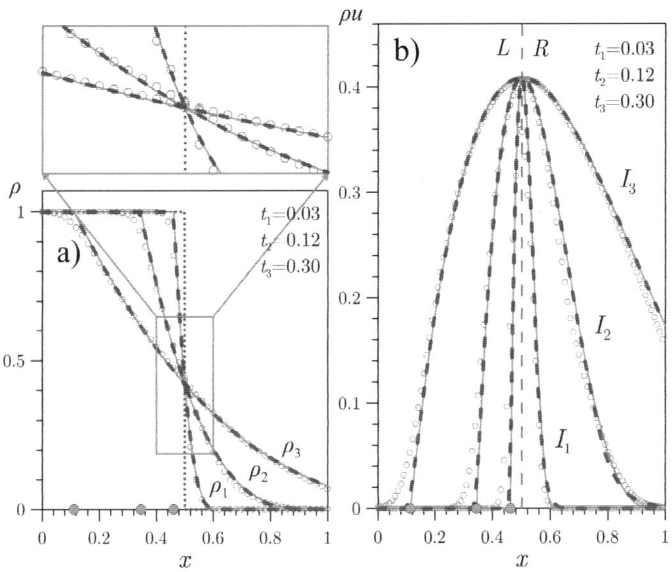

Figure 2. Gas outflow into the vacuum. The left shows the distribution of the density $\rho(x)$ at different points in time. CSPH - TVD for $N = 100$ (red circles), $N = 1000$ (blue dashed line). A solid line shows the exact solution. A dotted line indicates the initial density distribution. The inset shows the inflection point of the profile. The right shows the distribution of the specific momentum $I(x) = \rho u$ at different points in time. The dots mark the boundary of the gas flow area at different times.

one is 1.8×10^{-6}; approximation order of about 1.94; order of convergence is 2.04 and the calculation time is 21 second. The same values for the MUSCL scheme are the following: 3.48×10^{-6}; approximation order is 1.90; the order of convergence is 1.92; calculation time is 24 seconds.

Next, we consider a few astrophysical application of our numerical method: the interactions of strong shock waves, supersonic gas motion in a non-uniform gravitational field, dynamics gas – vacuum surface.

Consider a one-dimensional supersonic gas flow through a gravitational potential well. In our simulations, we adopted the gravitational potential in the following form: $\psi = \psi_0 \exp\left(-x^2/a^2\right)$, where a is the width and ψ_0 is the depth of our potential well. To verify our numerical results, we calculate the shock wave position also with an analytical approach (green points in Fig. 1). Figure 1 (left) shows the evolution of the flow from the beginning of the formation of a shock wave to the steady solution for gas inflow with the Mach number of $M_0 = 10$, $\psi_0 = -80$, $\gamma = 5/3$. At the initial moment of time, the homogeneous gas flow moves from left to right. The presence of the external force $-\partial\psi/\partial x$ initially leads to the formation of a shock wave at the outer edge of the potential well, however this solution is unstable, and the shock wave moves back towards the inner edge of the well where later a steady gas flow is established (Fig. 1, right). The results of our simulations demonstrate a high level of similarities of numerical solutions obtained with both CSPH-TVD and MUSCL schemes in the presence of non-homogenous gravitational fields.

One of the important issues in computational astrophysics, in particular in astrophysical hydrodynamics, is the correct treatment of open boundary between gas and vacuum. In order to test our algorithm on such a specific problem, we consider an outflow of gas into a vacuum. In Figure 2 from the left boundary we define the region filled with gas $\rho_L = 1$, $p_L = 1$, $u_L = 0$, on the right boundary we define the vacuum as follows: $\rho_R = 0$, $p_R = 0$, $u_R = 0$, assuming adiabatic index $\gamma = 5/3$. As a result, a rarefaction wave is formed, moving to the left; the outflow of gas into a vacuum occurs to the right. The density profiles at different points in time intersect at the inflection point shown in the sidebar. The rare-fraction wave propagates with sonic velocity and at the time $t_3 = 0.3$ it is located at $x_3 = 0.5$. Points on the x-axis depict the edge of the gas flow at different points in time. The results of numerical simulation are also in a good agreement with the exact solution of corresponded Riemann problem.

3. Parallel Algorithm Design

A parallel implementation of the numerical algorithm (4) – (8) for several GPUs was performed by using OpenMP-CUDA and GPU-Direct technologies, details of this approach are described in our work [8] for parallel SPH-algorithm.

The CSPH - TVD numerical algorithm consists of four main Global CUDA cores:

- The Hydrodynamics Force Computation (HFC) — this is the CUDA core for the calculation of hydrodynamic variables and external forces in (7) at the Lagrange stage. The kernel has two states {predictor, corrector }.
- The System Update Lagrange (SUL) — is the core of CUDA for updating of so-called characteristics of particles $(\mathbf{r}, \widetilde{\mathbf{U}})$ at the Lagrange stage, according to the equations (5) and (6). The kernel also has two states {predictor, corrector}.
- The Flux Gas Computation (FGC) — CUDA core which calculates the mass, momentum and energy flows ($\mathbf{F}^{n+1/2}_{i\pm1/2,j,k}$, $\mathbf{G}^{n+1/2}_{i,j\pm1/2,k}$ and $\mathbf{H}^{n+1/2}_{i,j,k\pm1/2}$) through the cell boundaries at the Euler stage at the moment $t_{n+1/2}$.
- The System Update Euler (SUE) — is the CUDA core for computation of the $\mathbf{U}$ at the Euler stage according to equations (4).

In Figure 3 we present the order of executing the Global CUDA cores described above.

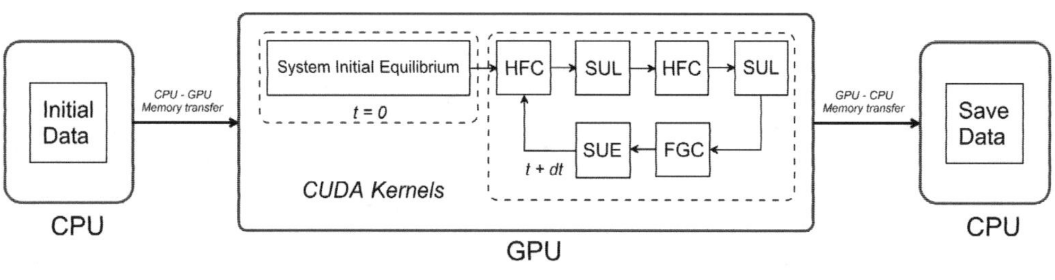

Figure 3. Flowchart for the calculation module.

The performance of the CSPH-TVD numerical algorithm was tested for various NVIDIA Tesla K20/40/80, P100 GPUs on the three-dimensional problem of supersonic gas flow through a gravitational potential well. The calculations showed that on Tesla P100 GPU, the computation

Journal of Physics: Conference Series **1392** (2019) 012041 doi:10.1088/1742-6596/1392/1/012041

time is lower by a factor of 3.8 compare to Tesla K80 and 4 times lower compare to Tesla K40 and 4.8 times correspond to the Tesla K20. The efficiency of parallelization of the algorithm on two and four Tesla K80 GPUs is 95% and 90%, respectively.

4. Conclusions

The results of numerical simulations using the CSPH-TVD method demonstrate its stability and efficiency in modelling gas-dynamic flows with various Mach numbers ($M \ll 1$, $M < 1$, $M \sim 1$, $M > 1$, $M \gg 1$).

Solutions based on CSPH-TVD and MUSCL converge in regions with large pressure gradients and non-homogeneous gravitational fields with strong discontinuities and non-steady "gas – vacuum" surfaces. The advantage of the proposed CSPH - TVD scheme is a good simulation accuracy and less computational costs compare to the Godunov-type numerical schemes. From another hand, CSPH - TVD scheme demonstrates a higher accuracy, lower dissipation rate, and better balanced compared to classical SPH methods.

We also report about parallel efficiency of our numerical scheme which allows calculating astrophysical flows in highly inhomogeneous gravitational fields taking into account non-steady "gas – vacuum" surfaces by using GPUs supercomputers.

One should note that there are various options for the implementation of the algorithm CSPH-TVD. In particular, different smoothing kernels can be used with different TVD - limiters, and methods for solving the Riemann problem. The technique can be easily extended to viscous hydrodynamics and non-stationary sources/sinks and gas self-gravity.

Acknowledgments

The work has been supported by the Ministry of Science and Higher Education (government task no. 2.852.2017/4.6 and 2019-0730), by the Russian Science Foundation (project no. 19-72-20089). The research is carried out using the equipment of the shared research facilities of HPC computing resources at Lomonosov Moscow State University supported by the project RFMEFI62117X0011.

References

[1] Busto S, Ferrín J L, Toro E F, Vázquez-Cendón M E 2018 *Journal of Computational Physics* **353** 169
[2] Chow A D, Rogers B D, Lind S J, Stansby P K 2018 *Computer Physics Communications* **226** 81
[3] Dakin G, Després B, Jaouen S 2019 *Journal of Computational Physics* **376** 339
[4] Ferracina L 2005 Monotonicity and boundedness in general Runge–Kutta methods *University of Leiden*
[5] Howard M.P., Panagiotopoulos A.Z., Nikoubashman A. Efficient mesoscale hydrodynamics: Multiparticle collision dynamics with massively parallel GPU acceleration // Computer Physics Communications, 2018, 230, 10-20
[6] Khrapov S S, Khoperskov A V, Kuz'min N M, Pisarev A V and Kobelev I A 2011 *Vychisl. Metody Programm* **12** 282
[7] Khrapov S S, Khoperskov S A, Khoperskov A V 2018 *Bulletin of the South Ural State University, Series: Mathematical Modelling, Programming and Computer Software* **11** 124
[8] Khrapov S S and Khoperskov A V 2017 *Communications in Computer and Information Science* **793** 266
[9] Khrapov S S, Pisarev A V, Kobelev I A, Zhumaliev A G, Agafonnikova E O, Losev A G and Khoperskov A V 2013 *Advances in Mech. Engineering* **5** 78708
[10] Monaghan J J 2005 *Report on Progress in Physics* **68** 1703
[11] Shen L, Lu C, Wu W, Xue S 2015 *Advances in Applied Mathematics and Mechanics* **7** 180
[12] Sokolichin A, Eigenberger G, Lapin A, Lubbert A 1997 *Chemical Engineering Science* **52** 611
[13] Titarev V A, Toro E F 2002 *Journal of Scientific Computing* **17** 609
[14] Toro E F 1999 Riemann solvers and numerical methods for fluid dynamics *Verlag: Springer* 624
[15] Toro E F 2012 Godunov Methods: Theory and Applications *Verlag: Springer* 1077
[16] Toro E, Eleuterio F 2018 *Book Series: Lecture Notes in Mathematics* **2212** 91
[17] van Leer B 1979 *J. Comput. Phys.* **32** 110

Mathematical modeling of free-surface flows using multiprocessor computing systems

V A Gushchin[1], V G Kondakov[2]

[1] Institute for Computer Aided Design of Russian Academy of Sciences, Moscow, Russia
[2] Nuclear Safety Institute of Russian Academy of Sciences, Moscow, Russia

E-mail: [1]gushchin@icad.org.ru,[2]kondakov@ibrae.ac.ru

Abstract. In this article the results of series of works concerning free-surface flows in liquid medium. Two models were considered: free-surface flows of homogeneous liquid and flows in density stratified liquid. For first model two problems were solved: fluid fluctuations in the container and the occurrence of an undular bore upon collision of a one-way flow with a wall. For second model also was performed test: collapse of homogeneously mixed spot of fluid in the stratified medium. For every test examples satisfactory results were obtained, comparable analysis was carried out whether it corresponds to analytic results.

1. Introduction

For first model of medium with free-surface flow were suggested following assumptions: liquid was viscous uncompressed medium with constant density, on free-surface boundary kinetic condition of disregard for capillary forces and absence of wind tension were made as boundary conditions. For second model uncompressed density-stratified medium was considered. Both models were considered in two-dimensional cases. With the presence of free-surface flow Navie-Stocks equations were transformed in curvilinear system of coordinates, nodes of this mesh were orthogonal and stationary. Full description of first model were given in [1, 2, 3]. For second model particularly with disregard for free-surface flow Navie-Stocks equations were supplemented with equation of continuity with the following assumption: density is not homogeneous and the incompressible fluid is assumed. The depiction of such mathematical setting is given in [4]. The aim of this work was development of new finite difference schemes based on CABARET method for wide series of problems concerning free-surface flows in non-homogeneous surface. The programs were written in Fortran90 language using MPI messaging protocol libraries. For test example of spot collapse parallel Laplas solver was used.

2. Mathematical model

A system of equations of non-homogeneously mixed viscous fluid (density-stratified) is considered. The system of equations of such medium are taken from [6]:

$$\frac{\partial \mathbf{v}}{\partial t} + (\mathbf{v}\cdot\nabla)\,\mathbf{v} = -\frac{1}{\rho}\nabla p + \frac{2}{\rho}\left(\nabla\cdot\mu\nabla\right)\mathbf{v} + \frac{1}{\rho}\nabla\times(\mu\nabla\times\mathbf{v}) + \mathbf{g}, \tag{1}$$

$$\frac{\partial \rho}{\partial t} + \nabla\cdot(\rho\mathbf{v}) = 0, \tag{2}$$

$$\nabla\cdot\mathbf{v} = 0. \tag{3}$$

It can be easily shown that in the right part of first equation with $\mu = const$ second and third addendum give the Laplas operator with multiplier $\frac{\mu}{\rho}$. In background mode without perturbation the distribution of density depended from height is assumed linear: $\rho(y) = \rho_0(1 + a \cdot y), a = \frac{1}{\rho_0}\left(\frac{\partial\rho}{\partial y}\right)_0 < 0$. Consequently the hydrostatic pressure is calculated as an integration of density over height:

$$p(y) = \int_y^{y_0} \rho(y) g dy = p(y_0) - \rho_0 g\left(y - \frac{a}{2}y^2\right) \tag{4}$$

with $p(y_0)$ is the pressure value in the high point of calculated domain. Now it is time to introduce a concept of deviation from hydrostatic pressure. Equation (1) can be rewritten as difference $\delta p = p - p(y)$. Substituting the expression (4) in equation of motion can give us following system of equations:

$$\frac{\partial \mathbf{v}}{\partial t} + (\mathbf{v} \cdot \nabla)\,\mathbf{v} = -\frac{1}{\rho}\nabla\delta p + \left(1 - \frac{\rho_0(1+ay)}{\rho}\right)\mathbf{g} + \frac{\mu}{\rho}\triangle\mathbf{v} \tag{5}$$

Equation (5) alongside (2,3) compose enclosed system of equations that can be solved using CABARET [5] method .

Furthermore plane non-stationary problem of collapsed spot (vertical collapse case) of homogeneously mixed fluid surrounded by stable and continuously stratified medium is considered: Solution of the problem can be found in the rectangular domain $\{(x, y) : -L_x \le x \le$

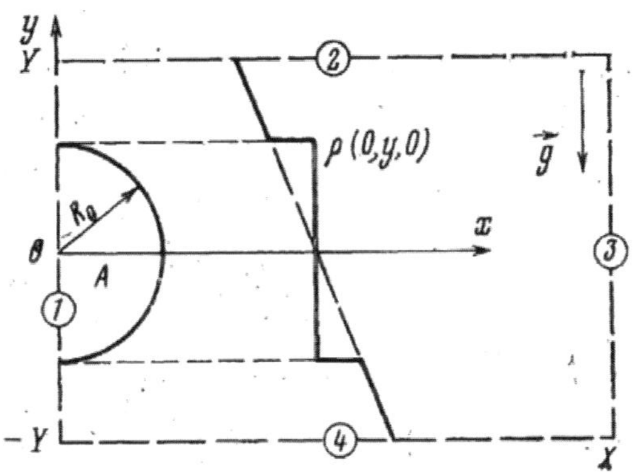

Figure 1. Spot in a stably-stratified liquid problem

$L_x, -L_y \le y \le L_y\}$. All four boundaries should be chosen on sufficiently enough distance form point of perturbation(from the location of the spot) in such a way that the operation of setting the boundaries on this particular boundaries in itself did not affect significantly on the motion of liquid. In current work the following boundary conditions of non-slip velocity and Neimann condition of pressure deviation were set:

$$\mathbf{v} = 0, \tag{6}$$

$$\frac{\partial \delta p}{\partial \mathbf{n}} = 0. \tag{7}$$

Choosing the radius of the spot R_0 at the initial moment of time as the characteristic linear size and the density inside the spot ρ_0 as the characteristic density at the characteristic time N^{-1}, where $N = \sqrt{|a|g}$ is the *Brunt−Väisälä* frequency, we can present equations non-dimensionally:

$$\tilde{f} = (\tilde{x}, \tilde{y}, \tilde{t}, \tilde{u}, \tilde{v}, \tilde{\delta p}, \tilde{\rho}),$$
$$x = \tilde{x}R_0, y = \tilde{y}R_0, t = \tilde{t}/N, u = \tilde{u}R_0N, v = \tilde{v}R_0N, \delta p = \tilde{\delta p}\rho_0 R_0^2 N^2, \rho = \tilde{\rho}\rho_0. \tag{8}$$

With Reynolds number $Re = \rho_0 R_0^2 N/\mu$. With Froude number $Fr = R_0 N^2/g$. Then the equations (1-3) can be presented in the following way:

$$\frac{\partial \mathbf{v}}{\partial t} + (\mathbf{v}\cdot\nabla)\,\mathbf{v} = -\frac{1}{\rho}\nabla\delta p + \left(1 - \frac{1-y\cdot Fr}{\rho}\right)\begin{pmatrix} 0 \\ -Fr^{-1} \end{pmatrix} + \frac{1}{\rho Re}\triangle \mathbf{v},$$
$$\frac{\partial \rho}{\partial t} + \nabla\cdot(\rho\mathbf{v}) = 0, \tag{9}$$
$$\nabla\cdot\mathbf{v} = 0.$$

Initial and boundary conditions were set as follows:

$$\begin{gathered} u=0, v=0, (x,y)\in R^2 \\ \rho = 1, (x,y)\in A, \\ \rho = 1 - y\cdot Fr, (x,y)\in R^2/A, \\ u=0, v=0, \tfrac{\partial \delta p}{\partial \mathbf{n}} = 0, (x,y)\in G. \end{gathered} \tag{10}$$

3. Finite difference scheme

In the rectangular domain$[-L_x;L_x][-L_y;L_y]$ orthogonal grid set with nodes is set:

$$\begin{aligned} x_i &= -L_x + 2L_x\tfrac{i}{N_x}, i = 0, N_x, \\ y_j &= -L_y + 2L_y\tfrac{j}{N_y}, j = 0, N_y. \end{aligned} \tag{11}$$

In the centers of cells the conservative velocity and density variables are introduced, in the centers of cell faces flux velocity and density variables are introduced. Conservative cells will be identified by half full layers of time, flux variables will be identified by full layers of time. In the initial moment conservative variables are initiated, further flux variables are calculated using any differential scheme of the first order (for example corner scheme) on the next full layer of time. The system of equations (9) can be fetched in the divergent form using continuity equations (2):

$$\frac{\partial \rho u}{\partial t} = -\nabla\cdot(\rho u\mathbf{v}) - \frac{\partial(\delta p)}{\partial x} + \frac{1}{Re}\triangle u, \tag{12}$$

$$\frac{\partial \rho v}{\partial t} = -\nabla\cdot(\rho v\mathbf{v}) - \frac{\partial(\delta p)}{\partial y} - (\rho - 1 + y\cdot Fr)\,Fr^{-1} + \frac{1}{Re}\triangle v. \tag{13}$$

Difference scheme can be rewritten in quadratures in the following way:

$$\begin{aligned} \tfrac{\rho^{n+1/2}-\rho^{n-1/2}}{\tau} &= -\nabla\cdot(\rho^n\mathbf{v}^n), \\ \tfrac{\rho^{n+1/2}\tilde{u}-\rho^{n-1/2}u^{n-1/2}}{\tau} &= -\nabla\cdot(\rho^n u^n\mathbf{v}^n) + \tfrac{1}{Re}\triangle u^{n-1/2}, \\ \tfrac{\rho^{n+1/2}\tilde{v}-\rho^{n-1/2}v^{n-1/2}}{\tau} &= -\nabla\cdot(\rho^n v^n\mathbf{v}^n) - \tfrac{1}{Fr}\left(\rho^{n-1/2} - 1 + y\cdot Fr\right) + \tfrac{1}{Re}\triangle v^{n-1/2}, \\ \nabla\left(\tfrac{1}{\rho^{n+1/2}\nabla\delta p^{n+1/2}}\right) &= \tfrac{\nabla\cdot\tilde{\mathbf{v}}}{\tau}, \\ \tfrac{\mathbf{v^{n+1/2}}-\tilde{\mathbf{v}}}{\tau} &= -\tfrac{1}{\rho^{n+1/2}}\nabla\delta p^{n+1/2}. \end{aligned} \tag{14}$$

Where the values with full indexes correspond to flux values, and variables with half full values correspond to conservative variables. We can consider fourth difference equation as a modified Laplas equation where the operator takes into the account the non-homogeneity of density. Current equation was solved using parallel method of conjugate gradients with preconditioners in the form of common Laplass operator with fixed density. Direct solver of Laplass operator inversion can be received using Furie transformation of unknown pressure variables in double decomposition and equating the corresponding decomposition component to the right side. Finally, the scheme ends with the calculation of flow variables at the new time layer in the following way:

$$\psi_S^{n+1} = 2\psi_C^{n+1/2} - \psi_{Sop}^n, \psi = \begin{pmatrix} \rho \\ u \\ v \end{pmatrix}, \tag{15}$$

Where S is the index of flow variable face, C is the index of adjacent cell from where the flow goes towars the S face, Sop is the face index of opposite face S and owned cell C.

4. Test

4.1. The fluctuations of liquid in a container

Let the rectangular container be partially filled with a liquid at rest. At the initial time t = 0, an instantaneous pressure pulse $\Phi = A\sigma(t)cos(kx)$ is applied to the free surface, after that the free surface starts to oscillate. From a linear analysis the shape of the free surface will be:

$$\eta(x,t) = -A\sqrt{\frac{\xi}{gH}}sin\left(\sqrt{\frac{\xi}{gH}}t\right)cos(kx), \xi = kHth(kH). \tag{16}$$

here $\xi = kHtanh(kH)$, $A = 11.78$, $k = 2\pi/\lambda$, the perturbation wavelength $\lambda = 80$, and the initial depth of the basin $H = 15$ in considered case. The shape of the free surface at some time moment, obtained using CABARET method, are shown in Fig. 2. Here dashed line – analytical solution, line with markers – our numerical results. The calculations were carried out for different numbers of grid points. The comparison of results obtained by SMIF and CABARET methods are shown in Fig. 3.

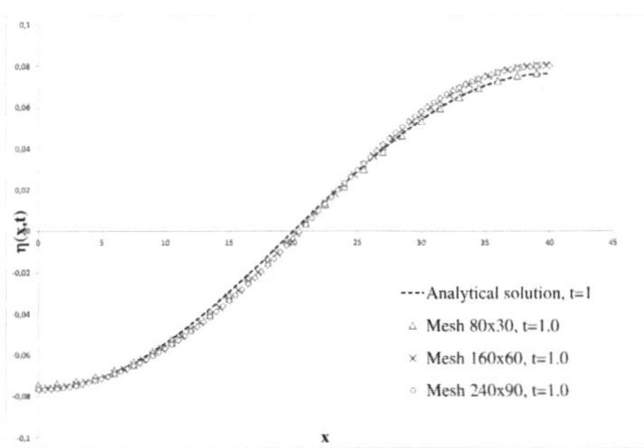

Figure 2. The shape of free surface. Comparison with the analytical solution (CABARET)

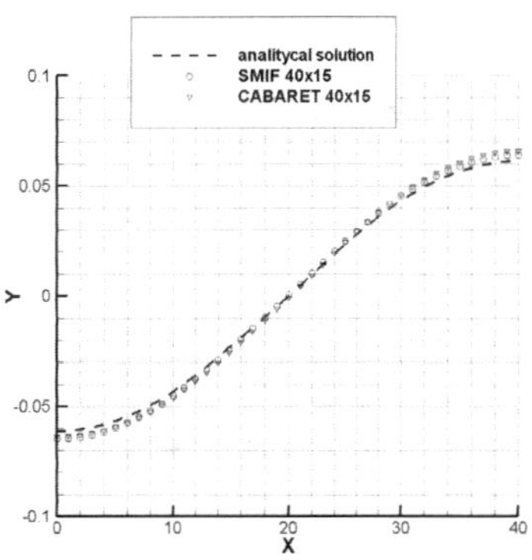

Figure 3. Comparison of SMIF and CABARET methods on mesh 40 × 15

4.2. The dynamics of a spot in a stably-stratified liquid

Time dependence of spot sizes is shown in Fig. 4–7. The isoclines of stream function for t=2, 4, 6, 8 are shown in Fig. 8–11 The first picture shows the emergence of a second vortex pair, the second clearly visible two developed vortex pairs, and by the time t=8 a third vortex pair arises. It should be noted that the vortices, located in the upper half-plane, have a greater intensity than those vortices from the lower half-plane.

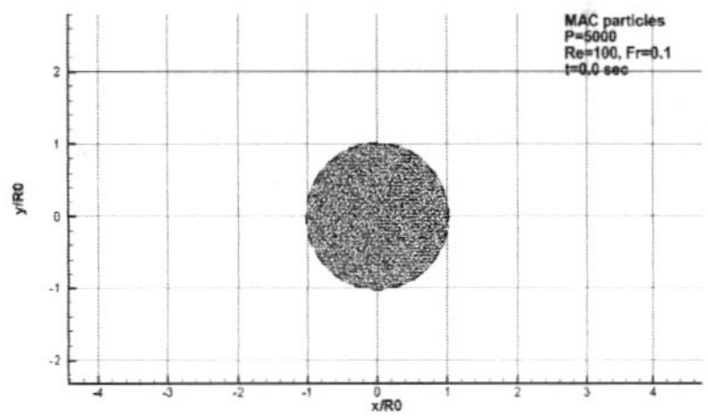

Figure 4. Initial markers distribution in circle of a spot.

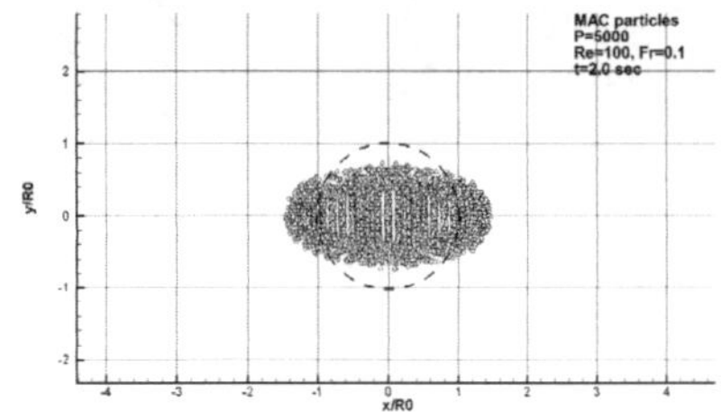

Figure 5. Markers at time t=2. First phase of spot dynamics.

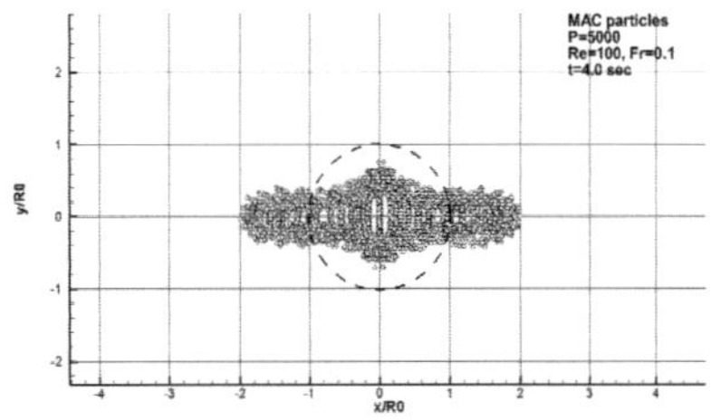

Figure 6. Markers at time t=4. Second phase of spot dynamics.

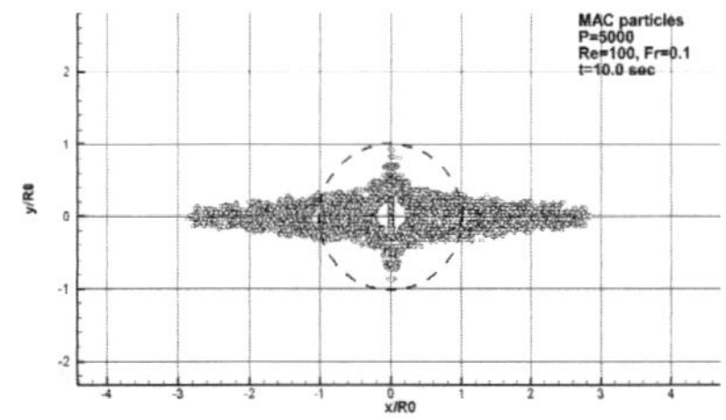

Figure 7. Markers at time t=10. Third phase of spot dynamics.

5. Conclusion

Using series of test examples it can be shown that suggested difference scheme correctly simulates flows with free-surface with absence of wind tension and disregard for capillary forces. In following setting there are following restriction: free-surface cant have breaks in breaking waves and splashing, i.e. the domain must be connected. With following restrictions fulfilled this model can used for simulations of wide range of problems such as internal waves, fluctuations of surface waves and interaction with vertical obstruction. In work [2] the following statement was shown, such as interaction of two pluming vortexes with free-surface and what oscillations occur. The test example with spot collapse has shown cylindrical spot in stratified medium. Using marker method [7], following horizontal and vertical dimensions of spot were obtained. Comparative analysis has shown good dynamics of spot size during first two phases and positive dynamic during third phase of spot flattening and disappearance. Using the results of calculations it can be concluded that suggested models are quite viable and can be useful for free-surface investigations in free-surface problems in liquid domains with non-homogeneous density.

Acknowledgments

The work of Gushchin V.A. was carried out in frame of the State task of ICAD RAS.

References

[1] Gushchin VA and Kondakov VG 2017 *Int. Multidisciplinary Scientific GeoConf.* (Sofia:Surveying Geology & mining Ecology Management) vol.17 pp.485–492
[2] Gushchin VA and Kondakov VG 2019 *J. Mathematical Models and Computer Simulations* **11** 499–508.
[3] Gushchin VA and Kondakov VG 2018 *AIP Conf. Proc.* (AIP Publishing) vol.2025 p.020003
[4] Gushchin VA and Kondakov VG 2019 *Int. Conf. on Finite Difference Methods, LNCS* (Cham:Springer) vol.11386 pp.274–280
[5] Faranosov GA et al. 2013 *J. Computer & Fluids* **88** 165–179
[6] Gushchin VA 1981 *USSR Computational Mathematics and Mathematical Physics* **21** 190–204
[7] Harlow FH, Welch JE 1965 *J. The Physics of Fluids* **8** 2182–89

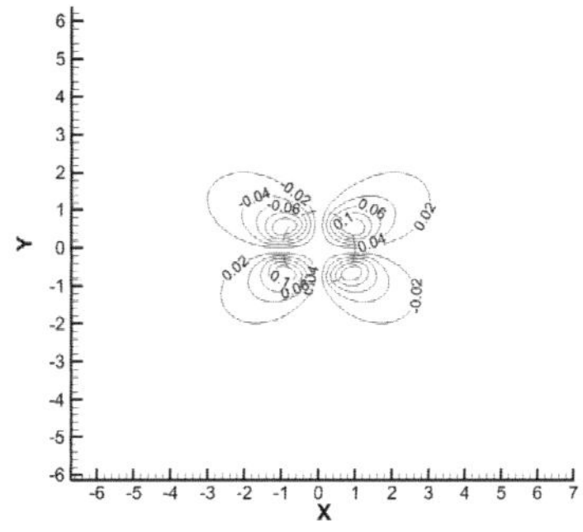

Figure 8. Isoclines of stream function t=2.

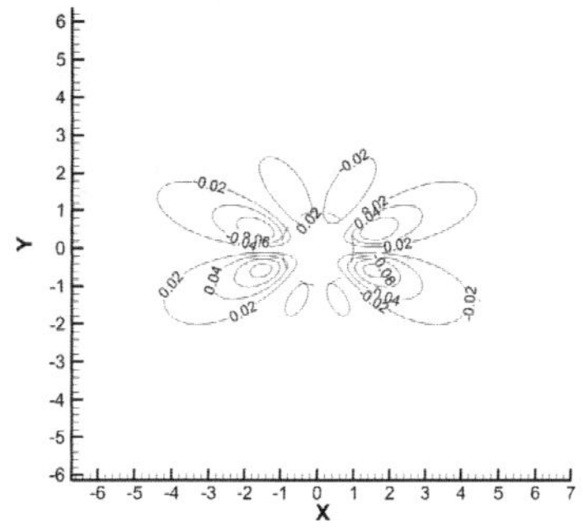

Figure 9. Isoclines of stream function t=4.

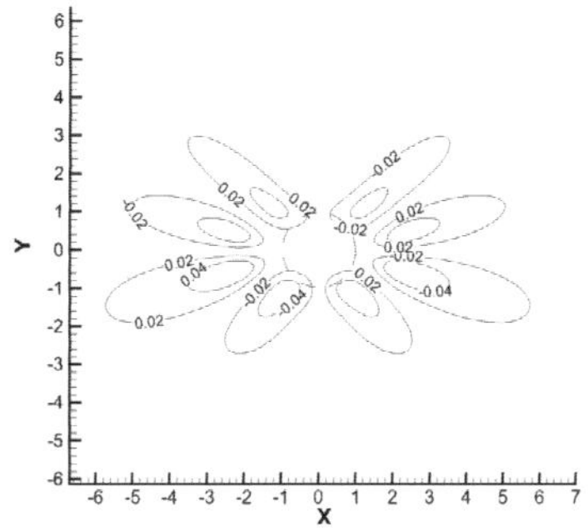

Figure 10. Isoclines of stream function t=6.

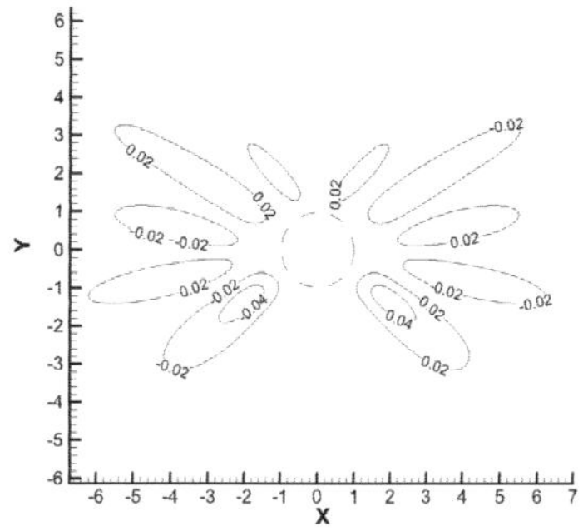

Figure 11. Isoclines of stream function t=8.

Calculation of current distribution in tungsten plate under exposure to a pulsed electron beam

G.G. Lazareva[1,3,4,5], **A.S. Arakcheev**[2,3,4], **V.A. Popov**[2,3], **A.G. Maksimova**[1,2,3] **and E.S. Kapina**[3]

[1]ICM&MG SB RAS 6, Ac. Lavrentieva Pr., Novosibirsk, 630090, Russian Federation
[2]BINP SB RAS 11, Ac. Lavrentieva Pr., Novosibirsk, 630090 Russian Federation
[3]Novosibirsk State University 1, Pirogova str., Novosibirsk, 630090 Russian Federation
[4]Novosibirsk State Technical University 20, Prospekt K. Marksa, Novosibirsk, 630073 Russian Federation
[5]Peoples' Friendship University of Russia (RUDN University) 6 Miklukho-Maklaya street, Moscow, 117198 Russian Federation

E-mail: lazareva@ssd.sscc.ru, asarakcheev@gmail.com, v.a.popov94@gmail.com, maksimova@oapmg.sscc.ru, katya_kapina@mail.ru

Abstract. In the mathematical model of melting of a tungsten plate exposed to a pulsed electron beam, the calculation of the current distribution in the medium is added. The model takes into account the heterogeneity of the resistivity. This enables modeling of non-uniform heated material. The current is expected to spread into the depth due to the increased resistance of the heated part. Changing the thickness of the tungsten plate, one can increase the current density in the melt. The calculation results for parameters corresponding to those used on the BETA facility at BINP SB RAS show that no current concentration occur. The Ampere force is not large enough for the rotation observed in the experiments.

1. Introduction

Currently, due to the development of thermonuclear fusion technologies, new practical problems of melting of refractory metals are emerging. To plan and analyze full-scale experiments it is necessary to have mathematical models that accurately take into account all the details and features of the process. The experimental facility Beam of Electrons for materials Test Applications (BETA), created at BINP SB RAS [1], yielded new results of evaporation of tungsten from the surface of a plate exposed to an electron beam. The full-scale experiment is constantly accompanied by the numerical one [2]. The tungsten heating model is based on solving the two-phase Stefan problem for temperature and electrodynamics equations in the sample region (solid metal and melt). The position of the phase boundary depends on discontinuous nonlinear coefficients. The condition at the free melt-solid boundary consists in the tungsten melting temperature of 3695K. The aim of the study is to simulate the erosion of the sample surface because of evaporation and penetration of heat flow deeply into the material, taking into account the rotation of the melt. Construction of a complete model of electron beam heating enables better assessment of the relevance of the simulation and determination of mechanisms that will occur in the case of plasma flow heating in modern plasma and future thermonuclear facilities.

2. Problem definition

In the mathematical model of melting in a tungsten plate exposed to a pulsed electron beam [3], the calculation of the current distribution in the medium is added. It is assumed that the characteristic time of change of the electric field is large as compared with the equilibration time of the electrodynamics equations in the scale of the problem. This means that the time derivatives are small and can be ignored; the time derivative in the charge continuity equation can also be neglected. These assumptions allow us to introduce the vector potential of the current: $\vec{j} = \nabla \times \vec{F}$. Because of the cylindrical symmetry, the current can be described solely with the function $F = F_\varphi$, which describes the "vorticity" of the current .

Consider the following heat equation with surface heating and current potential:

$$\begin{cases} c(T)\rho(T)\frac{\partial T}{\partial t} = \frac{1}{r}\frac{\partial}{\partial r} r\lambda(T)\frac{\partial T}{\partial r} + \frac{\partial}{\partial z}\lambda(T)\frac{\partial T}{\partial z}, \\ j_r = -\frac{\partial F}{\partial z}, \quad j_\varphi = 0, \quad j_z = -\frac{1}{r}\frac{\partial (rF)}{\partial r}, \\ \frac{\partial^2 F}{\partial r^2} + \frac{\partial^2 F}{\partial z^2} + \frac{\partial F}{\partial r}\left(\frac{1}{\rho_{el}}\frac{\partial \rho_{el}}{\partial r} + \frac{1}{r}\right) + \frac{\partial F}{\partial z}\frac{1}{\rho_{el}}\frac{\partial \rho_{el}}{\partial z} + F\left(\frac{1}{r\rho_{el}}\frac{\partial \rho_{el}}{\partial r} + \frac{1}{r^2}\right) = 0, \\ (n, \nabla T)|_\gamma = \frac{W(t,r)}{\lambda(T)}, \quad F|_\gamma = \frac{1}{r}\int\limits_0^r r' j_0(r')dr', \\ (n, \nabla T) = 0, \quad F = 0 \text{ at the other boundaries,} \\ T(r,z,t)\,|_{t=0} = T_0(r,z), \quad F(r,z,t)|_{t=0} = F_0(r,z). \end{cases} \tag{1}$$

where $T(r,z,t)$ is the temperature, $\vec{j} = (j_r, j_\varphi, j_z)$ is the current density, $F(r,z,t)$ is the current potential, $c(T)$ is the specific heat, $\rho(T)$ is the density, $\lambda(T)$ is the thermal conductivity, ρ_{el} is the electrical resistivity, $W(t,r)$ is the power of the heat flux on the surface γ, n is the normal to the surface, and T_0, j_0 and F_0 is the initial temperature, current and current potential respectively. The conditions on the free boundary are $[T]|_s = 0$ and $\left[\lambda(T)\frac{\partial T}{\partial t}\right]|_s = L_m v_n$, where L_m is the phase transition enthalpy, and v_n is the velocity of the phase transition boundary.

In the experiments on the BETA facility, samples of rolled tungsten were exposed to an axisymmetric electron beam. Electrons with an energy of $80 - -90keV$ heated the material in a layer thin as compared with the characteristic depth of heating of the material. The phase transitions inherent of the problem under consideration are included in the temperature coefficients in Eq. (1). The density, thermal conductivity, specific heat, and power loss are given as dependencies on the temperature of the material in the range of $300K \le T \le 8000K$. These functions have discontinuities or lose smoothness at the melting point $T_m = 3695K$. It is necessary to choose the function $\lambda(T)$ at large temperature values such that there be no division by zero. The thermal conductivity and heat capacity of solid tungsten are taken from [4]. The estimates for the thermal conductivity of liquid tungsten are taken from [5, 6]. The electrical resistance is taken from [7] (see Fig. 1):

$$\rho_{el}(T(\vec{x},t)) = \begin{cases} -0.968 + 1.9274\cdot 10^{-2}T + 7.826\cdot 10^{-6}T^2 - \\ \qquad - 1.8517\cdot 10^{-9}T^3 + 2.079\cdot 10^{-13}T^4, & 100 \le T \le 3695, \\ 135 - 1.855\cdot 10^{-3}(T - T_m) + 4.42\cdot 10^{-6}(T-T_m)^2, & 3695 \le T \le 6000. \end{cases} \tag{2}$$

The boundary condition for F was$F|_{z=z_0} = \frac{1-exp(-Ar^2)}{2Ar}$, which matches the spatial distribution of the heating power. Because of the linearity of the equations, we can omit the coefficient proportional to the peak current density $j_z(0,0)$.

It is more convenient to proceed in non-dimensional variables, for example, as follows:

$$r^* = \frac{r}{r_0}, \quad \lambda^* = \frac{\lambda}{\lambda_0}, \quad \rho^* = \frac{\rho}{\rho_0}, \quad c^* = \frac{c}{c_0}, \quad t^* = \frac{t}{t_0}, \quad T^* = \frac{T}{T_0}, \quad W^* = \frac{\lambda_0 T_0 W}{r_0} \tag{3}$$

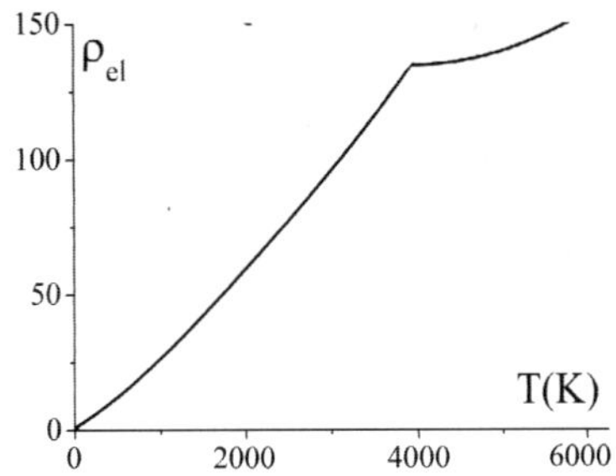

Figure 1. Graphs of temperature dependences of electrical resistance that were used in numerical simulation.

Table 1. Dimensional parameters of nondimensionalization.

Parameter	Typical value	Units
r_0	10^{-1}	mm
t_0	10^2	μs
λ_0	10^{-1}	$W/mm \cdot K$
ρ_0	10^{-5}	kg/mm^3
c_0	10^8	$W \cdot \mu s/kg\ K$
T_0	10^3	K
W_0	10^3	W/mm^2
j_0	10^4	A/m^2

3. Numerical simulation

For the melting process to be taken into account, the enthalpy of the phase transition L_m was introduced into the equation for temperature [8]:

$$\Big(c(T)\rho(T) + L_m\delta(T,\varepsilon)\Big)\frac{\partial T}{\partial t} = div\Big(\lambda(T)gradT\Big), \qquad (4)$$

where the melting point is $T_m = 3695K$.

$$\delta(T,\varepsilon) = \begin{cases} \frac{1}{2\varepsilon}, & |T - T_m| \le \varepsilon, \\ 0, & |T - T_m| > \varepsilon, \end{cases}$$

with the smoothing interval $[-\varepsilon, \varepsilon]$, $\varepsilon = 5K$ and the heat of fusion $L_m = 51.1 \cdot 10^5 \frac{W\mu s}{mm^3}$.

The process of evaporation at the boundary is taken into account via a given resultant energy flux $W_*(t,r) = W(t,r) - N(T|_\gamma)$, where $W(t,r)$ is the power of the heat flux (see (2)), $N(T|_\gamma) = L_e \frac{1}{S}\frac{dm}{dt}$ is the power loss (Figure 1d), $T|_\gamma$ is the boundary temperature, $P(T)$ is the saturated vapor pressure, and $\frac{1}{S}\frac{dm}{dt}$ is the mass evaporation rate. The power loss is calculated with the following parameter values:

$$L_e = 4.482 \cdot 10^{12}\frac{W\mu s}{kg},$$

$$\frac{1}{S}\frac{dm}{dt} = P(T|_\gamma)\sqrt{\frac{M}{2\pi RT|\gamma}} = \exp\left(26.19104 - \frac{83971.3K}{T|_\gamma}\right)\sqrt{\frac{0.184K}{2\pi 8.314T_\gamma}}10^{-12}\frac{kg}{mm^2 \cdot \mu s}. \qquad (5)$$

The derivation of Eq. (4) and justification of neglecting the surface temperature radiation in the energy balance are presented in more detail in [9]. The numerical implementation is based on the Douglas-Rachford scheme [10]. The Neumann boundary condition for the temperature and the Dirichlet boundary condition for the potential on the heated surface are used.

The power of the heat flux is

$$W(t, r) = W_{max}(t) \cdot \exp(-Ar^2), \quad A = 0.03088523. \quad (6)$$

At each time step in the numerical simulation, the variable $W_{max}(t)$ is taken from the experimental data file, individual for each heating pulse. The distribution of the heating power density over the surface was measured using X-ray visualization [11].

4. Simulation results

On the BETA facility, several photographs of the sample surface were taken during pulsed heating and cooling after exposure. The melt regions are experimentally registered, which can be used for estimation of the accuracy of calculations. It is shown that the evaporation model accurately describes the movement of the free boundary of the molten metal during cooling [1].

The heat absorbed by the surface propagates into the material. The current is expected to spread into the depth because of the increased resistance of the heated part. The model takes into account the heterogeneity of the resistivity, which enables simulation of inhomogeneously heated material. Changing the thickness of the sample, one can increase the current density in the melt. The sample for shot 688 has a $25mm \times 25mm$ cross section and a thickness of $4mm$. Since the sample is heated to a depth of several hundred microns in such a short time, the simulation region was a transverse section of the sample, a region of $12mm \times 4mm$. As for the time, the numerical simulation continued until the moment of the last measurement. Comparison of the calculated and experimental temperature on the surface of the plate is given in paper [12]. Calculation results for the radial temperature distribution over the sample surface are consistent with experimental data and analytical estimates.

Figures 2–5 shows isolines of temperature and isolines in the selected area, of current potential rF, module current and the normalized radial component of the current in the plane (r, z) calculated at $150\mu s$. For shot 688, the heating lasts for $119\mu s$. At $150\mu s$, cooling occurs and the surface temperature drops to 3000K. $3000K$.

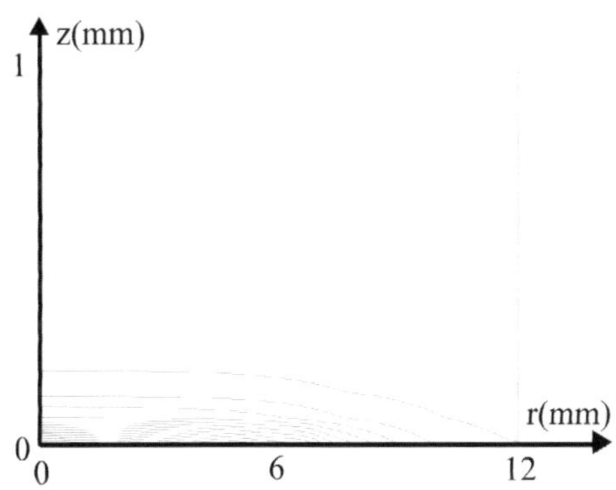

Figure 2. Isolines of temperature.

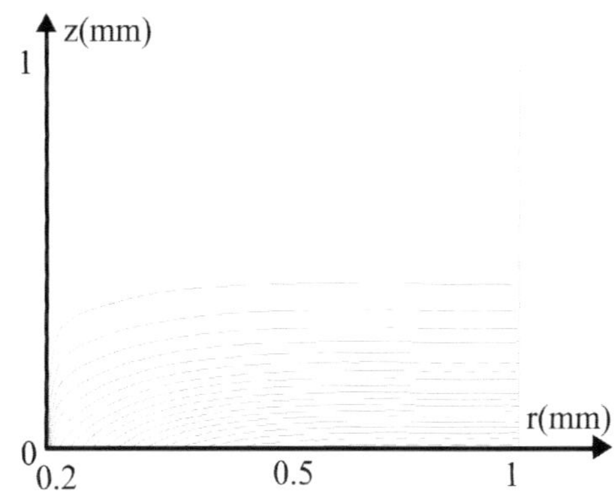

Figure 3. Current potential rF.

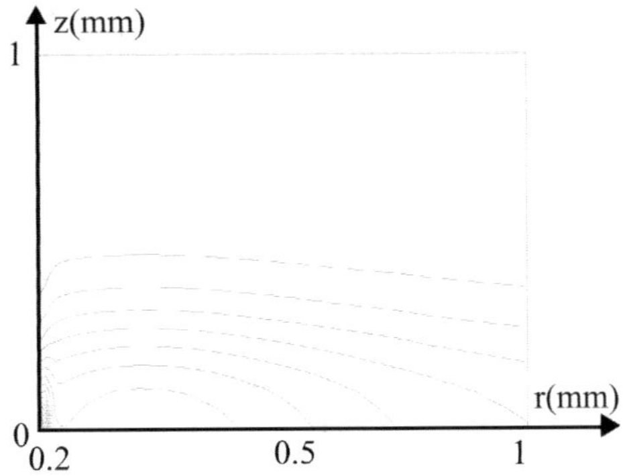

Figure 4. Module current.

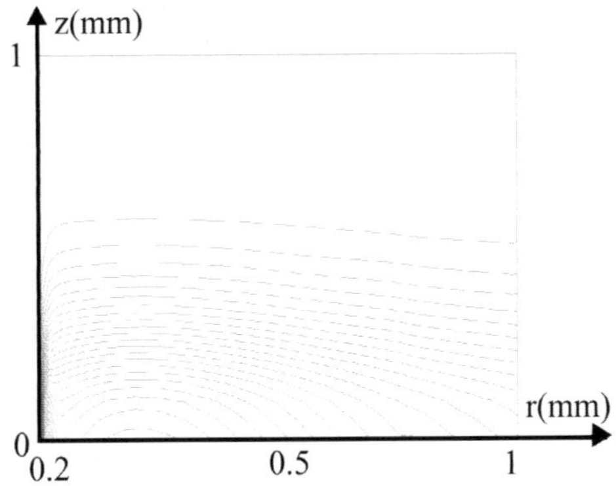

Figure 5. The normalized radial component of the .

Now we can study the effect of the Ampere force on the rotation. We know the current flow on the surface, but the rotation causes the radial component of the current in the melt, since it is perpendicular to the magnetic field. The density of the fluid depends on the temperature, but it can be estimated as $\rho = 1.6 \cdot 10^4 kg/m^3$. From the graph in Fig 2d, one can see that the maximum radial component in the fluid is $j_r \approx j_0$. The magnetic field in the setup can vary from shot to shot, and that time it was equal to $B = 1T$. On this basis, the acceleration can be calculated as $g = \frac{rj_r}{\rho} \approx 1m/s^2$. This is less than the acceleration of gravity, which means that we will not rotate the melt.

5. Conclusion

In the mathematical model of melting in a tungsten plate exposed to a pulsed electron beam, the calculation of the current distribution in the medium is added. The model takes into account the heterogeneity of the resistivity. This enables modeling of non-uniformly heated material. The current is expected to spread into the depth due to the increased resistance of the heated part. Changing the thickness of the tungsten plate, one can increase the current density in the melt. The results of the calculations coincide with the qualitative assessments.

The calculation results for the parameters corresponding to those used on the BETA installation show that no current concentration occur. The Ampere force is not large enough for the rotation observed in the experiments. This result is important for constructing a theoretical model of electron beam heating because there are no other forces that could rotate the melt. This may imply occurrence of current redistribution above the surface, which was not taken into account earlier. The constructed mathematical model is planned to be integrated with the calculation of gas dynamics over a tungsten plate and currents in the gas. The construction of a complete model of electron beam heating will enable better assessment of the relevance of the simulation and finding out which mechanisms will occur in the case of plasma flow heating in modern plasma and future thermonuclear installations.

Acknowledgements

The numerical results have been obtained under support of the RFBR (project No. 19-01-00422). Experimental data have been obtained under support of the Russian Science Foundation (project No 19-19-00272). Lazareva G.G. was partly supported by the "RUDN University Program 5-100".

References

[1] L. Vyacheslavov, A. Arakcheev, A. Burdakov, I. Kandaurov, A. Kasatov, V. Kurkuchekov, K. Mekler, V. Popov, A. Shoshin, D. Skovorodin, Y. Trunev, A. Vasilyev. Novel electron beam based test facility for observation of dynamics of tungsten erosion under intense ELM-like heat loads *AIP CP1771 (American Institute of Physics, Melville, NY, 2016)*, paper 060004.

[2] Arakcheev A.S., Apushkinskaya D.E., Kandaurov I.V., Kasatov A.A., Kurkuchekov V.V., Lazareva G.G., Maksimova A.G., Popov V.A., Snytnikov A.V., Trunev Yu.A., Vasilyev A.A., Vyacheslavov L.N.: Two-dimensional numerical simulation of tungsten melting under pulsed electron beam *Fusion Engineering and Design*, vol 132, pp. 13-17 (2018).

[3] Lazareva G.G., Arakcheev A.S., Vasilyev A.A., Maksimova A.G. Numerical simulation of tungsten melting under fusion reactor-relevant high-power pulsed heating *Smart Modeling for Engineering Systems. Proceedings of the Conference 50 Years of the Development of Grid-Characteristic Method.* Editors: Petrov, I.B., Favorskaya, A.V., Favorskaya, M.N., Simakov, S.S., Jain, L.C. Series: Smart Innovation, Systems and Technologies, Volume 133, 2019, 346 p., Springer International Publishing, P.41-51. doi : 10.1007/978-3-030-06228-6.

[4] J. W. Davis, P. D. Smith (1996) ITER material properties handbook *J. Nucl. Mater.* 233, 1593–1596.

[5] C.Y. Ho, R.W. Powell, and P.E. Liley (1972) Thermal conductivity of elements *Journal of Physical and Chemical Reference Data* 1, 279.

[6] G. Pottlacher (1999) Thermal conductivity of pulse-heated liquid metals at melting and in the liquid phase *Journal of Non-Crystalline Solids* 250, 177–181.

[7] P. Tolias, the EUROfusion MST1 Team Analytical expressions for thermophysical properties of solid and liquid tungsten relevant for fusion applications *Nuclear Materials and Energy* 13, 42-57 (2017) DOI: 10.1016/j.nme.2017.08.002

[8] A. A. Samarskii and P. N. Vabishchevich, Computational Heat Transfer *Editorial URSS*, Moscow, 2003, 784p. [in Russian]

[9] Lazareva G.G., Arakcheev A.S., Kandaurov I.V., Kasatov A.A., Kurkuchekov V.V., Maksimova A.G., Popov V.A., Snytnikov A.V., Trunev Yu.A., Vasilyev A.A., Vyacheslavov L.N. Computational experiment for solution of the Stefan problem with nonlinear coefficients *AIP Conference Proceedings* Vol. 2025, 080005 (2018).

[10] N. N. Yanenko, Method of Fractional Steps for Solving Multidimensional Problems of Mathematical Physics *Science. Sib. Deposition*, Novosibirsk, 1967, 196p. [in Russian]

[11] S. G. Taluts, Experimental study of the thermophysical properties of transition metals and alloys based on iron at high temperatures *abstract, PhD Thesis, code 01.04.14, Yekaterinburg: Publ. House UGGA*, p. 38 (2001). [in Russian]

[12] Lazareva G.G., Arakcheev A.S., Burdakov A.V., Kandaurov I.V., Kasatov A.A., Kurkuchekov V.V., Maksimova A.G., Popov V.A., Shoshin A.A., Snytnikov A.V., Trunev Yu.A., Vasilyev A.A. and Vyacheslavov L.N. Numerical model of high-power transient heating of tungsten with considering of various erosion effects *IOP Conf. Series: Journal of Physics: Conf. Series* 1103 (2018) 01200.

Journal of Physics: Conference Series **1392** (2019) 012044 doi:10.1088/1742-6596/1392/1/012044

Numerical model for calculating displacements near a crack

A.G. Maksimova[1,2,3], A.S. Arakcheev[2,3,4] and G.G. Lazareva[1,3,4,5]

[1]ICM&MG SB RAS 6, Ac. Lavrentieva Pr., Novosibirsk, 630090, Russian Federation
[2]BINP SB RAS 11, Ac. Lavrentieva Pr., Novosibirsk, 630090 Russian Federation
[3]Novosibirsk State University 1, Pirogova str., Novosibirsk, 630090 Russian Federation
[4]Novosibirsk State Technical University 20, Prospekt K. Marksa, Novosibirsk, 630073 Russian Federation
[5]Peoples' Friendship University of Russia (RUDN University) 6 Miklukho-Maklaya street, Moscow, 117198 Russian Federation

E-mail: maksimova@oapmg.sscc.ru, asarakcheev@gmail.com, lazareva@ssd.sscc.ru

Abstract. The paper presents a two-dimensional model of elastic deformations. An isotropic medium region in near a crack propagating along the surface is considered. The results correlate well with an analytical solution.

1. Introduction

The model of calculation deformations and mechanical stresses near a crack formed after a pulsed heat load on material is presented in paper [1]. The application of the model for description of experimental results is sufficiently limited by the assumption of plastic deformation absence. Plastic deformation can be taken into account in a two-dimensional numerical model, while the discussed model is based on analytical reduction of two-dimensional equations. The paper is aimed at development of the two-dimensional numerical model of deformations near a crack. The specific objective of the paper is the calculation of the deformations of a quarter space with forces in the form of a derivative of the delta function and comparison of the results with the analytical results [2], which are the basis of the model in [1, 3, 4].

2. Problem definition

We consider the problem of finding displacements around a crack at the surface of a rectangular sample. Let us assume that the crack is located along the axis x (Figure 1). The equations of the linear theory of elasticity of isotropic medium without volume forces and temperature changes have the following form [2]:

$$(1 - 2\nu)\triangle\mathbf{u} + \text{grad div}\mathbf{u} = 0,$$

where $\mathbf{u} = (u, v)$ is the displacement and ν is the Poisson ratio. We find a stationary solution for the system of equations of elliptic type via a steady-state solution of the following evolutionary system:

$$\begin{cases}(1-2\nu)\Big(\frac{\partial^2 u}{\partial y^2}+\frac{\partial^2 u}{\partial z^2}\Big)+\frac{\partial}{\partial y}\Big(\frac{\partial u}{\partial y}+\frac{\partial v}{\partial z}\Big)=\frac{\partial u}{\partial \tau},\\(1-2\nu)\Big(\frac{\partial^2 v}{\partial y^2}+\frac{\partial^2 v}{\partial z^2}\Big)+\frac{\partial}{\partial z}\Big(\frac{\partial u}{\partial y}+\frac{\partial v}{\partial z}\Big)=\frac{\partial v}{\partial \tau},\end{cases}$$

where τ is the parameter of the the steady state. In the calculation domain (Figure 2), let us set the location of the crack at the point z_0 on the axis z.

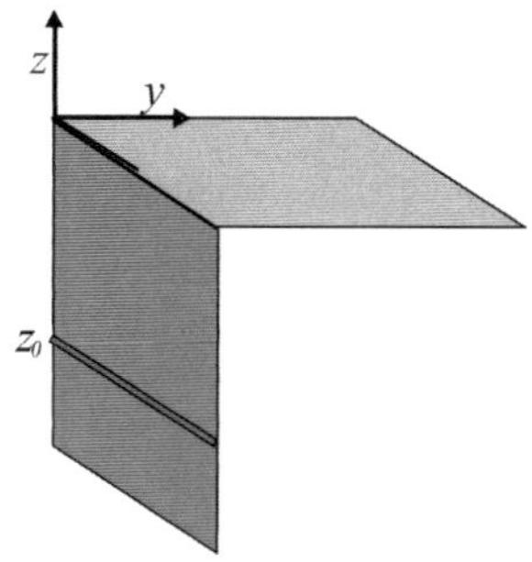

Figure 1. Quarter space filled with an elastic medium.

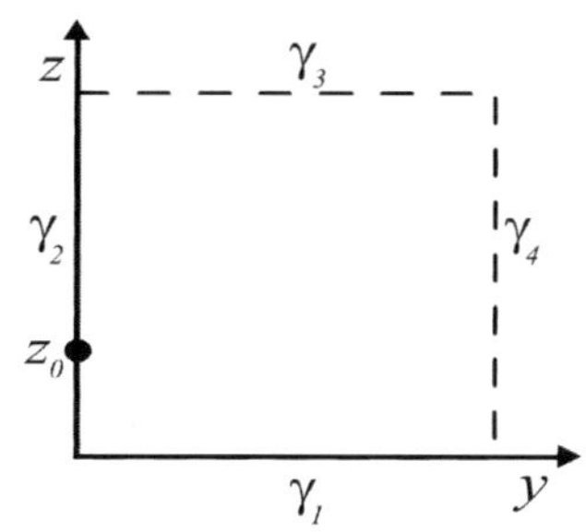

Figure 2. Coordinate system indicates to boundaries of computational domain.

3. Boundary condition

Let σ be the stress tensor with the components σ_{ij} and $f=(f_i)=(\sigma_{ij}n_i)$ be the surface forces, where the indexes correspond to the directions of the axes y, z. The relation between the stresses and displacements is expressed by the Hooke law:

$$\sigma_{ij}=E\Big(\frac{1}{1+\nu}\varepsilon_{ij}+\frac{\nu}{(1+\nu)(1-2\nu)}(\varepsilon_{yy}+\varepsilon_{zz})\delta_{ij}\Big),$$

where E is the Young modulus, δ_{ij} is the Kronecker delta, and ε_{ij} is the strain tensor:

$$\varepsilon_{ij}=\frac{1}{2}\Big(\frac{\partial u_i}{\partial x_j}+\frac{\partial u_j}{\partial x_i}\Big).$$

The Poisson ratio and the Young modulus fully characterize the elastic properties of the isotropic material.

We denote the boundaries as $\gamma_1, \gamma_2, \gamma_3, \gamma_4$, as in Figure 2. At the boundaries γ_1, γ_3 the surface forces are zero. The crack at the point z_0 is given through the surface forces on the boundary γ_3 by the derivative of the delta function. We fix the boundary γ_4 by the conditions $\mathbf{u}=0$. Eventually, the boundary conditions can be written as:

$$\gamma_1,\gamma_3:\begin{cases}\sigma_{yz}=0,\\ \sigma_{zz}=0,\end{cases}\Rightarrow\begin{cases}\frac{\partial u}{\partial z}+\frac{\partial v}{\partial y}=0,\\ \nu\frac{\partial u}{\partial y}+(1-\nu)\frac{\partial v}{\partial z}=0,\end{cases}$$

$$\gamma_2:\begin{cases}\sigma_{yz}=0,\\ \sigma_{yy}=\delta'(z-z_0),\end{cases}\Rightarrow\begin{cases}\frac{\partial u}{\partial z}+\frac{\partial v}{\partial y}=0,\\ (1-\nu)\frac{\partial u}{\partial y}+\nu\frac{\partial v}{\partial z}=\frac{(1-2\nu)(1+\nu)}{E}\delta'(z-z_0),\end{cases}$$

$$\gamma_4:\mathbf{u}=0\Rightarrow\begin{cases}u=0,\\ v=0.\end{cases}$$

4. Numerical simulation

We introduce in the calculation domain a square grid with a step h. Let us denote by u_{km}^n and v_{km}^n the values of u and v at the node (y_k, z_m), $k = \overline{1, N_y}$, $m = \overline{1, N_z}$, at the n-th iteration.

The numerical implementation is based on the Douglas–Rachford scheme and the run method [5]. The finite difference scheme for the presented system of equations has the following form:

$$\begin{cases} \frac{u_{km}^{n+1/2} - u_{km}^n}{\tau} = (1-2\nu)\left(\Delta_y^{n+1/2} u + \Delta_z^n u\right) + \Delta_y^{n+1/2} u + \Delta_{yz}^n v, \\ \frac{u_{km}^{n+1} - u_{km}^{n+1/2}}{\tau} = (1-2\nu)\left(\Delta_z^{n+1} u + \Delta_z^n u\right), \\ \frac{v_{km}^{n+1/2} - v_{km}^n}{\tau} = (1-2\nu)\left(\Delta_y^n v + \Delta_z^{n+1/2} v\right) + \Delta_{yz}^n u + \Delta_z^{n+1/2} v, \\ \frac{v_{km}^{n+1} - v_{km}^{n+1/2}}{\tau} = (1-2\nu)\left(\Delta_y^{n+1} v + \Delta_y^n v\right), \end{cases}$$

where

$$\Delta_y^n f = \frac{f_{k+1m}^n - 2f_{km}^n + f_{k-1m}^n}{h^2}, \quad \Delta_z^n f = \frac{f_{km+1}^n - 2f_{km}^n + f_{km-1}^n}{h^2},$$

$$\Delta_{yz}^n f = \frac{\left(f_{k+1m+1}^n - f_{k+1m-1}^n\right) - \left(f_{k-1m+1}^n - f_{k-1m-1}^n\right)}{4h^2}.$$

For the numerical implementation by the prediction method, the system of difference equations is rewritten in the canonical form. Each equation is supplemented by a pair of conditions at the boundary:

$$\begin{cases} u_{1m}^{n+1/2} = u_{2m}^{n+1/2} + \frac{\nu}{2(\nu-1)}\left(v_{2m-1}^n - v_{2m+1}^n\right) + \frac{(1-2\nu)(1+\nu)}{(1-\nu)E} f_m, \\ u_{N_y m}^{n+1/2} = 0. \end{cases}$$

$$\begin{cases} u_{k1}^{n+1} = u_{k2}^{n+1} + \frac{1}{2}\left(v_{k+12}^n - v_{k-12}^n\right), \\ u_{kN_z}^{n+1} = u_{kN_z-1}^{n+1} - \frac{1}{2}\left(v_{k+1N_z-1}^n - v_{k-1N_z-1}^n\right). \end{cases}$$

$$\begin{cases} v_{k1}^{n+1/2} = v_{k2}^{n+1/2} + \frac{\nu}{2(\nu-1)}\left(u_{k-12}^n - u_{k+12}^n\right), \\ v_{kN_z}^{n+1/2} = v_{kN_z-1}^{n+1/2} - \frac{\nu}{2(\nu-1)}\left(u_{k-1N_z-1}^n - u_{k+1N_z-1}^n\right). \end{cases}$$

$$\begin{cases} v_{1m}^{n+1} = v_{2m}^{n+1} + \frac{1}{2}\left(u_{2m+1}^n - u_{2m-1}^n\right), \\ v_{N_y m}^{n+1} = 0. \end{cases}$$

We used the following approximation f_m of the derivative of the delta function at the boundary:

$$f_m = \begin{cases} 1, & \text{if } m = m_z, \\ -1, & \text{if } m = m_z + 1, \\ 0, & \text{else}, \end{cases}$$

where the node number $m_z = \left[\frac{z_0}{h}\right]$ corresponds to the location of the crack.

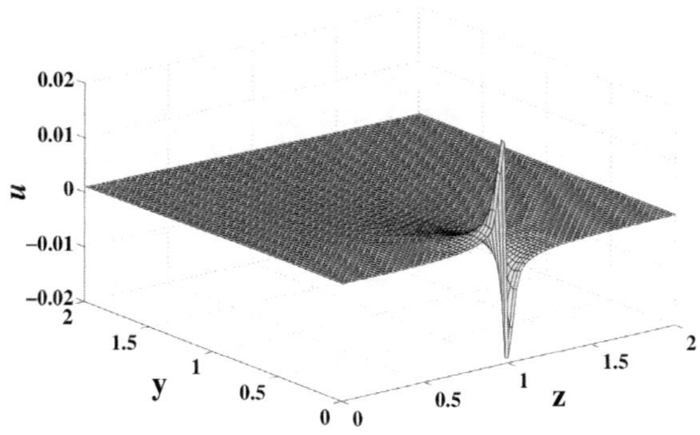

Figure 3. First component of displacement u.

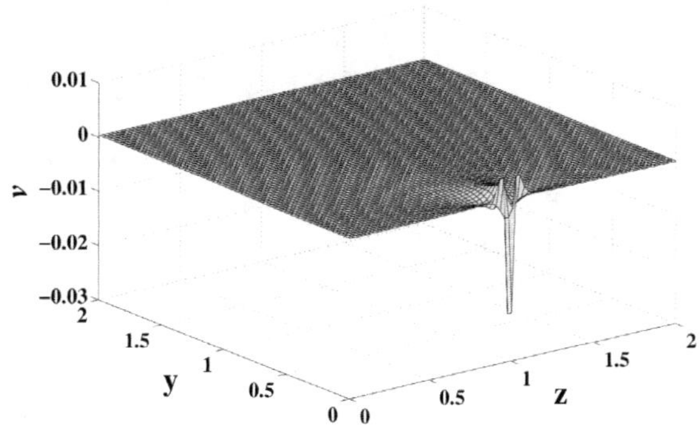

Figure 4. Second component of displacement v.

5. Simulation results

In the calculations for the problem with simplified conditions on the boundary ($f_0 =$ constant), the first order of convergence of the scheme was obtained.

The results of the calculation of the two components of the displacement vector in the vicinity of the crack are presented (Figure 3, 4). The calculations were carried out on a uniform grid of 401x401 nodes, with $h = 0.025$, $\tau = 0.01$. 15000 iterations were made. The crack location z_0 corresponds to nodes 40 and 41 at the boundary γ_2, which corresponds to 1.

Below is a comparison of the obtained calculations with the analytical solution [1, 6] for the u component on the boundary γ_2 (Figure 5) and v component on the boundary γ_1 (Figure 6). The calculations were made with the same parameters.

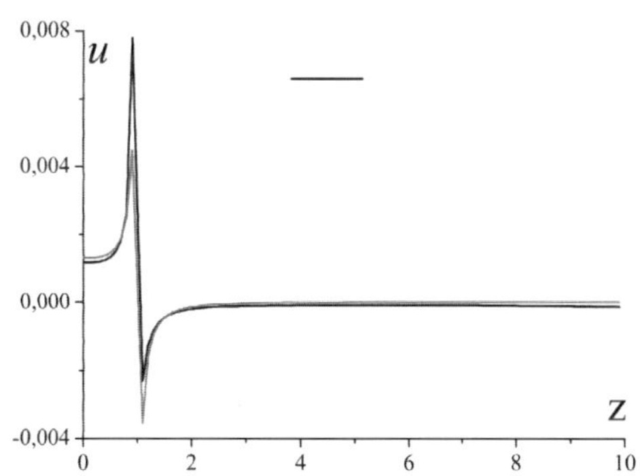

Figure 5. First component of displacement u on boundary γ_2.

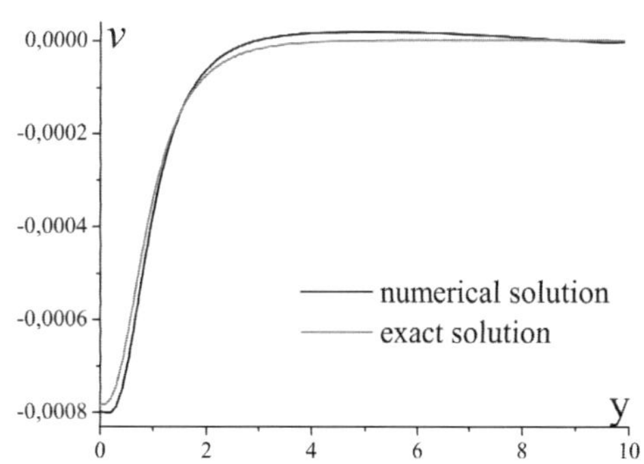

Figure 6. Second component of displacement v on boundary γ_1.

The small discrepancy between the numerical experiment and the analytical solution is presumably because of the distance h between the points of application of the force in calculations, while in reality this distance is infinitely small, and the fact that the infinite area is replaced by a finite square in the numerical calculation.

6. Conclusion

A model of elastic deformations was implemented, which describes displacements in the material around a crack propagating along the surface. The model problem in a two-dimensional formulation was solved numerically. The two components of the displacement vector were obtained. The calculated data correlate well with the analytical solution. In the future, it is planned to use the resulting model to calculate the conditions for the formation of cracks in metals under strong heat load.

Acknowledgements

The numerical results have been obtained within the framework of the budget project 0315-2016-0009 for ICM&MG SB RAS. The experimental data have been obtained under support of the Russian Science Foundation (project No 17-79-20203). Lazareva G.G. was partly supported by the "RUDN University Program 5-100".

References

[1] Arakcheev A.S., Apushkinskaya D.E., Kandaurov I.V., Kasatov A.A., Kurkuchekov V.V., Lazareva G.G., Maksimova A.G., Popov V.A., Snytnikov A.V., Trunev Yu.A., Vasilyev A.A., Vyacheslavov L.N. Two-dimensional numerical simulation of tungsten melting under pulsed electron beam *Fusion Engineering and Design* – July 2018. – V. 132. – P. 1317.

[2] Arakcheev A.S., Arakcheev S.A. Solution to Force Problem of Linear Elasticity Theory for Quarter Space with Edgeuniform Forces *J Apl Theol* 2(2) (2018): 5-12. doi: https://doi.org/10.24218/jatpr.2018.17.

[3] Arakcheev A.S., Huber A., Wirtz M. et al.Theoretical investigation of crack formation in tungsten after heat loads *Journal of Nuclear Materials.* – August 2015. – Vol. 463. – P. 246–249.

[4] Arakcheev A.S., Skovorodin D.I., Burdakov A.V. et al. Calculation of cracking under pulsed heat loads in tungsten manufactured according to ITER specifications *Journal of Nuclear Materials.* – December 2015. – Vol. 467. – P. 165–171.

[5] Yanenko N.N. Method of Fractional Steps for Solving Multidimensional Problems of Mathematical Physics *Novosibirsk: Science. Sib. Deposition*, 1967.

[6] Arakcheev S.A., Kandaurov I.V., Kasatovab A.A., Kurkuchekov V.V., Lazareva G.G., Maksimova A.G., Mashukov V.I., Popov V.A., Trunev Yu.A., Vasilyev A.A., Vyacheslavov L.N. On the mechanism of surface-parallel cracks formation under pulsed heat loads *Nuclear Materials and Energy*, Volume 20, August 2019, 100677. doi: 10.1016/j.nme.2019.100677.

Journal of Physics: Conference Series **1392** (2019) 012045 doi:10.1088/1742-6596/1392/1/012045

Thermal action of pulse radiation on the carbon conic shells loaded with internal pressure

A M Matveenko[1], A V Ostrik[2] and V N Bakulin[3]

[1] Moscow Aviation Institute,Volokolamsk Highway 4, 125993, Moscow, Russia
[2] Institute of Problems of Chemical Physics of the Russian Academy of Sciences, Academician Semenov Avenue 1, Chernogolovka, Moscow Region 142432, Russia
[3] Institute of Applied Mechanics of the Russian Academy of Sciences, Leningradsky prt 7, Moscow, 125040, Russia

E-mail: ostrik@ficp.ac.ru

Abstract. The new mechanism of constructions destruction as a result of thermal action of radiations and particles fluxes was considered. It is supposed that interruption of construction operability comes owing to non-stationary processes of deformation and destruction. Non-stationary deformation and destruction take place as a result of action of quasi-stationary working loadings at saltatory change of construction rigidity. Jump of rigidity arises at volume and impact heating of material by means of action of radiations and particles fluxes.

A numerical method for predicting of the thermal and tension-strain states of thin-walled orthotropic constructions having variable thickness under the action of energy fluxes of different physical nature are considered. Physical and chemical processes in result of heating of composite construction materials were taken into account. These processes are the binder pyrolysis, chemical reactions of carbon and air flow components, and the ablation of carbon residue. Non-stationary processes occurring in the shell has been numerically simulated. Model of layer-by-layer destruction is used in this simulation.

1. Introduction

Forecasting of consequences of interaction of radiations and particles fluxes (RPF) having various physical nature with composite structural elements of flight vehicles (VF) is of practical interest [1, 2, 3, 4]. In many cases thermal effect of RPF is had prevailing because on condition of sufficient remoteness of RPF source from VF mechanical action is not realized any more.

Vehicles can be irradiated in flight by RPF having spectrums which are characteristic for optical radiation and up to γ-quanta [4]. The main danger of thermal action to the thin-walled VF constructions consists in warming up materials and reduction of its thickness as a result of ablation in all this interval of quanta energy. Calculation of parameters of this danger (non-stationary temperature profile and distribution of thickness of the carried-away material) can be made by means of common physical and mathematical model of thermal action of RPF on multilayered composite targets [5, 6, 7] for all practically realized durations of irradiation and area densities of RPF energy. Such model was considered in quasione-dimensional statement [6] earlier. In the present work the model is generalized on a three-dimensional case.

Fast heating causes change of deformation and strength properties of materials and also change of thickness of the thin-walled construction. Then changes of these properties and thickness lead to development of non-stationary processes of deformation and destruction under the action of quasistationary flight loadings. Such mechanism of destruction can take place for thin-walled constructions of FV which are in flight. In particular nozzles of the working rocket engines of the top steps [8] are in severe heat-strength conditions. Thermal RPF action appears the largest in comparison with other s factors of powerful explosion (for example action of a shock wave) for the top steps. Orthotropic shell of rotation is made from a coal plastic can serve as nozzle model. This shell is jammed in the place of fastening and is free in the output section of a nozzle.

Now there is a representative set of numerical models of deformation of multilayered shells [9, 10, 11, 12, 13]. Unevenly heated shells having a variable thickness are among of such models. However problem calculation of non-stationary destruction for composite multilayered shells in non-stationary temperature fields are still under development. In present paper the numerical method [13] is used for a research of non-stationary deformation of composite multilayered shells. Calculations are made taking into account layer-by-layer destruction of thin-walled constructions at change of their stress-strain state.

Let's note that thermal and deformation parts of a problem can be considered independently. It is proved by the fact that change of a temperature profile can be neglected during process of non-stationary deformation and destruction of a shell (it is supposed that characteristic time of redistribution of heat is much more than the period of construction oscillation).

2. Temperature fields and thermal ablation

Calculation of temperature fields and distributions of ablation thickness in composite FV constructions is required for justification of their working capacity (it includes strength to flight loadings). Flight internal and external heat fluxes are axisymmetric in many settlement cases and the corresponding problem about a heat-mass transfer for such constructions is two-dimensional. However the regime of one-sided action is realized at irradiation of FV constructions by RPF and the axial symmetry of a problem does not take place.

Let's assume that the directions of axes of an orthotropy for a composite material coincide with the directions of axes of the coordinate system connected with a shell of rotation and temperature of the solid residue equals to temperature of gaseous products of thermal destruction. Then the equation of transfer of thermal energy in each layer of a shell is wrote in the form (the index of a layer is lowered)

$$\tilde{\rho}_s\tilde{c}_s\frac{\partial T}{\partial t}=\frac{\partial}{\partial z}\left(\tilde{\lambda}_z\frac{\partial T}{\partial z}\right)+\frac{\partial}{\partial s}\left(\tilde{\lambda}_s\frac{\partial T}{\partial s}\right)+\frac{\partial}{R\partial\varphi}\left(\tilde{\lambda}_\varphi\frac{\partial T}{\partial R\varphi}\right)+F, \tag{1}$$

$$F=\frac{sin\alpha}{R}\tilde{\lambda}_s\frac{\partial T}{\partial s}-\tilde{c}_g\tilde{G}_{gz}\frac{\partial T}{\partial z}+\frac{d\tilde{Q}_\Sigma}{dt},\tilde{\lambda}_z=a_sa_\varphi\lambda_z,\tilde{\lambda}_s=\frac{a_\varphi}{a_s}\lambda_s,\tilde{\lambda}_\varphi=\frac{a_s}{a_\varphi}\lambda_\varphi,\tilde{\rho}_s\tilde{c}_s=a_sa_\varphi\rho_sc_s,$$

$$\tilde{c}_g\tilde{G}_{gz}=a_sa_\varphi c_gG_{gz},\quad d\tilde{Q}_\Sigma/dt=a_sa_\varphi dQ_\Sigma/dt,\quad a_s=1-zd\alpha/ds,\quad a_\varphi=1+(z/R)cos\alpha,$$

where s,φ are longitudinal and angular coordinates of a curvilinear coordinates system connected with a shell surface, z is coordinate on a normal to a shell surface, T is temperature of a solid phase and gaseous products, $\alpha=\alpha(s)$ is the current corner between generating line and an axis of a shell of rotation (for a conic shell we have $\alpha=const$ is half corner of a cone); $R=R(s)$ is distance of generating line points from an axis, ρ_s, c_s are density and specific heat capacity of a composite, $\lambda_{z,s,\varphi}$ are total coefficients of heat conductivity for molecular and radiation (in a pores of a solid phase) transfers of heat in the corresponding directions, c_g, G_{gz} are specific heat and a z-component of a mass flux of gaseous products (components $G_{gs}, G_{g\varphi}$ are assumed by small); dQ_Σ/dt is total power of energy release per unit of volume at RPF absorption and the physical and chemical transformations (PCT).

The equation of heat transfer is complemented with relations for speeds of the motion of PCT boundaries, initial and boundary conditions. These conditions are formulated in rather common form [6] that allow to consider inputs and outputs of heat to construction surfaces. Heat contributions include losses from PCT and also convective and radiation supplies of energy from the high-temperature gases passing along an internal surface of a shell.

The solution of a problem is complicated by the fact that position and a form of boundaries of shell layers change as a result of PCT. New spatial variable is used for transforming of settlement region to area having motionless boundaries for each layer of a shell. This variable $\widetilde{z} = (z - z_l(t,s,\varphi))/\delta((t,s,\varphi)$ changes within each layer from zero to unit ($\delta = \delta(t,s,\varphi) = z_r(t,s,\varphi) - z_l(t,s,\varphi;\ z_l = z_l(t,s,\varphi), z_r = z_r(t,s,\varphi)$ are boundary surfaces of layers). The equation (1) for heat transfer is wrote in new variables

$$\widetilde{\rho}_s\widetilde{c}_s\frac{\partial T}{\partial t} = \frac{\partial}{\delta\partial\widetilde{z}}\left(\lambda_z^*\frac{\partial T}{\delta\partial\widetilde{z}}\right) + \frac{\partial}{\partial s}\left(\widetilde{\lambda}_s\frac{\partial T}{\partial s}\right) + \frac{\partial}{R\partial\varphi}\left(\widetilde{\lambda}_\varphi\frac{\partial T}{R\partial\varphi}\right) + \widetilde{F},$$

$$\widetilde{F} = C_{zs}\frac{\partial^2 T}{\delta\partial\widetilde{z}\partial s} + C_{z\varphi}\frac{\partial^2 T}{\delta\partial\widetilde{z}R\partial\varphi} + C_z\frac{\partial T}{\delta\partial\widetilde{z}} + C_s\frac{\partial T}{\partial s} + C_\varphi\frac{\partial T}{R\partial\varphi} + \frac{d\widetilde{Q}_\Sigma}{dt},$$

$$\lambda_z^* = \widetilde{\lambda}_z + \left(\frac{\partial\xi}{\partial s}\right)^2\widetilde{\lambda}_s + \left(\frac{\partial\xi}{R\partial\varphi}\right)^2\widetilde{\lambda}_\varphi, \qquad C_{zs} = -2\frac{\partial\xi}{\partial s}\widetilde{\lambda}_s, \qquad C_{z\varphi} = -2\frac{\partial\xi}{R\partial\varphi}\widetilde{\lambda}_\varphi,$$

$$\xi = \xi(t,\widetilde{z},s,\varphi) = (1-\widetilde{z})z_l(t,s,\varphi) + \widetilde{z}z_r(t,s,\varphi), \quad C_s = \frac{sin\alpha}{R}\widetilde{\lambda}_s - 2\frac{\partial\xi}{\partial s}\frac{\partial\widetilde{\lambda}_s}{\delta\partial\widetilde{z}}, \quad C_\varphi = -2\frac{\partial\xi}{R\partial\varphi}\frac{\partial\widetilde{\lambda}_s}{\delta\partial\widetilde{z}},$$

$$C_z = \widetilde{\rho}_s\widetilde{c}_s\frac{\partial\xi}{\partial t} - \widetilde{c}_g\widetilde{G}_{gz} - \frac{\partial\xi}{\partial s}\left(\frac{sin\alpha}{R}\frac{\partial\xi}{\partial s} + \frac{\partial^2\xi}{\delta\partial\widetilde{z}\partial s}\right)\widetilde{\lambda}_s - \frac{\partial\xi}{R\partial\varphi}\frac{\partial^2\xi}{\delta\partial\widetilde{z}R\partial\varphi}\widetilde{\lambda}_\varphi -$$

$$-\delta\frac{\partial}{\partial s}\left(\frac{\widetilde{\lambda}_s}{\delta}\frac{\partial\xi}{\partial s}\right) - \delta\frac{\partial}{R\partial\varphi}\left(\frac{\widetilde{\lambda}_\varphi}{\delta}\frac{\partial\xi}{R\partial\varphi}\right).$$

The problem formulated in new variables is solved by finite-difference method by means of the implicit integration scheme using splitting on spatial variables. Deviation from the the law of energy conservation is checked for all construction with the purpose of additional control of stability and accuracy of calculations during numerical integration.

3. Physical and chemical transformations

Temperature increase of a coal plastic at RPF action is followed by pyrolysis of binder and the corresponding losses of energy ($Q_{pyr} = 1...3$ kJ/g) in all volume of heated material. The expiring gas products of pyrolysis redistribute and carry away a part of energy from a construction. Change of density of material takes place owing to pyrolysis that leads to formation of the carbon rest finally. Velocity of pyrolysis is described by the simplest Arrhenius-type equation [14, 15, 16]

$$\frac{\partial\rho_s}{\partial t} = -K\rho_{s0}\left(\frac{\rho_s}{\rho_{s0}} - k_C\right)^\nu exp\left(\frac{E_{pyr}}{R_g T}\right), \tag{2}$$

where k_C is a mass fraction of the carbon rest (for coal plastic having phenolic binder that depend on mass contents and coke number of binder), K, ν, E_{pyr} are constants of pyrolysis kinetics ($K = (3...5) \times 10^6$ 1/s, $\nu \approx 1$, $E_{pyr} = 70...80$ kJ/mol that depend on a type of phenolic binder), R_g is a universal gas constant. The equation of pyrolysis kinetics (2) is solved numerically by means of the implicit scheme and together with the equation of transfer for thermal energy (1).

The most power-intensive process is ablation of the carbon rest. Mass velocity of ablation for the carbon rest (this velocity is included into boundary conditions for the equation (1)) calculates taking into account formation of molecular C_2 and C_3 compounds [5, 14, 15] $(C \to i=1, C_2 \to i=2, C_3 \to i=3)$

$$G_{ablC} = \sum_{i=1}^{3} \frac{\alpha_i(P_i^*(T) - P_i)}{\sqrt{2R_gT/\mu_i}} \qquad P_i = C_i \frac{\mu}{\mu_i} P, \qquad P_i^*(T) = K_i exp\left(-\frac{E_i}{R_gT}\right),$$

where G_{ablC} is mass velocity of ablation, α_i are accommodation coefficients ($\alpha_1 = 0,24, \alpha_2 = 0,5, \alpha_3 = 0,023$), μ, μ_i are molecular masses; $P_i^*(T), P_i, P$ are pressures of saturated steams, molecular compounds and mix of gases located at a surface, C_i are concentrations, K_i, E_i are constants for pressures of saturated steams.

The kinetics of mass losses for the carbon rest owing to heterogeneous chemical reactions of the carbon rest with five-component air mix is determined by the relations of papers [14, 17].

4. Deformation and destruction of a shell

The equations for the motion of a shell of rotation are wrote in the form [18]

$$\begin{gathered}
\frac{\partial RN_s}{\partial s} + \frac{\partial N_{\varphi s}}{\partial \varphi} - \gamma N_\varphi + k_s R Q_s = m_{sh} R \frac{\partial^2 u}{\partial t^2}, \\
\frac{\partial N_\varphi}{\partial \varphi} + \frac{\partial RN_{s\varphi}}{\partial s} + \gamma N_{s\varphi} + k_\varphi R Q_\varphi = m_{sh} R \frac{\partial^2 v}{\partial t^2}, \qquad (3) \\
\frac{\partial RQ_s}{\partial s} + \frac{\partial Q_{\varphi s}}{\partial \varphi} - R(k_s N_s + k_\varphi N_\varphi) + p(t,s,\varphi) = m_{sh} R \frac{\partial^2 w}{\partial t^2},
\end{gathered}$$

$$Q_s = \frac{\partial RM_s}{R\partial s} + \frac{\partial M_{\varphi s}}{R\partial \varphi} + \gamma \frac{M_\varphi}{R} - N_s\theta_s - N_{s\varphi}\theta_\varphi, \quad Q_\varphi = \frac{\partial M_\varphi}{R\partial \varphi} + \frac{\partial RM_{s\varphi}}{R\partial s} + \gamma \frac{M_{\varphi s}}{R} - N_\varphi\theta_\varphi - N_{\varphi s}\theta_s,$$

$$\gamma = \frac{dR}{ds} \quad k_s = \frac{d\theta}{ds}, \quad k_\varphi = \frac{cos\theta}{R}, \quad \theta = arctg\left(\frac{d^2X/ds^2}{d^2R/ds^2}\right), \quad \theta_s = k_s u - \frac{\partial w}{\partial s}, \quad \theta_\varphi = k_\varphi v - \frac{\partial w}{R\partial \varphi},$$

where u, v, w are displacements, $N_s, N_\varphi, N_{s\varphi}, N_{\varphi s}$ are intensities of the forces, $M_s, M_\varphi, M_{s\varphi}, M_{\varphi s}$ are intensities of the moments, Q_s, Q_{varphi} are intensities of shear forces, $X = X(s), R = R(s)$ is parametrical description for a form of a shell of rotation, $p(t,s,\varphi)$ is pressure. Stresses located in a layer of an orthotropic shell at z distance from the surface are wrote in the form of [19] (it is supposed that axes of an orthotropy coincide with the directions of axes of a curvilinear coordinate system)

$$\sigma_s = \frac{E}{1-\nu^2-\vartheta^2}\left[(1+\vartheta)(\varepsilon_s + \kappa_s z) + \nu(\varepsilon_\varphi + \kappa_\varphi z)\right],$$

$$\sigma_\varphi = \frac{E}{1-\nu^2-\vartheta^2}\left[(1-\vartheta)(\varepsilon_\varphi + \kappa_\varphi z) + \nu(\varepsilon_s + \kappa_s z)\right], \qquad \sigma_{s\varphi} = \sigma_{\varphi s} = \frac{E}{1+\psi}\left[\omega + \tau z\right], \qquad (4)$$

where Jung's module E , Poisson's coefficient ν and correction coefficients ϑ, ψ are determined by the known characteristics $E_s, E_\varphi, \nu_{s\varphi}, \nu_{\varphi s}, G_{s\varphi}$ ($E_s\nu_{\varphi s} = E_\varphi \nu_{s\varphi}$) for ortotropny material

$$E = \frac{2E_sE_\varphi}{E_s + E_\varphi}, \qquad \nu = \frac{E_s\nu_{\varphi s} + E_\varphi\nu_{s\varphi}}{E}, \qquad \vartheta = \frac{E_s - E_\varphi}{E_s + E_\varphi}, \qquad \psi = \frac{E}{2G_{s\varphi}} - 1.$$

It follows from (4) that parameters ν, ϑ, ψ have to satisfy to inequalities $\nu^2 + \vartheta^2 < 1$, $\psi > -1$. The parameters of deformations entering in relations (4) are determined by fields of displacements [18]

$$\varepsilon_s = \frac{\partial u}{\partial s} + k_s w + \frac{1}{2}\theta_s^2, \qquad \varepsilon_\varphi = \frac{\partial v}{R\partial \varphi} + k_\varphi w + \frac{1}{2}\theta_\varphi^2, \qquad \omega = \frac{\partial u}{r\partial \varphi} + R\frac{\partial}{\partial s}\left(\frac{v}{R}\right) + \theta_s\theta_\varphi,$$

$$\kappa_s = \frac{\partial\theta_s}{\partial s}, \qquad \kappa_\varphi = \frac{\partial\theta_\varphi}{R\partial\varphi} + \gamma\frac{\theta_s}{R} \qquad \tau = \frac{\partial\theta_s}{R\partial\varphi} + k_\varphi\frac{\partial v}{\partial s} - \gamma\frac{\theta_\varphi}{R},$$

Intensity of forces and the moments are determined by integration on thickness of a shell (we use (4))

$$N_s = \int \sigma_s(1 + k_\varphi z)dz = B^+_\vartheta\varepsilon_s + B_{\nu 2}\varepsilon_\varphi + S^+_\vartheta\kappa_s + S_{\nu 2}\kappa_\varphi,$$

$$N_\varphi = \int \sigma_\varphi(1 + k_s z)dz = B_{\nu 1}\varepsilon_s + B^-_\vartheta\varepsilon_\varphi + +S_{\nu 1}\kappa_s + S^-_\vartheta\kappa_\varphi,$$

$$M_s = \int \sigma_s(1 + k_\varphi z)zdz = S^+_\vartheta\varepsilon_s + S_{\nu 2}\varepsilon_\varphi + D^+_\vartheta\kappa_s + D_{\nu 2}\kappa_\varphi, \qquad (5)$$

$$M_\varphi = \int \sigma_\varphi(1 + k_s z)zdz = S_{\nu 1}\varepsilon_s + S^-_\vartheta\varepsilon_\varphi + +D_{\nu 1}\kappa_s + D^-_\vartheta\kappa_\varphi,$$

$$N_{s\varphi} = \int \sigma_{s\varphi}(1+k_\varphi z)dz = B_{\psi 2}\omega + S_{\psi 2}\tau, \quad M_{s\varphi} = \int \sigma_{s\varphi}(1+k_\varphi z)zdz = S_{\psi 2}\omega + D_{\psi 2}\tau \quad (s \leftrightarrow \varphi, 1 \leftrightarrow 2).$$

The rigidity characteristics entered in (5) for an orthotropic shell are calculated by integration on thickness in each point of a construction (calculation is necessary in each point as the shell has the variable thickness and uneven spatial distribution of temperature, $i = 1, 2$, $j = 0, 1, 2$)

$$B^\pm_\vartheta = I^\pm_0, \quad S^\pm_\vartheta = I^\pm_1, \quad D^\pm_\vartheta = I^\pm_2, \quad I^\pm_j = \int \frac{E(1 \pm \vartheta))}{1 - \nu^2 - \vartheta^2}(1 + k_\pm z)z^j dz, \quad (k_+ = k_\varphi, k_- = k_s,),$$

$$B_{\nu i} = J_{i0}, \qquad S_{\nu i} = J_{i1}, \qquad D_{\nu i} = J_{i1}, \qquad J_{ij} = \int \frac{E\nu}{1 - \nu^2 - \vartheta^2}(1 + k_i z)z^j dz, \qquad (6)$$

$$B_{\psi i} = K_{i0}, \qquad S_{\psi i} = K_{i1}, \qquad D_{\psi i} = K_{i1}, \qquad K_{ij} = \int \frac{E}{2(1 + \psi)}(1 + k_i z)z^j dz.$$

Let's note that intensities of forces and the moments (5) with rigidities (6) identically satisfy to the equation of the moments equilibrium for axis z: $N_{s\varphi} - N_{\varphi s} + k_s M_{s\varphi} - k_\varphi M_{\varphi s} = 0$.

The given relations are correct at elastic behavior of heated materials and when destructions are absent. Otherwise intensities of forces and moments are determined by numerical integration on relations (5) and calculation of stresses by means of model of layer-by-layer destruction [4, 6].

Initial conditions for the system of the equations (3) are wrote in form

$$u|_{t=0} = u_0(s), \quad v|_{t=0} = v_0(s), \quad w|_{t=0} = w_0(s), \quad \left.\frac{\partial u}{\partial t}\right|_{t=0} = \left.\frac{\partial v}{\partial t}\right|_{t=0} = \left.\frac{\partial w}{\partial t}\right|_{t=0} = 0, \qquad (7)$$

where $u_0(s)$, $v_0(s)$, $w_0(s)$ are displacements caused by action of flight loadings. These displacements formed from quasistationary and axisymmetric loadings are defined by means of the considered model of deformation and the relaxation method.

It is supposed that RPF action has the symmetry plane which passes through axis of a rotation shell. Then boundary conditions are

$$\varphi = \pm\frac{\pi}{2}: \quad v = \frac{\partial u}{\partial\varphi} = \frac{\partial w}{\partial\varphi} = \frac{\partial^3 w}{\partial\varphi^3}, \qquad s = 0: \quad u = v = w = \frac{\partial w}{\partial s} + k_s u = 0, \qquad (8)$$

$$s = L: \quad N_s = M_s = N_{s\varphi} + k_\varphi M_{s\varphi} = Q_s + \frac{\partial M_{s\varphi}}{R\partial\varphi} = 0. \qquad (9)$$

The implicit finite-difference scheme [4, 13] is used for numerical integration of the equations (3) of the shell motion with initial (7) and boundary (8), (9) conditions.

5. Conclusions

(i) The new mechanism of destruction of thin-walled constructions at a pulse supply of energy of RPF which are volume absorbed is offered. The essence of the mechanism consists in formation of non-stationary processes of deformation and destruction in a shell under the action of quasistationary flight loadings when rigidity of a construction changes practically instantly (time of energy supply is much less period of construction oscillation) as a result of pulse heating of materials.

(ii) The three-dimensional physical and mathematical model of thermal action of convective and radiation flight fluxes of heat together with RPF action is developed. Physical and chemical processes (there are pyrolysis of binder, chemical reactions with components of the running air flow and ablation of the carbon rest) taking place when heating composite materials of a construction are accepted in account. The algorithm of numerical realization of this multidimensional model of thermal action of fluxes on multilayered shell of variable thickness is offered.

(iii) The mechanical and mathematical model of deformation and layer-by-layer destruction of unevenly heated composite thin-walled construction of variable thickness is developed. The implicit finite-difference scheme for numerical integration of the motion equations for a shell was used.

Acknowledgments

Work is made with financial support from the Russian Foundation of Basic Researches (grant No. 18-08-00094-a, 19-08-00606-a).

References

[1] Ostrik A V 2003 *Thermomechanical action of x-ray radiation on multilayered heterogeneous barriers located in air* (Moscow: STC Information technique)
[2] Bakulin V N, Ostrik A V, Gribanov V M et al 2008 *Mechanical X-ray action on thin-walled composite constructions* (Moscow: Fizmatlit)
[3] Fortov V E, Gribanov V M, Ostrik A V, et al 2010 *Physics of nuclear explosion: In 5 vol. Vol. 2. Explosion action* (Moscow: Fizmatlit)
[4] Bakulin V N and Ostrik A V 2015 *Complex action of radiations and particles on the thin-walled constructions having heterogeneous sheetings* (Moscow: Fizmatlit)
[5] Polezhayev Yu V and Yurevich F B 1976 *Thermal protection* (Moscow: Energy)
[6] Ostrik A V and Slobodchikov S S 1995 *Mat. Modl.* **7**(10) 33–46
[7] Dimitriyenko Yu I 1997 *Mechanics of composite materials at high temperatures* (Moscow: Mech. engineering)
[8] Ellis R A et al 1997 *American Institute of Aeronautics and Astronautics Papers* **2672** 1–7
[9] Witmer E A, Balmer H A, Leech J W et al 1963 *AIAA J.* **1**(8) 1848–1857
[10] Volmir A S 1972 *Nonlinear dynamics of plates and shells* (Moscow: Science)
[11] Karmishin A V Lyaskovets V A Myachenkov V I et al 1975 *Statics and dynamics of thin-walled shell constructions* (Moscow: Mech. engineering)
[12] Andreyev A N and Nemirovsky Yu V 2001 *Multilayered anisotropic shells and plates: bend, stability, oscillations* (Novosibirsk: Science)
[13] Bakulin V N Bugay I V and Ostrik A V 2010 *Sea Int. Tech.* **2**(8) 3–6
[14] Grishin A M Parashin A D and Yakimov A S 1993 *Phys. of Comb. and Expl.* **29**(1) 87–95
[15] Dombrovsky L. A Yukina E P Kolpakov A V et al 1993 *Thermophys. of High Temp.* **31**(4) 619–625
[16] Tovstonog V A 2007 *Phys. of Comb. and Expl.* **24**(67) 15–34
[17] Zinchenko V I and Yakimov A S 1988 *Phys. of Comb. and Expl.* **24**(2) 141–149
[18] Ambartsumyan S A 1974 *Commonl theory of anisotropic shells* (Moscow: Science)
[19] Rabotnov Yu N 1988 *Mechanics of the deformed solid body* (Moscow: Science)

Journal of Physics: Conference Series **1392** (2019) 012046 doi:10.1088/1742-6596/1392/1/012046

Verification of the Cabaret schemes for the shallow water equations based on the conservation of angular momentum

Goloviznin V.M.[1], Solovjev A.V.[2] and Zalesny V.B.[3]

[1] Faculty of Computational Mathematics and Cybernetics, Moscow State University, Moscow, Russia

[2] Nuclear Safety Institute RAS, Moscow, Russia

[3] Institute of Numerical Mathematics RAS, Moscow, Russia

E-mail: gol@ibrae.ac.ru, solovjev@ibrae.ac.ru, zalesny@inn.ras.ru

Abstract. To solve the problems of large-scale dynamics of the atmosphere and the ocean, a system of single-layer two-dimensional shallow water equations on a sphere is often used [1, 2]. Such a system is the basis for solving complete baroclinic systems of prognostic equations and is used to assess the accuracy and efficiency of computational algorithms for solving direct and inverse problems. Important issues here are the choice of a suitable form of writing differential equations and effective algorithms for their numerical solution.
In [3], instead of using the equation for the conservation of momentum in a spherical coordinate system, the authors proposed a technique based on the use of the equation for the conservation of angular momentum in a Cartesian coordinate system. This technique allows us to get rid of the degeneracy of the coordinate system near the poles, while preserving only two equations of motion. The semi-discrete finite-volume scheme written in [3] should be supplemented with a method for determining flows on the edges of a grid. The authors used the Cabaret technique [4] for this purpose, which ensures time reversibility on flows where the characteristics of one family do not intersect.
This paper presents the original shallow-water equations in an integral form, a semi-discrete finite-volume scheme, and a brief description of the closure of this scheme using the Cabaret technique. The main attention is paid to the verification of the scheme on a series of traditional tests.

1. The scheme

Let a be the radius of a sphere centered at the origin of a Cartesian coordinate system. Let $h(\vec{r},t)=\zeta(\vec{r},t)-b(\vec{r}) \ll a$ be the thickness of the fluid at the point $\vec{r}$ of the sphere, ζ be the surface level, b be the topography and g be the gravitational acceleration. The equations describing the dynamics of such a thin layer are derived from the fundamental conservation laws. On the sphere, such laws are the law of conservation of mass and angular momentum.

$$\iint_G \frac{\partial h}{\partial t} dS + \oint_{\partial G} h(\vec{w}\cdot\vec{n})dL = 0 \quad (1.1)$$

$$\begin{gathered}\iint_G \frac{\partial}{\partial t}\left[h(\vec{r}\times\vec{w})\right]dS + \oint_{\partial G} h(\vec{r}\times\vec{w})(\vec{w}\cdot\vec{n})dL + \frac{g}{2}\oint_{\partial G}\zeta^2(\vec{r}\times\vec{n})dL = \\ = g\iint_G b(\vec{r}\times\nabla\zeta)dS + 2\iint_G h\cdot\vec{r}\times(\vec{\Omega}\times\vec{w})dS\end{gathered} \quad (1.2)$$

Here $\vec{w}$ is velocity vector, $\vec{n}$ is outer normal vector, dS is the area element of the sphere surface, dL is the length element of the sphere surface. The equation of motion written in this way is the basis for the so-called "well balanced" scheme

Consider an unstructured quadrangular grid on the sphere, the cell boundaries of which form arcs of large circles (geodesic grid). In [3], for equations (1.1) - (1.2), a finite-volume semi-discrete scheme was constructed, which has the following form:

$$\frac{\partial h_C}{\partial t} + \frac{1}{S_C}\sum_{m=1}^{4} h_m w_m^n \Delta L_m = 0 \quad (1.3)$$

$$\begin{gathered}\frac{\partial}{\partial t}\left[h(\vec{r}\times\vec{w})\right]_C + \frac{1}{S_C}\sum_{m=1}^{4} h_m w_m^n\left[w_m^n(\vec{r}_m\times\vec{n}_m)\Delta l_m + w_m^\tau(\vec{r}_m\times\vec{\tau}_m)\Delta L_m\right] + \\ + \frac{g}{2S_C}\sum_{m=1}^{4} h_m^2(\vec{r}_m\times\vec{n}_m)\Delta l_m = \frac{gb_C}{S_C}\sum_{m=1}^{4}\zeta_m(\vec{r}_m\times\vec{n}_m)\Delta l_m + \frac{\Phi_C}{S_C}\end{gathered} \quad (1.4)$$

Here, subscript C refers to conservative variables defined at the centers of the cells, and m refers to the flow variables specified at the centers of the faces, w_m^n and w_m^τ are the normal and tangential components of the flow velocity vector $\vec{w}_m$, ΔL_m and Δl_m are the length of the arc of the face and the length of the chord, based on the ends of the face. Φ_C is an approximation of the Coriolis force. For example:

$$2\int_S h\cdot\vec{r}\times(\vec{\Omega}\times\vec{w})ds \approx -2\cdot a\cdot\Omega\cdot h_C\cdot\vec{w}_C\cdot\sin\phi_C\cdot S_C = \Phi_C \quad (1.5)$$

The system of equations (1.3) - (1.4) describes the evolution of conservative variables defined in the cells. To close the system we need to determine the flows on the faces and the time discretization. For this, various approaches can be used (see, for example, [5]). In [3], the Cabaret method was used, according to which conservative variables are changed in accordance with the balance equations based on (1.3) - (1.4):

$$\frac{U_C^{t+1} - U_C^t}{\Delta t} + \frac{1}{S_C}\sum_{m=1}^{4} F_m^{t+1/2}\Delta L_m = Q \quad (1.6)$$

where $U_C = \left\{h_C, \left[h(\vec{r}\times\vec{w})\right]_C\right\}$, $F_m^{t+1/2} = \left(F_m^t + F_m^{t+1}\right)/2$. The form of the flows follows from (1.4). Balance equations in the form of (1.6) provide the reversibility of the scheme over time, provided that the calculation of the flow variables is reversible.

To find the flow variables on the time layer $t+1$, a characteristic approach is used, according to which local invariants are defined in the cells adjacent to the faces, which are extrapolated to the center of the face on the time layer $t+1$, and then the flow values of the variables $\left(h_m\right)^{t+1}, \left(w_m^n\right)^{t+1}, \left(w_m^\tau\right)^{t+1}$ are restored based on the values of the invariants.

2. Numerical tests

Below will be presented the results of several tests that are important for the verification of the Cabaret scheme.

2.1. *The stationary vortex*

As noted above, the Cabaret is time reversible in an area where the characteristic lines do not intersect. This means that in such areas the scheme is non-dissipative. To confirm this property, the problem of a stationary vortex with a radius $r_0 \ll a$ on a fixed sphere was solved. The initial data in spherical coordinates are as follows:

$$h(\lambda,\phi) = h_0 - \frac{\Theta^2}{4g\beta},\ u(\lambda,\phi) = \Theta\cdot\frac{\phi-\phi_0}{r_0},\ v(\lambda,\phi) = -\Theta\cdot\frac{\lambda-\lambda_0}{r_0},\ \Theta = \alpha\cdot\exp\left[\beta\left(1-\frac{r^2}{r_0^2}\right)\right] \quad (2.1)$$

Here h is depth, λ and ϕ are longitude and latitude, u and v are velocity components, $r = \left|\vec{r} - \vec{r}_C\right|$ is a distance to the vortex center, $\alpha > 0$ and $\beta > 0$ are some parameters. The calculation was carried out on the "cubosphere" grid with the cells number $6\cdot 90^2 = 48600$ up to a time corresponding to 50 vortex revolutions for points $r = r_0$. The vortex radius corresponds to the size of 3 cells, the background depth is $h_0 = 1000$, the depth in the center of the vortex is $h_{r=0} = 758.853$.

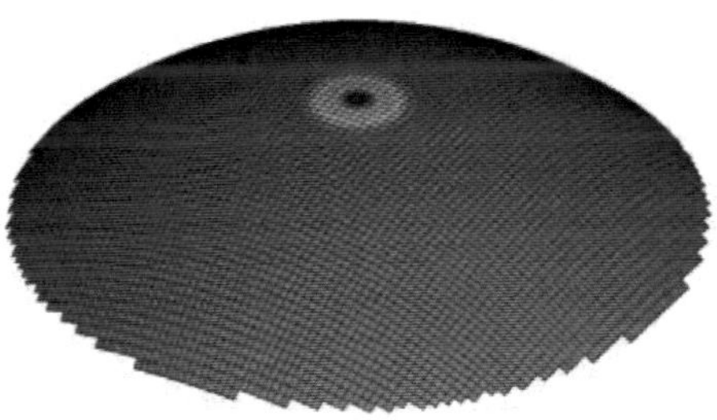

Figure 1a. Part of the grid and colorized initial value of the depth h

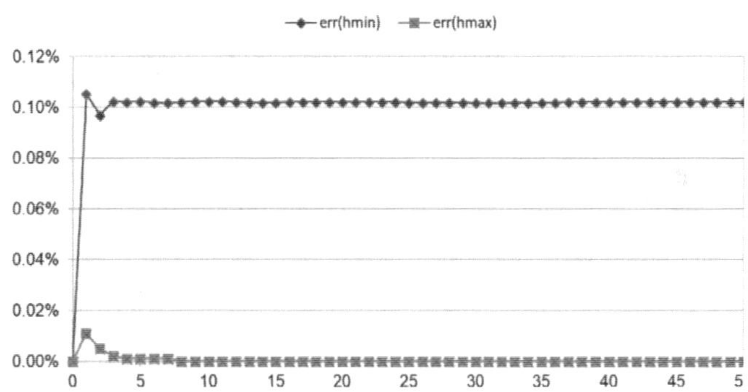

Figure 1b. $\varepsilon_{h\min}(t)$ and $\varepsilon_{h\max}(t)$

In the fig.1 the part of the grid and the initial distribution of the depth are shown, as well as the relative deviation of the minimum and maximum depth of the liquid from the initial one during the calculation time. On the horizontal axis of the fig. 1b marked the number of revolutions of the vortex, on the vertical - the value $\varepsilon = (h - h_{t=0}) / h_{t=0}$. From the above results it follows that during the first few turns the grid solution changes slightly, and then remains unchanged.

2.2. *Advection of Cosine Bell over the Pole*

This test is included in the set of tests proposed by Williamson in [6]. A incomplete system of equations is used. The continuity equation is left active, and the velocity field is assumed to be given and frozen so as to transfer the initial surface perturbation around the sphere without any distortions.:

$$u = u_0\left(\cos\phi\cos\alpha + \sin\phi\cos\lambda\sin\alpha\right),\ v = -u_0\sin\lambda\sin\alpha$$

The parameter $0 \le \alpha \le \pi/2$ is the angle between the axis rotation and the polar axis of the spherical coordinate system. The initial cosine bell pattern is given by

$$h(\lambda,\phi) = \begin{cases} (h_0/2)\left(1+\cos(\pi r/R)\right) & if\ r < R \\ 0 & if\ r \ge R \end{cases}$$

Following parameters were used:

$$u_0 = 2\pi a/(12\ days);\ h_0 = 1000\ m;\ R = a/3;\ a = 6.37122\cdot 10^6\ m.$$

The grid is "cubesphere" with the cell number $6\cdot 90^2 = 48600$. With these parameters, in the analytical case, the perturbation of the surface will move along the direction determined by the parameter α, and after 12 days will return to its original position. Below are the results of the transfer in directions $\alpha = 0$ and $\alpha = 0.05$. Due to the symmetry of the cubesphere grid, the results of calculations along directions $\alpha = \pi/2, \alpha = \pi/2 - 0.05$ are equal to $\alpha = 0, \alpha = 0.05$.

Figure 2. Surface level for t=12 days

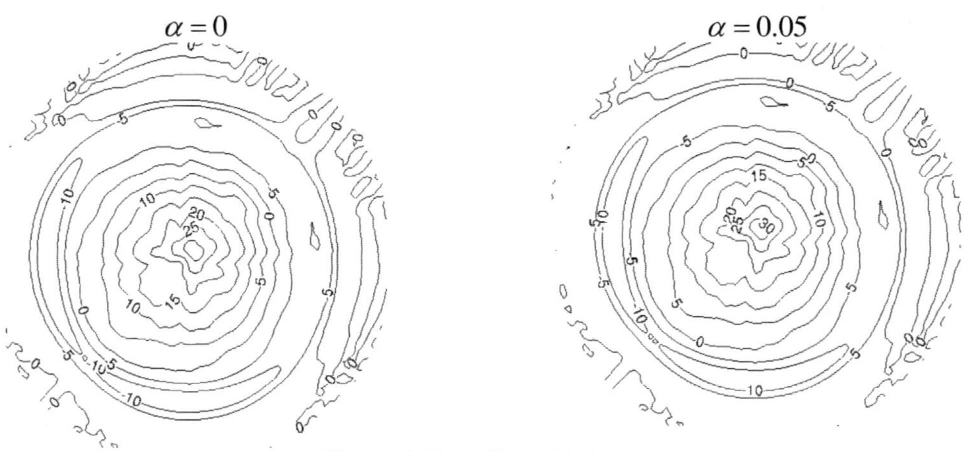

Figure 3. Error for t=12 days

Figure 2. shows the isolines of the surface level after one revolution of the perturbation around the sphere. The isolines are constructed in 100 m increments, the dotted lines correspond to the exact solution, the solid lines correspond to the result of the calculation for a time t = 12 days. Figure 3. shows the isolines of the error $\varepsilon = h_{t=0} - h_{t=12}$. The maximum error does not exceed 3%.

2.3. Development of instability in the zonal flow

This problem is taken from [7]. The problem is solved on a uniform rotating sphere. The background current in the analytical case is stationary. The speed is set by two components:

$$u(\phi) = \begin{cases} 0 & for \quad \phi \le \phi_0 \\ \dfrac{u_{\max}}{\exp\left[-4(\phi_1 - \phi_0)^2\right]} \cdot \exp\left[\dfrac{1}{(\phi - \phi_0)(\phi - \phi_1)}\right] & for \quad \phi_0 \le \phi < \phi_1 \\ 0 & for \quad \phi \ge \phi_1 \end{cases}$$

Here $\phi_0 = \pi / 7$, $\phi_1 = \pi / 2 - \phi_0$, $u_{\max} = 80\ m/s$. The depth is given by

$$h(\phi) = h_0 - \frac{1}{g}\int_0^{\phi} a u(\varphi)\left[2\Omega \sin(\varphi) + \frac{\tan(\varphi)}{a} u(\varphi)\right] d\varphi$$

The constant h_0 is defined so the middle depth is 10 km. $a = 6.37122 \cdot 10^6\ m$, $\Omega = 7.292 \cdot 10^5\ s^{-1}$, $g = 9.80616\ m/s$. The stationary flow is perturbed by adding small value to depth:

$$h'(\lambda, \phi) = \hat{h} \cdot \cos(\phi) \cdot \exp\left[-(\lambda / \alpha)^2\right] \cdot \exp\left[-(\phi_2 - \phi)^2 / \beta^2\right]$$

where $\phi_2 = \pi / 4$, $\alpha = 1/3$, $\beta = 1/15$, $\hat{h} = 120\ m$. This perturbation leads to flow instability. Below are the results of calculations by the Cabaret scheme on the G360 grid for a time of 144 hours in comparison with the results of calculations using the spectral method FSM-SWM on the T341 grid taken from [7]. From the comparison it is clear that the results are close, but the Cabaret circuit for this task generates less high-frequency noise.

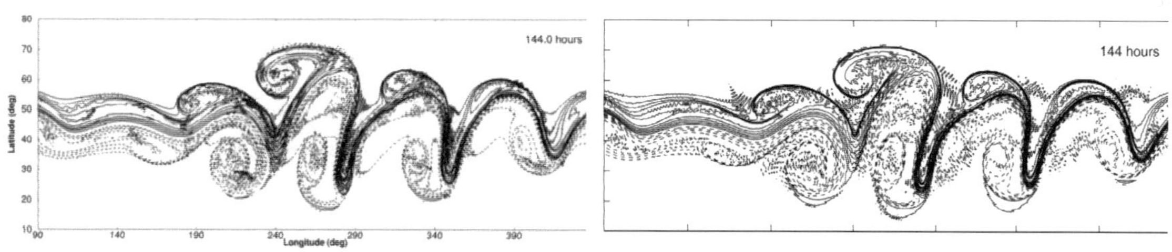

Figure 4. The vorticity field for the time 144 hours. Left – Cabaret scheme, right - FSM-SWM [7]. The contour interval is $2 \times 10^{-5}\ s^{-1}$. Negative contours are dashed. The zero level is skipped

2.4. Rossby-Haurwitz Wave

The problem is taken from [8]. The solution is a structure extending eastward without change, making one revolution in about 12 days. Below are the depth contours calculated by the Cabaret scheme on the G360 grid (left) and, for comparison, on the right, the results of the calculation from [8].

Comparing the results with [8], it is possible to conclude that the distortions are small and in general the flow calculated by the Cabaret scheme corresponds to the reference calculation.

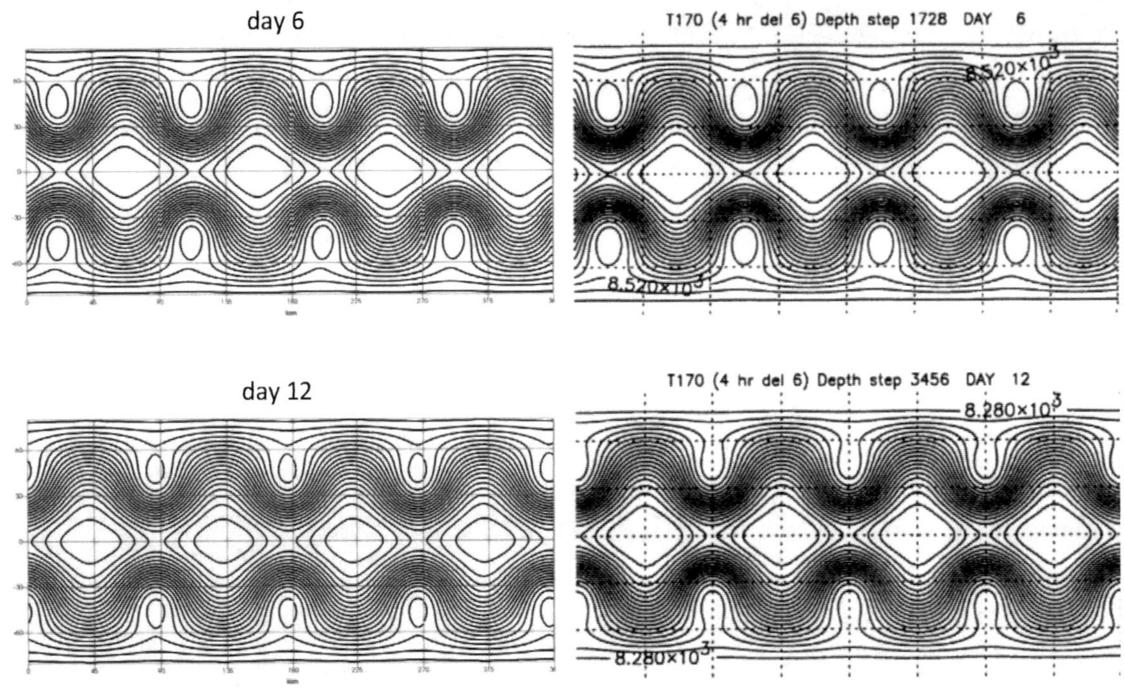

Figure 5. Isolines of density for 6 and 12 days. Left column – Cabaret scheme, right - [8].

Acknowledgments
The work is supported by the Russian Science Foundation under grant 18-11-00163.

References

[1] Mesinger F and Arakawa A. 1976 *GARP Publ. Ser.* **17** 1–64
[2] Arakawa A. and Lamb V R. "Computational design of the basic dynamical processes of the UCLA general circulation model."*Methods in Computational Physics* 17 (1977): 173-265.
[3] Goloviznin V M. et al 2018 J. *Phys.: Conf. Ser.*1128 012091. **doi** :10.1088/1742-596/1128/012091
[4] Goloviznin V, Zaytsev M, Karabasov S and Korotkin I 2013 Novel algorithms of computational hydrodynamics for multicore computing. Monograph *Publishing house of Moscow University*
[5] Goloviznin V.M., Chetverushkin B.N. New Generation Algorithms for Computational Fluid *"Computational Mathematics and Mathematical Physics"*, vol 58 (2018) 8, 1217-1225
[6] David L. Williamson, John B. Drake, James J. Hack, Rüdiger Jakob, Paul N. Swarztrauber, A standard test set for numerical approximations to the shallow water equations in spherical geometry, Journal of Computational Physics, Volume 102, Issue 1, 1992, 211-224, ISSN 0021-9991, doi:10.1016/S0021-9991(05)80016-6.
[7] Joseph Galewsky, Richard K. Scott & Lorenzo M. Polvani (2004) An initialvalue problem for testing numerical models of the global shallow-water equations, Tellus A: Dynamic Meteorology and Oceanography, 56:5, 429-440
[8] John Thuburn & Yong Li (2000) Numerical simulations of Rossby–Haurwitz waves, Tellus A: Dynamic Meteorology and Oceanography, 52:2, 181-189, DOI: 10.3402/ tellusa.v52i2.12258

Numerical simulation of a spatial – temporal model of epidemic distribution

T.S. Timofeeva[1] , **I.Farago, and E.L.Kim**
[1]North-Eastern Federal University, 48 Kulakovskogo str., Yakutsk 677000, Russia
E-mail: timofeevatc@mail.ru

Abstract. The problem of epidemiological situation forecasting is considered. The system of differential equations that describes the development of the epidemic process with regard to spatial dependence, proposed in works I.Farago and based on the Kermak-MacKendrick model. The paper presents the results of numerical calculations for different mathematical models of the spread of infectious diseases with different levels of detail.

1. Introduction

Infectious diseases have always been and still remain the most dangerous for the human body due to their ability to involve a large number of healthy people in a short period of time. There are many especially dangerous viruses, which are evaluated as a very effective means of waging biological warfare. Bioterrorism is an extremely serious threat to humanity.

Natural as well as social processes influence the rise and course of epidemic. Infectious diseases are among the diseases that significantly affect the indicators of health and life expectancy of the population. Therefore, it is very important to understand the mechanism of the epidemic and try to prevent its spread using such effective and affordable means as vaccination.

A timely and adequate prognosis is a prerequisite for planning the structure, scale, and timing of measures aimed at preventing epidemic and outbreak of diseases, as well as at reducing their negative consequences.

To model the rise, development and extinction of epidemic, systems of differential equations are often used, which, depending on the type of a model, determine the dynamics of the transition of individuals from one group to another. The most well-known is the SIR model, proposed by Kermack and McKendrick [1], as well as its generalizations [2, 3, 4]. The Kermack and McKendrick model is registered as a Cauchy problem for a system of ordinary differential equations:

$$\begin{aligned} \frac{dS}{dt} &= -aSI, \qquad t \in (0,T], \\ \frac{dI}{dt} &= aSI - bI, \qquad t \in (0,T], \\ \frac{dR}{dt} &= bI, \qquad t \in (0,T], \end{aligned} \tag{1}$$

where $S(t)$ is a number of people predisposed to a disease, $I(t)$ are infected individuals who are able to spread the disease, and $R(t)$ are individuals who had been infected and left the number of infected as a result of recovery or death. The disease coefficient a depends on the population density and the characteristic signs of the disease, and the recovery rate b depends inversely on the duration of the disease.

Using this model, we can calculate how many people should be immunized before the epidemic wave reaches a certain region, and we can stop the spread of the disease. Mathematical models can be used in the analysis and prediction of indicators of infectious diseases rate. Their use as a prediction tool is limited to modeling the spread of a single-type infection and its consequences. Modeling permits to predict the epidemiological situation, analyze the spatial spread of diseases, explore the mechanisms causing epidemic, develop programs to control and eradicate diseases, and analyze various preventive means. When building a model, it is necessary to take into account a number of limitations and assumptions about the nature of the disease and the structure of the population under consideration.

A number of models have been designed to detail the SIR and make possible their practical use. For example, in research [3], a SIR modification with dynamic parameters is used to predict the spread of HIV.

2. Spatial model of the disease distribution

Point models in their original form are not able to simulate the spatial movement of the disease. For example, they show the number of infected people depending on time, but do not give any information about their location. The location of infectious diseases and relations between neighboring countries have also played an important role in cases of an epidemic outbreak. Therefore, while modeling epidemics, it is necessary to take into account the spatial characteristics of the distribution of infection and the influence of geographical factors. Then the system of equations includes the mechanisms of movement of individuals between territories with a given intensity. As a result, we obtain the following one-dimensional initial-boundary problem for the system of partial differential equations:

$$
\begin{aligned}
\frac{\partial S}{\partial t} &= -S\Big(\delta I + \varphi\frac{\partial^2 I}{\partial x^2}\Big), \quad x \in (0,L), \quad t \in (0,T], \\
\frac{\partial I}{\partial t} &= S\Big(\delta I + \varphi\frac{\partial^2 I}{\partial x^2}\Big) - bI, \quad x \in (0,L), \quad t \in (0,T], \\
\frac{\partial R}{\partial t} &= bI, \quad x \in (0,L), \quad t \in (0,T]
\end{aligned}
\tag{2}
$$

with following initial and boundary conditions

$$
\begin{aligned}
S(x,0) = S_0(x), \quad I(x,0) = I_0(x), \quad R(x,0) = R_0(x), \quad x \in [0,L], \\
S(x,t) = S_0(x), \quad I(x,t) = I_0(x), \quad R(x,t) = R_0(x), \quad x = 0, L, \quad t \in (0,T],
\end{aligned}
\tag{3}
$$

where $S = S(x,t)$, $I = I(x,t)$, $R = R(x,t)$ depend on the spatial position and density of the relevant parts of the population. The initial - boundary problem 2 - 3 is called the sSIR spatial model.

For the numerical solution of the mathematical model 2 - 3, the finite difference method was used. A symmetric difference type of the Crank-Nicholson scheme was used, which provides a second order of accuracy with respect to $0(\tau^2 + h^2)$, where $\tau = T/M$ is the time grid step, $h = L/N$ is the spatial grid step. Since the difference scheme is implicit and nonlinear, the Picard iterations are used for numerical solution.

We present results of the numerical solution for following model parameters:

$$
L = 1, \quad T = 1240, \quad \delta = 0.01, \quad b = 0.03, \quad \varphi = \delta^3/12, \quad N = 100, \quad M = 620.
$$

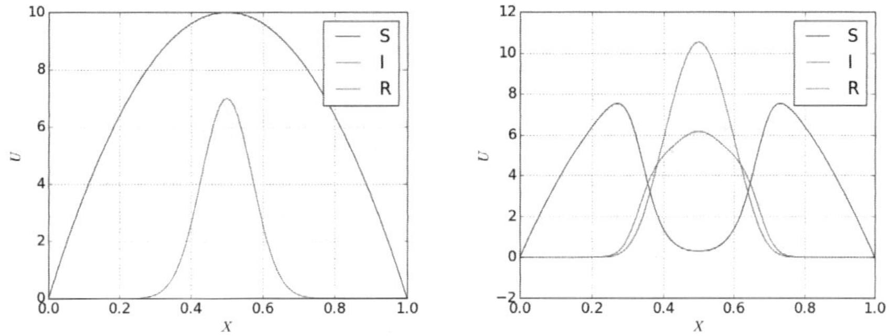

Figure 1: Initial contition (left) and numerical solution for $t = 40$ (right)

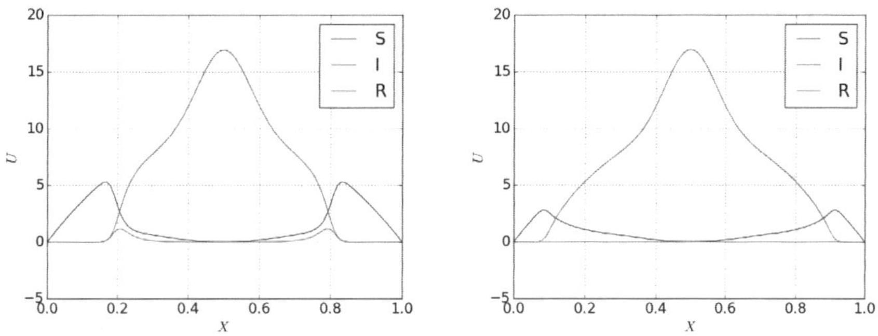

Figure 2: Results for $t = 240$ (left) and $t = 1240$ (right)

The mathematical model describes a situation when an infected individual infects more strongly than the uninfected, who are located at a distance less than δ. Initial conditions are given in the form:

$$S_0(x) = -40x(L-x), \quad I_0 = 7e^{100(x-L/2)^2}, \quad R_0 = 0, \quad x \in [0, L].$$

Fig. 1 - 2 show the distribution of S, I, R at the initial moment and at different moments in time. They are in good agreement with the results of the work [1], although it contains the calculations using explicit and linearized difference schemes. The number of iterations steadily decreases from 7 to 2 with increasing time. It should be noted that the exit from the iterative process was carried out when the following condition was fulfilled

$$\|R^{k+1} - R^k\| < 10^{-6}.$$

From the counting results it can be seen that over time the number of predisposed $S(x,t)$ to the disease decreases, the number of the infected $I(x,t)$, i.e. disease distribution also decreases, the number of the recovered$R(x,t)$ increases.

Next, we present the results of the numerical solution of a two-dimensional model problem with the following parameters:

$$L_1 = L_2 = 1, \quad T = 1240, \quad \delta = 0.01, \quad b = 0.03, \quad \varphi = \delta^3/12, \quad N = 100, \quad M = 620.$$

Initial conditions are given in the form:

$$S_0(x,y) = x(L-x)y(L-y)/(L/2)^4, \quad I_0(x,y) = 7e^{100[(x-L/2)^2+(y-L/2)^2}, \quad R_0(x,y) = 0, \quad x \in [0, L].$$

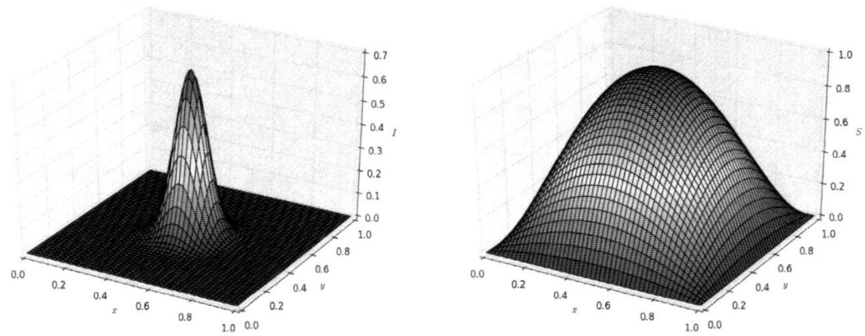

Figure 3: Initial condition for $I_0(x, y)$ (left) and $S_0(x, y)$ (right)

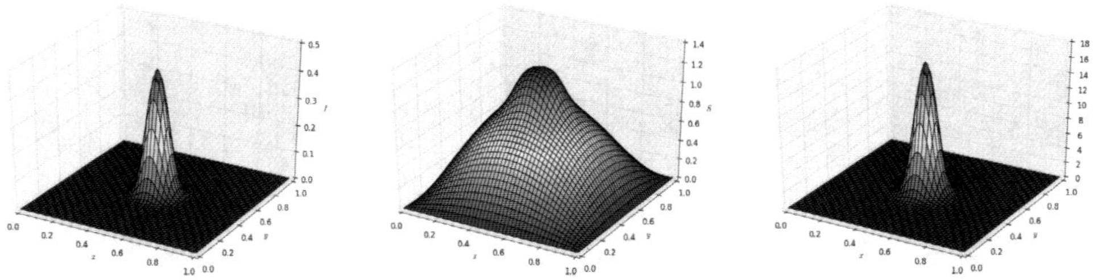

Figure 4: Numerical results for $I(x, y, T)$, $I(x, y, T)$ and $R(x, y, T)$ (from left to right)

The counting results of the two-dimensional model reproduce the counting results of one-dimensional model. The number of iterations on the time layers with time increase monotonously decreases from 14 to 2. Initial conditions and numerical results for $I(x, y, T)$, $I(x, y, T)$ and $R(x, y, T)$ are presented in Figs. 3 and 4.

3. Spatial model of the disease distribution with regard to fertility and mortality

When modeling "slow" infectious diseases characterized by long incubation and infectious periods (tuberculosis, HIV / AIDS), and when studying long-term circulation "rapid" infections, characterized by short incubation and infectious periods (measles, flu), demographic processes are taken into account. Therefore, it is necessary to take into account fertility and mortality in the mathematical models. In the classical spatial SIR model, we also included the case of the inflow of newborns into the class of susceptible individuals and the outflow of the dead from the group of susceptible, infected and insensitive individuals. Then the generalized model is recorded as follows:

$$
\begin{aligned}
\frac{\partial S}{\partial t} &= -S\Big(\delta I + \varphi \frac{\partial^2 I}{\partial x^2}\Big) + \mu(I + R), \quad x \in (0, L), \quad t \in (0, T], \\
\frac{\partial I}{\partial t} &= S\Big(\delta I + \varphi \frac{\partial^2 I}{\partial x^2}\Big) - (b + \mu)I, \quad x \in (0, L), \quad t \in (0, T], \\
\frac{\partial R}{\partial t} &= bI - \mu R, \quad x \in (0, L), \quad t \in (0, T],
\end{aligned} \tag{4}
$$

where b is a coefficient interpreted as the speed of recovery, μ is the death rate / fertility rate.

This mathematical model is supplemented by the initial and boundary conditions (3). For numerical solution of the mathematical model (4) - (3), we construct approximation using the finite difference method. Crank - Nicholson scheme was used for approximation by time and provide a second order of accuracy with respect to $0(\tau^2 + h^2)$. Since the discretization of the system of equations (4) represent a system of nonlinear equations, we used the Picard iterations for solution of the nonlinear discrete problem.

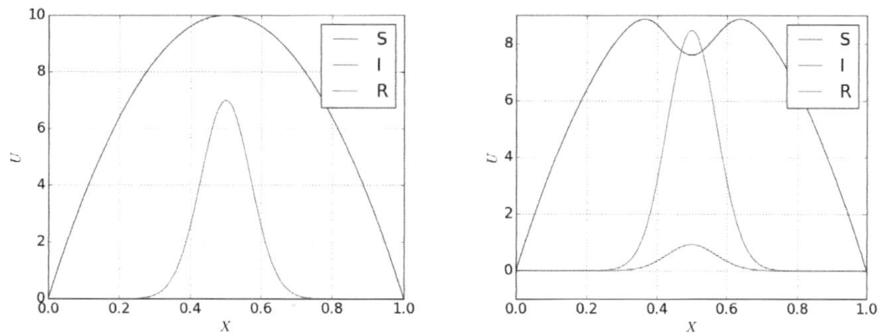

Figure 5: Initital condition (left) and numerical solution for $t = 40$ (right)

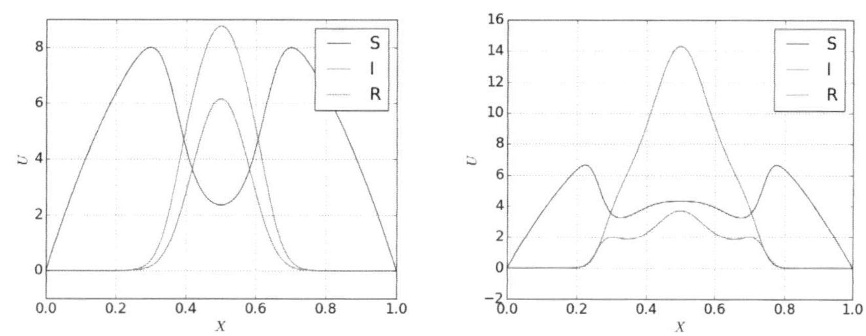

Figure 6: Numerical results for $t = 240$ (left) and $t = 1240$ (right)

We present the results for the constructed numerical implementation of the difference schemes. In the calculations for the spatial model of the disease distribution (4) - (3), we set:

$$L = 1, \quad T = 1860, \quad \delta = 0.01, \quad b = 0.03, \quad \varphi = \delta^3/12, \quad \mu = 0.00001, \quad N = 100, \quad M = 620.$$

We set the same initial conditions as in the previous model:

$$S_0(x) = -40x(L - x), \quad I_0 = 7e^{100(x-L/2)^2}, \quad R_0 = 0, \quad x \in [0, L].$$

Fig. 5 - 6 show the graphs of the distribution of S, I and R at the initial time and for different time of simulations. They are in good agreement with the results of the work of Professor I. Farago, although it carried out calculations using explicit and linearized difference schemes. The number of iterations monotonously decreases from 9 to 2 as time increases. It should be noted that the exit from the iteration process was carried out under the condition

$$\|R^{k+1} - R^k\| < 10^{-6}.$$

From the above results it can be seen that the accounting of mortality / fertility significantly slows down the healing process.

4. Conclusion

A mathematical spatial - temporal model of the infectious disease distribution was studied numerically. The model is an initial boundary value problem for the system of three partial differential equations for densities of healthy, infected and recovered populations. With the help of a numerical simulations, various situations of the epidemic in space were investigated.

References

[1] Kermack, W.O., McKendrick, A.G.: A contribution to the mathematical theory of epidemics // Proc. R. Soc. A Math. Phys. Eng. Sci. 1927. 115(772), 700–721.

[2] I.Farago, R.Horvath. On some qualitatively adequate discrete space-time models of epidemic propagation. J. of Computational and applied mathematics. 2016. 293, 45-54.

[3] I.Farago, R.Horvath. Qualitative properties of the finite difference solution of a space-time epidemic propagation model. Annales Univ. Sci. Budapest., Sect. Comp. 2016. 45, 157-168.

[4] I. Farago, R. Horvath, Qualitative properties of some discrete models of disease propagation, J. of Computational and applied mathematics. 2018. 340, 486–500.

Supercomputer Technologies in Mathematical Modelling | IOP Publishing
Journal of Physics: Conference Series **1392** (2019) 012048 doi:10.1088/1742-6596/1392/1/012048

An algorithm to identify text marked in rock core pictures with machine learning algorithm

Yunfeng Bai
Severnaya Dvina Emb. 17, Arkhangelsk, Russia
E-mail: yumengleme@126.com

Berezovsky,Vladimir
Severnaya Dvina Emb. 17,Arkhangelsk, Russia
E-mail: v.berezovsky@narfu.ru

Abstract. The aim of our project is to extract numerical descriptors or gather feature of rock samples from their images made under microscope with computer vision algorithm which can overcome the bottleneck of traditional method. It is impossible to solve it without getting the imformation of microscope magnification of images when they were made. That is why this paper is about an algorithm to identify the text used to mark microscope magnification information in rock core sample pictures. This is a sub project. As locations of text in these pictures are not fixed, font of text is not fixed, font size of text is not fixed, a method of two steps is designed to identify the text from the background. It works as follows: first, find the location of text; second, recognize the text with Convolution Neural Network. Experimental results demonstrate the potential value of the proposed approach to identify the magnification information in rock core sample pictures and to apply as a funtion to a program of our future project.

1. Introduction

The aim of our project is to extract numerical descriptors or gather feature of rock samples from their images made under microscope with computer vision algorithm, which can efficiently save time compared with traditional method. The first question we met in this project is to identify the text which used to mark microscope magnification information in rock core sample pictures. As it is impossible to extract numerical descriptors or gather feature of rock samples from their images without getting the imformation of microscope magnification of these images when they were made. A method of two steps is designed to identify the text from the background. It works as follows: first, find the location of text; second, recognize the text with Convolutional Neural Networks, which is a most popular machine learning algorithm these days.

The paper is organized as follows. Section 2 gives an overview of related works in the field of identification text in images. Section 3 makes an review of Convolutional Neural Networks. Section 4 describes Tensorflow which is a open source machine learning framework made by Google. Section 5 presents the individual steps of our approach to text localization and text regonition. Section 6 contains the experimental results obtained and discussion. Section 7 concludes the paper and outlines areas for future research.

2. Related works

In 2003, an approach was proposed by Julinda Gllavata, Ralph Ewerth and Bernd Freisleben to detect, localize and extract texts from color images with complex backgrounds [1]. It works as follows: Color images are first converted to grayscale images [1]. An edge image is generated using a contrast segmentation algorithm, which in turn uses the contrast of the character contour pixels to their neighboring pixels [1]. This is followed by the analysis of the horizontal projection of the edge image in order to localize the possible text areas [1]. After applying several heuristics to enhance the resulting image created in the previous step, an output image is generated that shows the text appearing in the input image with a simplified background [1]. These images are ready to be passed to OCR (Optical Character Recognition) system. Experimental results on a set of images demonstrated the performance of their approach [1].

In 2009, an Method was proposed by Weijuan Wen, Xianglin Huang, Lifang Yang, Zhao Yang and Pengju Zhang for text location and segmentation [2]. It works as follows: Firstly, the image is converted into wavelet transform domain [2]. Then three sub-band wavelet images of HxLy, LxHy, HxHy are binarizated and merged into a texture image by linear combination [2]. The texture image is processed by CRLA(constrained run length algorithm) and image smoothing to enhance the candidate text regions [2]. Finally, the text regions are located after 8-connected component growth and filtering out non-text regions [2]. Then the characters are segmented from text blocks based on the mean values of some slices of text-block are recognized by OCR [2]. Experimental results show that their method is robust to overlapped complex background [2].

In 2013, an algorithm was proposed by Anubhav Kumar and Neeta Awasthi for text localization and extraction for detection of both graphics and scene text in video images [3]. It works in four steps: Edge generation using Line edge detection mask, Text localization using projection profiles based method, Text segmentation and Text recognition [3]. The proposed technique gives better result than existing methods in terms of detection rate for large video image database with very few false alarms, reliable recall rate and precision rate [3].

In 2014, a connected component based approach was proposed by Kamrul Hasan Talukder and Tania Mallick to identify texts in images [4]. The text extraction process starts with conversion of the color image to gray scale image and then it converts the gray scale image into a binary image [4]. Then each text region is marked and the text is extracted from the image [4]. Finally, the extracted text is written into another gray scale image [4]. The experimental results demonstrate that the performance of the proposed method is valuable [4].

In 2017, Zhuoyao Zhong, Lei Sun and Qiang Huo proposed a LocNet to improve the localization accuracy of a Faster R-CNN based text detector [5]. Given a proposal generated by region proposal network (RPN), instead of predicting directly the bounding box coordinates of the concerned text instance, the proposal is enlarged to create a search region so that conditional probabilities to each row and column of this search region can be assigned, which are then used to infer accurately the concerned bounding box [5]. Experiments demonstrate that approach boosted the localization accuracy for Faster R-CNN based text detection significantly [5].

In 2015, an algorithm was proposed by T Kumuda and L Basavaraj for detecting and localizing text in natural scene images [6]. The entire work is divided into two stages. Text regions are detected in the first stage using texture features [6]. Discriminative functions are used to filter out non-text regions [6]. In the second stage the detected text regions are merged and localized [6]. An experimental results obtained showed that approach worked efficiently [6].

In 2014, an end-to-end text recognition system was presented by Michael O, Markus D, Stefan F, Florian K and Robert S [7]. The text recognition is done using a deep Convolution Neural Network (CNN) trained with backpropagation [7]. The system presented outperforms state of the art methods on the ICDAR 2003 dataset in the text-detection [7].

In 2016, Xinhao Liu, Takahito Kawanishi, Xiaomeng Wu and Kunio Kashino proposed a highly accurate character recognition model by utilizing the representational power of a specially designed Convolutional Neural Network (CNN) [8]. Based on the recognition model, they also developed an efficient post processing approach for error correction and hypothesis re-verification [8]. Character and word image recognition experiments on two public datasets show that the proposed approach provided superior or comparable results to the state-of-the-art techniques [8].

In 2018, a Rotational Region CNN was proposed by Yingying Jiang, Xiangyu Zhu, Xiaobing Wang, Shuli Yang, Wei Li and Hua Wang to estimate approximate text regions [9]. Their work follows three steps. First, a novel multi-task regression method was applied to support arbitrarily-oriented scene text detection [9]. Second, multiple ROIPoolings was introduced to address the scene text detection problem for the first time [9]. Third, an inclined Non-Maximum Suppression (NMS) was applied to post-process the detection candidates [9]. Experiments show that their method outperforms the state-of-the-art on standard benchmarks: ICDAR 2013, ICDAR 2015, COCO-Text and MSRA-TD500 [9].

3. Convolutional Neural Networks (CNN)

The CNN was first proposed by LeCun [10]. It simulates the processing system of human vision by using the local receptive field, shared weight, and subsampling [11]. Nowadays CNN is one of most popular artificial neural networks and widely used in computer vision, natural language processing, and speech recognition. A CNN is a multi-layered non-fully-connected neural network [11] composed by the input layers, the hidden layers and output layers. The hidden layer is of many neurons and connections between the input and output layers. The hidden layer can be composed of several convolutional layers, several pooling layers, and several full-connection layers in practice.

3.1. Convolutional layer

A convolutional layer is parametrized by the size and the number of the maps, kernel sizes, skipping factors, and the connection table [12]. Each layer has M maps of equal size (Mx, My). A kernel of size (Kx, Ky) is shifted over the valid region of the input image. The skipping factors Sx and Sy define how many pixels the filter/kernel skips in x and y direction between subsequent convolutions. The size of the output map is then defined as:

$$M_x^n = \frac{M_x^{n-1} - K_x^n}{S_x^n + 1} + 1 \tag{1}$$

$$M_y^n = \frac{M_y^{n-1} - K_y^n}{S_y^n + 1} + 1 \tag{2}$$

Where index n indicates the layer. Each map in layer Ln is connected to at most Mn-1 maps in layer Ln-1. Neurons of a given map share their weights but have different receptive fields.

3.2. Pooling layers

The purpose of the pooling layers is to achieve spatial invariance by reducing the resolution of the feature maps [13]. Each pooled feature map corresponds to one feature map of the previous layer [13]. There are two kinds of pooling layers: max pooling and subsampling.

The max pooling function

$$a_j = \max_{N*N}(\alpha_i^{n*n} u(n,n)) \tag{3}$$

applies a window function u(x;y) to the input patch, and computes the maximum in the neighborhood [13].

The subsampling function takes the average over the inputs, multiplies it with a trainable scalar β, adds a trainable bias b, and passes the result through the non-linearity [13].

In 2010, it has been found by Dominik Scherer, Adreas Muller and Sven Behnke that max-pooling can lead to faster convergence, select superior invariant features, and improve generalization [13].

3.3. Fully Connected Layers

The fully connected layer, as one of the important components of neural network, is mainly composed of two parts: linear operation and nonlinear operation. The linear transformation is a linear transformation from the perspective of the operation process. For an input vector $\vec{x} = [x_1, x_2, \cdots, x_n]^T$, it is transformed into $\vec{z} = [z_1, z_2, \cdots, z_n]^T$ by the matrix W, sometimes with an offset term $\vec{b} = [b_1, b_2, \cdots, b_n]^T$.

$$W\vec{x} + \vec{b} = \vec{z} \tag{4}$$

The linear part passes the summarized result to the nonlinear part, and the nonlinear part normalizes the obtained data. There will be problems when calculating from forward and reverse without the operation. Another important role of the nonlinear part is to break the previous linear mapping relationship. Suppose there is a two-layer fully connected neural network with no nonlinear layers, then for the first layer:

$$W^0\vec{x^0} + \vec{b^0} = \vec{z^1} \tag{5}$$

For the second layer:

$$W^1\vec{x^1} + \vec{b^1} = \vec{z^2} \tag{6}$$

Then:

$$W^1W^0\vec{x^0} + \vec{b^0} + \vec{b^1} = \vec{z^2} \tag{7}$$

$$W^1W^2\vec{x^0} + W^1\vec{b^0} + \vec{b^1} = \vec{z^2} \tag{8}$$

It is clear that the role of multi-layer network is no different from the function of a layer of network if it is without nonlinear part [14].

4. TensorFlow

TensorFlow is Google's second-generation machine learning algorithm implementation framework and Google chose to open up TensorFlow on GitHub in 2015 and released TensorFlow 1.0 in January 2017 [15]. It is a machine learning system that operates at large scale and in heterogeneous environments [16]. The front end of TensorFlow supports Python, C++, Go, Java and other development languages, and the back end is written in C++, CUDA, etc.

5. Proposed methods

Motivated by methods [1-9], an method is proposed to indentfy text which indicated microscope magnification in the picture of rock core samples. As there are 11 kinds of charecters need to be recognized and the code of this method will be one funtion of a program which is used to extract numerical descriptors or gather feature of rock samples from their images in future, this code should be as small as possible. Then a method was proposed. It includes two steps: first, find the location of text; second, recognize the text with Convolutional Neural Networks. For the first step, text regions are detected based on texture features. For the second step, CNN is composed of 5 layers which are include 2 convolutional layers, 2 pooling layers, 1 full-connection layers.

6. Experiment result and discussion

In our experiments, TensorFlow is used as a framework to implement CNN mentioned above. Weights parameters were initialized the before training networks with function of TensorFlow. Least squares method was considered as the cost function to minimize in this experiment. Gradient decent algorithm was used to calculate the gradient of the cost function and update the weights parameters by iterating.

The networks desinged were trianing with general-purpose GPU. The GPU used in the expriment is GeForce GTX 1050 Ti. It has 768 NVIDIA CUDA Cores and 4G GDDR5 memory which is of 7 Gbps Speed. The GPU Architecture is Pascal. The method was applied to rock image dataset which includes about 400 pictures. The correct rate is about 70%. It should be higher than this after optimization.

7. Conclusion

Experimental results demonstrate potential value of the proposed approach to identify the magnification information in rock core sample pictures. And it is small enough to apply as a funtion to program of our future project whose goal is to extract numerical descriptors or gather feature of rock samples from their images.

8. Acknowledgments

This work is supported by the Russian Foundation for Basic Research (RFBR) project No.16-29-15116 and China Scholarship Council.

References

[1] Gllavata J, Ewerth R and Freisleben B 2003 A robust algorithm for text detection in images *Proc. Int. Conf. on 3rd International Symposium on Image and Signal Processing and Analysis 2003* (IEEE)

[2] Weijuan W, Xianglin H, Lifang Y, Zhao Y and Pengju Z 2009 An Efficient Method for Text Location and Segmentation *Proc. Int. Conf. on 2009 WRI World Congress on Software Engineering* (IEEE)

[3] Anubhav K and Neeta A 2013 An efficient algorithm for text localization and extraction in complex video text images *Proc. Int. Conf. on 2013 2nd International Conference on Information Management in the Knowledge Economy* (IEEE)

[4] Kamrul T and Tania M 2014 Connected component based approach for text extraction from color image *Proc. Int. Conf. on 2014 17th International Conference on Computer and Information Technology (ICCIT)* (IEEE)

[5] Zhuoyao Z, Lei S and Qiang H 2017 Improved Localization Accuracy by LocNet for Faster R-CNN Based Text Detection *Proc. Int. Conf. on 2017 14th IAPR International Conference on Document Analysis and Recognition (ICDAR)* (IEEE)

[6] Kumuda T and Basavaraj L 2015 Detection and localization of text from natural scene images using texture features *Proc. Int. Conf. on 2015 IEEE International Conference on Computational Intelligence and Computing Research (ICCIC)* (IEEE)

[7] Michael O, Markus D, Stefan F, Florian K and Robert S 2014 End-to-End Text Recognition Using Local Ternary Patterns, MSER and Deep Convolutional Nets *Proc. Int. Conf. on 2014 11th IAPR International Workshop on Document Analysis Systems* (IEEE)

[8] Xinhao L, Takahito K, Xiaomeng W and Kunio K 2016 Scene text recognition with high performance CNN classifier and efficient word inference *Proc. Int. Conf. on 2016 IEEE International Conference on Acoustics, Speech and Signal Processing (ICASSP)* (IEEE)

[9] Yingying J, Xiangyu Z, Xiaobing W, Shuli Yang, Wei L and Hua W 2009 R^2CNN: Rotational Region CNN for Arbitrarily-Oriented Scene Text Detection *Proc. Int. Conf. on 2009 WRI World Congress on Software Engineering* (IEEE)

[10] Lcun Y, Bottou L, Bengio Y and Haffner P 1998 Gradient-based learning applied to document recognition (IEEE) 2278-324

[11] Min P, Chongyang W, Tong C and Guangyuan L 2016 NIRFaceNet: A Convolutional Neural Network for Near-Infrared Face Identification **7**

[12] Dan C, Ueli M, Jonathan M and Luca M 2011 Flexible, High Performance Convolutional Neural Networks for Image Classification *Proc. Int. Conf. on Proceedings of the Twenty-Second International Joint Conference on Artificial Intelligence* **2** 1237-42

[13] Dominik S, Adreas M and Sven B 2010 Evaluation of pooling operations in convolutional architectures for object recognition *Proc. Int. Conf. on International Conference on Artificial Neural Networks* (Springer) 92-101
[14] Chao F 2017 Deep learning in a easy way——learn core algorithms and visual practice (Beijing: Publishing House of Electronics Industry) 50
[15] Wenjian H and Yuan T 2017 TensorFlow combat (Beijing: Publishing House of Electronics Industry) 1-2
[16] Abadi M, Paul B, Jianmin Chen and Zhifeng C 2009 TensorFlow: A System for Large-Scale Machine Learning TensorFlow: A system for large-scale machine learning *Proc. Int. Conf. on 12th USENIX Symposium on Operating Systems Design and Implementation (OSDI16)* 265-84

Journal of Physics: Conference Series **1392** (2019) 012049 doi:10.1088/1742-6596/1392/1/012049

3-D electronic density maps for enzyme reaction intermediates: big data and high performance computing services in a virtualized environment

Anton Brekhov, Vladimir Mironov
Department of Chemistry, Lomonosov Moscow State University, Leninskie Gory 1/3, Moscow, 119991, Russian Federation

Alexander Moskovsky
RSC Group, Kutuzovsky prospect 36/23, Moscow, 121170, Russioan Federation

Dmitry Podgainiy, Oksana Streltsova, Dmitry Delyakov
Laboratory of Information Technologies, Joint Institute of Nuclear Research, ul. Joliot-Curie 6, Dubna, Moscow region, 141980, Russian Federation

E-mail: abrekhov@lcc.chem.msu.ru

Abstract. Enzymatic reactions play key role in metabolism of all living organisms. Recent advances in computational methods allow detailed study of enzyme-catalyzed processes, providing data on all intermediate states along reaction path. Furthermore, experimental data on the structure of the reaction intermediates become more available due to advancing of ultrafast X-Ray spectroscopy. These data describes electronic density of intermediates, sometimes even in time series. Currently, prominent Protein Data Bank (PDB) keeps records on the raw crystallographic data on electron density, but usually describes only enzymes or enzymes-inhibitor complexes. A database service where the both experimental and simulated data can be stored and analyzed simultaneously will be a valuable scientific tool for bringing theory and experiment together.

There are hundreds of enzymes known and one should expect large volumes of data on three-dimensional property like electronic density. At the same time, quantum chemistry simulations of large biomolecules require extensive computing resources. Thus, we turned into service model using Nautilus environment of Cognitive Hardware And Software Ecosystem Community Infrastructure (CHASE-CI) project as a blueprint. In that environment, each participating organization can donate their hardware resources, like storage or compute elements, while retaining flexible access control to running services and datasets. We have installed and tested a small-scale environment prototype using resources of LCC laboratory at Chemistry Department Lomonosov Moscow State University, Joint Institute of Nuclear Research (Dubna) and RSC Group company (Moscow). As it was efficiently implemented in Sloan Digital Sky Survey (SDSS) project, we identified basic search queries and structured database accordingly.

1. Introduction

In many fields of science the amount of accumulated data becomes enormous. The problem of effective storage and analysis of these data nowadays is becoming more and more important.

One of such research fields is study of protein-catalyzed reactions. The development of the ultrafast X-Ray spectroscopy allows to look in details on the intermediate reaction steps with extremely high time resolution. The resulting data are time series of three-dimensional electron density maps along the reaction path, i.e. the data are four-dimensional. One may understand the mechanism of the reaction by comparing the experimental electron density maps with those obtained in computer simulations. Today the amount of the experimental data is not too large and such analysis can be done manually. However, it will increase enormously in next years. Thus, the problem of its automatic analysis becomes more and more important.

The first and the most important problem is data storage. The ideal storage system should be large, extensible and reliable. Also, it is important to provide rapid access to the data for analysis. A prominent solution is to use cloud-base storage. In current study we tested distributed cloud-based solution deployed on the regular HPC servers, personal computers and even virtual machines. The reliability of data is provided by partial replication among the nodes. The environment is orchestrated by Kubernetes system.

2. The view on the service from the user's perspective

The design of the system assumes, that it would be used by the consortium of multiple organizations. Each member will provide the computational and storage resources for shared use. The resources can be geographically distributed and connected through the internet. Depending on needs of each member they will have different level of access to the shared resources. Currently the system is developed within the one organization, but we expect more when it comes to production.

Prior to developing of the system architecture it is important to identify probable use-cases to avoid possible failures at later steps. For example the developers of the Sloan Sky Digital Survey system recommend [1] to estimate around 20 requests. These requests can be obtained for example by surveying the future users. According to these recommendations we planned the design of the database which will be used to store 3D maps metadata.

The database will be filled in a (semi-)automatic way. For example, the computed data on the reaction intermediates will be published in the database after the successful simulation. The metadata will be used to search through the database. An example (incomplete) list of user requests are:

- Find all molecular structures available by the protein name, class, protein structure codes in Protein Data Bank (PDB);
- Find all intermediates and all known 3D maps (if any) by reaction type using enzyme code (EC) classification;

Next, we are going to implement indexing of 3D electron density maps basing on their similarity. There are few commonly used similarity metrics used to compare 3D data. Despite the fact, that the general problem of 3D data comparison is very complex, proteins related to the same enzyme class are alike each other. It allows to use relatively simple similarity metrics and comparison algorithms. For example, 3D electron density distributions are usually compared by constructing differential density map. To compute differential map, the input maps need to be properly aligned in 3D space. If needed, they are normalized and interpolated on the same 3D grid. Next, at each 3D grid point a difference between input maps is computed. This approach is commonly used in ultrafast X-Ray spectroscopy. It is planned to provide such analysis tools to users as they are needed.

3. Platform description

The platform comprises following services:

- Kubernetes system to orchestrate Docker containers.

- Distributed file system based on Ceph object storage. The object storage will be used to store large datasets like 3D maps.
- Database for metadata storage. We are going to use free open-source PostgreSQL database.
- High performance computing services. They will be used to run demanding calculations by user request.
- Web server for user access.

The structured metadata is stored directly in database tables. It includes the data which will be used in search queries, such is protein name, type, intermediate/transition state, etc. The metadata includes links to other databases, for example PDB codes. Semi structured data is stored in the JSON fields in the same database. The database also contain links to the bulky data in the object storage. Examples of these objects are density maps, calculation logs, protein structures.

4. Testing

To demonstrate the flexibility of the distributed platform we used three-node system, using the high-performance server provided by the RSC group, a desktop computed provided by the Department of Chemistry, and the virtual machine provided by the Joint Institution of Nuclear Research in Dubna, connected through the internet. To check the capabilities of this distributed system we used a series of tests for image processing. The dataset of 25,000 images was put to Ceph storage. To model the production workload we used computationally intensive job of image classification using ResNet network [2]. This test has I/O pattern similar to the production use. As a result, almost no performance degradation was obaseved when using this geographically distributed nodes setup.

5. Conclusion

This work is a part of a project aiming for developing an informational system for analysis of elementary steps of enzyme catalyzed reactions, including analysis of the electronic density. Using modern open-source cloud-based solutions it is possible to configure a geographically distributed informational system in a short time. The system features flexible sharing of the computational and storage resources between multiple users in different organizations. It can be used not only to store the data, but also to manage high-performance calculations. We plan to use the system as a tool for analysing enzyme reactions mechanisms.

Acknowledgement

This work is supported by Russian Foundation for Basic Research (project #18-29-13006).

References

[1] Szalay A.S., Gray J., Thakar A.R., Kunszt P.Z., Malik T., Raddick J., Stoughton C., vandenBerg J. The SDSS skyserver: public access to the sloan digital sky server data. In Proceedings of the 2002 ACM SIGMOD international conference on Management of data (SIGMOD '02). ACM, New York, NY, USA, 570-581. DOI: https://doi.org/10.1145/564691.564758.

[2] He K., Zhang X., Ren S., Sun J. Deep residual learning for image recognition. CoRR, abs/1512.03385. 2015.

Estimation of parameters of gas diffusion in models of microtransfactions of different type by the method of molecular dynamics

M.Yu. Antonov, A.V.Popinako, A.V.Grigoriev, I.N.Nikolaev

Type the author addresses here

mikhail@s-vfu.ru

Abstract. Estimation of gas transport parameters in organic nano-porous materials, such as kerogens, is important for understanding the fundamentals of the processes that co-produce natural gas. In this regard, the description and numerical parameterization of the processes of transfer of components of natural gas occurring at the micro level in organic nano-porous materials is of interest both from fundamental and practical points of view, including the search for possible ways to optimize the yield of natural gas and boost recoil. The purpose of this study is to study the methods of molecular modeling and the molecular dynamics of the mechanisms of gas transport in structures that mimic organic nano-porous materials, and the numerical evaluation of mass transfer parameters.

1. Introduction

Computer simulation methods have long shown their effectiveness in the study of the fundamental properties of physical systems, as well as in applied studies of mass transfer, diffusion / convection in complex systems [1–3]. To correctly solve diffusion / convection problems using computational physics methods based on solving differential equations for continuous media, as a rule, knowledge of such physical constants describing the properties of a simulated medium, such as diffusion coefficients, thermal expansion, etc., and their depending on the parameters of the system. These coefficients are usually obtained from experimental data. However, in some cases, obtaining experimental data is difficult or impossible. This applies, in particular, to the study of mass transfer processes in nanostructures at high pressures and temperatures, which correspond in particular to the natural conditions of gas mass transfer processes in kerogen. In this case, at the first step, other methods of computational physics can be applied, in particular, the molecular dynamics method (MD, [4]), which makes it possible to make sufficiently accurate estimates of the macroscopic parameters of gas diffusion processes on scales of the order of nanometers; n The use of MD allows one to analyze diffusion processes in structures and study the effect of dynamic mobility of molecules.

The obtained estimates of diffusion coefficients can be further used to refine the diffusion parameters of gas in nanostructures that mimic kerogen, for computer simulation of shale gas transport processes on a macroscopic scale, and to refine models describing shale gas transport processes. In particular, this applies to estimates of effective coefficients within the Pore-Network Model, for studying shale gas transport processes on a macroscopic scale, using grid methods (such as the finite element method).

Journal of Physics: Conference Series **1392** (2019) 012050 doi:10.1088/1742-6596/1392/1/012050

These models can be used in problems of rational nature management, in studying the development processes of shale deposits, many of which in the Russian Federation are concentrated in the Arctic zone. This applies, among other things, to the Kuonamian kolumplex of sediments in Yakutia, which, according to some experts, are of interest from the point of view of ease of exploration and production.

In the long term, due to the gradual exhaustion of traditional sources of hydrocarbons, there is an increase in the interest of the extractive industry to the deposits of shale hydrocarbons, many of which are located in the Russian Federation in the Arctic zone. Despite the fact that currently the development of shale deposits in the Russian Federation has no economic justification, monitoring and scientific research in the field of the extraction of hard-to-recover hydrocarbons are also necessary in order not to become in terms of technologies that may become relevant in the future in 20-50 years. This also applies to Yakutia, whose GRP is currently largely dependent on the extraction of minerals.

2. Materials and methods

2.1. Preparing of the structures

The original software BiLayer was used to build structures that simulate the gaps between the sheets of graphene, as well as structures that mimic the organic polymer. As a structure imitating the gap between the sheets of graphene, we used a structure with a size of 1.7 nm · 1.7 nm (in the X, Z plane) containing 836 carbon atoms. The structure of graphene is presented in Fig. 2. Periodic boundary conditions were used with regard to chemical bonds, and the MD OPLS force field was used [6–8].

The distance between the sheets (size of the periodic cell in the direction of the OY axis) varied from 0.5 nm to 19.5 nm (Figure 3, a). As a structure imitating the gap between the sheets of organic polymer, we used a structure with a size of 1.7 nm · 1.7 nm (in the X, Z plane), in which amino acid residues (a.k.o.) were added, supplemented with NH2 and C-OOH, while the position of the C-alpha atom was fixed in space to limit the drift of the molecule.

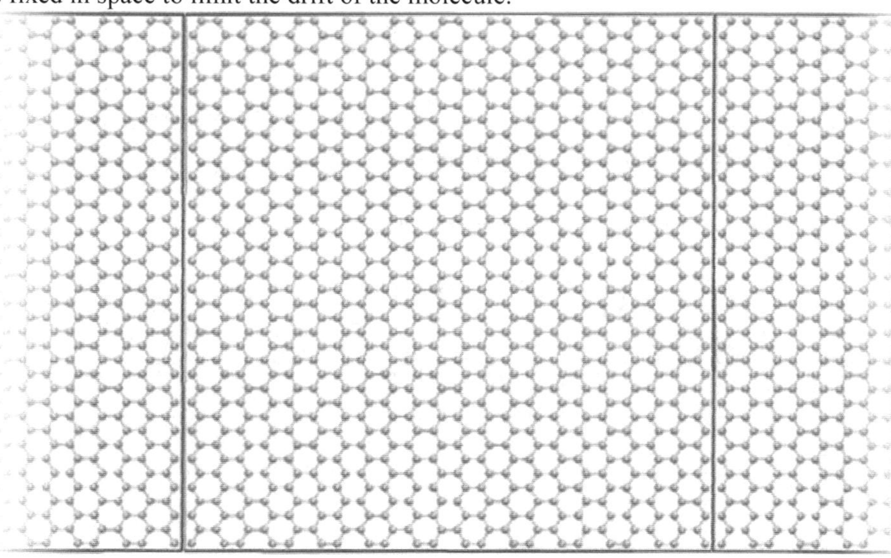

Figure 1. Graphene structure in periodic boundary conditions (blue lines).

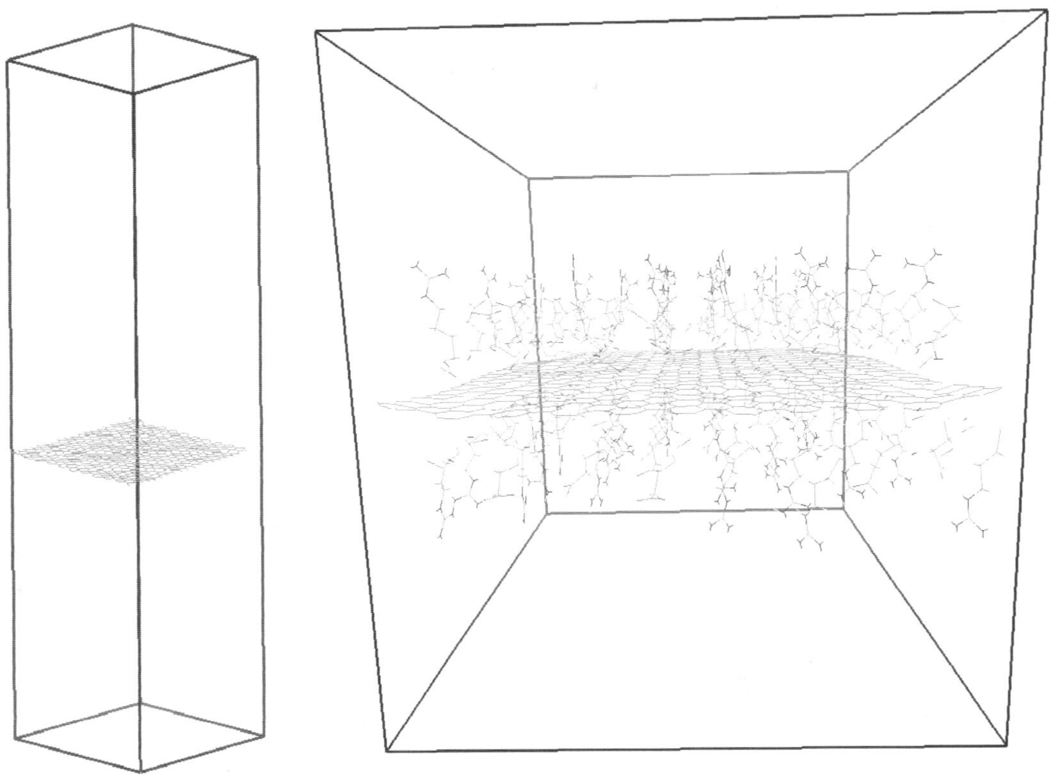

a b

Figure 2. Graphene with distance between the layers of 19.5 nm (a). System imitating organic polymer (Arg/Gly) with distance between the layers of 5 nm (b).

Three models were used that mimic the surface acid / alkaline, hydrophilic and hydrophobic. The residues Glu and Arg were chosen as the amino acid residues, and the acid a.k. and alkaline ako. respectively, Gln as a hydrophilic aco, Trp as a hydrophobic aco. (Figure 4). For each of the systems (Graphene, acid / alkaline, hydrophilic and hydrophobic surfaces), 38 structures were created with the distance between the walls from 0.5 nm to 19.5 nm with a step of 0.5 nm. The type of structure simulating the organic polymer Arg / Glu with a distance between the walls of 5 nm is shown in Figure 3, b. In this work, the molecules of methane and carbon dioxide included in the composition of natural gas were used as a model gas.

Since the presence of a graphene sheet does not have a significant effect on the diffusion of gas molecules, to accelerate the calculations, models simulating acid / alkaline, hydrophilic, and hydrophobic surfaces without the presence of a graphene sheet were also constructed. Amino acid residues were placed in a triangular grid, while the average area per 1 a.k. was 0.4 nm2. When preparing the structure, molecules a.k. aligned along the OY axis and rotated at a random angle in the XZ plane.

The prepared structures were then supplemented with structures obtained in the simulation of the gas phase (this process is described in more detail in Section 1.2). Figure 5 shows the structure model for the gap between the sheets of graphene. The periodic cell is highlighted in blue, supplemented with periodic images along the OX and OY axes.

The distance between the sheets (the size of the periodic cell in the direction of the OY axis) varied from 0.5 nm to 19.5 nm; in total, 38 structures were prepared for each system. During the simulation, amino acid residues supplemented with NH2 and C-OOH groups, imitating an organic polymer, were fixed in space due to the application of a harmonic potential with a stiffness constant of 1000 kJ / (mol · nm2) to C-alpha atoms a.k. .

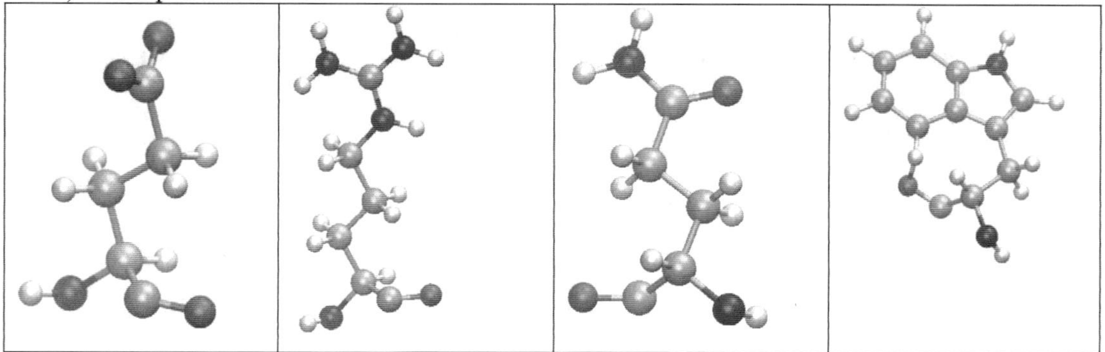

a, Glu	b, Arg	c, Gln	d, Trp

Figure 3. Models of molecules imitating organic polymer with acidic (a), basic (b), hydrophilic (c) and hydrophobic (d) properties.

2.2. Modelling of electrostatic interactions

Since the existence of electrons in the system is not explicitly taken into account in the MD method, the nonuniformity of the charge distribution is described by using partial charges for each atom and the Coulomb interaction energy between atoms is described by a standard expression. The calculation of the partial charges of atoms in the system is a separate fundamental point of molecular modeling, since these characteristics of atoms can have a significant impact on the dynamic behavior of the molecular system.

Different works use different partial charges corresponding to atoms of methane and carbon dioxide molecules, and the difference can reach 300% [6, 9, 10]. The problem is that when simulating gas transport under conditions that simulate natural conditions in kerogen, in this case, the parameters of the force fields selected for correct reproduction of the physical characteristics of the gas under normal conditions can give distortions. Therefore, to estimate the Coulomb interactions in the system, quantum chemical calculations were performed using the Firefly Gamess software [11]. The partial charges of atoms were estimated using the unrestricted Hartree-Fock method, the 6-31G * basis, and the Mulliken method. The calculation results are shown in table 1.

Table 1. Quantum-chemical calculations of the partial atom charges for methane and carbon dioxide molecules.

Molecule	Atom	Partial charge, e
CH4	C	0.660
	H	0.165
CO2	C	0.892
	O	-0.446

To prepare a periodic cell corresponding to the gas volume, molecular dynamics calculations were performed. Computational experiments were carried out in the Gromacs software package [12] version 4.6.7. The OPLS force field was used [6, 8], the minimization of the energy of the collected systems was carried out by the method of conjugate gradients. Then the system was subjected to multi-stage relaxation. The main parameters of modeling:

- Solving the equations of motion using the stochastic dynamics method with constant temperature control: τt = 1 ps;
- Integration step: 0.2-0.5 fs;

• Temperature: 300K and 353 K;
• Barostat: Berendsen with constant barostating τ = 5 ps;
• Pressure: 1 bar and 178.57 bar;
• Cutoff radius of non-valent interactions: 1.8 nm;
• The lengths of the connections and angles were not fixed.
• The initial atomic velocities were determined using a random number generator using the Maxwell distribution.

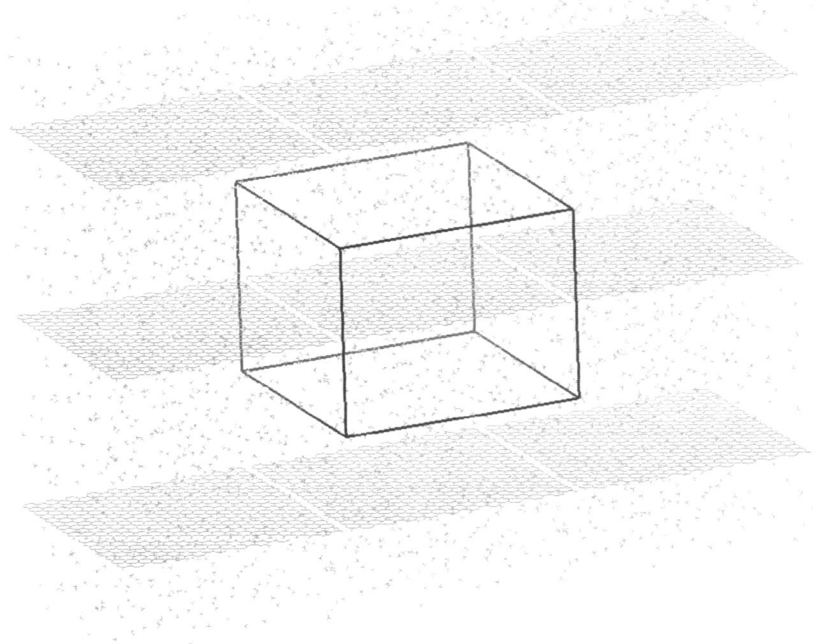

Figure 4. Model of the crack with methane gas between graphene layers under periodic boundary conditions (indicated with blue lines).

MD calculations were carried out for periodic cells containing 27,000 gas molecules under conditions of 300 K at T = 300 K, the results of the calculations are given in Table 2.

Table 2. Calculated diffusion coefficients in pure gas phase.

System	Calculated diffusion coefficient, cm^2/s
CH_4, Temperature 300K, Pressure 178 bar, OPLS forcefield.	83,9 ± 0,1 (• 10^{-5})
CH_4, Temperature 353K, Pressure 178 bar, OPLS forcefield.	153,9 ± 1,7 (• 10^{-5})

A total of 2 calculations were made with a total path length of 22 ns. For analysis, we used the data of the last 1 ns trajectories. As can be seen, with increasing temperature and pressure, the gas phase density and diffusion coefficient change. For further modeling of the conditions characteristic of the gas state in the kerogen, the simulation parameters were used: an integration step of 0.2 fs, a pressure of 178 bar, a temperature of 353 K. In the calculations of the diffusion of gases in model cracks, a model of methane was used.

3. Results

As a result of the calculations, an analysis was made of the diffusion coefficients of gas molecules in each system (Figure 6). As can be seen, with increasing the distance to the walls to about 0.8 nm, the graph reaches a plateau, and the difference in the diffusion coefficient in the thickness becomes insignificant and does not exceed the values of the standard deviation (on average 1.5 - 2.7 • 10-5 cm2 / s) however, it remains understated for the graphene system.

At the same time, the data near the border, on the contrary, are overestimated for the graphene system, and for the system from a.k. Trp simulating a hydrophobic organic polymer, the calculated diffusion coefficient is underestimated.

The results of the analysis of the energy of non-valent interactions of gas molecules with the walls are shown in Figure 7. As you can see, the total interaction energy decreases with the size of the system, and the interaction energy with the walls simulating a polymer is much stronger than in the graphene system to a distance of about 3 -5 nm, which may explain the underestimated values of the diffusion coefficient near the walls. As can be seen from Figure 8, a significant contribution to this effect is made by the partial Coulomb interaction potential, which is explained in particular by the fact that in the models of graphene used in the MD methods, the partial charge on the carbon atom is 0 and, thus, the Coulomb interaction between gas molecules and the wall is not taken into account. It can be seen that the expected contribution of Coulomb interactions is most significant in a system that mimics an acid / alkaline surface, and then decreases in the series: hydrophilic and hydrophobic surfaces.

It can also be noted here that in the case of using more polar molecules in the composition of the gas mixture (for example, carbon dioxide or water molecules), we can expect an additional enhancement of the role of Coulomb interactions on the rate of diffusion processes.

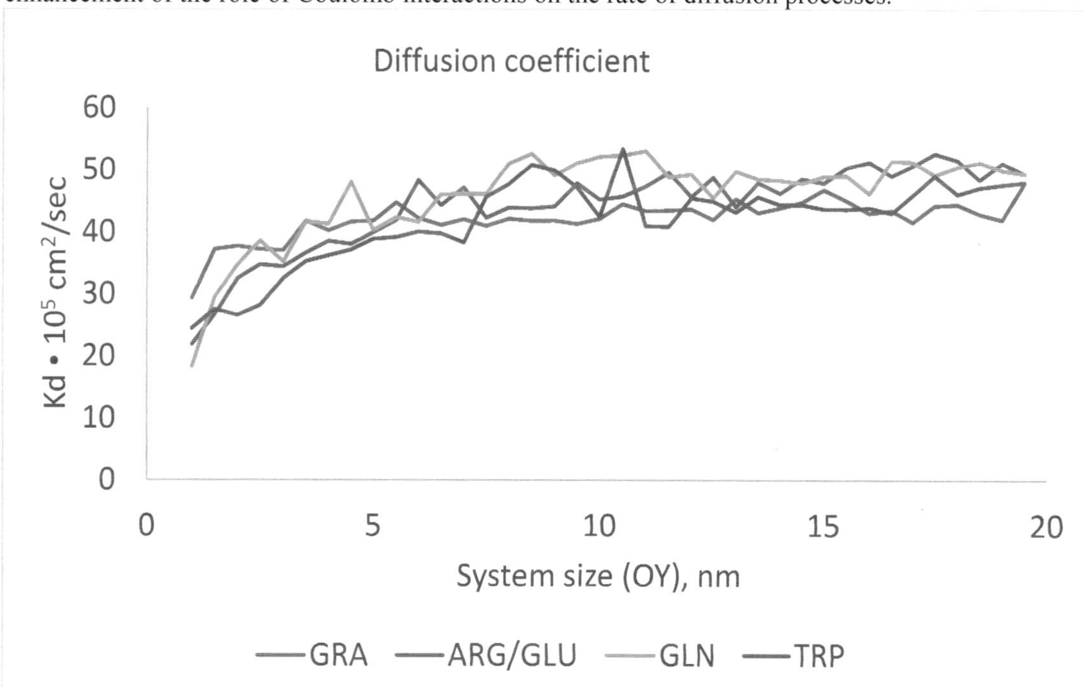

Figure 5. Estimated effective diffusion coefficient values for different systems.

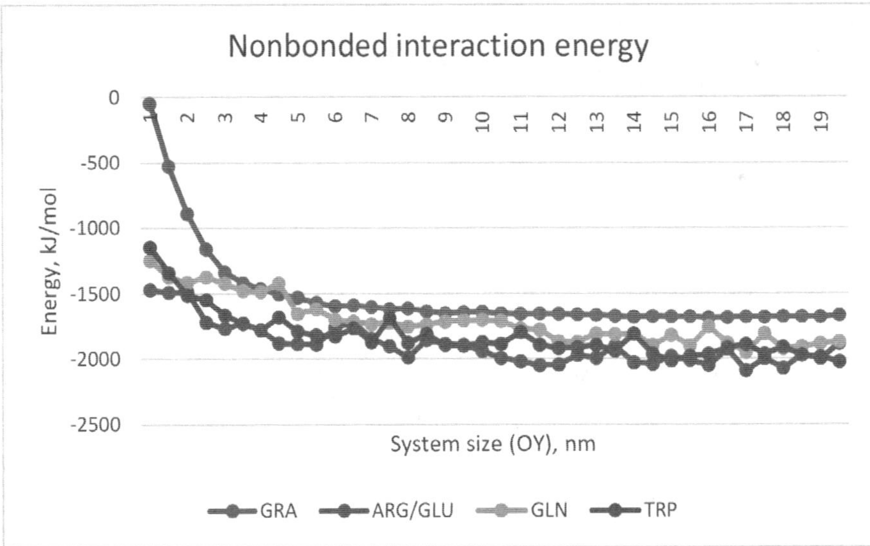

a

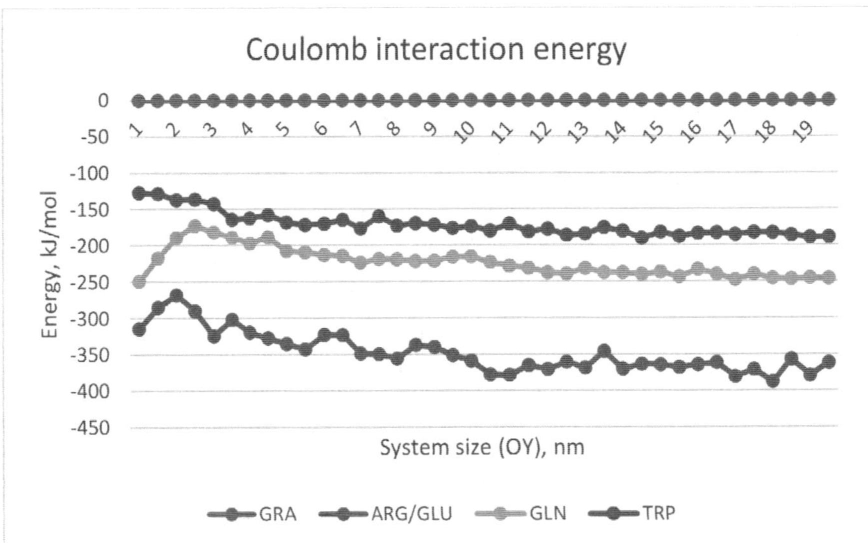

b

Figure 6. Calculated total energy of non-bonded interactions (a) and contribution of the Coulomb interaction energy (b).

4. Conclusion

Thus, in the course of the work, the structures of the gaps between the planes of pure carbon (graphene) and organic polymer were constructed, quantum-chemical calculations were performed to refine the force fields for non-standard system nodes, including methane and carbon dioxide molecules. Complexes of structures with a model gas were prepared (the methane model was used) and protocols were developed for simulating the diffusion of gases along the gap using the molecular dynamics method.

It is shown that in systems imitating an organic polymer, the processes of diffusion of gases near the system's boundary differ markedly from similar processes in the system, in which the crack boundary is described by the graphene model, which can be explained as a contribution of Lennard-Jones interactions with atoms other than carbon, and the contribution of Coulomb interactions.

The results can be used to refine macroscopic models of gas transport using grid methods (such as the Pore-Network Model and the finite element method) to study the transport processes of shale gas in kerogen.

Computational calculations were performed using the resources of the supercomputer complex of the NFCU “Arian Kuzmin”. The work on the preparation of three-dimensional structures was carried out with partial financial support from the Russian Foundation for Basic Research (Project No. 18-41-140005p_a, Antonov M.Yu.). Molecular calculations, analysis of the interaction energy and diffusion coefficients were carried out with support from the Ministry of Science and Higher Education of the Russian Federation (Grant No. FSRG-2017-0013).

References

[1] Akkutlu I.Y. et al. Multiscale model reduction for shale gas transport in a coupled discrete fracture and dual-continuum porous media // J. Nat. Gas Sci. Eng. 2017.

[2] Kou R., Alafnan S.F.K., Akkutlu I.Y. Multi-scale Analysis of Gas Transport Mechanisms in Kerogen // Transp. Porous Media. Springer Netherlands, 2017. Vol. 116, № 2. P. 493–519.

[3] Riewchotisakul S., Akkutlu I.Y. Adsorption-Enhanced Transport of Hydrocarbons in Organic Nanopores // SPE J. Society of Petroleum Engineers, 2016. Vol. 21, № 06. P. 1960–1969.

[4] Berendsen H.J.C. et al. Molecular dynamics with coupling to an external bath // J. Chem. Phys. 1984. Vol. 81, № 8. P. 3684.

[5] Antonov M.Ju. et al. Ocenka profilej potenciala srednej sily metodom umbrella sampling dlja transmembrannogo perenosa molekuly vody. // Matematicheskie zametki SVFU. 2014. Vol. 21, № 3. P. 95–105.

[6] Kaminski G.A. et al. Evaluation and Reparametrization of the OPLS-AA Force Field for Proteins via Comparison with Accurate Quantum Chemical Calculations on Peptides // J. Phys. Chem. B. American Chemical Society, 2001. Vol. 105, № 28. P. 6474–6487.

[7] Jorgensen W.L., Maxwell D.S., Tirado-Rives J. Development and Testing of the OPLS All-Atom Force Field on Conformational Energetics and Properties of Organic Liquids // J. Am. Chem. Soc. 1996. Vol. 118, № 45. P. 11225–11236.

[8] Jorgensen W.L., Tirado-Rives J. The OPLS [optimized potentials for liquid simulations] potential functions for proteins, energy minimizations for crystals of cyclic peptides and crambin // J. Am. Chem. Soc. 1988. Vol. 110, № 6. P. 1657–1666.

[9] Martin M.J., Siepmann J.I. Transferable Potentials for Phase Equilibria. 1. United-Atom Description of n-Alkanes. American Chemical Society, 1998.

[10] Sigfridsson E., Ryde U. Comparison of methods for deriving atomic charges from the electrostatic potential and moments // J. Comput. Chem. John Wiley & Sons, Ltd, 1998. Vol. 19, № 4. P. 377–395.

[11] Gordon M., Schmidt M. Advances in electronic structure theory: GAMESS a decade later. 2005. P. 1167–1189.

[12] Hess B. et al. GROMACS 4: Algorithms for Highly Efficient, Load-Balanced, and Scalable Molecular Simulation // J. Chem. Theory Comput. American Chemical Society, 2008. Vol. 4, № 3. P. 435–447.

Digital twins of multiscale 3D heterogeneous geological objects: 3D simulations and seismic imaging of faults, fractures and caves.

Vladimir Cheverda[1], Vadim Lisitsa[1], Maksim Protasov[1], Galina Reshetova[1], Boris Glinsky[2], Igor Chernykh[2], Anastasiya Merzlikina[3], Victoria Volyanskaya[4], Denis Petrov[5], Artjem Melnik[5] and Valery Shilikov[5]

[1] Institute of Petroleum Geology and Geophysics, 3 prosp. Koptyug, 630090, Novosibirsk, Russia

[2] Institute of Computational Mathematics and Mathematical Geophysics, 6 prosp. Lavrentiev, 630090, Novosibirsk, Russia

[3] IGIRGI, 25, Vavilova strt. bld.1, Moscow, Russia

[4] PJSC Rosneft

[5] RN KrasnoyarskNIPIneft Ltd., 65d, 9th May strt., Krasnoyarsk, Russia

E-mail: `cheverdava@ipgg.sbras.ru`

Abstract. Numerical simulation of seismic waves' propagation and imaging in a three-dimensional multiscale geological media is one of the main trends in the development of modern geophysics. We present the work flow for construction of the 3D digital twin of some real geological object. This object is cavernous fractured carbonate reservoir near the Riphean roof. One of the main features of this object is the presence of the family of faults filled with tectonic breccia. To simulate them, in particular, their interiors and surrounding damaged areas, we simulate paleotectonic processes with the use of 3D discrete elements technique. This simulation we do by GPU parallelization.

Next, when the fine digital geological model is done, we perform 3D finite-difference simulation of the seismic waves' propagation and get the full 3D synthetic data set. To do this simulation we use an original technique based on local grid refinement in time and space. To be able to deal with huge amount of input/output data we use High-Performance Computing (HPC) systems with parallel architecture and hybrid parallelization strategy by simultaneous use of MPI and OpenMP.

The final step is validation of our original multiscale algorithm for seismic imaging, which suppresses specular reflection, but accumulates weak scattered/diffracted waves. This approach opens the way to the reconstruction of subseismic geological objects, like fractures, fracture corridors and clusters of caves. To construct these images we use HPC with parallel architecture and MPI+OpenMP programming.

1. Introduction

Advances in a supercomputing technology make it feasible problems, which seemed impossible in the recent past but become a common practice today. In the geoscience, those have been opening new horizons for understanding the subsurface structures by getting 3D images and velocity models of high fidelity and microscale reservoir characterization. The forward seismic

Journal of Physics: Conference Series **1392** (2019) 012051 doi:10.1088/1742-6596/1392/1/012051

modeling and inversion, associated with field data processing and velocity estimation, may require terabytes of RAM and teraflops of computing power.

One of the principal direction of High-Performance Computing application in the Oil and Gas industry is the development of unique synthetic data sets for understanding the main peculiarities of the wave field's propagation in 3D heterogeneous multiscale media and approbation and validation of new data processing algorithms. These data must, to some extent, correspond to the real one, collected by the industrial acquisition systems.

In this research, we build the 3D multiscale digital model containing the main features inherent in geological objects at several licensed sites of Rosneft PJSC in the North of Eastern Siberia. They are:

(i) 3D geological faults, filled with tectonic breccia and surrounded by damage zones;
(ii) The systems of cracks confined to these faults as well as fracture corridors;
(iii) Areas of high cavernosity.

In this model we took into account a number of scales:
a) Macroscale, first tens of meters, obtained by 3D seismic study (seismic interfaces forming the skeleton of the model);
b) Mesoscale, first meters, obtained by various well log measurements;
c) Microscale, first tens of centimeters, captured by sonic logs, FMI fullbore formation microimager and analysis of core samples.

2. Model Building

2.1. Skeleton of the Model

The initial stage of building the model is the definition of the skeleton; in other words, a description of the totality of all interfaces known in the result of processing and interpretation of all available data - both surface seismic and VSP [1] as well as the entire spectrum of logs. We start with a mapping of all interfaces known by regular 3D seismic study, including 3D geological faults. Elastic parameters between interfaces are known on the base of a variety of additional data, like velocity analysis and well log. A general view of these surfaces one can see in Figure 1.

2.2. 3D Geological Faults

At present, the common knowledge is to treat a geological fault not like a slip surface, but as a complex 3D geological body [2]. Therefore, we consider faults as volumetric entities consisting of rocks deformed in the process of destruction caused by tectonic movements [1]. The products of this process are closely linked to a wide range of parameters, such as the tectonic regime, magnitude of fault displacement, and mechanical properties of the host rock. To perform numerical simulation of these complicated processes we apply the discrete elements technique with parameters calibration by comparing real observations and simulation results. Recall that when using this method the continuum medium is represented as a set of "discrete elements"/particles with simple geometric shape. In particular, we use their representation in the form of balls of different radii interacting according to a certain set of physical laws.

The main goal of 3D modeling is to determine and analyze the distribution of deformations in the horizontal direction along the fault at the macroscale level, especially for scenarios with displacement. The computational area was chosen in the form of a parallelepiped with dimensions of 500 meters vertically and 2000 meters in each horizontal direction. The size of the elements ranged from 0.5 to 15 meters with a uniform distribution. The modulus of

[1] Vertical Seismic Profile - measurement of seismic waves within within a borehole

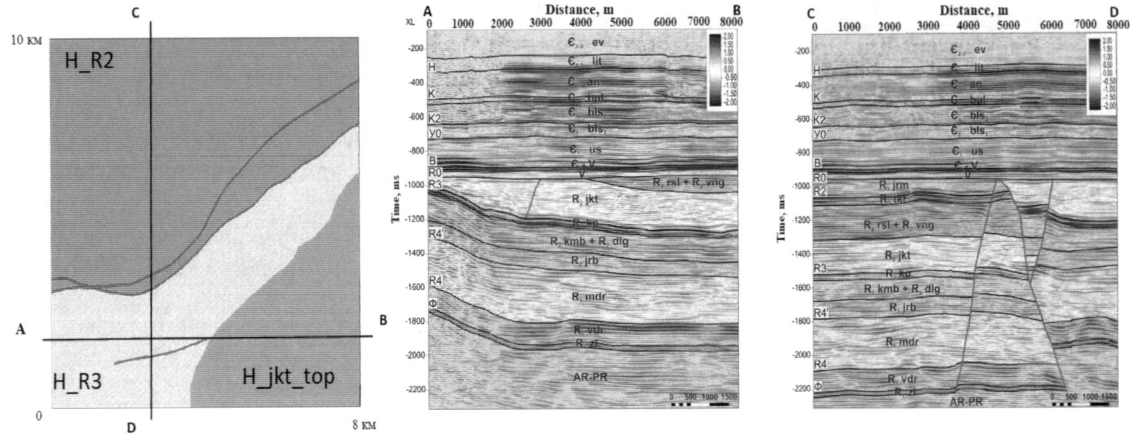

Figure 1. Skeleton of the model. Left - top view of the target layer (reservoir level), middle - vertical cross-section along line AB, right - vertical cross-section along line CD

stiffness of the elements was 16 GPa, regardless of which layer the element belonged to. To take into account the differences in the geomechanical properties of the layers, we varied the dynamic friction coefficient within them, which determines the intensity of tangential forces. The general scheme of the numerical experiments one can see in the Figure 2. These numerical

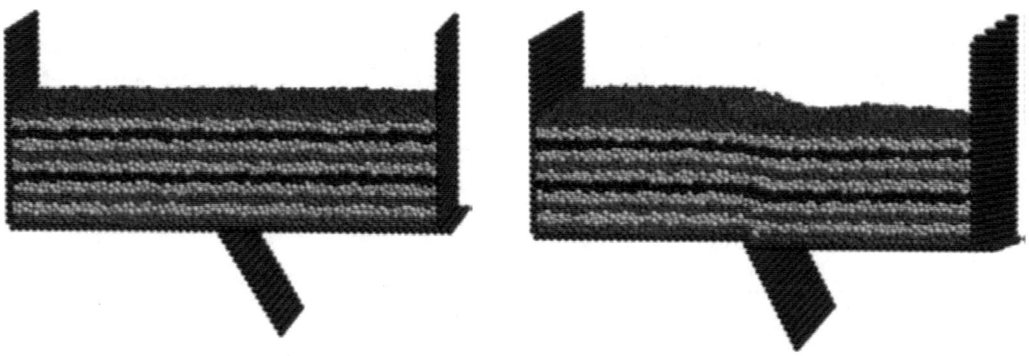

Figure 2. The general scheme of numerical simulation of faults formation with discrete elements. Left - initial stage, right - final stage

experiments give the spatial distribution of deformations. Next, with the use of the empirical curve describing connection of deformations with the change of elastic modules we reconstruct microscale heterogeneities in the damage areas around faults (see Figure 3). Finally we come to the full 3D elastic model of the target area with interfaces, faults and damage areas presented in the Figure4

3. 3D Multiscale Seismic Simulation

Let us present the formal parameters of the model and acquisition.

Model fills $8km \times 10km \times 6km$ parallelepiped with uniform cells of $5m \times 5m \times 5m$ everywhere out of localization of small-scale heterogeneities. Within areas with this kind of heterogeneities (damaged areas filled with tectonic breccia, clusters of caves and fractures and fracture corridors)

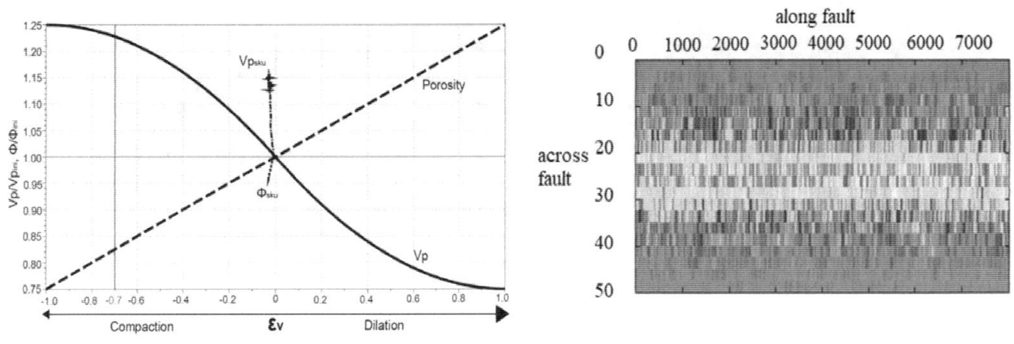

Figure 3. Left: The empirical dependence of perturbations of elastic modules of deformations. Right: Elastic parameters of tectonic breccia in the damage area around a fault

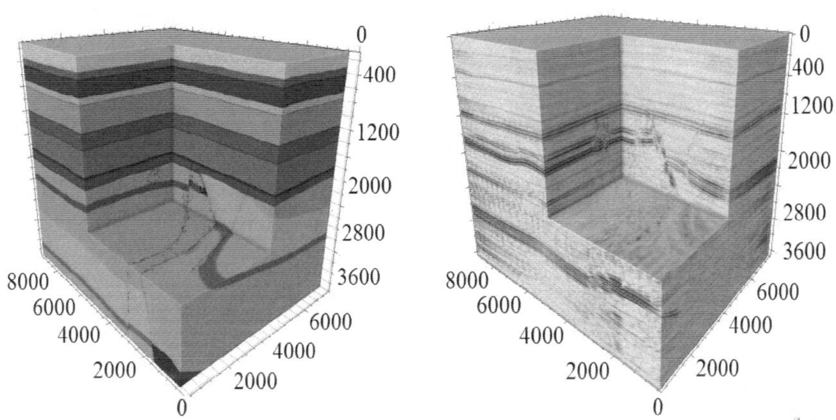

Figure 4. The full seismic model (left) and its seismic image (right). One can see 3D structure of the faults and their mapping on seismic wave imaging (right)

we use the locally refined grid with cells of 0.5 meters that allows us to describe the variability of elastic properties with reasonable accuracy.

Acquisition is the $8km \times 10km$ rectangular with:

- 3C receivers placed uniformly on the grid $25m \times 25m$;
- Vertical force point sources placed with the step of 50m along lines which are 300m from each other, 5280 sources at all;
- Source function is Ricker pulse with dominant frequency 40 Hz;
- Recording length for each receiver is 4 seconds with sample of 2 ms.

We perform numerical simulation of seismic wave's propagation in this 3D heterogeneous multiscale medium by explicit in time finite-difference technique with locally refined staggered grids in time and space [3].

4. Scattering Imaging

We use the synthetic data obtained and known velocity model to image small-scale heterogeneities. To do this we do asymmetric summation with the help of elastic Gaussian

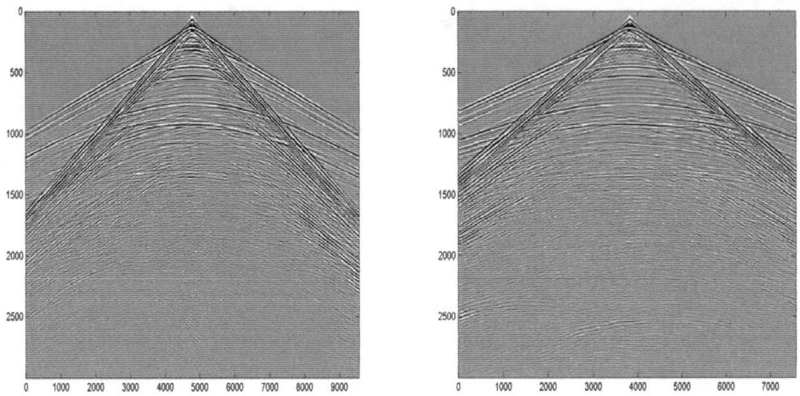

Figure 5. 3D seismograms for the source in the center of the acquisition. Left - along the short axis (8km), right - along the long one (10 km)

beams following paper [5]. The key feature of the method of scattering imaging is decomposition of full velocity model onto two constituent - smooth propagator given by smooth functions $\lambda_0(\vec{x})$, $\mu_0(\vec{x}), \varrho_0(\vec{x})$ and its sharp perturbations (reflector) described by sharp oscillating functions $\lambda_1(\vec{x})$, $\mu_1(\vec{x}), \varrho_1(\vec{x})$. The propagator does not return seismic energy back to the acquisition, but prescribes the time of wave's propagation between two distant points. In the contrast, reflectors do not change travel time, but turns the seismic wave towards the acquisition. In this paper we suppose the depth velocity model known and will image its perturbations, especially clusters of small-scale heterogeneities. To compute these images we use the 3C data $\vec{\phi}(x_r, y_r; x_s, y_s; \omega)$ produced by the sources (x_s, y_s) and acquired by receivers (x_r, y_r) and perform the asymmetric summation with weights computed by tracing two Gaussian beams from the target point towards sources and receivers (see details in [5]).

$$f(\vec{x}_i, \beta) = \int T^{gb_2}(x_s, y_s; \vec{x}_i; \gamma, \theta, \beta; \omega)\vec{T}^{gb_1}(x_r, y_r; \vec{x}_i; \gamma, \theta, \beta; \omega)\vec{\phi}(x_r, y_r; x_s, y_s; \omega) \times$$
$$\times k(\gamma, \theta, \beta) dx_r dy_r dx_s dy_s d\omega d\gamma d\theta \qquad (1)$$

The microlocal analysis done in [4] proves that left hand side of (1) is the local impedance and, hence, provides the variability of medium parameters at the imaging point.

In the Figure 6 one can see microscale heterogeneities in the model (left) and their image computed by the formula (1). There are all of them on the image and the intensity is proportional to the contrast of the impedance.

5. Conclusion

In this study we applied the variety of numerical methods which essentially use High-Performance Computations.

(i) Discrete elements based GPU simulation of the tectonic processes to estimate the elastic parameters of breccia filling geological faults;

(ii) Finite-difference technique on the base of 3D domain decomposition using huge high-performance computers and MPI/OpenMP parallel programming;

(iii) Original imaging technique on the base MPI/OpenMP parallel programs processing Terabytes of synthetic 3D seismic data (big data).

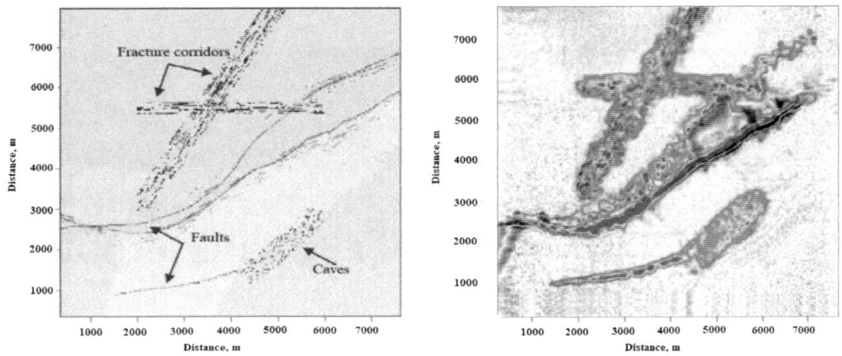

Figure 6. Small-scale heterogeneities in the model (left) and their scattering image (right)

Author contribution

Vladimir Cheverda proposed and together with Maksim Protasov justified the asymmetric summation with Gaussian beams to image small-scale heterogeneities. Dmitry Kolyukhin and Vadim Lisitsa applied discrete elements simulation to describe fault formation and estimate parameters of tectonic breccia. Maksim Protasov developed asymmetric summation for 3D imaging and constructed 3D images of small-scale heterogeneities; Galina Reshetova applied finite-difference technique with local grid refinement in time and space and computed 3D synthetic seismic wave fields. Boris Glinsky dealt with optimization of parallel programming, Igor Chernykh determined the technology of the workflow of high performance computations and Igor Kulikov learned the elasto-plastic processes within faults. Anastasiya Merzlikina developed 3D skeleton of the model, Victoriya Volyanskaya supervised simulation of tectonic motions creating faults, Denis Petrov is responsible for filling the skeleton with seismic parameters (density and wave propagation velocities), Valery Shilikov did interpretation of the data processing, Arjem Melnik developed faults geometry.

Funding

Vladimir Cheverda, Vadim Lisitsa, Maksim Protasov and Galina Reshetova have been supported by the Russian Science Foundation, project 17-17-01128. Boris Glinsky, Igor Chernykh and Igor Kulikov have been supported by Russian Foundation of Basic Research, project 16-29-15120.

Acknowledgements

Computations were done on the Siberian Supercomputer Center.

References

[1] Faulkner D.R., Jackson A.L., Lunn R.J., Schlische R.W., Shipton Z.K., Wibberly C.A.J., and M.O.Whitjack 2010 A review of recent developments concerning the structure, mechanics and fluid flow properties of fault zones *Journal of Structural Geology* **32** 1557 - 1575.

[2] Kolyukhin D.R, Lisitsa V.V., Protasov M.I., Qu D., Reshetova G.V., Tveranger J., Tcheverda V.A., Vishnevsky D.M. 2017 Seismic imaging and statistical analysis of fault facies models *Interpretation* **5**(4) 1 - 11.

[3] Kostin V., Lisitsa V., Reshetova G., Tcheverda V. 2015 Local time-space mesh refinement for simulation of elastic wave propagation in multi-scale media *Journal of Computational Physics* **281** 669 - 689

[4] Protasov M., Tcheverda V., Pravduhin A. 2019 3D-true amplitude anisotropic elastic Gaussian beam depth migration of 3D irregular data *Journal od Deismic Exploration* **28**(2) 121 - 146.

[5] Protasov M.I., Tcheverda V.A. and Reshetova G.V. 2016 Fracture detection by Gaussian beam imaging of seismic data and image spectrum analysis *Geophysical Prospecting* **64**(1) 68 - 82.

Journal of Physics: Conference Series **1392** (2019) 012052 doi:10.1088/1742-6596/1392/1/012052

Supercomputer technologies for solving problems of computational physics

Boris Glinskiy[1], **Yury Zagorulko**[2], **Igor Kulikov**[1] **and Anna Sapetina**[1]

[1] Institute of Computational Mathematics and Mathematical Geophysics SB RAS, Novosibirsk, Akad. Lavrentieva pr. 6, Russia

[2] A.P. Ershov Institute of Informatics System SB RAS, Novosibirsk, Akad. Lavrentieva pr. 6, Russia

E-mail: gbm@sscc.ru

Abstract. The issues of using supercomputer technologies developed by the authors to solve compute-intensive problems of mathematical physics are discussed. The technology of developing algorithms and software for supercomputers contains the three related stages: co-design, by which we understand the adaptation of the problem statement, the mathematical method and the computational algorithm to the parallel architecture of the supercomputer at all stages of the problem solution; studying the scalability of computational algorithms for the most promising supercomputers based on simulation modeling; evaluation of the energy efficiency of algorithms for various implementations on a given supercomputer architecture. It is proposed to further develop this approach with the use of the intelligence support for solving compute-intensive problems using the ontology of computational methods and algorithms for solving the problems of computational physics, the ontology of computational heterogeneous architectures and decision rules.

1. Introduction

The state-of-the-art in the of supercomputer development is characterized by the emergence of a variety of projects on the creation of an exascale-class supercomputer. Thus far, developments in the field of exascale supercomputers are conducted by different teams of developers in the United States. The collaboration in this direction is carried out, for example, by the national laboratories of the US Department of Energy: Sandia and Oak Ridge. In Europe, there are also similar programs; seven European countries have signed the declaration of the Joint Project EuroHPC aimed at the creation of exascale supercomputers. In Japan (the RIKEN institution), the development of a supercomputer has already begun, where it will be assembled and installed. There are numerous international projects to develop the system and the application software for exascale-class supercomputers with the participation of the United States, the countries of the European Union, Japan, China, Russia (IESP, G8 EXASCALE, CRESTA, etc.). In [1-3], a review and various approaches to the exascale scientific software design are given. In [4], the problems typical of exascale systems are listed, such as insufficient concurrent work available to maintain an extensive utilization of all resources, a time-distance delay intrinsic of parallel actions and resources on the critical execution path, which is not necessary in a sequential version, the delay due to the lack of availability of shared resources. In addition, in [4] various methods for overcoming the above-mentioned problems are discussed.

The current state of the computer technologies provides the researchers with a wide range of computational architectures with the ability to use high-performance parallel computing to solve their problems. It should be noted that along with classical computational clusters with MPP (massive parallel processing), SMP (symmetric multiprocessing) architectures, hybrid supercomputers equipped with accelerators of computing, are becoming increasingly popular. In the TOP 500 of the world's most powerful computers (`www.top500.org`) a supercomputer Summit with a capacity of 188 PFlops is in the first place. It includes 4068 computer nodes, each containing two 22-core processor IBM Power9 and six GPUs NVIDIA Tesla V100, linked by a dual-channel Mellanox EDR. Summit has also more than 10 PB of RAM combined with fast high-bandwidth channels for the efficient data transfer. Note that this supercomputer is optimized for the analysis of big data; nevertheless, intensive work on using the advantages of the V100 architecture for solving problems of plasma physics and accelerator physics is underway.

The performance of the most powerful Russian supercomputer Lomonosov-2 is 2.5 Pflops. However, in Russia, it is planned to build up the national supercomputer centers with a capacity of dozens PFlops. On such supercomputers, it is possible to solve problems of essential computational complexity.

In each subject area, where there is a set of methods for solving compute-intensive problems on modern multi-core computing systems, it is important to choose a method that would most effectively solve the problem in a reasonable time with a given accuracy. On the other hand, the representation of knowledge in the form of ontologies and inference rules for solving specific problems is becoming increasingly popular [5,6]. Note that ontologies (a system of concepts and relationships linking them) are the foundation of knowledge bases in complex areas that require systematization. The application of the ontologies facilitates the creation of generally available knowledge bases and the use of artificial intelligence methods thanks to a large set of freely available editors for creating ontologies and inference engines (reasoners) that make logical conclusions.

Thus, at present, the following situation is emerging: a researcher on the one hand, has a set of computational methods and algorithms that solve a complex computational problem, and, on the other hand – a set of computational architectures, libraries and software tools to solve the problem. Considering that it is difficult for a researcher to competently navigate both in computational methods for solving a problem and in existing means for high-performance computing (architectures, parallel languages and software tools), it is necessary to create a system of intelligence support (IS) for making decisions with allowance for a the features of the simulated environment to use a computing architecture.

It should be pointed out that numerical algorithms are being developed slower than hardware. Therefore the existing algorithms and programs for solving compute-intensive problems will be applied at the first stage of using petascale- and exascale supercomputers, containing hundreds of millions of cores.

Consequently, the creation of modern computing technologies for the parallel use of such a large number of computing cores is an urgent task.

2. A supercomputer technology for solving compute-intensive problems based on an integral approach

We have offered the integrated approach to the development of algorithms and software for petascale- and exascale-class supercomputers [7]. It contains the three stages.

The first stage is the co-design, which is based on the development of a parallel computational technology, with allowance for all aspects of parallelism. The co-design of parallel methods for solving large-scale problems is difficult to formalize. It is impossible to make a collection of recipes for the efficient solution of any problem. However, some general approaches can be proposed. The co-design approach concept consists of the following steps, with allowance for the target hardware/software platform: the formulation of physical statement of the problem, the mathematical formulation of the physical problem, the development of the numerical methods, the selection of data structures and parallel algorithms, the consideration of a supercomputer architecture, the usage of code optimization tools.

We use an extended definition of the co-design in contrast to the common conception which consists in the joint development of software and hardware. In such an approach, not only the comparison of the problem solution methods is becoming relevant but also the comparison of the efficiency of using various physical and mathematical statements [8, 9].

The second stage is the anticipated development of algorithms and software for the most promising exascale supercomputers. This stage is based on the simulation of the algorithm behavior within a certain supercomputer architecture. For the simulation of distributed systems it is best to use distributed simulation based on message passing.

The multi-agent approach [10] is used for the simulation of parallel programs run on a large number of cores due to such properties as decentralization, self-organization, and intelligent behavior. Among many multi-agent simulation platforms, the adaptable distributed simulation system AGNES [11] was chosen. It was successfully used in the scalability study of a series of parallel problems when executed on a large number of cores [12].

The third stage is estimating the energy efficiency of the algorithm with different implementations for a single architecture or for different supercomputer architectures. In this paper, the term "energy efficiency" for scientific HPC applications means the most efficient use of each core, a processor or a computational accelerator; the minimization of communications among computational nodes; a good workload balancing of the program. The minimization of communications enables a decrease in the idle standing time for processors and accelerators. The good workload balancing enables a uniformly load of a computational system. The most energy efficient algorithm gives the best FLOPS per Watts (Joules/sec) value. The efficiency of this approach is illustrated on some examples of complex computing problems in seismology, plasma physics and astrophysics.

Let us consider the problem of numerical simulation of the propagation of elastic waves in heterogeneous three-dimensional media with complex subsurface geometries, which is a difficult problem from the point of view of computation, which requires the use of effective methods of parallelizing and scaling algorithms.

Quite often, the topography of various real geophysical objects does not allow one to maintain an observational system. Therefore, constructing their 3D models requires solving the inverse problem by solving a set of direct problems: for different values of the elastic parameters of a heterogeneous medium; for various geometries of objects composing a model.

We apply the above-discussed approach to solving the problem of seismic wave propagation in a heterogeneous medium typical of magmatic volcanoes. Both active and sleeping volcanoes are potentially dangerous to the environment due to the possibility of sudden disaster eruptions. Using the methods of the active vibroseismic monitoring of magmatic structures will allow predicting a probable time of an eruption [13].

For the purpose of the co-design, we have made a comparison of the developed parallel implementations of solutions to the elastodynamic problem written in terms of the velocities of displacement and stress and in terms of displacements for the computational clusters, equipped with graphics cards. The simulation domain is considered to be an isotropic 3D-inhomogeneous elastic structurally complex medium which is a parallelepiped, one of whose sides is a free surface.

At the step of designing a numerical method, the most "flexible" and widespread technique for solving a three-dimensional elastodynamic problem is a finite difference method. Let us preliminarily notice that explicit finite difference schemes fit the architecture of graphics accelerator, because they are directly mapped to the topology of GPU architecture, and involve independent computations of values at each step and in each cell of the computational domain.

Journal of Physics: Conference Series **1392** (2019) 012052 doi:10.1088/1742-6596/1392/1/012052

In order to numerically solve elastodynamics equations in terms of the displacement velocities and the stress we apply the well-known Verrier finite difference scheme on a staggered grid. The calculation of its difference coefficients is based on integral conservation laws. To solve the problem in terms of displacements, we use a similar finite difference scheme [13]. We should note the main difference between the algorithms, which can be constructed on the above-mentioned finite difference schemes. The calculation of the displacement velocities and the stress requires a larger memory size (at least, 18 3D arrays with the unknowns should be stored), but requires a smaller number of float-in-point operations in total (57 operations for calculating the values in a cell for one time step). The calculation of the displacements requires a smaller memory size (at least, six 3D arrays with unknowns), but a larger number of operations (98 operations for one time step).

The numerical experiments have shown that the time of the calculations of the displacements and the time of the calculations of the displacement velocities and the stress at an equal number of nodes is roughly the same in spite of the fact that the displacements calculation is performed with a larger amount of floating point operations at each time step. Therewith, the displacement calculations requires almost half as much of the GPU memory size. Based on the results obtained we prefer using the approach proposed to calculating the displacements. Figure **??** shows a diagram of the vibroseis monitoring of the Elbrus volcano, its geophysical model and wave field propagation. The results of numerical experiments on the NCS-30T cluster have shown that it is possible to distinguish kinematic differences in the wave field and, approximately, restore the magma level in the output channel.

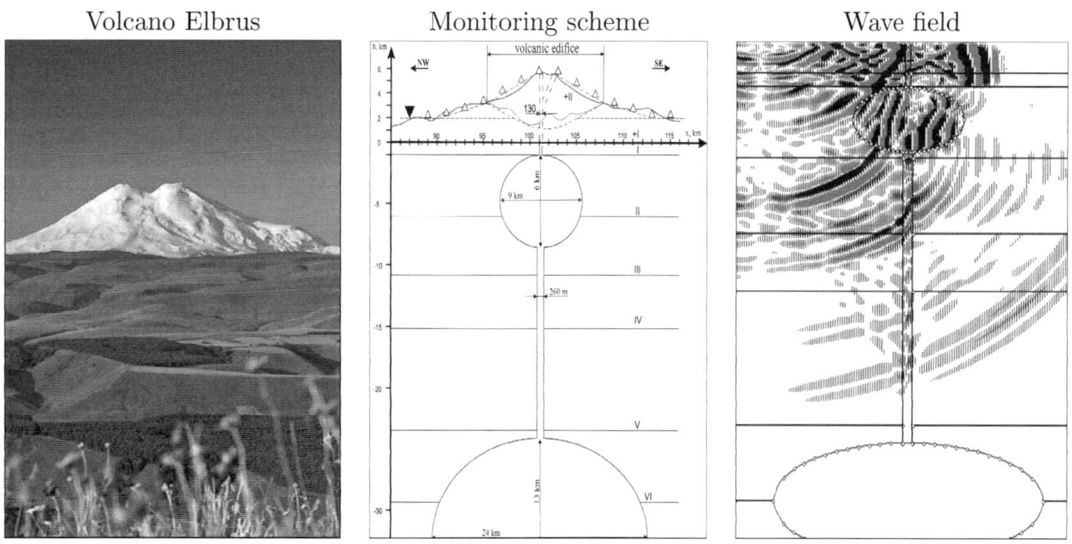

Figure 1. The scheme of vibroseismic monitoring of the Elbrus volcano and the result of the numerical experiment.

The scalability of the resulting algorithms has been tested using the AGNES simulation system. We have considered utilizing the hybrid supercomputers equipped with GPU or Xeon Phi architectures for solving the elastodynamic problem. The results of the research into the numerical simulation scalability of the elastic wave propagation problem in complex media are presented in Figure **??**. The two algorithms that solve this problem in two ways have been studied. From the figures it is clear that the scalability of the two approaches slightly differs, and both algorithms are suitable for execution on a large number of cores. The real calculations for the verification were performed with the Siberian Supercomputer Center.

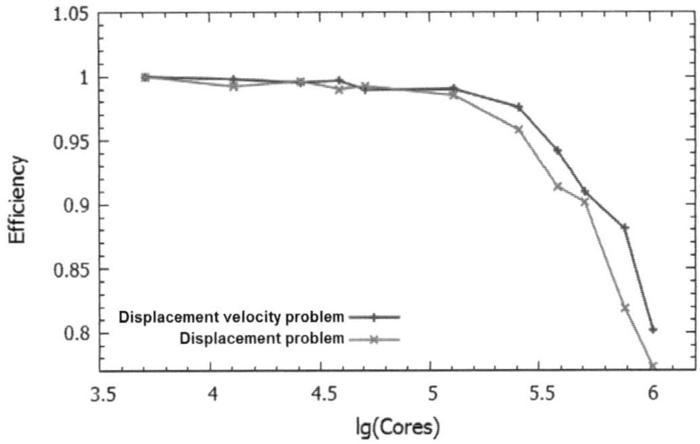

Figure 2. The scalability of numerical modeling of the elastic wave propagation problem in complex media based on the AGNES simulation.

The further research direction we see in the creation of methods and systems of intelligence support (IS) for solving compute-intensive problems on supercomputers.

3. The basic principles of building the IS system

Consider the principles of building the IS system for solving compute-intensive problems and its main blocks. Figure ?? shows a conceptual scheme of the IS system.

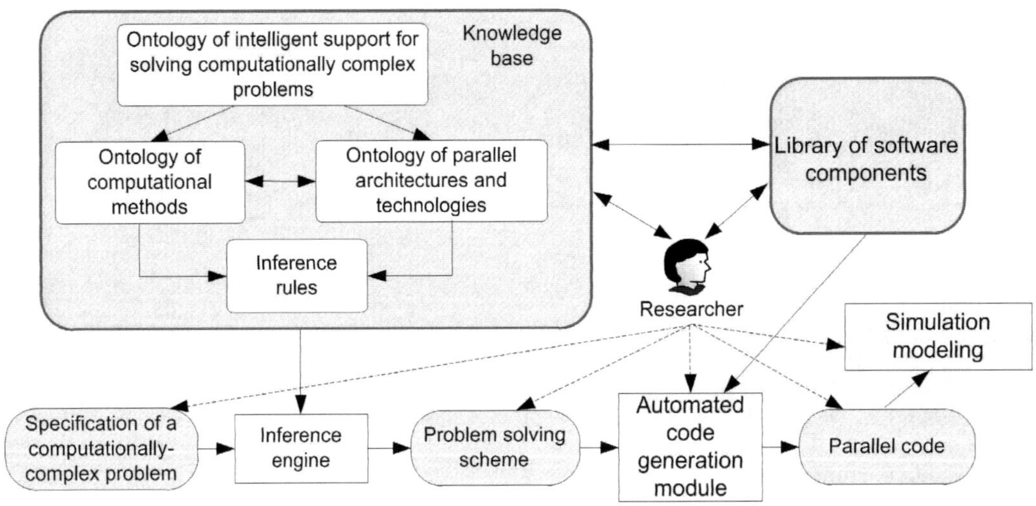

Figure 3. A conceptual scheme of the IS system for solving compute-intensive problems.

The core of such a system is a knowledge base, including a set of interconnected ontologies and inference rules that extend the logic of these ontologies. An important module of the IS system is an inference engine (a reasoner), which, using the knowledge base and the user specification of a problem, builds an optimal scheme for its solution.

Another component of the IS system, i.e. the automated code generation module, supports the creation of a parallel code that solves the problem set by the user. This module inserts the corresponding code fragments from the software component library (SC library) into a scheme of solving the problem. If there is no suitable component in the SC library, the user can substitute a necessary component by taking it from the standard library or writing a new one.

The IS system also includes a simulation modeling unit that evaluates the scalability of the resulting code on a given architecture before carrying out the final time-consuming calculations on a multi-core computing system. The main purpose of this unit is to select the optimal number of cores for the implementation of the algorithm by studying its behavior in the model time.

4. Conclusion

This paper considers the use of the integrated approach to the development of algorithms and software for the numerical simulations of seismic wave propagation in the media with complex subsurface geometries on a hybrid cluster. For the purpose of the co-design, we have made a comparison of the developed parallel implementations of solutions with the elastodynamic problem written in terms of the displacement velocities and the stress and in terms of displacements for the computational clusters, equipped with graphics cards. We have studied the execution time of the created parallel programs and their scalability (including the simulation of program execution on a large number of cores by using the simulation system AGNES).

In this paper, we have also presented the concept of the IS system for solving compute-intensive problems of mathematical physics. The paper shows how, having, on the one hand, the ontology of computational methods and the corresponding parallel algorithms, and on the other hand, the ontology of parallel architectures and technologies, it is possible to create parallel programs for solving problems as soon as possible.

Acknowledgments

This research was conducted within the framework of the budget project No. 0315-2019-0009 for ICMMG SB RAS and supported in part by the Russian Foundation for Basic Research [grant No. 19-07-00085].

References

[1] Reed D A and Dongarra J. 2015 *Commun. ACM* **58(7)** 56–68
[2] Keyes D E 2011 *C.R. Mechanique* **339** 70–77
[3] Asanovic K et al 2009 *Commun. ACM* **52** 56–67
[4] Sterling T 2013 *Adv. Parall. Comp* **24** 104–117
[5] Compton M et al 2012 *Web Semant. Sci. Serv. Agents World Wide Web* **17** 25–32
[6] Keet C M, Lawrynowicz A, dAmato C, Kalousis A, Nguyen P, Palma R, Stevens R and Hilario M H 2015 *Web Semant. Sci. Serv. Agents World Wide Web* **32** 43–53
[7] Glinskiy B, Kulikov I, Chernykh I, Snytnikov A, Sapetina A and Weins D 2017 *CCIS* **793** 278–289
[8] Glinskiy B, Kulikov I, Snytnikov A, Romanenko A, Chernykh I and Vshivkov V 2014 *Supercomputing Frontiers and Innovations* **1(3)** 88–98
[9] Glinsky B, Kulikov I, Chernykh I et al 2017 *LNCS* **10049** 342–353
[10] Bellifemine F L, Caire G and Greenwood D 2007 Developing Multi-Agent Systems with JADE (Chichester: Wiley)
[11] Podkorytov D, Rodionov A and Choo H 2012 *Proc. of the 6th Int. Conf. on Ubiquitous Information Management and Communication* (Kuala Lumpur: ACM) p 115
[12] Glinsky B M, Marchenko M A, Mikhailenko B G, Rodionov A S, Chernykh I G, Karavaev D A, Podkorytov D I and Vins D V 2013 Simulation Modeling of Parallel Algorithms for Exaflop Supercomputers (in Russian) *Inf. Technol. Comput. Syst.* **4** 3–14
[13] Boris Glinskiy, Anna Sapetina, Valeriy Martynov, Dmitry Weins, Igor Chernykh 2017 *CCIS* **753** 261–274

Numerical solution of the inverse coefficient problem of filtration in a multilayer reservoir

M. Kh. Khairullin[1], E. R. Badertdinova[2] and R. M. Khairullin[2]

[1] - Institute of Mechanics and Engineering, FRC Kazan Scientific Center, Russian Academy of Sciences, 2/31, Lobachevsky str., Kazan 420111 Russia
[2] - Kazan National Research Technological University, 68, Karl Marx street, Kazan 420015, Republic of Tatarstan, Russian Federation

E-mail: : khairullin@imm.knc.ru

E-mail: : badertdinova@yandex.ru

E-mail: : rustemkhairullin@yandex.ru

Abstract.
This work proposes a computational algorithm based on the regularization method for determining the hydraulic conductivity coefficient of a layered oil reservoir. The results of hydrodynamic investigations of vertical wells, opening a layered reservoir, are used as initial information.

1. Introduction

The majority of oil reservoirs has a layered structure due to features of the sedimentation process. If the ratio of the permeability coefficients of two neighboring layers is less than 10^{-3} then the Myatiev-Girinskii scheme is applicable [1].

One of the field hydrodynamic methods for determining the reservoir properties of a multilayer reservoir is the method of steady-state selection [2, 3]. The problem of determining the coefficient of hydraulic conductivity of a multilayered oil reservoir is among the class of inverse problems of underground hydromechanics. This work proposes a computational algorithm based on the regularization method for determining the hydraulic conductivity coefficient of a layered oil reservoir. In [4, 5] the problems of determining the filtration properties of a multilayer reservoir were considered.

2. Statement and solution of the coefficient inverse problem

According to the Myatiev-Girinskii scheme the problem of determining the pressure fields $p_1 = p_1(x,y)$, $p_2 = p_2(x,y)$, $\ldots$, $p_n = p_n(x,y)$ in a reservoir with nonpermeable roof and floor, separated by a weakly permeable layers under simultaneous separate exploitation, is the solving a system of partial differential equations in a multiconnected domain D with boundaries $\partial D = \Gamma + \Gamma_1$, Γ_1 is circle of radius $r_c = 0.1m$.

$$L_1 p_1 + \omega_1(p_1 - p_2) = 0,$$
$$L_2 p_2 + \omega_1(p_2 - p_1) + \omega_2(p_2 - p_3) = 0,$$

$$\cdots \tag{1}$$
$$L_n p_n + \omega_{n-1}(p_n - p_{n-1}) = 0$$

with boundary conditions

$$\int\limits_{\Gamma_1} \sigma_{2k-1}\frac{\partial p_k}{\partial n}dS = q_k \tag{2}$$

$$\left.\frac{\partial p_k}{\partial \tau}\right|_{\Gamma_1} = 0, \left.p_k\right|_{\Gamma} = 0, k = 1, 2, \cdots, n, \tag{3}$$

where $L_k p_k \equiv -div\,(\sigma_{2k-1} grad p_k)$, $\sigma_{2k-1}, H_{2k-1}\,(k = 1, 2, \cdots, n)$ is the hydraulic conductivity coefficient and thickness of well permeable layers; $\omega_k = \frac{\sigma_{2k}}{H_{2k}^2}$, σ_{2k}, H_{2k} is the hydraulic conductivity coefficient and thickness of weakly permeable layers.

The inverse problem is to find the hydraulic conductivity coefficient $\sigma = (\sigma_1, \sigma_2, \cdots, \sigma_{2n-1})$ of a multilayered oil reservoir. Its initial data are the flow rates q_k, the pressure functions on the boundary of the filtration domain, approximate values of the hydraulic conductivity coefficient, known from the technical documentation of the reservoir. Additionally, bottomhole pressures measured at a well are considered known.

$$p_k^{(z)} = \left.p_k\right|_{\Gamma_1}, k = 1, 2, \cdots, n \tag{4}$$

The solution of the inverse coefficient problem (1) - (4) is sought in the class of piecewise constant functions $\sigma_k = const$, $c_1 \le \sigma_k \le c_2$, c_1, c_2 are positive constants. This inverse problem generates a certain implicitly given nonlinear operator

$$A\sigma = P^* \tag{5}$$

where $\sigma = (\sigma_1, \sigma_2, \cdots, \sigma_{2n-1})$, $P^* = \left\{p_k^{(z)}\right\}_{k=1}^{n}$ is bottomhole pressure vector. The vector P^* is ordinarily known inexactly: $\|P^* - P_\delta^*\| \le \delta$, $\|\cdot\|$ is the norm in the Eulcidean space R^n ,δ is the error in measurement. Solution of the operator equation (5) with the approximate right side is realized on the basis of minimizing the smoothing functional A. N. Tikhonov [6,7]

$$M^\alpha(\sigma) = \|A\sigma - P_\delta^*\|^2 + \alpha\Omega(\sigma) \tag{6}$$

where $\Omega(\sigma) = \sum\limits_{i=1}^{2n-1}\left(\sigma_i - \sigma_i^0\right)^2$, $\alpha = \alpha(\delta)$ is the regularization parameter that agrees with the observation error, σ_i^0 is approximate values of the hydraulic conductivity coefficients.

Successive approximations σ^n are constructed in this manner: in the neighborhood of σ^n for a fixed value of the regularization parameter $\alpha = \alpha_n$ a nonlinear operator $A\sigma$ in the formula (6) is represented in the form

$$A\sigma = A\sigma^n + A'_\sigma(\sigma^n)(\sigma - \sigma^n) + o\|\sigma - \sigma^n\|$$

where $A'_\sigma(\sigma^n)(\sigma - \sigma^n)$ is the Frechet differential in σ^n. To find σ with a fixed α , the Gauss-Newton procedure is applied [7]. The functional (6) takes the form:

$$M^{\alpha_n}(\sigma) = \left\|A\sigma + A'_\sigma(\sigma^n)(\sigma - \sigma^n) - P_\delta^*\right\|^2 + \alpha_n\Omega(\sigma)$$

The Frechet differential is calculated on the basis of perturbation theory and has the form
$A'_\sigma(\sigma)(\sigma-\tilde\sigma) = (A_1, A_2, \cdots, A_n)$
where $A_i = (\delta\sigma_1 grad\tilde p_1, grad\tilde p_1^i) + \cdots + (\delta\sigma_{2n-1} grad\tilde p_n, grad\tilde p_n^i) + \sum_{l=1}^{n-1} \frac{\delta\sigma_{2l-2}}{H_{2l-2}^2}\Big((\tilde p_l - \tilde p_{l+1}, \tilde p_l^i) + (\tilde p_{l+1} - \tilde p_l, \tilde p_{l+1}^i)\Big)$, $(a,b) = \int_D a(x,y)\, b(x,y)\, dxdy$ is the scalar product, $\tilde p = (\tilde p_1, \tilde p_2, \tilde p_3, \cdots, \tilde p_n,)$ is solution of problem (1) - (4), when the hydraulic conductivity coefficient equals $\tilde\sigma$, $\tilde p_l^i$ are solutions conjugate problems:

$$\tilde L_1 \tilde p_1^i + \tilde\omega_1(\tilde p_1^i - \tilde p_2^i) = 0,$$
$$\tilde L_2 \tilde p_2^i + \tilde\omega_1(\tilde p_2^i - \tilde p_1^i) + \tilde\omega_2(\tilde p_2^i - \tilde p_3^i) = 0,$$
$$\cdots$$
$$\tilde L_n p_n + \tilde\omega_{n-1}(\tilde p_n^i - \tilde p_{n-1}^i) = 0$$

with boundary conditions

$$\int_{\Gamma_1} \sigma_{2k-1}\frac{\partial \tilde p_k^i}{\partial n} dS = \delta_{ki}, \frac{\partial \tilde p_k^i}{\partial \tau}\Big|_{\Gamma_1} = 0, \tilde p_k^i\Big|_{\Gamma} = 0, \quad i,k = 1,2,\cdots,n,$$

where $\tilde L_k \tilde p_k^i \equiv div\,(\sigma_{2k-1} grad\tilde p_k^i)$, δ_{ki} is Kronecker symbol.

3. The results of the numerical experiment

Two-layer oil reservoir $D = \{0 \le x, y \le 1000m\}$ is exploited by one well, pressure on the boundary Γ is zero. Reservoir parameters: fluid viscosity $\mu = 10mPa \cdot s$, permeability coefficients and thickness of layers $k_1 = 0.7\mu m^2$, $H_1 = 10m$, $k_2 = 0.0001\mu m^2$, $H_2 = 1m$, $k_3 = 0.35\mu m^2$, $H_3 = 8m$, $\sigma_i = \frac{k_i H_i}{\mu}$, $i = \overline{1,3}$.

Problem (1) - (3) and conjugate problems were solved by the finite difference method. In this case, the well is considered a point source with a power equal to the flow rate of the well [8]. The calculations were carried out with the grid step $h = 100m$ and the regularization parameter $\alpha = 200$. Initial permeability coefficient approximations: $k_1 = 0.5\mu m^2$, $k_2 = 0.0001\mu m^2$, $k_3 = 0.5\mu m^2$. At each step of the iterative process, the residuals were computed:

$$\|\Delta p^n\|^2 = \sum_{i=1}^{2}\sum_{j=1}^{3}\left(p_{ij}^{(N)} - p_{ij}^{(V)}\right)^2, \quad \|\sigma - \sigma^n\|^2 = \sum_{j=1}^{3}(\sigma_i - \sigma_i^n)^2, \quad i = 1,2,\cdots.$$

The calculations showed that the rate of convergence of the iterative process depends on the choice of the initial approximation of the permeability coefficient of the weakly permeable layer. Figure 1 and Figure 2 show the results of convergence of the iterative process with different regularization parameters.

The results of the calculations showed that the iterative process converges faster with the value of the regularization parameter $\alpha = 200$ and this regularization parameter is close to optimal according to the residual criterion.

4. Well testing N2046 of the Republic of Tatarstan

The well opened terrigenous Tula (in the range of 1101-1105 m.) and Bobrikov (in the range of 1112.5-1123 m) reservoir. Flow rate well $q = 13.9m^3/day$. Hydrodynamic investigations were carried out during simultaneous drainage of two layers, and during their separate exploitation.

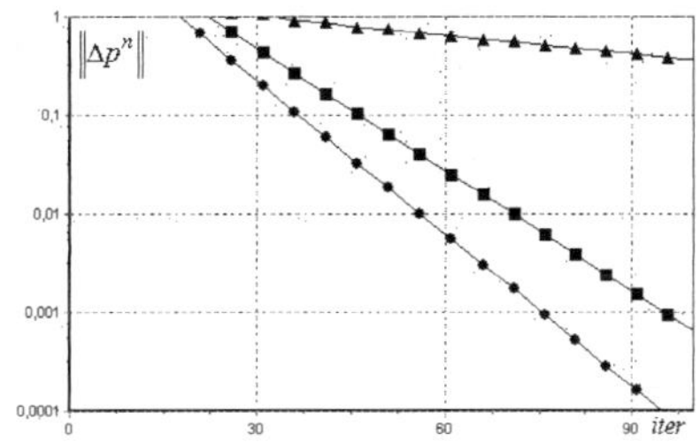

Figure 1. Decrease of the residual pressure functional.
△ – α = 190, ● – α = 200, ■–$\alpha = 250$.

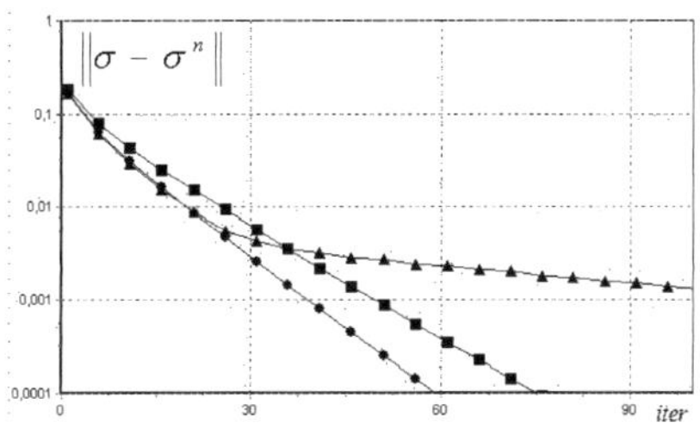

Figure 2. Convergence of the iterative process
△ – α = 190, ● – α = 200, ■–$\alpha = 250$.

The separation of the layers was carried out using the installation for simultaneous-separate exploitation.

An indicator card was constructed based on the results of field hydrodynamic investigations of a well operating only the Tula horizon. The estimates of the productivity coefficient $\eta_T = 0.027 m^3/MPa \cdot day$ and the flow rate $q_T \cong 11.1 m^3/day$, which gives the Tula horizon during the simultaneous exploitation of two layers, are obtained. The Table 1 shows the results of calculations. In the calculations, we used the scheme of "isolated" layers (1 line) and the Myatiev-Girinsky scheme (line 2).

Table 1. Calculation results.

P_{plB}, MPa	P_{plT}, MPa	P_B^z, MPa	P_T^z, MPa	$q_B, m^3/day$	$q_T, m^3/day$	$k_B, \mu m^2$	$k_T, \mu m^2$
6.123	6.193	2.339	2.433	2.8	11,1	0.018	0.213
6.156	6.156	2.339	2.433	2.8	11,1	0.025	0.254

Here P_{plB}, P_{plT} are reservoir pressure, P_B^z, P_T^z are bottomhole pressure, q_B, q_T are flow rates, k_B, k_T are permeability coefficients of Bobrikov and Tula horizons, respectively.

Calculations carried out both on model problems and on real data show that the proposed computational algorithm allows determining filtration properties of oil reservoirs.

References

[1] Polubarinova-Kochina P Ya 1977 *The Theory of Groundwater Movement* (Moscow : Science)) p 664
[2] Earlougher R ml 1977 *Hydrodynamic methods for well testing* (Moscow - Izhevsk :Inst. for Comp. Stud.) p 512.
[3] Diyashev R N 1984 *Joint development of oil reservoirs (Moscow : Science)* p 208
[4] Badertdinova E R and Khairullin M Kh 2006 Determination of filtration parameters of a layered bed using data of unsteady liquid inflow to a well *J. of Eng. Phys. and Thermophys.* **79** 3 553–556

[5] Khairullin M Kh , Shamsiev M N, Abdullin A I and Badertdinova E R 2014 Interpretation of the results of thermohydrodynamic studies of vertical wells that operate on multibed deposits *High Temperat.* **52** 5 703-707
[6] Tikhonov A N and Arsenin V Ya 1986 *Methods for solving ill-posed problems* (Moscow : Science)) p 288
[7] Khairullin M Kh 1996 Solution of inverse problems of filtration of multilayer beds by the regularization technique *J. DAN RAS* **347** 1 103-105
[8] Aziz H and Settari E 1982 *Mathematical modeling of reservoir systems* (Moscow : Science) p 407

Algorithm for restoration of friction power in drum brake device with a layer from the compositional material on temperature data

A S Kondakov

Institute of Oil and Gas Problems Siberian Branch of the Russian Academy of Sciences, Russia

E-mail: kondak_aleksey@mail.ru

Abstract. A quasilinear two-dimensional mathematical model of the thermal process in a drum braking device is considered, taking into account the rotational motion of the brake cylinder. The basic heat conduction equations and the algorithm for solving by method of conjugate gradients of the boundary inverse heat conduction problem to restore the friction power between the brake pad and the cylinder on temperature data are given.

Keywords: brake cylinder, brake pad, frictional heat generation, mathematical model, friction power, inverse problem, conjugate gradient method.

1. Introduction

The restoration of the total heat flux and, accordingly, the specific friction power resulting from frictional heat generation in a friction pair, according to additional temperature data, belongs to the class of boundary inverse heat transfer problems. An effective method for solving inverse heat transfer problems is an iterative regularization method based on gradient methods for minimizing the target residual functional [1]. Some similar problems for solving inverse problems of identification of frictional heat generation from temperature data for various friction pairs are given, for example, in [2–6].

In this paper, the problem of restoring the total heat flux in the friction zone of a drum braking device widely used in vehicles, for example, in the T-150K tractor and in VAZ-2101 and GAZ-3102 cars is considered [7]. Drum brake device consists of a pair of fixed brake pads, located inside the brake cylinder, fixed with the drive shaft of the wheel and rotating with it. The brake pads are pressed against the cylinder with a certain force during braking, ensuring that the rotation of the wheel axis stops. In the process of braking, the kinetic energy of a moving vehicle is converted into heat by the braking device, dissipated by its elements into the environment. Thermal analyzes carried out by means of temperature measurements for a friction pair pursue such important problems as estimating the power of frictional heat generation and determining the effective thermophysical properties of the materials from which the braking device elements are made [7]. In this regard, the estimation or restoration of the specific power of frictional heat generation from temperature measurements is an actual mathematical problem.

2. Problem statement

Consider a simplified design diagram of a drum brake device in Fig.1, consisting of a brake cylinder, rotating together with a drive shaft, and a fixed pair of brake pads. In the design scheme, the actual geometrical forms of the braking device are approximately idealized to describe the determining heat conduction equations in cylindrical coordinates. Heat sinks from the ends of the cylinder and pads will be considered negligible compared to the heat sinks from the side surfaces, which will allow us to apply the two-dimensional heat conduction equations in cylindrical coordinates. When braking, the initially rotating cylinder immediately does not stop, and its rotational speed will decelerate from the maximum initial value to zero during the time the vehicle stops. Therefore, we will consider the thermal model for a pair of brake pad-cylinder, taking into account the rotation of the cylinder, which has geometrical symmetry about the horizontal axis in the design diagram shown in Fig.1. Thus, we will set two-dimensional temperature problems for the brake pad and cylinder separately, in the most general case, taking into account the temperature dependence of the thermophysical properties of materials.

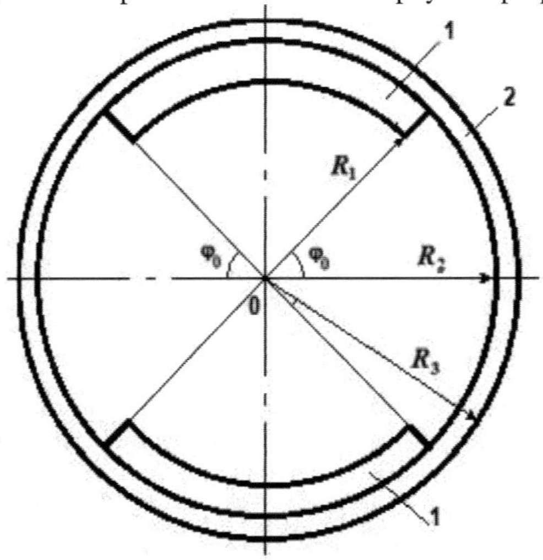

Figure1. Braking design: 1 – brake pads, 2 – cylinder

The temperature field $T(r,\varphi,t)$ in the brake pad is described by a quasilinear two-dimensional heat conduction equation in cylindrical coordinates with thermophysical coefficients $\boldsymbol{C_T = C_T(T)}$, $\lambda_T = \lambda(T)$:

$$C_T \frac{\partial T}{\partial t} = \frac{1}{r}\frac{\partial}{\partial r}\left(r\lambda_T \frac{\partial T}{\partial r} \right) + \frac{1}{r^2}\frac{\partial}{\partial \varphi}\left(\lambda_T \frac{\partial T}{\partial \varphi} \right), \tag{1}$$

$$R_1 < r < R_2, \ \ \varphi_0 < \varphi < \pi - \varphi_0, \ \ 0 < t \le t_m.$$

The temperature field $U(r,\varphi,t)$ in the cylinder is also described by a two-dimensional heat conduction equation with thermophysical coefficients $C_U = C_U(U)$, $\lambda_U = \lambda_U(U)$ and with a convective term that takes into account its rotational motion with an angular velocity $\Omega(t)$:

$$C_U \frac{\partial U}{\partial t} = \frac{1}{r}\frac{\partial}{\partial r}\left(r\lambda_U \frac{\partial U}{\partial r} \right) + \frac{1}{r^2}\frac{\partial}{\partial \varphi}\left(\lambda_U \frac{\partial U}{\partial \varphi} \right) + \Omega(t) C_U \frac{\partial U}{\partial \varphi}, \tag{2}$$

$$R_2 < r < R_3, \ \ 0 < \varphi < \pi, \ \ 0 < t \le t_m.$$

Conditions of non-ideal thermal contact are set in the contact zone of cylinder and brake pad:

$$\lambda_U \frac{\partial U}{\partial r}\bigg|_{r=R_2} - \lambda_T \frac{\partial T}{\partial r}\bigg|_{r=R_2} = q(\varphi,t), \ \ \varphi_0 < \varphi < \pi - \varphi_0 \tag{3}$$

$$U(R_2,\varphi,t) = T(R_2,\varphi,t), \ \ \varphi_0 < \varphi < \pi - \varphi_0. \tag{4}$$

The conditions of heat exchange according to Newton's law with the environment with a given temperature T_{rd} are set on the free surfaces of cylinder and brake pads:

$$\lambda_U \left.\frac{\partial U}{\partial r}\right|_{r=R_2} = \alpha_U (U|_{r=R_2} - T_{\text{rd}}),\ 0 \le \varphi \le \varphi_0,\ \pi - \varphi_0 \le \varphi \le \pi; \tag{5}$$

$$\lambda_U \left.\frac{\partial U}{\partial r}\right|_{r=R_3} = -\alpha_U (U|_{r=R_3} - T_{\text{rd}}),\ 0 \le \varphi \le \pi; \tag{6}$$

$$\left.\frac{\partial U}{\partial \varphi}\right|_{\varphi=0} = \left.\frac{\partial U}{\partial \varphi}\right|_{\varphi=\pi} = 0,\ R_2 \le r \le R_3; \tag{7}$$

$$\lambda_T \left.\frac{\partial T}{\partial r}\right|_{r=R_1,} = \alpha_T (T|_{r=R_1,} - T_{\text{rd}}),\ \varphi_0 \le \varphi \le \pi - \varphi_0; \tag{8}$$

$$\lambda_T \left.\frac{\partial T}{r\partial \varphi}\right|_{\varphi=\varphi_0} = \alpha_T (T|_{\varphi=\varphi_0} - T_{\text{rd}}),\ R_1 \le r \le R_2; \tag{9}$$

$$\lambda_T \left.\frac{\partial T}{r\partial \varphi}\right|_{\varphi=\pi-\varphi_0} = -\alpha_T (T|_{\varphi=\pi-\varphi_0} - T_{\text{rd}}),\ R_1 \le r \le R_2; \tag{10}$$

The initial temperature distribution in the elements of the braking device is assumed to be equal to the environment temperature

$$U(r,\varphi,0) = T(r,\varphi,0) = T_{\text{rd}}. \tag{11}$$

The direct problem (1) - (11) is solved by the finite difference method by splitting into spatial variables, consisting in a step-by-step solution by the one-dimensional difference equation sweep method with iteration over nonlinearity at each time step. In this case, the heat equation for a rotating cylinder with respect to the angular variable is solved using a monotonic difference scheme [8].

3. Algorithm for solving the boundary inverse problem

To solve the boundary inverse problem of identifying the total heat flux $q(\varphi,t)$ using temperature data, we use the iterative regularization method developed in [1] and successfully implemented in [2-6]. We present the basic relations necessary for the implementation of the method. Let the fixed point $(R_f,\varphi_i),\ i = 1,\dots,n,$ of the measurement of temperature over time be set in a fixed brake pad at a fixed distance $R_f,\ R_1 < R_f < R_2$ from the friction zone. The measured values of the temperature data are denoted by $f(\varphi_j,t)$.

The problem of identifying the heat flux function from the temperature data is to minimize the following residual functional:

$$J[q(\varphi,t)] = \frac{1}{2}\int_0^{t_m} \sum_{i=1}^{n} [T(R_f,\varphi_i t) - f(\varphi_i,t)]^2 dt, \tag{12}$$

on the solutions of system (1) - (11). Minimization of the functional (12) by the method of conjugate gradients is the most effective method for solving the inverse problem by iterative regularization.
To determine the gradient of the functional (12), the adjoint problem is considered with respect to variables $\Psi(r,\varphi,t),\ \Phi(r,\varphi,t)$ for brake pad and cylinder, respectively:

$$-C_T \frac{\partial \Psi}{\partial t} = \frac{\lambda_T}{r}\frac{\partial}{\partial r}\left(r\frac{\partial \Psi}{\partial r}\right) + \frac{\lambda_T}{r^2}\frac{\partial^2 \Psi}{\partial \varphi^2} + \frac{1}{r}\left[T(r,\varphi,t) - f(\varphi,t)\right]\delta(\varphi - \varphi_i), \tag{13}$$
$$R_1 < r < R_2,\ \varphi_0 < \varphi < \pi - \varphi_0,\ 0 < t \le t_m,$$

where $\delta(\varphi - \varphi_i) = 1$ at $\varphi = \varphi_i$, $\delta(\varphi - \varphi_i) = 0$ at $\varphi \ne \varphi_i$.

$$-C_U\frac{\partial\Phi}{\partial t}=\frac{\lambda_U}{r}\frac{\partial}{\partial r}\left(r\frac{\partial\Phi}{\partial r}\right)+\frac{\lambda_U}{r^2}\frac{\partial^2\Phi}{\partial\varphi^2}-\Omega(t)\cdot C_U\frac{\partial\Phi}{\partial\varphi} \qquad (14)$$

$$R_2<r<R_3,\ \ 0<\varphi<\pi,\ \ 0<t\le t_m,$$

$$\left.\left(\lambda_U\frac{\partial\Phi}{\partial r}\right)\right|_{r=R_2}-\left.\left(\lambda_T\frac{\partial\Psi}{\partial r}\right)\right|_{r=R_2}=0,\ \ \varphi_0\le\varphi\le\pi-\varphi_0 \qquad (15)$$

$$\Phi(r,\varphi,t)=\Psi(r,\varphi,t),\ \ \varphi_0\le\varphi|\le\pi-\varphi_0, \qquad (16)$$

$$\lambda_U\left.\frac{\partial\Phi}{\partial r}\right|_{r=R_2}=\alpha_U\left.\Phi\right|_{r=R_2},\ \ 0\le\varphi\le\varphi_0,\ \ \pi-\varphi_0\le\varphi\le\pi, \qquad (17)$$

$$\lambda_U\left.\frac{\partial\Phi}{\partial r}\right|_{r=R_3}=-\alpha_U\left.\Phi\right|_{r=R_3},\ \ 0\le\varphi\le\pi, \qquad (18)$$

$$\left.\frac{\partial\Phi}{\partial\varphi}\right|_{\varphi=0}=\left.\frac{\partial\Phi}{\partial\varphi}\right|_{\varphi=\pi}=0,\ \ R_2\le r\le R_3, \qquad (19)$$

$$\lambda_T\left.\frac{\partial\Psi}{\partial r}\right|_{r=R_1}=\alpha_T\left.\Psi\right|_{r=R_1},\ \ \varphi_0\le\varphi\le\pi-\varphi_0, \qquad (20)$$

$$\lambda_T\left.\frac{\partial\Psi}{r\partial\varphi}\right|_{\varphi=\varphi_0}=\alpha_T\left.\Psi\right|_{\varphi=\varphi_0},\ \ R_1\le r\le R_2, \qquad (21)$$

$$\lambda_T\left.\frac{\partial\Psi}{r\partial\varphi}\right|_{\varphi=\pi-\varphi_0}=-\alpha_T\left.\Psi\right|_{\varphi=\pi-\varphi_0},\ \ R_1\le r\le R_2, \qquad (22)$$

$$\Phi(r,\varphi,t_m)=\Psi(r,\varphi,t_m)=0. \qquad (23)$$

The formula for calculating the gradient of the functional (14) is determined by solving the adjoint boundary value problem (13) - (23)

$$J'\left[q(\varphi,t)\right]=\Psi(R_2,\varphi,t). \qquad (24)$$

The algorithm for minimizing the functional (12) using the conjugate gradient method is represented by the following sequence of transition from the k-th approximation $q^k(\varphi,t)$ to (k + 1) –th $q^{k+1}(\varphi,t)$:

$$q^k(\varphi,t)\to T_k(r,\varphi,t)\to\Psi(r,\varphi,t)\to J'[q^k]\to\gamma_k\to S^k(\varphi,t)\to V_k(r,\varphi,t)\to\beta_k\to q^{k+1}(\varphi,t),$$

Where

$$q^{k+1}(\varphi,t)=q^k(\varphi,t)-\beta_k S^k(\varphi,t),\ \ k=0,1,2,...;$$

$$S^k(\varphi,t)=J'[q^k]+\gamma_k S^{k-1}(\varphi,t),\ \ \gamma_0=0,\ \ \gamma_k=\frac{\int_0^{t_m}\int_{\varphi_0}^{\pi-\varphi_0}(J'[q^k(\varphi,t)])^2\,d\varphi dt}{\int_0^{t_m}\int_{\varphi_0}^{\pi-\varphi_0}(J'[q^{k-1}(\varphi,t)])^2\,d\varphi dt}.$$

The descent step β_k is determined from the expression

$$\beta_k=\frac{\int_0^{t_m}\int_{\varphi_0}^{\pi-\varphi_0}[T_k(R_f,\varphi,t)-f(\varphi,t)]V_k(R_f,\varphi,t)\,d\varphi dt}{\int_0^{t_m}\int_{\varphi_0}^{\pi-\varphi_0}V_k^2(R_f,\varphi,t)\,d\varphi dt},$$

where $V_k(r,\varphi,t)$ is determined by the solution of the boundary value problem for increments V and W of the temperature the brake pads and cylinder, respectively, at $\Delta q(\varphi,t)=S^k(\varphi,t)$:

$$\frac{\partial C_T V}{\partial t} = \frac{1}{r}\frac{\partial}{\partial r}\left(r\frac{\partial \lambda_T V}{\partial r}\right) + \frac{1}{r^2}\frac{\partial}{\partial \varphi}\left(\frac{\partial \lambda_T V}{\partial \varphi}\right),$$
$$R_1 < r < R_2,\ \varphi_0 < \varphi < \pi - \varphi_0,\ 0 < t \le t_m, \tag{25}$$

$$\frac{\partial C_U W}{\partial t} = \frac{1}{r}\frac{\partial}{\partial r}\left(r\frac{\partial \lambda_U W}{\partial r}\right) + \frac{1}{r^2}\frac{\partial}{\partial \varphi}\left(\frac{\partial \lambda_U W}{\partial \varphi}\right) + \Omega(t)\frac{\partial C_U W}{\partial \varphi},$$
$$R_2 < r < R_3,\ 0 < \varphi < \pi,\ 0 < t < t_m, \tag{26}$$

$$\left.\frac{\partial \lambda_U W}{\partial r}\right|_{r=R_2} - \left.\frac{\partial \lambda_T V}{\partial r}\right|_{r=R_2} = \Delta q(\varphi,t),\ \varphi_0 \le \varphi \le \pi - \varphi_0, \tag{27}$$

$$W(R_2,\varphi,t) = V(R_2,\varphi,t),\ \varphi_0 \le \varphi \le \pi - \varphi_0, \tag{28}$$

$$\left.\frac{\partial \lambda_U W}{\partial r}\right|_{r=R_2} = \alpha_U W\big|_{r=R_2},\ 0 \le \varphi \le \varphi_0,\ \pi - \varphi_0 \le \varphi \le \pi, \tag{29}$$

$$\left.\frac{\partial \lambda_U W}{\partial r}\right|_{r=R_3} = -\alpha_U W\big|_{r=R_3},\ 0 \le \varphi \le \pi, \tag{30}$$

$$\left.\frac{\partial W}{\partial \varphi}\right|_{\varphi=0} = \left.\frac{\partial W}{\partial \varphi}\right|_{\varphi=\pi},\ R_2 \le r \le R_3, \tag{31}$$

$$\left.\frac{\partial \lambda_T V}{\partial r}\right|_{r=R_1} = \alpha_T V\big|_{r=R_1},\ \varphi_0 \le \varphi \le \pi - \varphi_0, \tag{32}$$

$$\left.\frac{\partial \lambda_T V}{r\partial \varphi}\right|_{\varphi=\varphi_0} = \alpha_T V\big|_{\varphi=\varphi_0},\ R_1 \le r \le R_2, \tag{33}$$

$$\left.\frac{\partial \lambda_T V}{r\partial \varphi}\right|_{\varphi=\pi-\varphi_0} = -\alpha_T V\big|_{\varphi=\pi-\varphi_0},\ R_1 \le r \le R_2, \tag{34}$$

$$W(r,\varphi,0) = V(r,\varphi,0) = 0. \tag{35}$$

The problems (13) - (23) and (25) - (35) are solved numerically by the finite difference method in the same way as the direct problem (1)-(11).
In addition to the geometric dimensions and thermophysical properties of the materials from which the cylinder and brake pad are made, it is necessary to know the function of the angular velocity of rotation of the cylinder for the numerical realization of the problem. This angular velocity can be calculated by the method given in [7], based on the knowledge of the vehicle mass, its speed at the initial moment of deceleration, and the time of a complete stop.

4. Conclusion

1. An algorithm has been developed to solve the boundary inverse problem of heat transfer to restore the total heat flux, which characterizes the specific friction power between the brake pads and the cylinder according to temperature data.
2. The corresponding boundary nonstationary problems for conjugate variables and temperature increments are obtained, which are necessary for applying the conjugate gradient method to solve the boundary inverse problem.

5. References

[1] O.M. Alifanov, Ye.A. Artyukhin, S.V. Rumyantsev, 1988 *Ekstremal'nyye metody resheniya nekorrektnykh zadach* (Moscow: Nauka) p. 288

[2] A.I. Burya, B.I. Peleshenko, I.V. Rula, 2010 "Temperature and density determination for heat flux in the area of the shaft-bushing contact" *Journal of Friction and Wear.* **31(6)**, pp. 443-448

[3] Y.C. Yang, W.L. Chen, 2011 "A nonlinear inverse problem in estimating the heat flux of the disc in a disc brake system" *Appl. Thermal Engineering*, **31**, pp. 2439-2448

[4] N.P. Starostin, A.S. Kondakov, M.A. Vasilieva, 2013 "Identification of friction heat generation in sliding bearing by temperature data" *Invers Problem in Science and Engineering*, **21(2)**, pp. 298-313

[5] A.S. Kondakov, N.P. Starostin, M.A. Vasilieva, 2014 "Thermal Diagnostics of Friction in Sliding Bearings with Account for Propagation of Heat and Movement of Shaft" *Journal of Friction and Wear*, **35(1)**, pp. 48–54

[6] N.P. Starostin, A.S. Kondakov, M.A. Vasilieva, 2015 "Identification of friction torque in a sliding bearing by thermal data" *Inverse Problems in Science and Engineering*, **23(8)**, pp. 1388-1404

[7] V.A. Balakin, V.P. Sergiyenko, 1999 *Teplovyye raschoty tormozov i uzlov treniya* (Gomel': IMMS NANB) p. 220

[8] A.A. Samarskiy, 1977 *Teoriya raznostnykh skhem* (Moscow: Nauka) p. 656

Acknowledgments

The work has been accomplished within the framework of the State Order of the Federal Agency for Scientific Organizations of the Russian Federation (project No. AAAA-A17-117040710038-8 from 07.04.2017)

Journal of Physics: Conference Series **1392** (2019) 012055 doi:10.1088/1742-6596/1392/1/012055

Parallel computing applied to the molecular dynamics simulations

Victor L. Malyshev[1], **Elena F. Moiseeva**[1]

[1] Ufa State Petroleum Technological University, Ufa 450062, Russian Federation

E-mail: victor.l.malyshev@gmail.com

Abstract.
This paper discusses the use of high-performance algorithms for modeling the dynamics of dispersed systems using the method of molecular dynamics. Large-scale modeling, which makes it possible to determine various thermodynamic parameters and control the processes of phase transformations, requires consideration of a large number of particles, which imposes significant restrictions on the computational capabilities of the system. In work various problems of dynamics of disperse systems are considered. To solve the problems posed by the authors, a specialized data structure was developed, which reduces the computational complexity of the algorithm from quadratic to linear. The specificity of the molecular dynamics method makes it possible to achieve significant acceleration when using computations at heterogeneous stations equipped with a central processing unit (CPU) and graphic processing units (GPU). The authors showed that the presented technique can be effectively used for solving problems related to modeling processes occurring in areas with characteristic sizes of tens of nanometers on personal computers equipped with one or more GPUs.

1. Introduction

Molecular dynamics (MD) simulations is one of the best methods for solving the micro- and nanoscale problems. Despite the fact that the MD simulations are successfully applied for problems of nano-scale dynamics, the real systems investigation using MD seems very difficult because requires the consideration of enormous number of particles, and, thus, a lot of computational operations per minute. To speed up the calculations modern numerical algorithms and high performance programming can be used. The most used and simple in realization is the system consisting of monoatomic molecules which interact via Lennard-Jones (LJ) potential. As LJ potential function is short-range (for each atom only the influence of its nearest environment is significant), there is no need to calculate the interaction forces acting between each pair of molecules. The use of neighbour lists or data structures allows to simplify this procedure.

The first attempts in this direction were described by M.P. Allen and D.J. Tildesley [1], who built a list of neighbours. The main idea of this method is based on the fact that LJ potential has small interaction radius. Later hierarchical data structures were developed. Such structures are difficult to implement, but have the advantage of applying to long-range potentials. One of the algorithms for building such a structure can be found in [2, 3]. The implementation of MD simulations on systems with parallel computing capabilities can be attributed to high-performance hardware.

To date, there are a number of software packages for MD modeling, such as LAMMPS [4], DLPOLY [5], NAMD [6], GROMACS [7] and ESPResSO [8], which effectively implement the molecular dynamics method on clusters, however, only a few of them include the ability to implement computations on graphics processors.

Many authors have paid attention to the MD simulations on GPU [9, 10, 11, 12], but for the first time this method was fully implemented on the GPU in 2008 by Andersen et al [13]. In their work, they presented the first implementation of the MD simulations of monatomic molecules, in which all steps of the algorithm were performed on the GPU using NVIDIA CUDA technology. To reduce the complexity of the algorithm, they used neighbor lists. A similar implementation of the algorithm was presented by J.A. van Meel [14]. In 2011, Trott et al [15] presented the implementation of the molecular dynamics algorithm on a cluster of GPUs using the LAMMPS package. To speed up the calculations, the neighbor lists were also used in the algorithm. Calculations were carried out for different classes of materials (biomolecules, polymers, metals, semiconductors), and the speed of calculations was compared with existing implementations of MD-modeling on the GPU.

This paper describes the process of building a new data structure for non-polar molecules. The use of such a data structure, in conjunction with graphics processors, allows us to significantly speed up the calculation of short-range potentials in the MD method.

2. Model

The MD simulations is based on the idea that the considered system consists of molecules following the trajectories calculated from the Newtons second law and interacting with each other via the interaction potential. In a classical MD simulations, atomic positions evolve from their initial state according to Newtons equation of motion.

Except for the simplest cases, such equations are solved numerically, according to the chosen algorithm. For the integration of Newton's equations it is necessary to calculate the force, acting on atom, or the complex potential for the interaction of atom with the others.

The current paper represents the simulations of simple monoatomic fluids, which molecules interact via Lennard-Jones potential:

$$u_{LJ}(r) = 4\varepsilon \left[\left(\frac{\sigma}{r} \right)^{12} - \left(\frac{\sigma}{r} \right)^{6} \right], \tag{1}$$

where $r_{ij} = |r_i - r_j|$ is the distance between molecules i and j, ε is the depth of the potential energy well and σ is the distance at which the potential is zero.

The application of the MD simulations to study the systems with the size of tens and hundreds nanometers requires the consideration of millions molecules. For example, the cubic domain of $80\times80\times80$ nm^3 contains over 10 millions molecules. For most existing computational systems, the computations of the interactions between such a huge number of molecules is not a trivial problem. Thats why for such problems it is preferable to use modern computational methods and algorithms. In our simulations we use both algorithmic and hardware accelerations. As the MD simulations require fast and convenient data access, the data structures and efficient algorithms for their generation are of high priority.

3. Data structure

Efficient use of parallel algorithms requires convenient and fast data access. For this purpose, an efficient data structure was developed. We describe the main points of its construction. The algorithm consists of the following steps:

Preliminary calculations. Grid construction. Determination of the main parameters: the number of boxes, their linear dimensions, the order of numbering, the position of the center of each box. Building a list of neighbors for each box and centered vector of neighbors. These calculations are performed before the cycle. Such parts of the algorithm are sufficient to implement on the CPU, since their calculations are performed only once when generating the initial data.

Data structure generation. Since the modeling region is divided into boxes, the best way is to arrange the coordinates in such a sequence, when all the particles in the zero box are recorded first, then in the $1st$, and so on until the last box $N_{AllBox-1}$. To build a data structure, bucket sorting of atoms is used [16].

Converting positions to local coordinates. This operation is necessary for more convenient calculation of the distances between the particles - it allows you to avoid unnecessary branching of the algorithm, which has a bad effect on the performance of the GPU. Below is a detailed description of each of the presented items. All formulas are written for the three-dimensional case. For simplicity and clarity of the presentation of the material, we present figures for a two-dimensional data structure.

3.1. Number of boxes. Enumeration

Let us introduce two types of box indexing: pass-through and Cartesian. The pass-through index is a consecutive numbering of all boxes. Those end-to-end index traverses the curve filling the area (space filling curve). Cartesian method enumerates the box according to the shift along each of the axes relative to the origin (the lower left corner is selected by the reference point). The transition from one numbering system to another is carried out using simple formulas that use only the box numbering and the number of boxes in each direction.

3.2. Neighbour list construction

According to the selected algorithm, each atom located in the selected box can interact only with atoms located in the same box, as well as with atoms located in neighboring boxes. To find cells adjacent to the selected one, Cartesian numbering is used. Thus, having the index of the selected box in Cartesian numbering $[i, j, k]$, you can find 27 of its neighbors (including the selected cell itself): $[i-1, j-1, k-1]$, $[i, j-1, k-1]$, .., $[i+1, j+1, k+1]$. Then, for convenience, Cartesian indices can be transferred to the pass-through numbering system. The difficulty is in finding the neighbors for the boxes located on the edge of the simulated area. They are calculated in accordance with the selected boundary conditions. As a result, each box can be assigned a list of neighbors, consisting of 27 indices. The merged neighbor list will thus contain $27N_{AllBox}$ indices.

3.3. Data structure construction

To implement convenient data access, it is necessary to re-sort the particles in the system. Since the use of conventional sorting algorithms is not efficient on the GPU, a pseudo-sorting algorithm (bucket sorting) is used, as in the algorithm for generating a hierarchical data structure. It consists of the following steps:

(i) Constructing a placement histogram. It shows how many particles are in each box and the local index for each particle in the given box.

(ii) Parallel scanning to find "markers" that show the indices of the first and last atom in the specified box in the new sorted array of particles and the determination of the global particle index.

(iii) Ordering particles according to their global index.

3.4. Transformation of positions to local coordinates

When solving the Newton's equation, in order to calculate the interaction forces between atoms, it is necessary to know the distances between them. In order to unify the process of calculating the distances between particles and calculating the interaction forces under periodic boundary conditions, we transform the position of each particle into a local vector.

This reduces the computational complexity of calculating the interaction by reducing the excessive branching of the algorithm, which is detrimental to the effective implementation of the algorithm on the GPU.

3.5. Hardware acceleration using graphics processors

One of the most effective ways to speed up the calculation is to use graphical processes (GPU) for numerical calculations.

3.6. Application of GPU and data structures for molecular dynamics modeling

Let us consider the main features of the use of graphics processors for molecular dynamics simulations.

The algorithm for constructing the data structure described above assumes the existence of operations performed prior to the main time loop. Such calculations can be implemented once at the launch of the program code. Such operations include, for example, the neighbor list construction.

The initial data structure generation occurs on the main node. Next, data is split into nodes, and each node performs the same sequence of steps for different data sets corresponding to each node. From the flowchart it can be seen that at each time step, at each of the nodes, a data structure is checked, which reveals the number of particles that "flown" between the boxes. If necessary, the data structure is rebuilt, and the calculation continues. Upon completion of the execution of all steps of the algorithm, a transition to a new time step occurs.

3.7. Acceleration results

To demonstrate the acceleration obtained using a data structure and a GPU, a cubic region filled with liquid argon with a density of $\rho = 950$ kg/m^3 at a temperature of 85 K was considered. Particles are distributed throughout the simulation region. Potential cutoff radius is $r_{cutoff} = 3.5\sigma$, where σ is equal 3.405 Å. Calculations are performed on a heterogeneous workstation with two 6-core Intel Xeon 5660 2.8 GHz CPUs (12 physical cores and 12 virtual cores using Hyper-Threading technology), 12 GB RAM, and four NVIDIA Tesla C2050 GPUs/C2075 with 6 Gb RAM. The GPU code is written using NVIDIA CUDA technology; parallel operation of several GPUs implemented using OpenMP technology. Data is copied between GPUs using CUDA UVA (unified virtual addressing) technology, which allows data to be copied between GPUs directly, by passing the CPU. In the course of computational experiments, it was found that the use of NVIDIA CUDA UVA technology speeds up data exchange by about 30%. Calculations are made for double-precision floating point numbers.

Figure 1 presents the calculation time for one time step of the algorithm depending on the number of atoms in the system. The graph shows the calculation times by direct summation on the CPU and GPU (brute force). From the graph it can be seen that the use of the developed data structure makes it possible to reduce the computational complexity from quadratic to linear. Also on the graph you can see the performance comparison with the LAMMPS package (application package for molecular dynamics simulation).

The developed high-performance programming code for the MD simulations has been applied to several physical problems, such as cavitation in liquid argon (fig. 2) [17], surface bubble dynamics in a liquid flow (fig. 3) [18], particle movement by surface nanobubble (fig. 4) [19].

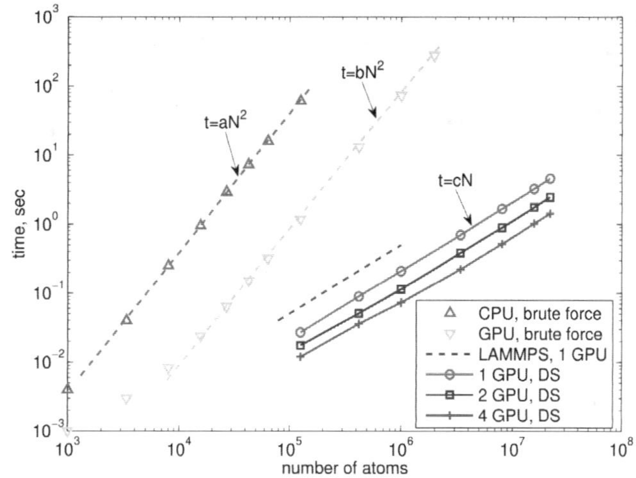

Figure 1. Full time calculation of one time step of the algorithm using a data structure. The dotted line shows the task time for a similar task obtained using the LAMMPS simulator. Graphs of calculation time are applied by direct summation on the CPU and GPU.

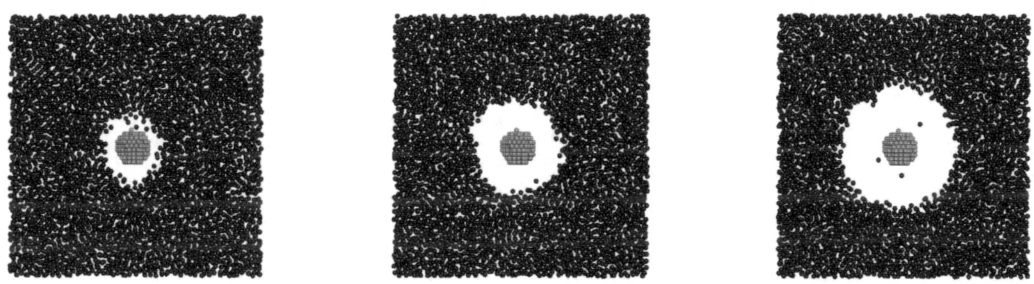

Figure 2. Cavitation in liquid argon near solid particle

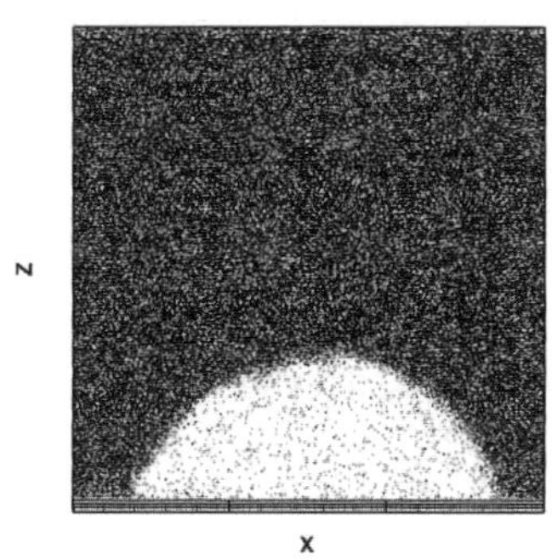

Figure 3. The initial molecules distribution (slice of the domain). Blue circles — Ar, yellow triangles — Ne, black squares — substrate. Number of particles is 1477852. Size of the domain is $40nm \times 40nm \times 40nm$.

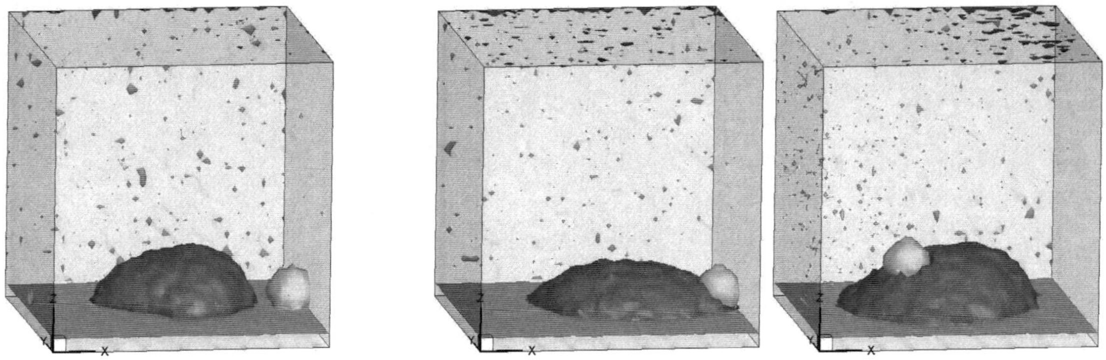

Figure 4. Surface nanobubble dynamics in liquid in presence of solid particle

4. Conclusion

The combination of the proposed data structure and graphics processors (GPU) for solving problems of the dynamics of dispersed systems using the molecular dynamics simulations reduces the computational complexity of the algorithm from $O(N^2)$ to $O(N)$.

The proposed approach allows to calculate one step of the algorithm for a system consisting of 20 million particles in less than 2 seconds. Therefore, the use of such algorithms is necessary for solving nano- and micro-scale problems such as phase transitions, cavitation, bubble dynamics, which require the large-scale modelling.

References

[1] Allen M and Tildesley D 1992 *Computer Simulation of Liquids* (Clarendon Press)
[2] Gumerov N A and Duraiswami R 2008 *Journal of Computational Physics* **227** 8290–8313
[3] Marin D, Malyshev V, Moiseeva E, Gumerov N, Akhatov I and Mikhailenko K 2013 *Numerical methods and programming* **14** 483–495
[4] Plimpton S 1995 *Journal of Computational Physics* **117** 1 – 19 ISSN 0021-9991
[5] Forester T and Smith W *DL POLY 2.0: A general-purpose parallel molecular dynamics simulation.*
[6] Phillips J C, Braun R, Wang W, Gumbart J, Tajkhorshid E, Villa E, Chipot C, Skeel R D, Kal L and Schulten K 2005 *Journal of Computational Chemistry* **26** 1781–1802 ISSN 1096-987X
[7] Berendsen H J C, Spoel D V D and Drunen R V 1995 *Comp. Phys. Comm* **91** 43–56
[8] Limbach H J, Arnold A, Mann B and Holm C 2006 *Computer Physics Communications* **174** 704–727
[9] Yang J, Wang Y and Chen Y 2007 *Journal of Computational Physics* **221** 799 – 804 ISSN 0021-9991
[10] Che S, Boyer M, Meng J, Tarjan D, Sheaffer J W and Skadron K 2008 *Journal of Parallel and Distributed Computing* **68** 1370 – 1380 ISSN 0743-7315
[11] Liu W, Schmidt B, Voss G and Mller-Wittig W 2008 *Computer Physics Communications* **179** 634 – 641 ISSN 0010-4655
[12] Moiseeva E, Mikhaylenko C, Malyshev V, Maryin D and Gumerov N 2012 *ASME International Mechanical Engineering Congress and Exposition, Proceedings (IMECE)* vol 7 pp 883–892
[13] Anderson J A, Lorenz C D and Travesset A 2008 *Journal of Computational Physics* **227** 53425359
[14] van Meel J, Arnold A, Frenkel D, Portegies Zwart S, and Belleman R 2008 *Molecular Simulation* **34** 259–266
[15] Trott C R, Winterfeld L and Crozier P S 2011 *http://arxiv.org/abs/1009.4330v2*
[16] Cormen T 2009 *Introduction to Algorithms, 3rd Edition:* (MIT Press) ISBN 9780262033848
[17] Malyshev V, Marin D, Moiseeva E, Gumerov N and Akhatov I 2015 *High Temperature* **53** 406–412
[18] Moiseeva E, Malyshev V, Marin D, Gumerov N and Akhatov I 2014 *ASME International Mechanical Engineering Congress and Exposition, Proceedings (IMECE)* vol 7
[19] Moiseeva E, Malyshev V, Marin D, Gumerov N and Akhatov I 2017 *Mathematical Models and Computer Simulations* **29** 131–140

Virtual stream computing systems with a fuzzy computing environment

Myshev A.V.
National Research Nuclear University MEPhI (IATE), Russia,Obninsk,Studgorodok 1
e–mail: mishev@iate.obninsk.ru

Abstract. The paper reveals the architecture (structure) of virtual stream computing systems with a fuzzy computing environment (closure conditions, constraints, exchanges, small scales and information uncertainty). The system assumes that hardware-software, computing and information-intellectual functions of such systems can actually be implemented and implemented in the form of hardware, software, brainware products for supercomputer technologies of mathematical modelling. Such products allow to obtain more adequate, reliable, meaningful and valuable results in the computational technologies of mathematical modeling. . This is due to the fact that the logic circuits of algorithms and procedures of computational technologies include an information system for referencing coordinates and checking the results of calculations in a fuzzy (diffuse) computing environment based on the principles of locality and specifying the model of the information environment for generating a result.

1. Introduction

The work relates to the field of digital computing, intelligent computing systems and intelligent information and measuring equipment with stream processing of information in a fuzzy and intermittent computing environment. It can be used both for creating intelligent virtual stream computing and information-measuring systems, computing environment and information processes in which are caused by model and algorithmic closure, physical environment limitations, fuzzy and interleaved information environment, energy and information exchange, information uncertainty, and for the development and implementation of artificial and computational intelligence systems, etc. The technical and technological result is the expansion of the functional and intellectual capabilities of computing and information-measuring systems, increasing the reliability of their information performance and obtaining adequate results in the conditions of isolation, constraints, exchange and uncertainty of a fuzzy computing environment. An intelligent virtual computational flow and / or information and measurement system with a fuzzy computing environment contains: a block of autonomously functioning processor elements (PE), the number of functioning PEs in a specific physical time quantum is determined by the conditions of the computational process algorithm or its result from the previous time quantum; a block of modules for preparing and generating data flow (measurement results, input data, intermediate and final calculation results) for stream processing and / or stream computing; a block of two-level virtual active memory for each PE and a group of PE, a block of total virtual active memory available for all PEs; information field of coordinate systems of information binding and verification of results (intermediate and final) of the computational process in the field of multi-level virtual active memory for each PE or group of PE; hardware – software interface for interfacing with the "external world".

Known types of architectures streaming and multi-pipeline computing and information-measuring systems are based on linear models of algorithms and procedures of calculations or measurements both in hardware implementation and in software without taking into account the facts of influence of the physical and information environment of formation and receipt of results. In practical implementation, these models formalize three ways of obtaining a result, and therefore they represent three classes of models: 1) additive models for obtaining a result; 2) multiplicative models of the result; 3) additive – multiplicative models for obtaining a result. The general principles of constructing functional and logical schemes for obtaining a result in such systems are based on the representation of the result either as a sum of deterministic and random variables, or as a product of deterministic and random variables, or as a superposition of the first two. Algorithmic aspects of the implementation designated models in the logical diagrams of the results formation procedures and the computing environment of intelligent streaming computing and information-measuring systems appear in certain limits in the same way. In intelligent computing systems, the computing and generating environment has a more complex logical organization and mobility than the same information measuring systems environment. The functional and logical structure of the procedures and technologies for obtaining the final results in the designated systems does not allow to take into account such factors of the computing environment as the information dynamics of the data flow of the computational processes and detection processes, the spatial – temporal scales of the physical environment, and the conditions of closure, constraints and exchange, information uncertainty and singular errors, virtualization and intellectualization, as well as much more.

It should be noted that the reasons for generating errors due to key quantities (accuracy and instability) of computational and information-measuring processes are based on the fact that traditionally practiced technologies of computation and measurement are very unstable. This instability is largely due to the fact that computational and measurement process technologies for streaming and scalar computations are implemented without taking into account the information dynamics of objects in a diffuse computing environment, which determines the eventfulness of the spatial – temporal and informational controllability conditions of computational processes.

2. Some features of structural organization of streaming computing systems with fuzzy computing environment

The architectures and ways of organizing computational and information-measuring systems of the new generation, which are currently offered and will be implemented in the near future, indicate that the principles of parallelism, virtualization and intellectualization, small space-time scales of the physical environment, isolation conditions, constraints, exchange, information uncertainty of the computing environment are the determining and dominant circuit design and system-engineering factors in the design of systems. It should be noted that the information dynamics of computational and information-measuring processes in such systems will evolve according to different physical and informational laws and on other spatial-temporal scales when spatial and dynamic uncertainty factors (informational, logical, and physical) dominate restrictions (spatial, temporal and informational), exchange (energy and information), closure (model and algorithmic). If the temporal evolution of the physical processes of flow calculations and measurements in such systems can be described by the equations of quantum mechanics (such as Schrödinger equations) or a similar type, then Boolean logic will have a different interpretation: fuzzy multi-valued logic, since due to the fact that information processes on elements of the indicated systems proceed on a nano-scale or less, state detection at the output with a given measurement scale (this is the number of significant symbols per unit of division and the detector relaxation time by orders of magnitude exceeds transients or phase transitions). And this means the following: on a given scale of measurement or obtaining the result of calculations and at a fixed relaxation time, within the framework of the currently available technical capabilities and proposed solutions, this problem is really insoluble. The semantic content of this fact means that the selection of the steady state of such a logical nano element of the system is an urgent problem both in physical and technical aspects, as well as in the intellectual and informational, technological, computing and measuring. This fact is partly noted by the developers of such nano elements (the first copies of which were developed in Russia, MIET). Here we also need new mathematics both to describe the flow processes at the physical level (apparently, the real elements use the formalism of quantum mechanics), and to describe the dynamics of computational and information processes in the virtual environment of such systems.

Intellectual capabilities and intellectual efficiency of the designated systems as an intelligent tool in the organization of computational and information and measurement processes, simulation and information processing in order to achieve adequate, accurate and stable results when the diffusion of information is localized at spatial and temporal scales with diffuse dynamics, were not considered and not investigated. Also open are such questions that are associated with the development of information models for obtaining the final results of stream computing and measurements under conditions of model and algorithmic closure, information exchange, limitations of the computing environment and information uncertainty.

Modern achievements in the relevant areas of natural science allow us to state with a high degree of confidence, on the one hand, that living intelligent systems process and analyze the flow of information in a fixed time quantum of perception. On the other hand, they implement their functions, including computational and information-measuring, under conditions of isolation (model and algorithmic), constraints (spatial, temporal and informational), exchange (energy and information) and uncertainty (informational, logical and physical). The designated conditions for the functioning of living systems are also characteristic of intelligent computing and measuring systems, which are currently not implemented and are not described in any model of obtaining final results for existing systems of calculations and measurements (real and theoretical).

3. Virtual stream computing system with fuzzy computing environment (construction method)

On the basis of the scheme developed by the author for constructing an intelligent virtual stream computing system and / or information-measuring system with a fuzzy computing environment, they include processor modules, memory modules, and data flow preparation modules. Such systems logically, functionally and informationally interact with each other according to the same principles as neurons in the network information model of a neural network, defines its organization and implementation in the form of a multilayer and multi-module streaming parallel processing system, where there are several layers, including a layer of processor elements (PE), a layer of training data streams, a layer of two-level virtual RAM for each PE and a module of the processor layer and a layer of total active virtual memory for the entire process weed layer, the processor layer is divided into virtual processor modules, and the total active virtual memory layer is divided into virtual memory modules, the processor modules have a network model for organizing information links and perform the functions of an interface both between the layer elements and the external environment, initialization of the computing system , routing information in the system, configuration of the architecture of a virtual streaming computing system under the structure of the information graph of the problem being solved, recovery of the result tat computational process in case of failure of the entire system, individual layers, modules or elements, and the modules of the virtual operating memory layer store functions of the attributes of modules, commands, storage segments of the local coordinate system of information binding and verification of results (intermediate and final) computing processes in the environment of the computations [1], the modules of the common active virtual memory are functionally and logically designed to store the global information system of the coordinates of checking and checking and reflecting on her results of the computational process at each step, and also performs buffer functions in tiling operations between modules of layers, functions of virtual ports and information gateways with the outside world.

According to the indicated method, the processor modules are logically and functionally combined and configured into a virtual computational structure with the geometry and connection topology of the network information model of the neural network and each processor module operates in its corresponding area of the information space of the coordinate system of reference and verification of the results, which ensures high reliability of the entire computing system and the reliability of the results obtained, and the basic and generating elements of the module virtual memory and shared virtual active memory are virtual cells of two types: active and passive virtual cells. The implementation of logical and arithmetic operations in processor modules is implemented according to the following computing principles: the principle of interaction of the computing environment operands with the information virtual environment [2, 3], the principle of virtual perspective in the design operator [4], the principle of connectedness of the results of calculations [1]: fractal and

informational .
On the other hand, with the help of the structure of a virtual intellectual streaming computational and / or information-measuring system that implements a new paradigm of computational intelligence based on the network information model of a neural network [2]. The network information model of a neural network describes the structure (architecture) of an intellectual stream computing and / or information-measuring system in the form of a global virtual information dynamic network in which the nature of interaction between elements at the control and information exchange levels is defined as virtual processes with local virtual interaction. An element of such a network, a formal neuron, is defined as a local virtual computing system that combines the resources of the processor elements and the memory of a real physical computing system with a rigid or reconfigurable (physical or virtual level) architecture, the logical and physical functions of which are as follows. First, it is a local virtual computer network, which defines and sets in the physical environment of a real streaming computing and / or information-measuring system a computing-measuring server-domain, i.e. the space of virtual address identifiers for its processor elements and their corresponding active memory segments. Secondly, a virtual computational – measuring server – domain specifies not only the area of information definition in the virtual address space of its active memory — its domain, but also defines in the information bounds of the address space of the memory domain as a common numerical scale for computing and / or information and measurement processes, and the scale division price (numerical and information accuracy) is its range. Elements of active memory are virtual cells of various types, differing in functional characteristics, logical organization of content. A neuron in the designated architecture (structure) of a virtual intelligent stream computing and / or information-measuring system is defined as its component, the purpose and function of which is as follows: 1) for a streaming parallel task implements its execution in a given range of possible values and a fixed quantum of physical time; 2) implementation of statistical synchronization of the computational and / or information-measuring process in a neural network within the allocated real-time quantum. Neurons are combined into a virtual computational and / or information and measurement structure that supports a computational and / or information and measurement process with a dynamically changing population of active neurons in a fixed and specific quantum of physical time. Active neurons can implement various computational and information-measuring processes within a specific quantum of physical time, if the interaction between them is parameterized by the task algorithm and the computing environment.
The memory area of a neuron is a local segment, the structural and logical organization of which is a model of active virtual memory [5] - passive cells (read only), which store the information coordinate system of a fuzzy virtual computing environment, and active cells form the RAM items. Geometric interpretation of the information coordinate system is that it determines the memory area of the segment, which reflects the dynamics of information processes of a fuzzy computing environment. The nodes of such a lattice in a broad sense are a figurative and symbolic reflection of the geometric and informational properties of the dynamics of the processes of a fuzzy computing environment [4, 6]. The informational essence of the geometry of nodes on a lattice defines it as a memory segment of virtual passive cells of fixed length and with unchanged information in them, i.e. this is a segment of constants in the form of character chains. This part of the segment defines the scope for determining the display of the computational and / or information measurement process in the information space of the active virtual memory of the computing environment — their domain and range of values is the range. The introduction of such a logical scheme of both the organization of memory and the construction of an algorithm for the computational and / or information-measuring process defines computational and / or information-measuring systems with such an architecture as intelligent virtual stream computing-measuring systems with variable domain-ranges. If the logic scheme of the computational process in a fuzzy neuron computing environment involves processes with local interaction, then interaction with other neurons of the neural network occurs. Computational processes in the designated systems are quantized by physical time. The duration of quanta can be arbitrary and is determined by the conditions of statistical synchronization and the computational process. In each quantum of physical time, the number of active neurons is determined by the result of the calculations in the previous quantum and the conditions of the problem. The virtual cells of the active memory of

the neuron, which do not belong to the coordinate system of information binding and verification, are active cells of variable length with dynamically changeable information in them, i.e. these are working virtual cells. The process of interaction of character chains stored in passive and active virtual cells is defined as an information process with local interaction. Any operation between cells of different types is specified in the form of two operands - interaction and design. A more detailed meaningful description of such processes is reflected in the author's works [1,3,4]. The indicated logical and algorithmic schemes allow us to build models of intellectual stream processing of information in a fuzzy computing environment.

4. Some conclusions

It should be noted here that in the ontology of the subject area of the creation of the designated computing and information and measuring systems, the flow of information or data in terms of content and meaning is logically and mathematically defined as follows. Firstly, in a fixed quantum of perception or detection time (physical and informational), the duration of which for computing systems is determined by the number of operations at one step of the computational process algorithm, and for measuring (analog or discrete) operations, the duration of physical processes of detection and relaxation, computational or measurement processes, the data stream is defined as a one-dimensional array or a linked list of data, by means of which on a scale of any nature (usually numerical) from azhaetsya finite set of quantum states calculated or measured value. Secondly, on an ordered set of designated time slices, the flow of information or data is defined as a two-dimensional array or a similar list of data, and the rows of such arrays are one-dimensional data streams of calculations or measurements for a specific time slice. This is the difference between streaming calculations or measurements from traditional ones, which in one step of an algorithm or a time slice only give one value of the result of a calculated or measured quantity, and in streaming - a data stream. At the functional and logical level, algorithms and procedures for the operations of synthesis, processing and analysis of data flows in such systems are formalized and described using the mathematical apparatus of streaming models.

5. References

[1] Myshev A V 2012 Metrological theory of the dynamics interacting objects in the information field of a neural network and a neuron *Information Technology* **4** 52–62 (in Russia)

[2] Myshev A V 2012 Information model of a neural network in computational intelligence technologies and forms of computing implementation *Information Technology* **1** 63–79 (in Russia)

[3] Myshev A V 2014 Virtual stream computing system based on information model of artificial neural network and neuron *Official bulletin "Inventions. Utility models"* **28** 30 pp. *Patent of invention №2530270* (in Russia)

[4] Myshev A V 2011 The method of virtual perspective in modeling fuzzy problems *Information Technology and computing systems* **11** 66–79 (in Russia)

[5] Myshev A V 2010 Model of active memory on binary fields in the virtualization technology of channels transmission and storage of information *Software products and systems* **1** 54–59 (in Russia)

[6] Myshev A V 2010 Dynamics of information processes in computing technologies of computer modeling *ISA RAN Press* **58** 137–148 (Moscow: Krasand)

Mathematical modeling of the dynamics of layered and block media with nonlinear contact conditions on supercomputers

I. S. Nikitin[1], N. G. Burago[2], V. I. Golubev[3] and A. D. Nikitin[1]

[1] Institute of Computer Aided Design of RAS, 2-ya Brestskaya st. 19/18, Moscow, 123056, Russia
[2] Ishlinski Institute for Problems in Mechanics of RAS, prospect Vernadskogo 101, b1, Moscow, 119526, Russia
[3] Moscow Institute of Physics and Technology (National Research University), Moscow,Russia

E-mail: i_nikitin@list.ru

Abstract. Continual models of solid media with a discrete set of slip planes (layered, block media) and with nonlinear slip conditions at the contact boundaries of structural elements are constructed. The constitutive equations of the resulting system of equations contain a small viscosity parameter in the denominator of nonlinear free terms. For a stable numerical solution of a system of differential equations, an explicit-implicit method is proposed with an explicit approximation of the equations of motion and an implicit approximation of the constitutive relations containing a small parameter. From implicit nonlinear difference approximations analytically, using the perturbation method, various effective formulas for the correction of stress components after an elastic time step are obtained. To calculate the elastic step, we used a grid-characteristic method on hexahedral grids, which allowed us to significantly increase the speed of calculations and simulate a non-stationary three-dimensional problem of generating a response from an oriented layered or block cracked cluster located in a homogeneous medium.

1. Introduction

Continual models of deformable solid media with a discrete set of slip planes (layered, block media) and with nonlinear slip conditions at contact boundaries can be obtained with the method of the asymptotic homogenization [1] or with the discrete variant of the slippage theory [2]. In all of these cases, in the constitutive system of equations with the non-linear free term and the small stress relaxation time are included. For the stable numerical solution of differential equations, the explicit-implicit method with the explicit approximation of motion equations and the implicit approximation of constitutive equations containing small parameter in the denominator of the free term was proposed. A set of effective formulas for the stress tensor correcting at the elastic step was obtained. Numerical simulations of the dynamic scattering process for subsurface layered and block objects in elastic 2D and 3D media were carried out using modern high-performance computing systems.

Journal of Physics: Conference Series **1392** (2019) 012057 doi:10.1088/1742-6596/1392/1/012057

2. Mathematical model

Non-linear interaction conditions between contacting structural elements may be formulated. In the Cartesian system x_i, $(i = 1, 2, 3)$. the unbounded elastic medium with the oriented system of periodic parallel slip planes is considered. The orientation is set with the unit vector of normal $\mathbf{n}$. The distance between slip planes is constant and equal to ε. The density of the material ρ, and Lame moduli λ and μ are known constants. The stress state of the medium is described with the stress tensor σ. The shear stress vector at the slip plane is equal to $\tau = \sigma \cdot \mathbf{n} - (\mathbf{n} \cdot \sigma \cdot \mathbf{n})\mathbf{n}$, the normal stress is equal to $\sigma_n = \mathbf{n} \cdot \sigma \cdot \mathbf{n}$. We define the slip velocity vector γ and the delamination velocity vector $\omega = \omega\mathbf{n}$, based on the discontinuity of tangential $[V_\tau]$ and normal $[V_n]$ velocity at contact boundaries: $\gamma = [\mathbf{V}_\tau]/\varepsilon$, $\omega = [V_n]/\varepsilon$.

We assume the presence of thick interlayers between elastic layers $d << \varepsilon$, however we will not take it into account explicitly, but using slip conditions at compressed layers boundaries. At the contact boundaries the special conditions are specified.

1) Piecewise linear condition of slip and weak delamination.

When $\sigma_n < 0$ (compressed contact): $\gamma = \dot{\tau}/k_\gamma$, $\omega = 0$.

When $\sigma_n \geq 0$ (weak delamination): $\gamma = \dot{\tau}/k_\gamma$, $\omega = \dot{\sigma}_n/k_\omega$. $k_\gamma/\mu << 1$, $k_\omega/\mu << 1$.

2) The slip condition for Coulomb friction with a low viscous additive and weak delamination.

When $\sigma_n < 0$ (compressed contact): $\tau = q\,|\sigma_n|\,(\gamma/\,|\gamma| + \eta\gamma)$ or

$$\gamma = \frac{1}{\eta}\frac{\tau}{|\tau|}\langle\frac{|\tau|}{q\,|\sigma_n|} - 1\rangle\ , \quad \omega = 0$$

When $\sigma_n \geq 0$ (weak delamination): $\gamma = \dot{\tau}/k_\gamma$, $\omega = \dot{\sigma}_n/k_\omega$. $k_\gamma/\mu << 1$, $k_\omega/\mu << 1$.

Here k_γ k_ω - coefficients of weak elastic tangential and normal bond of layers, η the viscosity coefficient, q - the Coulomb friction coefficient, $\langle F(y)\rangle = F(y)H(y)$, $H(y)$ the Heaviside function, $H(y) = 0$ if $y < 0$, $H(y) = 1$ if $y \geq 0$. The contact plane with the interaction condition defined is called the slip-delamination plane.

It should be noticed, that previously in paper [2] for the delamination regime the full delamination condition was used :$\Omega \geq 0$: $\tau = \sigma_n = 0$, that is the asymptotic case of the weak delamination when $k_\omega \to 0$, $k_\gamma \to 0$; $\Omega = [u_n]/\varepsilon$ the normalized discontinuity of normal displacements at the contact boundary, defining by the equation $\dot{\Omega} = \omega$. Also, for the case of the weak delamination $\omega = \dot{\sigma}_n/k_\omega$ inequalities $\sigma_n \geq 0$ and $\Omega \geq 0$ are equivalent.

From the numerical point of view, small parameters k_γ k_ω are regularizations that allow us to prevent the oscillations occurrence when sharp changing the compress boundary to the full delamination condition.

To construct the continuum model with a set of these slip-delamination planes we are going to deal with γ and ω as discontinuous functions of space and time. Also we will use main relationships from the slippage theory as many other authors. It allows us to take into account contributions from γ and ω into non-elastic strain tensors $\mathbf{e}^\gamma$ and $\mathbf{e}^\omega$ respectively:

$$\mathbf{e}^\gamma = (\mathbf{n} \otimes \gamma + \gamma \otimes \mathbf{n})/2, \quad \gamma \cdot \mathbf{n} = 0, \quad \mathbf{e}^\omega = (\mathbf{n} \otimes \omega + \omega \otimes \mathbf{n})/2 = \omega\mathbf{n} \otimes \mathbf{n}, \quad \omega = \omega\mathbf{n}$$

The full strain tensor equals to the sum of elastic and non-elastic parts:

$$\mathbf{e} = \mathbf{e}^e + \mathbf{e}^\gamma + \mathbf{e}^\omega, \quad \mathbf{e} = (\nabla\mathbf{v} + \nabla\mathbf{v}^T)/2$$

Here $\mathbf{v}$ is macroscopic velocity of medium particles, $\mathbf{e}^e$ is the elastic strain tensor accordingly to the Hookes law: $\dot{\sigma} = \lambda(\mathbf{e}^e : \mathbf{I})\mathbf{I} + 2\mu\mathbf{e}^e$

The final equation is the motion equation:

$$\rho\dot{\mathbf{v}} = \nabla \cdot \sigma$$

3. System of equations for a layered medium

In layered medium, containing a set of elastic layers, the only system of slip-delamination planes are possible with the normal **n**. If the normal to contact boundaries **n** is oriented along the coordinate x_2 then its components are $n_j = \delta_j^2$.

Through conditions for γ and ω, corresponding to the local contact conditions 1), are:

$$\gamma_j = \dot{\sigma}_{2j}/k_\gamma, \quad \omega = \dot{\sigma}_{22} H(\sigma_{22})/k_\omega$$

Through conditions for γ and ω, corresponding to the local contact conditions 2), are:

$$\gamma_j = \frac{1}{\eta}\frac{\sigma_{2j}}{|\tau|}\langle\frac{|\tau|}{q\,|\sigma_{22}|} - 1\rangle(1 - H(\sigma_{22})) + \dot{\sigma}_{2j} H(\sigma_{22})/k_\gamma$$

$$\omega = \dot{\sigma}_{22} H(\sigma_{22})/k_\omega, \quad |\tau| = \sqrt{\sum_{k\neq 2} \sigma_{2k}\sigma_{2k}}$$

Based on the chosen normal direction the final system of equations for this model is:

$$\rho\dot{v}_i = \sigma_{ij,j}, \quad \dot{\sigma}_{ii} \underset{i\neq 2}{=} \lambda v_{k,k} + 2\mu v_{i,i} - \lambda\omega, \quad \dot{\sigma}_{2j} \underset{j\neq 2}{=} \mu(v_{2,j} + v_{j,2}) - \mu\gamma_j$$

$$\dot{\sigma}_{22} = \lambda v_{k,k} + 2\mu v_{2,2} - (\lambda + 2\mu)\omega, \quad \dot{\sigma}_{ij} \underset{i,j\neq 2}{=} \mu(v_{i,j} + v_{j,i}), \quad i \neq j$$

4. System of equations for a block medium

The block medium consists of parallelepiped elastic elements with three possible slip-delamination planes. These planes are defined with normals $\mathbf{n}^{(s)}, \quad s = 1, 2, 3$. In this case the non-elastic strain tensors are:

$$\mathbf{e}^\gamma = \sum_{s=1}^{3} (\mathbf{n}^{(s)} \otimes \gamma^{(s)} + \gamma^{(s)} \otimes \mathbf{n}^{(s)})/2, \quad \gamma^{(s)} \cdot \mathbf{n}^{(s)} = 0$$

$$\mathbf{e}^\omega = (\mathbf{n}^{(s)} \otimes \omega^{(s)} + \omega^{(s)} \otimes \mathbf{n}^{(s)})/2 = \omega^{(s)}\mathbf{n}^{(s)} \otimes \mathbf{n}^{(s)}. \quad \omega^{(s)} = \omega^{(s)}\mathbf{n}^{(s)}$$

If three normals to slip-delamination planes are oriented along coordinate axis of the Cartesian system then $n_j^{(s)} = \delta_j^s$, where δ_j^s is Kronecker's symbol.

Through conditions for $\gamma^{(i)}$ and $\omega^{(i)}$, corresponding to the local contact conditions 1), are:

$$\gamma_j^{(i)} = \dot{\sigma}_{ij}/k_\gamma, \quad i \neq j, \quad \omega^{(i)} = \dot{\sigma}_{ii} H(\sigma_{ii})/k_\omega$$

Through conditions for $\gamma^{(i)}$ and $\omega^{(i)}$, corresponding to the local contact conditions 2), are:

$$\gamma_j^{(i)} = \frac{1}{\eta}\frac{\sigma_{ij}}{\left|\tau^{(i)}\right|}\langle\frac{\left|\tau^{(i)}\right|}{q\,|\sigma_{ii}|} - 1\rangle(1 - H(\sigma_{ii})) + \dot{\tau}_{ij} H(\sigma_{ii})/k_\gamma, \quad i \neq j$$

$$\omega^{(i)} = \dot{\sigma}_{ii} H(\sigma_{ii})/k_\omega. \quad \left|\tau^{(i)}\right| = \sqrt{\sum_{k\neq i} \sigma_{ik}\sigma_{ik}}$$

As for the layered medium, we can rewrite the main system in the suitable form:

$$\rho\dot{v}_i = \sigma_{ij,j}, \quad \dot{\sigma}_{jj} = \lambda v_{k,k} + 2\mu v_{j,j} - \lambda\sum_{l\neq j}\omega^{(l)} - (\lambda + 2\mu)\omega^{(j)}$$

$$\dot{\sigma}_{ij} = \mu(v_{i,j} + v_{j,i}) - \mu\gamma_j^{(i)} - \mu\gamma_i^{(j)}, \quad i \neq j$$

5. Numerical method

Both formulated systems are semi-linear hyperbolic systems and the numerical solution can be obtained with different explicit schemes. However, the slippage process switches on the non-linear free term with small viscosity parameter in the denominator. The system transforms into the form with small parameter and ordinary explicit schemes will not be stable. To overcome this problem, the usage of explicit-implicit method is proposed. The implicit approximation is used only for equations that contain small term in the denominator. All other equations are approximated with the explicit scheme.

Lets describe this approach for $\dot{\sigma}_{2j}$ for compressed contact case $\sigma_{22} < 0$ for the layered medium:

$$\dot{\sigma}_{2j} \underset{j\neq 2}{=} \mu(v_{2,j} + v_{j,2}) - \mu\sigma_{2j}\langle|\tau|/(q\,|\sigma_{22}|) - 1\rangle/(\eta\,|\tau|)$$

Implicit approximation with first order time approximation is:

$$(\sigma_{2j}^{n+1} - \sigma_{2j}^{n})\Big/\Delta t = \mu(v_{2,j}^{n+1} + v_{j,2}^{n+1}) - \mu\sigma_{2j}^{n+1}\langle\Sigma_e^{n+1}\Big/\left(q\left|\sigma_{22}^{n+1}\right|\right) - 1\rangle\Big/(\eta\Sigma_e^{n+1})$$

$$\Sigma_e^{n+1} = \sqrt{\left(\sigma_{12e}^{n+1}\right)^2 + \left(\sigma_{32e}^{n+1}\right)^2}$$

Here indices $n+1$ and n mean current and next time layers, Δt is the time step. We assume, that values v_i^{n+1} σ_{jj}^{n+1} were calculated with the explicit schemes for elastic equations. Solving this equation and also the same equation for normal tensions we can obtain correcting formulas. When $\sigma_{22e}^{n+1} < 0$ (compressed contact)

$$\sigma_{22}^{n+1} = \sigma_{22e}^{n+1}, \quad \sigma_{ii}^{n+1} = \sigma_{iie}^{n+1}, \quad \omega = 0, \quad i = 1,3$$

Whrn $\Sigma_e^{n+1} \geq q\left|\sigma_{22e}^{n+1}\right|$

$$\sigma_{i2}^{n+1} = q\left|\sigma_{22e}^{n+1}\right|\left(\sigma_{i2e}^{n+1}\big/\Sigma_e^{n+1}\right)\left(1 + \delta\Sigma_e^{n+1}\right)\Big/\left(1 + \delta q\left|\sigma_{22e}^{n+1}\right|\right)$$

$$\gamma_i = (\sigma_{i2e}^{n+1} - \sigma_{i2}^{n+1})/(\mu\Delta t), \quad i = 1,3$$

When $\Sigma_e^{n+1} < q\left|\sigma_{22e}^{n+1}\right|$

$$\sigma_{i2}^{n+1} = \sigma_{i2e}^{n+1}, \quad \gamma_i{}^{n+1} = 0, \quad i = 1,3, \quad \delta = \eta/(\mu\Delta t)$$

When $\sigma_{22e}^{n+1} \geq 0$ (delamination process)

$$\sigma_{22}^{n+1} = \frac{1}{(1+\beta)}(\beta\sigma_{22e}^{n+1} + \sigma_{22}^{n}), \quad \sigma_{ii}^{n+1} = \sigma_{iie}^{n+1} - \frac{1}{(1+\beta)}\frac{\lambda(\sigma_{22e}^{n+1} - \sigma_{22}^{n})}{(\lambda + 2\mu)}$$

$$\sigma_{i2}^{n+1} = \frac{(\alpha\sigma_{i2e}^{n+1} + \sigma_{i2}^{n})}{(1+\alpha)}, \quad \omega = \frac{(\sigma_{22e}^{n+1} - \sigma_{22}^{n})}{(1+\beta)(\lambda+2\mu)\Delta t}, \quad \gamma_i = \frac{(\sigma_{i2e}^{n+1} - \sigma_{i2}^{n})}{(1+\alpha)\mu\Delta t}, \quad i = 1,3$$

Coefficients for weak shear $\alpha = k_\gamma/\mu$ and stretching $\beta = k_\omega/(\lambda + 2\mu)$, $\alpha,\ \beta << 1$, ${\sigma_{ij}}_e^{n+1} = \sigma_{ij}^{n} + \left(\lambda v_{k,k}^{n+1}\delta_{ij} + \mu(v_{i,j}^{n+1} + v_{j,i}^{n+1})\right)\Delta t$ is the stress value after the elastic step. These formulas, in fact, is the adjustment of elastic stresses for the friction cone with viscous corrections. We used here the simplest formula for ${\sigma_{2j}}_e^{n+1}$ only to illustrate the derivation process.

In this research we used the grid-characteristic method on hexahedral meshes with the third order of the spatial approximation [3, 4, 5]. The monotonization was done based on the grid-characteristic monotonicity criterion, based on the internal structure of the linear transport equation. It decreases the approximation of the method to the second order on discontinuities. It allows us to increase significantly the precision and the robustness of the numerical simulation. The seismic response from the cluster of cracks and blocks in the homogenous medium was calculated in 2D and 3D cases.

6. Simulation results

On Fig. 1 two 2D wave fields of scattered seismic waves on the layered structure with different orientations for contact condition 2) are shown. The influence of the layer orientation on the

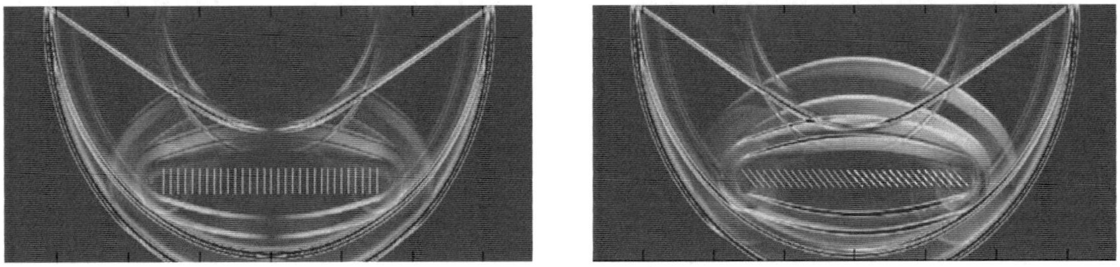

Figure 1. 2D-scattered seismic waves on the fractured cluster for various cracks orientation: vertical (left), under angle 45° (right).

scattered wave field is clearly seen. Seismic response is non-symmetric for non-vertical layers orientation. Based on this fact it is possible to make some assumptions about the main direction in the subsurface fractured medium.

On Fig. 2-3 3D wave fields of scattered seismic waves on layered and block structures for contact condition 1) are shown. The whole medium was a parallelepiped with sizes 10 x 10 x 3 km. P-wave velocity was 4500 m/s, S-wave velocity 2250 m/s, density 2500 kg/m3. The day surface condition was used on the upper side. In the center of the medium along OX and OY with the depth 50 m the point source with the 30 Hz Ricker time dependence function was applied. At the 2 km depth the object was set with sizes 3 x 3 x 0.2 km. Two different models were compared: block and layered with the normal vector along the axis OX. The parallelepiped mesh with 5 m cells, containing 2.4 billion of nodes was constructed. The time step was 1 ms, and totally 2000 steps were done. To visualize results we save each 40 time steps: 3 slices: 2 main vertical slices and one horizontal equals to the day surface. The total computation time was 5.5 hours on 100 computation cores for both mathematical models. As expected,

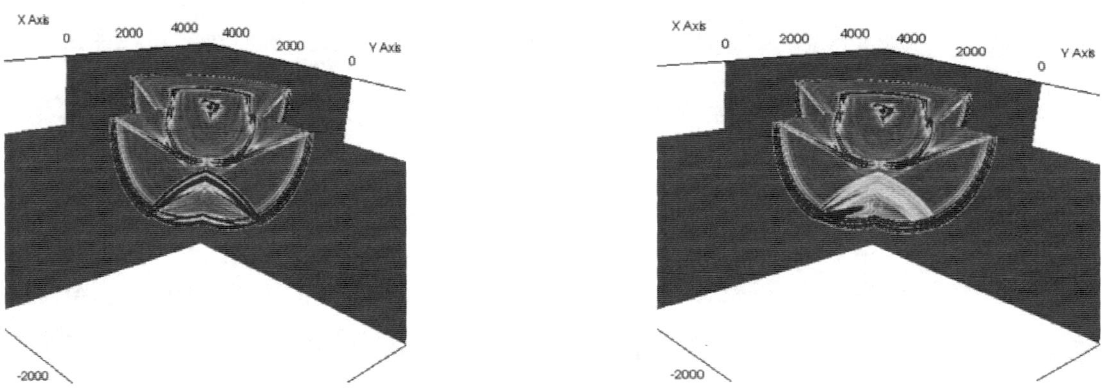

Figure 2. 3D - scattered seismic waves, t= 0.56 sec: in block (left), in layered cluster (right)

the amplitude of the seismic response for the block medium is significantly higher due to the existence of horizontal contact planes, that reflect the incident wave (see left Fig. 2, 3). Also, the 3D asymmetry of the response from the layered medium is clearly seen (see right Fig. 2, 3). It is explained by the presence of the preferred direction along the normal vector.

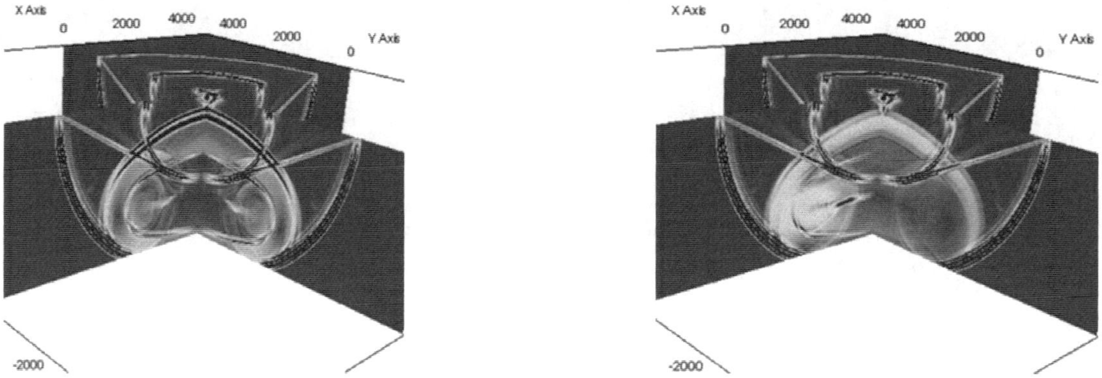

Figure 3. 3D - scattered seismic waves, t= 0.8 sec: in block (left), in layered cluster (right)

.

Conclusion

Continual models of solid media with a discrete set of slip planes (layered, block media) and with nonlinear slip conditions at the contact boundaries of structural elements were constructed. For a stable numerical solution of a system of differential equations, an explicit-implicit method is proposed with an explicit approximation of the equations of motion and an implicit approximation of the constitutive relations containing a small parameter in the denominator. A set of effective formulas for the stress tensor correcting at the elastic step was obtained. Numerical simulations of the dynamic scattering process for subsurface layered and block objects in elastic 2D and 3D media were carried out using modern high-performance computing systems.

Acknowledgments

The work was supported by State Contracts ICAD RAS AAAA-A19-119041590047-3, IPMech RAS AAAA-A17-117021310381-8. Numerical algorithms have been implemented using computing resources of the federal collective usage center Complex for Simulation and Data Processing for Mega-science Facilities at NRC Kurchatov Institute (http://ckp.nrcki.ru/).

References

[1] Burago N.G., Zhuravlev A.B., Nikitin I.S. Continuum model and method of calculating for dynamics of inelastic layered medium// Mathematical Models and Computer Simulations. 2019. Vol. 11. No 3. Pp. 59-74.

[2] Nikitin I.S. Dynamic models of layered and block media with slip, friction and separation. //Mechanics of Solids. Vol. 43. 4. 2008. Pp. 652-661.

[3] Golubev V.I., Petrov I.B., Khokhlov N.I., Shults K.I. Numerical computation of wave propagation in fractured media by applying the grid-characteristic method on hexahedral meshes // Computational Mathematics and Mathematical Physics, 2015, v. 55, 3, p. 509-518.

[4] Favorskaya, A., Golubev, V., Grigorievyh, D. Explanation the difference in destructed areas simulated using various failure criteria by the wave dynamics analysis // Procedia Computer Science, 2018, v. 126, pp. 1091-1099.

[5] Golubev, V., Khokhlov, N., Grigorievyh, D., Favorskaya, A. Numerical simulation of destruction processes by the grid-characteristic method // Procedia Computer Science, 2018, v. 126, pp. 1281-1288.

Mathematical model of decomposition of hydrates in a reservoir

Popov V.V.

677000 Yakutsk, 58, Belinsky str., Ammosov North-Eastern Federal University
677000 Yakutsk, 1 Oktyabrsky St., Institute of Oil and Gas Problems Sibirian Division RAS,

E-mail: imi.pm.pvv@mail.ru

Abstract. A mathematical model of the process of decomposition of hydrates in a reservoir by reducing pressure is implemented numerically. The dependence of results obtained on main parameters of the problem such as initial hydrate saturation, initial water saturation, depression on the formation, permeability coefficient, porosity has been studied. We compare results of the numerical implementation of the original model with results obtained after linearization of the heat equation, equations of gas flow continuity both separately and simultaneously. It is shown that with relatively low permeability, initial water saturation and depression one can use the model with linearized equations. The relative difference of the calculated hydrate saturation is a few percent. For the distribution of temperature and pressure, this difference is even smaller. The developed algorithm can easily be expanded for a multidimensional problem.

1. INTRODUCTION

The property of natural gas under certain conditions take in the earth's crust in a solid state and to form gas hydrate deposits, established by V.G. Vasiliev., Yu.F. Macogon, F.A. Trebin, A.T. Trofimuk and N.V. Chersky, was recognized as a discovery in [1]. Studies to determine the physical and mechanical properties of hydrates of natural gases and their fields are summarized in the monographs of Yu.F. Makogon [2], S.Sh.Byk, V.I.Fomina [3]. In the papers of N.V. Chersky and E.A.Bondarev [4], M.Selin and E.Sloyan [5], mathematical models of hydrate dissociation were constructed thermal effects on gas hydrate formation. The modeling of the process of dissociation of a gas hydrate due to a decrease in pressure was performed by Verigin H.H. et al. [6]. Here it was believed that the hydrate saturates the entire pore volume, and in addition it was assumed that the dissociation occurs under isothermal conditions. The mathematical model is reduced to the classical problem of Stefan. A self-similar solution of the linearized system of gas filtration equations was constructed for the case of gas extraction through the gallery. The first attempt at mathematical modeling of wet gas filtration with hydrate dissociation was undertaken in the monograph by E.A. Bondarev et al. [7]. In the above works, the phase transition occurs on a moving surface, as in problems like Stefan. Further development of these models were based on ideas of local thermodynamic equilibrium in the works of E.A.Bondarev, A.M.Maksimov, G.G.Tsypkina [8]. Also noteworthy are the works of Musakayev N. G., Borodin S. L., Belskikh D. S. [9], Khasanov M. K., Stolpovsky M. V., Kildibaeva S. R. [10] and Shagapov V.Sh., Khasanov M.K., Gimaltdinov I.K., Stolpovsky M.V. [11], where mathematical models are proposed and numerically implemented with a phase transition in an extended domain. This paper assesses the contribution of conductive and convective heat transfer to the process of decomposition of hydrates in near-well zone when extracting gas from gas hydrate deposits through the gallery.

2. PROBLEM STATEMENT

The heat equation contains a hydrate-gas term that takes into account the heat of the phase transition [8-11]:

$$C\frac{\partial T}{\partial t} = \frac{\partial}{\partial x}\left(\lambda\frac{\partial T}{\partial x}\right) + \frac{k(1-\nu)f_g}{\mu_g}\rho_g c_g\frac{\partial p}{\partial x}\frac{\partial T}{\partial x} + \frac{k(1-\nu)f_w}{\mu_w}\rho_w c_w\frac{\partial p}{\partial x}\frac{\partial T}{\partial x} + \phi\rho_h q_h\frac{\partial \nu}{\partial t}, \quad (1)$$

$$0 < x < l, \quad t > 0;$$

gas filtration equation [8-11]:

$$\phi\frac{\partial}{\partial t}\left((1-\nu-\sigma)\frac{p}{TR}\right) = \frac{\partial}{\partial x}\left(\frac{k(1-\nu)f_g}{\mu_g}\frac{p}{RT}\frac{\partial p}{\partial x}\right) - \phi\rho_h\varepsilon\frac{\partial \nu}{\partial t} \quad (2)$$

$$0 < x < l, \quad t > 0;$$

where k - permeability coefficient of the porous medium; μ_g - dynamic viscosity of gas, f_g - gas phase permeability:

$$f_g = \begin{cases} (1+3\sigma)\left(\frac{0.9-\sigma}{0.9}\right)^{3.5}, & 0 \le \sigma < 0.9, \\ 0, \quad 0.9 \le \sigma \le 1 \end{cases}$$

water filtration equation [8-11]:

$$\phi\frac{\partial \sigma}{\partial t} = \frac{\partial}{\partial x}\left(\frac{k(1-\nu)f_w}{\mu_w}\frac{\partial p}{\partial x}\right) - \phi\frac{\rho_h}{\rho_w}(1-\varepsilon)\frac{\partial \nu}{\partial t} \quad (3)$$

$$0 < x < l, \quad t > 0;$$

f_w - relative phase permeability of water; μ_w - dynamic viscosity of water;

$$f_w = \begin{cases} \left(\frac{\sigma-0.2}{0.8}\right)^{3.5}, & 0.2 < \sigma \le 1, \\ 0, \quad 0 \le \sigma \le 0.2 \end{cases}$$

equilibrium phase transition condition [2,3,7]:

$$T = A\ln p + B \quad (4)$$

$$0 < x < l, \quad t > 0;$$

Initial conditions:

$$T(0,x) = T_0, \; p(0,x) = A\ln p_0 + B, \; \nu(0,x) = \nu_0, \; \sigma(0,x) = \sigma_0 \quad (5)$$

Condition on the left border:

$$T(t,0) = T_g, \; p(t,0) = p_g, \quad (6)$$

Condition on the right border:

$$T(t,l) = T_0, \; p(t,l) = p_0, \nu(t,l) = \nu_0, \; \sigma(t,l) = \sigma_0 \quad (7)$$

3. SOLUTION ALGORITHM

In order to increase the efficiency of the numerical implementation of the mathematical model (1) - (7), we transform equations (2) and (3). From equation (2) we exclude $\frac{\partial \nu}{\partial t}$ using the equation (1):

$$\phi \frac{\partial}{\partial t}\left((1-\nu-\sigma)\frac{p}{TR}\right) = \frac{\partial}{\partial x}\left(\frac{k(1-\nu)f_g}{\mu_g}\frac{p}{RT}\frac{\partial p}{\partial x}\right) - C\frac{\varepsilon}{q_h}\frac{A}{p}\frac{\partial p}{\partial t} + \tag{8}$$

$$+\frac{\varepsilon}{q_h}\frac{\partial}{\partial x}\left(\lambda \frac{A}{p}\frac{\partial p}{\partial x}\right) + \frac{\varepsilon}{q_h}\frac{k(1-\nu)f_g}{\mu_g}\rho_g c_g \frac{\partial p}{\partial x}\frac{\partial T}{\partial x} + \frac{\varepsilon}{q_h}\frac{k(1-\nu)f_w}{\mu_w}\rho_w c_w \frac{\partial p}{\partial x}\frac{\partial T}{\partial x}$$

Hydrate saturation will be calculated from the equation

$$C\frac{\partial T}{\partial t} - \frac{\lambda}{K}\frac{A}{p}\phi\frac{\partial}{\partial t}\left((1-\nu-\sigma)\frac{p}{TR}\right) = \frac{\partial \lambda}{\partial x}\frac{\partial T}{\partial x} - \frac{\lambda}{K}\frac{A}{p}\frac{\partial K}{\partial x}\frac{\partial p}{\partial x} \tag{9}$$

$$+\frac{k(1-\nu)f_g}{\mu_g}\rho_g c_g \frac{\partial p}{\partial x}\frac{\partial T}{\partial x} + \frac{k(1-\nu)f_w}{\mu_w}\rho_w c_w \frac{\partial p}{\partial x}\frac{\partial T}{\partial x} + \phi\rho_h q_h \frac{\partial \nu}{\partial t} + \frac{\lambda}{K}\frac{A}{p}\phi\rho_h\varepsilon\frac{\partial \nu}{\partial t},$$

$$0 < x < l, \quad t > 0;$$

where

$$K = \frac{k(1-\nu)f_g}{\mu_g}\frac{p}{RT}$$

The equation of water flow continuity (3) is reduced to the form

$$\phi\frac{f_g}{\mu_g}\rho_g\frac{\partial \sigma}{\partial t} - m\frac{f_w}{\mu_w}\frac{\partial}{\partial t}((1-\nu-\sigma)\rho_g) + \phi\frac{f_g}{\mu_g}\rho_g\frac{\rho_h}{\rho_w}(1-\varepsilon)\frac{\partial \nu}{\partial t} + \frac{f_w}{\mu_w}m\rho_h\varepsilon\frac{\partial \nu}{\partial t} \tag{10}$$

$$= \rho_g\frac{f_g}{\mu_g}\frac{k}{\mu_w}\frac{\partial(1-\nu)f_w}{\partial x}\frac{\partial p}{\partial x} - \frac{f_w}{\mu_w}\frac{k}{\mu_g}\frac{\partial f_g \rho_g}{\partial x}\frac{\partial p}{\partial x}$$

obtained by lowering the order of the water filtration equation, (3) using the gas filtration equation (2).

We describe an algorithm for the numerical solution of problem (8) - (10), (4) - (7). Let the distributions of the sought functions on the n -th time layer to be known. To go to the $n+1$ th temporal layer, we first calculate the pressure distribution. We calculate the temperature distribution using the formula (4). Next, we calculate the distribution of hydrate saturation from equation (9). And at the end we calculate the distribution of water saturation from (10). And we perform simple iterations on nonlinearity.

4. NUMERICAL EXPERIMENTS

The assessment of the contribution of the terms taking into account convective and conductive heat transfer, both separately and jointly, has been carried out.For this, three more variants of equation (8) were considered:

excluding convective heat transfer

$$\phi \frac{\partial}{\partial t}\left((1-\nu-\sigma)\frac{p}{TR}\right) = \frac{\partial}{\partial x}\left(\frac{k(1-\nu)f_g}{\mu_g}\frac{p}{RT}\frac{\partial p}{\partial x}\right) - C\frac{\varepsilon}{q_h}\frac{A}{p}\frac{\partial p}{\partial t} + \tag{11}$$

$$+\frac{\varepsilon}{q_h}\frac{\partial}{\partial x}\left(\lambda \frac{A}{p}\frac{\partial p}{\partial x}\right)$$

without conductive heat transfer

$$\phi\frac{\partial}{\partial t}\left((1-\nu-\sigma)\frac{p}{TR}\right)=\frac{\partial}{\partial x}\left(\frac{k(1-\nu)f_g}{\mu_g}\frac{p}{RT}\frac{\partial p}{\partial x}\right)-C\frac{\varepsilon}{q_h}\frac{A}{p}\frac{\partial p}{\partial t}+ \tag{12}$$
$$+\frac{\varepsilon}{q_h}\frac{k(1-\nu)f_g}{\mu_g}\rho_g c_g\frac{\partial p}{\partial x}\frac{\partial T}{\partial x}+\frac{\varepsilon}{q_h}\frac{k(1-\nu)f_w}{\mu_w}\rho_w c_w\frac{\partial p}{\partial x}\frac{\partial T}{\partial x}$$

excluding both terms

$$\phi\frac{\partial}{\partial t}\left((1-\nu-\sigma)\frac{p}{TR}\right)=\frac{\partial}{\partial x}\left(\frac{k(1-\nu)f_g}{\mu_g}\frac{p}{RT}\frac{\partial p}{\partial x}\right)-C\frac{\varepsilon}{q_h}\frac{A}{p}\frac{\partial p}{\partial t} \tag{13}$$

and accordingly three versions of equation (9):

$$C\frac{\partial T}{\partial t}-\frac{\lambda}{K}\frac{A}{p}\phi\frac{\partial}{\partial t}\left((1-\nu-\sigma)\frac{p}{TR}\right)=\frac{\partial \lambda}{\partial x}\frac{\partial T}{\partial x}-\frac{\lambda}{K}\frac{A}{p}\frac{\partial K}{\partial x}\frac{\partial p}{\partial x} \tag{14}$$
$$+\phi\rho_h q_h\frac{\partial \nu}{\partial t}+\frac{\lambda}{K}\frac{A}{p}\phi\rho_h\varepsilon\frac{\partial \nu}{\partial t},$$
$$0<x<l,\quad t>0;$$

$$C\frac{\partial T}{\partial t}=\frac{k(1-\nu)f_g}{\mu_g}\rho_g c_g\frac{\partial p}{\partial x}\frac{\partial T}{\partial x}+\frac{k(1-\nu)f_w}{\mu_w}\rho_w c_w\frac{\partial p}{\partial x}\frac{\partial T}{\partial x}+\phi\rho_h q_h\frac{\partial \nu}{\partial t}, \tag{15}$$
$$0<x<l,\quad t>0;$$

$$C\frac{\partial T}{\partial t}=\phi\rho_h q_h\frac{\partial \nu}{\partial t} \tag{16}$$

$$0<x<l,\quad t>0;$$

Also carried out the linearization of equation (2):

$$\phi\left((1-\nu_0-\sigma_0)\frac{1}{T_0R}\right)\frac{\partial p}{\partial t}=\frac{k(1-\nu_0)f_g(\sigma_0)}{\mu_g}\frac{p_0}{RT_0}\frac{\partial^2 p}{\partial x^2}-\phi\rho_h\varepsilon\frac{\partial \nu}{\partial t} \tag{17}$$

$$0<x<l,\quad t>0;$$

The calculations were carried out at the following values of the parameters:
$c_s = 700$ J/(kg ·K; $\rho_s = 2650 kg/m^3$; $c_h = 3210$ J/(kg·K; $\rho_h = 910 kg/m^3$;
$c_w = 4.19\cdot 10^3$ J/(kg·K); $\rho_w = 1000 kg/m^3$; $c_g = 2093$ J/(kg·K);
$\lambda_s = 2$Wt/(m·K); $\lambda_h = 2,11$Wt/(m·K); $\lambda_w = 0,58$Wt/(m·K); $\lambda_g = 0.0034$Wt/(m·K);
$q_h = 5.1\cdot 10^5$J/kg; $\mu_w = 1.8\cdot 10^{-3}$ Pa·s; $\mu_g = 1.3\cdot 10^{-5}$ Pa·s $\varepsilon = 0.129; \phi = 0.1-0.3$; $k = 10^{-13}-10^{-15} m^2$.

The results of numerical calculations are shown in Tables 1-2. and in Figures 1a-1b. Table 1 presents the relative differences in the values of hydrate saturation (in absolute value in percent) obtained as a result of the numerical implementation of the mathematical model under consideration (let's call model 1) and numerical solution of the following problems: (11), (14), (10), (4) - (7) (model 2); (12), (15), (10), (4) - (7) (model 3); (13), (16), (10), (4) - (7) (model 4) at $t = 5$ $\nu_0 = 0.5$, $\sigma_0 = 0.3$, $T_0 = 280$, $p_g = 3$ Mpa, $\phi = 0.1$.

The table shows that the effect of conductive heat transfer is relatively high at low permeabilities, and decreases sharply at high. And the effect of convection is more stable and increases with increasing permeability.

Similar results are presented in Table 2, where the difference between the results obtained by the implementation of models 1 and 4 is dependent on the porosity. Note that at high initial pressures, the values shown in the figures and tables are much smaller.

The effects of initial hydrate saturation and water saturation on the results of numerical calculations are small.

k	10^{-13}	$5 \cdot 10^{-14}$	10^{-14}	$5 \cdot 10^{-15}$	$2.5 \cdot 10^{-15}$	10^{-15}
model 1-2	0.11	0.12	0.51	1.04	2.11	5.52
model 1-3	3.54	3.37	2.90	2.70	2.52	2.38
model 1-4	3.55	3.42	3.39	3.74	4.66	7.95

Table 1: Effect of conductive and convective heat transfer.

ϕ	0.1	0.15	0.2	0.25	0.3
model 1-4	7.95	3.5	2.58	2.06	1.69

Table 2: Dependence on porosity at $k = 10^{-13}$

Figure 1a shows the distribution of hydrate saturation obtained by solving the original problem and problem (16), (17), (10), (4) - (7) with $t = 5$ days, $\nu_0 = 0.5$, $\sigma_0 = 0.3$, $T_0 = 285$, $p_g = 3$ Mpa, $k = 10^{-15}$, $\phi = 0.1$. Figure 1b shows a similar distribution of hydrate saturation at $p_g = 4$ Mpa.

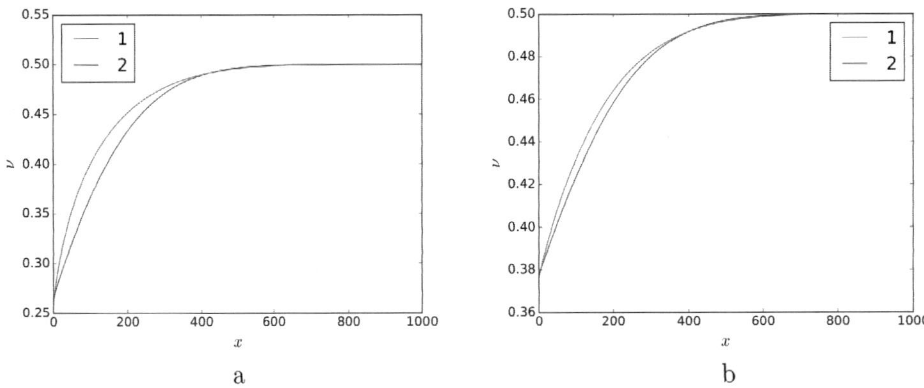

Fig. 1 Distribution of hydrate saturation. at $t = 5$ $\nu_0 = 0.5$, $\sigma_0 = 0.3$, $T_0 = 285$, $k = 10^{-13}$ $p_g = 3$ Mpa – (a), $p_g = 4$ Mpa – (b)

5. CONCLUSION

As shown by numerical studies, conformal and conductive heat transfer does not greatly affect the calculation results. In the case of medium and high penetrability, in the energy conservation equation, the terms taking into account thermal conductivity and convective heat transfer can be neglected. Numerical experiments have shown that the heat capacity coefficient can be considered constant. Then for the approximate calculation of the distribution of hydrate saturation, you can use the formula:

$$\nu = \nu_0 + \frac{C_0}{\phi \rho_h q_h}(T - T_0) \tag{18}$$

6. ACKNOWLEDGMENTS

The work was partially supported by the Russian National Science Foundation (project 19-11-00230 and 18-45-140035 p_a), as well as within the state order of the Ministry of Science and Higher Education of the Russian Federation 0377-2018-0002.

References

[1] *Vasiliev V.G., Makogon Yu.F., Trebin FA, Trofimuk A.A. , Chersky N.V.* The property of natural gases to be in the earth bark in the solid state and form gas hydrate deposits // Collection discoveries of the USSR, 1968-1969. - M .: TSNIIPI. 1970.

[2] *Makogon Yu. F.* Hydrates of natural gases. - M.: Nedra. 1974.

[3] *Byk S.Sh., Makogon Yu.F., Fomina V.I.* Gas hydrates [in Russian], Moscow, Chimiya, (1980).

[4] *Chersky N.V., Bondarev E.A.* On the thermal method of development of gas hydrate deposits // Dokl. Academy of Sciences of the USSR. 1972. T.203. No 3. S. 550- 552.

[5] *Selin M.S. , Sloan E.D.* Heat and mass transfer during the dissociation of hydrates in porous media // A I C h E Journal. 1989. V.35. No 6. Pp.1049-1052

[6] *Verigin H.H., Khabibullin I.L., Khalikov G.A.* Linear problem of decomposition of gas hydrates in a porous medium // Izv. Academy of Sciences of the USSR. Ser. fur. Jew. and gas. 1980. No 1. p. 174-177.

[7] *Bondarev E.A., Babe G.D., Groysman A.G. and Kanibolotsky M.A.* Mechanics of hydrate formation in gas flows. -Novosibirsk: Science, 1976.

[8] *Bondarev E.A., Maksimov A.M. , Tsypkin GG* To the mathematical modeling of the dissociation of gas hydrates // Report Academy of Sciences of the USSR. 1989. T. 308. No. 3. p. 575-577.

[9] *Musakayev N. G., Borodin S. L., Belskikh D. S.* "Mathematical model and algorithm for solving the problem of non-isothermal gas filtration in the reservoir whith consider of hydrate decomposition [in Russian]," Bulletin of the South Ural State University, Mathematics Series, Mechanics, Physics, **9** , No. 2, 22-29 (2017).
DOI: 10.14529/mmph170203

[10] *Khasanov M. K., Stolpovsky M. V., Kildibaeva S. R.* Numerical simulation of the formation of a gas hydrate in a porous medium during gas injection // Bulletin of the Bashkir University. 2013. V. 18. No4 C. 969-972.

[11] *Shagapov V.Sh., Khasanov M.K., Gimaltdinov I.K., StolPovsky M.V.* "Numerical simulation of gas hydrate formation in a porous reservoir of the finite length at purging it with gas [in Russian]," Prikl. mechanics and tech. physics, **52**, No. 4, 116-126, (2011).

Mathematical technology for solving the Kardar-Parisi-Zhang equation with a source

A E Rassadin[1] and A V Stepanov[2]

[1] Laboratory of infinite-dimensional analysis and mathematical physics, Lomonosov Moscow State University, Faculty of Mechanics and Mathematics, 1 Leninskiye Gory, Main Building, Moscow 119991, GSP-1, Russia

[2] Department of Engineering, Chuvash State Agricultural Academy, 29 K. Marx str., Cheboksary 428000, Russia

E-mail: brat_ras@list.ru

Abstract. The description of an algorithm for numerical solution of the two-dimensional Kardar-Parisi-Zhang equation with fractal initial condition and spatially inhomogeneous stochastic source of particles has been presented. Using the Cole-Hopf substitution, the Kardar-Parisi-Zhang equation is reduced to an auxiliary linear parabolic equation with a variable coefficient. This reduction provides both compatibility of the algorithm with the concept of mathematical technology and possibility of its practical realization on supercomputer using the MPI parallel programming technology and the ROOT framework. The required profile of the evolving surface is restored by the inverse substitution by solving the auxiliary equation, Opportunity of estimation of its autocorrelation function and its spectral power density, as well as the time dynamics of Shannon surface information, have been discussed.

1. Introduction

One of the key questions under implementation of nanotechnologies into practice is the question about quality of the solid state surface. To clarify this question models for temporal evolution of crystal surface under different technological processes are required. One of these models is the Kardar-Parisi-Zhang (KPZ) equation, at present the KPZ-equation being the most popular phenomenological equation for modelling of crystal growth (see [1] and references therein). This equation was derived in paper [2] to describe the process of growth of the surface of a solid with epitaxial technology. In accordance with this paper the KPZ-equation looks as follows:

$$\frac{\partial h}{\partial t} = c + \frac{c}{2}\left[\left(\frac{\partial h}{\partial x}\right)^2 + \left(\frac{\partial h}{\partial y}\right)^2\right] + \nu\left(\frac{\partial^2 h}{\partial x^2} + \frac{\partial^2 h}{\partial y^2}\right) + Q(x,y)\,, \tag{1}$$

where $h(x,y,t)$ is the height of the surface under investigation, x and y are spatial Cartesian coordinates, c is the rate of growth and ν is coefficient of surface diffusion, function $Q(x,y)$ being additional source of sputtering substance in the vicinity of the surface [1].

Equation (1) ought to be provided by initial condition corresponding to initial shape of the surface under investigation:

$$h(x,y,0) = h_0(x,y)\,, \qquad (x,y) \in \Omega \subset R^2\,. \tag{2}$$

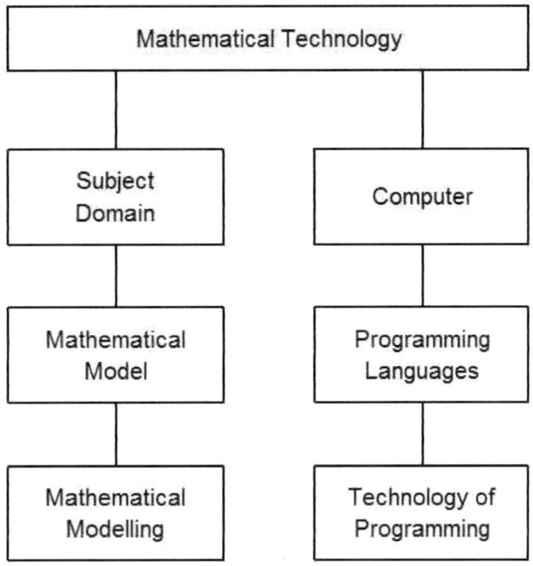

Figure 1. Structural scheme of mathematical technology concept

If $Q(x,y)$ is stochastic field then input equation (1) becomes stochastic partial differential equation.

There is a number of important values connected with solution of equation (1). The first of them is autocorrelation function of height:

$$K(\xi,\eta,t) =< (h(x,y,t)- < h(x,y,t) >)(h(x+\xi,y+\eta,t)- < h(x+\xi,y+\eta,t) >) > , \quad (3)$$

where angular brackets denote statistical averaging of function in them.

The two-dimensional spatial Fourier-transform from function (3) is known to be spectral power density of roughness of the surface:

$$S(p,q,t) = \int\int K(\xi,\eta,t)\, \exp(-i\,p\,\xi - i\,q\,\eta)\, d\xi d\eta\,. \quad (4)$$

Function (4) is significant because at every moment of time $t > 0$ it can be extracted from measurement of bistatic cross-section of monochromatic visible light with fixed polarization on this surface [3]. If λ is wavelength of this light source then for validity of this extraction the following inequality ought to be true [3]: $|h(x,y,t)- < h(x,y,t) > | \ll \lambda$. Other conditions for validity of this extraction namely $\left|\frac{\partial h(x,y,t)}{\partial x}\right| \ll 1$ and $\left|\frac{\partial h(x,y,t)}{\partial y}\right| \ll 1$ that is conjecture about small angles of the surface [3] coincide with conditions of validity for input equation (1) [2].

Another important value which can be constructed via the solution $h(x,y,t)$ of the Cauchy problem (1) and (2) is the Shannon information for the surface:

$$I(t) = \int_\Omega |h(x,y,t)|\, \log_2 |h(x,y,t)|\, dxdy\,. \quad (5)$$

At last often initial shape of the surface (2) proves to be fractal (see [4] and references therein). Moreover there is a variety of methods for estimation of fractal dimension D of network approximation for function $h(x, y, t)$ [5]. And this value depends on time due to temporal evolution of the solid state surface in accordance with the KPZ-equation (1): $D = D(t)$.

Thus subject domain for the KPZ-equation (1) and values (3)-(5) connected to them is quite wide. It means that one ought to use under solution of this series of connected problems on supercomputer the concept of mathematical technology [6]. This concept includes two chains namely the first of them contains subject domain¡ mathematical models of considered subject domain and their mathenatical modelling and the second of them contains computer, programming languages and technology of programming (see Figure 1).

The rest of the article is organized as follows: in Section 2 we construct algorithm for numerical solution of the KPZ-equation (1). Section 3 deals with derivation of fractal initial shapes (2). In section 4 methods of numerical calculation of values (3)-(5) related to the KPZ-equation (1) are considered. In final section we summarize the results obtained and discuss perspectives of further investigations.

2. The algorithm for numerical solution of the KPZ-equation

Let us introduce a new unknown function $\varphi(x, y, t)$ as follows:

$$h(x, y, t) = c\,t + \frac{2\,\nu}{c}\,\ln\varphi(x, y, t)\,. \tag{6}$$

The expression (6) is known to be the Cole-Hopf substitution [2]. By means of this formula, one can reduce the nonlinear equation (1) to a linear parabolic equation with a variable coefficient:

$$\frac{\partial\varphi}{\partial t} = \nu\left(\frac{\partial^2\varphi}{\partial x^2} + \frac{\partial^2\varphi}{\partial y^2}\right) + \frac{c\,Q(x, y)}{2\,\nu}\,\varphi\,. \tag{7}$$

In accordance with (6) initial condition for equation (7) is equal to:

$$\varphi(x, y, 0) = \exp\left[\frac{h_0(x, y)}{2\,\nu}\right], \qquad (x, y) \in \Omega \subset R^2\,. \tag{8}$$

To solve numerically the auxiliary linear Cauchy problem (7)-(8) at first let us suppose that domain Ω in (8) is a rectangular domain. Further one ought to segment the rectangular domain Ω into M rectangular subdomains Ω_m ($\bigcup_{m=1}^{M}\overline{\Omega}_m = \overline{\Omega}$) and to define on each Ω_m the simple problem in the framework of the method of fractional steps [7]:

$$\frac{\varphi_{ij}^{k+\frac{1}{2}} - \varphi_{ij}^{k}}{\tau/2} + \hat{\Lambda}_x\,\varphi_{ij}^{k+\frac{1}{2}} + \hat{\Lambda}_y\,\varphi_{ij}^{k} = \frac{c\,Q_{ij}}{2\,\nu}\,\varphi_{ij}^{k}\,, \tag{9}$$

$$\frac{\varphi_{ij}^{k+1} - \varphi_{ij}^{k+\frac{1}{2}}}{\tau/2} + \hat{\Lambda}_x\,\varphi_{ij}^{k+\frac{1}{2}} + \hat{\Lambda}_y\,\varphi_{ij}^{k+1} = \frac{c\,Q_{ij}}{2\,\nu}\,\varphi_{ij}^{k+\frac{1}{2}}\,, \tag{10}$$

where τ and Δ are steps on time and space respectively, φ_{ij}^k and Q_{ij} are network functions corresponding to continuous functions $\varphi(x, y, t)$ and $Q(x, y)$ and

$$\hat{\Lambda}_x\,\varphi_{ij}^k \equiv -\nu\,\frac{\varphi_{i+1,j}^k - 2\,\varphi_{ij}^k + \varphi_{i-1,j}^k}{\Delta^2}\,, \quad \hat{\Lambda}_y\,\varphi_{ij}^k \equiv -\nu\,\frac{\varphi_{i,j+1}^k - 2\,\varphi_{ij}^k + \varphi_{i,j-1}^k}{\Delta^2}\,.$$

To join solution of the simple problem (9)-(10) on subdomain Ω_m with solutions on neighboring subdomains one should impose exchange boundary conditions [6].

After that one can restore network function h_{ij}^k using discrete analog of formula (6).

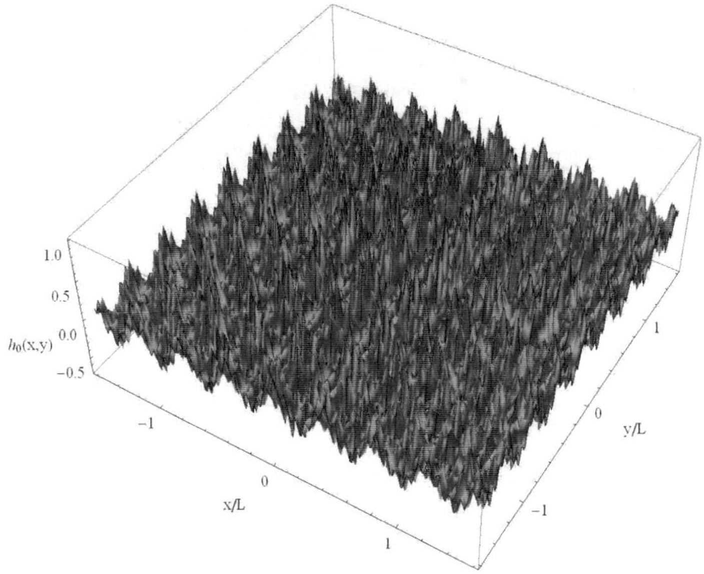

Figure 2. Regularized fractal initial shape

3. Regularization of fractal initial condition

Fractal functions are known to be nowhere differentiable (see [5] and references therein). For instance the well-known Weierstrass function

$$W(x,\vec{\mu}) = \sum_{n=1}^{\infty} a^n \cos\frac{\pi\, b^n\, x}{L}\,, \tag{11}$$

where $\vec{\mu} = (a, b, L)$ is vector of its parameters under $0 < a < 1$ and $a\, b > 1$ has no derivative but possesses by fractal dimension $D = 2 + \ln a / \ln b$ [5].

From function (11) it is easy to construct two-dimensional fractal surface:

$$W(x, y, \vec{\mu}) = W(x, \vec{\mu})\, W(y, \vec{\mu})\,. \tag{12}$$

On the other hand for parabolic equations existence and uniqueness theorems require that initial conditions (2) and (8) for the KPZ-equation (1) or auxiliary equation (7) must be quite smooth [8].

One can overcome this obstacle by means of some regularization of function (12) namely let us consider truncated Weierstrass function:

$$W_N(x,\vec{\mu}) = \sum_{n=1}^{N} a^n \cos\frac{\pi\, b^n x}{L}\,. \tag{13}$$

It is not difficult to check that direct product of functions (13) obeys to the next inequality:

$$|W(x, y, \vec{\mu}) - W_N(x, y, \vec{\mu})| \leq \frac{2\, a^{N+2}}{(1-a)^2}\,, \tag{14}$$

where

$$W_N(x, y, \vec{\mu}) = W_N(x, \vec{\mu})\, W_N(y, \vec{\mu})\,. \tag{15}$$

Thus inequality (14) means that for any fixed accuracy smooth function (15) can approximate fractal function (12) under quite large number N of terms in sum (13) hence regularized fractal function (15) can be chosen as initial condition (2) $h_0(x, y)$ for the KPZ-equation (1). The result of such approximation under $a = 0.5$, $b = 5.0$ and $N = 11$ is presented on Figure 2.

It is obvious that the above described approach can be generalized on unequal numbers of terms and unequal vectors of parameters under taking direct product on x and y of truncated Weierstrass functions (13).

4. Estimation of values connected with the KPZ-equation

In order to calculate numerically autocorrelation function (3) it is necessary to describe a procedure for statistical averaging of different functions of height $h(x, y, t)$.

The most trivial algorithm of averaging is the following one. At first one ought to solve the simple problem (9)-(10) on each rectangular subdomain $\overline{\Omega}_m$ for different realizations of source $Q_\alpha(x, y)$, $\alpha = 1, 2, \ldots, A$. After that average value of function f is equal to:

$$< f >= \frac{1}{A} \sum_{\alpha=1}^{A} f_\alpha \,, \tag{16}$$

where value f_α corresponds to realization $Q_\alpha(x, y)$.

Further spectral power density of roughness of the surface (4) it is convenient to estimate as discrete Fourier transform because of there are parallel algorithms to perform this operation [9].

The Shannon information for the surface (5) is additive function of domain Ω:

$$I(t) = \sum_{m=1}^{M} I_m(t) \,, \tag{17}$$

where

$$I_m(t) = \int_{\Omega_m} |h(x, y, t)| \log_2 |h(x, y, t)| \, dxdy \,. \tag{18}$$

To calculate integrals (18) on the basis of network functions φ_{ij}^k one ought to use different cubature formulae [10].

In addition, a triple autocovariation function can be calculated on individual lines as a characteristic of the non-Gaussianity of the process.

5. Conclusion

This article presents the algorithm of numerical solution of the two-dimensional KPZ-equation with a time-independent random surface source of particles and a regularized fractal initial condition adjusted to the concept of mathematical technology [6]. Such a reanimation of the concept of mathematical technology is currently extremely relevant in anticipation of the emergence of supercomputer exaflop performance [11].

This algorithm is expected to be realized on the Lomonosov supercomputer [12] using the MPI parallel programming technology and the ROOT object oriented data analysis framework [9].

The elaborated algorithm is very important for practice because of the study of the properties of solutions of the KPZ-equation is essential for understanding the processes of nanoengineering [13].

Acknowledgments

This work was supported by the Russian Foundation for Basic Research, grant number 18-08-01356-a.

References

[1] Gubinelli M and Perkowski N 2017 *Communications in Mathematical Physics* **349:1** 165
[2] Kardar M, Parisi G and Zhang Y C 1986 *Phys. Rev. Lett.* **56** 889
[3] Ishimaru A 1978 *Wave Propagation and Scattering in Random Media. Volume 2* (New York, San Francisco, London: Academic Press)
[4] Kulikov D A, Potapov A A, Rassadin A E and Stepanov A V 2017 *IOP Conf. Ser.: Mater. Sci. Eng.* **256** 012026
[5] Potapov A A, Gulyaev Yu V, Nikitov S A, Pakhomov A A and German V A 2008 *The Modern Methods of Image Processing* (Moscow: FIZMATLIT)
[6] Yanenko N N, Karnachuk V I and Konovalov A N 1977 *Numerical methods for mechanics of continuous media* **8** 129
[7] Yanenko N N 1967 *Method of fractional steps for solution of multidimensional problems of mathematical physics* (Novosibirsk: Nauka)
[8] Tikhonov A N and Samarskii A A 1966 *Equations of mathematical physics* (Moscow: Nauka)
[9] Bruna R and Rademakers F 1996 *Nucl. Inst. and Meth. in Phys. Res.* A **389** 81
[10] Sobolev S L 1974 *Introduction into the theory of cubature formulae* (Moscow: Nauka)
[11] Gergel V P and Linev A V 2012 *Vestnik of the UNN n.a. N I Lobachevsky* **3** 189
[12] Sadovnichy V, Tikhonravov A, Voevodin Vl and Opanasenko V 2013 *Contemporary High Performance Computing: From Petascale to Exascale* 283
[13] Rassadin A E, Sazanova T S, Stepanov A V and Fomin L A 2018 *IOP Conf. Ser.: Mater. Sci. and Eng.* **443** 012027

Journal of Physics: Conference Series **1392** (2019) 012060 doi:10.1088/1742-6596/1392/1/012060

Solution of two-phase nonisothermal fluid flow problem with nonlinear filtration law on heterogeneous computing systems

A V Tsepaev

Institute of Mechanics and Engineering, FRC Kazan Scientific Center, Russian Academy of Sciences, Lobachevskii ul. 2/31, Kazan, Tatarstan, 420111 Russia

E-mail: alex_tsepaev@mail.ru

Abstract. The work is devoted to the methods of solving two-phase fluid flow problems in the reservoirs. It is assumed that in domain with high velocities Darcy's law is violated and a nonlinear filtration law is used. To determine the temperature use the law of conservation of energy (first law of thermodynamics). The temperature of the fluid and of the skeleton are considered the same. The fluid viscosities are taken by the temperature functions. The methods for solving two-phase flow problems in porous medium with a nonlinear filtration law based on the decomposition methods are proposed. The proposed methods are implemented on the heterogeneous computing systems.

1. Introduction

Mathematical models of fluid flow in porous media are systems of coupled nonlinear unsteady partial differential equations [1], [2]. An important equation in these systems is the law of filtration - the dependence between fluid velocity and pressure. Most fluid flows are described by the Darcy's filtration law, which expresses a linear dependence between velocity and pressure gradient. A lot of works are devoted to the verification and study of the limits of the applicability of the Darcy's law. In the course of these studies, it is shown that there are two main reasons for the deviation from the Darcy's law. The Darcy's law is violated in cases when there are high flow velocities (the upper limit of the applicability of the Darcy's law) or too low flow velocities caused by the emergence of non-Newtonian rheological properties of the liquid [3]. In this paper we consider the first case related to the action of inertial forces.

In the development of oil and gas reservoirs the fluids contained in them can reach a temperature different from the reservoir temperature. The temperature change in the reservoir can occur due to the action of thermodynamic effects during the movement of fluids in a porous medium [3]: the barothermal effect, the effect of phase transformations, the injection into the reservoir of various displacers with a temperature different from the initial(cold or hot water, steam), the implementation of various thermochemical oxidation processes. In this paper, the model problem of hot water injection into the reservoir was considered. To determine the temperature used the law of conservation of energy (first law of thermodynamics). The temperature of the fluid and of the skeleton was considered to be the same. The viscosity of the fluid was taken as functions of temperature. The solution algorithm based on the decomposition method is proposed. The problem was solved using decomposition methods described in [4], [5] on the heterogeneous computing systems [6].

2. Problem statement

We will consider the three-dimensional confined reservoir D. In case of violation of the Darcy's law, the flow is subject to the nonlinear Forchheimer's law [3]: $\mathbf{q}_\alpha = -(A_\alpha + Bq_\alpha)^{-1}\text{grad } p$, where $\alpha = \{o, w\}$ is the index corresponding to oil or water, $A_\alpha = 1/K_\alpha$, $K_\alpha = kf_\alpha/\mu_\alpha$ is the phase permeabilities, f_α is the relative permeabilities, k is the absolute permeability, μ_α is the viscosities, q_α is the module of the fluid velocity vector, B is the constant of the porous medium. For the solution domain D introduce the following notations: $\bar{D} = \bar{D}_0 \cup (\bigcup\limits_{l=1}^{N} \bar{D}_l)$, $D_i \cap D_j = \emptyset$, $i \neq j$, $\bar{D}_0 \cap \bar{D}_l = \gamma_l$, where D_0 is the domain outside of the well region, D_l is the well region. It is required to determine the pressure p, the temperature T and the saturations S_α from the system of equations:

$$\text{div}(\mathbf{q}_w + \mathbf{q}_o), \tag{1}$$

$$\mathbf{q}_\alpha = -(A_\alpha + Bq_\alpha)^{-1}\text{grad } p \quad \text{in} \quad G_{l\alpha}, \quad l = 1, ..., N, \quad \alpha = \{o, w\}, \tag{2}$$

$$\mathbf{q}_\alpha = -K_\alpha \text{grad } p \quad \text{in} \quad D \backslash (\bigcup_{l=1}^{N} G_{l\alpha}), \quad \alpha = \{o, w\}, \tag{3}$$

$$\frac{\partial}{\partial t}(\sum_\alpha \rho_\alpha S_\alpha U_\alpha + (1-m)\rho_s C_s T) + \text{div}(\sum_\alpha \rho_\alpha \mathbf{q}_\alpha H_\alpha) - \text{div}(\kappa_T \text{grad } T) = 0, \tag{4}$$

$$\sum_\alpha S_\alpha = 1, \tag{5}$$

$$\text{div}(\mathbf{q}_w) + m\partial(S_w)/\partial t = 0. \tag{6}$$

Boundary conditions:

$$p = p_\Gamma \quad \text{on} \quad \Gamma_1, \quad \mathbf{q}_w = \mathbf{q}_{\Gamma n} \quad \text{on} \quad \Gamma_2, \quad p|_{\partial V_l} = P_l, \quad l = 1, ..., N, \tag{7}$$

$$T = T_{w_l}, \quad l = 1, ..., M, \quad T = T_\Gamma \quad \text{on} \quad \Gamma, \tag{8}$$

$$S = S_{w_l}, \quad l = 1, ..., M, \quad S = S_{w_\Gamma} \quad \text{on} \quad \Gamma_3. \tag{9}$$

Initial conditions:

$$T = T^0 \quad \text{in} \quad D, \tag{10}$$

$$S = S_w^0 \quad \text{in} \quad D. \tag{11}$$

Here ρ_α is the densities of phase α, $U_\alpha = C_\alpha T$ is the internal energies of phase α, C_α is the phases heat capacity, ρ_s the density of the skeleton, C_s is the heat capacity of the reservoir, $H_\alpha = U_\alpha + p/\rho_\alpha$ is the enthalpies of the phases, κ_T is the thermal conductivity of the reservoir, $\Gamma_1 + \Gamma_2 = \Gamma$ is the external boundary surface of the domain D, Γ_3 is the part of the surface Γ through which the liquid enters the reservoir, ∂V_l is the surface interval of opening of the l-th well, P_l is the set pressure on the well, p_Γ is the set pressure on the boundary Γ_1, $G_{l\alpha} = \{x \in D_l | Re(q_\alpha) > Re_{cr}\}$ is the domains with high flow rates, $\mathbf{q}_{\Gamma n}$ is the set value of the normal component of the fluid velocity at the boundary Γ_2, N is the number of the wells, M is the number of the injection wells ($M < N$), S_{wl} is the set value saturation on the injection wells.The boundaries of domains $G_{l\alpha}$ are not known in advance and must be determined in the decision process. In the problem statement $G_{l\alpha} \subset D_l$. The problem will be solved numerically with a given error ϵ on a grid that condenses to the well opening intervals. The reservoir D is represented by a multi-connected region, the inner surfaces of which are determined by the surfaces of wells in the intervals of reservoir opening.

3. The solution methods

At each time step the systems of equations to determine the pressure p, temperature T and the systems of equations to determine saturation S_α are solved separately. The nonlinear problem (1)-(4), (7), (8), (10) by definition the pressure p and the temperature T is solved iteratively. In the first iteration, the functions p^1 and T^1 are defined from the system of equations:

$$\mathrm{div}(\mathbf{q}_w^0 + \mathbf{q}_o^0),$$

$$\mathbf{q}_\alpha^0 = -K_\alpha \mathrm{grad}\, p^1,$$

$$\frac{\partial}{\partial t}(\sum_\alpha \rho_\alpha S_\alpha U_\alpha(T^1) + (1-m)\rho_s C_s T^1)) + \mathrm{div}(\sum_\alpha \rho_\alpha \mathbf{q}_\alpha^0 H_\alpha(p^1, T^1)) - \mathrm{div}(\kappa_T \mathrm{grad}\,(T^1)) = 0,$$

with the conditions (7), (8), (10). Next, the module of fluid velocity vector q_α^1 on each element is calculated by the formula $q_\alpha^1 = |\mathrm{K}_\alpha \mathrm{grad}\; p^1|$. Subdomains $G_{l\alpha}^1 \subset D_l$ are defined by the elements on which the condition $Re(q_\alpha^1) > Re_{cr}$ is satisfied. Let the functions p^i, q_α^i, T^i and the subdomains $G_{l\alpha}^i$ be known. The functions p^{i+1} and T^{i+1} are determined from the solution of the equations:

$$\mathrm{div}((F_o^i + F_w^i)\mathrm{grad}\, p^{i+1}) = 0, \quad \text{in} \quad D, \tag{12}$$

$$\mathbf{q}_\alpha^i = -F_\alpha^i \mathrm{grad}\, p^{i+1} \quad \text{in} \quad D, \quad \alpha = \{o, w\}, \tag{13}$$

$$\frac{\partial}{\partial t}(\sum_\alpha \rho_\alpha S_\alpha U_\alpha + (1-m)\rho_s C_s T^{i+1})) + \mathrm{div}(\sum_\alpha \rho_\alpha \mathbf{q}_\alpha^i H_\alpha) - \mathrm{div}(\kappa_T grad\; T^{i+1}) = 0, \tag{14}$$

where $U = U(T^{i+1})$, $H = H(p^{i+1}, T^{i+1})$, $\mathbf{q}_\alpha^i = -K_\alpha \mathrm{grad}\, p^{i+1}$ on the elements in the subdomains $D \backslash (\bigcup_{l=1}^{N} G_{l\alpha}^i)$ and $\mathbf{q}_\alpha^i = -(A_\alpha + Bq_\alpha^i)^{-1} \mathrm{grad}\; p^{i+1}$ on the elements in the subdomains $G_{l\alpha}^i$. The module of the flow velocity vector q_α^{i+1} is defined by equality $q_\alpha^{i+1} = |K_\alpha \; \mathrm{grad} p^{i+1}|$ in the domain $D \backslash (\bigcup_{l=1}^{N} G_{l\alpha}^i)$. On the elements of the grid of the domains $G_{l\alpha}^i$ the module of the fluid velocity is defined from the condition $|\mathrm{grad} p^{i+1}| = (A_\alpha + Bq_\alpha^{i+1})q_\alpha^{i+1}$. The subdomains $G_{l\alpha}^{i+1}$ are defined by the elements on which the condition $Re(q_\alpha^{i+1}) > Re_{cr}$. The system of equations (1)-(4), (7), (8), (10) to determine the pressure and temperature is solved, if the system (12) -(14) with the conditions (7), (8), (10) is solved with a given tolerance ϵ and $G_{l\alpha}^{i+1} = G_{l\alpha}^i$.

To determine the saturations S_α is solved a system consisting of equations (5), (6), (9), (11), in which the pressure p, the temperature T and the domains with high flow rates $G_{l\alpha}$ are considered constant values. Approximation of the system is carried out by the method of control volumes [7] with an implicit scheme of time "upflow".

Two different decomposition methods were used to solve the problem of two-phase fluid flow without taking into account high flow rates in [4], [5]. The decomposition method for solving difference equations for determining the pressure based on the independent solution of systems of algebraic equations for the fine grid segments in the subdomains D_l and a new type of coupling these solutions with the solution on the coarse grid. To take into account the high flow rates in the proposed decomposition method, each system for the fined grid segments must be solved by the algorithm described above. The decomposition method for determining the saturation field is based on a combination of elements of explicit and implicit schemes [8].

4. Numerical results

The proposed algorithm was tested in solving of two-phase flow problems with a different number of vertical producing and injection wells. The reservoir consisted of 10 layers (1 km×1 km×0,018 km) with the layers thicknesses $d_1 = 1$ m, $d_2 = 1$ m, $d_3 = 3$ m, $d_4 = 1$ m, $d_5 = 1$ m, $d_6 = 1$ m, $d_7 = 2$ m, $d_8 = 1$ m, $d_9 = 2$ m, $d_{10} = 5$ m and absolute permeability's $k_1 = 10^{-3}$ darcy, $k_2 = 10^{-2}$

darcy, $k_3 = 25 \times 10^{-3}$darcy, $k_4 = 10^{-2}$ darcy, $k_5 = 10^{-3}$ darcy, $k_6 = 10^{-2}$ darcy, $k_7 = 5 \times 10^{-2}$ darcy, $k_8 = 10^{-2}$ darcy, $k_9 = 10^{-3}$ darcy, $k_{10} = 15 \times 10^{-3}$ darcy respectively. On the top of the reservoir was considered impermeable, saturation on the side surface $S_w = 0$, saturation on the bottom surface $S_o = 0$. The initial saturation $S_o = 1$. The viscosity of the oil at a given temperature can be estimated by knowing the viscosity at two other temperatures $\mu_o = \mu_1 \exp[(1-T_1/T)(T_2/(T_2-T_1))\ln(\mu_2/\mu_1)]$. The water viscosity is given by the Bingham formula $\mu_w = \left(0.021482(T-8.435)+(80784+(T-8.435)^2)^{0.5}\right)^{-1}$. Oil density $\rho_o = 0.882\text{g/cm}^3$, water density $\rho_w = 1\text{g/cm}^3$. Specific heat capacity $C_w = 4.182kJ/(kgK)$, $C_o = 0.92kJ/(kgK)$. Thermal conductivity of the reservoir $\kappa_T = 1.7W/(mK)$, initial temperature $T^0 = 300K$. Each autopsy interval was simulated in a circular cylinder with base radius $r = 0.1$ m and is closed at the top and bottom spherical surfaces of radius $r = 0.1$ m. Thus, for each point on the surface of the opening intervals of the normal vector uniquely determined. Due to the lack of experimental data, the values of the constant B were chosen from the condition of proximity of the modules of flow velocity q_α, calculated using Darcy's laws and the Forchheimer's law in the transition from the linear to the nonlinear law. To calculate Reynolds numbers we used the formula [3] $Re(q) = 4\sqrt{2\kappa}q\rho/(m^{1.5}\mu)$, where $\kappa = k\mu/(\rho g)$, μ is the viscosity, ρ is the density, g is the acceleration of gravity, m is the porosity. For various porous media Re_{cr} according to [3] lies within 0.0085–3.4. Two types of solutions were considered: the solution of the system without domain decomposition and the solution with domain decomposition. As previously noted, the coefficient B was chosen in such a way that the velocities calculated using Darcy and Forchheimer laws were close at the boundary of the transition from the linear law to the nonlinear one. Numerical experiments were carried out in cases where these velocities differed by 1%, 0.1%, 0.01%. For each of these cases, table 1 shows the values of the coefficient B, at which numerical testing was carried out.

Table 1. Value of the coefficient B.

$\Delta q(\%)$	$Re_{cr} = 0.3$	$Re_{cr} = 0.01$
1%	$2.07 \cdot 10^{-5}$	$6.22 \cdot 10^{-4}$
0.1%	$2.05 \cdot 10^{-6}$	$6.17 \cdot 10^{-5}$
0.01%	$2.05 \cdot 10^{-7}$	$6.16 \cdot 10^{-6}$

In table 2 the time of solving problems with different number of wells by methods without decomposition and with decomposition with Reynolds number $Re_{cr} = 0.3$ is given. The results of the solution showed the superiority in time of the speed of the solution of decomposition methods in comparison with the methods without decomposition.

The proposed algorithm was tested on a cluster consisting of 4-core computing nodes with Intel Core i7 2600 processors and equipped with NVIDIA GTX 560 TI graphics accelerators when solving a model three-dimensional problem of three-phase fluid flow with vertical production and injection wells. Each well area contained about 25000 nodes. The total number of nodes for the gathering of 200 wells was 5×10^6. Systems of nonlinear equations for determining the saturations were solved by the implicit Newton method [9]. The algorithms are parallelized using MPI processes, OpenMP and CUDA technologies. We used C++ language and Visual Studio 2008 application development environment. When solving the problem, each MPI process is allocated an equal number of thickening portions of the grid. To solve problems in the thickening areas

Table 2. Solution time with decomposition and without decomposition.

Number of fine grid segments	Number of nodes	Solution time without decomposition (min)	Solution time with decomposition (min)
1	11744	7	6
50	365524	301	147
100	726721	722	523

corresponding to a single MPI process, streams are generated using OpenMP technology, which distribute these tasks to the processor cores and graphics devices. In this case, the tasks are distributed dynamically, that is, as they are solved. The problem was solved on two computing nodes with a total number of 8 cores (4 cores per node) and 2 GPU accelerators. The tasks were started in the mode of one MPI process per node, each MPI process corresponds to one GPU device. The following options for running tasks were considered: 1) On all available cores using a single MPI process and without using GPU accelerators. 2) On two cores using two MPI processes and two GPU devices. In this case, each MPI process had one GPU device and one core. 3) On all available cores using MPI processes and GPU devices. In this case, all available cluster resources(2 GPUs and 8 cores) were used.

The tables 3–4 shows the results of the solution. The comparison was made with the solution of the problem on one core without using GPU devices.

Table 3. Solution time acceleration without GPU devices.

Number of cores	Acceleration
1	1
2	1.3
3	1.9
4	2.3

Table 4. Solution time acceleration with GPU devices.

Number of GPU devices	1 MPI(1 core)	1 MPI(4 core)	2 MPI(2 core)	2 MPI(8 core)
1	17.3	20.2	–	–
2	–	–	30.1	36.3

The table 4 shows the acceleration of the problem solution time when GPU devices was used. The efficiency of use of GPU devices reaches 80%.

5. Conclusion

The methods to solve the two-phase nonisothermal fluid flow problem with fine grid segments based on the methods of decomposition are constructed. In the case of high flow velocities (the Reynolds number is greater than the critical value), the nonlinear Forchheimer law was applied. Algorithms for solving the problem on heterogeneous computing systems are constructed on the basis of the proposed decomposition methods. The high efficiency of using multiprocessor systems built on the basis of graphics processors is shown.

References

[1] Chen Z, Huan G and Ma Y 2006 *Computational methods for multiphase flows in porous media* (SIAM) p 569
[2] Aziz Kh and Settari 1982 *A Petroleum Reservoir Simulation* (Nedra, Moscow) p 407
[3] Basniev K S, Vlasov A M, Kochina I M and Maksimov V M 1986 *Podzemnaya gidravlika* (Moscow, Nedra) p 303
[4] Mazurov P A and Tsepaev A V 2006 *Vychislitel'nye metody i programmirovanie* **7** 115-23
[5] Tsepaev A V 2013 *Uchenye Zapiski Kazanskogo Universiteta. Seriya Fiziko-Matematicheskie Nauki* **155** 71-9
[6] Voevodin V V and Voevodin Vl V 2002 *Parallel'nye vychisleniya* (BHV-Petersburg) p 609
[7] Mazzia A and Putty M 2002 *Mixed-finite element and finite volume discretization for heavy brine simulations in groundwater* (J. of Computational and Applied Mathematics) 191-213
[8] Kalitkin N N 1978 *CHislennye metody* (Moscow, Nauka) p 512
[9] Samarskii A A and Gulin A V 1989 *Chislennye metody* (Moscow, Nauka) p 432
[10] Chao TungYangChih LinHuangCheng FangLin 2011 *Computer Physics Communications* **182** 266-9

Propagating large open quantum systems towards their asymptotic states: cluster implementation of the time-evolving block decimation scheme.

Valentin Volokitin[1], Ihor Vakulchyk[2,3], Evgeny Kozinov[1], Alexey Liniov[1], Iosif Meyerov[1], Michail Ivanchenko[1], Tatyana Laptyeva[1], Sergey Denisov[4]

[1]Lobachevsky State University of Nizhny Novgorod, Russia
[2]Institute for Basic Science, Daejeon, Korea
[3]Korea University of Science and Technology, Daejeon, Korea
[4]Oslo Metropolitan University, N-0130 Oslo, Norway

E-mail: `valentin.volokitin@itlab.unn.ru`

Abstract. Many-body quantum systems are subjected to the Curse of Dimensionality: The dimension of the Hilbert space $\mathcal{H}$, where these systems live in, grows exponentially with number of their components. However, with some systems, it is possible to escape the curse by using a low-rank tensor approximation known as "matrix-product state/operator (MPS/O) representation" in the quantum community and "tensor-train decomposition" among applied mathematicians. Motivated by recent advances in computational quantum physics, we consider chains of N spins coupled by nearest-neighbor interactions. The spins are subjected to an action coming from the environment. Spatially disordered interaction and environment-induced decoherence drive systems into non-trivial asymptotic states. The dissipative evolution is modeled with a Markovian master equation in the Lindblad form. By implementing the MPO technique and propagating system states with the time-evolving block decimation scheme, which allows keeping the length of the state descriptions fixed, it is in principle possible to reach the asymptotic states. We propose and realize a cluster implementation of this idea. The implementation on four nodes allowed us to resolve the asymptotic states of the model systems with $N = 128$ spins (total dimension of the Hilbert space $\dim\mathcal{H} = 2^{128} \approx 10^{39}$).

1. Introduction

Many-body systems are at the focus of the current research in theoretical and experimental quantum physics. In addition to their fundamental importance for quantum thermodynamics and information [1], these systems are perspective from the technological point of view; e.g., all manufactured (by now) quantum computers are based on arrays of interacting superconducting qubits [2].

All real-life quantum systems are open, meaning that they interact – to a different extent – with their environments [3]. This 'action from outside', termed "decoherence" or "dissipation", works together with the unitary evolution stemming from system's Hamiltonians and, on large time scales, these joint efforts result in the creation of an asymptotic stationary state. The evolution of an open quantum system towards its asymptotic states is usually modeled with a Markovian master equation, which describes the dynamics of the system density operator

Published under licence by IOP Publishing Ltd

$\varrho(t)$, $\dot{\varrho}(t) = \mathcal{L}\varrho(t)$ [3]. Formally, similar to the Schroedinger equation used to describe unitary evolution of an isolated quantum system, this is a linear differential equation which can be solved numerically, e.g., by diagonalizing generator of evolution $\mathcal{L}$.

However, computational studies of many-body quantum systems are limited by the so-called Course of Dimensionality: the total length L of description (number of parameters required to specify a state) of an isolated quantum system consisting of N components (spins, qubits, ions, etc.), each one with d degrees of freedom, scales as $L(N) \sim d^N$. To specify an *arbitrary* state of a system of 50 qubits one needs $2^{50} \approx 10^{15}$ complex-valued parameters. This exceeds the memory capacity of the supercomputer "Titan"[4]. In the case of open quantum systems, the complexity squares: to describe a density operator one needs $L(N) \sim d^{2N}$ real-valued parameters.

This is a famous problem in modern data science – manipulations (or even simply storing) with data tensors becomes impossible when the data are sorted in high-dimensional spaces. The attempts to break the curse led to the development of a variety of low-rank tensor approximation algorithms [5]. These algorithms are used now in signal processing, computer vision, data mining, and neuroscience [6]. The most robust algorithms are based on Singular Value Decomposition (SVD), and one particularly efficient for multilinear algebra manipulations is the so-called Tensor-Train (TT) decomposition [7]. In physical literature, it is commonly referred to as Matrix Product State (MPS) [or Matrix Product Operator (MPO)] representation [8]. While these two names are used simultaneously (though in different fields), the underlying mathematical structure is the same [9]. The MPS/MPO/TT approach allows to reduce descriptions of *some* many-body states to a linear scaling $L(N) \sim N$[7].

The MPS/MPO representation allows for effective propagation of quantum many-body systems in time by using the so-called Time-Evolving Block Decimation (TEBD) scheme [10]. In short, this is a procedure to reduce the description of the state, obtained after every propagation step, to a given fixed length L_{cut}. The accuracy of the propagation is controllable through L_{cut}: If the information is thrown out after the restriction is substantial, the used TEBD propagation is bad and leads to a wrong description. Otherwise, it is good. Some many-body systems 'behave' well during the TEBD propagation and so the amount of the neglected information is tolerable (we are not going to discuss physical properties underlying such a 'good behavior' and refer the reader to an extensive literature on the subject; see. e.g., Ref. [8]). Important is that the MPO/TT-TEBD scheme can be used to propagate open systems [11] and thus get in touch with the corresponding asymptotic states [12, 13]. It is crucial therefore to estimate computational resources needed for the realization of this program. Here we report the results of our studies in these directions.

2. The algorithm

2.1. Tensor-Train Decomposition and Time-Evoling Block Decimation Propagation

Here we mainly follow works [7] and [8]; for more details, we refer the interested reader to them.

We start with the TT representation of a N-dimensional complex-valued tensor $A^{i_1,i_2\ldots i_N}$ with $i_k = 1, 2, \ldots M$ [7],

$$A\left[i_1, i_2 \ldots i_N\right] = \sum_{\alpha_1 \ldots \alpha_{N-1}} \Gamma^{[1]i_1}_{\alpha_0,\alpha_1} \lambda^{[1]}_{\alpha_1} \Gamma^{[2]i_2}_{\alpha_1,\alpha_2} \lambda^{[2]}_{\alpha_2} \ldots \Gamma^{[N-2]i_{N-1}}_{\alpha_{N-2},\alpha_{N-1}} \lambda^{[N-1]}_{\alpha_{N-1}} \Gamma^{[N]i_N}_{\alpha_{N-1},\alpha_N}. \tag{1}$$

One may interpret this structure as a "train" of Γ's that encode local structure in each dimension, and λ's that quantify correlations between them. Each $\Gamma^{[k]}$ is an array of M matrices $r_{k-1} \times r_k$ with restrictions $r_j \leq M \max(r_{j-1}, r_{j+1})$ with boundary conditions $r_0 = r_N = 1$. Thus, the dimensions of the matrices are $1 \times M$, $M \times M^2$, $M^2 \times M^3 \ldots M^2 \times M$, $M \times 1$, which corresponds to the full representation with M^N complex parameters. The construction of the TT representation invlove SVD operations (their number is proportional to the number

Journal of Physics: Conference Series **1392** (2019) 012061 doi:10.1088/1742-6596/1392/1/012061

Algorithm 1 : TEBD method implementation

1: **upload**: system & method parameters (N, $T_j[i_j, i'_j]$, $T_{j,j+1}[i_j, i'_j, i_{j+1}, i'_{j+1}]$, dt, $T_{\max}$, R), initial state ($\Gamma^{[j]i_j}_{\alpha_{j-1}\alpha_j}$, $\lambda^{[j]}_{\alpha_j}$)
2: **for** $t = 0$ **to** $T_{\max}$ **do**
3: propagate all $\Gamma^{[j]i_j}_{\alpha_{j-1}\alpha_j}$ on $[t; t + dt/2]$
4: propagate $\Gamma^{[j]i_j}_{\alpha_{j-1}\alpha_j}$, $\lambda^{[j]}_{\alpha_j}$, $\Gamma^{[j+1]i_{j+1}}_{\alpha_j\alpha_{j+1}}$ with odd j on $[t; t + dt/2]$ ▷ 4-6 do with hard cutoff of the local bond dimension
5: propagate $\Gamma^{[j]i_j}_{\alpha_{j-1}\alpha_j}$, $\lambda^{[j]}_{\alpha_j}$, $\Gamma^{[j+1]i_{j+1}}_{\alpha_j\alpha_{j+1}}$ with even j on $[t; t + dt]$
6: propagate $\Gamma^{[j]i_j}_{\alpha_{j-1}\alpha_j}$, $\lambda^{[j]}_{\alpha_j}$, $\Gamma^{[j+1]i_{j+1}}_{\alpha_j\alpha_{j+1}}$ with odd j on $[t + dt/2; t + dt]$
7: propagate all $\Gamma^{[j]i_j}_{\alpha_{j-1}\alpha_j}$ on $[t + dt/2; t + dt]$
8: **end for**
9: **save** results
10: **release** memory

of the system components). When SVD is performed, one can keep only certain singular values based on the approximation criterion. One possibility is to discard all values smaller than a fixed number. An alternative approach is to introduce a so-called **bond dimension** R, a cut-off value such that on each bound i only singular values $\lambda^{[i]}_j$, $j \leq R$, are kept and the rest are truncated. We use the latter option. Each such local approximation on the set of singular value $\{\lambda^{[i]}\}$ introduces a truncation error, $E_i(R) = \sum_{j>R} \left(\lambda^{[i]}_j\right)^2$.

The TT representation provides a basis for an approximate tensor propagation algorithms. Here we use Time-Evolving Block Decimation (TEBD) scheme [14, 11], which was specifically designed for quantum systems but applicable also in the general case. Consider a tensor flow governed by an evolution generator consisting only of operations acting on one or two adjacent dimensions,

$$\frac{d}{dt}A[i_1 \dots i_N] = \sum_j \sum_{i_j} T^{[1]j}_{i_j} A[i_1 \dots i_j \dots i_N] + \sum_{j_1,j_2} \sum_{i_{j_1},i_{j_2}} T^{[2]j}_{i_{j_1},i_{j_2}} A[i_1 \dots i_{j_1} i_{j_2} \dots i_N], \quad (2)$$

where operations $T^{[1]}$ and $T^{[2]}$ act only on the i-th component and a pair of components, respectively. We use standard time discretization to iteratively integrate this equation (starting from some initial tensor). In terms of operations the solution reads

$$A(t + dt) = L(dt)A(t) = = \exp\left[\left(\sum_j \hat{T}^{[1]j} + \hat{T}^{[2]j}\right) dt\right] A(t). \quad (3)$$

As T operators generally do not commute, we have to approximate the matrix exponents. To minimize the error, it is convinient to separate the operators into groups as large as possible and such that all the operators belonging to one group commute with each other. All one-dimension operators commute by default, and two-dimension acting on odd/even pairs commute within their oddity groups. We use modified second order Suzuki-Trotter decomposition [8]. As all operators are commuting by construction, corresponding computation can be parallelized. Each two-index operator may include a cut-off if after the reorthogonalizationm, the number of singular value exceeds bound dimension R. Corresponding accumulated truncation error $E(t, R)$ is then calculated as a sum of local errors over all the operation during evolution up to time t.

Computations are dominated by SVD, so resulting complexity is $\mathcal{O}(NR^3)$, where R is the bond dimension. With $\mathcal{O}(N)$ cores available, it becomes $\mathcal{O}(R^3)$ and thus the computational task is perfectly scalable.

2.2. Lindblad Equation

We apply both TT and TEBD methods to evolve numerically many-body open quantum models. The state of such systems is described by a density matrix $\varrho(t)$ of the size $M^N \times M^N$, where N is the number of particles/spins and M is number of the local states, which we put to $M = 2$ for a 1/2-spins that we consider in the paper. Evolution of a quantum system in contact with the environment is governed by Lindblad equation [3]

$$\dot{\varrho}(t) = \mathcal{L}\varrho(t) = \mathcal{L}_H\varrho(t) + \mathcal{L}_{\text{dis}}\varrho(t) = -i\,[H, \varrho(t)] + \sum_{s=1}^{M} \gamma_s \left[D_s\varrho(t)D_s^\dagger - \frac{1}{2}\{D_s^\dagger D_s, \varrho(t)\} \right], \quad (4)$$

where $\mathcal{L}$ is the Lindblad superoperator consisting of conservative $\mathcal{L}_H$ and dissipative $\mathcal{L}_{\text{dis}}$ parts, H is Hamiltonian, D_s are dissipation operators, and γ_s are corresponding dissipation rates. There is a stationary state solution for any Lindblad superoperator $\mathcal{L}\varrho(\infty) = 0$ which is unique (aside of special cases of symmetries which we do not address here). Finally, many-body density operator ϱ can be represented as an $2N$-dimensional tensor $\varrho\,[i_1, i_2 \dots i_N; i'_1, i'_2 \dots i'_N]$ where every pair of indexes i_j, i'_j (each one runs from 1 to 2) correspond to the j-th qubit/spin.

3. Implementation

The method described in Section II is implemented as shown in Algorithm 1. The algorithm is implemented using the C++ programming language. We found, that the matrix operations (mainly SVD) are the most time-consuming parts of the algorithm. In this regard, we employ the Armadillo software library integrated with highly optimized mathematical routines from the Intel Math Kernel Library to improve performance. Finally, Armadillo/MKL routines take about 50-80% of computation time during the propagation step depending on the current system state.

The algorithm assumes performing a set of integration operations for individual components of the system at every time step. These operations are not independent but can be ordered according to their dependencies for the organization of parallel computations. In particular, all one- and two-particle interactions can be performed completely in parallel.

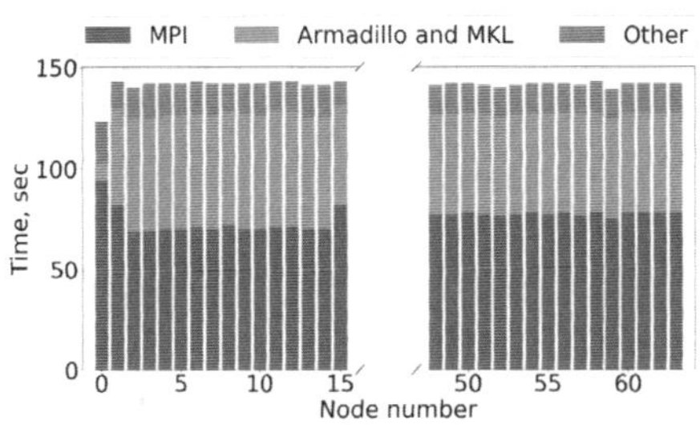

Figure 1. Distribution of computational and communication functions run time. 64 MPI-processes were executed on four nodes of the cluster.

The cluster parallelization is done by using the MPI technology. We apply the classic master-worker scheme for parallelization of the algorithm. For that, the single managing MPI-process (master) forms separate tasks for one– and two–particle interactions, monitors their dependencies from each other and readiness, distributes tasks to all other processes (workers) and accumulates the results.

All computational experiments have been done on the Lobachevsky cluster with a 2×8-core Intel Xeon CPU E5-2660, 2.20GHz, 64 GB RAM, Infiniband QDR interconnect. The code was compiled with the Intel C++ Compiler, Intel Math Kernel Library and Intel MPI from the Intel Parallel Studio XE suite of development tools and the Armadillo library.

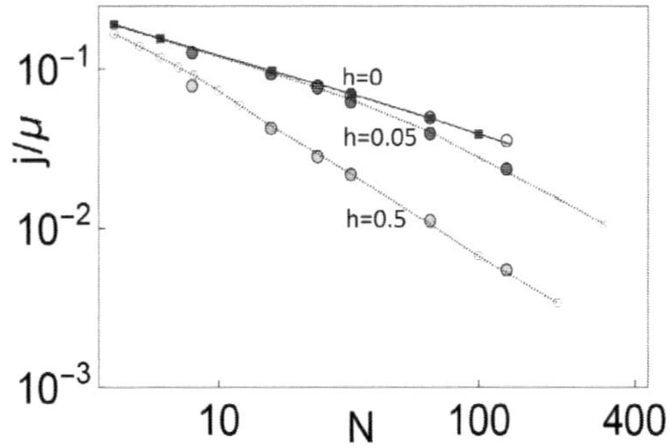

Figure 2. Scaling of the spin current j through a disordered spin chain with N spins for different values of disorder strength h. Our results (big colored circles) are plotted on top of the results reported in Ref. [12]). The maximal size of the model system used in our simulations is $N = 128$. For every set of parameters, we performed averaging over 20 disorder realizations. The propagation time step $dt = 0.1$ and bond dimension $R = 50$.

4. Results

As test-beds we used two models of open spin chains with next-neigbor couplings [12, 13].

We integrate model from Ref. [12] with following parameters: $N = 128$, $R = 50$, $T_{\max} = 50$, $dt = 0.1$. Parallel code was run on four computational nodes of the cluster (1 MPI-process per CPU core, 64 MPI-processes overall). Total computation time was 143 s. The resulted diagram for the distribution of computational and communication functions run time is presented in Fig. 1. It is shown that the calculations are fairly well balanced, which is an undoubted advantage of the parallelization scheme. However, MPI communications take a significant part of the computation time, while further increasing the number of cluster nodes used will not significantly speed up the calculations, which is a limitation of the scheme. Computational efficiency (ratio of computation time to total execution time) was 47%.

We find that it is possible to reproduce - with high accuracy – the results reported in Ref. [12] by using bond-dimension $R = 50$. On Fig. 2 we present a comparison of the results of the sampling we perform with our code (big circles; yellow, red and green) with the results by Žnidarič and his co-authors. We use propagation step $dt = 0.1$ and propagate every system up to $t = 10^4$, irrespectively of its size. For every value of N and disorder strength h, we

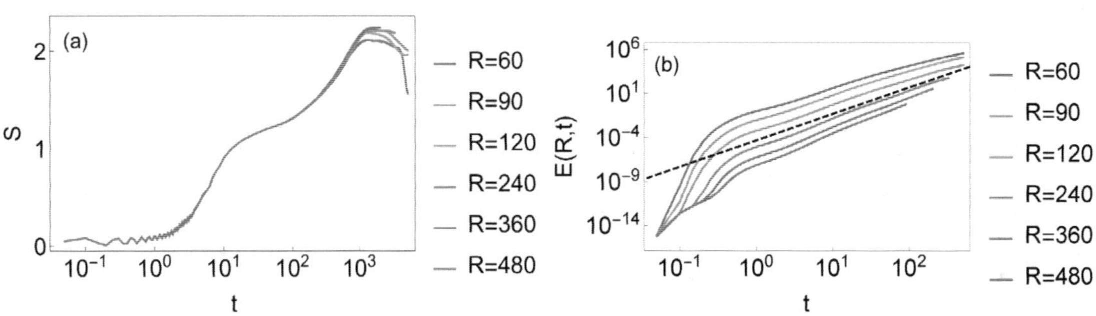

Figure 3. (a) Evolution of the operator entanglement entropy S for a single disorder realization for the model from Ref. [13], for different values of bond dimension R. The propagation time step $dt = 0.1$ and the system size is $N = 128$. Note that for $R = 480$ we did not reach the asymptotic 'plateau' because it was not possible to numerically propagate system further (we hit the two-week limit). (b) Increase of the accumulated truncation error in time.

additionally performed averaging for 20 disorder realizations. Each realization took from 2 minutes to 2 hours depending on the size of the system and involved up to two nodes (for large system sizes, $N > 64$).

For the model from Ref. [13] we study the dynamics of the operator entanglement entropy (for a fixed disorder realization) for different bond dimensions as shown on Fig. 3a. We found that $R_c = 360$ constitutes a threshold value after which the asymptotic entropy does not change upon further increase of the bond dimension. The calculation time for this value of bond dimension was four weeks of continuous propagation on four computing nodes. We also analyze also the evolution of accumulated error $E(R, t)$ in this case. The operator entropy, which signals the arrival to the asymptotic state, is not accompanied by the saturation of the error. The latter continues to grow in a power-law manner, see the dashed line in Fig. 3b. This means that MPO states – even with $R = 480$ – are still different from the genuine asymptotic state of the model (which is the zero-value eigenelement of the corresponding Lindbladian).

5. Conclusions

We presented a parallel implementation of the MPO-TEBD algorithm to propagate many-body open quantum systems. Parallelization is performed using the MPI technology and employs the master-worker scheme for computational tasks distribution. High-performance implementations of linear algebra from the Intel MKL were used to better utilize computational resources of modern hardware.

The performance tests on the Lobachevsky cluster demonstrated that 64 MPI processes running on four computational nodes is the optimal configuration for the model systems with $N = 128$ spins. As a next step, we plan to explore the possibility of further improvements of the parallelization by reducing the communications and increasing the efficiency of using computational resources. After that, we hope to reach the limit $N \simeq 400$ with the test-bed models.

6. Acknowledgments

The authors acknowledge support of the Russian Foundation for Basic Research and the Government of the Nizhni Novgorod region of the Russian Federation, grant # 18-41-520004. IV acknowledges support by the Institute for Basic Science, Project Code (IBS-R024-D1), and by the Korea University of Science and Technology Overseas Training program.

References

[1] J. Eisert, M. Friesdorf, and C. Gogolin, Nature Phys. **11**, 124 (2015).
[2] J. M. Gambetta, J. M. Chow, and M. Steffen, npj Quant. Inf. **3**, 2 (2017).
[3] Breuer, H.-P., and F. Petruccione, *The Theory of Open Quantum Systems* (Oxford University Press, Oxford, 2002).
[4] https://en.wikipedia.org/wiki/Titan (supercomputer)
[5] T. G. Kolda and B. W. Bader, SIAM Rev. **51**, 455 (2009).
[6] A. Cichocki, *Era of Big Data processing: a new approach via Tensor Networks and Tensor Decompositions*, https://arxiv.org/abs/1403.2048.
[7] I. V. Oseledets, SIAM J. Sci. Comput. **33**, 2295 (2011).
[8] U. Schollwoeck, Ann. of Phys. **326**(1) (2010).
[9] J. Haegeman, C. Lubich, I. Oseledets, B. Vandereycken, and F. Verstraete, Phys. Rev. B **94**, 165116 (2016).
[10] G. Vidal, Phys. Rev. Lett. **91**, 147902 (2003).
[11] M. Zwolak and G. Vidal, Phys. Rev. Lett. **93**, 207205 (2004).
[12] M. Žnidarič, A. Scardicchio, and V. K. Varma, Phys. Rev. Lett. **117**, 040601 (2016).
[13] I. Vakulchyk, I. Yusipov, M. Ivanchenko, S. Flach, and S. Denisov, Phys. Rev. B **98**, 020202(R).
[14] G. Vidal, Phys. Rev. Lett. **91**(14) (2003)
[15] T. Prosen and I. Pižorn, Phys. Rev. A. **76**, 032316 (2007).

Journal of Physics: Conference Series **1392** (2019) 012062 doi:10.1088/1742-6596/1392/1/012062

High Throughput Calculations as an Elevator on the Way from Chemical Structure to Novel Materials

I.D. Yushina, G.I. Makarov, Y.V. Matveychuk, E.V. Bartashevich

Nanotechnology Research & Education Centre, South Ural State University, Chelyabinsk, Russia

E-mail: `iushinaid@susu.ru`

Abstract. Development of new materials and drugs presently requires trial chemical methods that are time consuming and that depends on the chance more than we would have liked. We aim to describe, how the high throughput calculations can be applied for computational predictions of the organic crystalline structure and its properties. In this work we tried to illustrate the examples of techniques that were applied to accelerate the design of iodine-contained organic materials with nonlinear optical properties, to explain the plasticity of maleate amino acids crystals by means of analysis of hydrogen bonds orientation, to clarify the biological activity of traditional antibiotics using molecular dynamics modeling

1. Introduction

Nowadays, the high throughput calculations have penetrated and infiltrated all main sections of chemistry. Advances in the field of physical chemistry are unthinkable without solving the problems of structure and properties predicting, such as optical, elastic, electrical, thermodynamic, thermochemical properties. In turn, the properties predictions are based on virtual models of the molecules, crystals, polymers and solids, as well as on our understanding how fragments of multicomponent systems are bound to each other.

Nowadays due to the diffusion of supercomputer technologies in our current knowledge, the new disciplines are born at the close junctions of various fundamental sciences. At least two prominent examples can confirm this fact: the formation of quantum crystallography [1, 2] and the relevance of skills that chemoinformatics [3] forms. Quantum crystallography is a research area exploiting the fact that parameters of quantum-mechanically valid electronic model of a crystal can be derived from the accurately measured set of X–ray coherent diffraction structure factors [1]. On the one hand, the chemistry describes how atoms and molecules interact via chemical bonds to form new chemical compounds; on the other hand, the informatics involves the practice of information processing and the engineering of information systems. Thus, the chemoinformatics is aimed to reproduce the data into information and information into knowledge for making the better and faster decisions in material science and drug design.

The purpose of our present work is to illustrate two important aspects: 1) how the understanding the nature of chemical bonds helps us in the structure modeling and the properties predictions; 2) how the digital tools of computational chemistry inextricably linked with the modern approaches of the drug design.

Pursuing the goals, firstly, we consider how the concept of halogen bonding and the methods of evolutionary algorithms [4] could be applied to accelerate design of organic materials with nonlinear optical properties. Secondly, we describe an example in which quantitative analysis of the hydrogen bonds orientation helped to explain the plasticity of amino acid maleates crystals. Thirdly, we touch the case study based on molecular dynamics modeling that returned the attention and interest to the traditional antibiotics.

2. Computational design strategy for NLO properties of iodine–containing organic crystals

One of our specific tasks was to develop and deploy methodology for modeling of novel materials based on the modification of the molecular and crystal structure of iodine–derived compounds bound by the iodine–iodine halogen bonds [5, 6] and the other electrostatically driven interactions. Our interest to iodine–containing materials is justified by the high polarizability of iodine atoms and their ability to build the multiple non–covalent interactions, forming infinite chains and nets in crystals.

Recently the computational design strategy for iodine–containing organic crystals was proposed [7]. In general, the strategy is based on lattice energy minimization for the crystal packing predictions using evolutionary algorithms implemented in USPEX program [8]. The protocol of structure and properties prediction for iodine–containing organic crystals [9] is illustrated on Fig. 1A

Suggested protocol allowed us to predict the unknown polymorphs of iodine-derived benzenes, with the higher efficiency of second harmonic generation. We obtained the justification for NLO properties impact on hyperpolarizabilities of crystals with iodine halogen bonding [10]. Next, we illustrate the results of our modeling and NLO properties predictions for 2–iodo–3–hydroxypyridin, which crystalline structure was described earlier [11]. Here we deepen our understanding of the role of halogen and hydrogen bonding as well as focus on several computational methodology observations in wider context.

Periodic Kohn-Sham calculations were carried out in CRYSTAL14 [12] with PBE0 exchange–correlation functional with D2 dispersion correction and localized atomic basis sets. The vibrational frequencies were computed via coupled perturbed Kohn–Sham analytical approach

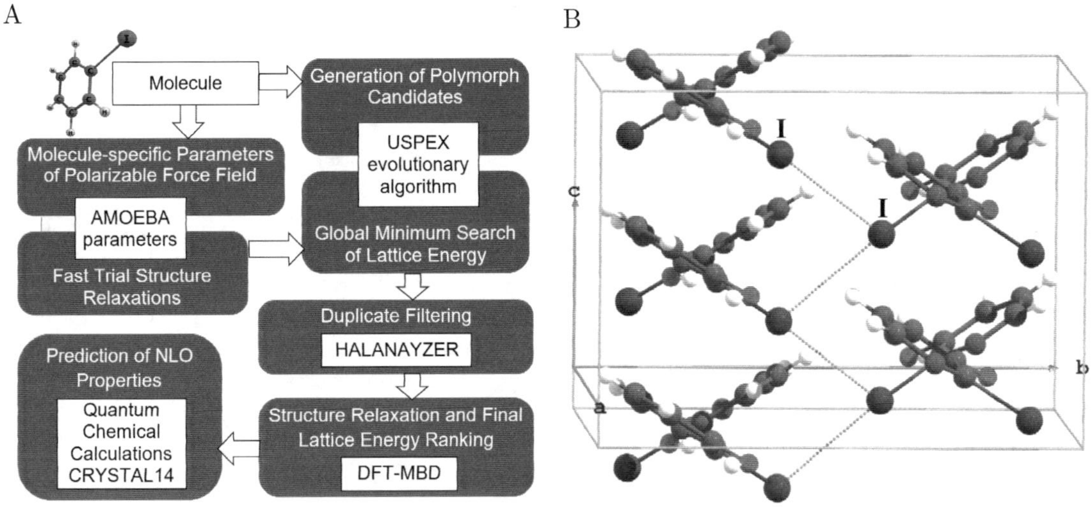

Figure 1. A: The protocol of structure and properties prediction for iodine–containing organic crystals. B: 2–Iodo–3–hydroxypyridine molecules interact via L–shaped I. . . I halogen bonds.

at G point of the Brillouin zone in the harmonic approximation in order to control localization of optimized experimental and generated structures in local energy minima.

Crystal structure of 2–iodo–3–hydroxypyridin has two significant structure–determining motives: the N...H—O hydrogen bond that organizes molecules in anti–parallel way and is directed mainly in ***a*** crystallographic axis direction or the I...I halogen bond that forms L–shaped zigzags along ***c*** crystallographic axis direction (Fig. 1B).

Thus, the main task of crystal packing prediction and reproduction in this case was to preserve these structural features.

Besides analysis of sustainability of main packing motives, it is essential to control space groups that are obtained after generation of modelled structures and their optimization. The study of second order NLO properties, which are commonly expressed as the ability for second harmonic generation, one usually takes into consideration only noncentrosymmetric space groups in which second order hyperpolarizabilities β have nonzero values.

As we have previously showed [10], we succeeded in reproducing the crystal packing of experimental 2–iodo–3–hydroxypyridin structure and obtained values of $\chi(2)_Z$ — the only nonzero direction in tensor of second order susceptibilities $\chi(2)$ according to symmetry restrictions of $Pna2_1$ space group. Moreover, due to the usage of generic algorithms is was possible to obtain model structure in Cc space group that due to absence of symmetry limitations has also nonzero component $\chi(2)_X$ in addition to $\chi(2)_Z$, thus resulting higher overall values of NLO activity.

Combination of hydrogen and halogen bonding features allows analyzing the influence of both of them on NLO properties.

The effect of hydrogen bonding on hyperpolarizability values is mostly manifested in the direction of ***a*** crystallographic axis. Only in a direction there is a clear trend: the shorter is ***a*** axis value, the shorter is N...H—O distance and the higher is the value of total hyperpolarizability. Other crystallographic directions are less sensitive to the changes of N...H—O distance, although the values of hyperpolarizability change drastically.

Thus, in case of 2–iodo–3–hydroxypyridine hydrogen bonding has greater impact on the stabilization of experimental and generated structures while halogen bonding is more significant for the emergence and increasing of NLO properties.

3. Effect of the H–bonds orientation the features of crystal structure and elastic properties

The numerical description of crystal mechanical properties was offered for a long time in the form of stiffness and compliance fourth rank tensors. However, the solutions describing the anisotropy of crystalline materials and following directly from these tensors were represented only during last decade. These are general elastic-anisotropy index A* [13], universal elastic anisotropy index A^U [14], log–Euclidean anisotropy index A^L [15]. The influence of the anisotropy of intermolecular interactions on crystal mechanical properties was studied by Desiraju et al. [16]. However, the complex comparison of H–bonds orientation with elastic moduli anisotropy in combination with A^U calculation was presented in our previous work [17]. In this approach the quantitative analysis of H–bonds orientation and stiffness tensor anisotropy was made for L–alanine crystal (**A**) [18], L–isoleucinium hydrogen maleate hemihydrate (**IL**) and L–leucinium hydrogen maleate (**L**) [19]. The structure relaxation and the stiffness tensor calculation were carried out on B3LYP/6–31G(d,p) level of theory. The universal elastic anisotropy indexes A^U and the spatial dependences of elastic moduli can be calculated using ELATE software [20].

The anisotropy of Young's and shear moduli is approximately identical for the **IL** and **A** crystals, as well as the differences of minimum and maximum values of shear modulus. The corresponding values for **L** crystal are qualitatively bigger. The calculations of A^U have quantitatively confirmed this fact. The values of the universal elastic anisotropy index are

1.38 (**IL**), 3.76 (**L**) and 2.36 (**A**). As the A^U value is one of key indicators of mechanical flexibility/brittleness of a crystal, we can assume that the **L** crystal can show flexibility in some directions, but other samples do not demonstrate such property. This assumption has been confirmed experimentally for **L** and **IL** earlier [19].

The comparison of the stiffness tensor properties with the structural features of these crystals has allowed establishing the main reason of differences in mechanical properties. The main structure–forming interactions in crystals of amino acids and their salts are intermolecular H–bonds.

The lack of molecular layers in **A** crystals assumes the small probability of the elastic properties manifestation caused mainly by the shift of these layers. The calculated anisotropy of elastic moduli of **A** crystals is generally small and corresponds to such assumption. The different situation is observed with layered **L** and **IL** crystals. The smallest resistance to shift in these crystals is observed in the direction parallel to layers, what is caused by lack of strong interactions between layers. However, in **IL** crystals the possibility of layers shift is complicated by action of other factors, mainly steric. That is, the layered structure is not a sufficient condition for possibility of crystal flexibility manifestation in certain directions. The calculation of stiffness tensor in this case gives the chance to reduce the series of the crystals capable to have such unique properties.

4. Molecular dynamics simulation method in the study of bacterial ribosome function

The modeling of biopolymers by the method of classical molecular dynamics (MD) enables to investigate their conformational mobility, to estimate the stability of intramolecular non–covalent interactions, for example, hydrogen bonds, and to describe their interaction with low–molecular ligands, ions and water molecules. The MD method consists of the evolution of molecular systems during time; in the method of classical MD, the movements of molecules is described by classical mechanics model, so that molecular systems are represented as a set of material points interacting with each other. The potential energy of the system is given by the sum of the potentials describing some or other covalent and non–covalent interactions. We use GROMACS [21] software to perform MD simulations. Biopolymers were modeled with the parm99sb force field [22], while small ligands are modeled with the GAFF [23].

An important example of the application of classical MD is the simulations of a bacterial ribosome [24]. MD simulations enables to investigate consistent conformational alterations in ribosomal RNA [25], which is especially important in the context of multiple evidence of the developed allosteric relationships regulating the function of the ribosome [26]. Methods of MD enable to simulate the interaction of antibiotics with a ribosome, explaining the differences in their activity [27] and the mechanism of their action.

The ribosome is a very large ribonucleoprotein complex, so that its simulation requires a huge expenditure of computational resources. However, when studying the binding of low molecular ligands to the ribosome, it is possible to model a sufficiently large fragment of the ribosome, completely including the binding site and its nearest vicinity. Such an approach preserves the local conformational mobility of biopolymers and excludes the global one, allowing to considerably economize required computational resources. We use cubic fragment of core area of large subunit of *E. coli* ribosome. This fragment includes all amino acid and nucleotide residues were set apart, as well as K^+ and Mg^{2+} in which at least one atom fell into a cubic area with 7 nm edge including the entire nascent peptide exit tunnel and peptidyl transferase center in such a way that the center of this region was located at the nascent peptide exit tunnel aligned along an imaginary applicate axis. During simulation this area is centered in a cubic cell with 8.8 nm edge filled with water molecules so that the system edges were covered by a 0.9 nm thick solvent layer.

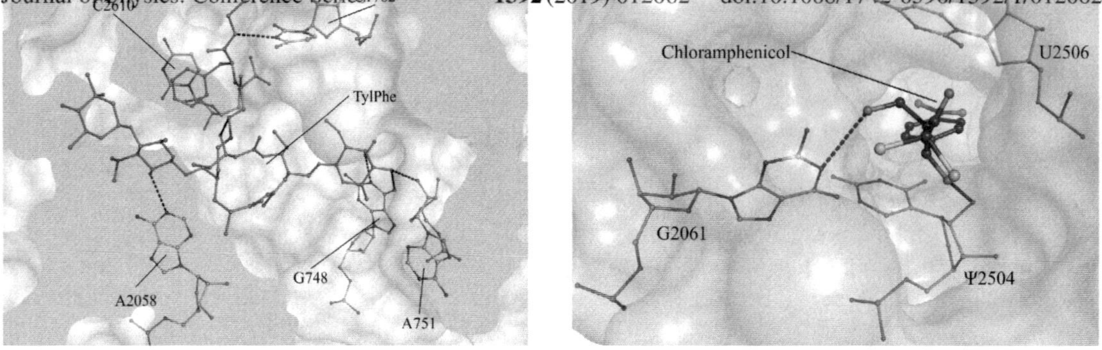

Figure 2. A: Conformation of TylPhe phenylalanyl substituent modeled by MD methods; the hydrogen bonds are shown by dashes. B: Interaction between chloramphenicol and *E. coli* ribosome, staying in canonical A,A/P,P–state, modeled by MD methods. Hydrogen bond is shown by black dashes.

In our previous work [28] we have modeled the structure of the complex of bacterial ribosome with TylPhe antibiotic obtained from a tylosin. This compound is superior in activity to the tylosin, despite the fact that it is unable to covalently bind to 23S rRNA. We established the conformation of the phenylalanyl substituent, which explains this effect (Fig. 2A) . Moreover, the molecular dynamics method allows the structure of antibiotic complexes to be simulated with a certain functional state of the ribosome: for example, recently we modeled the structure of the chloramphenicol complex with the E.coli ribosome stalled in the canonical A,A/P,P–state, which explains the results of biochemical studies the binding of chloramphenicol [29] (Fig. 2B) .

5. Conclusion

Thus, in this way we have illustrated how the global challenges in material science and drug design can be solved with the help of high throughput calculations and supercomputer technologies applied as an elevator in modern structural chemistry. We have demonstrated the examples how the following problems can be solved: to expand the possibilities in the organic crystal structure modeling; to deepen the understanding of the chemical bonds and non-covalent interactions nature; to develop methods for the nonlinear optical and elastic properties predictions. Such studies based on the time-consuming calculations [30]are of direct importance for the development and improvement of novel materials and antibacterial drugs necessary to maintain the modern quality of life.

Acknowledgments

This work was supported by the Ministry of Education and Science of the Russian Federation, grant 4.1157.2017/4.6, and by the Government of the Russian Federation, Act 211, contract No. 02.A03.21.0011. Quantum–chemical calculations have been performed using computational possibilities of TORNADO supercomputer, South Ural State University.

References

[1] Tsirelson V 2018 *Journal of Computational Chemistry* **39** 1029–1037

[2] Massa L, Huang L and Karle J 1995 *International Journal of Quantum Chemistry* **56** 371–384

[3] Madzhidov T, Baskin I, Antipin I and Varnek A 2015 *Introduction to chemoinformatics (series of benefits)* vol 10–2 (Kazan) pp 198–200

[4] Zhu Q, Oganov A R, Glass C W and Stokes H T 2012 *Acta Crystallographica Section B* **68** 215–226

[5] Desiraju G, Ho P, Kloo L, Legon A, Marquardt R, Metrangolo P, Politzer P, Resnati G and Rissanen K 2013 *Pure Appl. Chem.* **85** 1711–1713

[6] Bartashevich E and Tsirelson V 2014 *Russian Chemical Reviews* **83** 1181–1203

[7] Glass C W, Oganov A R and Hansen N 2006 *Computer Physics Communications* **175** 713 – 720

[8] Masunov A E, Torres K, Dyakov A A, Yushina I D and Bartashevich E V 2018 *The Journal of Physical Chemistry C* **122** 22622–22631

[9] Tkatchenko A, DiStasio R A, Car R and Scheffler M 2012 *Phys. Rev. Lett.* **108** 236402

[10] Yushina I D, Masunov A E, Lopez D, Dyakov A A and Bartashevich E V 2018 *Crystal Growth & Design* **18** 5069–5079

[11] Saha B K, Nangia A and Nicoud J F 2006 *Crystal Growth & Design* **6** 1278–1281

[12] Dovesi R, Orlando R, Erba A, Zicovich-Wilson C M, Civalleri B, Casassa S, Maschio L, Ferrabone M, De La Pierre M, D'Arco P, Noël Y, Caus M, Rérat M and Kirtman B 2014 *International Journal of Quantum Chemistry* **114** 1287–1317

[13] Ledbetter H and Miglioria A 2006 *Journal of Applied Physics* **100** 063516

[14] Ranganathan S I and Ostoja-Starzewski M 2008 *Phys. Rev. Lett.* **101** 055504

[15] Kube C M 2016 *AIP Advances* **6** 095209

[16] Mishra M K, Ramamurty U and Desiraju G R 2016 *Current Opinion in Solid State and Materials Science* **20** 361 – 370

[17] Matveychuk Y V, Bartashevich E V and Tsirelson V G 2018 *Crystal Growth & Design* **18** 3366–3375

[18] Simpson Jnr H J and Marsh R E 1966 *Acta Crystallographica* **20** 550–555

[19] Arkhipov S G, Rychkov D A, Pugachev A M and Boldyreva E V 2015 *Acta Crystallographica Section C* **71** 584–592

[20] Gaillac R and Coudert F X ELATE: Elastic tensor analysis. http://progs.coudert.name/elate

[21] van der Spoel D, Lindahl E, Hess B and Kutzner C 2008 *J. Chem. Theory Comput.* **4** 435–447

[22] Hornak V, Abel R, Okur A, Strockbine B, Roitberg A and Simmerling C 2006 *Proteins: Structure, Function, and Bioinformatics* **65** 712–725

[23] Wang J, Wolf R M, Caldwell J W, Kollman P A and Case D A 2004 *J. Comput. Chem.* **25** 1157–1174

[24] Makarov G I, Makarova T M, Sumbatyan N V and Bogdanov A A 2016 *Biochemistry (Moscow)* **81** 1579–1588

[25] Makarov G I, Golovin A V, Sumbatyan N V and Bogdanov A A 2015 *Biochemistry (Moscow)* **80** 1047–1056

[26] Makarova T and Bogdanov A 2017 *Biochemistry (Moscow)* **82** 1557–1571

[27] Shishkina A V, Makarova T M, Tereshchenkov A G, Makarov G I, Korshunova G A and Bogdanov A A 2015 *Biochemistry (Moscow)* **80** 1500–1507

[28] Makarov G I, Sumbatyan N V and Bogdanov A A 2017 *Biochemistry (Moscow)* **82** 925–932

[29] Makarov G and Makarova T 2018 *Biochim. Biophys. Acta* **1862** 2940 – 2947

[30] Kostenetskiy P and Semenikhina P 2018 SUSU supercomputer resources for industry and fundamental science *2018 Global Smart Industry Conference (GloSIC)* (IEEE) pp 1–7

Multiscale simulation of the heat and mass transfer with Brinkman model

Valentin Alekseev[1], Maria Vasilyeva[2], Vasily Vasiliev[3]

[1]Multiscale model reduction Laboratory, North-Eastern Federal University, Yakutsk, Russia
[2]North-Eastern Federal University, Yakutsk, Russia, Institute for Scientific Computation, Texas A&M University, College Station, TX, USA
[3]North-Eastern Federal University, Yakutsk, Russia

E-mail: alekseev.valen@mail.ru

Abstract. In this work, we consider a heat and mass transfer problem in heterogeneous media. Mathematical model is described by a system of equations for temperature, pressure and velocity. For simulations of the flow problem in domain with low permeable obstacles, we use a Brinkman model. Fine grid approximation is constructed based on the discontinuous Galerkin (IPDG) method for heat transfer and flow problems. For coarse grid approximation of the flow problem, we use a Generalized Multiscale Discontinuous Galerkin Method. Numerical results are presented for a two-dimensional model problem with different number of the multiscale basis functions.

1. Problem formulation

In this work, we consider the transport and flow problems in domain with a low permeable obstacles $\Omega = \Omega_1 \cup \Omega_2$, where Ω_1 is the free flow domain and Ω_2 is the obstacles subdomain. Transport problem simulate a heat transfer in the heterogeneous domains with free flow in the subdomain Ω_1. For the simulation of the flow in the heterogeneous domains with the free flow in subdomain Ω_1 and Darcy's flow in low permeable obstacles, we use a Brinkman model (Darcy-Stokes-Brinkman model) [1, 2, 3, 4]. Such mathematical models are used, for example, to describe a complex processes in the porous media at the pore-level [5, 6, 7, 8, 9].

We consider flow and heat transfer problem in heterogeneous domain $\Omega = \Omega_1 + \Omega_2$. The flow problem is described by the Brinkman model

$$\begin{aligned} \mu \Delta u - \nabla p + A u = 0, &\quad x \in \Omega, \\ \nabla \cdot u = 0, &\quad x \in \Omega, \end{aligned} \tag{1}$$

with following boundary conditions

$$u = g, \quad x \in \Gamma_L, \quad (\mu \nabla u - p\mathcal{I}) \cdot n = 0, \quad x \in \Gamma_R, \quad u = 0, \quad x \in \Gamma_T \cup \Gamma_B,$$

where μ is the viscosity, p is the pressure, $A = k^{-1}$. Here permeability k in the obstacles domain Ω_2 is defined using Carman-Kozeny equation

$$A = \begin{cases} 0, & x \in \Omega_1, \\ -C\dfrac{(1-\epsilon)^2}{\epsilon^3 + b}, & x \in \Omega_2, \end{cases}$$

where C is the constant accounting for the mushy-region morphology, ϵ is the porosity, b is a constant introduced to avoid division by zero.

The heat transfer equation is described by a convection-diffusion equation with given velocity field u

$$\alpha \left(\frac{\partial T}{\partial t} + u \cdot \operatorname{grad} T \right) - \operatorname{div} \cdot (\lambda \operatorname{grad} T) = 0, \quad x \in \Omega \tag{2}$$

with initial condition $T = T_0$ and boundary conditions

$$T = T_1, \quad x \in \Gamma_L, \quad -\lambda \frac{\partial T}{\partial n} = 0, \quad x \in \Gamma_R \cup \Gamma_T \cup \Gamma_B,$$

where $\partial\Omega = \Gamma_L \cup \Gamma_R \cup \Gamma_T \cup \Gamma_B$.

For the coefficients of the equation, we have the following relationships

$$\alpha = \begin{cases} \rho_w c_w, & x \in \Omega_1, \\ \rho_{sc} c_{sc}, & x \in \Omega_2, \end{cases} \quad \lambda = \begin{cases} \lambda_w, & x \in \Omega_1, \\ \lambda_{sc}, & x \in \Omega_2, \end{cases}$$

where ρ_w, c_w, λ_w and ρ_{sc}, c_{sc}, λ_{sc} are density, specific heat capacity and thermal conductivity of water and grains, respectively.

2. Fine grid approximation

To approximate the problem with respect to spatial variables, we use the discontinuous Galerkin method (IPDG) for both heat transfer and flow equations (2) and (1). For a classic approximation, we construct a sufficiently fine grid that resolve all small scale heterogeneities on the grid level. Let $\mathcal{T}^h$ be a triangulation of the computational domain Ω and Γ^h be the set of all faces between the elements $\mathcal{T}^h$. On the edge $E \in \Gamma^h$ between the elements K_1 and K_2, we define an average and the jump of the function u

$$\{u\} = \frac{u|_{K_1} + u|_{K_2}}{2}, \quad [u] = u|_{K_1} - u|_{K_2}.$$

The fine scale velocity space $V^h = \{v \in L^2(\Omega) : v|_K \in (\mathbb{P}_1(K))^2, \forall K \in \mathcal{T}^h\}$ contains functions which are piecewise linear in each fine-grid element K and discontinuous across coarse grid edges. For the pressure, we use the space of piecewise constant functions Q^h. We have following variational formulation for the flow problem (1): find $(u,p) \in V^h \times Q^h$ such that

$$\begin{aligned} a_{DG}(u,v) + b_{DG}(p,v) = 0, \quad \forall v \in V^h \\ b_{DG}(u,q) = 0, \quad \forall r \in Q^h, \end{aligned} \tag{3}$$

where

$$a_{DG}(u,v) = \sum_{K \in \mathcal{T}^h} \int_K (\mu \nabla u \cdot \nabla v + A\, u\, v) dx - \sum_{E \in \Gamma_h} \int_E \left([\mu\, u \cdot n]\, \{v\} - \{\mu\, v \cdot n\}\, [u] - \{\mu\} \frac{\gamma_u}{h}\, [u]\, [v] \right) ds,$$

$$b_{DG}(u,p) = \sum_{K \in \mathcal{T}^h} \int_K p \nabla u dx + \sum_{E \in \Gamma_h} \int_E \{p\} [u \cdot n]\, ds,$$

where γ_u is the penalty parameter and n is the unit normal to the edge E.

The variational formulation of the transport equation (2) using the discontinuous Galerkin method (IPDG, Interior Penalty Discontinuous Galerkin) is defined as follows: find $T^{n+1} \in W^h$ such that

$$\sum_{K\in\mathcal{T}^h} \int_K \left(\alpha \frac{T^{n+1} - T^n}{\tau} r + \alpha\, u \cdot \nabla T^{n+1} r + \lambda \nabla T^{n+1} \cdot \nabla r \right) dx + \sum_{E\in\Gamma_h} \int_E [T^{n+1} u \cdot n][r]\, ds$$
$$- \sum_{E\in\Gamma_h} \int_E \left(\{\lambda \nabla r\} \cdot [T^{n+1} n] - \{\lambda \nabla T^{n+1}\} \cdot [r\, n] + \{\lambda\} \frac{\gamma_T}{h} \cdot [T^{n+1} n]\, [r\, n] \right) ds = 0, \quad \forall r \in W^h, \tag{4}$$

where γ_T is the penalty parameter and $W^h = \{v \in L^2(\Omega) : v|_K \in (\mathbb{P}_1(K))^2, \forall K \in \mathcal{T}^h\}$. Here, we use an implicit scheme for approximation by time with time step τ.

Next, we describe construction of the coarse grid approximation using Generalized finite element method for flow problem. Note that, we solve a transport problem on the fine grid.

3. Multiscale method

In this section, we describe construction of the coarse grid approximation using Generalized finite element method for flow problem. Note that, we solve a transport problem on the fine grid. Let $\mathcal{T}^H$ be a coarse grid of domain Ω with mesh size H and $\mathcal{E}^H$ be the set of all facets of the coarse grid in $\mathcal{T}^H$. For the pressure approximation, we use the piecewise constant function space Q^H over the coarse cells. We define V^H as the multiscale velocity space, which contains a set of basis functions supported in each coarse block K. [10, 11, 12, 13].

We construct a multiscale space for the velocity

$$V^H = \text{span}\{\psi_i\}_{i=1}^{N_u}, \quad Q^H = \{r \in L^2(\Omega) : r|_K \in P^0(K), \forall K \in \mathcal{T}_H\}$$

and for the pressure, we use the space of piecewise constant functions over the coarse triangulation, $N_u = \dim(V^H)$ is the number of basis functions and $N_p = \dim(Q^H)$ is equal to the number of coarse grid cells.

For the coarse grid approximation, we use a Discontinuous Galerkin(DG) approach and have following variational formulation: find $(u_H, p_H) \in V^H \times Q^H$ such that

$$\begin{aligned} a_{\text{DG}}(u_H, v) + b_{\text{DG}}(p_H, v) &= f(v) \quad \forall v \in V^H, \\ b(u_H, q) &= 0, \quad \forall q \in Q^H. \end{aligned} \tag{5}$$

For construction of the multiscale space for the velocity, we solve a spectral problem to select dominant modes of the snapshot space. We construct local snapshot basis in each coarse block $K_i, (i = 1, \cdots, N)$, where N is the number of coarse blocks in Ω. The local snapshot space consists of functions which are solutions $\phi_l^i \in V^h(K_i)$ of

$$\begin{aligned} a_{\text{DG}}(\phi_l^i, v) + b_{\text{DG}}(\eta, v) &= 0 \quad \forall v \in V^h(K_i), \\ b(\phi_l^i, q) &= \int_{K_i} c\, q\, dx, \quad \forall q \in Q^h(K_i). \end{aligned} \tag{6}$$

with $\phi_l^i = \delta_i^l$ on ∂K_i $(l = 1, \cdots, J_i)$, where J_i is the number of fine grid nodes on the boundary of K_i, and δ_i^l is the discrete delta function defined on ∂K_i. Here, c is chosen by the compatibility

Journal of Physics: Conference Series **1392** (2019) 012063 doi:10.1088/1742-6596/1392/1/012063

condition, $c = \frac{1}{|K_i|}\int_{\partial K_i} \delta_i^l \cdot n\, ds$. Next, we form a local snapshot space in K_i using the all local solutions $V^{i,\text{snap}} = \{\phi_l^i : 1 \leq l \leq J_i\}$ and define projection matrix to the snapshot space $R_{i,\text{snap}} = \left[\phi_1^i, \ldots, \phi_{J_i}^i\right]$.

To reduce size of the snapshot space, we solve following local spectral problem

$$A^{i,\text{snap}}\psi_k^{i,\text{snap}} = \lambda_k S^{i,\text{snap}}\psi_k^{i,\text{snap}}, \tag{7}$$

where $A^{i,\text{snap}} = R_{i,\text{snap}} A^i R_{i,\text{snap}}^T$, $S^{i,\text{snap}} = R_{i,\text{snap}} S^i R_{i,\text{snap}}^T$. Here A^i is the matrix representation of the bilinear form $a_i(u, v)$ and S^i is the matrix representation of the bilinear form $s_i(u, v)$

$$a_i(u, v) = \int_{K_i} \mu \nabla u \cdot \nabla v \, dx, \quad s_i(u, v) = \int_{\partial K_i} u \cdot v \, dx.$$

We arrange the eigenvalues in increasing order and choose the first eigenvectors corresponding to the first M_i the smallest eigenvalues $\psi_k^i = R_{i,\text{snap}}\psi_k^{i,\text{snap}}$ as the basis functions ($k = 1, ..., M_i$)

$$V^H = \text{span}\{\psi_k^i : 1 \leq i \leq N_e, \ 1 \leq k \leq M_i\},$$

This multiscale space will be used as the approximation space for the velocity.

4. Numerical results

We consider numerical simulation using the proposed models in heterogeneous computational domain $\Omega = [0, 1]^2$ (see Figure 1). The fine grid contains 89778 vertices, 30127 facets and 59562 triangular cells. We consider two coarse grids: (1) 5×5 grid with 36 vertices, 60 facets and 25 cells; (2) 10×10 grid with 121 vertices, 220 facets and 100 cells. We set $\mu = 1.0$, $C = 1.0$, $\epsilon = 10^{-5}$, $b = 10^{-6}$, $\rho_w = 1000$, $c_w = 4200$, $\lambda_w = 560$, $\rho_{sc} = 1260$, $c_{sc} = 840$, $\lambda_{sc} = 1100$, $g = (1.0, 0.0)$, $T_0 = 1$ and $T_1 = 10$. We simulate for $t = 0.5$ with 50 time steps.

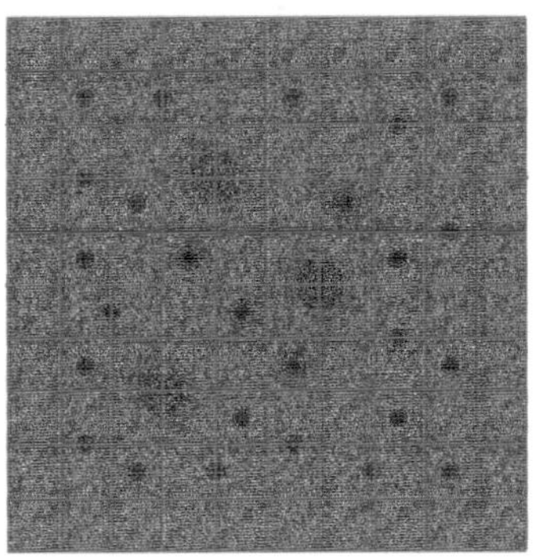

Figure 1. Computational domain with coarse and fine grids

To compare the results, we use the relative L_2 and H_1 errors between fine-scale solutions v_f and multiscale solutions v_{ms} for velocity ($e_{L_2}^u$ and $e_{H_1}^u$). For pressure, we use a L_2 errors on the fine and coarse grids ($e_{L_2}^p$ and $e_{L_2}^{\bar{p}}$). For temperature, we calculate L_2 error on the fine grid using fine scale and multiscale velocity ($e_{L_2}^T$).

The results of calculations are shown in Figure 2, where we present the X and Y components of velocity, pressure and temperature distributions for fine grid and multiscale solutions.

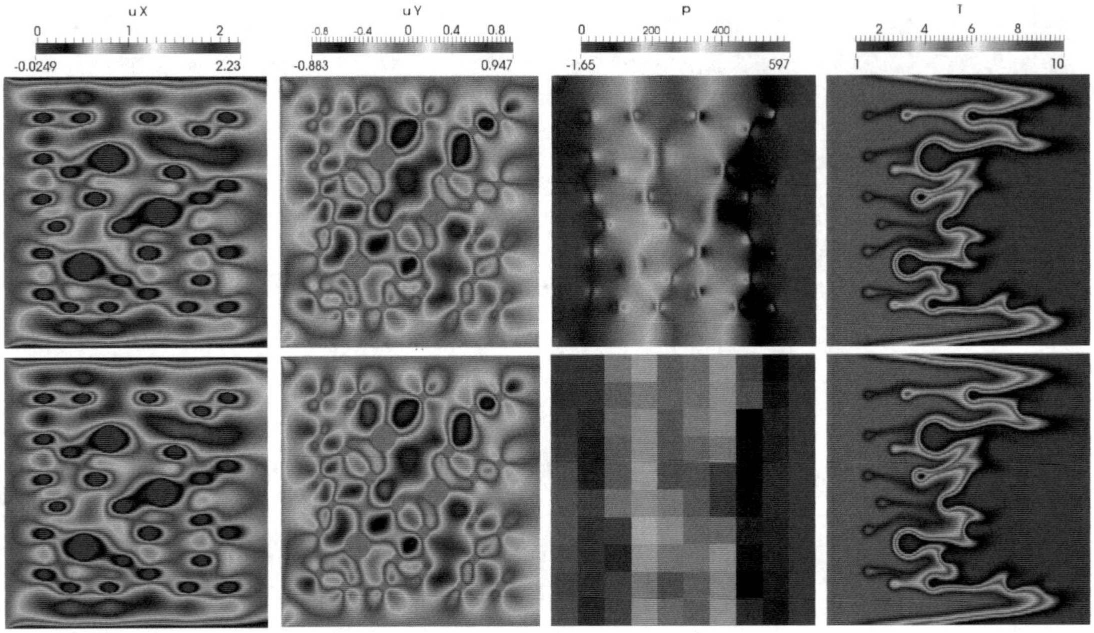

Figure 2. The distribution of the velocity (u_x and u_y), pressure and temperature (from left to right). Top: Fine grid solution. Bottom: Multiscale solution.

5 × 5						
M	DOF_c	$e^u_{L_2}$ (%)	$e^u_{H_1}$ (%)	$e^p_{L_2}$ (%)	$e^{\overline{p}}_{L_2}$ (%)	$e^T_{L_2}$ (%)
10	275	43.059	78.449	732.669	41.409	28.847
20	525	15.426	39.407	38.187	0.19	8.598
30	775	8.424	29.258	20.21	0.026	5.610
40	1025	5.317	24.173	16.595	9.1e-03	3.251
60	1525	3.044	18.645	14.723	1.79e-03	1.976
10 × 10						
10	1100	22.24	51.489	91.398	4.079	12.705
20	2100	6.081	30.154	20.233	0.054	3.792
30	3100	3.683	23.081	15.025	0.013	2.712
40	4100	2.673	19.833	13.572	9.04e-03	1.888
60	6100	2.154	17.056	12.604	4.87e-03	1.497

Table 1. Relative errors for velocity and pressure with different number of multiscale basis functions for 5x5 and 10x10 meshes

Calculation was performed using 60 multiscale basis functions for the velocity filed. Temperature distribution is presented at final time for solution on the fine grid with fine scale and multiscale velocity field. Relative errors are presented in Table 1 for 5×5 and 10×10 coarse grids. We observe that the velocity L_2 error reduce from 43% to 3% and the pressure coarse grid error reduce from 41% to 0.00179% for 5×5 grids, when we use 10 and 60 multiscale velocity basis functions respectively. On the 10×10 coarse grid, the velocity L_2 error reduce from 22% to 2% and the pressure coarse grid error reduce from 4% to 0.00487%, when we use 10 and 60 multiscale

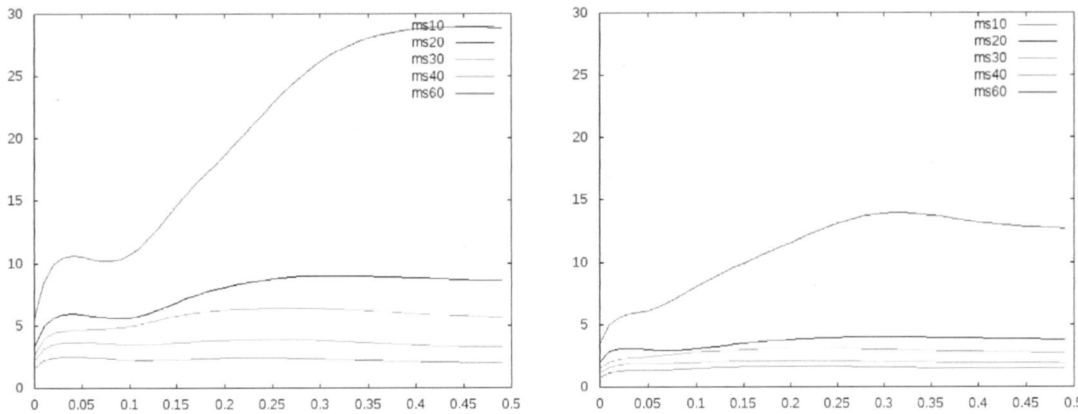

Figure 3. The L_2 error between the fine-scale and multiscale solution of the temperature. (Left: Error for 5x5 meshes. Right: Error for 10x10 meshes.)

velocity basis functions respectively. In Figure 3, we present dynamic of the relative L_2 error for temperature, where we obtain a good solutions using sufficient number of the multiscale basis functions for accurate approximation of the velocity field.

5. Acknowledgments

The work are supported by the mega-grant of the Russian Federation Government N14.Y26.31.0013 and RSF N19-11-00230.

References

[1] Young T. J., Vafai K. Convective flow and heat transfer in a channel containing multiple heated obstacles *International Journal of Heat and Mass Transfer (Vol. 41, No. 21, pp. 3279-3298)* 1998.

[2] Hsu C. T., Cheng P. The Brinkman model for natural convection about a semi-infinite vertical flat plate in a porous medium *International Journal of Heat and Mass Transfer (Vol. 28, No. 3, pp. 683-697)* 1985.

[3] Lesinigo M., DAngelo C., Quarteroni A. A multiscale DarcyBrinkman model for fluid flow in fractured porous media *Numerische Mathematik (Vol. 117, No. 4, pp. 717-752)* 2011.

[4] Khaled A. R. A., Vafai K. The role of porous media in modeling flow and heat transfer in biological tissues *International Journal of Heat and Mass Transfer (Vol. 46, No. 26, pp. 4989-5003)* 2003.

[5] Chung E. T., Vasilyeva M., Wang Y. A conservative local multiscale model reduction technique for Stokes flows in heterogeneous perforated domains *Journal of Computational and Applied Mathematics (Vol. 321, pp. 389-405)* 2017.

[6] Chung, E. T., Efendiev, Y., Li, G., Vasilyeva, M. Generalized multiscale finite element methods for problems in perforated heterogeneous domains. *Applicable Analysis (Vol. 95, No. 10, pp. 2254-2279)* 2016.

[7] Spiridonov, D., Vasilyeva, M., Leung, W. T. A Generalized Multiscale Finite Element Method (GMsFEM) for perforated domain flows with Robin boundary conditions. *Journal of Computational and Applied Mathematics (Vol. 357, pp. 319-328)* 2019.

[8] Vasilyeva, M., Chung, E. T., Leung, W. T., Wang, Y., Spiridonov, D. Upscaling method for problems in perforated domains with non-homogeneous boundary conditions on perforations using Non-Local Multi-Continuum method (NLMC). *Journal of Computational and Applied Mathematics (Vol. 357, pp. 215-227)* 2019.

[9] Alekseev, V., Gavrileva, U., Spiridonov, D., Tyrylgin, A., Vasilyeva, M Numerical simulation of the transport and flow problems in perforated domains using generalized multiscale finite element method *AIP Conference Proceeding AIP Publishing (Vol. 2025, No. 1, pp. 100001)* 2018.

[10] Chung, E. T., Efendiev, Y., Vasilyeva, M., Wang, Y. A multiscale discontinuous Galerkin method in perforated domains. *In Proceedings of the Institute of Mathematics and Mechanics (Vol. 42, No. 2, pp. 212-229)* 2016.

[11] Chung, E. T., Vasilyeva, M., Wang, Y. A conservative local multiscale model reduction technique for Stokes flows in heterogeneous perforated domains. *Journal of Computational and Applied Mathematics (Vol. 321, pp. 389-405)* 2017.

[12] Stepanov S., Vasilyeva M., Vasilev V. I. Generalized multiscale discontinuous Galerkin method for solving the heat problem with phase change *Journal of Computational and Applied Mathematics (Vol. 340, pp. 645-652)* 2018.

[13] Tyrylgin A., Vasilyeva M., Brown D. Generalized Multiscale Finite Element Method for Poroelasticity Problems in Heterogeneous Media *International Conference on Finite Difference Methods Springer, Cham (pp. 566-573)* 2018.

[14] Gavrileva U. et al. Generalized Multiscale Discontinuous Galerkin Method for Helmholtz Problem in Fractured Media *International Conference on Finite Difference Methods Springer, Cham (pp. 250-257)* 2018.

Journal of Physics: Conference Series **1392** (2019) 012064 doi:10.1088/1742-6596/1392/1/012064

Quasi-cellular nets based models of flow systems

Anton O. Aristov
National university of science and technology "MISiS", Moscow, Russia
E-mail: batan-87@mail.ru

Abstract. It considered specific type of discrete structures quasi-cellular nets (designed by author). The basic idea is space division to discrete areas cells, which distributed through the directions of flows in different subject areas (transport, industry, social systems, life systems etc.). Each cell has computer memory view as data structure that consists of data about cell coordinates, flow object, and changeable parameters of cell. Distribution of flow being considered as transition of flow between neighboring cells. Quasi-cellular nets are models that will be use for implementation of flow dualism (micro- and macrolevel of simulations) for different subject areas.

1. Introduction

There are many systems in science, technologies, technics, engineering, medicine, social and other areas based on flows which distributed in the limited space. Flow simulation considered as common science method of researches. The feature of flow simulations is dualism of simulation suggesting flow as set of interacting objects on micro level of simulation; flow as object on macrolevel[1][2].

Existing models of flow systems designed for concrete subject areas and levels (micro- or macro). Decision of described problems is designing of new type of discrete structures that allows simulation on micro- and macrolevel.

2. Basics of quasi-cellular nets

Quasi cellular nets based on ideas of space division into areas cells. Cells situated along the flow (in different subject areas) distribution directions. More often structure of flow based system sets as graph (Figure 1, Figure 2) that simulates space of flow distribution (road nets, network, blood vessels, routes of people flows motion etc.).

Set cells (circle areas with constant radius) along the edges of graph. Cells touch each other and make sampling of graph edges (Figure 3, Figure 4). Set each cell with data set about area state, that consists of:

- coordinates of cell as area of space;
- parameters of flow object that located of in this cell;
- cell parameters that changed under influence of flow that move through the cell.

Define discrete simulation time. Each time step flow based object transit to neighboring cell. Flow distribution define as flow based object transition between neighboring cells for each time step.

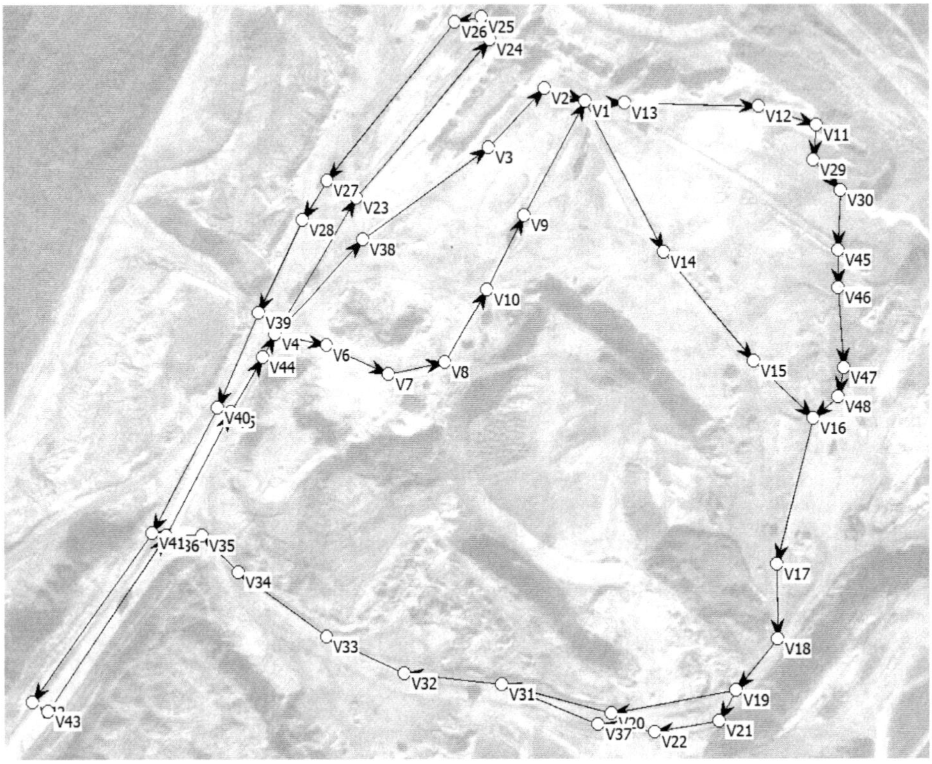

Figure 1. Structure of quarry logistical routes

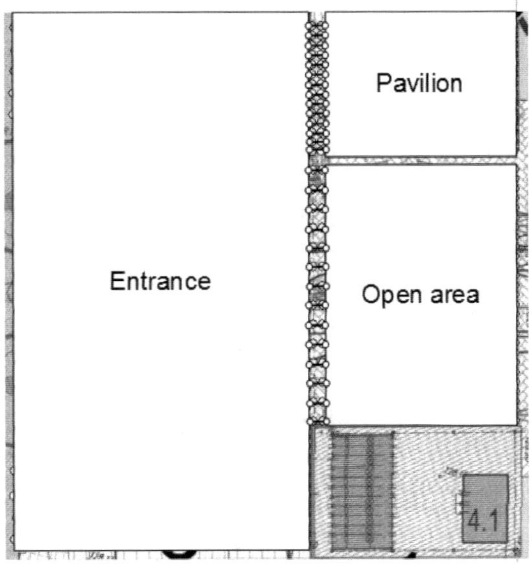

Figure 2. Structure of place of mass stay of people

Quasi-cellular nets simulation based on calculations of cell parameters form cell data structure for each time moment(time step). Simulation on macrolevel based on calculations by cells of subset of quasi-cellular net based on parameters of data about cell[3]. Also structure of quasi-

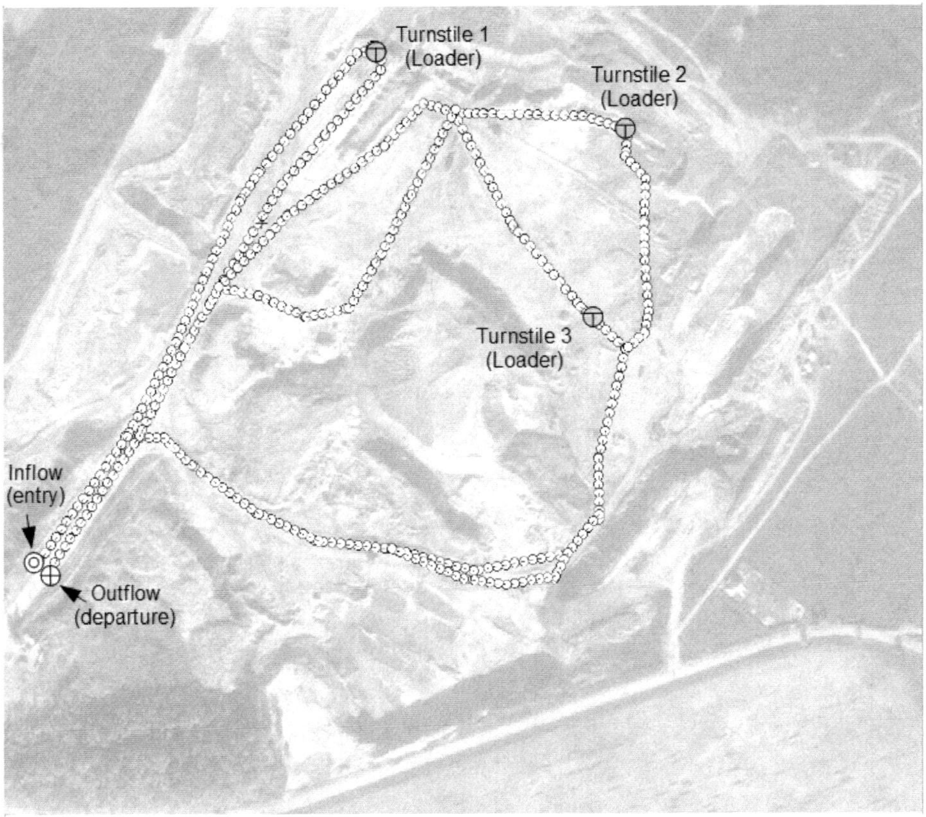

Figure 3. Quasi-cellular net for quarry logistics

cellular net includes cell of flow object delay (turnstile), flow generator (inflow), outflow, counter, which influence to flow distributions.

3. Applications of quasi-cellular nets

Practical application of quasi-cellular nets based on common scheme of designing subject area usings. It consist of several stages[4]:

- definitions of flows in flow system, design graph model;
- synthesis of quasi-cellular net structure;
- definition of cell structure;
- set extension elements (counters, delay cells, inflows, outflows)

Let's describe design of models on examples : quarry transport logistics[4] and people flow simulation[3]. At the next stage let's create quasi-cellular nets based on graphs (Figures 1,2). Simulation with quasi-cellular nets allows to calculate micro level parameters such as :

- coordinates of each object in flow;
- flow intensity for area of quasi-cellular net;
- duration of passage for flow object;
- flow value through the cell

etc.

Macrolevel parameters :

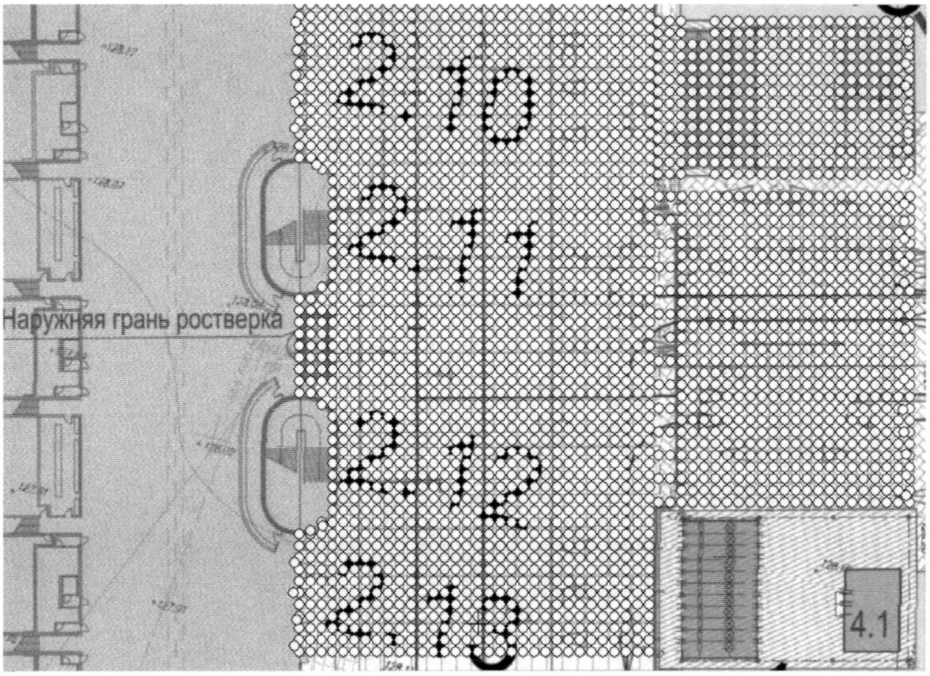

Figure 4. Quasi-cellular net for place of mass stay of people

- average waiting time for queue;
- average quantity of objects in the space area (subnet)
- flow value for subset of quasi-cellular net.

Quasi-cellular nets also may be used for visualization of data that distributed through the space. Samples of visualization are temperature maps which shows stress on parts of quasi-cellular net[5] (Figure 5, Figure 6). Described features (micro- and microlevel simulation and visualization) implemented under the same discrete structure.

4. Conclusion

Area of quasi-cellular nets using is design and development of computer simulation tools of flow systems for different subject areas. Simulation of micro- and macrolevel implemented under the same discrete structure. Quasi-cellular nets also used for implementation of flow dualism (micro- and macrolevel).

5. References

[1] Eremin V. Badalyan A. 2007 *Computer simulation of conflict situation for evaluation to assess the level of traffic safety on two-lane highways* (Moscow: IKF-Catalog)

[2] Fedorov N.V. 2015 *Simulation of complex systems* (Moscow: MSIU)

[3] Aristov A. 2018 Theoretical and applied baseline for application of quasi-cellular nets in computer modeling and visualization of transport logistics in open pit sand mine *MINING INFORMATIONAL AND ANALYTICAL BULLETIN (SCIENTIFIC AND TECHNICAL JOURNAL)* 2018 1

[4] Aristov A. 2014 Quasicellular networks and their application for simulation of visitor flow in public spaces *Computer research and modeling* 2014 2

[5] Aristov A. 2014 Geometrical modeling and visualization of simulation results in quasi cellular nets *J. Scientific Visualization* 2014 5

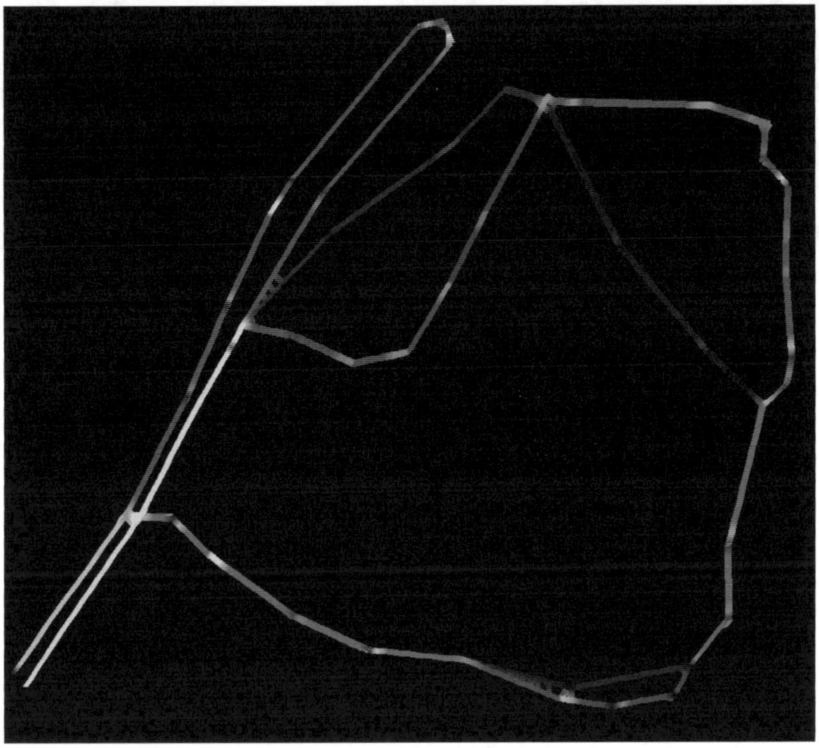

Figure 5. Temperature map for logistical routes

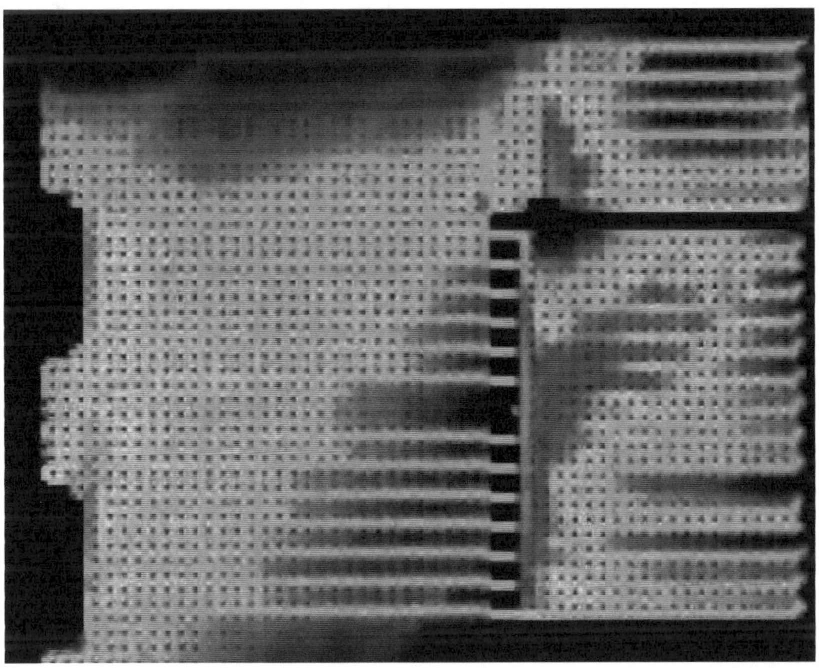

Figure 6. Temperature map for place of mass stay of people

Block finite-element approach to building refined models of layer-by-layer analysis of the stress-strain state of three-layer irregular shells

V N Bakulin[1,2]

[1]Moscow Aviation Institute (National Research University), Volokolamskoye highway 4, Moscow, 125993,Russia
[2]Institute of Applied Mechanics of the Russian Academy of Sciences, Leningradsky avenue 7, Moscow, 125040, Russia

E-mail: vbak@yandex.ru

Abstract. This work offers a block finite element approach to building refined models of layer-by-layer analysis of stress-strain state of three-layer generally irregular cylindrical shells.

1. Introduction

Three-layer shells are being increasingly used for rocket and space, aircraft and other modern equipment structures [1, 2]. The application of three-layer shells is constrained among others by insufficiently developed models dedicated to various exposures to individual layers and real conditions of layers fixing.

The building of models and finite element calculations of heterogeneous sandwich and three-layer shells including those made of composites, are described in works [3, 4, 5]. Three-layer cylindrical shells involve application of broken line hypothesis [3, 5]. Sandwich shells in most cases involve Timoshenko layer packet hypothesis [4]. If the layer packet hypothesis is applied, the number of degrees of freedom a finite element model does not depend on the number of layers. However, such approach restricts the range of loadings and layer fixing conditions as well as results in simplified models or models for narrow class of problems, which often meet the computational accuracy requirements.

The issues of layer-by-layer analysis of heterogeneous sandwich and three-layer shells have not been sufficiently studied yet. The layer-by-layer analysis approach is that the layer wall including the filler, if necessary, are divided into layers in thickness, which are afterwards linked together. With such approach, depending on stress-strain properties, layer thickness and problem situation, we can create and apply various models to calculate layers using analytic and numerical methods.

2. Formulation of problem

This article involves the layer-by-layer analysis-based building of finite element models for a refined study of stress-strain state of three-layer generally irregular cylindrical shells taking into account heterogeneity including filler layers, application of loads to different layers and different layer fixing conditions, lateral strains and stress, momental state of carrier layers, three dimensional stress state in filler layer. The models shall enable to make calculations with variable geometrical and stress-strain properties, and consider the properties changes and stress-strain state parameters not only in meridian and circumferential coordinates but also in filler layer thickness, which are currently taken with insufficient accuracy, and some of peculiarities are not taken into account at all.

The considered layer-by-layer analysis model of stress-strain state of three-layer irregular shells is based on building and following blocking of two different types of natural curved finite elements. Thin and rigid carrier layers are simulated with two-dimensional finite elements of momental carrier layers, and filler layer is simulated with three-dimensional finite elements of thick-wall layer. Thickness and stress-strain properties of a filler layer vary greatly.

However, this approach involves significantly increasing degrees of freedom of finite element model (it depends on the number of shell layers; the number of finite elements used to simulate layers including in thickness, and the number of finite element degrees of freedom). Thus, calculation errors, required computer resources and calculation time increase too. It much complicates and often makes impossible the study of such structures with appropriate accuracy. In this regard there is a problem of reducing the size of finite element models of heterogeneous sandwich generally irregular shells.

The building of finite element models usually involves shape functions obtained through approximation of displacement fields by first and second-order polynomials [4] it often results in intricate finite element models. In this regard it is important to build efficient models with shape functions which ensure increased velocity of convergence and hence result in reduced order of equation systems as against known approaches. This is especially important for model building based on layer-by-layer analysis.

The most efficient are shape functions based on analytical solutions of the shell theory problems [5]. Such approach has been successfully developed as to axisymmetric three-layer cylindrical shells only.

Efficient shape functions can be obtained through the approximation of generalized strains [6] followed by satisfying the equation of strain compatibility. This approach involving building efficient models of irregular three-layer cylindrical shells is described in [3, 7, 8, 9, 10, 11, 12].

3. Finite-element model of carrier layers of three-layer cylindrical generally irregular rotational shell

Basic relations.

If carrier layers of the three-layer shell are sufficiently thin and rigid, then stress-strain state simulation shall involve two-dimensional natural curved finite elements developed from general theory of shells based on Kirchhoff-Love hypotheses. Let us index layers (fig.1a), beginning from the shell inner surface: $i = 1, 2, 3$; assign index c to carrier layers, and index f to filler layers.

Displacement of carrier layer points are defined by displacement of middle surface points $\delta_i^c = \{u, v, w\}^T$ and angular displacements of normal to middle surface against axes $x, y(\varphi)$ $\vartheta_i^c = \{\vartheta_x, \vartheta_y\}^T$ (fig. 1b).

Hereinafter, we shall not indicate vector coefficients and matrix indices corresponding to a layer index.

Generalized strain vector $\varepsilon_i^c = \{\varepsilon_1, \varepsilon_2, \gamma, æ_1, æ_2, \chi\}^T$ hereinafter referred to as strain vector is linked to displacement vector δ_i^c through Cauchy relations [13]:

$$\varepsilon_1 = \frac{\partial u}{\partial x}; \quad \varepsilon_2 = \frac{1}{R}\frac{\partial v}{\partial \varphi} + \frac{w}{R}; \quad \gamma = \frac{1}{R}\frac{\partial u}{\partial \varphi} + \frac{\partial v}{\partial x}$$

$$æ_1 = -\frac{\partial^2 w}{\partial x^2}; \quad æ_2 = \frac{1}{R^2}\frac{\partial v}{\partial \varphi} - \frac{1}{R^2}\frac{\partial^2 w}{\partial \varphi^2}; \quad \chi = \frac{1}{R}\frac{\partial v}{\partial x} - \frac{1}{R}\frac{\partial^2 w}{\partial x \partial \varphi} \quad (1)$$

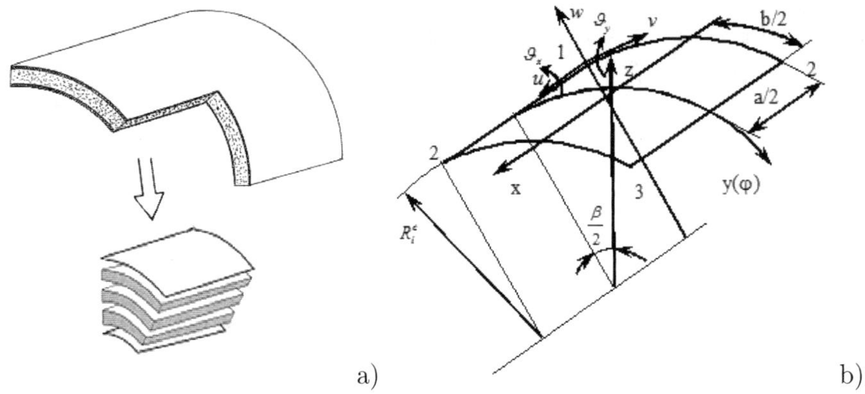

Figure 1. a – Finite element block example, b – Carrier layer finite element.

Rigid displacement is defined by integrating relations (1) with zero strain.

The selection of displacement approximation functions is tied to the selection of nodal displacements and the number of nodes. Carrier layer finite element nodal displacements involve u, v, w and angular displacements $\vartheta_x^c, \vartheta_y^c$. Thus, finite element of carrier layers of three-layer generally irregular cylindrical rotational shell has twenty degrees of freedom.

Displacement and strain approximation functions shall be written using twenty coefficients α, which number is equal to the number of finite element degrees of freedom. From among twenty indefinite coefficients α six are used for rigid displacement $\delta_c^0 = \left\{u^0, v^0, w^0\right\}^T$

$$\delta_c^0 = T_c^0 \alpha_c^0; \quad \alpha_c^0 = \{\alpha_1, \ldots, \alpha_6\}^T ,$$

where $\alpha_1, \ldots, \alpha_6$ are constants of integration used here as indefinite coefficients, T_c^o is coefficient matrix at $\alpha_1, \ldots, \alpha_6$.

Finite element of carrier layers of three-layer generally irregular cylindrical shell.

The finite element (fig. 1b) is generated through the intersection of two planes perpendicular to the center line and two planes passing through the shell center line. The nodes are located in the finite element corners and marked 1 - 4 (fig. 1b).

Unlike most works dedicated to finite element method involving displacement function approximation, this work describes generalized strain function approximation. This has significantly increased the velocity of convergence of numerical procedures and enabled to reduce the required number of finite elements, which is very important for layer-by-layer analysis models application.

Since finite element of carrier layers of three-layer generally irregular cylindrical rotational shell has twenty degrees of freedom, six of which are responsible for rigid displacement, then the rest fourteen degrees of freedom are responsible for shell strain displacement.

Let us distribute the remaining fourteen indefinite coefficients between strain vector coefficients (the formula in brackets is added to satisfy the equations of continuity of deformations) [3, 7, 8].

$$\begin{aligned} &\varepsilon_1 = \alpha_7 + \alpha_8\varphi; \quad \varepsilon_2 = \alpha_9 + \alpha_{10}x - \left(\alpha_{12} + \tfrac{\alpha_{13}x}{3} + \alpha_{14}\varphi + \tfrac{\alpha_{15}x\varphi}{3}\right)\tfrac{x^2}{2R}; \gamma = \alpha_{11}; \\ &æ_1 = \alpha_{12} + \alpha_{13}x + \alpha_{14}\varphi + \alpha_{15}x\varphi \;\; ; \quad æ_2 = \alpha_{16} + \alpha_{17}x + \alpha_{18}\varphi + \alpha_{19}x\varphi; \\ &\chi = \alpha_{20} + \left(-\tfrac{1}{R^2}\alpha_8 x + \tfrac{1}{R}\alpha_{14}x + \tfrac{1}{2R}\alpha_{15}x^2 + R\alpha_{17}\varphi + \tfrac{R}{2}\alpha_{19}\varphi^2\right). \end{aligned} \tag{2}$$

$$\text{or} \quad \varepsilon_i^c = \Omega_i^c \alpha_i^c.$$

where $\alpha_i^c = \{\alpha_1 \dots \alpha_{20}\}^T$ is an indefinite coefficient vector, Ω_{ic} (6×20) is a strain function approximation matrix.

To define displacement approximation functions caused by element strain, let us integrate Cauchy relations (1) using strain expressions of (2).

Displacement vector δ_i^c of finite element of shell carrier layers is a sum of rigid δ_c^0 and strain δ_c^d displacements $\delta_c^d = \{u_d, v_d, w_d\}^T$. The displacement approximation function obtained in this manner shall be written using twenty indefinite coefficients

$$u = \alpha_2 R\cos\varphi + \alpha_4 R\sin\varphi + \alpha_5 + \alpha_7 x + \alpha_8 x\varphi + \alpha_{11} R\varphi - \alpha_{17}\frac{1}{2}R^3\varphi^2 +$$
$$+\alpha_{19}R^3\varphi\left(1 - \frac{1}{6}\varphi^2\right) - \alpha_{20}R^2\varphi$$
$$v = (\alpha_1 + \alpha_2 x)\sin\varphi - (\alpha_3 + \alpha_4 x)\cos\varphi + \alpha_6 - \frac{\alpha_8}{2R}x^2 + \alpha_{16}R^2\varphi + \alpha_{17}R^2 x\varphi +$$
$$+\frac{1}{2}\alpha_{18}R^2\varphi^2 + \alpha_{19}R^2 x\left(\frac{\varphi^2}{2} - 1\right) + \alpha_{20}Rx$$
$$w = -(\alpha_1 + \alpha_2 x)\cos\varphi - (\alpha_3 + \alpha_4 x)\sin\varphi + R\alpha_9 + R\alpha_{10}x - \frac{1}{2}x^2\alpha_{12} - \frac{1}{6}\alpha_{13}x^3 -$$
$$-\frac{1}{2}\alpha_{14}x^2\varphi - \frac{1}{6}\alpha_{15}x^3\varphi - R^2\alpha_{16} - R^2\alpha_{17}x - R^2\alpha_{18}\varphi - R^2\alpha_{19}x\varphi$$

or as a matrix

$$\delta_i^c = T_i^c \alpha_i^c, \tag{3}$$

where T_i^c (3×20) is a matrix of carrier layer finite element displacement approximation functions. By inserting nodal points coordinates into relations (3) and into expressions of normal angular displacement around axes x and φ, written through indefinite coefficients in terms of relations (3), we shall obtain relations, which link finite element nodal displacement vector q_i^c and vector α_i^c:

$$q_i^c = C_i^c \alpha_i^c, \tag{4}$$

where C_i^c (20×20) is a vector α_i^c and vector q_i^c constraint matrix.

Using the expression (4) let us find vector

$$\alpha_i^c = \left(C_i^c\right)^{-1} q_i^c \tag{5}$$

By rearranging the expressions (3) in terms of relations (5), we shall obtain shape functions of finite element of three-layer cylindrical shell carrier layers

$$\delta_i^c = T_i^c \left(C_i^c\right)^{-1} q_i^c$$

By obtaining expressions of generalized strains as well as force and moment (stresses) we shall calculate finite element stiffness matrix using algorithms described in works [3, 7, 8], and build a finite element model of carrier layers.

Journal of Physics: Conference Series **1392** (2019) 012065 doi:10.1088/1742-6596/1392/1/012065

4. Finite-element model of filler layer of three-layer cylindrical generally irregular rotational shell.

Finite element for simulating stress-strain state in a filler layer.

The application of two and three-dimensional finite elements for simulating momental carrier layers and filler layer respectively, shall lead to the generalized displacement discontinuity upon contact areas of these finite elements, if they are built using different displacement approximation functions.

To avoid errors caused by such generalized displacement discontinuity, a filler layer finite element on cylindrical surfaces should involve as many nodes as carrier layer finite element has; the generalized displacements and approximation functions similar to those of carrier layer finite element should be used as nodal unknowns and approximation functions.

The finite element is generated through intersection of two planes perpendicular to the center line and two planes passing through the shell center line. The finite element nodes are located in the corners, four on internal (node numbers 1 - 4) and external (node numbers 5 - 8) cylindrical surfaces.

The degrees of freedom in the filler layer finite element node conjugated in internal and external cylindrical surfaces with carrier layer finite elements include three linear displacements u,v,w and angular displacement of the normal around axes x and $\varphi(y)$. Thus, the filler layer finite element has generally forty degrees of freedom. When the filler layer finite elements are located one by one over the shell thickness, there is no meaning in the normal angular displacement being taken into account. In this case corresponding generalized displacements are eliminated by a computational procedure involved in boundary conditions.

By reducing degrees of freedom of the filler layer finite element we may switch from one filler model to another.

The filler layer simulation is done using meridional, circumferential and normal coordinates to the shell surface.

Approximation displacement functions of the filler layer finite elements. The local coordinate system of the filler layer finite elements is located on the finite element middle surface with its origin positioned at the intersection of diagonals.

Since the generalized displacements of the filler layer finite elements is related to the middle surface, when the finite elements of carrier and filler layers are joined, we shall transfer from the carrier layer middle surface to the filler layer surface using transfer matrix like [4, 14].

As is noted before, the filler layer finite element displacement fields approximation functions are built on internal and external cylindrical shells of these finite elements. Besides, the carrier layer-dedicated functions are used here.

Since the development finite element model enables to reproduce any law of stress-strain state properties variation through simulation with a required number of finite elements, these finite elements involve the linear law of displacement variation over the radial coordinate like in the work [15].

By inserting the obtained expressions of approximation functions of the displacement fields of the filler layer finite element being a thick-wall cylindrical shell into Cauchy relations, which link strains with rigid displacements in curvilinear coordinates (see work [16]), we shall obtain dependencies of generalized strains written using indefinite coefficient vector.

Using the obtained dependencies of generalized strains, we shall write down expressions of stress based on a physical law [17, 18].

Using strain and stress expressions we can calculate the filler layer finite element stiffness matrix using algorithms similar to those described in works [19, 20].

Block finite-element model of layer-by-layer analysis of the stress-strain state of three-layer generally irregular cylindrical rotational shells

By joining the developed filler finite elements over thickness we shall build the finite element block, like in works [21, 22], for simulating stress-strain state in the filler layer.

By joining over thickness the built filler finite element block with the reviewed carrier layer finite elements like in works [21, 22], we shall build finite element block for simulating stress-strain state in the layers of three-layer generally irregular shells.

Acknowledgments

The work has been performed under the state order. Official registration number: **AAAA - A** 19-119012290177-0.

References

[1] Bakulin V N, Obraztsov I F and Potopakhin V A 1998 *Dynamic problems of Nonlinear Theory of Multilayer Shells. The Effect of Intense Thermoforce Loads, Concentrated Energy Flows* (Moscow: Fizmatlit) 464

[2] Bakulin V N, Ostrik A V 2015 *Complex action of radiation and particles on thin-walled structures with heterogeneous coatings* (Moscow: Fizmatlit) 280

[3] Bakulin V N 1985 *The Finite-Element Method for Studying the Stress-Strain State of Three-Layer Cylindrical Shells* (Moscow: TsNII Informatsii) 140

[4] Bakulin V N, Rassokha A A 1987 *The Finite-Element Method and Holographic Interferometry in the Mechanics of Composites* (Moscow: Mechanical Engineering) 312

[5] Bakulin V N2004 *Approximation for Simulation of Stress-Strain State of Sandwich Cylindrical Shells Mat. Model.* **16** 6 101-105

[6] Ashwell D G, Sabir A B 1972 New Cylindrical Shell Finite Element, Based on Simple Independent Strain Functions *Int. J. of Mech.Sc.* **14** 3 171-183

[7] Bakulin V N, Demidov V I 1978 Three-Layered Natural Curved Finite Element *Izv. Vyssh. Uchebn. Zaved. Mashinostroenie* 5 5–10

[8] Bakulin V N 2006 Construction of Approximations for Simulation of Stress-Strain State of Carrier Layers and Filler Layers of Three-Layered Non-Axisymmetric Cylindrical Shells *Mat. Model.* **18** 8 101-110

[9] Bakulin V N 2007 Effective Models in the Refined Analysis of The Strain State for Three-Layer Non-Axisymmetric Cylindrical Shells *Dok. Ph.* **52** 6 330-334

[10] Bakulin V N 2007 Construction of Approximations and Models for Studying Strain-Stress State of Three-Layered Non-Axisymmetric Cylindrical Shells *Mat. Model.* **19** 12 118-128

[11] Bakulin V N 2009 Testing of Finite-Element Model Designed for Studying Stress-Strain State of Irregular Sandwich Shells *Mat. Model.* **21** 8 121–128

[12] Bakulin V N 2018 An Efficient Model for Layer-by-Layer Analysis of Sandwich Irregular Cylindrical Shells of Revolution *Dok. Ph.* **63** 1 23-27

[13] Balabukh L I, Kolesnikov K S, Zarubin V S et al 1969 *The Fundamentals of Rocket Structural Mechanics* (Moscow: High school) 494

[14] Bakulin V N, Krivtsov V S, Rassokha A A 1983 Algorithm for Anisotropic Shell Finite Element Stiffness Matrix *Izv. VUZov. Aviats. Tekh.* 4 14-18

[15] Bakulin V N 2017 A Corrected Model of Level-by-Level Analysis of Three-Layer Irregular Conical Shells *Dok. Ph.62* 1 37- 41

[16] Vasil'ev V V 1988 *Mechanics of Structures from Composite Materials* (Moscow: Mechanical Engineering) 272

[17] Bakulin V N 2019 Layer- by-Layer Analysis of Stress-Strain State of Three-Layer Shells with Cutouts *Izv. RAN, Mekhanika Tverdogo Tela* 2 111-125

[18] Bakulin V N 2019 Model of Refined Analysis of Stress-Strain State of Three-Layer Conical Irregular Shells *Prik. Mat. i Meh.* **83** 2 315-327

[19] Bakulin V N 2006 Finite Element Model for Analysis of Stress-Strain State of Three-Layer Shells *Mat. Mod.* **18** 1 3-9

[20] Bakulin V N, Repinskii V V 2001 Finite-Element Strain Models of One-Layer and Three-Layer Conical Shells *Mat. Mod.* **13** 6 39–46

[21] Bakulin V N 2018 Block Finite Element Model for Layer-by-Layer Analysis of Stress-Strain State of Three-Layer Irregular Structured Shells *Izv. RAN, Mekhanika Tverdogo Tela* 4 65-72

[22] Bakulin V N 2019 Block Finite-Element Model of Layer-by-Layer Analysis of the Stress-Strain State of Three-Layer Generally Irregular Shells of Double-Curvature Revolution *Dok.Ph.* **64** 1 9-13

Journal of Physics: Conference Series **1392** (2019) 012066 doi:10.1088/1742-6596/1392/1/012066

Homogenization of the elasticity problem with periodically located cracks

Alena A.Egorova, Natalia V.Neustroeva, Nadezhda M.Afanaseva
42 Kulagovskogo str, Yakutsk, Republic of Sakha (Yakutia), Russia, 677027
E-mail: alena.egorova@gmail.com, nnataliav@mail.ru, afanasieva.nm@gmail.com

Abstract. In this paper we study the nonlinear problem of elasticity with periodically located cracks. On the edges of these cracks non-penetration conditions are given, which leads to a variational inequality. The period of distribution of cracks, as well as their sizes, depends on the small parameter.

In this work the corresponding varational problem and its homogenized problems are presented. We derive the necessary estimates for the theorem about convergence of the solutions of the variational inequality to the solution of the homogenised problem. For the first corrector of the classical asymptotic expansion we construct a penalty equation and a linear iterative equation in integral form. The convergence theorems for corresponding problems are formulated.

1. Introduction

In this work we consider the elasticity problem on a domain with periodically located cracks. It is supposed, that opposite edges of cracks can not penetrate into each other. The period of distribution of cracks, as well as their sizes, depends on the small parameter $\varepsilon << 1$. The problem is non-linear, it results in a variational inequality on some convex closed set of admissible displacements. The behavior of the solution $\mathbf{u}^\varepsilon$ is determined by the first two terms $\mathbf{u}^0$, $\mathbf{u}^1$ of the asymptotic expansion, where $\mathbf{u}^0(x)$ is the solution of the elasticity problem in a domain without cracks, and the first corrector $\mathbf{u}^1(x,y)$ is the solution of the variational inequality for a given function $\mathbf{u}^0(x)$ on a periodicity cell (see [1, §6 chapter VI]). The convergence of the $\mathbf{u}_\varepsilon$ to the $\mathbf{u}^0(x)$ in L^2-sense was proved in [2]. In this paper we analyze complete convergence, using the periodic unfolding method. The basics of this method are given in [3], as well as its applications in [4] - [6]. The main ideas of this method are the introduction of the unfolding operator and the decomposing of function to micro-macro parts. The macroscopic part does not capture the oscillations of order ε, while the microscopic part is designed to do so. The unfolding operator doubles the dimension of spaces and the weak convergence of unfolding functions in L^2-sence in such space is equivalent to the two-scale convergence, which simplifies the proofs.

For the local problem on a cell with one crack we construct a penalty problem. We formulate the theorem about convergence of the solutions of the penalty problem to the solution of the problem on the cell, when the small regularization parameter tends to zero. Also we present the iteration equation, which approximate the nonlinear penalty problem. The rate of convergence is given in the Theorem 3. The results can be applied for numerical implementation.

[1] The authors were supported by the Government of Russian Federation Megagrant No. 14.Y26.31.0013

2. The periodic unfolding operator in cracked domain

Let consider a bounded domain $\Omega \subset \mathbb{R}^2$ with Lipschitz boundary $\partial\Omega$. In the following $Y = (0,1)^2$ is a unit cell. Let the crack Γ is a smooth curve, strictly included in Y, and Y_Γ is open set $Y \setminus \Gamma$;

Denote $\Xi_\varepsilon = \{z \in Z^2 : \varepsilon(\bar{Y} + z) \subset \Omega\}$ and $\widehat{\Omega}_\varepsilon = \{\bigcup_{z\in\Xi_\varepsilon}(\varepsilon z + \varepsilon\overline{Y})\}$ is the largest union of $\varepsilon(z+\overline{Y})$ cells, included in Ω, while $\Lambda_\varepsilon = \Omega \backslash \widehat{\Omega}_\varepsilon$ is the subset of Ω, containing parts from $\varepsilon(z+\overline{Y})$ cells, intersecting the boundary $\partial\Omega$ (see Fig.2)

Then the union of all the cracks is denoted $\Gamma_\varepsilon = \cup_{z\in\Xi_\varepsilon}\varepsilon(\Gamma + z)$, strictly included in Ω. So, the set Γ_ε can be considered as a system of periodically located cracks (with the period εY). Denote Ω_ε a cracked domain $\Omega_\varepsilon = \Omega \cap \{x : \frac{x}{\varepsilon} \in Y_\Gamma\}$. The boundary of Ω_ε consists of an outer boundary $\partial\Omega$ and an inner boundary Γ_ε with two edges Γ_ε^+, Γ_ε^-. Assume, that $\partial\Omega$ does not intersect with the system Γ_ε (see Fig.1).

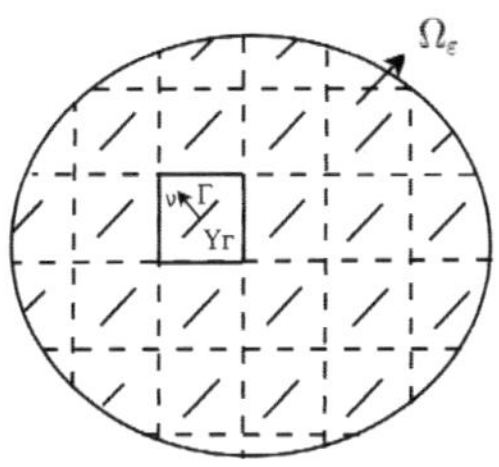

Figure 1. The domain with periodically located cracks

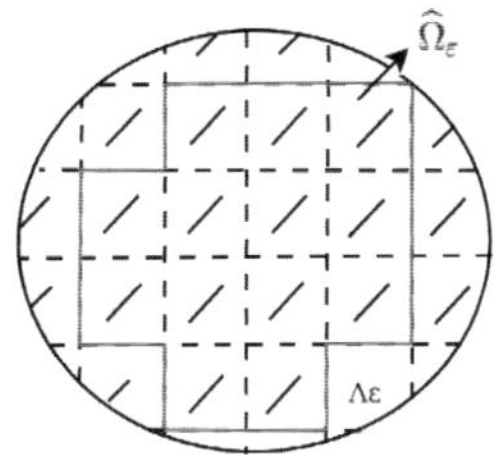

Figure 2. The domain $\widehat{\Omega}_\varepsilon$

In the periodic setting, almost every point $z \in \mathbb{R}^N$ can be written as $z = [z]_Y + \{z\}_Y$, $[z]_Y \in Z^2$, $\{z\}_Y \in Y$. Then, for almost every $z \in \mathbb{R}^2$, there exists a unique element in $\mathbb{R}^2$, denotes by $\left[\frac{x}{\varepsilon}\right]_Y$, such that $x - \varepsilon\left[\frac{x}{\varepsilon}\right]_Y = \varepsilon\left\{\frac{x}{\varepsilon}\right\}_Y$, where $\left\{\frac{x}{\varepsilon}\right\}_Y \in Y$.

Definition 1. Let $p \in [1, +\infty]$. For all $\phi \in L^p(\Omega_\varepsilon)$, the unfolding operator $\mathcal{T}_\varepsilon \in L^p(\Omega \times Y_\Gamma)$ is defined as follows:

$$\mathcal{T}_\varepsilon(\phi)(x,y) = \phi(\varepsilon\left[\frac{x}{\varepsilon}\right] + \varepsilon y) \text{ for a.e. } (x,y) \in \widehat{\Omega}_\varepsilon \times Y_\Gamma, \;\; \mathcal{T}_\varepsilon(\phi)(x,y) = 0 \text{ for a.e. } (x,y) \in \Lambda_\varepsilon \times Y_\Gamma.$$

The main properties of the unfolding operator for the fixed domain, proved in [3], can be adapted for a domain with cracks.

3. Problem on equilibrium of an elastic body with periodically located cracks

In our model it is supposed, that the edges of the cracks can not penetrate each other. Therefore nonlinear boundary condition is given in the form of the following inequality: $[\mathbf{u}]\nu = [\mathbf{u}_\nu] \geq 0$ on Γ, where $[\mathbf{u}] = \mathbf{u}^+ - \mathbf{u}^-$ denotes the jump of the vector $\mathbf{u}$ on Γ, and $\mathbf{u}^\pm$ correspond to the values $\mathbf{u}$ on the edges $\Gamma^\pm$.

The equation of state is $\sigma_{ij} = a_{ijkl}e_{kl}(\mathbf{u})$, where $e(\mathbf{u}) = \{e_{kl}(\mathbf{u})\}$ is the strain tensor, $e_{kl}(\mathbf{u}) = \frac{1}{2}\left(\frac{\partial u_k}{\partial x_l} + \frac{\partial u_l}{\partial x_k}\right)$, σ_{ij} is a stress tensor, $i,j,k,l = 1,2$; the tensor of elastic modulus a_{ijkl} has the usual properties of symmetry and positive definiteness: $a_{ijkl} = a_{jikl} = a_{klij}$, $a_{ijkl}\xi_{kl}\xi_{ij} \geq C_0|\xi|^2 \;\forall \xi_{ij} = \xi_{ji}$, $C_0 = \text{const} > 0, i,j,k,l = 1,2$. Here and below repeated indices are summed. For derivative we use the notation $\frac{\partial v}{\partial x_i} = v_{,i}$.

We define a convex set of the admissible displacement $K_\varepsilon = \{\mathbf{v} \in H^1_{\partial\Omega}(\Omega_\varepsilon) \mid [\mathbf{v}_\nu] \geq 0 \text{ on } \Gamma_\varepsilon\}$, where $H^1_{\partial\Omega}(\Omega_\varepsilon) = \{\mathbf{v} = (v_1, v_2) \in H^1(\Omega_\varepsilon) \mid \mathbf{v} = 0 \text{ on } \partial\Omega\}$. There is only one solution $\mathbf{u}^\varepsilon$ in K_ε of the variational inequality:

$$\mathbf{u}^\varepsilon \in K_\varepsilon, \quad \int\limits_{\Omega_\varepsilon} \sigma_{ij}(\mathbf{u}^\varepsilon)e_{ij}(\mathbf{v}-\mathbf{u}^\varepsilon) \geq \int\limits_{\Omega_\varepsilon} \mathbf{f}(\mathbf{v}-\mathbf{u}^\varepsilon) \quad \forall \mathbf{v} \in K_\varepsilon. \tag{1}$$

The strong formulation of the static problem (1) is the following: Find $\mathbf{u}^\varepsilon = (u_1^\varepsilon, u_2^\varepsilon)$ and $\sigma_{ij}(\mathbf{u}^\varepsilon)$, $i,j = 1,2$:

$$-\sigma_{ij,j} = \mathbf{f}, \;\; i,j = 1,2 \;\text{ in }\; \Omega_\varepsilon, \tag{2}$$

$$\mathbf{u}^\varepsilon = 0, \;\text{ on }\; \partial\Omega, \tag{3}$$

$$[\mathbf{u}_\nu^\varepsilon] \geq 0, \;\; [\sigma_\nu] = 0, \;\; \sigma_\nu \cdot [\mathbf{u}^\varepsilon]\boldsymbol{\nu} = 0 \;\text{ on }\; \Gamma_\varepsilon, \tag{4}$$

$$\sigma_\nu \leq 0, \;\; \boldsymbol{\sigma}_\tau = 0 \;\text{ on }\; \Gamma_\varepsilon^\pm, \tag{5}$$

where $\mathbf{f} = (f_1, f_2) \in L^2(\Omega)$ are given external forces, σ_ν, $\boldsymbol{\sigma}_\tau$ are normal and tangent components of the stress tensor at the boundary respectively, $\sigma_\nu = \sigma_{ij}\nu_j\nu_i$, $\boldsymbol{\sigma}_\tau = \sigma\boldsymbol{\nu} - \sigma_\nu \cdot \boldsymbol{\nu}$, $\boldsymbol{\sigma}_\tau = (\sigma_{\tau 1}, \sigma_{\tau 2})$, $\sigma\boldsymbol{\nu} = (\sigma_{1j}\nu_j, \sigma_{2j}\nu_j)$. The models of the form (2) - (5) were studied in the works [1], [2], [7] and so on.

4. Homogenization process

Below we give a result related to the unfolding method. Let $\mathbf{u}_\varepsilon$ is a solution of the problem (1).

Definition 2. Let be ω an open set, strictly included in Ω. For every $\varepsilon \leq \dfrac{dist(\omega, \partial\Omega)}{4}$ and for every $\forall\, \boldsymbol{\phi} \in L^p(\Omega_\varepsilon)$, $p \in [1, +\infty]$, we define $Q_\varepsilon \in W^{1,\infty}(\omega; L^p(Y_\Gamma))$:

for $\xi \in \Xi_\varepsilon$, $Q_\varepsilon(\phi)(\varepsilon\xi, y) = \mathcal{T}_\varepsilon(\phi)(\varepsilon\xi, y)$ for $y \in Y_\Gamma$,

for $x \in \omega$, $Q_\varepsilon(\phi)(x,y)$ is the Q_1 interpolate of $Q_\varepsilon(\phi)$ at the vertices of the cell $\left(\varepsilon\left[\dfrac{x}{\varepsilon}\right]_Y + \varepsilon Y\right) \times \{y\}$ for $y \in Y_\Gamma$.

Lemma. Suppose $\omega \Subset \Omega$. One has

$$\|Q_\varepsilon(e(\mathbf{u}^\varepsilon))\|_{L^2(\omega, H^1(Y_\Gamma))} \leq C\|\mathbf{f}\|_{L^2(\Omega)}, \; \|Q_\varepsilon(e(\mathbf{u}^\varepsilon)) - \mathcal{T}_\varepsilon(e(\mathbf{u}^\varepsilon))\|_{L^2(\omega; H^1(Y_\Gamma))} \leq C\varepsilon\|\mathbf{f}\|_{L^2(\Omega)}. \tag{6}$$

The constants do not depend on ε (it depends on ω).

Proof. For any function $\phi \in L^2(\Omega_\varepsilon)$ with support strictly included in Ω and for ε small enough denote $\Delta_k(\phi) = \phi(\cdot + \varepsilon\mathbf{k}) - \phi$, where $\mathbf{k} = \mathbf{e_i}$, $i = 1,2$. Then $\Delta_k(\phi) \in L^2(\Omega_\varepsilon)$. Let $dist(\omega, \partial\Omega) = \delta > 0$. There exists a function $\rho_\omega \in \mathcal{D}(\Omega)$ such that $0 \leq \rho_\omega(x) \leq 1 \; \forall x \in \Omega$ and $\rho_\omega = 0$, when $dist(x, \partial\Omega) \leq \frac{\delta}{4}$, $\rho_\omega = 1$, when $dist(x, \partial\Omega) \geq \frac{\delta}{2}$, $x \in \Omega$. Function ρ_ω satisfies:

$$\|\frac{\partial\rho_\omega}{\partial x_i}\|_{L^\infty(\Omega)} \leq \frac{C}{\delta}, \; \|\Delta_k \frac{\partial\rho_\omega}{\partial x_i}\|_{L^\infty(\Omega)} \leq \frac{C\varepsilon}{\delta^2}, \tag{7}$$

where $i,j = 1,2$, the constant C depends on $\partial\Omega$.

Let $\mathbf{v} \in K_\varepsilon$. In (1) choose as a test function $\mathbf{v}^\varepsilon = \mathbf{u}^\varepsilon + \rho_\omega(\mathbf{v} - \rho_\omega\mathbf{u}^\varepsilon) \in K_\varepsilon$. By entering the notations $\bar{\mathbf{u}}^\varepsilon = \rho_\omega\mathbf{u}^\varepsilon$ and $\bar{\mathbf{f}}^\varepsilon = \rho_\omega\mathbf{f}^\varepsilon$ one can obtains

$$\mathbf{a}^\varepsilon(\bar{\mathbf{u}}^\varepsilon, \mathbf{v} - \bar{\mathbf{u}}^\varepsilon) \geq (\bar{\mathbf{f}}, \mathbf{v} - \bar{\mathbf{u}}^\varepsilon) + \mathbf{b}_\varepsilon(\mathbf{u}^\varepsilon, \mathbf{v} - \bar{\mathbf{u}}^\varepsilon) \;\forall\, \mathbf{v} \in K_\varepsilon \tag{8}$$

where $\mathbf{a}^\varepsilon$ is a bilinear form, defined by the left hand of inequality (1) and $\mathbf{b}_\varepsilon$ is a bilinear form, which is equal to

$$\mathbf{b}_\varepsilon(\mathbf{u}^\varepsilon, \mathbf{w}) = \int\limits_{\Omega_\varepsilon} a_{ijkl}\left(\frac{\partial\rho_\omega}{\partial x_i}\mathbf{u}^{\varepsilon^j} e_{kl}(\mathbf{w}) - e_{ij}(\mathbf{u}^\varepsilon)\frac{\partial\rho_\omega}{\partial x_k}\mathbf{w}^l\right) dx.$$

For a small enough ε the set $(\varepsilon\mathbf{k}+\varepsilon\Gamma_\varepsilon)\cap\omega \subset \Omega$ and $\bar{\mathbf{u}}^\varepsilon(\cdot\pm\varepsilon\mathbf{k}) \in H_\varepsilon(\Omega)\cup K_\varepsilon$. Further, substituting the $\mathbf{v} = \bar{\mathbf{u}}^\varepsilon(\cdot \pm \varepsilon\mathbf{k})$ in (8) and adding to each other the resulting inequalities we can obtain

$$\mathbf{a}^\varepsilon(\Delta_k\bar{\mathbf{u}}^\varepsilon, \Delta_k\bar{\mathbf{u}}_\varepsilon) \leq (\Delta_k\bar{\mathbf{f}}, \Delta_k\bar{\mathbf{u}}^\varepsilon) + \mathbf{b}_\varepsilon(\Delta_k\mathbf{u}^\varepsilon, \Delta_k\bar{\mathbf{u}}^\varepsilon) \tag{9}$$

Notice, that

$$\|\mathbf{u}^\varepsilon\|_{H^1(\Omega_\varepsilon)} \leq C\|\mathbf{f}\|_{L^2(\Omega)}. \tag{10}$$

Besides,

$$\|\Delta_{-k}\Delta(\bar{\mathbf{u}}^\varepsilon)\|_{L^2(\Omega_\varepsilon)} \leq C\varepsilon\|\nabla(\Delta_k(\bar{\mathbf{u}}^\varepsilon))\|_{L^2(\Omega_\varepsilon)} \leq C\varepsilon\|e(\Delta_k(\bar{\mathbf{u}}^\varepsilon))\|_{L^2(\Omega_\varepsilon)}. \tag{11}$$

Therefore, using the above estimates (7), (10) and (11) one can show that

$$|\mathbf{b}_\varepsilon(\Delta_k\mathbf{u}^\varepsilon, \Delta_k\bar{\mathbf{u}}^\varepsilon)| \leq C\varepsilon\|\mathbf{f}\|_{L^2(\Omega)}\|e(\Delta_k\bar{\mathbf{u}}^\varepsilon)\|_{L^2(\Omega_\varepsilon)} + C\varepsilon^2\|\mathbf{f}\|^2_{L^2(\Omega)}. \tag{12}$$

Further, neglecting the terms of the order ε^2 in (12) and using the positive definiteness of the tensor a_{ijkl}, we obtain the following estimate:

$$\|e(\Delta_k\bar{\mathbf{u}}^\varepsilon)\|_{L^2(\Omega_\varepsilon)} \leq C\varepsilon\|\mathbf{f}\|_{L^2(\Omega)}, \tag{13}$$

where the constant C does not depend on ε, but it depends on ω.

By definition of Q_1- interpolate, the operator Q_ε reaches its maximum at some $\varepsilon\xi$. So, the first estimate in (6) follows from the properties of the unfolding operator $\mathcal{T}_\varepsilon$ (see [3],[4]) and the estimate (10). The second one is a consequence of the properties of the operator $\mathcal{T}_\varepsilon$ and the estimate (13).

Let be O a bounded domain in with Lipschitz boundary. Denote $W(O) = \{\mathbf{v} \in (H^1(O))^2 \mid \int_O \mathbf{v}(x)\mathbf{r}(x)dx = 0 \text{ for all } \mathbf{r} \in R\}$, where R is the space of rigid displacements $R = \{\mathbf{r} = (r_1, r_2) \mid \mathbf{r}(x) = a + b\wedge x,\ (a,b) \in \mathbb{R}^2\times\mathbb{R}^2\}$. The space $W(O)$ is the orthogonal of R in $(H^1(O))^2$ for the scalar product $\ll \mathbf{u},\mathbf{v} \gg = \int_O e(\mathbf{u})e(\mathbf{v})dx + \int_O \mathbf{u}\mathbf{v}dx$.

Let $L^2(\Omega; K_{per}(Y_\Gamma)) = \{\mathbf{v}_1 \in L^2(\Omega; H^1_{per}(Y_\Gamma))^2 |\ \mathbf{v}_1(x,\cdot) \in K_{per}(Y_\Gamma) \text{ for } x \in \Omega\}$, where $K_{per}(Y_\Gamma) = \{\phi \in H^1_{per}(Y_\Gamma)^2 \cap W(Y_\Gamma) \mid [\phi_\nu] \geq 0 \text{ on } \Gamma\}$.

Now, for every open set ω with Lipschitz boundary, satisfying $\omega \Subset \Omega' \Subset \Omega$, due to estimates (6) and the compact embedding theorems, one obtains

$$Q_\varepsilon(e(\mathbf{u}^\varepsilon)) \to e(\mathbf{u}^0) + e_y(\mathbf{u}^1) \text{ weakly in } H^1(\omega; H^1(Y_\Gamma))^{2\times2}; \tag{14}$$

$$Q_\varepsilon(e(\mathbf{u}^\varepsilon)) \to e(\mathbf{u}^0) + e_y(\mathbf{u}^1) \text{ strongly in } L^2(\omega\times Y_\Gamma)^{2\times2}; \tag{15}$$

$$\mathcal{T}_\varepsilon(e(\mathbf{u}^\varepsilon)) \to e(\mathbf{u}^0) + e_y(\mathbf{u}^1) \text{ strongly in } L^2(\omega\times Y_\Gamma)^{2\times2}. \tag{16}$$

Theorem 1. Let $\mathbf{u}^\varepsilon$ is a solution of the problem (1).

There exists a subsequence, still denoted $\{\mathbf{u}^\varepsilon\}$, $\mathbf{u}^0 \in H^1_\Gamma(\Omega)^2$, and $\mathbf{u}^1 \in L^2(\Omega; K^1_{per}(Y_\Gamma))$ such that the following convergences hold:

$$\mathcal{T}_\varepsilon(\mathbf{u}^\varepsilon) \to \mathbf{u}^0 \text{ strongly in } L^2(\Omega; H^1(Y_\Gamma))^2,$$
$$\mathcal{T}_\varepsilon(e(\mathbf{u}^\varepsilon)) \to e(\mathbf{u}^0) + e_y(\mathbf{u}^1) \text{ strongly in } L^2(\Omega\times Y_\Gamma)^{2\times2}.$$

Besides, the pair $(\mathbf{u}^0, \mathbf{u}^1) \in (H^1_0(\Omega))^2 \times L^2(\Omega; K_{per}(Y_\Gamma)$ is a unique solution of the following variational inequality problem:

$$\int_\Omega\int_Y a_{ijkl}(e_{kl}(\mathbf{u}^0) + e_{y,kl}(\mathbf{u}^1))(e_{ij}(\mathbf{v}-\mathbf{u}^0) + e_{y,ij}(\mathbf{w}-\mathbf{u}^1))dxdy \geq \int_\Omega f(\mathbf{v}-\mathbf{u}^0)dx$$

$\forall\ (\mathbf{v},\mathbf{w}) \in (H^1_0(\Omega))^2 \times L^2(\Omega; K_{per}(Y_\Gamma))$.

The proof is based on the above convergences (14)-(16) and the standart substitution of test functions.

5. Iterative method for equation with penalty

We introduce spaces $H_{Y\Gamma} = \{\mathbf{v} = (v_1, v_2) : v_i \in H^1(Y_\Gamma), \quad v_i - 1 \text{ periodic by } y, \ i = 1,2\}$, $\tilde{H}_{Y\Gamma} = \{\mathbf{v} \in H_{Y\Gamma} : \bar{\mathbf{v}} = 0\}$, where $\bar{\mathbf{v}} = \int\limits_Y \mathbf{v} dy$, and the sets of functions on the cell Y_Γ :

$$K_{Y\Gamma} = \{\mathbf{v} \in H_{Y\Gamma} : \ [\mathbf{v}_\nu^1] \geq 0 \ \text{ on } \Gamma\}, \ \ \tilde{K}_{Y\Gamma} = \{\mathbf{v} \in K_{Y\Gamma} : \ \bar{\mathbf{v}} = 0\}.$$

Consider functions $\mathbf{u}^\varepsilon$ and $\mathbf{v} = \mathbf{v}^\varepsilon$ in the form

$$\mathbf{u}^\varepsilon = \mathbf{u}^0(x) + \varepsilon \mathbf{u}^1(x,y) + \varepsilon^2 \mathbf{u}^2(x,y) + ..., \ , \tag{17}$$

and $\mathbf{v}^\varepsilon = \mathbf{v}^0(x) + \varepsilon \mathbf{v}^1(x,y) + \varepsilon^2 \mathbf{v}^2(x,y) + ..., \quad y = \frac{x}{\varepsilon}, \quad x \in \Omega$. Then,using inequality (1), one can obtained problems for functions $\mathbf{u}^0$, $\mathbf{u}^1$ from the formal expansion (17), as it is shown in [1, §6 chapter VI]. The function $\mathbf{u}^0$ is a solution of the following variational problem: Find $\mathbf{u}^0 \in (H_0^1(\Omega))^2$ such that

$$\int\limits_\Omega \int\limits_Y a_{ijkl} \left(\frac{\partial u_k^0}{\partial x_l} + \frac{\partial u_k^1}{\partial y_l} \right) \frac{\partial}{\partial x_j} (v_i^0 - u_i^0) dx dy - \int\limits_\Omega \mathbf{f}(\mathbf{v}^0 - \mathbf{u}^0) dx = 0 \ \ \forall \mathbf{v} \in (H_0^1(\Omega))^2.$$

This problem is equivalent to the following boundary problem: Find $\mathbf{u}^0 \in (H_0^1(\Omega))^2$ such that

$$-\bar{\sigma}_{ij,j}(\mathbf{u}^0) = \mathbf{f} \ \text{ in } \ \Omega, \ \mathbf{u}^0 = 0 \ \text{ on } \ \partial\Omega \tag{18}$$

where $\bar{\sigma}_{ij} = \int\limits_Y \sigma_{ij}(\mathbf{u}^0) dy, \ \ \sigma_{ij}(\mathbf{u}^0) = a_{ijkl} \left(\frac{\partial u_k^0}{\partial x_l} + \frac{\partial u_k^1}{\partial y_l} \right), \ i,j = 1,2.$

Now consider a local problem on Y.

For fixed x and $\mathbf{u}^0$ the function $\mathbf{u}^1(x,y) \in K_{Y\Gamma}$ satisfies the following variational inequality

$$\int\limits_Y a_{ijkl} \left(\frac{\partial u_k^0}{\partial x_l} + \frac{\partial u_k^1}{\partial y_l} \right) \frac{\partial}{\partial y_j} (w_i - u_i^1) dy \geq 0 \ \ \forall w_i \in K_{Y\Gamma}. \tag{19}$$

Note, that the inequality (19) includes only gradiends in the y variable of the functions $\mathbf{u}^1$, $\mathbf{w}$. Therefore we can add a constant to each function and the left hand of the inequality (19) does not change. So we can use spaces $\tilde{H}_{Y\Gamma}$ and $\tilde{K}_{Y\Gamma}$ instead of $H_{Y\Gamma}$, $K_{Y\Gamma}$.

Introduce the following bilinear and linear forms in $H_{Y\Gamma}$:

$$a_Y(\mathbf{v}^1, \mathbf{w}) = \int\limits_Y a_{ijkl} \frac{\partial v_k^1}{\partial y_l} \frac{\partial w_i}{\partial y_j} dy, \ \ L(\mathbf{w}) = \int\limits_Y a_{ijkl} \frac{\partial u_k^0}{\partial x_l} \frac{\partial w_i}{\partial y_j} dy.$$

Then we can rewrite the variational inequality (19) in the form: Find a function $\mathbf{u}^1(x,y) \in \tilde{K}_{Y\Gamma}$ such that

$$a_Y(\mathbf{u}^1, \mathbf{w} - \mathbf{u}^1) + L(\mathbf{w} - \mathbf{u}^1) \geqslant 0 \tag{20}$$

for all $\mathbf{w} \in \tilde{K}_{Y\Gamma}$. The local problem (20) has an unique solution on $\tilde{K}_{Y\Gamma}$ (see [1, Theorem 6.1, §7 chapter VI]). Convergence of the solution $\mathbf{u}^\varepsilon$ of the variational inequality (1) to the solution $\mathbf{u}^0$ of the homogenised problem (18) is proved in [2, Theorem 4, §2].

Let $H(Y)$ be the space $(\tilde{H}_{Y\Gamma})^2$, with norm $\|v\|_{H(Y)}^2 = \|v\|_{L^2(Y_\Gamma)}^2 + \sum_{i=1}^2 \|v_{,i}\|_{L^2(Y_\Gamma)}^2$. Then we introduce the penalty operator $\beta(\mathbf{u}^1) : H(Y) \to H(Y)^*$

$$\langle \beta(\mathbf{u}^1), \mathbf{v} \rangle = - \int\limits_\Gamma [\mathbf{u}_\nu^1]_- [\mathbf{v}_\nu] d\Gamma,$$

where angle brackets denote the duality between the spaces $H(Y)$ and $H(Y)^*$ and lower minus sign is a negative part of a function, i.e.: $[\mathbf{u}^1_\nu]_- = 0$, if $[\mathbf{u}^1_\nu] \geq 0$; and $[\mathbf{u}^1_\nu]_- = -[\mathbf{u}^1_\nu]$, if $[\mathbf{u}^1_\nu] < 0$.

Consider the nonlinear penalty equation with the parameter $\delta > 0$

$$a_Y(\mathbf{u}^{1\delta}, \mathbf{w}) + \frac{1}{\delta}\langle \beta(\mathbf{u}^{1\delta}), \mathbf{w}\rangle = -L(\mathbf{w}) \;\; \forall\, \mathbf{w} \in H(Y). \tag{21}$$

Theorem 2. Let $\mathbf{u}^{1\delta}$ is a solution of the penalty problem (21) and $\mathbf{u}^1$ is a solution of the problem (20). Then

$$\mathbf{u}^{1\delta} \to \mathbf{u}^1 \text{ strongly in } H(Y) \text{ with } \delta \to 0.$$

Let introduce scalar product $(\mathbf{u}, \mathbf{w}) = a_Y(\mathbf{u}, \mathbf{w}) + \int\limits_\Gamma [\mathbf{u}_\nu][\mathbf{w}_\nu] d\Gamma$ and equivalent norm $\|\mathbf{u}\|^2_a = (\mathbf{u}, \mathbf{u})$ in the given space $H(Y)$.

To solve the nonlinear penalty problem (21) approximately we construct the following iterative procedure for fixed δ and for a given function $\mathbf{u}^{1\delta,0} \in H(Y)$, $n = 0, 1, 2, ..$:

$$a_Y(\mathbf{u}^{1\delta,n+1}, \mathbf{w}) + \frac{1}{\delta}(\mathbf{u}^{1\delta,n+1} - \mathbf{u}^{1\delta,n}, \mathbf{w}) = -L(\mathbf{w}) - \frac{1}{\delta}\langle \beta(\mathbf{u}^{1\delta,n}), \mathbf{w}\rangle \;\; \forall\, \mathbf{w} \in H(Y). \tag{22}$$

There exists a solution $\mathbf{u}^{1\delta,n+1} \in H(Y)$ of the problem (22).

Theorem 3. There is convergence $\mathbf{u}^{1\delta,n+1} \to \mathbf{u}^{1\delta}$ strongly in $H(Y)$ with $n \to \infty$. Moreover,

$$\|\mathbf{u}^{1\delta,n+1} - \mathbf{u}^{1\delta}\|^2_a \leq (1 + 2C\delta)^{-(n+1)} \|\mathbf{u}^{1\delta,0} - \mathbf{u}^{1\delta}\|^2_a.$$

Proofs of Theorem 2 and Theorem 3 are given in [8].

References

[1] Sanchez-Palencia E. 1984 *Non-Homogeneous Media and Vibration Theory* (Moscow, Mir)
[2] Pastukhova S. E. 2000 On homogenization of a variational inequality for an elastic body with periodically distributed fissures *Sb. Math.* **191** No.2 291–306
[3] Cioranescu D., Damlamian A. and Griso G. 2008 The Periodic Unfolding Method in Homogenization *SIAM J. Math. Anal. Society for Industrial and Applied Mathematics* **40** No. 4 1585–620
[4] Cioranescu D., Damlamian A. and Orlik J. 2013 Homogenization via unfolding in periodic elasticity with contact on closed and open cracks *Asymptotic Analysis* **82** 201–32
[5] Griso G., Migunova A. and Orlik J. 2016 Homogenization via unfolding in periodic layer with contact *Asymptotic Analysis. IOS Press* **99** 23–52
[6] Griso G., Orlik J. Homogenization of contact problem with Coulomb's friction on periodic cracks arXiv:1811.06615 [math.AP]
[7] Khludnev A.M. 2010 *Elasticity Problems in Nonsmooth Domains* (Moscow, Fizmathlit)
[8] Neustroeva N.V., Afanaseva N.M.,Egorova A.A. 2019 Variational problem for an elastic body with periodically located cracks *Mat.Zamet.SVFU* **26** No. 2

Multiscale Finite Element Method for scattering problem in heterogeneous domain

Uygulaana Gavrilieva[1], **Maria Vasilyeva**[2], **Isaac Harris**[3], **Eric T. Chung**[4], **Yalchin Efendiev**[5]

[1]Multiscale model reduction laboratory, North-Eastern Federal University, Yakutsk, Republic of Sakha (Yakutia), Russia, 677980
[2]Institute for Scientific Computation, Texas A&M University, College Station, TX 77843 & Department of Computational Technologies, North-Eastern Federal University, Yakutsk, Republic of Sakha (Yakutia), Russia, 677980
[3]Department of Mathematics, Purdue University, West Lafayette, Indiana USA
[4]Department of Mathematics, The Chinese University of Hong Kong (CUHK), Hong Kong SAR
[5]Department of Mathematics & Institute for Scientific Computation (ISC), Texas A&M University, College Station, Texas, USA

E-mail: `lanasemna@mail.ru`[1], `vasilyevadotmdotv@gmail.com`[2], `tschung@math.cuhk.edu.hk`[4], `efendiev@math.tamu.edu`[5]

Abstract. In this paper, we consider the wave scattering problem in heterogeneous domain. The mathematical model is described by the Helmholtz equation for time-harmonic wave propagation with absorbing boundary condition. We construct a coarse grid approximation for the efficient numerical solution on coarse grids using the Multiscale Finite Element Method (MsFEM), where we construct a multiscale space using solution of some local problems in each local domain. Numerical results are presented to illustrate the performance of the method.

1. Introduction

In this paper, we consider time harmonic wave scattering in heterogeneous media. These problems occur in many real world applications, for example, in engineering design, where the wave propagation through composite materials is used to understand particular microstructure behavior and construct materials with special properties [3]. For problems in periodic media, macroscale mathematical model is constructed by the homogenization theory [4, 3, 2, 1]. For the wave propagation in non-periodic media, macroscale models are developed based on the numerical homogenization or using multiscale methods [6, 7, 8, 10, 5, 9]. In this work, we present a Multiscale Finite Element Method (MsFEM) for solution of wave propagation in heterogeneous media. For the construction of the coarse grid approximation, we solve local problems to construct multiscale basis functions. We test two approaches for construction of the local basis functions based on the two types of local problems using elliptic part of operator and using the Helmholtz operator. We present numerical results for two-dimensional model problem and study the performance of the proposed method for various wave numbers.

The paper is organized as follows. In Section 2, we present the mathematical model and fine grid approximation of the wave propagation problem in heterogeneous media. In Section 3, we present a coarse grid approximation based on the Multiscale Finite Element Method. Numerical results for two-dimensional model problem is presented in Section 4.

2. Problem formulation and fine grid approximation

We consider the comlex-valued Helmholtz problem in heterogeneous domain Ω

$$\nabla \cdot (A(x)\nabla u_s) + k^2 n(x)\, u_s = f(x), \quad x \in \Omega, \tag{1}$$

where the wave number k is constant and assumed to be positive, $u = u_i + u_s$ is the total field, u_s is the scattered field and

$$\Omega = \Omega_0 \cup \sum_{j=1,2,...} D_j.$$

with

$$n(x) = \begin{cases} 1 & x \in \Omega_0 \\ \alpha_j & x \in D_j, j = 1,2,.. \end{cases}, \quad A(x) = \begin{cases} \mathcal{I} & x \in \Omega_0 \\ A_j & x \in D_j, j = 1,2,.. \end{cases}$$

and non-zero right hand side in subdomains D_j

$$f(x) = \nabla \cdot ((I - A(x))\nabla u_i) + k^2(1 - n(x))u_i,$$

where $u_i = e^{ikx\cdot d} = \cos(kx \cdot d) + i\sin(kx \cdot d)$ is the given incident field and d is the incident wave direction.

We consider equation with Sommerfield radiation condition

$$\frac{\partial u_s}{\partial n_b} - iku_s = 0, \quad x \in \partial\Omega,$$

where n_b is the unit normal on boundary.

For approximation of the Helmholtz equation on the fine grid, we use finite element method and have following variational formulation: find $u \in V_h = H_1(\Omega)$ such that

$$a(u_s, v) - k^2 m(u_s, v) - ib(u_s, v) = l(v), \quad v \in V_h, \tag{2}$$

where

$$a(u,v) = \int_\Omega A\nabla u \cdot \nabla v\, dx, \quad m(u,v) = \int_\Omega n\, u\, v\, dx,$$

$$b(u,v) = \int_{\partial\Omega} k\, u\, v\, ds, \quad l(v) = -\int_\Omega f\, v\, dx,$$

and $u_s = Re(u_s) + iIm(u_s)$.

We can write the complex valued problem in matrix form

$$(K_h - k^2 M_h - iB_h)\, U = F_h, \tag{3}$$

where $U = Re(U) + iIm(U)$, $(F_h)_j = l(\phi_j)$, and complex valued matrices M_h, B_h and K_h are given by

$$(M_h)_{rj} = m(\phi_r, \phi_j), \quad (B_h)_{rj} = b(\phi_r, \phi_j), \quad (K_h)_{rj} = a(\phi_r, \phi_j),$$

and $u_s = \sum_j U_j \phi_j$, ϕ_j is the linear basis functions for fine grid approximation.

3. Coarse grid approximation using MsFEM

For coarse-scale approximation, we use Multiscale Finite Element Method with following computational algorithm:

- the construction of the multiscale basis functions in local domain ω_i and
- the construction and solution of the coarse grid system on multiscale space.

We define the coarse grid $\mathcal{T}_H$ in domain Ω, $\mathcal{T}_H = \cup_{i=1}^{N_c} \omega_i$ with mesh sizes $H \gg h > 0$ where ω_i is obtained by the combining all the coarse cells K around one vertex of the coarse grid (local domain), N_c is number of cells. Let V_H be a finite dimensional function space, which consists of fuctions that are smooth on each local domain related to the coarse mesh cell. We construct a local multiscale space $V_H^{\omega_i}$ by solution of the local problems. The global multiscale space V_H is then defined as the linear span of all $V_H^{\omega_i}$, $\omega_i \in \mathcal{T}_H$.

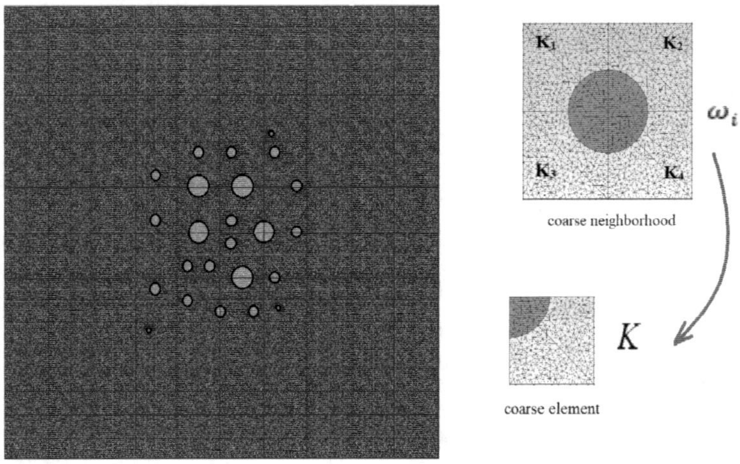

For approximation on the coarse grid, we use a continuous Galerkin (CG) approach and have following variational formulation: find $u_H \in V_H$:

$$a(u_H, v) - k^2 m(u_H, v) - ib(u_H, v) = l(v), \quad v \in V_H. \tag{4}$$

For construction of the multiscale basis functions in local domains, we consider following local problems in $K_j \in \omega_i$

- Case 1 with elliptic operator

$$\nabla \cdot (A(x) \nabla \phi_i) = 0, \quad x \in K_j, \quad \forall K_j \in \omega_i, \tag{5}$$

  with linear boundary condition on ∂K_j.

- Case 2 with full operator

$$\nabla \cdot (A(x) \nabla \phi_i) + k^2 n(x) \phi_i = 0, \quad x \in K_j, \quad \forall K_j \in \omega_i, \tag{6}$$

  with linear boundary condition on ∂K_j.

Then multiscale space V_H is defined as follow

$$V_H = \text{span}\{\phi_i\}.$$

For construction of the coarse grid system, we use a projection approach and define the projection operator

$$R = (\phi_1, ..., \phi_{N_c})^T,$$

and therefore Galerkin projection 4 can be written in the following matrix form

$$(K_H - k^2 M_H - iB_H)U_H = F_H, \quad (7)$$

where

$$K_H = RK_hR^T, \quad M_H = RM_hR^T, \quad B_H = RB_hR^T, \quad F_H = RF_h,$$

and the fine-scale representation of the multiscale solution is given by $U_{ms} = R^T U_H$.

4. Numerical results

We present a numerical results for the fine-scale solution and the coarse-scale solution using multiscale finite element method. Computational domain with heterogeneous obstacles is presented in Figure 1 and have dimensions $\Omega = [-2, 2] \times [-2, 2]$. For the computational mesh we consider fine mesh, 10×10 and 20×20 coarse meshes. Fine mesh contains 100802 vertices and 200482 triangle elements, structured coarse 10×10 mesh with 121 vertices, 100 triangle elements and 20×20 mesh 441 vertices, 400 triangle elements. We set $k = 0.1, 1, 2$ and $d = (1, 0)$. For background media, we set $\alpha = 1$ and $n = 1$ and for circle inclusions $\alpha = 10$ and $n = 0.2$.

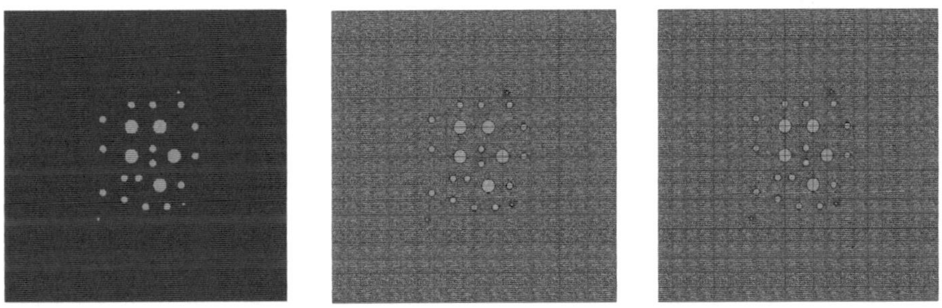

Figure 1. Heterogeneous domain with fine mesh that contains 100802 vertices and 200482 triangle elements. Coarse mesh with 121 vertices and 100 triangle elements (middle). Coarse mesh with 441 vertices and 400 triangle elements (right).

In Figure 2, we present the fine-scale and coarse-scale solution for the real part, imaginary part and absolute value with wave number $k = 1$. We obtain solution with good accuracy for reduced order model using MsFEM. In Table 1, we show a multiscale solver errors for Case 1 and Case 2 for the coarse grids with 121 and 441 local domains. We observe error reduction for a finer coarse grid. Furthermore, we have larger errors for higher values of the wave number k. Comparing the errors from Case 1 and Case 2, we see that the case with full operator provide smaller errors. Our results show that the presented method give good approximation of the solution and reduce size of system for scattering problems with the small wave numbers.

5. Acknowledgements

The paper is supported by the mega-grant of the Russian Federation Government N14.Y26.31.0013. The research of Eric Chung is partially supported by the Hong Kong RGC

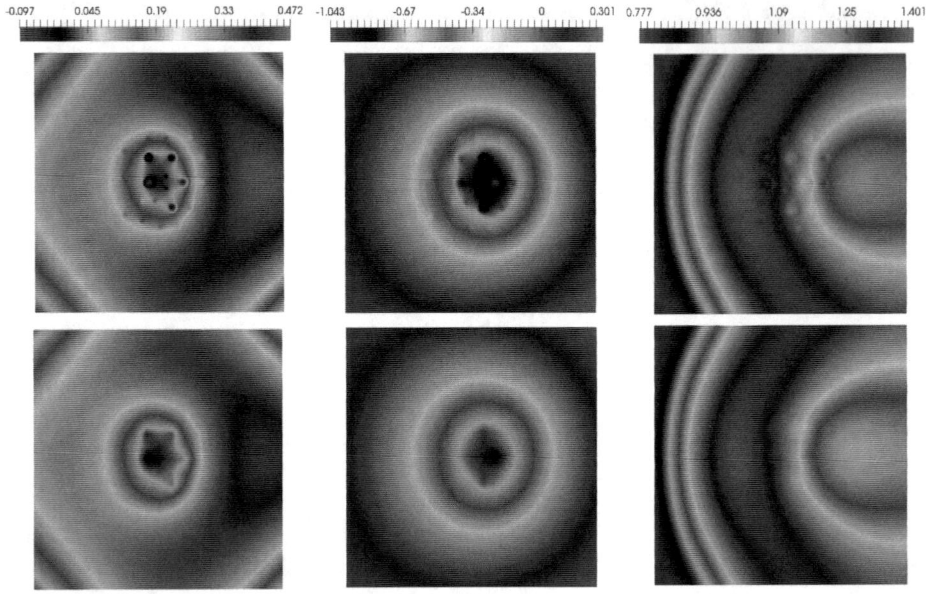

Figure 2. Fine-scale (top) and coarse-scale (bottom) solutions for heterogeneous domain with $k = 1.0$. Re(u), Im(u), Abs(u) (from left to right).

k	$\|\|e_{u_s^{Re}}\|\|_{L_2}$	$\|\|e_{u_s^{Im}}\|\|_{L_2}$	$\|\|e_{u_s^{Abs}}\|\|_{L_2}$
	10×10		
0.1	8.085	4.594	0.015
1	10.767	10.209	2.963
2	40.894	41.064	17.316
	20×20		
0.1	6.124	4.437	0.012
1	6.331	6.058	1.707
2	26.305	27.304	10.756

k	$\|\|e_{u_s^{Re}}\|\|_{L_2}$	$\|\|e_{u_s^{Im}}\|\|_{L_2}$	$\|\|e_{u_s^{Abs}}\|\|_{L_2}$
	10×10		
0.1	8.075	4.591	0.015
1	6.486	6.427	1.686
2	28.912	28.928	11.681
	20×20		
0.1	6.117	4.436	0.012
1	4.216	4.320	1.090
2	18.796	19.536	7.448

Table 1. Relative L_2 errors in %. Coarse mesh with 121 (10×10) and 441 local domains (20×20). Case 1 (left) and Case 2 (right).

General Research Fund (Project numbers 14304217 and 14302018) and CUHK Faculty of Science Direct Grant 2018-19.

References

[1] Fioralba Cakoni, David Colton, and Houssem Haddar. The interior transmission problem for regions with cavities. *SIAM Journal on Mathematical Analysis*, 42(1):145–162, 2010.

[2] Fioralba Cakoni, Bojan B Guzina, and Shari Moskow. On the homogenization of a scalar scattering problem for highly oscillating anisotropic media. *SIAM Journal on Mathematical Analysis*, 48(4):2532–2560, 2016.

[3] Fioralba Cakoni, Houssem Haddar, and Isaac Harris. Homogenization approach for the transmission eigenvalue problem for periodic media and application to the inverse problem. *arXiv preprint arXiv:1410.3729*, 2014.

[4] Fioralba Cakoni and Isaac Harris. The factorization method for a defective region in an anisotropic material. *Inverse Problems*, 31(2):025002, 2015.

[5] Eric T Chung, Yalchin Efendiev, Richard L Gibson, and Maria Vasilyeva. A generalized multiscale finite element method for elastic wave propagation in fractured media. *GEM-International Journal on Geomathematics*, 7(2):163–182, 2016.
[6] Eric T Chung, Yalchin Efendiev, and Wing Tat Leung. Generalized multiscale finite element methods for wave propagation in heterogeneous media. *Multiscale Modeling & Simulation*, 12(4):1691–1721, 2014.
[7] Shubin Fu, Eric Chung, and Guanglian Li. Edge multiscale methods for elliptic problems with heterogeneous coefficients. *Journal of Computational Physics*, 2019.
[8] Shubin Fu, Guanglian Li, Richard Craster, and Sebastien Guenneau. Wavelet-based edge multiscale finite element method for helmholtz problems in perforated domains. *arXiv preprint arXiv:1906.08453*, 2019.
[9] U Gavrilieva, V Alekseev, and M Vasilyeva. Numerical homogenization for wave propagation in fractured media. In *AIP Conference Proceedings*, volume 2025, page 100002. AIP Publishing, 2018.
[10] M Vasilyeva, JD De Basabe, Y Efendiev, and RL Gibson Jr. Multiscale model reduction of the wave propagation problem in viscoelastic fractured media. *Geophysical Journal International*, 217(1):558–571, 2019.

Neural networks for multicontinuum models

A. Grigorev, S. Stepanov, Dj. Nikiforov

Ammosov North-Eastern Federal University, Yakutsk, Russia

E-mail: `re5itsme@gmail.com`, `cepe2a@inbox.ru`, `dju92@mail.ru`

Abstract.
Neural networks are developed for multicontinuum models. Multicontinuum models [1] assume that processes divided into several continuums and they are connected by some exchange member. In problems connected with PDE not only solution, but also coefficients can depends on time. First we construct LSTM neural network for exchange coefficient that depend on time. We compare the results obtained on the basis of the reference coefficient and on the basis of the neural network. Next we apply same neural network on the solutions. The proposed approach involves the use of a neural network for the approximate generation of a solutions and the further adaptation of the solution of a neural network for generate coarse grid solution based on DG finite elements.

1. Introduction

Nowadays neural networks [2] are very popular technology for wide range of problems. Neural networks are subclass of machine learning methods. About last 5 years neural networks demonstrate better results than other methods in many applications — language analysis, image processing, artificial intellect, etc. There are several reasons for this — big amount of data (internet, data generation methods, kaggle, etc), high-performance computations (parallel calculations, supercomputers, GPU computing), modern algorithms (convolutional neural networks, recurrent neural networks, open-source libraries like tensorflow). For every kind of problem we need to apply special neural network. For example for image classification we need to use CNN. In problems connected with PDE not only solution, but also coefficients can depends on time. We will use LSTM [3] (one effective recurrent neural network) for coefficient predictions on time and expand this approach on solution prediction.

2. Neural network approach

Most of neural networks methods can be expressed as optimization algorithm to setup model parameters as weights and bias values for minimize error functional. A recurrent neural network (RNN) is a class of artificial neural network where connections between nodes form a directed graph along a temporal sequence. This allows it to exhibit temporal dynamic behavior. Unlike feedforward neural networks, RNNs can use their internal state (memory) to process sequences of inputs. This makes them applicable to tasks that connected with sequences. Let's discuss idea of LSTM. An LSTM network is a special type of RNN capable of learning long-term dependencies. LSTM networks were introduced [in Hochreiter and Schmidhuber, 1997]. Such networks do an excellent job with many tasks and are widely used at this time. LSTM networks are designed specifically to solve the problem of long-term dependencies. Storing information for long periods of time is their default behavior. All RNN have the form of a chain of repeating modules. The repeating module of a standard RNN has a very simple structure, for example, a single tanh layer (the activation function is the hyperbolic tangent). An LSTM network is a similar chain but a repeating module has a different structure. Instead of one layer it contains four layers which interact in a special way. A key feature of the LSTM network is the cell state. The state of the cell can be compared with a conveyor belt. It runs through the entire chain with only minor linear interactions. The LSTM network has the ability to remove and add information to the cell state. This process is governed by special structures called gateways. A gate is a mechanism for allowing information to be passed selectively. It consists of a sigmoid layer (activation function sigmoid) and a pointwise multiplication operation. The output of the sigmoid layer is a number from 0 to 1 which determines the level of transmission. The cell has three gates controlling its state. At the first stage it is necessary to decide what information should be removed from the state of the cell. This decision is made by a sigmoid layer called the forget gate. The next step is to decide what new information should be written to the cell state. This stage is divided into two parts. First the sigmoid layer called the input gate (input gate) decides which values need to be updated. The tanh layer then creates a vector of candidate values that can be added to the cell state. It's time to update the previous state of the cell to the current state. At the previous stages it was already decided what needs to be done now it remains only to accomplish this. Finally you need to decide what to send out. The output will be the filtered state of the cell. First the sigmoid layer decides which elements of the cell state should be transferred to the output. The state of the cell is then converted using a tanh layer to the interval from -1 to 1 and multiplied by the output of the sigmoid layer in order to output only what was decided to output.

3. Preprocessing problem of exchange coefficient

In many work researchers interested in exchange mechanism between continua (e.g. [4]). In our case we suppose that exchange coefficient depends on time. We generated CSV file that consist values of exchange coefficient that depends on time. This data builded with step 1 month. For simulate real data we add noise See Figure 1.

- Let's suppose exchange coefficient depends on time
- Let's suppose one-phase filtration flow not so deep underground
- Let's suppose the soil degrades over time so the exchange coefficient grows
- Period: 1999 Jan – 2028 Dec

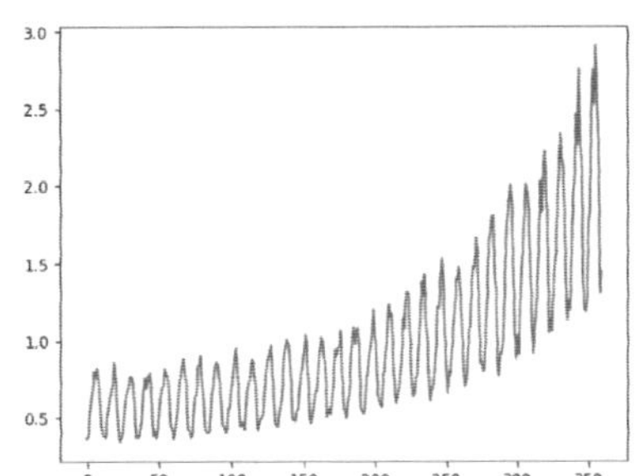

Figure 1. Exchange coefficient on time

We using open source libraries on Laptop (i7-8750H + Nvidia GeForce GTX 1050 Ti). Moreover for realise neural network part on GPU version of Tensorflow. TensorFlow is a free and open-source software library for dataflow and differentiable programming across a range of tasks. It is a symbolic math library, and is also used for machine learning applications such as neural networks.

We applying LSTM network based on next algorithm:

- We using SKlearn preprocessor to normalize data (0, 1)
- We divide generated data to train part and test part
- We build up Sequentional model using Keras library
- We add LSTM model to Sequentional with 4 extra layers
- Parameters of NN: epochs = 100, batch = 1

Comparing results

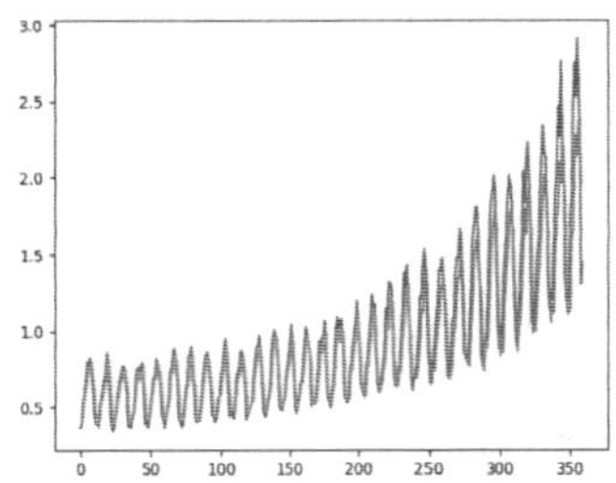

Noise 10% (Training 67%)

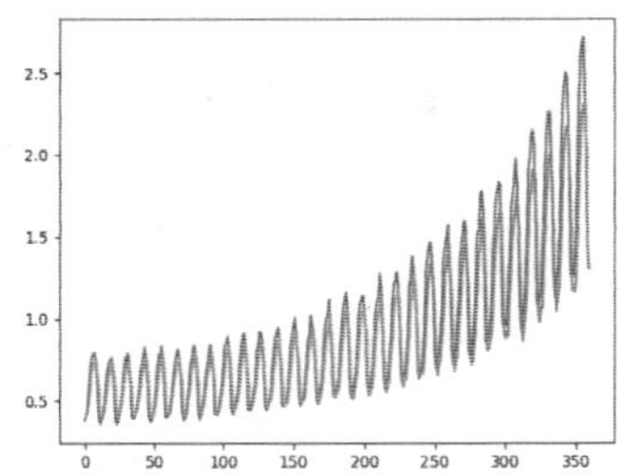

Noise 5% (Training 67%)

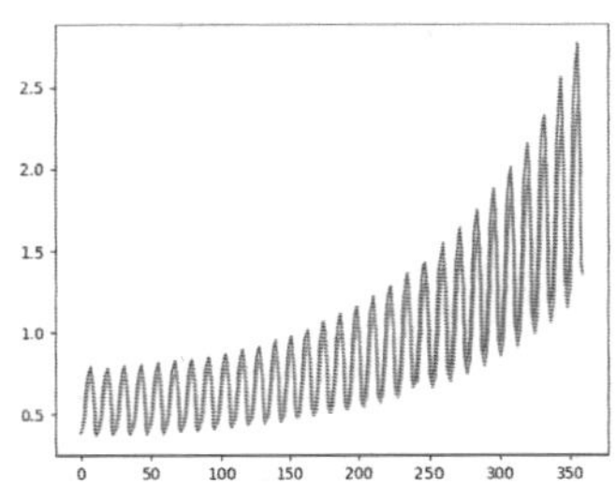

Noise 1% (Training 67%)

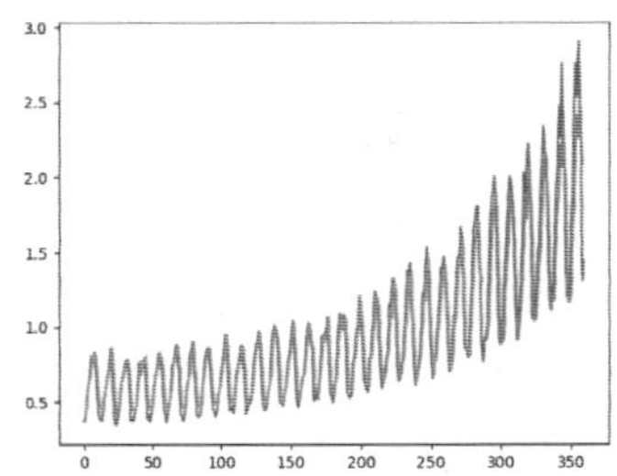

Noise 10% (Training 80%)

Figure 2. Blue – reference, orange – training, green - prediction

4. Model problem

Let's consider model problem with constant coefficient case excluding exchange coefficient:

$$c\frac{\partial u_1}{\partial t} - \nabla \cdot \nabla u_1 + r(t)(u_1 - u_2) = 0,$$

$$\frac{\partial u_2}{\partial t} - d\nabla \cdot \nabla u_2 - r(t)(u_1 - u_2) = 0.$$

We want to compare not only coefficient values but also results of modeling that based on reference coefficient and coefficient that predicted by neural network (last 2 year). We using

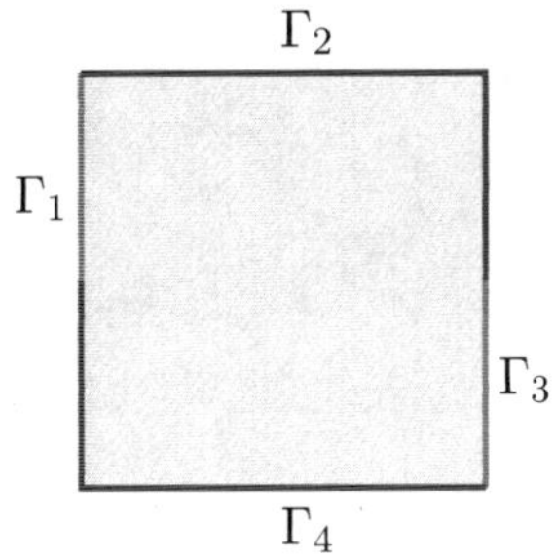

Figure 3. Unit rectangle

next computation domain Ω ($\Gamma_D = \Gamma_1 \cup \Gamma_3$, $\Gamma_N = \Gamma_2 \cup \Gamma_4$):
We applying next boundary conditions:

$$u_\alpha(\boldsymbol{x}, t) = 1 - \exp(-10t), \quad \boldsymbol{x} \in \Gamma_1,$$

$$u_\alpha(\boldsymbol{x}, t) = 0, \quad \boldsymbol{x} \in \Gamma_3,$$

$$\frac{\partial u_\alpha}{\partial \boldsymbol{n}}(\boldsymbol{x}, t) = 0, \quad \boldsymbol{x} \in \Gamma_2 \cup \Gamma_4,\, t \in (0,\, T],$$

and initial conditions:

$$u_\alpha(\boldsymbol{x}, 0) = 0, \quad x \in \Omega, \alpha = 1,\, 2.$$

Solving method

We implement finite element method using Fenics library [5]: CG 1st order ($\tau = 0.015$, Structured triangular mesh $32 \times 32 \times 2$). And on time we use fully implicit scheme:

$$c \int_\Omega \frac{y_1^{n+1} - y_1^n}{\tau} v_1 d\boldsymbol{x} + \int_\Omega \nabla y_1^{n+1} \nabla v_1 d\boldsymbol{x} + r \int_\Omega (y_1^{n+1} - y_2^{n+1}) v_1 d\boldsymbol{x} = 0,$$

$$\int_\Omega \frac{y_2^{n+1} - y_2^n}{\tau} v_2 d\boldsymbol{x} + d \int_\Omega \nabla y_2^{n+1} \nabla v_2 d\boldsymbol{x} + r \int_\Omega (y_2^{n+1} - y_1^{n+1}) v_2 d\boldsymbol{x} = 0.$$

On the model two-dimensional problem the dependence of the accuracy of an NN based solution on reference solution is investigated (See Figures 4, 5). Before we can see (Figure 2) LSTM neural network even can predict noise behavior of coefficient but also L_2 errornorm of solution is not exceed coefficient discrepancy maximum value (For example Figure 4).

Also we solution at the last moment of time which demonstrate adequate solution based on NN predicted coefficient (See Figure 6).

Next idea is we constructing continuation on the solution. We can apply the same LSTM neural network already on solutions that we obtain by simulating based coeficient predicted by LSTM. In this easier case LSTM neural network demonstrate much better results. And by predicting in set of points of domain we can construct coarse grid solution. On Figure 7 we demonstrate constructing coarse grid solution by pulling pointwise LSTM solution on piecewise constant DG finite elements [6].

Conclusions

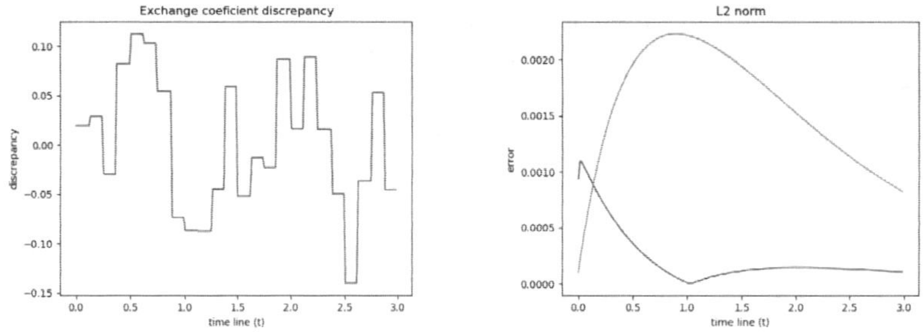

Coeficient discrepancy blue — fractures, orange — matrix

Figure 4. Comparing reference solution and NN coeficient based

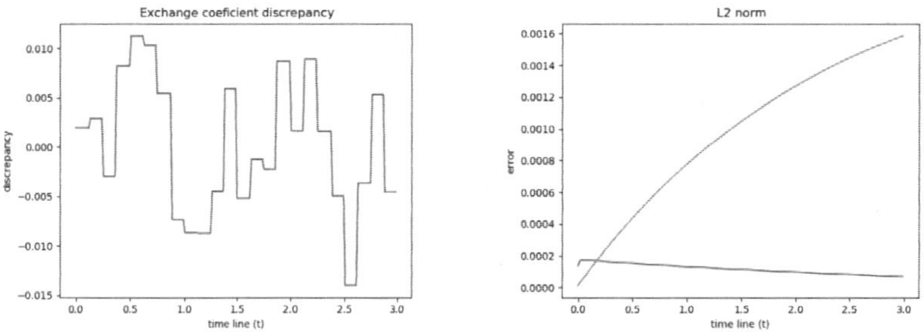

Coeficient discrepancy blue — fractures, orange — matrix

Figure 5. Comparing reference solution and NN coeficient based (×0.1)

- We apply LSTM network for coefficient preprocessing problem. Results demonstrate that neural network can recreate noise behaviour of time dependent sequences.
- We extended predictions on solutions of model problem. And this approach let us construct coarse grid solution. This method provide high accuracy of NN predictions for pointwise solution.

4.1. Acknowledgments
The work was supported by RSCF grant 17-71-10106 for A. Grigorev, Stepanov's work is supported by grant of RFBR N 18-31-00318

References
[1] Barenblatt, G. I., Zheltov, I. P., Kochina, I. N. (1960). Basic concepts in the theory of seepage of homogeneous liquids in fissured rocks [strata]. Journal of applied mathematics and mechanics, 24(5), 1286-1303.
[2] Haykin, S. (2009). Neural networks and learning machines. vol. 3 Pearson. Upper Saddle River, NJ, USA.
[3] Hochreiter, S., Schmidhuber, J. (1997). LSTM can solve hard long time lag problems. In Advances in neural information processing systems (pp. 473-479).
[4] Vabishchevich, P. N., Grigoriev, A. V. (2016). Numerical modeling of fluid flow in anisotropic fractured porous media. Numerical Analysis and Applications, 9(1), 45-56.
[5] Logg, A., Mardal, K. A., Wells, G. (Eds.). (2012). Automated solution of differential equations by the finite element method: The FEniCS book (Vol. 84). Springer Science & Business Media.
[6] Arnold, D. N., Brezzi, F., Cockburn, B., Marini, L. D. (2002). Unified analysis of discontinuous Galerkin methods for elliptic problems. SIAM journal on numerical analysis, 39(5), 1749-1779.

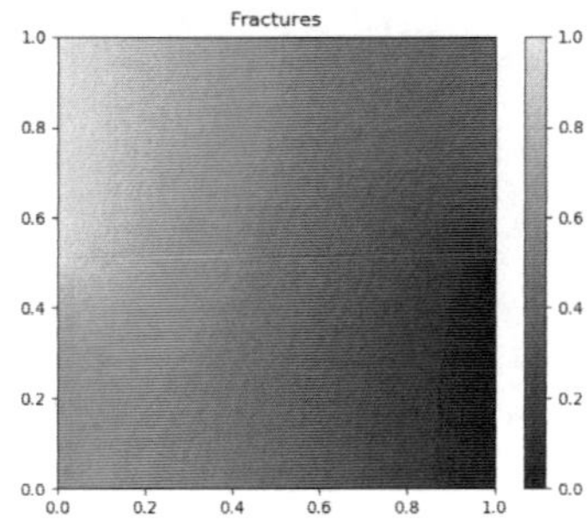

Ref. fractures

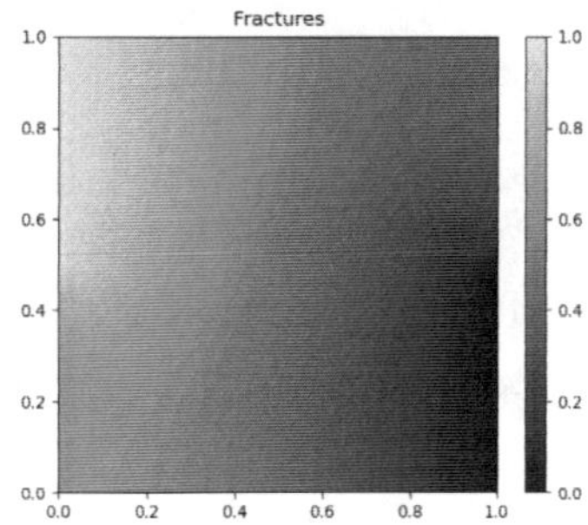

NN based fractures

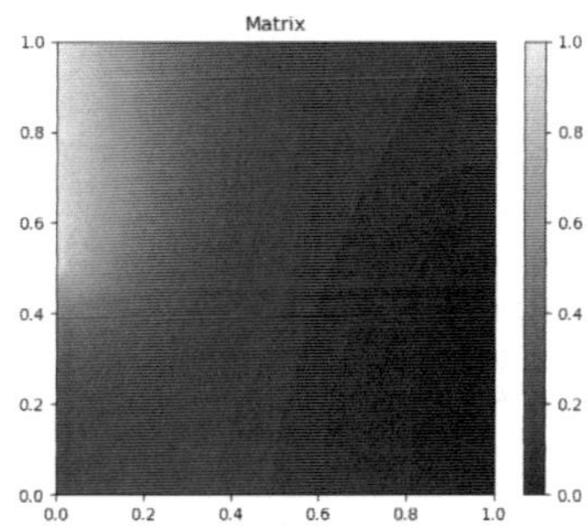

Ref. matrix

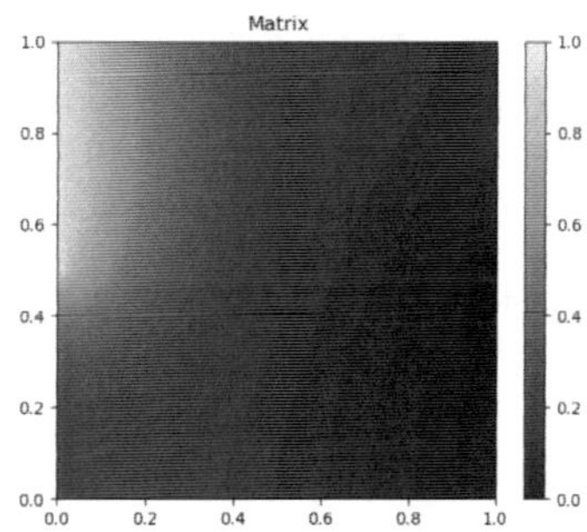

NN based matrix

Figure 6. Solutions at last moment of time

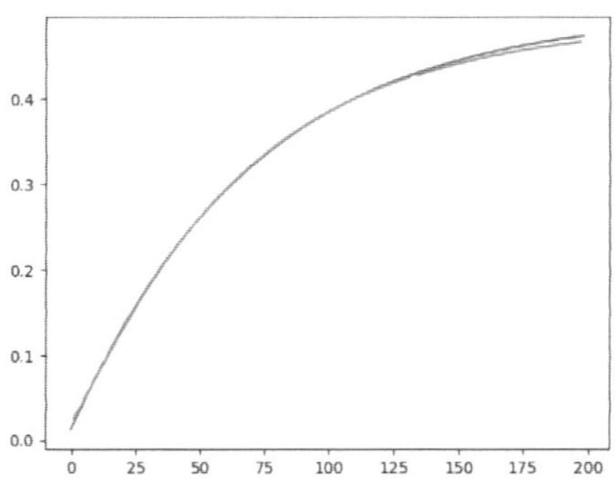

Point solution prediction

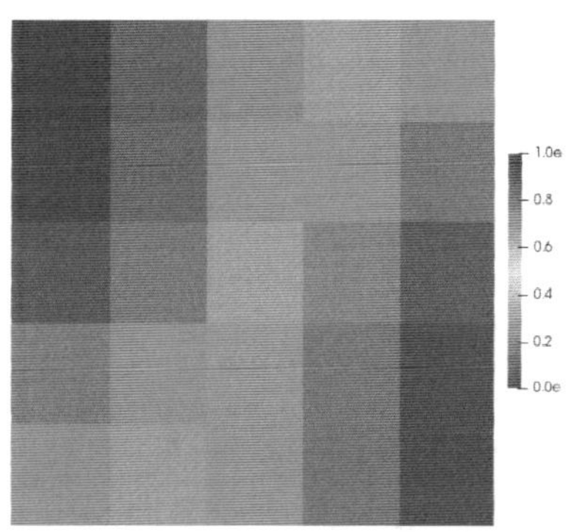

coarse grid solution (DG 0)

Figure 7. Constructing coarse grid solution

Effective calculation of thermophysical properties of composite materials with multiple configurations by asymptotic homogenization technique

Vasiliy Grigoriev[1], Petr Zakharov[1] and Mir Akimov[2]

[1] International research laboratory "Multiscale model reduction". North-Eastern Federal University, 42 Kulakovskogo st, Yakutsk, 677891, Russian Federation.
[2] Department of Applied Mathematics and Computer Science, North-Eastern Federal University, 48 Kulakovskogo st, Yakutsk, 677891, Russian Federation.

E-mail: d_alighieri@rambler.ru

Abstract. In this work a three-scale asymptotic homogenization technique is considered for calculating the thermophysical properties of composite materials with multiple periodic configurations. Heterogenieties of porous structures were considered by the periodic distribution of single cells at the microscale and mesoscale levels. Two effective thermal conductivity coefficients are calculated to obtain an average solution on the global structure. A three-scale algorithm is also compared with the classical two-scale homogenization, where microscale allows all heterogenieties. We provide a numerical experiments which show the efficiency and accuracy for the calculation of heat transfer of composite material with multiple configurations.

1. Introduction

Porous materials are widely used in engineering and manufacturing because of their good thermal stability, high thermal resistance and low relative density. In particular, in the conditions of the Far North, the issue of thermal insulation is in the first place.

Currently, the materials used in thermal insulation, consist of alternating volumes of substances with different properties. A significant part of such materials are composite materials with a periodic structure.

Averaging methods describe global behaviors by reducing the original equations with fast oscillating coefficients to equations with equivalent coefficients that can not only save computational resources, but also guarantee good computational accuracy [1–9].

In this paper, we thoroughly investigated the process of heat transfer in porous materials with multiple configurations, considered as periodic layers of single cells at the microscale and mesoscale, as illustrated in Figure 1. A level separation approach is used, namely, a three-scale asymptotic homogenization algorithm and comparison of the results obtained with finite element approximation. Also for the interest and completeness of the numerical experiment, the method of classical asymptotic averaging was implemented, where the region is divided into only two scales: micro and macro.

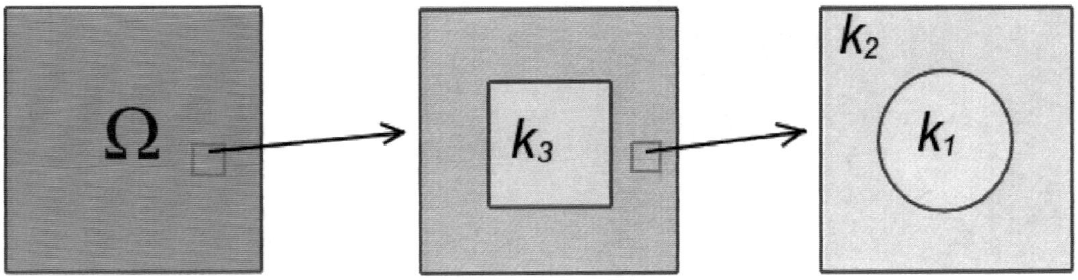

Figure 1. The scheme of the material with multiple configurations. From left to right: macro scale, meso scale, micro scale

2. Three-scale asymptotic homogenization method

In this work, we apply the scale-separation approach, namely a three-scale asymptotic homogenization method, and compare the result with finite element approximation.

Consider the following family of linear elliptic partial differential equations with a periodically oscillating coefficient $a_\varepsilon(x)$ where $\varepsilon \to 0$

$$-\nabla \cdot (a_\varepsilon \nabla u_\varepsilon) = f(x), x \in \Omega \subset \mathbb{R}^2, \tag{1}$$

$$u_\varepsilon(x) = 0, \quad x \in \partial\Omega, \tag{2}$$

where the period of oscillations $a_\varepsilon(x)$ is εL, where L — is domain lenght i.e.,

$$a_\varepsilon(x + k(\varepsilon L)\mathbf{e}_i) = a_\varepsilon(x), \quad \forall k \in \mathbb{Z}, \quad i = \{1, 2\}, \tag{3}$$

where $\mathbf{e}_i$ being the canonical basis $\mathbb{R}^2$. We assume that $f(x)$ is a given function, such that $f \in H^{-1}(\Omega)$.

Now consider the three-scale asymptotic homogenization method. We are considering three different scales, namely d_1, d_2 and L (micro, meso and macro scales), which characterize different structural sizes, i.e.

$$\varepsilon_1 = \frac{d_1}{L} \ll 1 \text{ and } \varepsilon_2 = \frac{d_2}{d_1} \ll \varepsilon_1. \tag{4}$$

Using the relation (4) two formally independent variables are introduced:

$$y = \frac{x}{\varepsilon_1} \text{ and } z = \frac{x}{\varepsilon_2} \tag{5}$$

Moreover, we assume that each property of a material is y and z periodic functions in Y and Z, respectively. Thus, we assume that the desired solution can be represented as a power series in terms of small parameters ε_1 and ε_2, namely

$$u_\varepsilon(x, y, z) = \tilde{u}(x) + \sum_{i=1}^{\infty} u_i(x, y)\varepsilon_1^i + \sum_{i=1}^{\infty} u_i^*(y, z)\varepsilon_2^i \tag{6}$$

Since the quantities involved vary in scale y and z, the following cell averaging operators are defined

$$\langle \bullet \rangle_y = \frac{1}{|Y|} \int_Y \bullet \, dy \text{ and } \langle \bullet \rangle_z = \frac{1}{|Z|} \int_Z \bullet \, dz,$$

where $|Y|$ and $|Z|$ represent periodic cell sizes.

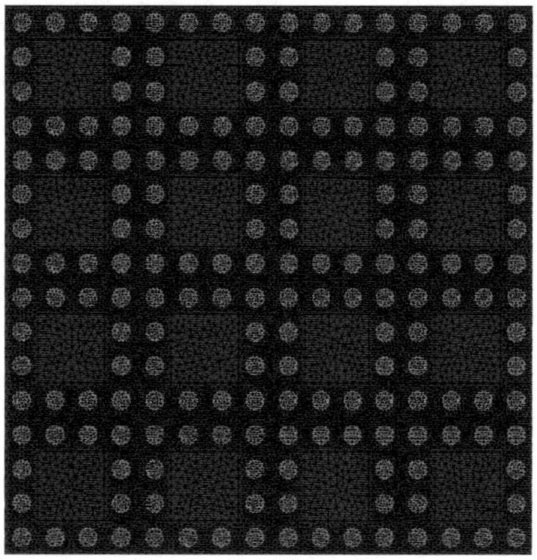

Figure 2. Overview of the finite element approximation mesh.

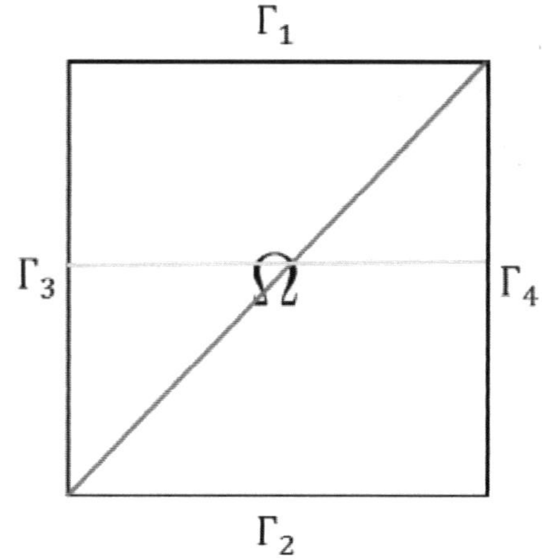

Figure 3. Computational domain.

3. Numerical experiment

Due to the fact that it is too difficult to obtain a theoretical solution to this problem (1),we will replace our exact solution with the finite element approximation obtained on a very fine mesh. Were selected linear triangular elements for the original problem on a very fine mesh for the reference solution and the same elements for the coarse mesh. For this study, as a coarse mesh was chosen unit area broken into rectangles, which in turn are divided into four elements.

To demonstrate the thermal conductivity properties in porous materials with multiple configurations, the effective thermal conductivity coefficients were calculated, which are equal for the classical two-scale asymptotic homogenization

$$\mathcal{A}^H = \begin{pmatrix} 0.265618330939 & 5.6792373e-05 \\ 5.6792373e-05 & 0.265617481342 \end{pmatrix}$$

and three-scale asymptotic homogenization

$$\mathcal{A}_Y^H = \begin{pmatrix} 0.160221931759 & 7.4758127e-05 \\ 7.4758127e-05 & 0.160221880602 \end{pmatrix},$$

$$\mathcal{A}_Z^H = \begin{pmatrix} 0.269216950836 & 0.000212804424 \\ 0.000212804424 & 0.269216950836 \end{pmatrix}$$

micro and meso scales respectively.

We also calculated L^2 and H^1 error norms for two-scale and three-scale asymptotic homogenization technique, which are presented in the tables 1 and 2.

For numerical studies were selected case where $k_1 = 0.1$ W/(m·K), $k_2 = 1.0$ W/(m·K), $k_3 = 10.0$ W/(m · K), $T(\mathbf{x}) = 0.0$, $\mathbf{x} \in \Gamma_i, i = 1, 2, 3, 4$ and with heat source $f = 8.0$ W/m^3. Calculated temperature fields are shown in Figures 5, 6 and 7. For clarity, the contour lines are also presented in the same place for all three solutions compared to each other in the Figure 4.

In the case of two-scale asymptotic homogenization, we get 2.4% of relative error, for three-scale asymptotic homogenization, the relative error equals 3.7%. This is explained by the fact that for asymptotic homogenization, the periodicity of the cells is important, i.e.

$$\lim_{\varepsilon \to 0} ||\mathcal{E}_{rel}||_{L^2} \longrightarrow 0$$

Table 1. Absolute and relative error norms for two-scale asymptotic homogenization

Coarse grid	$\|\|\mathcal{E}_{rel}\|\|_{L^2}$ (%)	$\|\|\mathcal{E}_{rel}\|\|_{H^1}$ (%)
4x4	16.012	126.913
8x8	7.744	88.818
16x16	4.764	70.670
32x32	3.280	55.001
64x64	2.584	28.264
128x128	2.513	18.747
256x256	2.491	12.193
512x512	2.488	11.051

Table 2. Absolute and relative error norms for three-scale asymptotic homogenization

Coarse grid	$\|\|\mathcal{E}_{rel}\|\|_{L^2}$ (%)	$\|\|\mathcal{E}_{rel}\|\|_{H^1}$ (%)
4x4	16.723	128.977
8x8	8.412	90.395
16x16	5.452	72.112
32x32	4.252	56.540
64x64	3.732	30.512
128x128	3.680	21.924
256x256	3.665	16.898
512x512	3.663	15.947

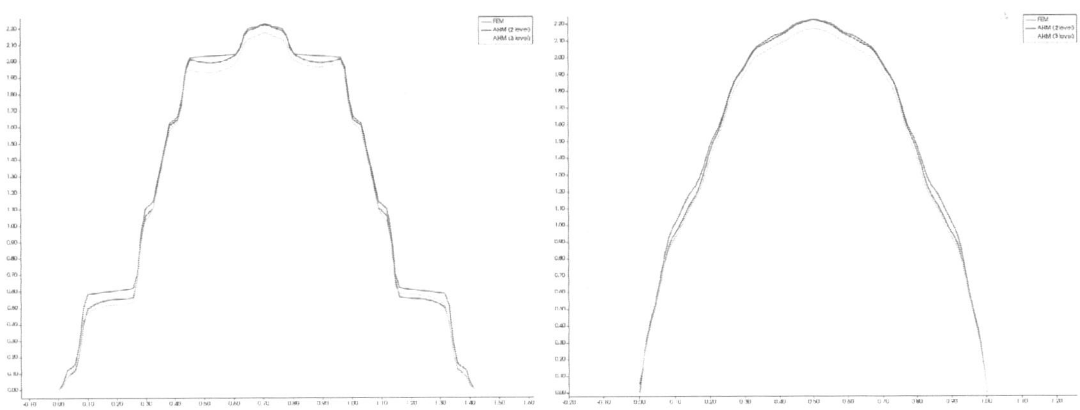

Figure 4. Diagonal and transverse isolines

Details about this can be found in the following papers [10–14]. In this paper, we have chosen a small periodicity. With large periodicities, both of these methods show good accuracy.

4. Conclusion

It is obvious that the use of the usual finite-element approximation on a grid that resolves the entire heterogeneity of the material is very expensive in terms of computing resources. In

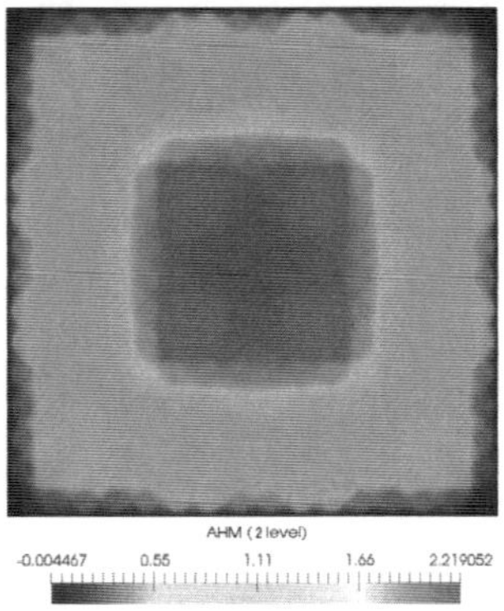

Figure 5. Two-scale asymptotic homogenization.

Figure 6. Three-scale asymptotic homogenization.

Figure 7. Finite element approximation

the case of a three-scale asymptotic homogenization, the scale separation provides undeniable advantages in reducing computational complexity. Also, this approach is a useful tool not only for calculating the thermophysical properties of composite materials, but also for studying biomechanical and engineering applications in which there are several length scales. Thus, the scale separation is the first step to computationally executable multiscale modeling of complex hierarchical materials.

Acknowledgments
The work was supported by the mega-grant of the Government of the Russian Federation (N 14.Y26.31.0013).

References

[1] Bensoussan A, Lions J L and Papanicolaou G 2011 *Asymptotic analysis for periodic structures* vol 374 (American Mathematical Soc.)
[2] Cioranescu D and Donato P 1999
[3] Oleinik O A, Shamaev A and Yosifian G 2009 *Mathematical problems in elasticity and homogenization* vol 2 (Elsevier)
[4] Jikov V V, Kozlov S M and Oleinik O A 2012 *Homogenization of differential operators and integral functionals* (Springer Science & Business Media)
[5] Temizer I and Wriggers P 2011 *Journal of the Mechanics and Physics of Solids* **59** 344–372
[6] Li Z H, Ma Q and Cui J 2016 *Journal of Computational Physics* **314** 712–748
[7] Zhang H, Zhang S, Bi J Y and Schrefler B 2007 *International Journal for Numerical Methods in Engineering* **69** 87–113
[8] Fish J, Yu Q and Shek K 1999 *International journal for numerical methods in engineering* **45** 1657–1679
[9] Asakuma Y, Kanazawa Y and Yamamoto T 2014 *International Journal of Heat and Mass Transfer* **73** 97–102
[10] Lukkassen D, Nguetseng G and Wall P 2002 *International Journal of Pure and Applied Mathematics* **2** 35–86
[11] Alexandre R 1997 *Proceedings of the Royal Society of Edinburgh Section A: Mathematics* **127** 441–455
[12] Allaire G 1992 *SIAM Journal on Mathematical Analysis* **23** 1482–1518
[13] Allaire G 1994 *PITMAN RESEARCH NOTES IN MATHEMATICS SERIES* 1–1
[14] Trucu D, Chaplain M and Marciniak-Czochra A 2012 *Applicable Analysis* **91** 1351–1373

GPU-accelerated discrete element modeling of geological faults

V. Lisitsa[1,2], **V. Tchebverda**[1], **V. Volianskaia**[3]

[1] Institute of Petroleum Geology and Geophysics SB RAS, Novosibirsk, Russia
[2] Sobolev Institute of Mathematics SB RAS, Novosibirsk, Russia
[3] PAO "Rosneft", Moscow, Russia

E-mail: `lisitsavv@ipgg.sbras.ru`

Abstract. In this paper, we present an original algorithm for the numerical simulation of tectonic movements and the related formation of geological faults. The approach is based on the use of Discrete Element Method, where the geological media is represented as an agglomeration of discrete particles which interact as elastic, visco-elastic, or elasto-plastic bodies. This approach naturally allows accounting for large deformations and discontinuities in the geological media; thus, allows simulation of faults formation. Implementation of the algorithm is based on the use of Graphical Processor Units. The Discrete Element Method requires a high number of floating point operations and logical operations per a single particle (degree of freedom) per time step; whereas the number of memory access operations is relatively low. Thus, the use of GPUs decreases the computational time substantially.

1. Introduction

In seismic processing and interpretation, geological faults are considered and represented as discontinuities in seismic signal phase and further discontinuities in sedimentary, metamorphic or magmatic rock bodies. Thus, no physical properties are assigned to a fault. However, real geological faults have a complex structure which includes the main fault body ("fault core") and fractured or damage zones around [1], [2]. Damage zone may be highly fractured, thus, permeable especially for carbonates [3], or it can be an impermeable due to the presence of deformation bonds which is typical for the sandstones [4]. Such differences of the local permeability near faults may strongly affect the reservoir performance [5]. Thus an accurate representation of the fault and damage zone is required for efficient oil and gas exploration.

Direct observation and analysis of fault structures are troublesome due to some natural reasons. Thus, numerical simulation is a reliable and efficient way to investigate the peculiarities of the structure's forming and tectonic movement process. There are several approaches to simulate faults and fractures formation. These includes grid-based methods [6], [7], [8] and meshless approaches also known as discrete elements method (DEM) [9], [10]. The letter is preferred because no predefined crack or fault geometry is needed for simulation. However, particle-based methods are more computationally intense [10]. Despite this, the particle-based methods are incredibly flexible and can be used to generate multiple statistical realizations of the fault zones and study statistical features of the strongly deformed and highly-distorted zones. Meshless methods of geological faults formation simulations can be used to generate faults geometries in realistic environments. After that simulated faults can be introduced in geological models which are used for seismic modeling and imaging [11], [12], moreover use advanced simulation techniques such as local mesh refinement [13], [14] allow studying seismic responses of the fine structure of near-fault damage zones.

2. Discrete Elements Method

Following [15] and [16], to simulate the tectonic movements causing finite deformations and geological fault formation in the Earth's crust, we use the discrete element method. In this approach, the media is represented as an assembly of individual particles with a particular geometry and physical properties. Each particle is characterized by the coordinate of its center $\vec{x}^j$, radius R_j, repulsion and attraction bulk moduli K_r^+ and K_r^- respectively, tangential sliding stiffness K_s, and two friction coefficients μ_s is the static one and μ_d is the dynamic friction coefficient. Having set these parameters, one may define the interaction forces between two adjoint particles.

2.1. Computation of inter-particle forces

Consider two particles with the numbers i and j, with the coordinates $\vec{x}^i$ and $\vec{x}^j$ and radii R^i and R^j respectively. Particle j acts on particle i with the normal elastic forces:

$$\vec{F}_n^{ji} = \begin{cases} K_r^-(R^i + R^j - \|\vec{X}^{ji}\|)\vec{n}^{ji}, & R^i + R^j - \|\vec{X}^{ji}\| > 0, \quad \textit{repulsion}, \\ K_r^+(R^i + R^j - \|\vec{X}^{ji}\|)\vec{n}^{ji}, & 0 \le R^i + R^j - \|\vec{X}^{ji}\| \le r_0, \quad \textit{active bond}, \\ 0, & R^i + R^j - \|\vec{X}^{ji}\| > r_0, \quad \textit{no bond}, \end{cases} \tag{1}$$

where r_0 is the bond length, vector $\vec{X}^{ji} = \vec{x}^i - \vec{x}^j$ connects the centers of the particles and directed from particle j to particle i, vector $\vec{n}^{ji} = \vec{X}^{ji}/\|\vec{X}^{ji}\|$ is the unit vector directed from the centers of particle j to the center of particle i or normal vector.

Additionally frictional forces are taken into account if two particles are in a contact [16]:

$$\vec{F}_t^{ji} = \begin{cases} -K_s\delta_t\vec{t}^{ji}, & K_s\delta_t \le \mu^s\|\vec{F}_n^{ji}\|, \quad \textit{static friction}, \\ -\mu^d\|\vec{F}_n^{ji}\|\vec{t}^{ji}, & K_s\delta_t > \mu^s\|\vec{F}_n^{ji}\|, \quad \textit{dynamic friction}, \end{cases} \tag{2}$$

where K_s is the tangential sliding stiffness, usually considered to be equal to bulk modulus; i.e., $K_s = K_r$, vector $\vec{t}^{ji}$ is the unitary tangential vector directed along the projection of the relative velocity onto the contact plane of two particles; i.e.,

$$\vec{t}^{ji} = \frac{\vec{v}^{ji} - (\vec{v}^{ji}, \vec{n}^{ji})\vec{n}^{ji}}{\|\vec{v}^{ji} - (\vec{v}^{ji}, \vec{n}^{ji})\vec{n}^{ji}\|}, \quad \vec{v}^{ji} = \vec{v}^i - \vec{v}^j. \tag{3}$$

In this notations $\vec{v}^{ji}$ is the relative velocity of the particle i with respect to particle j. Parameter δ^{ji} denotes the tangential displacement of the contact point from its initial position. Tangential forces provided by formula (2) satisfy the Coulombs law; i.e., the static friction governs the particles interaction if the forces as below a critical value. If the tangential forces exceed the critical dynamical friction proportional to normal force is applied. Typically the static friction is much higher than the dynamical one.

Additionally, potential external forces are considered, and an artificial dissipation is introduced in the system to prevent elastic waves from propagating through the model and ensuring the media to remain stable at infinite instants.

Having computed all external forces acting at j-th particle one may recompute its position using classical mechanics principles:

$$M^i\frac{d^2\vec{x}^i}{dt^2} = \vec{F}^i\left(t, \vec{x}^i, \vec{x}^j, \frac{d\vec{x}^i}{dt}, \frac{d\vec{x}^j}{dt}\right), \tag{4}$$

where dissipative $\vec{F}_d^i$ and frictional forces $\vec{F}_t^{ji}$ explicitly depend on the particles velocities $\vec{v}^i = \frac{d\vec{x}^i}{dt}$.

To numerically resolve the system of equations (4), we use the Verlet-like scheme with the velocity half-step [17].

2.2. Boundary conditions

Proper implementation of the boundary conditions is a challenging task for the particles-based methods. In our research, we deal with two types of boundary conditions. First, we impose the rigid boundary condition; i.e., the surface Γ_s is fixed, or its movement is prescribed. Moreover, it is stiff; thus, the particles cannot penetrate through it. Formally, this type of boundary condition can be stated as follows. Assume a boundary $\Gamma_s = \{\vec{x}|x_2 = x_2^B\}$. If a particle is close enough to the boundary; i.e., if for the j-th particle $|x_2^j - x_2^b| \leq R^j$, then $F_2^{jB} = K_r^-(R^j - |x_2^j - x_2^b|)$.

However, the numerical implementation of this condition requires extra conditional operators. Thus it is worth implementing stiff-boundary as a series of particles, to make the simulation uniform either in the interior of the domain or near the boundary. To do so, we introduced the "boundary" particles with the same physical properties as those of the interior particles. However, we do not compute the forces acting on the "boundary" particles but allow the "boundary" particles to move according to a prescribed law.

The second type of boundary conditions is $P_{over} = const$. This condition ensures constant overburden pressure. Note that, condition P_{over} assumes that external forces act at the upper boundary of the domain $\Gamma_p(t)$ along the normal direction to the boundary. This boundary is flexible, and it evolves in time; thus, to impose the boundary condition, we need to follow the elements which form the upper boundary. This can be done, for example, by computing Voronoi diagrams for upper elements. However, such procedures are computationally intense. To overcome this difficulty, we suggest using the flexible membrane at the upper boundary [18], [19]. The idea of the approach is to introduce a layer of discrete elements so that the membrane elements are affected only by the normal forces.

If two adjoin membrane elements are interacting

$$\vec{F}_n^{m,m\pm1} = K_r(R^{m\pm1} + R^m - \|\vec{X}^{m,m\pm1}\|)\vec{n}^{m,m\pm1}, \tag{5}$$

if membrane element interacts with other elements

$$\vec{F}_n^{mi} = K_r(R^i + R^m - \|\vec{X}^{mi}\|)\vec{n}^{mi}, \quad R^i + R^m - \|\vec{X}^{mi}\| > 0. \tag{6}$$

It means that the adjoint membrane elements are bonded, and these bonds never bake, however no bonds of friction are considered when membrane elements interact with elements of other types. The membrane elements are ordered; thus, it is easy to approximate constant pressure condition. If a membrane element with number m is considered then additional force, due to the pressure is

$$\vec{F}_p^m = 2PR^m\vec{n}, \tag{7}$$

where $\vec{n}$ is the vector normal to the boundary, which can be computed as:

$$\vec{n} = (x_2^{m-1} - x_2^{m+1}, x_1^{m-1} - x_1^{m+1})^T,$$

the direction of the normal vector is defined uniquely due to the ordering of the membrane elements.

2.3. Output parameters
Numerous parameters can be obtained as a result of discrete element simulations. If rock properties are studied using uniaxial and triaxial stress tests, then the primary attention is paid to the distribution of the braked bonds [20], [21], stresses, and normal forces distribution [22] However, at the scale of the geological bodies a reliable parameter to determine fault zones is the strains distribution [23], [15], [17], [24], [25]. These strains can be further translated to the changes of physical parameters of rocks using the experimental laboratory measurements.

To estimate the strains distribution, one may compute the relative displacements element-wise, after that the components of the strain tensor components can be computed and interpolated to a regular grid. A detailed discussion of the estimation of the strain can be found in [23].

3. Implementation of the Algorithm
According to the general formulation of the particle-based methods, one has to compute the forces affecting each particle due to the interaction with all other particles. However, in geomechanical modeling by the discrete element method, for each particle, only a small number of neighboring particles directly contact the considered one. The adjacency matrix is sparse, but it can evolve. Thus, two related problems should be solved. First, organizing the process of adjacency matrix construction (approximation). Second, computing forces and applying time stepping.

To construct the adjacency matrix, we suggest using the lattice method. As it follows from the equations (1) and (2), only directly contacting particles affect each other; thus, for each particle, the domain of dependence does not exceed $2R_{max} + r_0$, where R_{max} is the maximal radius of the particles. Also, due to the stability criterion of the Verlet scheme, a single particle cannot move more than $0.1R_{min}$ per a single time step, where R_{min} is the minimum radius over all particles. Thus, we can introduce a grid with the lattice size equal to $2R_{max} + r_0$, so that each particle and all its neighbors belong to the same lattice of directly adjoint lattice. Now we can state the rule of adjacency matrix approximation - for each particle, all the particles belonging to the same or directly adjoint lattices are neighbors. In this case, we overestimate the number of connected particles but strictly simplify the process of the matrix construction.

The initial assignment of the particles to the lattices is performed by a sequential code by CPU. It is implemented particle-by-particle so that we determine the lattice number for considered particle and add the particle number to the list of particles for this lattice. This procedure is inapplicable under OMP of CUDA parallelization. Thus, the GPU implementation of the reassignment of the particles to the lattices is done lattice-by-lattice. The lattices are large enough, so that after one time step a particle may either stay in the same lattice or move to a directly adjoint lattice. Thus, to update the list of particles for each lattice, we need to check the particles which previously belonged to this lattice or the directly adjusted one. Similar ideas are used in the molecular dynamics and lattice Boltzmann methods but with different principles of lattices construction [26].

Computation of the forces and the numerical solution of the equation of motion is implemented on GPU. The parallelization is applied particle-by-particle so that a GPU core compute forces for one particle at a time.

4. Numerical experiments
We use the algorithm to simulate geological fault position for different directions of tectonic displacements in 2D. The computational domain was 4000 m by 500 m. The boundary conditions introduced the displacements. We consider five scenarios with the displacement direction equal to 0° (vertical displacement), 15°, 30°, 45°, and 60° from vertical direction pointing downward. So, in all cases, these are normal dip-slip faults. The vertical displacement was 100 m. Radii

of the elements were homogeneously distributed from 0.75 to 1.5 m. The dynamic friction coefficient $\mu_d = 0.3$. We provide the results of simulation - distribution of the horizontal strains ε_{xx} in figure 1. Note, that the sub-vertical displacements lead to the formation of a wide damage zone, which has a form of a wedge. The closer the displacements to the horizontal direction, the narrower is the wedge with higher deformations in the fault body. Moreover, the main direction of the fault; i.e., the dip angle tends to about 30°, which is governed by the properties of the media, rather than displacement direction.

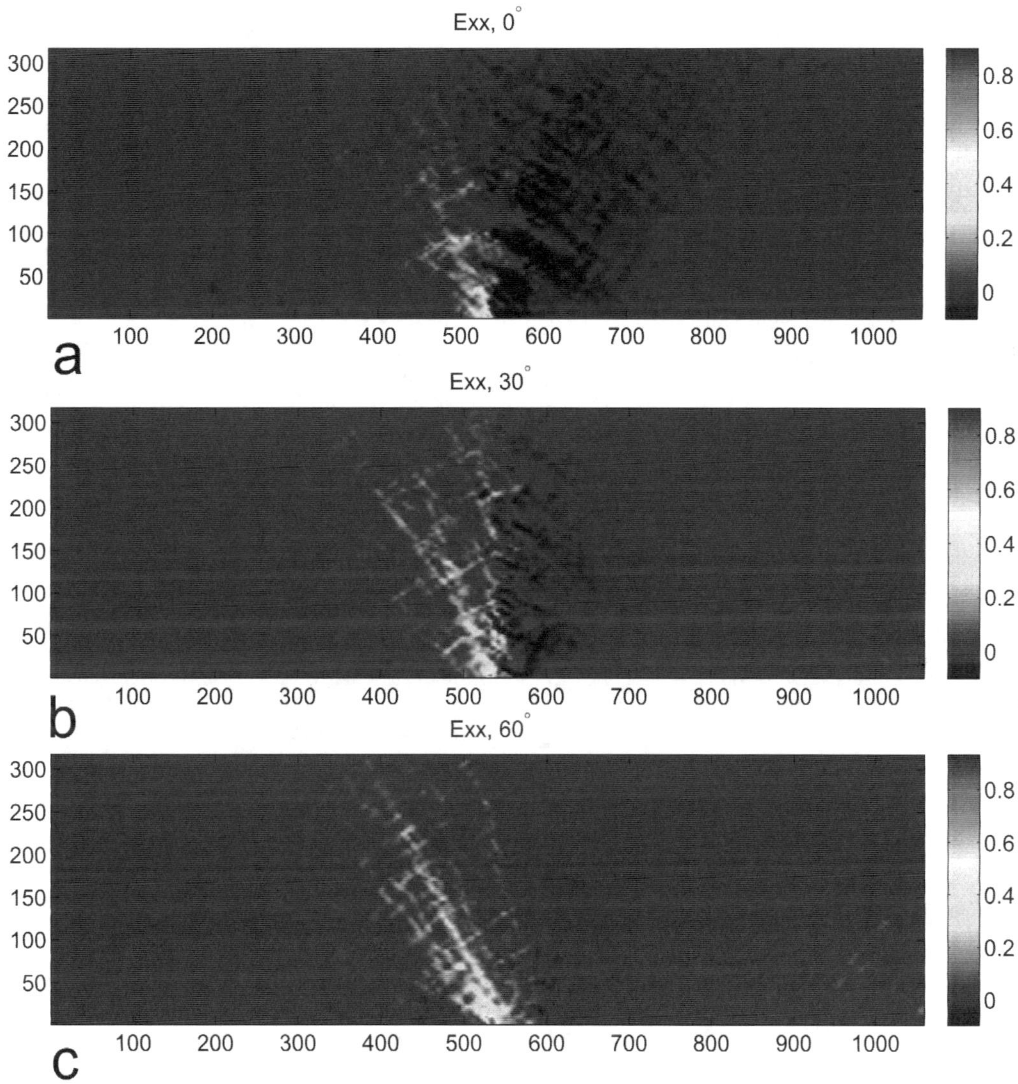

Figure 1. Distribution of horizontal strains ε_{xx} for the displacement directions 0° (a), 30° (b), and 60° (c)

5. Conclusions

We presented an algorithm for numerical simulation of the geological fault geometry and position. The approach is based on the Discrete Elements Method, which allows modeling of finite deformations and structural discontinuities in the Earth's crust. We implement the algorithm using CUDA technology to simulate a single statistical realization of the model, whereas MPI is used to parallelize for different statistical realizations. The presented numerical experiments illustrate the formation of the V-shaped fault damage zone. The obtained strains distributions can be used further to estimate the mechanical and transport properties of the fault damage zone.

Acknowledgments
V. Lisitsa did development of the mathematical basis of the algorithm in IPGG SB RAS under financial support of the Russian Precedent agency grant MD-20.2019.5. Parallel implementation of the algorithm was done by V. Lisitsa in IM SB RAS under the support of the Russian Science Foundation grant no. 19-77-20004. Numerical experiments were carried out by V. Tcheverda under the support of the Russian Science Foundation grant no. 17-17-01128. V. Volianskaia made the geological interpretation. The research is carried out using the equipment of the shared research facilities of HPC computing resources at Lomonosov Moscow State University and cluster NKS-30T+GPU of the Siberian supercomputer center.

References

[1] Choi J H, Edwards P, Ko K and Kim Y S 2016 *Earth-Science Reviews* **152** 70–87
[2] Peacock D C P, Dimmen V, Rotevatn A and Sanderson D J 2017 *Journal of Structural Geology* **102** 179–192
[3] Hausegger S, Kurz W, Rabitsch R, Kiechl E and Brosch F J 2010 *Journal of Structural Geology* **32** 1349–1362
[4] Fossen H, Schultz R A, Shipton Z K and Mair K 2007 *Journal of the Geological Society* **164** 755–769
[5] Bense V F, Gleeson T, Loveless S E, Bour O and Scibek J 2013 *Earth-Science Reviews* **127** 171–192
[6] González G, Gerbault M, Martinod J, Cembrano J, Carrizo D, Allmendinger R and Espina J 2008 *Journal of Structural Geology* **30** 791–808
[7] Guiton M L E, Sassi W, Leroy Y M and Gauthier B D M 2003 *Journal of Structural Geology* **25** 1317–1330
[8] Resor P G and Pollard D D 2012 *Journal of Structural Geology* **41** 98–109
[9] Gray G G, Morgan J K and Sanz P F 2014 *Journal of Structural Geology* **59** 19–36
[10] Lisjak A and Grasselli G 2014 *Journal of Rock Mechanics and Geotechnical Engineering* **6** 301–314
[11] Kolyukhin D R, Lisitsa V V, Protasov M I, Qu D, Reshetova G V, Tveranger J, Tcheverda V A and Vishnevsky D M 2017 *Interpretation* **5** SP71–SP82
[12] Vishnevsky D M, Kolyukhin D R, Lisitsa V V, Protasov M I, Reshetova G V, Tcheverda V A, Qu D and Tveranger J 2017 *Doklady Earth Sciences* **473** 477–481
[13] Kostin V, Lisitsa V, Reshetova G and Tcheverda V 2015 *Journal of Computational Physics* **281** 669–689
[14] Lisitsa V, Reshetova G and Tcheverda V 2012 *Computational Geosciences* **16** 39–54
[15] Hardy S and Finch E 2005 *Basin Research* **17** 507–520
[16] Luding S 2008 *European Journal of Environmental and Civil Engineering* **12** 785–826
[17] Hardy S, McClayc K and Munozb J A 2009 *Marine and Petroleum Geology* **26** 232–248
[18] Wang Y and Tonon F 2009 *Journal of Engineering Mechanics* **135** 1029–1037
[19] Bono J d, Mcdowell G and Wanatowski D 2012 *Géotechnique Letters* **2** 199–203
[20] Cundall P A and Strack O D L 1979 *Géotechnique* **29** 47–65
[21] Li Z, Wang Y H, Ma C H and Mok C M B 2017 *Acta Geotechnica* **12** 987–1002
[22] Duan K, Kwok C Y and Ma X 2017 *Acta Geotechnica* **12** 495–510
[23] O'Sullivan C, D Bray J and Li S 2003 *International Journal for Numerical and Analytical Methods in Geomechanics* **27** 859–877
[24] Abe S, Gent H v and Urai J L 2011 *Tectonophysics* **512** 12–21
[25] Botter C, Cardozo N, Hardy S, Lecomte I and Escalona A 2014 *Marine and Petroleum Geology* **57** 187–207
[26] Alpak F O, Gray F, Saxena N, Dietderich J, Hofmann R and Berg S 2018 *Computational Geosciences* **22** 815–832
[27] Calinski T and Harabasz J 1974 *Communications in Statistics* **3** 1–27
[28] Rousseeuw P J 1987 *Journal of Computational and Applied Mathematics* **20** 53–65

Journal of Physics: Conference Series **1392** (2019) 012071 doi:10.1088/1742-6596/1392/1/012071

GMsFEM on unstructured grids for single-phase flow in fractured porous media

Djulustan Nikiforov[1], **Maria Vasilyeva**[3,4], **Yalchin Efendiev**[2,3] **and Vasily Vasil'ev**[4]

[1] Multiscale model reduction laboratory, North-Eastern Federal University, Yakutsk, Republic of Sakha (Yakutia), Russia, 677980

[2] Department of Mathematics, Texas A&M University, College Station, Texas, USA.

[3] Institute for Scientific Computation, Texas A&M University, College Station, TX 77843-3368, USA

[4] Department of Computational Technologies, North-Eastern Federal University, Yakutsk, Republic of Sakha (Yakutia), Russia, 677980.

E-mail: dju92@mail.ru, vasilyevadotmdotv@gmail.com, efendiev@math.tamu.edu, vasvasil@mail.ru

Abstract. In this work, we consider an unstructured Generalized Multiscale Finite Element Method (GMsFEM) for solution of the filtration problem in a fractured media. The basic idea is that coarse grid blocks are formed as sets of fine grid triangular cells and, thus, can be of an almost arbitrary polygonal shape. In this approach, when constructing the fine grid, the faces of coarse blocks are not considered. Discrete fractured model is used to describe a flows in a fractured media.

1. Introduction

In numerical solution of the applied problems using multiscale methods, a fine-scale grid is constructed by taking into account coarse blocks, which is a rather complicated process. In order to avoid difficulties with coarse grid definition, various modifications of the multiscale methods are developed. For example, in [1] and [2] a dynamic a grid coarsening strategy was developed for a mixed flow, where the grid should separate high-flow regions from low-flow regions. In [3], an unstructured coarse grids are used in Multiscale Finite Folume Method (MSFVM) for simulation of the flow processes in the fractured porous media. Generalazed Multiscale Finite Element Method with discontinuous Galerkin coupling on the unstructured coarse grid was presented in [11] for solution of the problems in perforated domains. GMsFEM for flow in fractured porous media was presented in [10] for structured coarse grids.

In this work, we consider a Generalized Multiscale Finite Element Method on unstructured coarse grids for solution of the filtration problem in fractured porous media. In Section 2, we present a problem formulation and fine grid approximation using discrete fracture model. In Section 3, we present a multiscale algorithm for solution of the problem on unstructured coarse grids based on the GMsFEM. Numerical results are presented in Section 4.

2. Model problem and fine grid approximation

In this work we consider a single-phase fluid flow in fractured porous media [10]

$$\begin{aligned} c_m \frac{\partial p_m}{\partial t} - \operatorname{div} \cdot \left(\frac{k_m}{\mu} \operatorname{grad} p_m \right) + \sigma_{mf}(p_m - p_f) = 0, \quad x \in \Omega, \\ c_f \frac{\partial p_f}{\partial t} - \operatorname{div} \cdot \left(\frac{k_f}{\mu} \operatorname{grad} p_f \right) - \sigma_{fm}(p_m - p_f) = 0, \quad x \in \gamma, \end{aligned} \tag{1}$$

where Ω is the porous media domain, γ is the lower-dimensional domain for fractures, p_m and p_f are the porous matrix and fractures pressures. Here μ is the fluid viscosity, $c_\alpha = \phi_\alpha c_{R,\alpha}$, ϕ_α, $c_{R,\alpha}$ and k_α are the porosity, compressibility and permeability for continuum α ($\alpha = m, f$). Note that the flow of an incompressible fluid in an elastically deformable porous medium is considered and k_f is much larger than k_m (high - permeable fractures).

System of equations (1) is considered with following initial and boundary conditions

$$\begin{aligned} & p_\alpha(x, 0) = p_0(x), \quad x \in \Omega, \\ & p_\alpha = p_1, \quad x \in \Gamma_L, \quad -\frac{k_\alpha}{\mu} \frac{\partial p_\alpha}{\partial n} = 0, \quad x \in \partial\Omega / \Gamma_L. \end{aligned} \tag{2}$$

for $\alpha = m, f$, n is the outward unit normal to the boundary $\partial\Omega$, Γ_L is left boundary.

The finite element method and discrete fracture model are used to discretize the system on a fine grid, where fractures are represented on the fine grid as the facets of finite element mesh ([4] – [6]). Using superposition principle and suppose that $p = p_f = p_m$ on the fractures $p = p_m$ on porous matrix domain, we obtain following approximation in the matrix form

$$M \frac{p^{n+1} - p^n}{\tau} + A p^{n+1} = 0, \tag{3}$$

where

$$A = [a_{i,j}], \quad a_{i,j} = \int_\Omega \left(\frac{k_m}{\mu} \operatorname{grad} \phi_i, \operatorname{grad} \phi_j \right) dx + \int_\gamma \left(\frac{k_f}{\mu} \operatorname{grad} \phi_i^f, \operatorname{grad} \phi_j^f \right) dx,$$

$$M = [m_{i,j}], \quad m_{i,j} = \int_\Omega c_m \phi_i \, \phi_j \, dx + \int_\gamma c_f \phi_i^f \, \phi_j^f \, dx.$$

and ϕ_i, ϕ_i^f are the linear basis functions in domain Ω and γ. Here for approximation by time, we used an implicit scheme with time step size τ, superscripts n and $n+1$ denote previous and current time levels, respectively.

3. Coarse grid approximation using GMsFEM

In this section, we give a brief description of the unstructured GMsFEM for flow problems in fractured media [7] – [11]. We start with the construction of the coarse grids. We define unstructured grid as a partition of the fine grid by the mesh partitioning library SCOTCH [12]. We emphasize the use of ω_i to denote a coarse neighborhood, and K to denote a coarse block ($i = 1, ..., N$, N is the number of coarse neighborhoods).

To construct the multiscale space for pressure approximation, we solve following spectral problem

$$A^{\omega_i} \Psi_j^i = \lambda_j^i S^{\omega_i} \Psi_j^i, \tag{4}$$

where A^{ω_i} is the restriction of the global matrix A in local domain ω_i and

$$S^{\omega_i} = \left[s_{i,j}^{\omega_i} \right], \quad s_{i,j}^{\omega_i} = \int_{\omega_i} \frac{k_m}{\mu} \phi_i \, \phi_j \, dx + \int_\gamma \frac{k_f}{\mu} \phi_i^f \, \phi_j^f \, dx.$$

Next, we choose the eigenvectors corresponded to the smallest M_i eigenvalues.

For construction of the conforming multiscale space, we define a multiscale partition of unity functions χ_i as follows

$$A^K \chi_i = 0, \quad \forall K \in \omega_i, \tag{5}$$

where A^K is the restriction of the global matrix A in coarse partitioning K. We solve eq. (5) with linear boundary conditions $\chi_i = g_i$ on ∂K, where g_i is a continuous function on ∂K and is linear on each edge of ∂K.

To form a multiscale space and projection matrix, we multiply the partition of unity functions χ_i by the eigenfunctions Ψ^i_j

$$V_{\text{ms}} = \text{span}\{\psi_{i,j} = \chi_i \Psi^i_j : 1 \leq i \leq N, 1 \leq j \leq M_i\}, \quad R = [\psi_{1,1}, \ldots, \psi_{N,M_N}]^T,$$

and define a projection matrix $R^T = [\psi_1, \ldots, \psi_N]$, where ψ_i are used to denote the nodal values of each basis function defined on the fine grid.

Finally, we have following coarse grid system

$$M_c \frac{p_c^{n+1} - p_c^n}{\tau} + A_c p_c^{n+1} = 0, \tag{6}$$

where $A_c = RAR^T$, $M_c = RMR^T$ and $p_{ms}^{n+1} = R^T p_c^{n+1}$.

4. Numerical results

In this section, we present numerical results to show the performance of the presented multiscale method on unstructured coarse grids. For coefficients representing matrix and fracture properties, we set $\phi_m = 0.4$, $c_{R,m} = 10^{-9}\text{Pa}^{-1}$, $k_m = 10^{-15}\text{m}^2$, $\phi_f = 1$, $c_{R,f} = 10^{-9}\text{Pa}^{-1}$, $k_f = 10^{-10}\text{m}^2$, $\mu = 2 \cdot 10^{-3}\text{Pa} \cdot \text{s}$, $p_0 = 10\text{MPa}$, $p_1 = 20\text{MPa}$. Final time is 30 months with 30 time steps. We construct three coarse grids (Figure 4). Fine grid contains 70297 elements and 35471 vertices. Coarse grid 1 contains 25 elements and 52 vertices. Coarse grid 2 contains 50 elements and 102 vertices. Coarse grid 3 contains 100 elements and 202 vertices. To compare the results, we use relative errors in L^2_a and H^1_a norms.

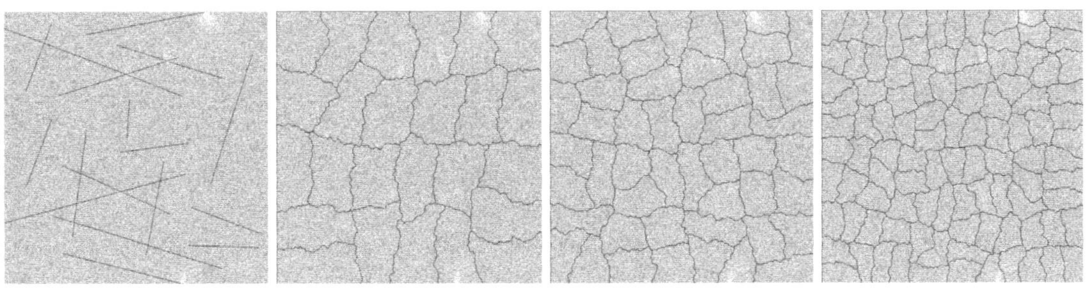

Figure 1. Fine grid and three unstructured coarse grids. First: Fine grid with 70297 elements and 35471 vertices. Second: Coarse grid 1 with 25 elements and 52 vertices. Third: Coarse grid 2 with 50 elements and 102 vertices. Fourth: Coarse grid 3 with 100 elements and 202 vertices.

The fracture geometry with several connected fractures is illustrated in Figure 4. The fine grid (reference) and multiscale solutions on coarse grid 2 with $M = 4$ at different times are shown in Figure 4. We observe that the coarse-scale solution provide a good approximation of the reference solution. The relative L^2_a and H^1_a relative errors are shown in Table 4. We see that the H^1_a relative error and L^2_a relative error on coarse grid 3 are 0.04 % and 2.3 % for $M = 32$ on the coarse grid 1. We obtain a smaller error for finer coarse grid and error reduce with increasing number of the multiscale basis functions.

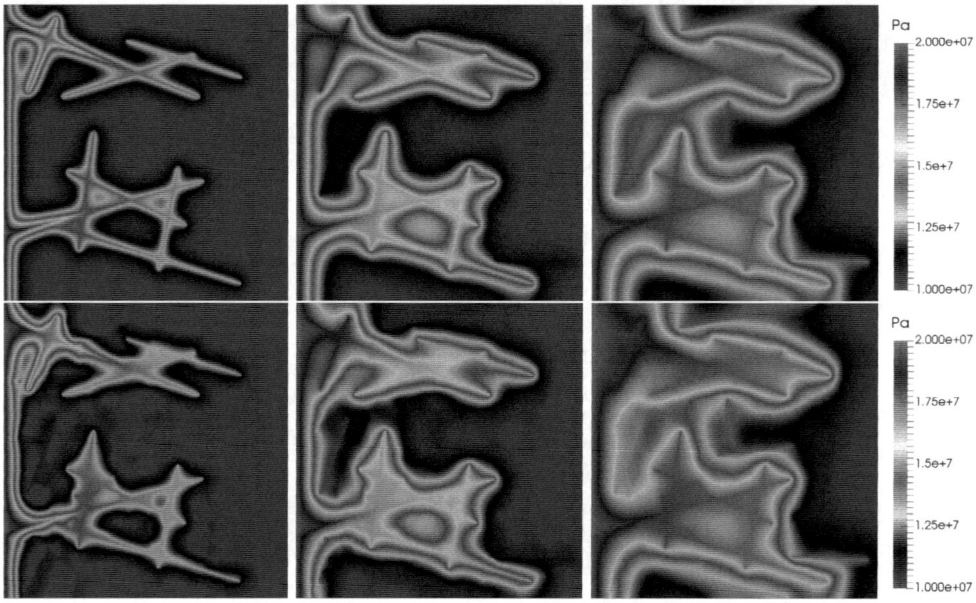

Figure 2. Fine-scale solution (top) and multiscale solutions on coarse grid 2 with $M_{\text{off}} = 4$ (bottom) for three time steps t_m, $m = 4, 13, 30$.

M	Coarse grid 1		Coarse grid 2		Coarse grid 3	
	L_a^2	H_a^1	L_a^2	H_a^1	L_a^2	H_a^1
1	12.6	81.7	10.9	76.6	5.9	63.8
2	5.1	52.1	4.1	46.5	2.45	40.1
4	1.6	23.8	1.5	22.9	0.78	19.6
8	0.8	15.1	0.3	10.2	0.16	7.8
16	0.1	5.2	0.08	4.3	0.04	3.7
32	0.04	2.3	0.02	1.9	0.01	1.9

Table 1. Relative errors in % at the final time for different number of multiscale basis functions $M = M_i$, $i = 1, ..., N$.

5. Conclusion

In this paper, we presented an unstructured multiscale method for single phase flow problem with discrete fracture model. Multiscale basis functions are constructed in the offline stage via local spectral problems. To represent our approach we solve the flow problem on three different coarse grids. Numerical results are presented and demonstrate that GMsFEM can provide accurate solution on the unstructured coarse grids.

6. Acknowledgements

Work is supported by the mega-grant of the Russian Federation Government N14.Y26.31.0013. MV's work is supported by the grant RSF N17-71-20055.

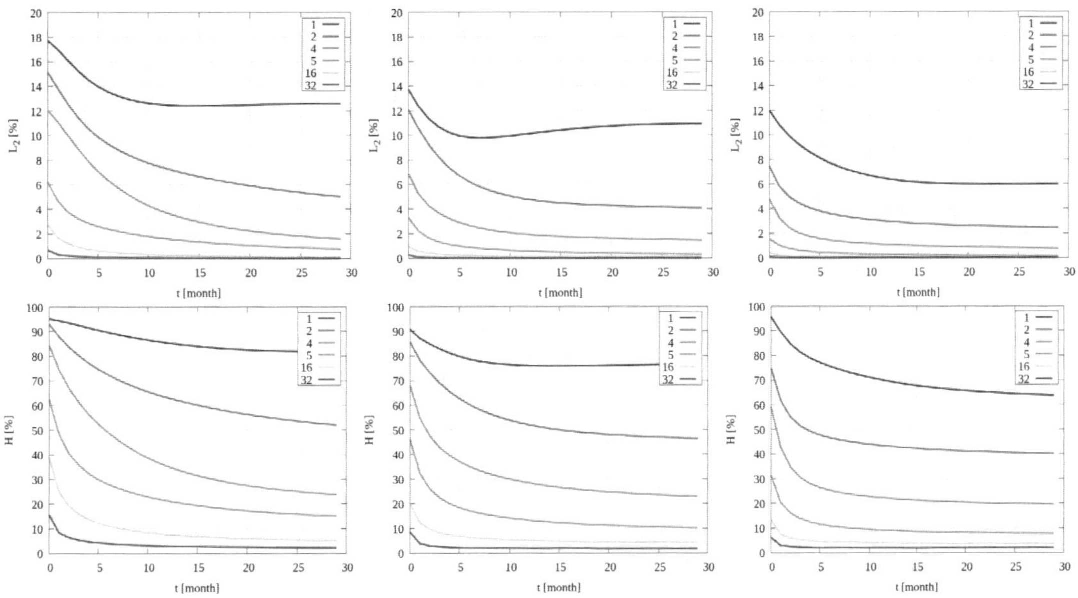

Figure 3. Relative errors L^2_a and H^1_a in % by time (top and bottom roes). Coarse grid 1, 2 and 3 (from left to right)

References

[1] Aarnes, J. E., Hauge, V. L., Efendiev, Y. (2007). Coarsening of three-dimensional structured and unstructured grids for subsurface flow. Advances in Water Resources, 30(11), 2177-2193.

[2] Aarnes, J. E., Efendiev, Y., Hou, T. Y., Jiang, L. (2010). Mixed multiscale finite element methods on adaptive unstructured grids using limited global information. Multiscale Methods: Bridging the Scales in Science and Engineering, 1.

[3] Bosma, S., Hajibeygi, H., Tene, M., Tchelepi, H. A. (2017). Multiscale finite volume method for discrete fracture modeling on unstructured grids (MS-DFM). Journal of Computational Physics, 351, 145-164.

[4] Vasil'eva, M. V. E., Vasiliev, V. I., Krasnikov, A. A., Nikiforov, D. U.Y. (2017). Numerical simulation of single-phase fluid flow in fractured porous media. Uchenye Zapiski Kazanskogo Universiteta. Seriya Fiziko-Matematicheskie Nauki, 159(1), 100-115.

[5] Vasiliev, V. I., Vasil'eva, M. V. E., Gladkikh, V. S., Il'in, V. P., Nikiforov, D. U. Y., Perevozkin, D. V. E., Prokop'ev, G. A. (2018). Numerical solution of a fluid filtration problem in a fractured medium by using the domain decomposition method. Sibirskii Zhurnal Industrial'noi Matematiki, 21(4), 15-27.

[6] Nikiforov, D. Y., Stepanov, S. P. (2019). Numerical simulation of the embedded discrete fractures by the finite element method. In Journal of Physics: Conference Series (Vol. 1158, No. 3, p. 032038). IOP Publishing.

[7] Efendiev, Y., Galvis, J., Hou, T. Y. (2013). Generalized multiscale finite element methods (GMsFEM). Journal of Computational Physics, 251, 116-135.

[8] Chung, E. T., Efendiev, Y., Leung, T., Vasilyeva, M. (2017). Coupling of multiscale and multi-continuum approaches. GEM-International Journal on Geomathematics, 8(1), 9-41.

[9] Akkutlu, I. Y., Efendiev, Y., Vasilyeva, M., Wang, Y. (2017). Multiscale model reduction for shale gas transport in a coupled discrete fracture and dual-continuum porous media. Journal of Natural Gas Science and Engineering, 48, 65-76.

[10] Akkutlu, I. Y., Efendiev, Y., Vasilyeva, M. (2016). Multiscale model reduction for shale gas transport in fractured media. Computational Geosciences, 20(5), 953-973.

[11] Chung, E. T., Efendiev, Y., Vasilyeva, M., Wang, Y. (2016, January). A multiscale discontinuous Galerkin method in perforated domains. In Proceedings of the Institute of Mathematics and Mechanics, 42 (2), 212-229

[12] http://www.labri.fr/perso/pelegrin/scotch.

Journal of Physics: Conference Series **1392** (2019) 012072 doi:10.1088/1742-6596/1392/1/012072

Scaling in Super Element Model of Petroleum Reservoirs

Alexandr Mazo, Konstantin Potashev

Institute of Mathematics and Mechanics, Fluid Mechanics department,
Kazan Federal University, Russian Federation, Kazan, Kremlevskaya st., 35.

KPotashev@mail.ru

Abstract. A feature of the super element two-phase flow model in an oil reservoir is the construction of a solution on computational grids with a spatial step of the order of the distance between wells (hundreds of meters) horizontally and of the order of the total thickness of the reservoir (tens of meters) vertically. To maintain the acceptable accuracy of the solution, the upscaling procedures are performed during the transition from a detailed geological to a large super element computational grid and downscaling during a reverse transition at the local refinement of the solution. The field of absolute permeability and functions of relative phase permeability are subject to upscaling. The downscaling procedure is applied to the grid saturation function on the super element grid to formulate the initial conditions for local refinement of the solution on the detailed computational grid. The paper describes ways to simplify and implement these procedures, taking into account the characteristic features of the geological structure and the process of oil reservoir development.

1. Introduction

To reduce the dimension of computational grids in modeling of the oil reservoir, the averaging technique (upscaling) [1, 2, 3, 4] and the use of detailed grids in local subregions [5, 6, 7, 8] are usually used. This paper describes the functions of upscaling and downscaling with two-stage super element modeling [9, 10, 11, 12]. This approach allows not only to reduce the computational costs by orders of magnitude, but also to improve the accuracy of calculations in comparison with many traditional methods.

2. Basic equations of the super element model

For simplicity, we will neglect the compressibility of fluids, capillary and gravitational forces. In normalized variables, the mathematical total flow model describing two-phase flow in an oil reservoir contains two equations for pressure p and water saturation s [13]:

$$\beta\frac{\partial p}{\partial t}+\operatorname{div}\mathbf{u}=0,\quad \mathbf{u}=-\sigma\,\operatorname{grad}p, \tag{1}$$

$$m\frac{\partial s}{\partial t}+\operatorname{div}\left(f\left(s\right)\mathbf{u}\right)=0, \tag{2}$$

$$\mathbf{u}=-\sigma(s)\nabla p,\ \ \sigma=k\,\varphi(s),\ \ \varphi(s)=k_w(s)+\mathrm{K}_\mu k_o(s),\ \mathrm{K}_\mu=\frac{\mu_w}{\mu_o},$$
$$f(s)=\frac{k_w(s)}{\varphi(s)},\ k_w(s)=\bar{s}^n,\ \ k_o(s)=(1-\bar{s})^n,\ \bar{s}=\frac{s-s_*}{s^*-s_*}\in[0,1], n=1\div4. \tag{3}$$

Here t is the time; s_*, s^* – saturation range; m_0, k – porosity and normalized absolute permeability of the formation; $\mathbf{u}$ is the normalized total flow rate; f is the proportion of water in the stream; φ is the mobility of the mixture; μ_w, μ_o – viscosity of the water and oil phases; k_w, k_o – relative phase permeability functions; β – elastic capacity.

Integration of the initial equations (1), (2) over an arbitrary super element (SE) V using the Gauss-Ostrogradsky theorem gives [9, 10, 11]

$$\beta|V|\frac{\partial\langle p\rangle}{\partial t}+Q_\Gamma+Q_\gamma=0, \tag{4}$$

$$|V|\langle m\rangle\frac{\partial\langle s\rangle}{\partial t}+Q_\Gamma^W+Q_\gamma^W=0, \tag{5}$$

where $|V|$ is the volume of the super element V, and the average values and fluid flows through the outer and inner boundaries of the SE are defined by the formulas

$$\langle m\rangle=m,\ \ \langle p\rangle=p,\ \ \langle s\rangle={ms}/{m},\ \ (\ )=\frac{1}{|V|}\int_V(\ )dV, \tag{6}$$

$$Q_\Gamma=\int_\Gamma u_n\,\mathrm{d}\Gamma,\ \ Q_\gamma=\int_\gamma u_n\,\mathrm{d}\gamma,\ \ Q_\Gamma^W=\int_\Gamma f\,u_n\,\mathrm{d}\Gamma,\ \ Q_\gamma^W=\int_\gamma f\,u_n\,\mathrm{d}\gamma. \tag{7}$$

Here Γ is the outer boundary of the FE, represented by a set of its faces Γ_j, and γ is the inner boundary, which is the perforated part of the surface of the wells inside V. Included in the definition of flow (7) u_n is determined by the formula

$$x,y,z\in\gamma,\Gamma:\ u_n=-\sigma\frac{\partial p}{\partial n}.$$

For the closure of a model, the values u_n, f, σ are expressed in terms of averaged $\langle p\rangle, \langle s\rangle$:

$$Q_\Gamma=\sum_j|\Gamma_j|u_n\big|_{\Gamma_j},\ Q_\gamma=|\gamma|u_n\big|_\gamma,\ Q_\Gamma^W=\sum_j|\Gamma_j|fu_n\big|_{\Gamma_j},\ Q_\gamma^W=|\gamma|fu_n\big|_\gamma, \tag{8}$$

where $(\ )\big|_{\Gamma_j}=\frac{1}{|\Gamma_j|}\int_{\Gamma_j}(\ )d\Gamma,\ \ (\ )\big|_\gamma=\frac{1}{|\gamma|}\int_\gamma(\ )d\gamma.$

To calculate the averaged two-phase flow rates in the SEM, relations similar to formulas (3) are used. The set of values $\langle p\rangle$ in the super elements determines the field of average pressure in the reservoir. The use of one or another procedure of filling the grid function $\langle p\rangle$ allows determining the gradient of the average pressure field and using a ratio for the averaged field of the flow rate, similar to the Darcy law:

$$\mathbf{U}=-\boldsymbol{\sigma}\nabla\langle p\rangle,\ \ \boldsymbol{\sigma}=\mathbf{K}\langle\varphi\rangle,\ \ \langle\varphi\rangle\equiv\Phi(\langle s\rangle)=K_w(\langle s\rangle)+\mathrm{K}_\mu K_o(\langle s\rangle). \tag{9}$$

Here $\mathbf{K}$ is the effective absolute permeability tensor in the SE; K_w, K_o are functions that play the same role as the RP functions in the original equations. The definition of functions K_w, K_o is the task of upscaling of RP; tensor $\mathbf{K}$ is constructed as a result of solving the problem of absolute permeability upscaling.

It should be noted that the entry (9) contains the gradient from the discrete grid function $\langle p\rangle$ and is introduced only formally to obtain approximations of average integral flows:

$$u_n \approx U_n = \mathbf{U}\cdot\mathbf{n} = -\mathbf{k}\,\Phi\left(\langle s\rangle\right)\nabla\langle p\rangle = -|\mathbf{k}|\,\Phi\left(\langle s\rangle\right)\frac{\partial\langle p\rangle}{\partial \mathrm{k}},\quad \mathbf{k}=\mathbf{n}\cdot\mathbf{K};$$
$$f\,u_n \approx \langle f\rangle U_n,\quad \langle f\rangle \equiv F\left(\langle s\rangle\right)=\frac{K_w\left(\langle s\rangle\right)}{\Phi\left(\langle s\rangle\right)}. \tag{10}$$

The vector $\mathbf{k}=\mathbf{n}\cdot\mathbf{K}$ can be called the *full permeability vector* on the face of the SE, and $\partial/\partial\mathrm{k}$ is the derivative in the direction of this vector, which is constructed by the method of colocations:

$$U_n = -|\mathbf{k}|\langle\varphi\rangle\frac{\partial\langle p\rangle}{\partial\mathrm{k}} \approx -\overline{k\varphi}\frac{P_2-P_1}{h_1+h_2},\quad \overline{k\varphi}=\frac{|\mathbf{k}_1||\mathbf{k}_2|\langle\varphi\rangle_1\langle\varphi\rangle_2\left(h_1+h_2\right)}{|\mathbf{k}_2|\langle\varphi\rangle_2 h_1+|\mathbf{k}_1|\langle\varphi\rangle_1 h_2}. \tag{11}$$

Through the values of P_1 and P_2 the average pressure on the lines of directions k_1 and k_2, which in turn are calculated by linear interpolation of the average pressures $\langle p\rangle$ in the nearest SE, surrounding the face Γ. Thus, the colocation algorithm is essentially a way of differentiating the average field $\langle p\rangle$ by nodal values that do not require explicit completion of this function.

3. Absolute permeability upscaling

The task of local upscaling of absolute permeability is the definition of matrices of effective tensors **K** in each SE [3]. The coefficients of these tensors are found from the condition of the best approximation of average normal velocities (10), while the "exact" solution for the velocities $u_{n\,j}$ on the faces Γ_j is a quantity in which the pressure p is obtained as a solution of problem (1), (2) with the initial coefficient $\sigma(\mathbf{x})$ of relatively small grid in the area V of this super element.

To reproduce the various variants of the filtration flow through the FE, linear boundary conditions are set.

$$p^b\Big|_{\mathbf{x}\in\Gamma} = \sum_{i=1}^{3}\delta_i^b x^i,\quad b=1..3, \tag{12}$$

where x^i are the Cartesian coordinates of the points of the outer boundary of the FE, δ_i^b is the Kronecker symbol. When upscaling performed for the SE containing the well γ, in addition to the conditions (12), the characteristic value of flow rate on the well is set:

$$\int_\gamma u_n\,\mathrm{d}\gamma = q. \tag{13}$$

The components K^{ij} of the effective tensor matrix **K** for the SE V are determined from the condition of minimizing the functional of the residual

$$J\left(K^{ij}\right)=\frac{1}{3}\sum_{b=1}^{3}\rho^b \to \min_{K^{ij}},\quad \rho^b\left(K^{ij}\right)=\frac{1}{M}\sum_{m=1}^{M}\left(|\Gamma_m|\frac{u_{nm}{}^b-U_{nm}^b\left(K^{ij}\right)}{\max\limits_{l=1..M}\left(|\Gamma_l|\left|u_{nm}{}^b\right|\right)}\right)^2, \tag{14}$$

where M is the number of faces of the SE. The values $U_{nm}^b\left(K^{ij}\right)$ on each face Γ_m for each variant b of the boundary conditions are calculated according to the scheme (11), in which the value P_Γ is calculated in the center of the face Γ_m from the boundary conditions (12), and the value P_1 is determined by interpolation according to the values of pressure on the faces and the average pressure $\langle p \rangle$ in the SE, which is calculated by numerical integration of the pressure obtained on a fine grid.

We note that the formulated scheme for calculating the coefficients K^{ij} does not impose a symmetry condition on the matrix of the effective permeability tensor.

4. Relative phase permeability functions upscaling

The functions K_w, K_o should provide the best approximation for average velocities of the total flow and the aqueous phase across all faces of the SE. Just as for the initial RP, we will use for them power dependences (modified relative permeability functions, MRP)

$$K_w(\langle s \rangle) = S^A, \quad K_o(\langle s \rangle) = (1-S)^B, \quad S = \frac{\langle s \rangle - s_{\min}}{1-s_{\min}}, \tag{15}$$

$$A(S) = A_0 + A_1 S + A_2 S^2, \quad B(S) = B_0 + B_1 S + B_2 S^2,$$

where $s_{\min}$, $\mathbf{A} = \{A_i\}$, $\mathbf{B} = \{B_i\}$, $i = 0,1,2$ are the desired constants. The value $s_{\min}$ is the value of the average water saturation in the FE, before reaching which the total flow through the "output" face of the SE does not flooding.

The upscaling of RP is performed after rescaling the absolute permeability as follows [4].

1) In the area D containing the actual SE and surrounding elements around it, equations (1), (2) are solved on a fine grid, functions $p(t)$, $s(t)$ are constructed, and average velocities $u_n(t), f\,u_n(t)$ on the faces of the actual SC and corresponding flows (8) are calculated.

2) For each SE, the average $\langle p \rangle(t), \langle s \rangle(t)$ are determined by formulas (6) and with their help, using formulas (10), at fixed coefficients $A_i, B_i, s_{\min}$, average velocities $U_n(t)$, $\langle f \rangle U_n(t)$ on the outer and inner faces of the SE are calculated.

3) The desired 7 coefficients of MRP (15) are sought as a solution to the problem of minimizing the functional

$$J(\mathbf{A},\mathbf{B},s_{\min}) = \frac{1}{T}\int_0^T \left(w_1 \rho_\Gamma + w_2 \rho_\Gamma^w + w_3 \rho_\gamma + w_4 \rho_\gamma^w\right)dt; \; \rho_\Gamma = \left(|\Gamma| U_n|_\Gamma - Q_\Gamma\right)^2,$$

$$\rho_\Gamma^w = \left(|\Gamma| U_n^w|_\Gamma - Q_\Gamma^w\right)^2, \; \rho_\gamma = \left(|\gamma| U_n|_\gamma - Q_\gamma\right)^2, \; \rho_\gamma^w = \left(|\gamma| U_n^w|_\gamma - Q_\gamma^w\right)^2, \tag{16}$$

where T is the flooding time; w_i are weights. The first two terms in integral (16) are responsible for the total and water flow through the outer faces of the SE, and the last – for flows through the well inside the SE.

The implementation of the above-described general upscaling technique is rather complex, since it involves solving three-dimensional unsteady flow problems in the vicinity of each SE with an individual porosity and permeability distribution. Therefore, it is advisable to introduce two assumptions that greatly simplify this procedure.

First, for an oil reservoir that has a predominantly layered structure, all the parameters of each layer can be considered constant within the SE. Secondly, all the super elements can be divided into three groups, which differ in the presence and mode of operation of the well. For a SE with a production well, a radially symmetric fluid inflow to the well and its flooding is characteristic; similarly for a solar cell with an injection well – radially symmetric displacement of oil by water and flooding of the flow on the outer faces. For a SE without a well, the filtration flow can be considered as plane-parallel. Accordingly, for each SE it is sufficient to consider one of the three listed watering scenarios, each of which is reduced to solving a two-dimensional problem.

1) Radially symmetric filtration in the FE with a production well in its center is modeled by equations (1) - (4) in a cylindrical coordinate system (r, z) in the region $r_w < r < 1$, $0 < z < H$,

$0<t<T$. At the initial moment $s=0,\ p=0$; the condition is set on the well $p_w=-1$, and the values $p=0$, $s=1$ are supported on the contour $r=1$.

2) Radially symmetric filtration in an ESS with an injection well in its center is described by the same system of equations, but with different boundary conditions: conditions $p_w=1,\ s=1$ are set at the well, and pressure $p=0$ is maintained at the outer boundary.

3) Planar-parallel filtration through the FE without a well is modeled by equations (1) - (4) in the Cartesian coordinate system (*x*, *z*) in the region $0<x<1,\ 0<z<H$, $0<t<T$. At the initial moment $s=0,\ p=0$; on the left boundary, conditions $p_w=1$, $s=1$ are given, and pressure $p=0$ is maintained on the right.

In work [14], a method was proposed for taking into account the lateral heterogeneity of the flooding of the near-wellbore zone of the formation, enclosed in the volume of one grid block containing a production well. The algorithm is designed for large-block finite volume simulation of two-phase filtration in the development of an oil reservoir and can significantly improve the accuracy of the calculation of the water-cut and flow rate of the well, as well as the average saturation in the near-well zone. At the same time, as the rescaled functions of relative phase permeabilities, modified functions are used obtained from solving a reference, upscaling problem on a given grid unit, which does not require significant computational costs under the assumption of radial symmetry of the flow under uniform boundary conditions. The effectiveness of the proposed method is demonstrated by the example of a local problem for a single well in a reservoir with heterogeneous boundary conditions for saturation and by the example of super element modeling of a section of a reservoir with a seven-point development system at various flow rates of wells.

Note that the time required to build a model on a super element grid consists of three components – performing upscaling of the absolute permeability, upscaling of the RP and directly modeling the reservoir flooding. And the cost of modeling is less than 1%, since upscaling procedures require solving a large number of auxiliary tasks on detailed grids. On the other hand, upscaling is performed only once when constructing a grid of super elements; therefore, the use of super element model has significant advantages when performing multivariate design calculations on large oil fields. The greatest computational costs (more than 80%) fall on the RP upscaling procedure, which requires solving two-phase flow problems for each SE. One of the options for reducing the number of local upscaling tasks is to allocate areas of similarity of geological structure. Another variant of the principle acceleration of the solution of this problem is the use of an artificial neural network apparatus and machine learning using the statistical parameters of the local distribution of reservoir properties of the reservoir as input data [15].

5. Saturation downscaling

The super element grid allows describing the behavior of average saturation, but does not provide a detailed picture of the saturation distribution along the interlayers necessary to set the initial condition on the detailed computational grid. For this, downscaling of the saturation field is performed.

As one of the saturation downscaling algorithms that gives the correct results under the assumption of a stratified structure, the following is proposed:

1. vertical averaging of the reservoir and hydroconductivity is performed;
2. one builds the solution of the steady-state problem for pressure with average values of fluid flow rates for the period (t_1-t_0) of modeling on the fine grid:

$$-\nabla\left(\bar{\sigma}\nabla p\right)=\bar{Q},\quad \bar{Q}=\frac{1}{T}\int_{t_0}^{t_1}Q(t)\,\mathrm{d}t\,;$$

3. stream lines are constructed in a two-dimensional reservoir;
4. along each stream line the time of flight (TOF) is calculated;
5. saturation field on the detailed grid is built as

$$s_h = \begin{cases} s^0, \; k_h / m_h \tau > \tau^*; \\ \dfrac{k_h}{m_h} \dfrac{\tau}{\tau^*}(s_c - 1) + 1, \; k_h / m_h \tau \le \tau^*; \end{cases}$$

the threshold value $\tau^*(s_H)$ is determined from the conservative for the aqueous phase condition:

$$R(\tau^*) \to \min, \quad R = \sqrt{\frac{1}{N}\sum_{i=1}^{N}\left(m_H^i s_H^i V_H^i - \sum_{k\in M^i} m_h^k s_h^k V_h^k \right)^2}, \quad s_h^k = \frac{1}{|V_h^k|}\int_{V_h^k} s_h dV,$$

where is the index of the super element, is the set of indexes of the blocks of the detailed grid in the SE.

Calculations showed that over time, the agreement between saturation fields and, accordingly, in well performance indicators, obtained from the local refinement model and detailed model of the entire reservoir, is preserved.

Acknowledgments
This work was funded by the subsidy allocated to Kazan Federal University for the state assignment in the sphere of scientific activities 9.9786.2017/8.9.

References
[1] Durlofsky L 1998 *Computational Geosciences* **2** 73-92
[2] Belyaev A 2000 *Averaging in problems of filtration theory* (*in russian*) (Moscow: Nauka) 200
[3] Mazo A and Potashev K 2018 *Math. Models and Comp. Simulations* **10** 26-35
[4] Mazo A and Potashev K 2017 *Math. Models and Comp. Simulations* **9** 570-579
[5] Aarnes J, Kippe V and Lie K-A 2005 *Adv. in Water. Res.* **28** 257-271
[6] Efendiev Y, Ginting V, Hou T and Ewing R 2006 *J. Comput. Phys.* **220** 155-174
[7] Jenny P, Lee S and Tchelepi H 2006 *J. Comput. Phys.* **217** 627-641
[8] Pergament A, Semiletov V and Tomin P 2011 *Math. Models and Comp. Simulations* **3** 365-374
[9] Mazo A and Bulygin D 2011 *Neft. Gas. Novacii* (*in russian*) **11** 6-8
[10] Mazo A, Potashev K, Kalini E and Bulygin D 2013 *Matematicheskoe modelirovanie* **25** 51-64
[11] Mazo A, Potashev K and Kalinin E. 2015 *Proc. Earth and Planetary Sci.* **15** 482-487
[12] Mazo A and Potashev K 2017 *Georesources* **19** 323-330
[13] Barenblatt G, Entov V and Ryzhik V 1990 *Theory of Fluid Flow Through Natural Rocks* (London: Kluwer Academic Publishers)
[14] Potashev K and Abdrashitova L 2017 *Uchenye Zapiski Kazanskogo Universiteta - Seriya Fiziko-Matematicheskie Nauki* **159** 116-129
[15] Potashev K 2017 *Lobachevskii J. of Mathematics* **38** 843-848

Finite-element modeling of critical loads of buckling of structures considering of nonlinearity of pre-buckling state

V V Repinskii [1], V N Bakulin[1,2]

[1]Moscow Aviation Institute (National Research University), Volokolamskoye highway 4, Moscow, 125993,Russia
[2]Institute of Applied Mechanics of the Russian Academy of Sciences, Leningradsky avenue 7, Moscow, 125040, Russia

E-mail: repinskii.vadim@yandex.ru

E-mail: vbak@yandex.ru

Abstract. An approach is proposed to determine the critical loads of buckling, taking into account the geometric nonlinearity of the subcritical stress-strain state of structures. The first stage involves the structure strain-stress state in the context of geometrical non-linearity at some preset loading level. Afterwards the found solution neighborhood is supplemented with linearized non-linear equation system and the solution for the obtained linearized system is found. Following is the second linearization of the initial equation system in the linearized system solution neighborhood, obtained at the previous stage. The condition of equation systems degeneracy obtained after the second linearization causes a generalized eigenvalue problem. The approach is implemented using finite element models for the calculation of shell structural elements.

1. Introduction

The procedure of finite-element-based finding buckling load involves two methods of increasing accuracy of a final finite-element solution with necessary computer resources being reduced at the same time. The first method is a development of refined approaches and algorithms as to finding buckling loads. The second one is a creation of new, more efficient finite-element models.

2. Formulation of problem

This work offers an algorithm of defining buckling loads, which enables to consider a geometrical non-linearity pre-buckling state of structure elements at issue. The first stage involves the structure strain-stress state in the context of geometrical non-linearity at some preset loading level. Afterwards the found solution neighborhood is supplemented with linearized non-linear equation system and the solution for the obtained linearized system is found. Following is the second linearization of the initial equation system in the linearized system solution neighborhood, obtained at the previous stage. The condition of equation systems degeneracy obtained after the second linearization causes a generalized eigenvalue problem. The algorithm is implemented on the basis of finite element models for the calculation of shells.

3. Approach to finding buckling loads in the context of stress-strain state geometrical non-linearity

Let us consider a shell exposed to statistical loads and temperature with set pre-stress σ_0 and initial strain ε_0. The physical law is assumed as linear [1]:

$$\sigma = \sigma_0 + D\left(\varepsilon + \varepsilon_0 - \alpha T\right), \tag{1}$$

where σ is a stress vector, ε is a strain vector, σ_0 and ε_0 are pre-stress and initial strain vectors corresponding to the initial state, α is a linear thermal expansion coefficients vector, T is a temperature counted from the initial state, D is a symmetric positive definite matrix of material elastic constants.

A relation between nodal displacements $\boldsymbol{q}$ and strains $\boldsymbol{\varepsilon}$ is assumed as quadratic:

$$\varepsilon_i = \varepsilon_i^0 + \varepsilon_i^n = B_{ij}q_j + C_{ijk}q_jq_k. \tag{2}$$

Here and further, the indices for the vectors $\sigma, \varepsilon, \alpha$ and for the matrix D correspond to the components of the stress and strain vectors, the indices for the vectors q correspond to the components of the vector of nodal displacements.

The type of C matrix is defined by a certain definite element type, and matrix B is a linear strain matrix. Its type is specified by reviewing certain finite elements. The formula (2) and further assume a summation over repeated dummy index. Strain energy density w is found through (1) as stresses $\boldsymbol{\sigma}$ acting on strains $\boldsymbol{\varepsilon}$:

$$w = \int_0^{\varepsilon} \sigma^{\mathrm{T}} d\varepsilon = \frac{1}{2}\varepsilon^{\mathrm{T}} D\varepsilon + \sigma_0^{\mathrm{T}}\varepsilon + \varepsilon_0^{\mathrm{T}} D\varepsilon - T\alpha^{\mathrm{T}} D\varepsilon.$$

By integrating over the volume using (2) and subtracting external forces energy R at displacements q, we shall find a potential energy. Using Lagrange variation principle, we shall obtain equilibrium equations:

$$\begin{aligned}\frac{\partial \Pi}{\partial q_i} \equiv 0 = -R_i + \int_V \sigma_{0k} B_{ki} dV + \int_V \varepsilon_{0k} D_{kl} B_{li} dV - \int_V T\alpha_k D_{kl} B_{li} dV + \int_V B_{ki} D_{kl} B_{lj} dV q_j + \\ + \int_V \sigma_{0k}\left(C_{kij} + C_{kji}\right) dV q_j + \int_V \varepsilon_{0k} D_{kl}\left(C_{lij} + C_{lji}\right) dV q_j - \int_V T\alpha_k D_{kl}\left(C_{lij} + C_{lji}\right) dV q_j + \\ + \int_V B_{li} D_{lm} C_{mjk} dV q_j q_k + \int_V B_{lk} D_{lm}\left(C_{mij} + C_{mji}\right) dV q_j q_k + \frac{1}{2}\int_V C_{lij} D_{lm} C_{mnk} dV q_j q_k q_n + \\ + \frac{1}{2}\int_V C_{lji} D_{lm} C_{mnk} dV q_j q_k q_n + \frac{1}{2}\int_V C_{lkj} D_{lm} C_{mni} dV q_j q_k q_n + \frac{1}{2}\int_V C_{lkj} D_{lm} C_{min} dV q_j q_k q_n\end{aligned}$$

Let us introduce the following notations:

$$\begin{aligned}&\textstyle\int_V \sigma_{0k} B_{ki} dV = R_i^{\sigma}, \int_V \varepsilon_{0k} D_{kl} B_{li} dV = R_i^{\varepsilon}, \int_V T\alpha_k D_{kl} B_{li} dV = R_i^{a}, \int_V B_{ki} D_{kl} B_{lj} dV = C_{ij}, \\ &\textstyle\int_V \sigma_{0k}\left(C_{kij} + C_{kji}\right) dV = C_{ij}^{\sigma}, \int_V \varepsilon_{0k} D_{kl}\left(C_{lij} + C_{lji}\right) dV = C_{ij}^{\varepsilon}, \\ &\textstyle\int_V T\alpha_k D_{kl}\left(C_{lij} + C_{lij}\right) dV = C_{ij}^{a}, \int_V \left(B_{li} D_{lm} C_{mjk} + B_{lk} D_{lm} C_{mij} + B_{lj} D_{lm} C_{mki}\right) dV = Q_{ijk}, \\ &\textstyle\int_V \left(C_{nij} D_{mn} C_{mlk} + C_{nkj} D_{mn} C_{mil}\right) dV = X_{ijkl},\end{aligned} \tag{3}$$

As matrix C_{ijk} is symmetric with respect to indices 2 and 3 ($C_{ijk} = C_{ikj}$), matrix Q_{ijk} is symmetric with respect to any pair of indices:

$$Q_{\mathrm{ik}} = Q_{\mathrm{iki}} = Q_{\mathrm{ik}} = Q_{\mathrm{ik}} = Q_{\mathrm{ki}};$$

matrix X_{ijkl} is symmetric with respect to a pair of indices (i,k) and (l,j); the equilibrium equations shall take the following form:

$$R_i + R_i^a - R_i^\varepsilon - R_i^\sigma = C_{ij}q_j + C_{ij}^\sigma q_j + C_{ij}^\varepsilon q_j - C_{ij}^a q_j + Q_{ijk}q_j q_k + X_{ijkl}q_j q_k q_l.$$

Let us assume the shell stressing as one-parameter and linearly dependent on parameter λ:

$$R_i = \overline{R}_i + \lambda \hat{R}_i, \quad \sigma_i = \overline{\sigma}_i + \lambda\hat{\sigma}_i, \quad \varepsilon_i = \overline{\varepsilon}_i + \lambda\hat{\varepsilon}_i, \quad T_i = \overline{T}_i + \lambda\hat{T}_i. \tag{4}$$

In this regard the equilibrium equations shall take the following form:

$$\overline{R}_i + \overline{R}_i^\alpha - \overline{R}_i^\varepsilon - \overline{R}_i^\sigma + \lambda\left(\hat{R}_i + \hat{R}_i^\alpha - \hat{R}_i^\varepsilon - \hat{R}_i^\sigma\right) = \left(C_{ij} + \overline{C}_{ij}^\sigma + \overline{C}_{ij}^\varepsilon - \overline{C}_{ij}^\alpha\right) q_j +$$
$$+ \lambda\left(\hat{C}_{ij}^\sigma + \hat{C}_{ij}^\varepsilon - \hat{C}_{ij}^\alpha\right) + Q_{ijk}q_j q_k + X_{ijkl}q_j q_k q_l \tag{5}$$

The solution of non-linear equations system (5) defines the displacements of nodes of the shell finite-element model, which correspond to its equilibrium state. At some λ the system (5) degenerates, thus being equivalent to the shell critical equilibrium.

Let us assume the critical values λ are sufficiently small, and linearize the system (5) in its solution neighborhood corresponding to $\lambda = 0$. Suppose

$$\overline{R}_i + \overline{R}_i^\alpha - \overline{R}_i^\varepsilon - \overline{R}_i^\sigma = \left(C_{ij} + \overline{C}_{ij}^\sigma + \overline{C}_{ij}^\varepsilon - \overline{C}_{ij}^\alpha\right) q_j^* + Q_{ijk}q_j^* q_k^* + X_{ijkl}q_j^* q_k^* q_l^*,$$

where q_j^* is a solution of non-linear equation system (5) at $\lambda = 0$.

By linearizing the system (5) in neighborhood q_j^* and neglecting the last summand we shall obtain the following:

$$\hat{R}_i = \overline{C}_{ij}\hat{q}_j + \hat{C}_{ij}q_i^* + \lambda\hat{C}_{ij}\hat{q}_j + \left(Q_{ijk}q_k^* + Q_{ikj}q_k^*\right)\hat{q}_j + \lambda Q_{ijk}\hat{q}_j\hat{q}_k,$$

where

$$\overline{C}_{ij} = C_{ij} + \overline{C}_{ij}^\sigma + \overline{C}_{ij}^\varepsilon - \overline{C}_{ij}^\alpha; \hat{C}_{ij} = \hat{C}_{ij}^\sigma + \hat{C}_{ij}^{\tilde{\varepsilon}} - \hat{C}_{ij}^\alpha; \hat{R}_i = \hat{R}_i + \hat{R}_i^\alpha - \hat{R}_i^\varepsilon - \hat{R}_i^\sigma;$$

$\hat{q}_j$ is an unknown increment of nodal displacement vector.

With $\lambda = 0$ let us find the following:

$$-\hat{C}_{ij}q_j^* + \hat{R}_i = \left(\overline{C}_{ij} + Q_{ijk}q_k^* + Q_{ikj}q_k^*\right)\hat{q}_j,$$

Now suppose $\tilde{q}_j^* = q_j^* + \lambda\hat{q}_j$, where q_i^* is a solution of the system (5) at $\lambda = 0$. Let us linearize the system (5) in neighborhood of vector q_i^* by introducing the following notations:

$$R_i = \overline{R}_i + (\lambda + \mu)\hat{R}_i; \quad q_j = \tilde{q}_j^* + \mu\tilde{q}_j. \tag{6}$$

From (5) and (6) taken into account we shall obtain the following:

$$\overline{R}_i + (\lambda+\mu)\hat{R}_i = \overline{C}_{ij}\left(\tilde{q}_j^* + \mu\tilde{q}_j\right) + (\lambda+\mu)\hat{C}_{ij}\left(\tilde{q}_j^* + \mu\tilde{q}_j\right) +$$
$$+ Q_{ikj}\left(\tilde{q}_j^* + \mu\tilde{q}_j\right)\left(\tilde{q}_k^* + \mu\tilde{q}_k\right) + X_{ijkl}\left(\tilde{q}_j^* + \mu\hat{q}_j\right)\left(\tilde{q}_k^* + \mu\tilde{q}_k\right)\left(\tilde{q}_l^* + \mu\tilde{q}_l\right). \tag{7}$$

Let us differentiate the relation (7) with respect to μ:

$$\hat{R}_i = \overline{C}_{ij}\tilde{q}_j + \hat{C}_{ij}\left(\tilde{q}_j^* + \mu\tilde{q}_j\right) + (\lambda+\mu)\hat{C}_{ij}\tilde{q}_j + Q_{ikj}\tilde{q}_j\left(\tilde{q}_k^* + \mu\tilde{q}_k\right) + Q_{ikj}\tilde{q}_k\left(\tilde{q}_j^* + \mu\tilde{q}_j\right) + \ldots$$

(the highest degrees q_j are neglected).

At $\mu = 0$ we shall obtain the following:

$$\left\{\left[\overline{C}_{ij} + (Q_{ijk}q_k^* + Q_{ikj}q_k^*)\right] + \lambda\left[\hat{C}_{ij} + (Q_{ijk}\hat{q}_k + Q_{ikj}\hat{q}_k)\right]\right\}\tilde{q}_j = \hat{R}_i - \hat{C}_{ij}q_j^* - \lambda\hat{C}_{ij}\hat{q}_j. \quad (8)$$

The degeneracy condition of the system (8) has the following form:

$$\det\left\{\left[\overline{C}_{ij} + (Q_{ijk}q_k^* + Q_{ikj}q_k^*)\right] + \lambda\left[\hat{C}_{ij} + (Q_{ijk}\hat{q}_k + Q_{ikj}\hat{q}_k)\right]\right\} = 0 \quad (9)$$

Thus, the problem reduces to defining eigenvalues of a pair of symmetric matrixes.

Now we shall note some aspects of selecting the load constant component in equations (8). As is clear from the method of reducing the non-linear system (5) to problem (9), the results being obtained are the more trustworthy the less the absolute value λ is, while the value sign is not important. Thus, it is useful to choose constant components close to expected critical values. If the solution of non-linear system (5) is not extremely expensive in terms of a computer-based system resources cost, the iterative improvement of buckling loads is essentially possible.

On the other hand, there is a number of cases of practical importance, when a shell deforms in a linear manner to the extent of reaching critical equilibrium. Such cases include, among others, the axial compression of a cylindrical shell. In this situation the result does not depend on the selection of load intercept.

4. Finite-element implementation of defining buckling loads as illustrated by rotational shell of Gaussian zero curvature

Let us assume that structure deformation with a reasonable degree of accuracy is described by models based on the Kirchhoff-Love hypotheses. As an example of defining the type of matrixes C, B and Q (2,3) we shall consider a Kirchhoff-Love hypotheses-based two-dimensional finite element designed to simulate thin conical and cylindrical shells. The element is a rectangular-plan quadrinodal finite element of a thin anisotropic rotational shell of Gaussian zero curvature [1,2]. Radial stress is approximated with incomplete bicubic polynomial, and tangential displacement is approximated with bilinear polynomial. Besides, the finite element functions explicitly include statements, which describe it as a rigid-body displacement, that significantly improves the solution convergence [3-5].

Let us write a strain vector for a conical shell as follows: $\varepsilon = \{\varepsilon_1, \varepsilon_2, \varepsilon_{12}, \varkappa_1, \varkappa_2, \varkappa_{12}\}^{\mathrm{T}}$, where $\varepsilon, \varkappa$ are tangential strains and curvatures. Taking into account non-linear components the strain vector components are linked to displacement through the following equations [6]:

$$\begin{aligned}
\varepsilon_1 &= \tfrac{\partial u}{\partial x} + \tfrac{1}{2}\left[e_{11}^2 + e_{12}^2 + e_{13}^2\right], \\
\varepsilon_2 &= \tfrac{\partial v}{r\partial\beta} + \tfrac{u}{r}\sin\gamma + \tfrac{w}{r}\cos\gamma + \tfrac{1}{2}\left[e_{21}^2 + e_{22}^2 + e_{23}^2\right], \\
\varepsilon_{12} &= \tfrac{\partial u}{r\partial\beta} - \tfrac{v\sin\gamma}{r} + \tfrac{\partial v}{\partial x} + e_{11}e_{21} + e_{22}e_{12} + e_{13}e_{23}, \\
\varkappa_1 &= -\tfrac{\partial^2 w}{\partial x^2}, \\
\varkappa_2 &= -\tfrac{\partial^2 w}{r^2\partial\beta^2} + \tfrac{\partial v}{r^2\partial}\cos\gamma - \tfrac{\partial w}{r\partial x}\sin\gamma, \\
\varkappa_{12} &= -\tfrac{\partial^2 w}{r\partial x}\cos\gamma + \tfrac{\partial w}{r^2\partial\beta}\sin\gamma - \tfrac{v}{r^2}\sin\gamma\cos\gamma,
\end{aligned} \quad (10)$$

where u, v, w are displacements of the shell reference surface,γ is a cone angle, $r = R_2\cos\gamma$ [1],

$e_{11}, e_{12}, e_{13}, e_{21}, e_{22}, e_{23}$ for conical shell are defined as follows:

$$e_{11} = \frac{\partial u}{\partial x}, \quad e_{12} = \frac{\partial v}{\partial x}, \quad e_{13} = \frac{\partial w}{\partial x}, \quad e_{21} = \frac{\partial u}{r\partial\beta} - \frac{v\sin\gamma}{r},$$

$$e_{22}=\frac{\partial v}{r\partial\beta}+\frac{u}{r}\sin\gamma+\frac{w}{r}\cos\gamma,\quad e_{23}=\frac{\partial w}{r\partial\beta}-\frac{v\cos\gamma}{r}.$$

As is seen from equations (10), non-linear components are taken into account only in case of tangential strains.

Let us take a quadrangular finite element of a thin anisotropic rotational shell of Gaussian zero curvature with nodes in vertexes, which is generated through the intersection of two planes perpendicular to the center line and two planes passing through the center line. Thus, the element has 4 nodes and 20 degrees of freedom - 5 in each node: $u_{\mathrm{i}}, v_{\mathrm{i}}, w_{\mathrm{i}}, \vartheta_{\mathrm{xi}}, \vartheta_{\beta\mathrm{i}}$, where $\vartheta x=\frac{\partial w}{r\partial B}-\frac{v\cos\gamma}{r}$, is an angle of normal rotation around axis x, $\vartheta_\beta=-\frac{\partial w}{\partial x}$ is an angle of normal rotation around axis β.

Let us define the element displacement vector:

$$u=u^0+u^{\mathrm{d}}$$

where

$\mathrm{u}^0=\{u_0,v_0,w_0\}^{\mathrm{T}}$ is a rigid element displacement vector, obtained in the similar manner $[4,5,7-12]$,
$u^{\mathrm{d}}=\{u_{\mathrm{d}},v_{\mathrm{d}},w_{\mathrm{d}}\}^{\mathrm{T}}$ is a displacement vector, caused by shell straining:
$u_{\mathrm{d}}=a_7x+a_8\beta+a_9x\beta,$
$v_{\mathrm{d}}=a_{10}\beta+a_{11}x\beta,$
$w_{\mathrm{d}}=a_{12}x\beta+a_{13}x^2+a_{14}\beta^2+a_{15}x^2\beta+a_{16}x\beta^2+a_{17}x^3++a_{18}\beta^3+a_{19}x^3\beta+a_{20}x\beta^3,$

where $a_i, i=1,2,\ldots,20$ are indefinite coefficients linked to nodal displacement through the following equation: $a=Aq$, where $a=\{a_1,\ldots,a_{20}\}^{\mathrm{T}}$ is an indefinite coefficient vector, $q=\{q_1,\ldots,q_{20}\}^{\mathrm{T}}$ is a nodal displacement vector, A is a $20-\text{by}-20$ matrix.

As matrix the displacement vector shall be written as follows:

$$u=N(x,\beta)a,$$

where $u=(u,v,w)^{\mathrm{T}}$, $N(x,\beta)$ is a basis function matrix.

By rearranging the latter equation in terms of nodal displacements we shall obtain the following:

$$u=\boldsymbol{N}(x,\beta)Aq. \tag{11}$$

By substituting (11) in Cauchy equations (10), we shall obtain a strain vector non-linear component expression arranged using nodal displacement vector as follows:

$$\varepsilon^l=\tilde{N}Aq,$$

where $\tilde{N}=\Omega N(x,\beta)$, Ω is an operators matrix [1], and for strain vector non-linear component:

$$\varepsilon_1^n=\frac{1}{2}e_x^{\mathrm{T}}e_x=\frac{1}{2}q^{\mathrm{T}}A^{\mathrm{T}}N_x^{\mathrm{T}}N_xAq,\varepsilon_2^n\ =\frac{1}{2}e_\beta^{\mathrm{T}}e_\beta=\frac{1}{2}q^{\mathrm{T}}A^{\mathrm{T}}N_\beta^{\mathrm{T}}N_\beta Aq,$$
$$\varepsilon_{12}^n=e_x^{\mathrm{T}}e_\beta=q^{\mathrm{T}}A^{\mathrm{T}}N_x^{\mathrm{T}}N_\beta Aq,$$

where N_x, N_β are matrixes obtained from the following equations:

$$e_{\mathrm{x}}=N_{\mathrm{x}}(x,\beta)a, e_\beta=N_\beta(x,\beta)a,$$

where $e_x=\{e_{11},e_{12},e_{13}\}^{\mathrm{T}}$, $e_\beta=\{e_{21},e_{22},e_{23}\}^{\mathrm{T}}$. Hence we can obtain matrix C_{ijk} (2), linking nodal displacements with non-linear strains:

$$C_{1\mathrm{ij}}=1/2A^{\mathrm{r}}N_{\mathrm{x}}^{\mathrm{r}}N_{\mathrm{x}}A, C_{2\mathrm{ij}}=1/2A^{\mathrm{T}}N_\beta^{\mathrm{r}}N_\beta A, C_{3\mathrm{ij}}=A^{\mathrm{T}}N_{\mathrm{x}}^{\mathrm{r}}N_\beta A,$$

$C_{ijk}=A^T\overline{C}_{kij}A$. Thus, matrix Q_{ijk} (7) shall have the following form:

$$Q_{ijk}=\iiint_V \tilde{N}_{lp}D_{lm}\overline{C}_{mrq}dV\left(A_{pi}A_{rj}A_{qk}+A_{pk}A_{ri}A_{qj}+A_{pj}A_{rk}A_{qi}\right).$$

5. Conclusion

An approach, methodology and algorithms for determining critical loads of buckling taking into account the geometric nonlinearity of the subcritical stress-strain state of structures, have been developed. The proposed approach is implemented in the form of finite-element models for determining the critical loads of buckling of shells of revolution. The developed approach, methodology and algorithms make it possible to obtain refined values of critical loads of buckling in those cases in which the geometric nonlinearity of the subcritical stress-strain state is realized (the presence of cutouts, areas of significant weakening of the rigidity of the structure, etc.).

Acknowledgments

The work has been performed under the state order. Official registration number: AAAA-A19-119012290177-0.

References

[1] Bakulin V N, Repinskii V V 2001 Finite-Element Strain Models of One-Layer and Three-Layer Conical Shells *Mat. Model.* **13** 6 39
[2] Repinskii V V 1997 Effective Finite Elements for Calculation of stability of Thin Anisotropic Rotationl Shells *Vopr. Oboron. Tekhniki Ser. 15* 117 3
[3] Cantin G., Glagh R W 1968 A curved, cylindrical shell finite element. *AIAA J.* **6** 6 1057-1062
[4] Bakulin V N, Demidov V I 1978 Three-Layered Natural Curved Finite Element. *Proc. of High. Ed. Inst. ch. Build.* 5 5-10
[5] Bakulin V N 1985 The Finite-Element Method for Studying the Stress-Strain State of Three-Layer Cylindrical Shells *Proc. of High. Ed. Inst. ch. Build.* 5 5-10
[6] Novozhilov V V 1948 The Fundamentals of Non-Linear Elasticity Theory (Leningrad: Gostekhizdat) 211
[7] Bakulin V N 2006 Construction of Approximations for Simulation of Stress-Strain State of Base Layers and Core Layers of Three-Layered Non-Axisymmetric Cylindrical Shells *Mat. Model.* **18** 8 101-110
[8] Zheleznov L P, Kabanov V V 1990 Displacement Functions of Rigid Finite Elements of Rotational Shell *Izv. RAN. MT* 1 131
[9] Bakulin V N 2007 Effective Models in the refined Analysis of the Strain State for Three-Layer Non-Axisymmetric Cylindrical Shells*Doklady Physics* **52** 6 330-334
[10] Bakulin V N 2007 Construction of Approximations and Models for Studying Strain-Stress State of Three-Layered Non-Axisymmetric Cylindrical Shells *Mat. Model.* **19** 12 118
[11] Bakulin V N 2009 Testing of Finite-Element Model Designed for Studying Stress-Strain State of Irregular Sandwich Shells *Mat. Model.* **21** 8 121
[12] Bakulin V N 2018 An Efficient Model for Layer-by-Layer Analysis of Sandwich Irregular Cylindrical Shells of Revolution *Dokl. Ph.* **63** 1 23-27

Numerical homogenization of elastoplastic deformations of composite material with small proportion of inclusions

Petr V. Sivtsev[1], Aleksandr E. Kolesov[1], Petr E. Zakharov[1] and Ying Yang[2]

[1]Research Laboratory "Multiscale Matheatical Modeling and Computing", Ammosov North-Eastern Federal University, 58, Belinskogo, 677000 Yakutsk, Russia

[2]Hunan Key Laboratory for Computation and Simulation in Science and Engineering, School of Mathematics and Computational Science, Xiangtan University, Xiangtan 411105, Hunan, China.

Abstract. Numerical simulation of the stress-strain state of a composite material may be difficult due to large computational complexity associated with a grid resolution of a large number of inclusions. To overcome the problem one may use the homogenization method. But for material with plastic properties, proper modeling of yield stress and hardening may be overcomplicated. In this work we use some simplification associated with a small proportion of inclusions and restriction of stress values by matrix material strength. As model problem, we use deformation of concrete deep beam reinforced with steel or basalt fiber inclusions. For the numerical solution, the finite element method was applied using the FEniCS computing platform.

1. Introduction

In the concrete industry one of the most innovative reinforcements are inclusions in the form of fibers, which can significantly enhance the strength characteristics of concrete [1]. In recent decades, interest in the study of composite reinforcement by basalt plastic for the subsequent replacement of metal rods in load-bearing reinforced concrete structures in construction has been increased [2]. One of the innovative trends in the development of the North of Russia are technologies related to the production of basalt continuous fiber and composite materials based on it. The presence of huge stocks of raw materials and the relatively low cost of fibers produced from it makes it possible to consider their prospects in economic terms.

When modeling the problem of calculating the stress-strain state of a composite material with inclusions in the form of fibers, there is a problem associated with a large quantity of the cells needed for the grid resolution of each fiber. One way to solve this problem is to use a multiscale method or a method of numerical homogenization. However, in real applications, the number of fibers can be so large that the solution to the problem of constructing local bases for the classical multiscale method will also be quite complex. Of course, one can use the method of numerical homogenization on the representative element and solve the problem at the macro level using effective coefficients [3, 4]. But for the material with plastic properties, proper modeling of yield stress and hardening may be overcomplicated. In this work, we use

Material	Young modulus E, GPa	Poisson ratio ν
Concrete	40	0.15
Basalt	89	0.26
Steel	200	0.3

Table 1: Elastic parameters of concrete matrix, basalt and steel fibers

some simplification associated with a small proportion of inclusions. As model problem, we use deformation of concrete deep beam reinforced with steel or basalt fiber inclusions [5].

For a numerical solution, we have used the implementation of the finite element method using the FEniCS computing platform [6–8]. Results of comparison of deformations between a solution with a full grid resolution of fibers and a solution with effective coefficients are presented. Two different variants of elastic properties of inclusions in concrete were used: ones for basalt and steel material.

2. Problem Statement

Let us consider mathematical model describing stress-strain state of composite concrete with basalt or steel fiber inclusions

$$\operatorname{div} \boldsymbol{\sigma} = 0, \quad \boldsymbol{x} \in \Omega = \Omega_1 \cup \Omega_2, \tag{1}$$

where Ω_1 is subdomain of concrete, Ω_2 is subdomain of fibers, $\boldsymbol{\sigma}$ is stress tensor.

For plasticity model [9, 10] strain tensor consists of elastic and plastic parts:

$$\varepsilon = \begin{pmatrix} \frac{\partial u_1}{\partial x_1} \\ \frac{\partial u_2}{\partial x_2} \\ \frac{\partial u_2}{\partial x_1} + \frac{\partial u_1}{\partial x_2} \end{pmatrix} = \varepsilon^e + \varepsilon^p. \tag{2}$$

For convenience, we use Voight notation. There is also a linear relation between stress and elastic strain tensors.

$$\boldsymbol{\sigma} = \begin{pmatrix} \sigma_{11} \\ \sigma_{22} \\ \sigma_{12} \end{pmatrix} = \boldsymbol{C}\varepsilon^e, \quad \boldsymbol{C} = \begin{pmatrix} C_{1111} & C_{1122} & C_{1112} \\ C_{2211} & C_{2222} & C_{2212} \\ C_{1211} & C_{1222} & C_{1212} \end{pmatrix},$$

here $\boldsymbol{C}$ is elastic tensor, which is represented as follows

In the case of isotropic body, elastic tensor can be uniquely determined by Young modulus E and Poisson coefficient ν. Elastic parameters of materials under the investigation are presented in Table 1 [11, 12]. Hardening parameter and yielding stress equal to 21.5 GPa and 30.0 MPa, respectively

Concrete is modeled as plastic material with isotropic hardening, which can be represented as figure 1.

In this article, we limit the maximum value of stress that occurs in concrete around the destruction values. So we can model inclusion materials as pure elastic.

Equation of solid body is supplemented with boundary conditions related to the fixation on the left side and tangent stress on the right side, which is illustrated in figure 2. These conditions can be defined with following Dirichlet and Neumann boundary conditions:

$$\boldsymbol{u}_Y = 0, \quad \boldsymbol{x} \in \Gamma_D, \quad \boldsymbol{\sigma}_{\boldsymbol{n}} = (0, -P), \quad \boldsymbol{x} \in \Gamma_N. \tag{3}$$

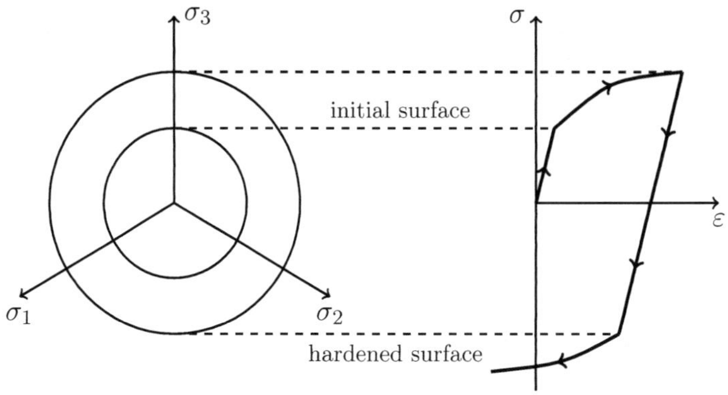

Figure 1: Isotropic hardening

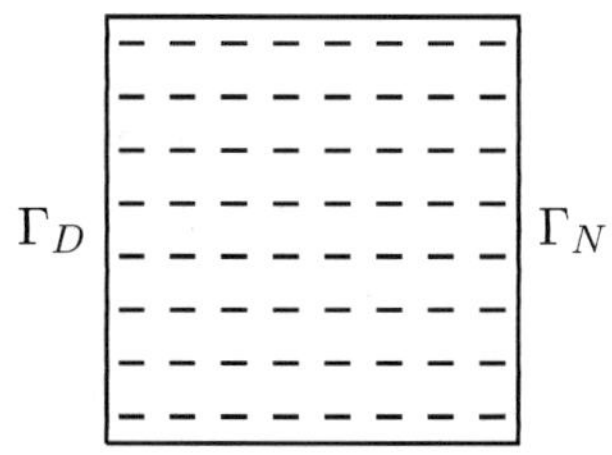

Figure 2: Problem definition. There are 64 fiber inclusions with height to length ratio equal to 1:8 located evenly on volume.

Pressure stress grows from 0 to 10 MPa. At the final moment of time, pressure value is high enough to produce stresses more than tensile strength of the material, which is roughly equal to 100 MPa.

To approximate the space, the finite element method is used. The original differential problem is reduced to a discrete variational problem. The numerical implementation is performed using the FEniCS computing platform.

3. Homogenization Method

For homogenized plasticity, we use homogenized elastic tangent calculated using the asymptotic homogenization method for elastic coefficients. Plastic parameters, such as yield stress and hardening coefficient, remain the same as matrix material, i.e. concrete. We assume that such approximation can be suitable for stress-strain computation of composites with a low volumetric concentration of inclusions limited by stress values equal to matrix material strength. Without such restrictions, one must use multiple yielding points with tensor values for the homogenization parameter, which could describe the anisotropic nature of composite material.

The usage of the asymptotic homogenization method is associated with the computation of an effective elastic tensor. Let denote an average value of function ψ by the following notation

$$\langle \psi \rangle = \frac{\int_\omega \psi \, \mathrm{d}\boldsymbol{x}}{\int_\omega \mathrm{d}\boldsymbol{x}},$$

where ω is a periodic domain. Thus, an average stress tensor can be expressed by

$$\langle \boldsymbol{\sigma} \rangle = \langle \boldsymbol{C} \, \varepsilon \rangle = \boldsymbol{C}^* \langle \varepsilon \rangle,$$

where $\boldsymbol{C}^*$ is the effective elastic tensor. To compute components of the effecitve elastic tensor, we consider three problems in the periodic domain ω with the following force sources:

(i) $\boldsymbol{f} = -\nabla \cdot \boldsymbol{C}\,\varepsilon((x_1, 0))$,
(ii) $\boldsymbol{f} = -\nabla \cdot \boldsymbol{C}\,\varepsilon((0, x_2))$,
(iii) $\boldsymbol{f} = -\nabla \cdot \boldsymbol{C}\,\varepsilon((x_2/2, x_1/2))$.

To ensure the uniqueness of the solution we fix displacement to $(0, 0)$ in the middle point of the domain ω.

4. Comparative Analysis

In figures 4, 3, comparison of the maximum displacement magnitude between plastic, elastic, homogenized plastic and homogenized elastic solutions over time steps are presented for concrete composite with steel and basalt fiber inclusions. According to these results, our approximation provides good accuracy for the case of basalt fiber inclusions. Our approximation for composite with steel inclusions provides results with bigger error, but still tolerable. It is confirmed by a comparison of the displacement results. Its error for steel fiber inclusions is presented in figures 5–8.

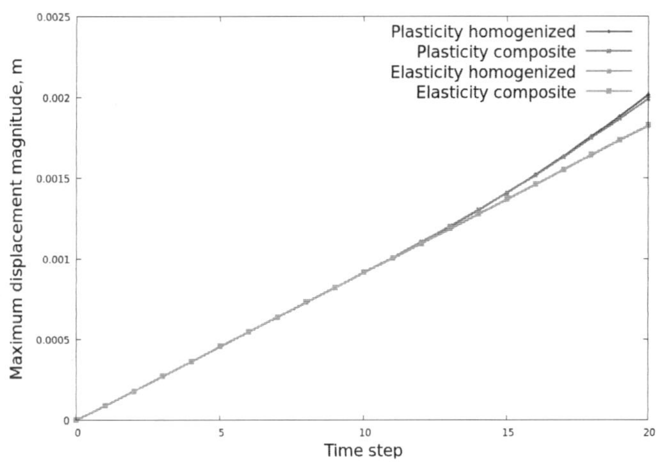

Figure 3: Comparison of the maximum displacement magnitude between plastic, elastic, homogenized plastic and homogenized elastic solutions for concrete composite with steel fiber inclusions over time steps.

5. Conclusion

We have made a solution for the plasticity problem for a composite material with fiber inclusion using the homogenization method. Results demonstrate that the homogenized plasticity model with homogenized elasticity tangent, yield stress and hardening values equal to ones of the matrix can be applied to a model stress-strain state of the composite material in the whole stress range until it reach tensile strength, in other words, the values are high enough to produce cracks and fractures.

Acknowledgments

The authors would like to thank the referees for the helpful suggestions. The work was supported by mega-grant of the Russian Federation Government (N 14.Y26.31.0013), National Natural

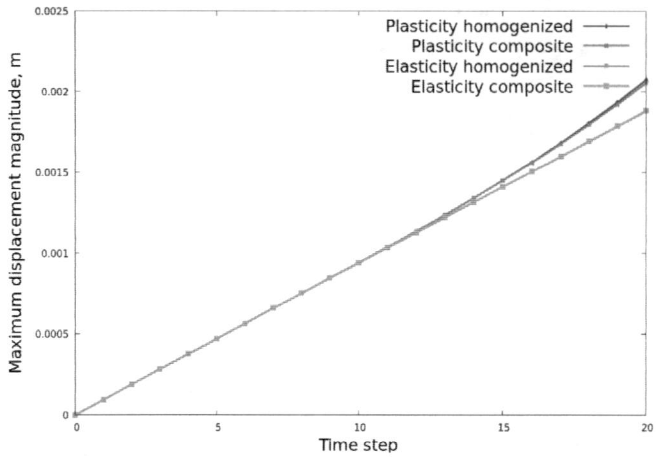

Figure 4: Comparison of the maximum displacement magnitude between plastic, elastic, homogenized plastic and homogenized elastic solutions for concrete composite with basalt fiber inclusions over time steps.

Figure 5: Horizontal displacement value at the last time step of plasticity solution for concrete composite with steel fiber inclusions over time steps.

Figure 6: Vertical displacement value at the last time step of plasticity solution for concrete composite with steel fiber inclusions over time steps.

Science Foundation of China Project (No. 11671342), Project of Scientific Research Fund of Hunan Provincial Science and Technology Department, China (No.2018JJ2374, 2018WK4006).

Journal of Physics: Conference Series **1392** (2019) 012074 doi:10.1088/1742-6596/1392/1/012074

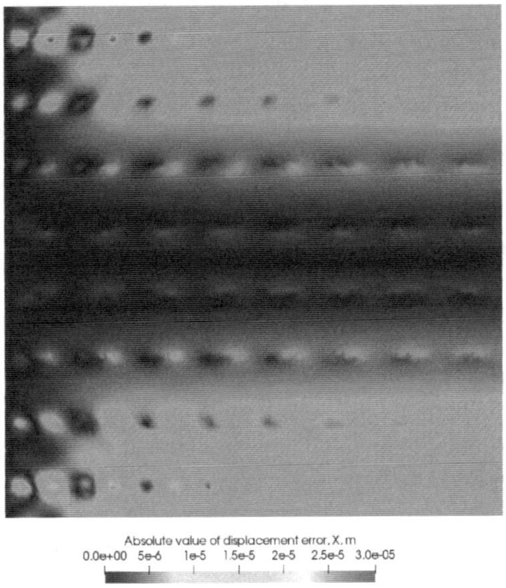

Figure 7: Horizontal displacement error at the last time step of homogenized plasticity solution for concrete composite with steel fiber inclusions over time steps.

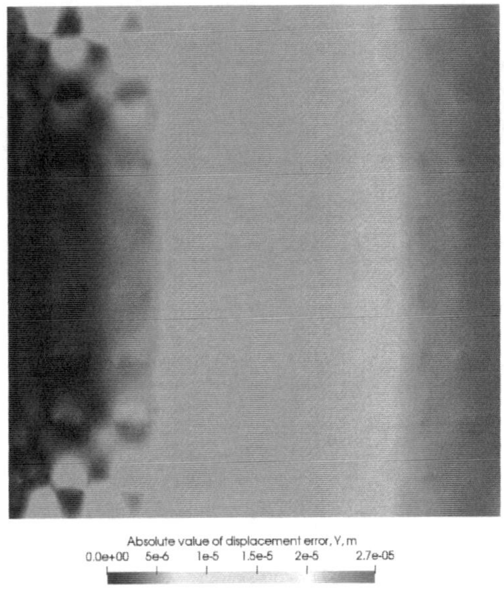

Figure 8: Vertical displacement error at the last time step of homogenized plasticity solution for concrete composite with steel fiber inclusions over time steps.

References

[1] Smarzewski, P.: Influence of basalt-polypropylene fibres on fracture properties of high performance concrete. Composite Structures 209, 23-33 (2019)

[2] Sivtseva, A.V., Sivtsev, P.V.: Numerical Simulation of Deformations of Basalt Roving. International Conference on Finite Difference Methods, 501–508 (2018).

[3] Zakharov, P.E., Sivtsev, P.V.: Numerical calculation of the effective coefficient in the problem of linear elasticity of a composite material. Mathematical notes of NEFU 24.2, 75-84 (2017).

[4] Stepanov, S.P., Vasilyeva, M.V., Vasil'ev, V.I.: Generalized multiscale discontinuous Galerkin method for solving the heat problem with phase change. Journal of Computational and Applied Mathematics 340, 645-652 (2018).

[5] Kolesov A. E. et al.: Numerical Analysis of Reinforced Concrete Deep Beams. International Conference on Numerical Analysis and Its Applications. – Springer, Cham, 414-421 (2016).

[6] Logg, A., Mardal, K.A., Wells, G.N.: Automated solution of differential equations by the finite element method. Springer (2012).

[7] Antonov M. Y., Grigorev A. V., Kolesov A. E.: Numerical modeling of fluid flow in liver lobule using double porosity model. International Conference on Numerical Analysis and Its Applications. – Springer, Cham, 187-194 (2016).

[8] Avvakumov A. V. et al.: Solution of the 3D neutron diffusion benchmark by FEM. International Conference on Large-Scale Scientific Computing. – Springer, Cham, 435-442 (2017).

[9] Simo J. C., Hughes T. J. R.: Computational inelasticity. Springer Science & Business Media, 2006.

[10] de Souza Neto E. A., Peric D., Owen D. R. J.: Computational methods for plasticity: theory and applications. John Wiley & Sons, 2011.

[11] Valentino P. et al.: Mechanical characterization of basalt fibre reinforced plastic with different fabric reinforcements–Tensile tests and FE-calculations with representative volume elements (RVEs). Convegno IGF XXII (2013).

[12] Smarzewski P.: Processes of Cracking and Crushing in Hybrid Fibre Reinforced High-Performance Concrete Slabs. Processes, 44 (2019).

Embedded fracture model in numerical simulation of the fluid flow and geo-mechanics using Generalized Multiscale Finite Element Method

Aleksei Tyrylgin [1], **Maria Vasilyeva**[2], **Eric T. Chung**[3]

[1]Multiscale model reduction laboratory, North-Eastern Federal University, Yakutsk, Republic of Sakha (Yakutia), Russia, 677980.

[2]Institute for Scientific Computation, Texas A&M University, College Station, TX 77843-3368 & Department of Computational Technologies, North-Eastern Federal University, Yakutsk, Republic of Sakha (Yakutia), Russia, 677980.

[3]Department of Mathematics, The Chinese University of Hong Kong (CUHK), Hong Kong SAR.

E-mail: aa.tyrylgin@mail.ru

Abstract. In this work, we consider a pororelasticity problem in fractured porous media. Mathematical model contains a coupled system of equations for pressure and displacements, for which we use an embedded fracture model. The fine grid approximation is constructed based on the finite volume approximation for the pressure in fractured media and finite element method for the displacements. Multiscale approximation is developed using a structured coarse grid and is based on the Generalized Multiscale Finite Element Method for pressures and displacements. The performance of the method is tested using a two-dimensional model problem with different number of the multiscale basis functions.

1. Introduction

Effective numerical simulations of the problems in fractured porous media is important for many real world applications, for example, in oil and gas reservoirs, geothermal fields and underground waste disposal [1, 2]. Mathematical models of the flow problems in fractured porous media are constructed based on the mixed dimensional formulations [3, 4]. Fine grid approximation depends on the mesh construction, where the discrete fracture model is used for conforming fracture and porous matrix grid [5, 6, 7], and in the case of separate independent construction of the fracture and porous matrix grids, an embedded fracture model is used [8, 9, 10]. Mixed dimensional formulation of the flow problem is similar to the dual continuum approaches [11].

For accurate numerical solution of the poroelasticity problem using embedded fracture model, we should use a fine grid which leads to the large discrete system of equations. To reduce the size of the discrete system, multiscale methods are used [12, 13, 14, 15]. In this paper, we develop a Generalized Multiscale Finite Element method for solution of the poroelasticity problems in fractured media with embedded fracture model. We construct multiscale basis functions for pressures and displacement via solution of the local spectral problems. We numerically investigate our method using a two-dimensional model problems in fractured and heterogeneous poroelastic media.

The paper is organized as follows. In Section 2, we present the mathematical model and fine grid approximation of the poroelasticity problem with embedded fracture model. In Section 3, we construct a multiscale coarse grid solver using Generalized Multiscale Finite Element Method. Numerical results for two - dimensional model poroelasticity problems are presented in Section 4.

2. Mathematical model and fine grid approximation

Let Ω is computation domain for the porous matrix and γ is the lower dimensional fracture domain. We consider a mathematical model of flow and geo-mechanics in fractured poroelastic medium that described by a following system of equations for displacements, pressure in porous matrix and fractures

$$\begin{aligned} -\operatorname{div}\sigma(u) + \alpha \operatorname{grad} p = 0, \quad x \in \Omega, \\ c_m \frac{\partial p_m}{\partial t} - \operatorname{div}(k_m \operatorname{grad} p_m) + r_{mf}(p_m - p_f) = f_m, \quad x \in \Omega, \\ c_f \frac{\partial p_f}{\partial t} - \operatorname{div}(k_f \operatorname{grad} p_f) - r_{fm}(p_m - p_f) = f_f, \quad x \in \gamma, \end{aligned} \tag{1}$$

with a linear relation between stress σ and strain ε tensors

$$\sigma(u) = \lambda \varepsilon_v \mathcal{I} + 2\mu\varepsilon(u), \quad \varepsilon(u) = 0.5(\nabla u + (\nabla u)^T),$$

where λ, μ are the Lames parameters, u is the displacements, p is the pressure, α is the Biot coefficient, f is the source term, $k_\alpha = \kappa_\alpha/\nu$, ν is the viscosity, κ_α is the permeability for $\alpha = m, f$, $r_{\alpha\beta} = \eta_{\alpha\beta} r$, r is the transfer coefficient, $\eta_{\alpha\beta}$ is the geometric factors, c_α is the compressibility for $\alpha = m, f$. Note that, we neglect the gravitational forces, suppose that effect of the mechanics to the flow is relatively small and consider effect of the porous matrix pressure to the displacements.

In this work, we use the two-dimensional problem for illustration of the presented method and consider an implicit scheme for approximation of time with given time step τ. Let $\mathcal{T}_h = \cup_i \varsigma_i$ be a fine scale finite element partition of the domain Ω and $\mathcal{E}_\gamma = \cup_l \iota_l$ is the fracture mesh. N_f^m is the number of cells in $\mathcal{T}_h$, N_f^f is the number of cell for fracture mesh $\mathcal{E}_\gamma$.

For approximation of the flow problem, we use a finite volume approximation on the structured fine girds and obtain the following discrete system

$$\begin{aligned} c_m \frac{p_{m,i} - \check{p}_{m,i}}{\tau} |\varsigma_i| + \sum_j T_{ij}(p_{m,i} - p_{m,j}) + \sum_l q_{il}(p_{m,i} - p_{f,l}) = f_m |\varsigma_i|, \quad \forall i = 1, N_f^m \\ c_f \frac{p_{f,l} - \check{p}_{f,l}}{\tau} |\iota_l| + \sum_n W_{ln}(p_{f,l} - p_{f,n}) - \sum_i q_{il}(p_{m,i} - p_{f,l}) = f_f |\iota_l|, \quad \forall l = 1, N_f^f \end{aligned} \tag{2}$$

where $T_{ij} = k_m |E_{ij}|/\Delta_{ij}$ ($|E_{ij}|$ is the length of interface between cells ς_i and ς_j, Δ_{ij} is the distance between mid point of cells ς_i and ς_j), $W_{ln} = k_f/\Delta_{ln}$ (Δ_{ln} is the distance between points l and n), $|\varsigma_i|$ and $|\iota_l|$ is the volume of the cells ς_i and ι_l, $q_{il} = r$ if $\varsigma_i \cap \iota_l \neq 0$ and equals zero otherwise. Here $(\check{p}_m, \check{p}_f)$ are solutions from the previous times step.

For displacement, we use Galerkin method with linear basis functions.

$$\int_\Omega \sigma(u) : \varepsilon(v)\, dx - \int_\Omega \alpha p_m \mathcal{I} \cdot \varepsilon(v)\, dx = 0, \tag{3}$$

Therefore, we have a following computational algorithm for solution on the fine grid in the matrix form

- Solve pressure system for $p = (p_m, p_f)^T$:

$$\left(\frac{1}{\tau}M + A\right) p = F, \tag{4}$$

where

$$M = \begin{pmatrix} M_m & 0 \\ 0 & M_f \end{pmatrix}, \quad F = \begin{pmatrix} F_m + \frac{1}{\tau}M_m\check{p}_m \\ F_f + \frac{1}{\tau}M_f\check{p}_f \end{pmatrix}, \quad A = \begin{pmatrix} A_m + Q & -Q \\ -Q & A_f + Q \end{pmatrix},$$

$$M_m = \{m_{ij}^m\}, \quad m_{ij}^m = \begin{cases} c_m|\varsigma_i| & i = j, \\ 0 & i \neq j \end{cases}, \quad M_f = \{m_{ln}^f\}, \quad m_{ln}^f = \begin{cases} c_f|\iota_l| & l = n, \\ 0 & l \neq n \end{cases},$$

and A_m, A_f are the transmissibility matrices, Q is the tranfer term matrix between porous matric and fractures, $F_m = \{f_i^m\}$, $f_i^m = f_m|\varsigma_i|$, $F_f = \{f_l^f\}$, $f_i^m = f_f|\iota_l|$.

- Solve displacements system for $u = (u_x, u_y)$

$$Du = B, \tag{5}$$

where D is the elasticity stiffness matrix

$$D = \begin{pmatrix} D_{xx} & D_{xy} \\ D_{xy} & D_{yy} \end{pmatrix}, \quad B = \begin{pmatrix} B_x \\ B_y \end{pmatrix},$$

and $D_{xx} = [d_{ij}^{xx}] = \int_\Omega \sigma_x(\psi_i) : \varepsilon_x(\psi_j)\, dx$, $D_{yy} = [d_{ij}^{yy}] = \int_\Omega \sigma_y(\psi_i) : \varepsilon_y(\psi_j)\, dx$, $D_{xy} = [d_{ij}^{xy}] = \int_\Omega \sigma_x(\psi_i) : \varepsilon_y(\psi_j)\, dx$, $B_x = [b_j^x] = \int_\Omega \alpha p_m \cdot \varepsilon_x(\psi_j)\, dx$, $B_y = [b_j^y] = \int_\Omega \alpha p_m \cdot \varepsilon_y(\psi_j)\, dx$ with linear basis functions ψ_i.

3. Coarse grid approximation using GMsFEM

For coarse grid approximation of the poroelasticity problems in fractured porous media, we use the Generalized Multiscale Finite Element Method (GMsFEM). GMsFEM contains following steps: (1) construction of the coarse and fine meshes; (2) generation of the projection matrix using local multiscale basis functions; (3) construction of the coarse grid system using projection matrix; (4) solution of the coarse scale problem and reconstruction of the fine grid solution.

For construction of the multiscale basis functions for pressures and displacements, we solve a spectral problem in the local domain ω_i

- pressures, $\phi_j^{\omega_i} = (\phi_{m,j}^{\omega_i}, \phi_{f,j}^{\omega_i})$

$$A^{\omega_i}\phi_j^{\omega_i} = \lambda_{p,j} S^{\omega_i}\phi_j^{\omega_i},$$

- displacements, $\Phi_j^{\omega_i} = (\Phi_{x,j}^{\omega_i}, \Phi_{y,x}^{\omega_i})$

$$D^{\omega^i}\Phi_j^{\omega_i} = \lambda_{u,j} G^{\omega_i}\Phi_j^{\omega_i},$$

where D^{ω_i} and A^{ω_i} are the restrictions of the global matrices D and A to the local domain ω_i. Here

$$S^{\omega_i} = \begin{pmatrix} S_m^{\omega_i} & 0 \\ 0 & S_f^{\omega_i} \end{pmatrix}, \quad G^{\omega_i} = \begin{pmatrix} G_x^{\omega_i} & 0 \\ 0 & G_y^{\omega_i} \end{pmatrix},$$

Journal of Physics: Conference Series **1392** (2019) 012075 doi:10.1088/1742-6596/1392/1/012075

$$S_m^{\omega_i} = \{s_{ij}^{\omega_i,m}\}, \quad s_{ij}^{\omega_i,m} = \begin{cases} k_m|\varsigma_i| & i=j, \\ 0 & i \neq j \end{cases}, \quad S_f^{\omega_i} = \{s_{ln}^{\omega_i,f}\}, \quad s_{ln}^{\omega_i,f} = \begin{cases} k_f|\iota_l| & l=n, \\ 0 & l \neq n \end{cases},$$

and $G_x^{\omega_i} = G_y^{\omega_i} = [g_{ij}] = \int_{\omega_i} (\lambda + 2\mu)\psi_j dx$.

We form the multiscale spaces for pressures and displacements using eigenvectors $\{\phi_1^{\omega_i}, \phi_2^{\omega_i}, ..., \phi_{L_p}^{\omega_i}\}$ and $\{\Phi_1^{\omega_i}, \Phi_2^{\omega_i}, ..., \Phi_{L_u}^{\omega_i}\}$ corresponding to the first smallest L_p and L_u eigenvalues, where $\lambda_{p,1} \leq \lambda_{p,2} \leq ... \leq \lambda_{p,L_p}$ and $\lambda_{u,1} \leq \lambda_{u,2} \leq ... \leq \lambda_{u,L_u}$. Furthermore, for obtaining conforming basis functions we use linear partition of unity functions χ^{ω_i}. We construct transition matrices R_u and R_p from a fine grid to a coarse grid and use it for reducing the dimension of the problem

$$R_u = \{\chi^{\omega_1}\Phi_1^{\omega_1}, \chi^{\omega_1}\Phi_2^{\omega_1}, ..., \chi^{\omega_1}\Phi_{L_u}^{\omega_1}, ..., \chi^{\omega_{N_c}}\Phi_1^{\omega_{N_c}}, \chi^{\omega_{N_c}}\Phi_2^{\omega_{N_c}}, ..., \chi^{\omega_{N_c}}\Phi_{L_u}^{\omega_{N_c}}\}$$

$$R_p = \{\chi^{\omega_1}\phi_1^{\omega_1}, \chi^{\omega_1}\phi_2^{\omega_1}, ..., \chi^{\omega_1}\phi_{L_p}^{\omega_1}, ...\chi^{\omega_{N_c}}\phi_1^{\omega_{N_c}}, \chi^{\omega_{N_c}}\phi_2^{\omega_{N_c}}, ..., \chi^{\omega_{N_c}}\phi_{L_p}^{\omega_{N_c}}, \}.$$

where χ^{ω_i} is linear partition of unity functions, L_p and L_u are the number of basis functions for pressure and displacements, N_c is the number of vertices of a coarse grid.

Then the system of equations can be translated into a coarse grid, and we have following computational algorithm in the matrix form

- Solve pressure system for $p_c = (p_{c,m}, p_{c,f})^T$:

$$\left(\frac{1}{\tau}M_c + A_c\right) p_c = F_c, \tag{6}$$

  where $M_c = R_p M R_p^T$, $A_c = R_p A R_p^T$ and $F_c = R_p F$.
- Solve displacements system for $u = (u_x, u_y)$

$$D_c u_c = B_c, \tag{7}$$

  where $D_c = R_u D R_u^T$ and $B_c = R_u B$.

After obtaining of a coarse-scale solution, we can reconstruct fine-scale solution $u_{ms} = R_u^T u_c$ and $p_{ms} = R_p^T p_c$. Next, we present numerical results for a test problems and compare multiscale solutions for pressures and displacements (p_{ms} and u_{ms}) with the reference fine grid solutions (p and u).

4. Numerical results

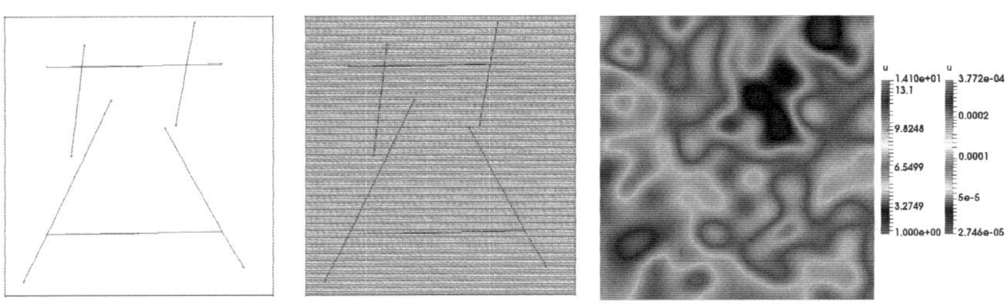

Figure 1. Fracture distribution, fine grid, heterogeneous permeabilities K and elasticity parameter E (from left to right).

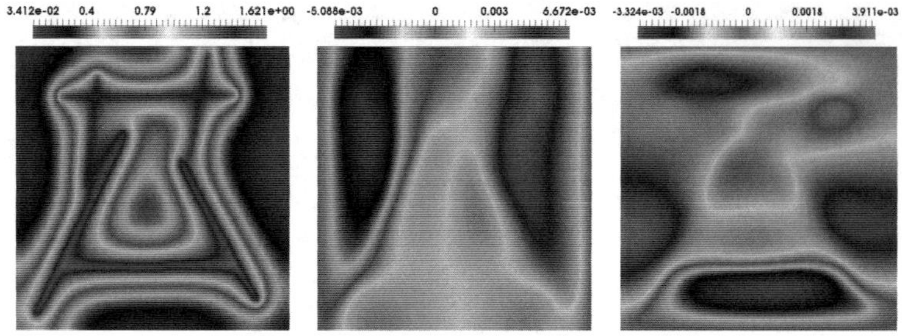

Figure 2. Distribution of pressure, displacement along X and Y directions at the last moment of time for 8 multiscale basis functions for *Test 1*.

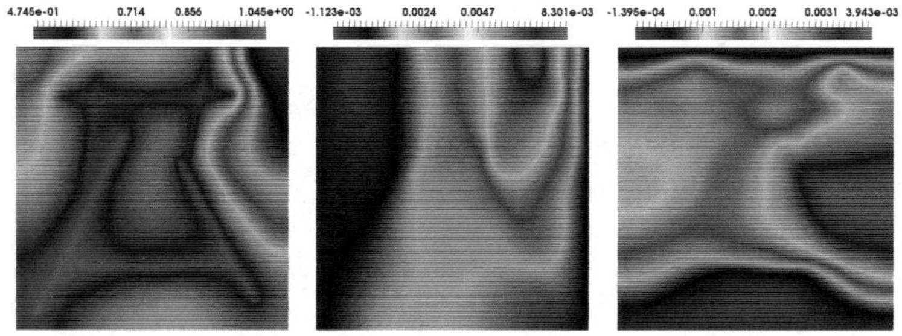

Figure 3. Distribution of pressure, displacement along X and Y directions at the last moment of time for 8 multiscale basis functions for *Test 2* .

	10×10		20×20	
L	e^u	e^p	e^u	e^p
1	4.850	1.840	1.794	0.896
2	1.933	1.332	0.535	0.550
4	1.120	0.648	0.372	0.292
8	0.395	0.488	0.136	0.195
12	0.289	0.330	0.077	0.144

	10×10		20×20	
L	e^u	e^p	e^u	e^p
1	13.461	10.781	3.195	4.065
2	4.285	7.350	1.037	2.477
4	3.850	4.208	1.082	1.278
8	1.103	3.502	0.274	0.992
12	1.009	1.628	0.126	0.552

Table 1. Relative errors in % for displacement and pressure with different numbers of multiscale basis functions for *Test 1* (left) and *Test 2* (right). $L_p = L_u = L$.

In this section, we consider poroelasticity problem in fractured porous media and consider two test problems. We set parameters of model problem $E = 10$, $\nu = 0.3$, $k_m = 10^{-5}$, $k_f = 1$ for *Test 1* . For *Test 2*, we use a heterogeneous coefficients, where elasticity modulus and heterogeneous permeability are presented in Figure 1. We set $\alpha = 1$, $\sigma = k_m$, $c_m = 0.1$ and $c_f = 0.01$. The calculation is performed by $T_{max} = 100$ with time step $\tau = 10$. Computational grids and fracture distribution are presented in Figure 1. Fine grid contains 40401 vertices and 40000 cells. For study of the presented multiscale method, we consider two coarse grids 10×10 and 20×20.

In Figures 2 and 3, distribution of pressure and displacement along X and Y directions at final time for 8 multiscale basis functions in homogeneous and heterogeneous media are presented. In

Table 1, we present the relative L_2 errors for different number of the multiscale basis functions on the coarse grid 10×10 and 20×20 between multiscale and fine grid solutions. Presented results of the poroelasticity problem show that proposed multiscale method can provide good accuracy for a small number of multiscale basis functions for embedded fracture model.

5. Acknowledgements

MV's and TA's works are supported by the mega-grant of the Russian Federation Government N14.Y26.31.0013, RSF N17-71-20055 and RFBR N19-31-90120. YE's work is supported by the mega-grant of the Russian Federation Government N14.Y26.31.0013. The research of Eric Chung is partially supported by the Hong Kong RGC General Research Fund (Project numbers 14304217 and 14302018) and CUHK Faculty of Science Direct Grant 2018-19.

References

[1] I Yucel Akkutlu, Yalchin Efendiev, Maria Vasilyeva, and Yuhe Wang. Multiscale model reduction for shale gas transport in a coupled discrete fracture and dual-continuum porous media. *Journal of Natural Gas Science and Engineering*, 48:65–76, 2017.

[2] I Yucel Akkutlu, Ebrahim Fathi, et al. Multiscale gas transport in shales with local kerogen heterogeneities. *SPE journal*, 17(04):1–002, 2012.

[3] Vincent Martin, Jérôme Jaffré, and Jean E Roberts. Modeling fractures and barriers as interfaces for flow in porous media. *SIAM Journal on Scientific Computing*, 26(5):1667–1691, 2005.

[4] Luca Formaggia, Alessio Fumagalli, Anna Scotti, and Paolo Ruffo. A reduced model for darcy's problem in networks of fractures. *ESAIM: Mathematical Modelling and Numerical Analysis*, 48(4):1089–1116, 2014.

[5] I Yucel Akkutlu, Yalchin Efendiev, Maria Vasilyeva, and Yuhe Wang. Multiscale model reduction for shale gas transport in poroelastic fractured media. *Journal of Computational Physics*, 353:356–376, 2018.

[6] Liyong Li, Seong H Lee, et al. Efficient field-scale simulation of black oil in a naturally fractured reservoir through discrete fracture networks and homogenized media. *SPE Reservoir Evaluation & Engineering*, 11(04):750–758, 2008.

[7] Mohammad Karimi-Fard, Luis J Durlofsky, Khalid Aziz, et al. An efficient discrete fracture model applicable for general purpose reservoir simulators. In *SPE Reservoir Simulation Symposium*. Society of Petroleum Engineers, 2003.

[8] Seong H Lee, MF Lough, and CL Jensen. Hierarchical modeling of flow in naturally fractured formations with multiple length scales. *Water resources research*, 37(3):443–455, 2001.

[9] Liyong Li, Seong Hee Lee, et al. Efficient field-scale simulation for black oil in a naturally fractured reservoir via discrete fracture networks and homogenized media. In *International oil & gas conference and exhibition in China*. Society of Petroleum Engineers, 2006.

[10] Matei Tene, MS Al Kobaisi, H Hajibeygi, et al. Algebraic multiscale solver for flow in heterogeneous fractured porous media. In *SPE Reservoir Simulation Symposium*. Society of Petroleum Engineers, 2015.

[11] GI Barenblatt, Iu P Zheltov, and IN Kochina. Basic concepts in the theory of seepage of homogeneous liquids in fissured rocks [strata]. *Journal of applied mathematics and mechanics*, 24(5):1286–1303, 1960.

[12] Y. Efendiev, J. Galvis, and T. Hou. Generalized multiscale finite element methods. *Journal of Computational Physics*, 251:116–135, 2013.

[13] Y. Efendiev and T. Hou. Multiscale Finite Element Methods: Theory and Applications. 4, 2009.

[14] Patrick Jenny, Seong H Lee, and Hamdi A Tchelepi. Adaptive multiscale finite-volume method for multiphase flow and transport in porous media. *Multiscale Modeling & Simulation*, 3(1):50–64, 2005.

[15] Hadi Hajibeygi, Giuseppe Bonfigli, Marc Andre Hesse, and Patrick Jenny. Iterative multiscale finite-volume method. *Journal of Computational Physics*, 227(19):8604–8621, 2008.

Numerical calculation of spectral problems in SP_3 approximation by FEM

A. O. Vasilev[1,2], A. V. Avvakumov[3], V. F. Strizhov[2] and P. N. Vabishchevich[2]

[1]North-Eastern Federal University, 58, Belinskogo, Yakutsk, Russia
[2]Nuclear Safety Institute, Russian Academy of Sciences, 52, B. Tulskaya, Moscow, Russia
[3]National Research Center *Kurchatov Institute*, 1, Sq. Ac. Kurchatov, Moscow, Russia

E-mail: haska87@gmail.com

Abstract. The SP_3 approximation of the neutron transport equation allows improving the accuracy for both static and transient simulations for reactor core analysis compared with the neutron diffusion theory. Besides, the SP_3 calculation costs are much less than higher order transport methods (S_N or P_N). Another advantage of the SP_3 approximation is a similar structure of equations that is used in the diffusion method. Therefore, there is no difficulty to implement the SP_3 solution option to the multi-group neutron diffusion codes. In this work, the application of the SP_3 methodology based on solution of the λ- and α-spectral problems has been tested for reactor benchmarks. The FEM is chosen to achieve the geometrical generality. The results calculated with the diffusion and SP_3 methods are compared with the reference calculation results.

1. Introduction

The diffusion approximation of the neutron transport equation is widely used in nuclear reactor analysis allowing whole-core calculations with reasonable accuracy. The SP_3 [1] method can be considered as an improved approximation of the neutron transport equation compared with the diffusion method. In this regard, it will be very useful to compare the spectral parameters, calculated by both the diffusion and SP_3 methods.

To characterize the reactor steady-state conditions or dynamic behavior, some spectral problems are considered [2, 3]. The steady-state condition is usually described by solution of a spectral problem (λ-eigenvalue problem); the fundamental eigenvalue (the largest eigenvalue) is called k-effective of the reactor core. The reactor dynamic behavior can naturally be described on the basis of the approximate solution expansion in time-eigenvalue of α-eigenvalue problem [2]. At large times, one can talk about the asymptotic behavior of a neutron flux, whose amplitude is $\exp(\alpha t)$ [4]. Previously the complex eigenvalues and eigenfunctions were found in the spectral problems for some numerical tests [5].

2. Problem statement

Lets consider the symmetric form of the SP_3 equation for the neutron flux [6]. The neutron dynamics is considered in the limited convex two-dimensional or three-dimensional area Ω ($\boldsymbol{x} = \{x_1, ..., x_d\} \in \Omega$, $d = 2, 3$) with boundary $\partial\Omega$. The neutron transport is described by the system of equations

$$\begin{aligned}
\frac{1}{v_g}\frac{\partial \phi_{0,g}}{\partial t} - \frac{2}{v_g}\frac{\partial \phi_{2,g}}{\partial t} - \nabla\cdot D_{0,g}\nabla\phi_{0,g} + \Sigma_{r,g}\phi_{0,g} - 2\Sigma_{r,g}\phi_{2,g} = \\
=(1-\beta)\chi_{n,g}S_n + S_{s,g} + \chi_{d,g}S_d, \\
-\frac{2}{v_g}\frac{\partial \phi_{0,g}}{\partial t} + \frac{9}{v_g}\frac{\partial \phi_{2,g}}{\partial t} - \nabla\cdot D_{2,g}\nabla\phi_{2,g} + (5\Sigma_{t,g} + 4\Sigma_{r,g})\phi_{2,g} - 2\Sigma_{r,g}\phi_{0,g} = \\
= -2(1-\beta)\chi_{n,g}S_n - 2S_{s,g} - 2\chi_{d,g}S_d,
\end{aligned} \tag{1}$$

where

$$S_n = \sum_{g'=1}^{G} \nu\Sigma_{f,g'}\phi_{g'}, \quad S_{s,g} = \sum_{g\neq g'=1}^{G} \Sigma_{s,g'\to g}\phi_{g'}, \quad S_d = \sum_{m=1}^{M}\lambda_m c_m,$$

$$\phi_{0,g} = \phi_g + 2\phi_{2,g}, \quad D_{0,g} = \frac{1}{3\Sigma_{tr,g}}, \quad D_{2,g} = \frac{9}{7\Sigma_{t,g}}, \quad g = 1, 2, ..., G.$$

Here G — number of energy groups, $\phi_g(\boldsymbol{x},t)$ — scalar flux, $\phi_{0,g}(\boldsymbol{x},t)$ — 0th moment of angular flux, $\phi_{2,g}(\boldsymbol{x},t)$ — 2th moment of angular flux, $\Sigma_{t,g}(\boldsymbol{x},t)$ — total cross-section, $\Sigma_{tr,g}(\boldsymbol{x},t)$ — transport cross-section, $\Sigma_{r,g}(\boldsymbol{x},t)$ — removal cross-section, $\Sigma_{s,g'\to g}(\boldsymbol{x},t)$ — scattering cross-section, χ_g — spectra of neutrons, $\nu\Sigma_{f,g}(\boldsymbol{x},t)$ — generation cross-section, $c_m(\boldsymbol{x},t)$ — density of sources of delayed neutrons, λ_m — decay constant of sources of delayed neutrons, M — number of types of delayed neutrons.

The density of sources of delayed neutrons is described by the equations

$$\frac{\partial c_m}{\partial t} + \lambda_m c_m = \beta_m S_n, \quad m = 1, 2, ..., M, \tag{2}$$

where β_m is the fraction of delayed neutrons of m-type, and $\beta = \sum_{m=1}^{M}\beta_m$. The Marshak-type conditions are set at the boundary of the area $\partial\Omega$

$$\begin{bmatrix} J_{0,g}(\boldsymbol{x}) \\ J_{2,g}(\boldsymbol{x}) \end{bmatrix} = \begin{bmatrix} \frac{1}{2} & -\frac{3}{8} \\ -\frac{3}{8} & \frac{21}{8} \end{bmatrix} \begin{bmatrix} \phi_{0,g}(\boldsymbol{x}) \\ \phi_{2,g}(\boldsymbol{x}) \end{bmatrix}, \quad J_{i,g}(\boldsymbol{x}) = -D_{i,g}\nabla\phi_{i,g}(\boldsymbol{x}), \quad i = 0, 2. \tag{3}$$

System of equations (1) and (2) is supplemented with boundary conditions (3) and corresponding initial conditions

$$\phi_g(\boldsymbol{x},0) = \phi_g^0(\boldsymbol{x}), \quad g = 1, 2, ..., G, \quad c_m(\boldsymbol{x},0) = c_m^0(\boldsymbol{x}), \quad m = 1, 2, ..., M. \tag{4}$$

Let's write the boundary problem (1)–(4) in operator form. The vectors $\boldsymbol{u}_1 = \{\phi_{0,1}, \phi_{0,2}, \cdots, \phi_{0,G}\}$, $\boldsymbol{u}_2 = \{\phi_{2,1}, \phi_{2,2}, \cdots, \phi_{2,G}\}$, $\boldsymbol{c} = \{c_1, c_2, ..., c_M\}$ and matrices are defined as follows

$$V = (\mathrm{v}_{gg'}), \quad \mathrm{v}_{gg'} = \frac{1}{v_g}\delta_{gg'}, \quad B = (b_{gg'}), \quad b_{gg} = -2\Sigma_{r,g}, \quad b_{gg'} = 2\Sigma_{s,g'\to g},$$

$$\begin{aligned}
A_1 = (a_{gg'}), \quad a_{gg} = -\nabla\cdot D_{0,g}\nabla + \Sigma_{r,g}, \quad a_{gg'} = -\Sigma_{s,g'\to g}, \\
A_2 = (a_{gg'}), \quad a_{gg} = -\nabla\cdot D_{2,g}\nabla + 5\Sigma_{tr,g} + 4\Sigma_{r,g}, \quad a_{gg'} = -4\Sigma_{s,g'\to g}, \\
F = (f_{gg'}), \quad f_{gg'} = \chi_{n,g}\nu\Sigma_{f,g'}, \quad E = (e_{gm}), \quad e_{gm} = \chi_{d,g}\lambda_m, \\
\Lambda = (\lambda_{mm'}), \quad \lambda_{mm'} = \delta_{mm'}\lambda_m, \quad Q = (q_{mg}), \quad q_{mg} = \beta_m\nu\Sigma_{f,g},
\end{aligned}$$

where $\delta_{gg'}$ is the Kronecker symbol. We shall use the set of vectors $\boldsymbol{u_1}, \boldsymbol{u_2}$, whose components satisfy the boundary conditions (3). Using the set definitions, the system of equations (1) and (2) can be written as following

$$\begin{gathered}
V(\frac{\partial \boldsymbol{u}_1}{\partial t} - 2\frac{\partial \boldsymbol{u}_2}{\partial t}) + A_1\boldsymbol{u}_1 + B\boldsymbol{u}_2 = (1-\beta)F(\boldsymbol{u}_1 - 2\boldsymbol{u}_2) + E\boldsymbol{c},\\
V(-2\frac{\partial \boldsymbol{u}_1}{\partial t} + 9\frac{\partial \boldsymbol{u}_2}{\partial t}) + A_2\boldsymbol{u}_2 + B\boldsymbol{u}_1 = -2(1-\beta)F(\boldsymbol{u}_1 - 2\boldsymbol{u}_2) - 2E\boldsymbol{c},\\
\frac{\partial \boldsymbol{c}}{dt} + \Lambda\boldsymbol{c} = Q(\boldsymbol{u}_1 - 2\boldsymbol{u}_2).
\end{gathered} \tag{5}$$

The Cauchy problem is formulated for equations (5) when

$$\boldsymbol{u}_1(0) = \boldsymbol{u}_1^0, \quad \boldsymbol{u}_2(0) = \boldsymbol{u}_2^0, \quad \boldsymbol{c}(0) = \boldsymbol{c}^0, \tag{6}$$

where $\boldsymbol{u}_1^0 = \{\phi_{0,1}^0, \phi_{0,2}^0, ..., \phi_{0,G}^0\}$, $\boldsymbol{u}_2^0 = \{\phi_{2,1}^0, \phi_{2,2}^0, ..., \phi_{2,G}^0\}$ and $\boldsymbol{c}^0 = \{c_1^0, c_2^0, ..., c_M^0\}$.

3. Spectral problems

The spectral problem, which is known as the λ-spectral problem, is usually considered. For the system of equations (5), we have

$$L\boldsymbol{\varphi} = \lambda^{(k)} M\boldsymbol{\varphi}, \tag{7}$$

where

$$\boldsymbol{\varphi} = \{\boldsymbol{\varphi}_1, \boldsymbol{\varphi}_2\}, \quad L = \begin{pmatrix} A_1 & B \\ B & A_2 \end{pmatrix}, \quad M = \begin{pmatrix} F & -2F \\ -2F & 4F \end{pmatrix}.$$

The minimal eigenvalue is used for characterisation of neutron field, thus $k = 1/\lambda_1^{(k)}$ is the effective multiplication factor (k-effective). The value $k = 1$ is related to the critical state of the reactor, and the corresponding eigenfunction $\boldsymbol{\varphi}^{(1)}(\boldsymbol{x})$ is the stationary solution of the Eq (5). At $k > 1$, one can speak about supercriticality, at $k < 1$ — about subcriticality.

The spectral problem (7) cannot directly be connected with the dynamic processes in a nuclear reactor. The eigenvalues of the multiplication factor of the reactor and the corresponding eigenfunctions do not depend on the time delay for the emission of delayed neutrons. At the best, we can get only the limiting case — the stationary critical state. The more acceptable spectral characteristics for the non-stationary equation (5) are related the spectral problem

$$\begin{gathered}
L\boldsymbol{\varphi} - (1-\beta)M\boldsymbol{\varphi} - I\boldsymbol{s} = \lambda^{(\alpha)} W\boldsymbol{\varphi},\\
\Lambda\boldsymbol{s} - R\boldsymbol{\varphi} = \lambda^{(\alpha)}\boldsymbol{s}.
\end{gathered} \tag{8}$$

where

$$I = \begin{pmatrix} E \\ -2E \end{pmatrix}, \quad R = \begin{pmatrix} Q & -2Q \end{pmatrix}, \quad W = \begin{pmatrix} V & -2V \\ -2V & 9V \end{pmatrix}$$

The fundamental eigenvalue $\alpha = \lambda_1^{(\alpha)}$ is called the α-eigenvalue or the period eigenvalue, because it is inversely related to the reactor period. If $\alpha = 0$, then the reactor is critical; if $\alpha < 0$ — supercritical state; if $\alpha > 0$ — subcritical state.

The problem of the period eigenvalues essentially takes into account the contribution of delayed neutrons. In particular, the long lifetime of the predecessors of delayed neutrons makes a large contribution to the slowly decreasing eigenfunctions of the reactor period, and this does not occur when only prompt neutrons are taken into account.

The asymptotic behaviour of Cauchy problem solution (5)-(6) at large times can be connected with the eigenvalue α. In this regular mode, the reactor behaviour is described by the function $\exp(-\alpha t)\boldsymbol{\varphi}^{(1)}(\boldsymbol{x})$.

4. IAEA-2D with reflector

We consider the test IAEA-2D with reflector. The two-group model ($G = 2$) and one group of delayed neutrons ($M = 1$) are used. Description and coefficients are presented here [7].

The method of finite elements [8] on triangular calculation grids is used for the approximate solution of the spectral problems. The standard Lagrangian finite elements are used. The software has been developed using the library FEniCS [9]. SLEPc has been used for numerical solution of the spectral problems. We used a Krylov-Schur algorithm with an accuracy of 10^{-15}. The following parameters were varied in the calculations: n — the number of triangles per one assembly; p — the order of finite element.

In this work we compare the SP_3 calculations with the previous diffusion model calculations [10].

4.1. Solution of Lambda Modes spectral problem

As a reference solution we take the solution obtained using the MCNP4C code [11]. The maximum difference in assembly power between two models is about 2 percent for the rodded assemblies (material 3 [7]).

Table 1. The effective multiplication factor.

n	p	k_{dif}	Δ_{dif}	δ_{dif}	t	k_{sp_3}	Δ_{sp_3}	δ_{sp_3}	t
	1	1.01041	418	14.34	0.01	1.01159	536	14.14	0.02
6	2	1.00623	0	1.95	0.04	1.00711	88	2.19	0.14
	3	1.00558	65	0.70	0.10	1.00636	13	0.35	0.47
	1	1.00699	76	4.82	0.03	1.00792	169	4.96	0.13
24	2	1.00561	62	0.77	0.21	1.00640	17	0.42	0.96
	3	1.00551	72	0.70	0.56	1.00626	3	0.17	2.97
	1	1.00591	32	1.47	0.19	1.00671	48	1.42	0.85
96	2	1.00552	71	0.73	1.14	1.00626	3	0.18	6.37
	3	1.00551	72	0.72	3.49	1.00625	2	0.18	20.34
Ref.		1.00623				1.00623			

Table 2. The eigenvalues $k_i = 1/\lambda_i^{(k)}$ for $p = 3, n = 96$.

i	Diffusion	SP_3
1	1.005509678491	1.006244515310
2	0.996489969484	0.997253707156
3	0.996489969416	0.997253707121
4	0.976790579591	0.977758816863
5	0.976790579352	0.977758816817
6	0.958683528959	0.959895076934
7	0.928979605001	0.930969293068
8	0.924186320247	0.925931354406
9	0.904788471277	0.907348873881
10	0.904788471274	0.907348873799

The comparison of the calculated effective multiplication factors is shown in Table 1. The following notation is used: k_{dif} — effective multiplication factor for the diffusion model; k_{sp_3} —

effective multiplication factor for the SP_3 model; Δ — absolute deviation from the reference value in pcm (10^{-5}); δ — the standard deviation of the relative power in percent, t — calculation time in sec. These data demonstrate the convergence of the computed eigenvalues with refinement of the calculation mesh and increase polynomial degree.

The results of the first 10 eigenvalues for $p = 3, n = 96$ are presented in Table 2. The power distributions and calculation errors for $p = 3, n = 96$ using the diffusion model are shown in Fig 1 and using the SP_3 model are shown in Fig 2.

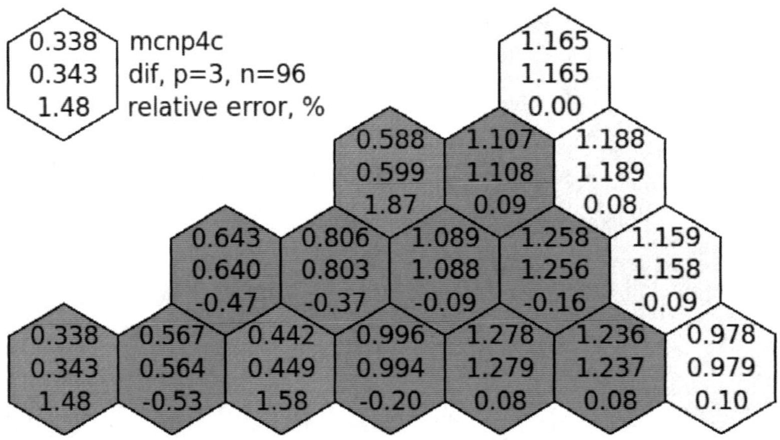

Figure 1. Power and error distributions using the diffusion model.

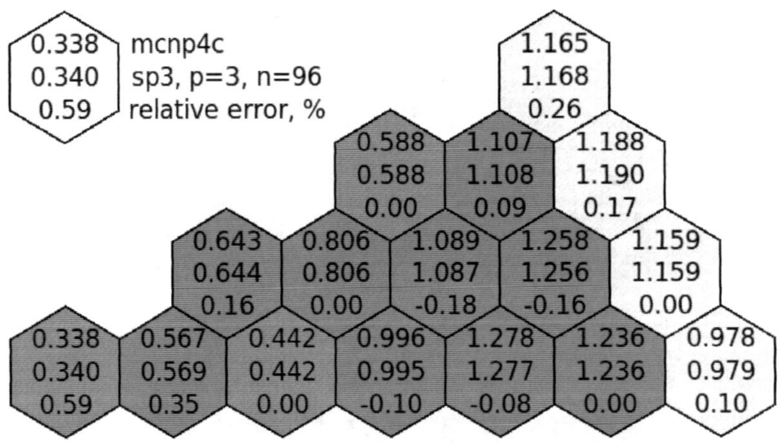

Figure 2. Power and error distributions using the SP_3 model.

4.2. Solution of α-spectral problem with delayed neutrons

As a reference solution, we use the fine mesh solutions obtained using the transport SP_3 model ($p = 3, n = 96$). The α-spectral problem results are shown in Table 3.

The calculation results for the first 10 eigenvalues are shown in Table 4. The eigenvalues are well separated. The fundamental eigenvalue is negative and therefore the main harmonic will increase, while all others will attenuate. A regular mode of the reactor is thereby defined. The value $\alpha = \lambda_1^{(\alpha)}$ determines the amplitude of neutron flux and is connected directly with reactor period in the regular mode.

Table 3. The α-eigenvalues.

n	p	α_{dif}	Δ_{dif}	t	α_{sp3}	Δ_{sp3}	t
	1	−68.2268	−66.8893	0.01	−88.9461	87.6086	0.02
6	2	−1.2810	0.0565	0.05	−11.1554	9.8179	0.19
	3	−0.4506	0.8869	0.18	−1.8063	0.4688	0.65
	1	−9.0267	−7.6892	0.05	−25.1658	23.8283	0.17
24	2	−0.4686	0.8689	0.37	−1.9832	0.6457	1.54
	3	−0.4202	0.9173	1.26	−1.3787	0.0412	5.10
	1	−0.7018	0.6357	0.33	−4.9794	3.6419	1.36
96	2	−0.4225	0.9769	2.66	−1.3994	0.0619	11.65
	3	−0.4184	0.9191	9.01	−1.3375	–	38.16
Ref.		−1.3375			−1.3375		

Table 4. The eigenvalues $\alpha_i = \lambda_i^{(\alpha)}$ for $p = 3, n = 96$.

i	Diffusion	SP_3
1	−0.418414021	−1.337480417
2	0.028108057	0.023799162
3	0.028108075	0.023804256
4	0.062814035	0.062218273
5	0.062814041	0.062220971
6	0.069514636	0.069228497
7	0.073730817	0.073541211
8	0.074126208	0.073987939
9	0.075346220	0.075210662
10	0.076266017	0.076175218

Acknowledgements
This work was supported by the grant of the Russian Federation Government (# 14.Y26.31.0013), the Russian Foundation for Basic Research (# 18-31-00315) and the Russian Science Foundation (# 19-71-00008).

References

[1] Brantley P S and Larsen E W 2000 *Nuclear Science and Engineering* **134** 1-21
[2] Bell G I and Glasstone S 1970 Nuclear Reactor Theory *Van Nostrand Reinhold Company*
[3] Stacey W M 2007 Nuclear Reactor Physics. *Wiley*
[4] Avvakumov A V et al. 2017 *Matematicheskoe Modelirovanie* **29(7)** 44-62
[5] Avvakumov A V, Strizhov V F, Vabishchevich P N and Vasilev A O 2017 *Annals of Nuclear Energy* **99** 68-79
[6] Ryu E H and Joo H G 2013 *Annals of Nuclear Energy* **56** 194-207
[7] Chao Y A and Shatilla Y A 1995 *Nuclear Science and Engineering* **121** 210-225
[8] Brenner S C and Scott L R 2008 The mathematical theory of finite element methods *Springer*
[9] Logg A, Mardal K A and Wells G 2012 The FEniCS book *Springer Science & Business Media*
[10] Avvakumov A V, Vabishchevich P N and Vasilev A O 2014 *Vestnik of the NEFU* **11** 7-18
[11] Bahabadi M H, Pazirandeh A and Mitra A 2016 *Annals of Nuclear Energy* **98** 74-80

Journal of Physics: Conference Series **1392** (2019) 012077 doi:10.1088/1742-6596/1392/1/012077

Analytical matrix solutions of linear ordinary differential equations with constant coefficients

Y I Vinogradov[1], **D B Molchanov**[1], **V N Bakulin**[2,3]

[1]Bauman State Technical University, 2nd Bauman str. 5, Moscow, 107005, Russia
[2]Moscow Aviation Institute (National Research University), Volokolamskoye highway 4, Moscow, 125993,Russia
[3]Institute of Applied Mechanics of the Russian Academy of Sciences, Leningradsky avenue 7, Moscow, 125040, Russia

E-mail: vbak@yandex.ru

Abstract. The article puts forward a modified finite element method based on decomposition and analytical solution techniques. The algorithm is as follows. A complex structure is divided into simple form sub-regions which involve partial differential equations. Next, the equations are decomposed. The decomposed equation solutions are written using analytical solution formulae. Meanwhile, the finite element size of the method proposed is defined only by the value of an averaging interval of required functions, since ordinary differential equation formulae are analytical. The algorithm has been tested by solving rectangular plate and bicurved shallow shell bending problems. The results proved proper convergence to precise values with increasing number of finite elements.

1. Introduction

The applied problem of continuum mechanics involves the analysis of thin structure stresses, e.g. aircraft.

Meanwhile, the mathematical problem involves the controllable error construction and analysis of mathematical models of shell and plate deformation mechanics, i.e. in an analytical manner [1-6].

The mathematical models of plate and shell mechanics are generally based on partial differential equations. Such equations reduce to ordinary differential expressions and are solved using the Fourier method (separation of variables) among others. After the separation of variables, the complete system of ordinary differential equations as to the required values describing she'll section state is written down as first order equation system

$$\frac{dy_1}{dx} = a_{11}y_1 + a_{12}y_2 + \ldots + a_{1n}y_n + f_1,$$
$$\frac{dy_2}{dx} = a_{21}y_1 + a_{22}y_2 + \ldots + a_{2n}y_n + f_2,$$
$$\frac{dy_n}{dx} = a_{n1}y_1 + a_{n2}y_2 + \ldots + a_{nm}y_n + f_n,$$

and represented as matrix

$$\boldsymbol{y}'(x) = A(x)\boldsymbol{y}(x) + \boldsymbol{f}(x), \tag{1}$$

where $\|A(x)_{ij}\| \quad (i = 1, 2, \ldots, 8; j = 1, 2, \ldots, 8)$ is a matrix of variable and constant coefficients of the equation system, $\|\boldsymbol{y}_{ii}\| \quad (i = 1, 2, \ldots, 8; j = 1)$ is a shell section properties column.

2. Matrix solution of boundary value problems

To solve the boundary value problem, the given boundary conditions are written down in a matrix form

$$\begin{aligned} U\boldsymbol{y}\left(x_0\right) &= \boldsymbol{u} \\ V\boldsymbol{y}\left(x_n\right) &= \boldsymbol{v} \end{aligned} \tag{2}$$

where

$$\|U_{ij}\| \quad (i = 1, 2, \ldots, 4; j = 1, 2, \ldots, 8), \|V_{ij}\| \quad (i = 1, 2, \ldots, 4; j = 1, 2, \ldots, 8),$$

$$\|\boldsymbol{u}_{ij}\| \quad (i = 1, 2, \ldots, 4; j = 1), \quad \|\boldsymbol{v}_{ij}\| \quad (i = 1, 2, \ldots, 4; j = 1).$$

The solution of the ordinary differential equation (1) includes the general solution of the corresponding homogeneous differential equation and particular solution of non-homogeneous equation. The solution of the corresponding homogeneous equation

$$\boldsymbol{y}'(x) = A(x)\boldsymbol{y}(x), \tag{3}$$

in $\left(x_{(1}, x_n\right)$ is obtained through Picard method and can be generally written as follows

$$\boldsymbol{y}(x) = K_{x_{\mathrm{p}}}^{x_a}(A(x))\boldsymbol{y}_0,$$

where $K_{x_0}^{x_n}(A(x))$ is a Cauchy-Krylov function matrix. Cauchy-Krylov function matrix has a meaning of a matrix exponential, thus, it has the multiplicative property. If the matrix exponential solution interval is divided into sub-intervals, then the product of sub-interval matrix exponentials is equal to a complete interval matrix exponential

$$K_{x_0}^{x_n}(A(x)) = K_{x_{n-1}}^{x_n} K_{x_{n-2}}^{x_{n-1}} \ldots K_{x_0}^{x_1}.$$

If the differential equation coefficients are constant (e.g. cylindrical shell), Cauchy-Krylov function matrix is found as follows

$$K_{x_0}^{x_n}(A) = \sum_{m=0}^{m=\infty} \frac{(A\Delta x)^m}{m!}, \Delta x = x_n - x_0. \tag{4}$$

In case of variable coefficients (conical, spherical shells) the integration segment is divided into n sub-intervals, where the differential equation coefficients (1) can be averaged. Then, each sub-interval can involve formula (4) followed by the multiplicative property of Cauchy-Krylov function matrix

$$K_{x_0}^{x_n}(A(x)) = \prod_{i=n}^{i=1} \sum_{m=0}^{m=\infty} \frac{\left[A\left(\tau_i\right) \Delta x_i\right]^m}{m!},$$

where $A\left(\tau_i\right)$ are averaged coefficient matrixes of the equation (3).

The partial solution of the differential equation is defined by an integral of so-called Cauchy matrix. In an arbitrary interval (x_{i-1}, x_i) it is defined as follows

$$\boldsymbol{y}_{x_{i-1}}^{*x_i} = K_{x_{i-1}}^{x_i}\left(A_i\right) \int_{x_{i-1}}^{x_i} \left[K_{x_{i-1}}^{\tau_i}\left(A_i\right)\right]^{-1} \boldsymbol{f}_i d\tau_i. \tag{5}$$

Or:

$$\boldsymbol{y}_{x_{i-1}}^{*x_i} = e^{A_i \Delta x_i} \int_{x_{i-1}}^{x_i} e^{-A_i \Delta \tau_i} \boldsymbol{f}_i d\tau_i, \ \left[e^{A_i \Delta \tau_i}\right]^{-1} = e^{-A_i \Delta \tau_i}$$

$$\Delta x_i = x_i - x_{i-1}, \quad \Delta \tau_i = \tau_i - x_{i-1}, \tau_i \in [x_{i-1}, x_i]$$

The elements of column $\boldsymbol{f_i}$ are averaged and the column is taken out of the integral. The matrix exponential under the integral is replaced with a matrix series. Then integration is performed. Then

$$\boldsymbol{y}_{x_{i-1}}^{*x_i} = e^{A_i \Delta x_i} \overline{T}_i \boldsymbol{f}_i \Delta x_i, \quad \overline{T}_i = E - \frac{A_i \Delta x_i}{2!} + \frac{(A_i \Delta x_i)^2}{3!} - \frac{(A_i \Delta x_i)^3}{4!} + \ldots \tag{6}$$

By multiplying series of the matrix exponential and matrix $\overline{T}_i$, we shall obtain a formula of a particular solution in the arbitrary interval (x_{i-1}, x_i)

$$\boldsymbol{y}_{x_{i-1}}^{*x_i} = T_i f_i \Delta x_i, T_i = E + \frac{A_i \Delta x_i}{2!} + \frac{(A_i \Delta x_i)^2}{3!} + \frac{(A_i \Delta x_i)^3}{4!} + \ldots \tag{7}$$

Formula (7) defines the particular solution of the equation (1).

The simple algorithm of stable calculation dedicated to the boundary value problem (1)(2), lies in the fact that the main interval with intermediate points $x_1, x_2, \ldots, x_i, \ldots, x_{n-1}$ is divided into stable calculation intervals. The differential equation solution is defined in each interval with a controllable error under one of the formulae. In fact, the solution establishes a link between the interval ends, i.e. between the parameters of she'll section states

$$\boldsymbol{y}(x_i) = K_{x_{i-1}}^{x_i}(A(x)) \boldsymbol{y}(x_{i-1}) + \boldsymbol{y}_{x_{i-1}}^{*x_i}. \tag{8}$$

Let us write down the equations (8) for each interval as follows

$$\begin{array}{l} -K_{x_0}^{x_1}(A(x)) y(x_0) + E y(x_1) = \boldsymbol{y}_{x_0}^{*x_1}, \\ -K_{x_1}^{x_2}(A(x)) y(x_1) + E y(x_2) = \boldsymbol{y}_{x_1}^{*x_2}, \\ \ldots\ldots\ldots\ldots\ldots \\ -K_{x_{n-1}}^{x_n}(A(x)) y(x_{n-1}) + E y(x_n) = \boldsymbol{y}_{x_{n-1}}^{*x_n}. \end{array}$$

The obtained equations are one less than the section number including extremes. Then we supplement the system with boundary conditions (2). The algebraic equation system derived in such a manner is represented as matrix

$$\left\| \begin{array}{ccccccc} U & 0 & 0 & 0 & \ldots & 0 & 0 \\ -K_{x_0}^{x_1} & E & 0 & 0 & \ldots & 0 & 0 \\ 0 & -K_{x_1}^{x_2} & E & 0 & \ldots & 0 & 0 \\ \ldots & \ldots & \ldots & \ldots & \ldots & \ldots & \ldots \\ 0 & 0 & 0 & 0 & \ldots & -K_{x_{n-1}}^{x_n} & E \\ 0 & 0 & 0 & 0 & \ldots & 0 & V \end{array} \right\| \times \left\| \begin{array}{c} \boldsymbol{y}(x_0) \\ \boldsymbol{y}(x_1) \\ \boldsymbol{y}(x_2) \\ \ldots \\ \boldsymbol{y}(x_{n-1}) \\ \boldsymbol{y}(x_n) \end{array} \right\| = \left\| \begin{array}{c} \boldsymbol{u} \\ \boldsymbol{y}_{x_0}^{*x_1} \\ \boldsymbol{y}_{x_1}^{*x_2} \\ \ldots \\ \boldsymbol{y}_{x_{n-1}}^{*x_n} \\ \boldsymbol{v} \end{array} \right\| \tag{9}$$

The boundary value problem solution ends with a solution of the obtained matrix algebraic equation system.

3. Decomposition method

However, the class of problems restricted by the potential of the Fourier method is rather narrow. Besides, the method of solving ordinary differential equations mentioned above can be applied to a wide variety of ordinary differential equations. Thus, the creation of finite element modified method with analytical solution of element-based equations requires a more general method of reducing partial differential equations to ordinary ones. The modification of such method with a high degree of generality is a decomposition method. The application of the decomposition method to solving some particular problems of structural mechanics is described in works by Pshenichnov G.I. Before this the method was described in works by Rozin L.A. named as 'Split Method'.

$$L(z) = f(\alpha, \beta),$$

let us represent the differential operator as h summands

$$L = \sum_{k=1}^{h} L_k,$$

and represent the right part as a sum h of unknown functions

$$f(\alpha, \beta) = \sum_{k=1}^{k_k} f^{(k)}(\alpha, \beta).$$

By associating the summand of the right part with each summand of the differential operator the original equation can be replaced with a system of simpler equations

$$L_k\left(z^{(k)}\right) = f^{(k)}(\alpha, \beta),$$

and equations of constraints

$$z^{(1)}(\alpha, \beta) = z^{(k)}(\alpha, \beta). \tag{10}$$

The decomposition method can be more clearly demonstrated using the plate bending equation

$$\frac{\partial^4 w}{\partial x^4} + 2\frac{\partial^4 w}{\partial x^2 \partial y^2} + \frac{\partial^4 w}{\partial y^4} = \frac{q_z}{D}. \tag{11}$$

By representing the right part as three summands (unknown functions)

$$\frac{q_z}{D} = f^{(1)}(x, y) + f^{(2)}(x, y) + f^{(3)}(x, y), \tag{12}$$

let us associate the summand of the right part with each summand of the differential operator

$$\begin{cases} \frac{\partial^4 w}{\partial x^4} = f^{(1)}(x, y) \\ \frac{\partial^4 w}{\partial y^4} = f^{(2)}(x, y) \\ 2\frac{\partial^4 w}{\partial x^2 \partial y^2} = f^{(3)}(x, y) \end{cases}. \tag{13}$$

It is easy to show that using the system of equations, by exclusion, the original equation can be easily reached. Next, the system equations can be solved through analytical solution formulae mentioned above. Let us assume that the first equation solution resulted in $w^{(1)}\left(f^{(1)}(x, v)\right)$, the second resulted in $w^{(2)}\left(f^{(2)}(x, y)\right)$ and the third resulted in $w^{(3)}\left(f^{(3)}(x, y)\right)$. Now we can write down the equations of constraint

$$\begin{cases} w^{(1)} = w^{(2)} \\ w^{(2)} = w^{(3)} \end{cases}. \tag{14}$$

Equations, appear as a consistent system to define unknown functions

$$f^{(1)}(x,y), f^{(2)}(x,y), f^{(3)}(x,y).$$

After defining unknown functions and inserting them in $w^{(1)}, w^{(2)}, w^{(3)}$, the solution of the original equation is determined completely.

However, the constraint equations solution involves a fundamental difficulty. The unknown functions are in the right parts of the equations of the system . Thus, they appear under the integral in the partial solution formula . Example for $w^{(1)}$:

$$w^{(1)} = K_{x_0}^{x}(A_x)\,w_{x_0}^{(1)} + \int_{x_0}^{x} K(x,\tau)f^{(1)}(\tau)d\tau.$$

Thus, the constraint equations are not algebraic but integral equations. The simplest way of overcoming the difficulty is dividing the integration interval (x_0, x) into smaller intervals, where functions $f^{(1)}(x,y), f^{(2)}(x,y), f^{(3)}(x,y)$, can be averaged as it is done for external load in formula (6). In other words, a piecewise constant approximation of unknown functions should be used. Then the averaged functions can be taken out of the integral, so the integral equations turn to algebraic. A priori errors cannot be controlled in this case.

Improvement of constraint equation solution and development of a priori error estimate method involve the opportunity of improving and developing of the method at issue.

The mentioned method is naturally extended from partial differential equations into partial differential equation systems. Thus, decomposition can be applied to the shell resolving equation systems. The transition to systems does not lead to any crucially new difficulties, it only results in more intricate calculations.

4. Modified method of finite elements

Decomposition and analytical solution methods make the basis for the modified method of finite elements. The method algorithm is as follows: Complex structure is divided into simple form sub-regions with partial differential equations available; Equations are decomposed; After the decomposition the equation solutions are written using analytical solution formulae; Elements are converged; Global system of linear algebraic equations is produced (9). The size of the finite element of the proposed method is defined only by the value of unknown function averaging interval $f^{(1)}(x,y), f^{\cdot(2)}(x,y), f^{(3)}(x,y)$, since ordinary differential equation formulae are analytical. Thus, the improvement of solving constraint equations results in reducing number of finite elements. The algorithm has been tested by solving rectangular plate and bicurved shallow shell bending problems. The results proved proper convergence to precise values with increasing number of finite elements.

References

[1] Vinogradov Y I, Matveenko A M 2015 Opportunities and Prospects of Generating Finite Element Analysis*Proc. of 19th Int.Conf. on Comp. Mech. and Appl. Syst.* (Moscow: MAI) 47-49

[2] Vinogradov Y I, Bakulin V N 2016 Experimental and Analytical Investigation of the Stressed- Strained State of a Cylindrical Shell Under the Action of Concentrated Radial Forces *Mat. Ph. and Mech.* **26** 1 49-52

[3] Bakulin V N, Vinogradov Y I 2017 Analytical and Asymptotic Solution of Boundary Value Problems in the Mechanics of Deformed Shells Under Concentrated Loading *Rus. Aer.* **60** 1 13-20

[4] Bakulin V N, Vinogradov Y I, Men'kov G B 2018 The Stressed State of a Stiffened Conical Shell with Thermal Protective Coating with Temperature-Dependent Properties *Rus. Aer.* **61** 2 156-164

[5] Bakulin V N, Repinskii V V 2018 Comparison of Finite Element Solution with Analytical Solution in Spherical Shell Deformation Mechanics Problems *Proc. of the XII Int. Conf. on Appl.Math. and Mech. in the Aer. Ind.* (Moscow: MAI) 338-340

[6] Vinogradov Y I, Molchanov D B 2018 Basics of the Modified Finite Element Method for Solving Boundary Problems of Structural Mechanics of Thin-Walled Structures *Prikl. Mat. i Mech.* **82** 4 498-511

Journal of Physics: Conference Series **1392** (2019) 012078 doi:10.1088/1742-6596/1392/1/012078

Comparison of discrete fiber and asymptotic homogenization methods for modeling deformations of fiber-reinforced materials

Petr E. Zakharov, Petr V. Sivtsev

Ammosov North-Eastern Federal University, 58, Belinskogo, 677000 Yakutsk, Russia

E-mail: zapetch@gmail.com

Abstract.
The main issue in a numerical simulation of fiber-reinforced materials deformation is high computational complexity. Exact resolution of fibers thickness leads to a huge number of degrees of freedom. In this work, we compare discrete fiber and asymptotic homogenization methods for modeling deformations. The discrete fiber method considers fibers as one-dimensional lines and the asymptotic homogenization method averages elasticity coefficients. We investigate the accuracy of these methods for a large range of models and approximation parameters.

1. Introduction

Fiber-reinforced materials are recommended as most of the strongest composite materials, e.g. concrete with basalt-polypropylene fiber inclusions [1]. Another example of fiber-reinforced material is load-bearing concrete structure reinforced with basalt plastic or steel [2, 3]. One of the popular trends in development technologies is related to the production of basalt fiber and composite materials based on it.

In numerical modeling of the stress-strain state of fiber-reinforced materials, the main issue is related to a huge quantity of the mesh cells to take into account. There are numerical methods, which reduce the initial problem to a more simplified problem, e.g. homogenization or multiscale methods [4, 5]. The discrete fiber method, which is similar to the discrete fracture method [6, 7] in porous media simulation, can be used for fiber-reinforced material simulation as well.

In this work, we compare two approaches of fiber-reinforced material modeling: the discrete fiber and the asymptotic homogenization methods. In the discrete fiber method (DFM) we assume that fibers are one-dimensional and we take account only stress-strain state along the fiber length. In this simplification, we can reduce the number of mesh cells, i.e. degrees of freedom. The asymptotic homogenization method (AHM) [9–12] is convenient for the domain containing a large amount of repeated subdomains. With AHM, we compute averaged elastic tensor and solve the problem on a coarse mesh. These two approaches have advantages and, in some cases, one of them can be more convenient than the other. We make a numerical investigation of the cases, in which the methods show better accuracy. Numerical implementations are performed using finite element package FEniCS [8].

2. Problem Statement

We consider a problem in the unit square $\Omega = (0,1)^2$, where Ω_1 is the main material and $\Omega_2 = \cup_{i=1}^{K} \phi_i$ is the subdomain of K fiber inclusions. The stress-strain state of fiber-reinforced material is described by the following equation for the displacement $\boldsymbol{u} = (u_1, u_2)$

$$\nabla \cdot \boldsymbol{\sigma} = \boldsymbol{f}, \quad \boldsymbol{x} \in \Omega, \tag{1}$$

where $\boldsymbol{\sigma} = \boldsymbol{C}\varepsilon$ is the stress tensor, $\boldsymbol{C}$ is the elastic tensor and ε is the strain tensor, $\boldsymbol{f}$ is the force source. To describe the numerical methods we use Voight notation

$$\boldsymbol{\sigma} = \begin{pmatrix} \sigma_{11} \\ \sigma_{22} \\ \sigma_{12} \end{pmatrix}, \quad \boldsymbol{C} = \begin{pmatrix} C_{1111} & C_{1122} & C_{1112} \\ C_{2211} & C_{2222} & C_{2212} \\ C_{1211} & C_{1222} & C_{1212} \end{pmatrix}, \quad \varepsilon = \begin{pmatrix} \varepsilon_{11} \\ \varepsilon_{22} \\ \varepsilon_{12} \end{pmatrix} = \begin{pmatrix} \frac{\partial u_1}{\partial x_1} \\ \frac{\partial u_2}{\partial x_2} \\ \frac{\partial u_2}{\partial x_1} + \frac{\partial u_1}{\partial x_2} \end{pmatrix}.$$

We assume that the main material and fibers are isotropic. Then the elastic tensor can be expressed by Lame parameters λ_i, μ_i depending on the subdomain

$$\boldsymbol{C}_i = \begin{pmatrix} \lambda_i + 2\mu_i & \lambda_i & 0 \\ \lambda_i & \lambda_i + 2\mu_i & 0 \\ 0 & 0 & \mu_i \end{pmatrix}, \quad \boldsymbol{x} \in \Omega_i, \quad i = 1, 2.$$

Lame parameters are expressed by Young modulus E_i and Poisson coefficient ν_i

$$\lambda_i = \frac{E_i \, \nu_i}{(1+\nu_i)(1-2\nu_i)}, \quad \mu_i = \frac{E_i}{2(1+\nu_i)}, \quad \boldsymbol{x} \in \Omega_i, \quad i = 1, 2.$$

Equation (1) is supplemented with boundary conditions. On the left border Γ_L, we fix the displacement using Dirichlet condition

$$\boldsymbol{u} = (0, 0), \quad \boldsymbol{x} \in \Gamma_L.$$

And on the right border Γ_R, we apply the force using Neumann condition

$$\boldsymbol{\sigma_n} = \boldsymbol{g}, \quad \boldsymbol{x} \in \Gamma_R.$$

where $\boldsymbol{\sigma_n} = \boldsymbol{\sigma}\, \boldsymbol{n}$ and $\boldsymbol{n}$ is the normal vector to the border.

3. Finite Element Mehtod

As a reference solution, we use the solution of finite element approximation where fibers are taken into account in full size. We get the following variational statement for our problem: find the function $\boldsymbol{u} \in V$, which satisfies $a(\boldsymbol{u}, \boldsymbol{v}) = L(\boldsymbol{v}), \forall \boldsymbol{v} \in \widehat{V}$, where $a(\boldsymbol{u}, \boldsymbol{v})$ is the bilinear form, $L(\boldsymbol{v})$ is the linear form

$$a(\boldsymbol{u}, \boldsymbol{v}) = \int_{\Omega_1} \boldsymbol{C}_1 \, \varepsilon(\boldsymbol{u}) : \varepsilon(\boldsymbol{v}) \, \mathrm{d}\boldsymbol{x} + \int_{\Omega_2} \boldsymbol{C}_2 \, \varepsilon(\boldsymbol{u}) : \varepsilon(\boldsymbol{v}) \, \mathrm{d}\boldsymbol{x}, \tag{2}$$

$$L(\boldsymbol{v}) = \int_{\Omega} \boldsymbol{f} \, \boldsymbol{v} \, \mathrm{d}\boldsymbol{x} + \int_{\Gamma_R} \boldsymbol{g} \, \boldsymbol{v} \, \mathrm{d}\boldsymbol{s}, \tag{3}$$

and V, $\widehat{V}$ are the trial and test function spaces

$$V = \widehat{V} = \{\boldsymbol{v} \in H^1(\Omega) : \boldsymbol{v} = (0,0), \boldsymbol{x} \in \Gamma_L\},$$

$H^1(\Omega)$ is the Sobolev function space. We use first order Lagrange elements as finite elements.

Journal of Physics: Conference Series **1392** (2019) 012078 doi:10.1088/1742-6596/1392/1/012078

4. Discrete Fiber Method

In the discrete fiber method similar to the discrete fracture method, we consider fibers as one-dimensional. We represent subdomain Ω_2 by set of lines and denote by $\Gamma_2 = \cup_{i=1}^K \gamma_i$, where γ_i is the fiber line. In this case, we get the following bilinear and linear forms for our problem

$$a(\boldsymbol{u}, \boldsymbol{v}) = \int_{\Omega} \boldsymbol{C}_1 \, \varepsilon(\boldsymbol{u}) : \varepsilon(\boldsymbol{v}) \, \mathrm{d}\boldsymbol{x} + \int_{\Gamma_2} d \, (\lambda_2 + 2\mu_2 - \lambda_1 - 2\mu_1)(\nabla \boldsymbol{u}_{\boldsymbol{\tau}} \, \boldsymbol{\tau})(\nabla \boldsymbol{v}_{\boldsymbol{\tau}} \, \boldsymbol{\tau}) \mathrm{d}\boldsymbol{s},$$

$$L(\boldsymbol{v}) = \int_{\Omega} \boldsymbol{f} \, \boldsymbol{v} \, \mathrm{d}\boldsymbol{x} + \int_{\Gamma_R} \boldsymbol{g} \, \boldsymbol{v} \, \mathrm{d}\boldsymbol{x},$$

where d is the thickness of fibers, $\boldsymbol{u}_{\boldsymbol{\tau}} = \boldsymbol{u} \, \boldsymbol{\tau}$ and $\boldsymbol{\tau}$ is the tangent vector to a fiber line. In this approximation, we take into account only compression and tension of fibers along to fiber line and neglect other effects. Function spaces and finite elements are the same as in the finite element approximation.

5. Asymptotic Homogenization Method

We assume that our composite material can be divided into equal squares with fibers on the same locations. Then we can compute an effective elastic tensor using well known asymptotic homogenization method. Let denote an average value of function ψ by following notation

$$\langle \psi \rangle = \frac{\int_{\omega} \psi \, \mathrm{d}\boldsymbol{x}}{\int_{\omega} \mathrm{d}\boldsymbol{x}},$$

where ω is a periodic domain. Thus, an average stress tensor can be expressed by

$$\langle \boldsymbol{\sigma} \rangle = \langle \boldsymbol{C} \, \varepsilon \rangle = \boldsymbol{C}^* \langle \varepsilon \rangle,$$

where $\boldsymbol{C}^*$ is the effective elastic tensor. To compute corresponding components of the effecitve elastic tensor, we consider three problems in the periodic domain ω with the following force sources:

(i) $C^*_{ij11} = \langle \sigma_{ij} \rangle, \quad ij = 11, 22, 12$, from $\boldsymbol{f} = -\nabla \cdot \boldsymbol{C} \, \varepsilon((x_1, 0))$,
(ii) $C^*_{ij22} = \langle \sigma_{ij} \rangle, \quad ij = 11, 22, 12$, from $\boldsymbol{f} = -\nabla \cdot \boldsymbol{C} \, \varepsilon((0, x_2))$,
(iii) $C^*_{ij12} = \langle \sigma_{ij} \rangle, \quad ij = 11, 22, 12$, from $\boldsymbol{f} = -\nabla \cdot \boldsymbol{C} \, \varepsilon((x_2/2, x_1/2))$.

To ensure the uniqueness of the solution, we fix displacement to $(0, 0)$ in the middle point of the domain ω. To solve the periodic problems, we use the same variational forms (2), (3) as in the finite element approximation. We change only function spaces

$$V = \widehat{V} = \{\boldsymbol{v} \in H^1(\omega) : \boldsymbol{v}(\boldsymbol{x}) = \boldsymbol{v}(\boldsymbol{x} + \boldsymbol{p}), \boldsymbol{p} = (0, l), (l, 0)\},$$

where l is the side of square domain ω.

After computing the average elastic tensor, we find the coarse solution using the next variational statement: find the function $\boldsymbol{u} \in V$, which satisfies $a(\boldsymbol{u}, \boldsymbol{v}) = L(\boldsymbol{v}), \forall \boldsymbol{v} \in \widehat{V}$, where

$$a(\boldsymbol{u}, \boldsymbol{v}) = \int_{\omega} \boldsymbol{C}^* \, \varepsilon(\boldsymbol{u}) : \varepsilon(\boldsymbol{v}) \, \mathrm{d}\boldsymbol{x}, \quad L(\boldsymbol{v}) = \int_{\Omega} \boldsymbol{f} \, \boldsymbol{v} \mathrm{d}\boldsymbol{x} + \int_{\Gamma_R} \boldsymbol{g} \, \boldsymbol{v} \, \mathrm{d}\boldsymbol{s}.$$

Function spaces and finite elements are the same as in the finite element approximation. We compare only a coarse solution and do not compute a higher order solution of the asymptotic homogenization method. In this case, for a coarse solution, we can use the same number unknowns as in periodic problems, i.e. a finer mesh. Also, it gives less difference with a reference solution.

6. Numerical Experiments

For numerical comparison, we consider simple square domain Ω with $n \times n$ equal subdomains ω. Each subdomain ω contains uniformly distributed $k = K/n^2$ fibers with size $l \times d$, where $l = \frac{1}{2n}$ is the lenght and d is the thickness. In fig. 1 we show domain with $n \times n = 4 \times 4$ subdomains ω with $k = 2$ fibers. We consider two variants of subdomains: ω_{ahm} with full size fibers and ω_{dfm} with one-dimensional fibers. To discretize domain Ω, we use a uniform triangular mesh with grid step h. The grid size of subdomain ω is $n \times n$ less than the grid size of domain Ω.

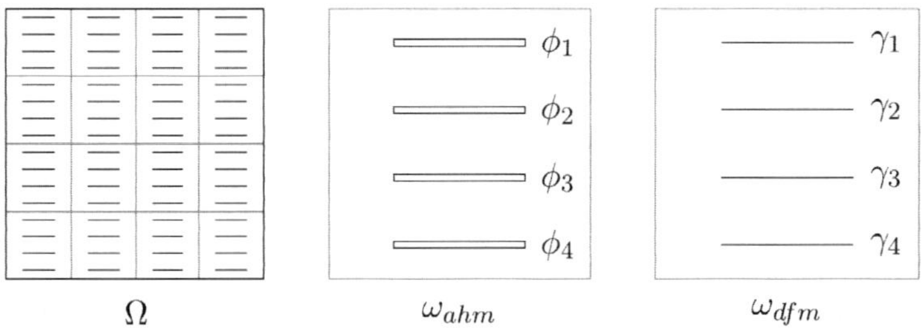

Figure 1. Problem domain Ω, subdomain ω_{ahm} for AHM and ω_{dfm} for DFM

In our numerical investigation we use the following model parameters: $E_1 = 2 \cdot 10^{10}, E_2 = \alpha E_1, \nu_1 = \nu_2 = 0.3, \boldsymbol{g} = (0, -10^5)$, where α is the ratio of Young modulus.

Let denote two norms of relative errors by

$$\epsilon^i_{L_\infty} = \frac{\|\boldsymbol{u}^i - \boldsymbol{u}^{fem}\|_{L_\infty}}{\|\boldsymbol{u}^{fem}\|_{L_\infty}}, \quad \epsilon^i_{L_2} = \frac{\|\boldsymbol{u}^i - \boldsymbol{u}^{fem}\|_{L_2}}{\|\boldsymbol{u}^{fem}\|_{L_2}}, \quad i = dfm, ahm,$$

where $\boldsymbol{u}^{fem}$ is the FEM reference solution, $\boldsymbol{u}^{dfm}$ is the DFM solution and $\boldsymbol{u}^{ahm}$ is the AHM solution. We compare DFM and AHM implementations in 5 tests. In each test we vary one of this parameters: k, d, α, n, h, and all other parameters are fixed.

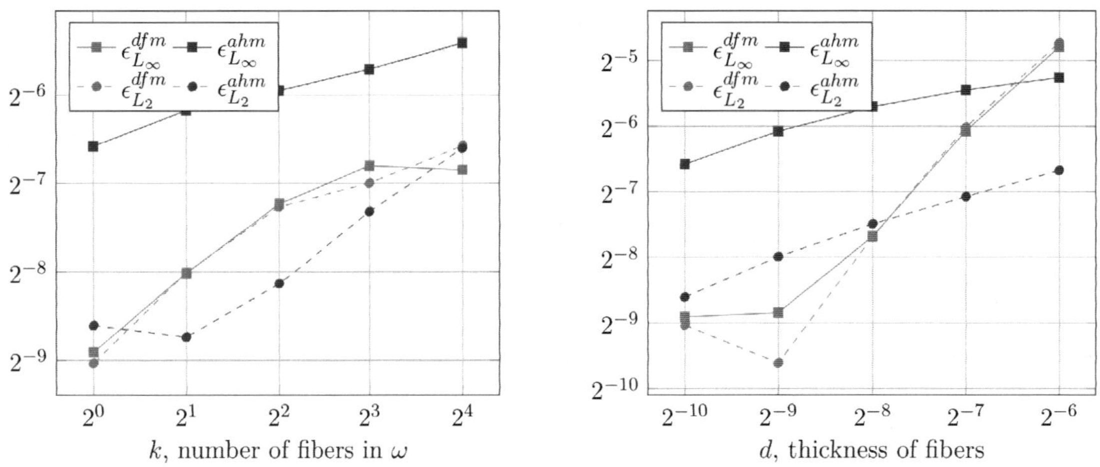

Figure 2. Errors depending on number and thickness of fibers

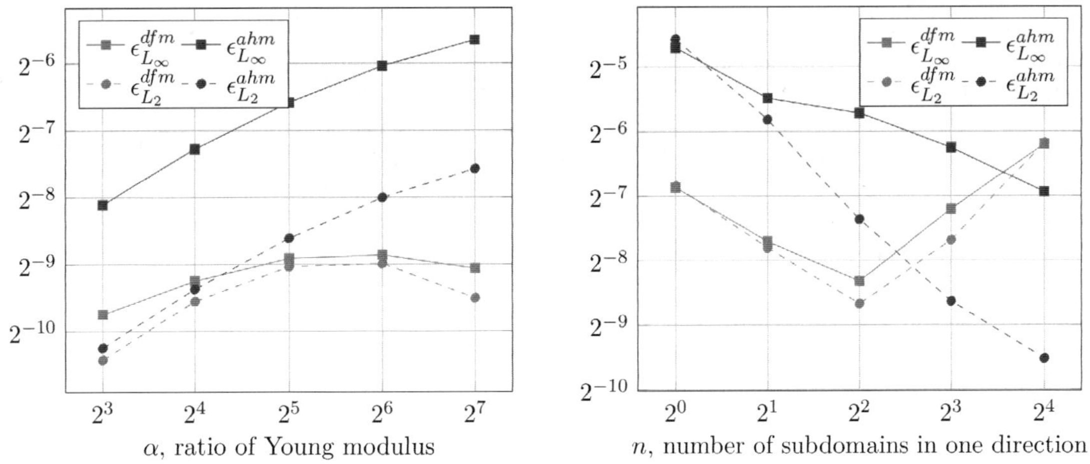

Figure 3. Errors depending on ration of Young modulus and number of subdomains

In fig. 2 we show a comparison of relative errors for different number k and thickness d of fibers. For these calculations, we fix the next parameters: $n = 4, d = 1/1024, \alpha = 32, h = 1/2048$. As we can see, both relative errors of DFM are almost the same. And the AHM L_∞ norm of error is much higher than L_2 due to the localization of the error near fibers. Both methods show that the relative errors increase with a number of fibers. For all cases k, the error L_2 norms are similar and differ less than 2 times. And DFM error L_∞ norm is better more than 2 times comparing to AHM. DFM solution is more sensitive to the thickness of fibers d and the relative error grows faster than AHM error. But for fibers with a small thickness, i.e. with a high ratio of length and thickness, DFM gives a more precise solution than AHM.

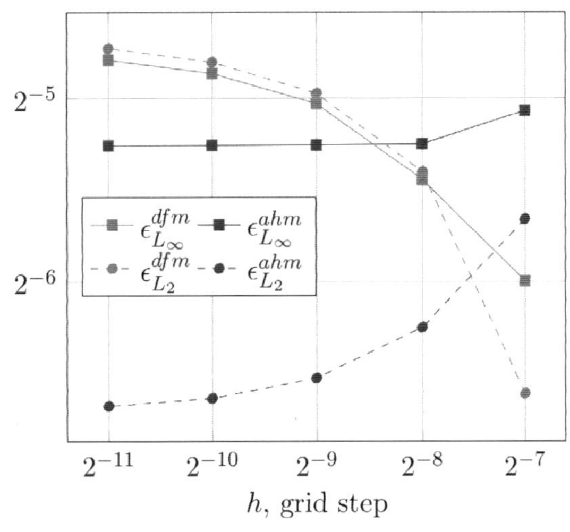

Figure 4. Error depending on grid step

In the next two tests, we investigate dependence on α ration of Young modulus and n number of subdomains. Calculated relative erros are plotted in fig. 3. For simulations with α variation, we use the following parameters: $n = 4, d = 1/1024, k = 1, h = 1/2048$ and for n variation: $d = 1/1024, k = 1, h = 1/(128\,n)$. DFM solution shows better accuracy for high α ratio of

Young modulus and the error is almost constant for different α. AHM error increses with α. However, AHM error decreases on increasing n number of subdomains. For large n AHM accuracy is better than DFM and DFM error increases linearly with number of subdomains.

The last test is conducted to compare errors depending on h grid step, see fig. 4. To neglect FEM error, we solve the reference solution on the finest grid with step $h = 1/2048$. Due to the variation of grid steps, we choose the large thickness of fibers $d = 1/64$, i.e length and thickness ratio $l/d = 8$, and DFM error is larger than AHM error. Other parameters: $n = 4, k = 1, \alpha = 32$. For the fine grids, the errors are close to methods of approximation errors and are almost constant. AHM error increases for the coarse grids and this corresponds to the theory. DFM error decreases for the coarse grids due to opposite signs of FEM (in DFM) and DFM errors, which neglect each other.

7. Conclusion

We have numerically investigated two approaches of deformation modeling of fiber-reinforced materials. DFM showed better accuracy for a large ration of Young modulus between main material and fibers comparing to AHM. Moreover, DFM is more convenient for thick fibers, which have a large ration of length and thickness. AHM solution is better for domains containing a large number of equal subdomains. And one of the most important advantages of DFM is the use of more coarse meshes, where fibers are represented as lines, which reduces degrees of freedom.

References

[1] Smarzewski, P.: Influence of basalt-polypropylene fibres on fracture properties of high performance concrete. Composite Structures 209, 23-33 (2019)

[2] Sivtseva, A.V., Sivtsev, P.V.: Numerical Simulation of Deformations of Basalt Roving. International Conference on Finite Difference Methods, 501–508 (2018).

[3] Kolesov A. E. et al.: Numerical Analysis of Reinforced Concrete Deep Beams. International Conference on Numerical Analysis and Its Applications. – Springer, Cham, 414-421 (2016).

[4] Zakharov, P.E., Sivtsev, P.V.: Numerical calculation of the effective coefficient in the problem of linear elasticity of a composite material. Mathematical notes of NEFU 24.2, 75-84 (2017).

[5] Stepanov, S.P., Vasilyeva, M.V., Vasil'ev, V.I.: Generalized multiscale discontinuous Galerkin method for solving the heat problem with phase change. Journal of Computational and Applied Mathematics 340, 645-652 (2018).

[6] Vasil'ev V. I. et al.: Numerical solution of a fluid filtration problem in a fractured medium by using the domain decomposition method. Journal of Applied and Industrial Mathematics 12(4), 785-796 (2018).

[7] Akkutlu I. Y. et al.: Multiscale model reduction for shale gas transport in a coupled discrete fracture and dual-continuum porous media. Journal of Natural Gas Science and Engineering 48, 65-76 (2017).

[8] Logg, A., Mardal, K.A., Wells, G.N.: Automated solution of differential equations by the finite element method. Springer (2012).

[9] Pinho-da-Cruz J., Oliveira J. A., Teixeira-Dias F. Asymptotic homogenisation in linear elasticity. Part I: Mathematical formulation and finite element modelling. Computational Materials Science 45(4), 1073-1080 (2009)

[10] Oliveira J. A., Pinho-da-Cruz J., Teixeira-Dias F. Asymptotic homogenisation in linear elasticity. Part II: Finite element procedures and multiscale applications. Computational Materials Science 45(4), 1081-1096 (2009)

[11] Pellegrino C., Galvanetto U., Schrefler B. A. Numerical homogenization of periodic composite materials with non-linear material components. International Journal for Numerical Methods in Engineering 46(10), 1609-1637 (1999)

[12] Yuan Z., Fish J. Toward realization of computational homogenization in practice. International Journal for Numerical Methods in Engineering 73(3), 361-380 (2008)

Development of algorithms for the formation of steady-state modes based on the topology of electric power systems

D S Akhmetbaev[1], **A R Dzhandigulov**[2]

[1] (Mathematics Faculty), Seifullin Kazakh Agro Technical University, 62 Zhenis avenue, 010011 Nur-Sultan, Kazakhstan
[2] Department of Algebra and Geometry, L.N. Gumilyov Eurasian National University, 13 Kazhymukan St,010008 Nur-Sultan, Kazakhstan

E-mail: `axmetbaev46@mail.ru`

Abstract. Implementation of the new algorithm for the formation of the steady state mode based on the distribution coefficients of the driving currents in complex electrical networks. is proposed.

The advantage of the proposed approach, in contrast to previously known ones, is that it is not necessary to calculate all 2-trees, as is done in previously known topological methods, which reduces the number of operations by n^2 times.

The results of calculations by the proposed method on various test examples 99% coincided with the results of calculations obtained using the industrial software RastrWin3.

1. Introduction.

In this paper, an algorithm for the formation of the steady state of complex electrical networks using the distribution coefficients of driving currents is implemented.

In the paper [1], a comparative review of known topological methods for studying electric power networks is presented and a new algorithm for calculating current distribution coefficients in complex electric networks is proposed.

The steady state mode of electric power networks is described by a system of nonlinear equations, which can be solved by numerical methods. The difficulties of ensuring the scheduling and speed of iterative processes in the analysis of high-dimensional systems have led to the development of many programs that implement various methods.

Currently, various methods of calculating the steady state are used. Each of them has its advantages and disadvantages [2, 3, 4, 5, 6].

One of the authors of this work has developed an improved topological method, where the complex voltage of an arbitrary node of a complex electric power grid circuit is determined from the matrix equation [1, 7, 8]:

$$U = U_0 + C^T Z_d C \overline{U}^{-1} \overline{S}, \qquad (1)$$

where C is a rectangular complex matrix of current distribution coefficients; Z_d is a diagonal matrix of branch resistances; U is the complex-valued vector of nodal voltages; S is a complex-valued power vector of nodal loads, generators and transverse branches. The dash above means complex conjugation, and the "T" sign indicates transposition of the matrix.

From the expression (1) it can be seen that, mathematically, the properties of electrical network circuits are described by the natural parameters of the branches and the generalized parameter by the matrix of the distribution coefficients of the driving currents. Consequently, the formation of nodal voltages of a complex electric power grid for given generators and loads is reduced to the determination of the distribution coefficients of nodal currents. These current distribution coefficients are calculated on the basis of expressions obtained using matrix methods or methods of network circuit topology.

2. Formulation of the problem.

An electrical network (city-wide, country-wide) is represented as a replacement scheme with m - nodes and n - branches. It is required to determine the voltage in the nodes of the circuit at steady state, according to the formula (1).

3. Decision.

System (1) is solved by an iterative method. Moreover, the implementation of algorithms for calculating nodal voltages depends on the form of representation of the nodal parameters of the electric power network scheme. If the node parameters are specified by active and reactive powers, then the node voltage of the i–th node for the k-th iteration is defined as [8, 9]:

$$U_i^{'(k)} = U_0 + \sum_{j=1}^{n} Z_{ij} U_j^{-1(k-1)} \left(P_j \cos(\delta_j^{k-1} + \psi_{ij}) + Q_j \sin(\delta_j^{k-1} + \psi_{ij}) \right),$$

$$U_i^{"(k)} = \sum_{j=1}^{n} Z_{ij} U_j^{-1} \left(P_j \sin(\delta_j^{k-1} + \psi_{ij}) - Q_j \cos(\delta_j^{k-1} + \psi_{ij}) \right),$$

where $U_i^k = \sqrt{(U_i^{'(k)})^2 + (U_i^{"(k)})^2}$- voltage module of the i-th node in the k-th iteration;

$\delta_i^k = arctg \frac{U_i^{"(k)}}{U_i^{'(k)}}$ - the voltage phase of the i–th node in the k-th iteration;

$\underline{Z}_{ij} = \sqrt{(Re \sum_{j=1}^{m} \underline{C}_{ij}^t \underline{Z}_j \underline{C}_{ji})^2 + (Im \sum_{j=1}^{m} \underline{C}_{ij}^t \underline{Z}_j \underline{C}_{ji})^2}$ - the module of mutual knot resistance;

$\psi_{ij} = arctg \frac{Im \sum_{j=1}^{m} \underline{C}_{ij}^t \underline{Z}_j \underline{C}_{ji}}{Re \sum_{j=1}^{m} \underline{C}_{ij}^t \underline{Z}_j \underline{C}_{ji}}$ - the phase of mutual knot resistance.

If the node parameters are set by active powers and voltages, then for reactive power of the i-th node, for k-th iteration, the expression [10, 11] is true:

$$Q_i^k = \frac{U_i^2 \cos \delta_i^{k-1} - U_0 U_i - Z_{ii} P_i \cos(\delta_i^{k-1} + \psi_{ii})}{Z_{ii} \sin(\delta_i^{k-1} + \psi_{ii})} -$$

$$\frac{\sum_{j \neq i, j=1}^{n} Z_{ij} U_i U_j^{-1(k-1)} \left(P_j \cos(\delta_j^{k-1} + \psi_{ij}) + Q_j \sin(\delta_j^{k-1} + \psi_{ij}) \right)}{Z_{ii} \sin(\delta_i^{k-1} + \psi_{ii})}.$$

The imaginary part of the voltage of the given node, respectively, is equal to:

$$U_i^{"(k)} = \sum_{j=1, j \neq i}^{n} Z_{ij} U_j^{-1(k-1)} \left(P_j \sin(\delta_j^{k-1} + \psi_{ij}) - Q_j \cos(\delta_j^{k-1} + \psi_{ij}) \right) +$$

$$Z_{ii} U_i^{-1} \left(P_i \sin(\delta_i^{k-1} + \psi_{ii}) - Q_i^k \cos(\delta_i^{k-1} + \psi_{ii}) \right).$$

From here, the phase value of the i-th node voltage for the k-th iteration is:

$$\delta_i^k = \arcsin \frac{U_i^{''(k)}}{U_i}.$$

Further, the calculations are carried out similarly to obtain results for a given accuracy.

In the present work, the algorithm described above is implemented. The developed program was tested on the example of the design scheme of a dedicated part of the network of a real power system of the Republic of Kazakhstan with a voltage of 220 kV (Figure 1).

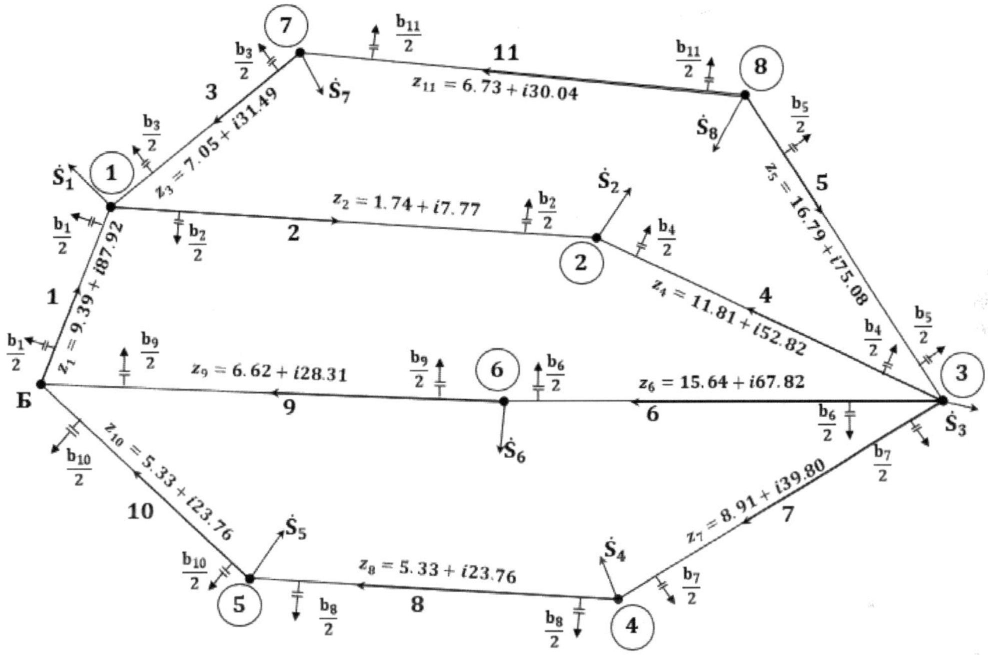

Figure 1. Substitution scheme

In order to verify the correctness of obtained results, we calculated voltages in the network's nodes at steady state using both the proposed method and industrial software RastrWin3. The results of comparison are presented in Table 1.

The deviation of the calculated values by the proposed method from using the industrial program RastrWin3 is less than 1%.

The counting time of our program was 10^{-1} seconds. This means that there is a real possibility to eliminate the main drawback of the existing software systems - the lack of the ability to dynamically analyze the power system.

4. Conclusion.

The paper implements an algorithm for modeling the steady-state mode of complex electric power networks based on the distribution coefficients of the driving currents. It is shown that the software implementation of the algorithm for the formation of a steady state significantly increases the efficiency of research as the number of nodes and branches of the power grid's electric network increases.

Table 1. Results of comparative calculations of node voltages.

	Calculations using matrix C		RASTR verifications calculations		Deviations		
Node number	U, kV	$\delta, grad$	U, kV	$\delta, grad$	$\Delta U, kV$	$\Delta U, \%$	$\Delta\delta, grad$
1	241.43	-4.8315	240.65	-4.79	0,78	0,32	-0,04
2	241.53	-4.8433	240.78	-4.81	0,75	0,31	-0,03
3	240.89	-4.3086	240.36	-4.28	0,53	0,22	-0,03
4	237.38	-3.0902	237.1	-3.08	0,28	0,12	-0,01
5	235.60	-1.8466	235.45	-1.84	0,15	0,06	-0,01
6	236.27	-1.9230	236.12	-1.92	0,15	0,06	0,00
7	239.60	-6.2297	238.33	-6.17	1,27	0,53	-0,06
8	239.12	-6.5359	238.03	-6.49	1,09	0,46	-0,05

5. References.

[1] D S Akhmetbayev A D A and Berdygozhin A S 2018 *Elektrichestvo. Moskva* 18–27
[2] PI Bartolomey SK Okulovskiy A A and Yaroslavtsev A 1982 *Elektrichestvo* 1–5
[3] V A Venikov B I Golovitsyn M S L and Unarokov A A 1976 *Izvestiya ANSSSR. Energetika i transport* 39–49
[4] Akhmetbayev D S 2010 *Elektrichestvo. Moskva* 23–27
[5] Akhmetbayev D S 2010 *Elektrooborudovaniye: ekspluatatsiya i remont. Moskva* 19–26
[6] Akhmetbayev D S 2010 *Elektrooborudovaniye: ekspluatatsiya i remont. Moskva* 15–23
[7] Akhmetbayev D and et al 2017 *Networks, Results in Physics* 1644–1649
[8] D Akhmetbayev A Akhmetbayev A S and Kolcun M 2017 Modeling the set mode of complex power grid, based on infeed coefficients *Proceedings of the 9th International Scientific Symposium, Elektroenergetika, 2017* pp 12–14
[9] D Akhmetbayev A A and Aidarova A 2018 Determination of rational transformation coefficients transformers distribution networks *E3S Web of conferences 25.04003 (2018)*
[10] D Akhmetbayev A Akhmetbayev S Z and Zhakishev B 2018 New modeling of steady-state modes of complex electrical grids of power systems matec *Web of Conferences 155, 01043 (2018)*
[11] D Akhmetbayev A A and Zhantlessova A 2018 Formation of the z-form of equations of steady-state modes of energy systems complex electric networks *Web of Conferences 58,02021 (2018)*

Journal of Physics: Conference Series **1392** (2019) 012080 doi:10.1088/1742-6596/1392/1/012080

Data space reflectivity full waveform inversion

Kirill Gadylshin[1], **Vladimir Tcheverda**[1]

[1] Institute of Petroleum Geology and Geophysics, Novosibirsk, Russia

E-mail: GadylshinKG@ipgg.sbras.ru

Abstract. The full waveform inversion of seismic data aroused the hope to perform simultaneously and in automated way tomography and imaging by solving non-linear least-squares optimization problem. As it has been recognized early, brute force minimization by classical methods is hopeless if low time frequencies are absent in the data. The paper develops a reliable numerical technique for smooth velocity reconstruction via model space decomposition. We present realistic synthetic examples for validating presented algorithm.

1. Introduction

The velocity model building in the depth domain is necessary to guarantee the correct travel-times of wave propagation and therefore is a key element of the up-to-date seismic data processing. As early as the middle of 80th of the last century A. Tarantola introduced the Full Waveform Inversion (FWI) based on the matching the observed and the synthetic seismograms. The L2 norm is widely used for such matching, though other criteria are also considered. To minimize the misfit function and to find the elastic parameters of the subsurface, iterative gradient-based algorithms are usually applied. Such approach to FWI proposed originally by Lailly [1] and Tarantola [2] has been developed and studied in a great number of publications (see [3] and the references therein). However, the straightforward application of FWI reconstructs reliably only the reflectivity component of the subsurface but fails to provide a smooth velocity model. The smooth component could not be recovered stably without the presence of extremely low time frequencies. The matter is the shape of the data misfit functional differs a lot with respect to various velocity components it is nearly quadratic with respect to reflectors, but perturbations of the smooth velocity component (propagator) lead to a very complicated and non-linear behavior (see e.g. [4]). Heuristically it is explained by the so-called "cycle-skip" problem when phase shifts between the recorded and synthetic data produce local minima. In the paper [5] authors introduce some new inversion strategy based on the use of increasing time frequencies: they start inversion with the lowest available ones and at each subsequent iteration they increase the frequency and use the results from the previous step as the initial guess. But they perform this inversion in time domain, hence have to use at each step the low-pass band filtering of data in the time domain which reduces the efficiency of this technique. The next step in implementation of the frequency-domain FWI with increasing time frequencies was done by G. Pratt with co-authors in [6]. They use minimization in time frequency domain and very naturally proceed sequentially from low to high frequencies. The key moment in the successful implementation of this approach is effective solver for the stationary Helmholtz/Elastic waves equation. Unfortunately, this approach did not resolve the problem of stable recovery of a macrovelocity when there is a lack of low time frequencies in the data for both surface and well

(Vertical Seismic Profile) acquisitions ([4],[7]). Of course, on this base they enhance the role of existing low time frequencies and in some sense improve stability of a propagator recovery, but cannot propose the stable solution of the problem. The alternative to this standard formulation was proposed in the paper [8]. The main message of this alternative is modification of the data misfit functional on the basis of the decomposition of a model space into two subspaces:

- subspace of smooth propagators (macrovelocity components);
- subspace of spatial reflectors, which, in turn, is the image of the subsurface.

This approach was modified for the frequency domain in [9] and was applied for reconstruction of well-known Marmousi model for realistic frequency bandwidth and offsets. In this paper, implementation of this approach for elastic Full Waveform Inversion is presented.

2. Method and theory

2.1. Data Space Reflectivity FWI reformulation

Seismic inverse problem is equivalent to solution of the nonlinear operator equation:

$$F(m) = d, \tag{1}$$

where $F : M \to D$ is a nonlinear forward operator, associated with the elastic waves equation, which maps model space M into data space D. As in acoustic MBTT case [8] the essence of this formulation is to decompose the velocity model into two constituents: smooth propagators p and oscillating reflector r. Since we search for both V_p and V_s velocity models, in fact p is a vector $p = (pV_p, pV_s)$, where pV_p – P-velocity propagator and pV_s – S-velocity propagator. In turn, reflector is treated as the result of true amplitude migration applied to the part of the data called data-space reflectivity (DSR) – preimage of the depth reflector in data space:

$$m = p + r = M(p)s. \tag{2}$$

The key moment of this decomposition is propagator-reflector interrelation $r = M(p)s$ with operator $M(p)$ being some kind of true-amplitude prestack migration/linearized inversion. In particular, reweighted version of adjoint operator based migration:

$$M(p)s = W \circ \Re\{DF^* s\} \tag{3}$$

Here DF is Frechet derivative of the full nonlinear map F, * means adjoint operator and W is some linear operator providing true amplitudes imaging/migration. This kind of the model space decomposition leads to the following modified data misfit functional:

$$E(p, s) = \|F(p + M(p)s) - d\|_D^2. \tag{4}$$

Minimization with respect to propagator p and data-space reflectivity s is performed independently and by turn. We start with admitting $s = d$ and do search for some intermediate value of propagator p. After stabilization of this process, the search is switched to data-space reflectivity s and so on.

The search for propagator is implemented using Gauss-Newton method [6]. Since the dimension of propagator space is small enough with respect to the dimension of the whole model space M, we can calculate partial SVD (only highest singular vectors which correspond the highest singular values) of the approximate Hessian H_a on the first iteration using matrix-free methods (such as Krylov-subspace based methods, for numerical implementation see [10]).

Then we use r-pseudoinverse of this operator [11] as a preconditioner in Conjugate-Gradient method during propagator minimization:

$$p_{k+1} = p_k + \mu_k S_k,\ S_0 = \nabla_0, \tag{5}$$

$$S_k = -P\nabla_k - \frac{< P\nabla_k, \nabla_k - \nabla_{k-1} >_M}{< P\nabla_k, \nabla_{k-1} >_M} S_{k-1}, \tag{6}$$

where $p_k = (pV_p, pV_s)$ – propagator on k-th iteration, $P = (H_a)_r^\dagger$ – pseudoinverse of approximate Hessian. The gradient ∇_k with respect to the propagator unknown is calculated as follows:

$$\nabla_k = \Re\{DF^*\delta d_k + (D^2F(p_k)W^*DF^*\delta d_k)^* s\} \tag{7}$$

here $\delta d_k = F(m_k) - d$ is data residual on current iteration, DF – first derivative of forward map calculated at point $m_k = p_k + M(p_k)s$ and D^2F – second derivative calculated at point p_k. Incorporation into inversion process Hessian H_a helps to mitigate crosstalk between V_p and V_s models. Minimization of the data-space reflectivity s is done via steepest descent method. The corresponding gradient ∇_s is calculated as follows:

$$\nabla_s = M^*(p)DF^*\delta d, \tag{8}$$

where $M^*(p)$ – adjoint to the migration operator.

3. Numerical results

3.1. Experiment 1

We use Gullfaks South [12] model to validate the approach based on propagator/reflector decomposition. 2D elastic model one can see in the Fig. 1 (left column). We would like to pay the attention that $\gamma = \frac{V_p}{V_s}$ is not a constant, that is there no dependence between V_p and V_s. Input data are generated for the set of 19 uniformly sampled frequencies in the range $5 - 20Hz$. The acquisition system has 39 vertical point force sources and 200 2C-receivers located at surface $z = 0m$ with a lateral spacing of 100m and 20m respectively. We start with depth dependent initial model (1, middle column) which is far from the real velocities. 2D B-Splines of order 3 were used as the basis of the smooth propagator space. Doing minimization, finally, we come to the result which fairly well coincides with original model for all three parameters: V_p, V_s and γ (1, right column).

The data comparison is demonstrated on Fig. 2. Synthetics are generated for the shot located at $x = 700m$, source wavelet is Ricker with dominant frequency $10Hz$. As it can be clearly observed the reconstructed velocity model desribe main part of the data for both vertical and horizontal receiver components.

3.2. Experiment 2

Finally, we consider the realistic example based on the Marmousi2 elastic synthetic velocity model [13]. Input data are synthesized in frequency domain for the set of twenty uniform frequencies in the range $5 - 15Hz$. We choose the lowest temporal frequency equal to $5Hz$ following some publications and discussions with colleagues dealing with real seismic data. The acquisition system has 46 volumetric sources and 460 receivers located at the top of the model with lateral spacing of 200m and 20m respectively. As initial guess for propagator, we use the vertically heterogeneous model (see Fig. 3). As an initial guess for the time reflectivity variable are used the observed data itself. The results of DSR FWI we present in Fig. 3. As in previous example, we search for the propagator in the space of 2D B-Splines functions of order 3. This guarantees the smoothness of the solution and reasonable size of the model space. To minimize

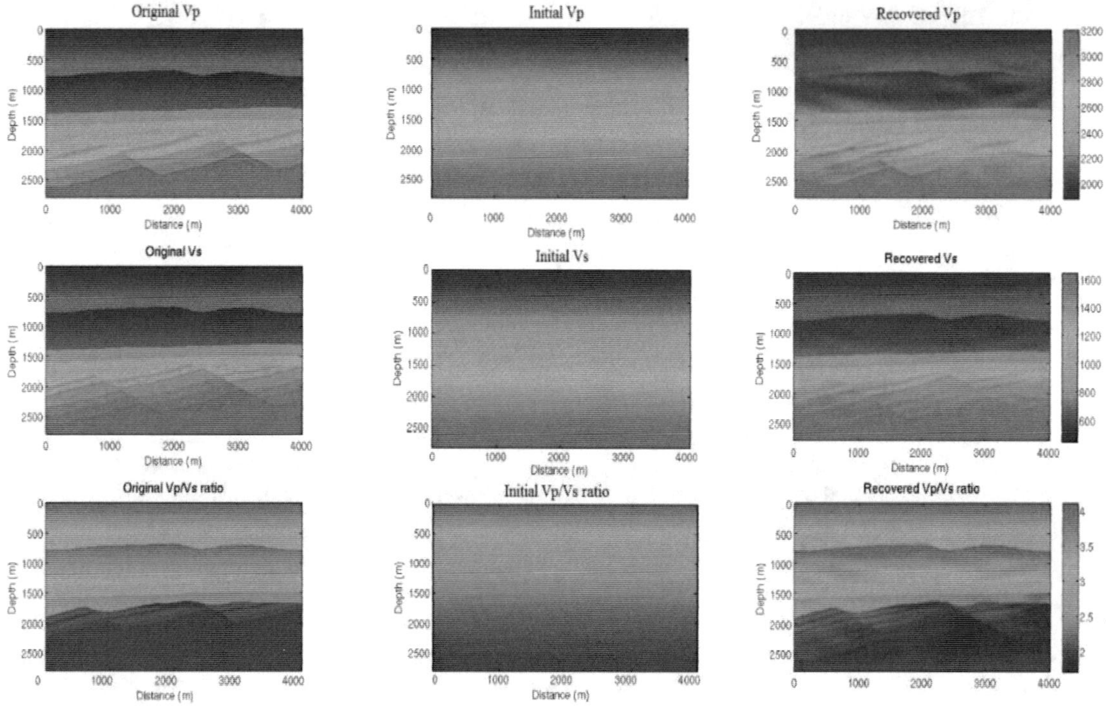

Figure 1. Gullfaks velocity model. Left column from top to bottom: original Vp,Vs and Vp/Vs models. Middle column: initial Vp,Vs and Vp/Vs models. Right column: reconstructed Vp,Vs and Vp/Vs velocity models after DSR FWI

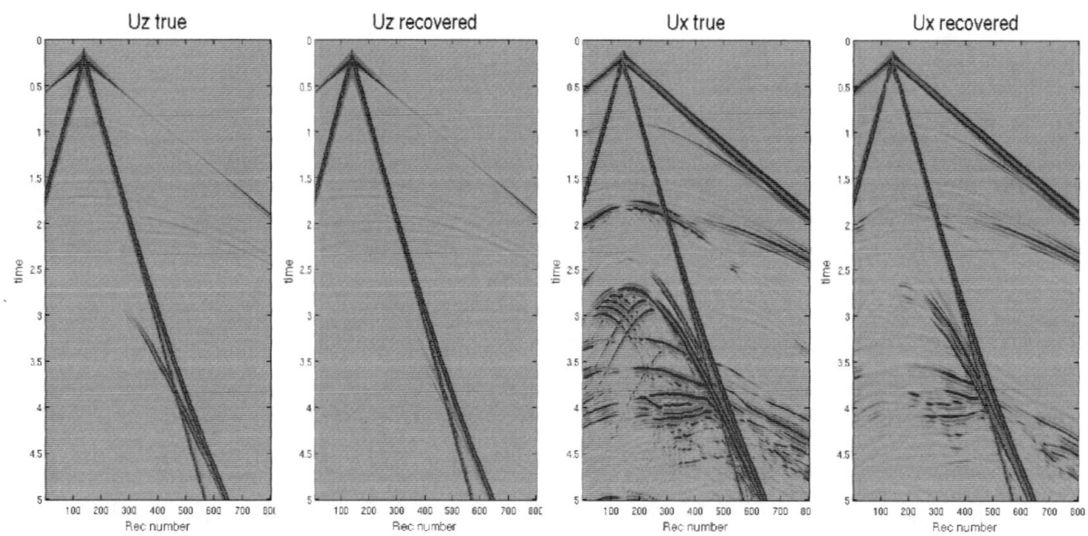

Figure 2. Synthetics generated for shot located at position $x = 700m$. Source wavelet Ricker with dominant frequency $10Hz$. From the left to the right: vertical component for true model and for recovered model, horizontal component for true model and for recovered model

the cost function we use the projected conjugate gradient method, with orthogonal L_2-projector onto smooth space (B-splines). To compute true-amplitude migration M we apply the imaging procedure on the base of inverse generalized Radon transform with representation of the Greens function as decomposition of Gaussian beams. As one can see, local reconstruction of Marmousi velocity model is good enough.

Next, to evaluate the quality of the macrovelocity (propagator) reconstruction we generate common image gathers (CIGs) in the offset-depth domain (see Fig. 4). These CIGs are computed for the initial model and for the model reconstructed via DSR FWI. Reconstructed model significantly improve the flatness of the events in the CIGs with respect to the initial model CIGs. This proves the reliability of the velocity model reconstruction.

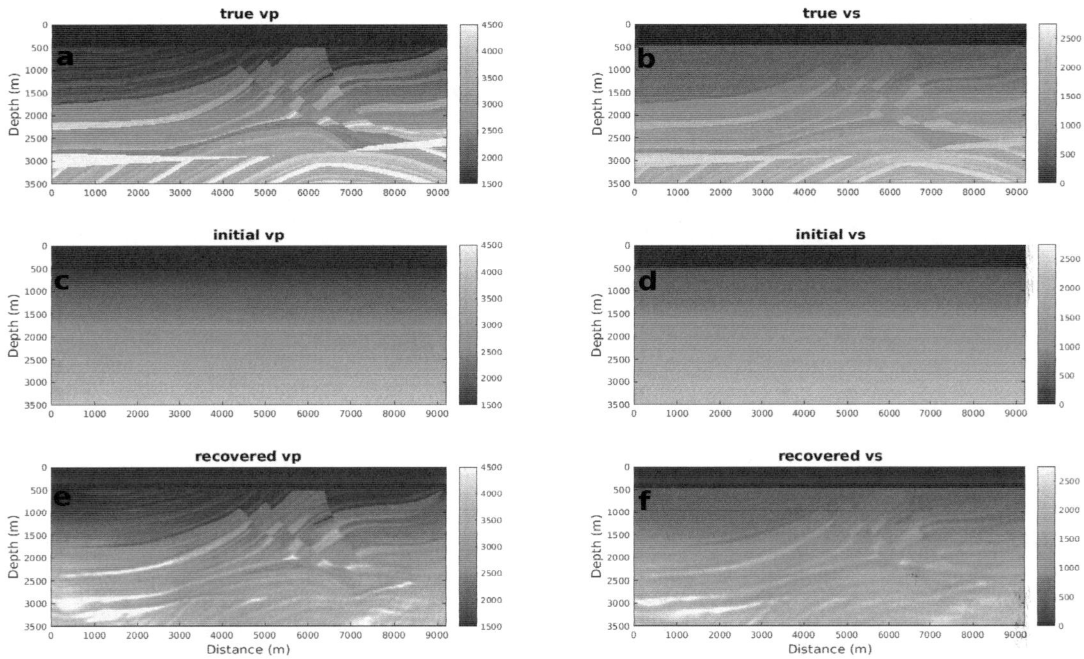

Figure 3. a),b) exact Marmousi2 Vp and Vs models; c),d) initial guess for Vp, and Vs; e),f) reconstructed Vp and Vs models through DSR FWI algorithm

4. Conclusions

We introduce DSR FWI modified data misfit functional on the base of propagator/reflector decomposition of the model space. This decomposition provides alternative way to perform full waveform inversion. We do inversion separately for the search of smooth macrovelocity (reduced model dimension) and sharp reflectors (classical true amplitude prestack migration for known macrovelocity). By this approach we succeeded to reconstruct Gullfaks South model and the Marmousi2 model without low time frequencies we used frequencies from $5Hz$ to $20Hz$ only and do not use huge offsets.

Acknowledgments

Statement of the problem was done by Prof. Tcheverda (funded by Russian Science Foundation project number 17-17-01128). The numerical implementation was done by Kirill Gadylshin who is partially supported by RFBR grant 18-35-00253 and Russian Government Grant MK-670.2019.5.

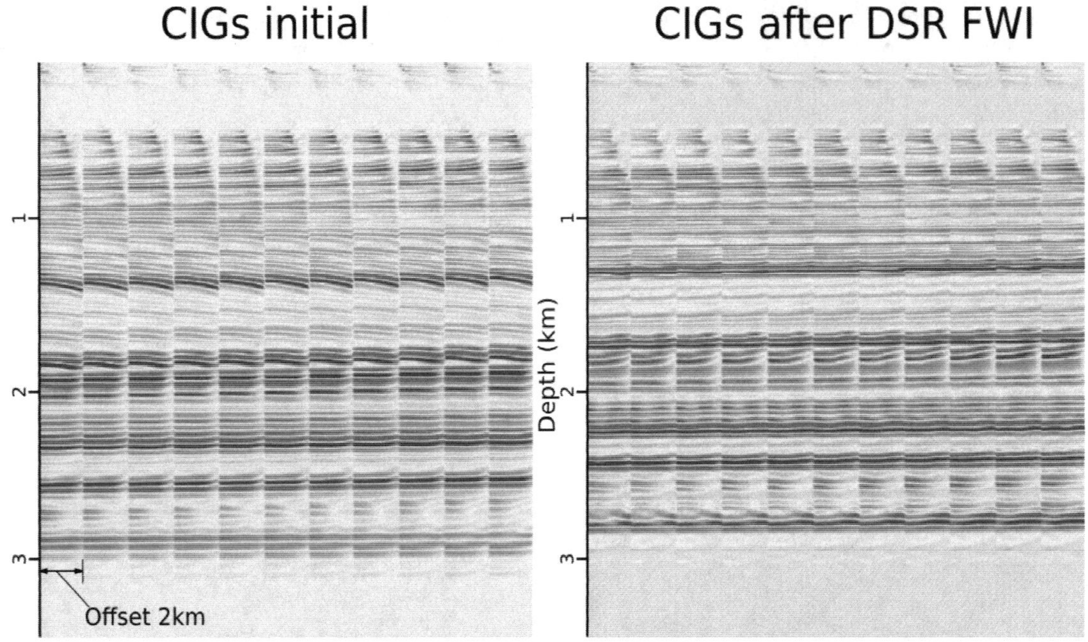

Figure 4. Common image gathers (CIGs) at positions X=[3.5:0.01:3.6] km from the initial guess (left) and model obtained via DSR FWI

References

[1] Lailly P 1983 *The seismic inverse problem as a sequence of before stack migrations. Conf. on Inverse Scattering: Theory and Application. SIAM.* pp 206–220

[2] Tarantola A 1984 *Inversion of seismic reflection data in the acoustic approximation. Geophysics*

[3] Virieux J and Operto, S 2009 *An overview of full-waveform inversion in exploration geophysics.*

[4] Sirgue L 2006 *The importance of low frequencies and large offset in waveform inversion.* 68th EAGE Technical Conf. and Exhibition, A037

[5] Bunks C, Saleck, F M, Zaleski S and Chavent G 1995 *Multiscale seismic inversion. Geophysics* vol 60, iss.5. pp 1457–1473

[6] Pratt G, Shin C and Hicks G J 1998 *Gauss-Newton and full Newton methods in frequency-space seismic waveform inversion Geophysical Journal International* vol 133, iss. 2. pp 341–362.

[7] Silvestrov I, Neklyudov D, Kostov C and Tcheverda V 2013 *Full-Waveform Inversion for Macro Velocity Model Reconstruction in Look-Ahead Offset Vertical Seismic Prfile: Numerical Singular Value DecompositionBased Analysis. Geophysical Prospecting,* 61(6) pp 1099–1113

[8] Clement F, Chavent G and Gomez S 2001 *Migration-based traveltime waveform inversion of 2-D simple structures: A synthetic example Geophysics* vol 66. pp 845–860

[9] Tcheverda V Chavent G and Gadylshin K 2016 *Macrovelocity reconstruction by reflection FWI, 78th EAGE Conference and Exhibition,* Vienna, Austria, 30 May2 June 2016. Th SRS2 06.

[10] Hernandez V, Roman J E and Vidal V 2005 *SLEPc: A scalable and flexible toolkit for the solution of eigenvalue problems ACM Trans.* Math. Software vol 31, iss. 3. pp 351–362.

[11] Cheverda V A and Kostin V I 1995 *Rpseudoinverses for compact operators in hilbert spaces: existence and stability Journal of Inverse and Illposed Problems.* vol 3, iss. 2. pp 131–148

[12] Thompson M, Arntsen B and Amundsen L 2003 *Acquisition geometry versus 4C image quality: A study from Gullfaks South SEG Technical Program Expanded Abstracts.* vol 22, iss. 1. pp 793–796

[13] Martin G S, Wiley R, Marfurt K J 2006 *Marmousi2: An elastic upgrade for Marmousi The Leading Edge.* vol 25. pp 156–166.

Journal of Physics: Conference Series **1392** (2019) 012081 doi:10.1088/1742-6596/1392/1/012081

Numerical method for recovering the piecewise constant right-hand side function of an elliptic equation from a boundary overdetermination data

D Kh Ivanov[1], A E Kolesov[1] and P N Vabishchevich[1,2]

[1]North-Eastern Federal University, 48 Kulakovskogo str., Yakutsk 677000, Russia
[2]Nuclear Safety Institute of RAS, 52 B. Tulskaya str., Moscow 115191, Russia

E-mail: `djulus.ivanov@yandex.ru`

Abstract. An elliptic problem in an open set Ω is considered with Dirichlet boundary condition on piecewise smooth boundary $\partial\Omega$. The inverse problem is to recover piecewise-constant source term, which means an identification function of an unknown subset $D \subset \Omega$. An additional information is taken as Neumann boundary condition, which leads us to consider the inverse problem under Cauchy boundary data. In this work a new computational algorithm for recovery the unknown source term is proposed. The main idea is that the identification function of subset D is replaced by a Heaviside function over the solution of an auxiliary elliptic equation. Then we formulate a minimization problem of a residual for overdetermination data, when a control is taken as the right-hand side of the auxiliary equation. Numerical implementation is based on the finite element method applying the open-source computing platform `FEniCS` and its package `dolfin-adjoint`. The capabilities of the given computational algorithm are shown by results of numerical solutions of 2D test problems.

Keywords: coefficient inverse problem, right-hand side function, elliptic equation, finite element method, FEniCS, dolfin-adjoint

1. Introduction

Inverse problems for partial differential equations are known as non-classical and often belong to the class of ill-posed problems. When considering such problems we are focusing on investigations of issues of the existence, the uniqueness and the stability of the solution over an input data, and also to developments of numerical algorithms.

Among inverse problems for elliptic equations one can derive a problem of identification of the right-hand side (RHS) of an elliptic equation by given boundary observations. This problem relates with the potential theory, for which a lot of interesting results have been obtained. The earliest work is thought to be classical Novikov's work [1] about an uniqueness of the solution of the inverse problem of potential theory in a star shaped domain, when a constant density of mass distribution is known. In later works, e.g. [2], existence and uniqueness problems were considered in the case, where the RHS does not depend on one of the spatial variables. In [3], a wider class of uniqueness was established, when the RHS can include two functions with fewer variables.

General methods of solving inverse problems of reconstruction of the RHS function in an elliptic equation one can find in following works [4, 5, 6] and references therein. For considered class of inverse problems the popular method is Tikhonov's regularization in various modifications. For instance, in [7], an optimal control problem was considered for an identification of the RHS of an elliptic equation from known additional boundary measurements with representing the solution (RHS function) by a finite system of basis functions.

In this work an inverse problem of an identification of the piecewise-constant RHS function of a second order elliptic equation. An unknown RHS function is considered as the identification function of a set of non-overlapping bounded star-shaped domains. A new approach is given to approximate the solution of the considered inverse problem, when objective RHS function is sought by the Heaviside function over the solution of an auxiliary elliptic equation. An optimal control problem is posed to find the minimum of the residual functional for the boundary overdetermination, and the RHS of the auxiliary elliptic problem is taken as the control. For discretization in space the finite element method is used. The gradient iterative method is used for solving the optimal control problem.

2. Problem statement

Let's consider a bounded open region $\Omega \subset \mathbb{R}^2$ with sufficiently smooth boundary $\partial\Omega$, and $\boldsymbol{x} = (x_1, x_2) \in \Omega$. A *direct problem* is stated as follows: find a function $u(\boldsymbol{x})$ s.t. it satisfies the solution of a Poisson problem with homogeneous Dirichlet boundary condition

$$\begin{aligned} -\Delta u &= \rho(\boldsymbol{x}), \quad \boldsymbol{x} \in \Omega, \\ u(\boldsymbol{x}) &= 0, \quad \boldsymbol{x} \in \partial\Omega, \end{aligned} \tag{1}$$

where $\rho(\boldsymbol{x}) \geq 0$.

We consider an *inverse problem* of finding the RHS, $\rho(\boldsymbol{x})$, from a priori assumption that it has next form:

$$\rho(\boldsymbol{x}) = \begin{cases} 1, & \boldsymbol{x} \in D, \\ 0, & \boldsymbol{x} \in \Omega \backslash D, \end{cases} \tag{2}$$

where $D \subset \Omega$, $\partial D \cap \partial\Omega = \emptyset$. Under an assumption that D consists of a finite number of star-shaped subdomains an additional information is taken as a Neumann condition on the whole boundary $\partial\Omega$:

$$\frac{\partial u}{\partial n} = g(\boldsymbol{x}), \quad \boldsymbol{x} \in \partial\Omega, \tag{3}$$

where $\boldsymbol{n}$ is outer normal of the boundary $\partial\Omega$. So, the inverse problem can be uniquely solved, and under such conditions it was investigated, for example, in [6], see section 4.1.

From (2) we can notice that $\rho(\boldsymbol{x}) = \chi_D(\boldsymbol{x})$, which is an identification function of region D. Therefore the inverse problem is basically to find unknown region D or more precisely the boundary of D. There are a lot of works related to such kind of inverse interface problems based on the level set method, e.g. [8, 9]. In this paper we introduce a new method to solve the inverse problem (1)–(3) that differs from the level set method, and it is based on minimization of a residual functional:

$$\min_{\rho} J(\rho) = \int_{\partial\Omega} \left(\frac{\partial u}{\partial n}(\boldsymbol{x}; \rho) - g(\boldsymbol{x}) \right)^2 \mathrm{d}s. \tag{4}$$

3. Computational algorithm

An approximate solution, ρ, of the inverse problem (1)–(4) is sought using the Heaviside function over some auxiliary smooth function $c(\boldsymbol{x})$:

$$\chi_D(\boldsymbol{x}) \longrightarrow H(c(\boldsymbol{x})), \quad \boldsymbol{x} \in \Omega.$$

The auxiliary function $c(\boldsymbol{x})$ is defined from a solution of next boundary-value problem:

$$\begin{aligned} -\gamma\Delta c + c = f(\boldsymbol{x}), \quad & \boldsymbol{x} \in \Omega, \\ c = 0, \quad & \boldsymbol{x} \in \partial\Omega, \end{aligned} \tag{5}$$

with parameter $\gamma = \text{const} > 0$. Thus the objective RHS function (2) has the following form

$$\rho(\boldsymbol{x}) = H(c(\boldsymbol{x})) = \begin{cases} 1, & c(\boldsymbol{x}) > 0, \ \boldsymbol{x} \in \Omega \\ 0, & c(\boldsymbol{x}) \le 0, \ \boldsymbol{x} \in \Omega. \end{cases} \tag{6}$$

Notice that D can be defined from the function $c(\boldsymbol{x})$, where it takes positive values.

The introduction of the auxiliary problem is justified by the fact that the piecewise-constant RHS function of (1) is approximated by using the smooth function $c(\boldsymbol{x})$ over all computational domain. The auxiliary function smoothness is controlled by a term $-\gamma\Delta c$, whereas γ has a meaning of regularization parameter. From (1)–(3) and (5)–(6) it's seen that one can minimize the functional (4) over the RHS function $f(\boldsymbol{x})$ of (5):

$$J(f) = \int_{\partial\Omega} \left(\frac{\partial u}{\partial n}(\boldsymbol{x}; f) - g(\boldsymbol{x}) \right)^2 \mathrm{d}s \longrightarrow \min_f . \tag{7}$$

The latter can be found by using gradient based method. In this case, one need to find derivative of the residual w.r.t. the control function. It can be made by deriving tangent linear and adjoint models of the inverse problem.

Computational implementation is performed by using the finite element method. Let's introduce a Hilbert space $L_2(\Omega)$ with standard scalar product $(\cdot,\cdot)$ and norm $\|\cdot\|$. Let $V^h \subset H^1(\Omega)$ and $\tilde{V}^h = \{v \in V^h : v(\boldsymbol{x}) = 0, \ \boldsymbol{x} \in \partial\Omega\}$. Elements of V^h are represented by a system of pyramid-like piecewise-linear basis functions $\xi_i(\boldsymbol{x})$:

$$v(\boldsymbol{x}) \approx \sum_{i=1}^{N_h} v_i \xi_i(\boldsymbol{x}), \quad v_i = v(\boldsymbol{x}_i),$$

where $\{\boldsymbol{x}_i\}_{i=1}^{N_h}$ are triangulation points of the region Ω. Now we introduce finite element analogues for aforesaid functions

$$u_h(\boldsymbol{x}), \ c_h(\boldsymbol{x}), \ \rho_h(\boldsymbol{x}) \in \tilde{V}^h \quad \text{and} \quad f_h(\boldsymbol{x}) \in V^h.$$

Taking into account finite element approximations and notations we introduce an Algorithm 1 for the solution of the inverse problem (1)–(3), (5)–(7).

4. Numerical experiments

Let computational region be a unit square with 64×64 uniform grid on it. For exact values of observation data we use more refined grid as 256×256. The smoothing parameter γ is set to 0.02 as computation results not included in this work had given us that it should be not so large — an extreme smoothness case, and not so small — an almost discrete case.

Numerical solutions of direct and inverse problems and corresponding PDEs are evaluated in `python3` by using an open source library `FEniCS` (version 2018.1.0) [10]. For finding the minimum of the residual we use the package `dolfin-adjoint` (version 2018.1.0) [11] with `SciPy` minimization (`L-BFGS-B` method [12]).

Algorithm 1 Solution of the inverse problem.

input g, f_h^0, γ, M, ε_1, ε_2 {M is a large number}
output ρ_h
for $k = 0$ **to** M **do**
  $c_h^k \leftarrow$ the solution of the variational problem of (5)

$$\gamma(\nabla c_h^k, \nabla v) + (c_h^k, v) = (f_h^k, v), \quad \forall v \in \tilde{V}^h$$

  $\rho_h^k \leftarrow H(c_h^k)$
  $u_h^k \leftarrow$ the solution of the variational problem of the direct problem (1)

$$(\nabla u_h^k, \nabla v) = (\rho_h^k, v), \quad \forall v \in \tilde{V}^h$$

  $J_k \leftarrow \int_{\partial\omega} (\nabla u_h^k \cdot \boldsymbol{n} - g)^2 \, \mathrm{d}s$
  $J_{f,k} \leftarrow \dfrac{\mathrm{d}J_k}{\mathrm{d}f_h^k}$ {by solving adjoint problem, `dolfin-adjoint`}
  if $|J_{k-1} - J_k| < \varepsilon_1$ **or** $\|J_{f,k}\|_\infty < \varepsilon_2$ **then**
    $\rho_h \leftarrow \rho_h^k$
    stop
  end if
  $f_h^{k+1} \leftarrow f_h^k + d^k$ {step d^k is calculated by `SciPy` minimizer}
end for
$\rho_h \leftarrow \rho_h^k$

Initial guess for the unknown region D is taken as a resulting circle D_0 with radius r and center $\boldsymbol{x}_0 = (X_1, X_2)$, which are found from observation data:

$$r = \left(-\frac{1}{\pi}\int_{\partial\Omega} g \, \mathrm{d}s\right)^{\frac{1}{2}}, \qquad X_1 = \frac{\int_{\partial\Omega} x_1 g \, \mathrm{d}s}{\int_{\partial\Omega} g \, \mathrm{d}s}, \qquad X_2 = \frac{\int_{\partial\Omega} x_2 g \, \mathrm{d}s}{\int_{\partial\Omega} g \, \mathrm{d}s}.$$

Then the initial guess for control function, RHS of (5), is taken as a smooth function:

$$f_0(\boldsymbol{x}) = \begin{cases} \cos^2\left(\dfrac{\pi}{2r}\int_\Omega |\boldsymbol{x} - \boldsymbol{x}_0| \, \mathrm{d}x\right), & \boldsymbol{x} \in D_0, \\ 0, \ \boldsymbol{x} \in \Omega/D_0. \end{cases}$$

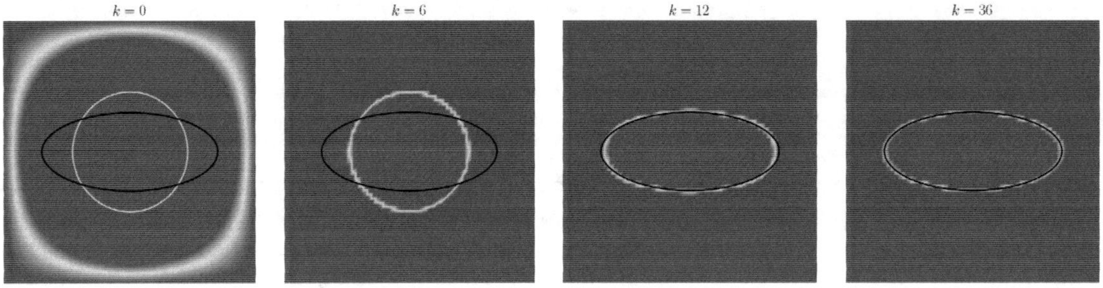

Figure 1. Approximation results of the ellipse shape at some iterations.

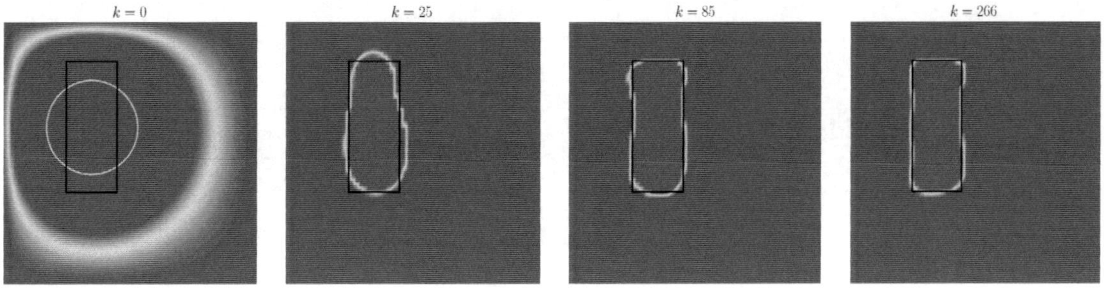

Figure 2. Approximation results of the rectangle shape at some iterations.

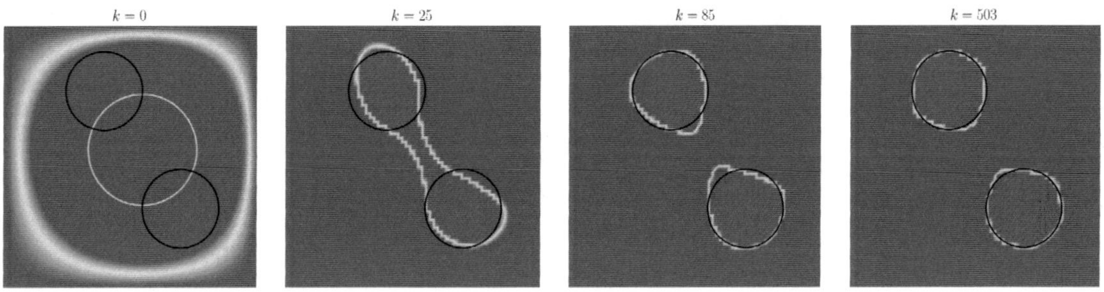

Figure 3. Approximation results of the two circles shape at some iterations.

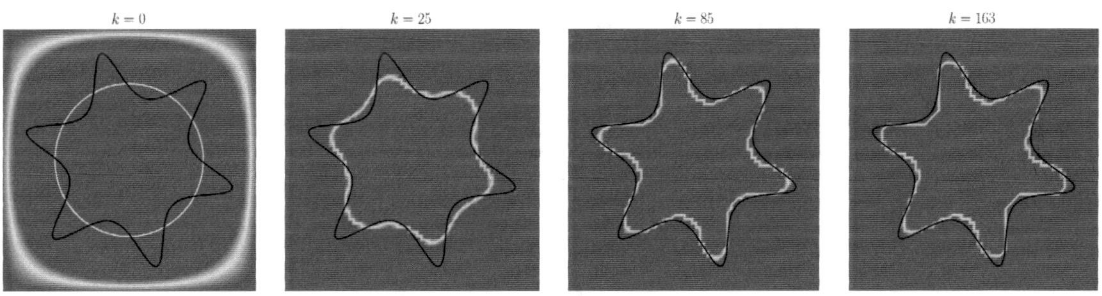

Figure 4. Approximation results of the star shape at some iterations.

The choice of the initial guess for $f(\boldsymbol{x})$ should be done carefully, otherwise the algorithm may fail or terminate earlier if the auxiliary function $c(\boldsymbol{x})$ becomes strictly negative at some iteration.

The Heaviside function is approximated by a following smooth function

$$H_\eta(c) = \begin{cases} 0, & c < 0, \\ 0.5 - 0.5\cos\left(\pi\dfrac{c}{\eta}\right), & 0 \leq c < \eta, \\ 1, & c \geq \eta \end{cases} \tag{8}$$

with a small parameter η, in our case $\eta = 0.01$. Notice, due to the statement of the auxiliary problem (5) and the assumption that $\partial D \cap \partial\Omega = \emptyset$ the Heaviside function should vanish at zero.

Table 1. Details of shapes for the unknown region D.

Shape form	Description
Ellipse	$((x_1 - 0.5)/0.35)^2 + ((x_2 - 0.5)/0.15)^2 \leq 1$
Rectangle	$\lvert x_1 - 0.35\rvert \leq 0.1 \ \cap \ \lvert x_2 - 0.6\rvert \leq 0.25$
Two circles	$(x_1 - 0.7)^2 + (x_2 - 0.3)^2 \leq 0.15^2 \ \cup \ (x_1 - 0.4)^2 + (x_2 - 0.75)^2 \leq 0.15^2$
Star	$((x_1 - 0.5)^2 + (x_2 - 0.5)^2)^2(1 + 0.8 \sin(6 \tan^{-1}((x_2 - 0.5)/(x_1 - 0.5)))) \leq 0.005$

In this work, true measurement data is derived from evaluation of the direct problem in the refined grid. We simulate test problems with different forms of the unknown region D for RHS of the equation (1), see Figures 1–4 and Table 1 for more details. In the first pictures, an initial approximation along with an initial circle D_0 for control function (yellow solid line) are drawn. The exact contour of unknown region is drawn by black solid line. The second and third pictures represent approximations at some intermediate iterations, the last picture is for the exit iteration, when the algorithm terminates with $\varepsilon_1 = 10^{-14}$ and $\varepsilon_2 = 10^{-12}$.

5. Conclusion

In this work we presented new method of finding the piecewise-constant right-hand side function of the elliptic equation from total boundary observation data. Capabilities of the method are demonstrated by 2D test problems for different shapes of unknown region D. Notice that this method can be easily applied for the other coefficients of elliptic equations.

Acknowledgments

The work was supported by RFBR grant 17-01-00689 and Mega grant of Russian Government #14.Y26.31.0013.

References

[1] Novikov PS 1938 On the uniqueness of the solution of the inverse potential problem *Dokl. AN SSSR* **18**(3) 165–8

[2] Prilepko AI 1967 Inverse problems of potential theory *Diff. Eq.* **3**(1) 30–44

[3] Vabishchevich PN 1982 Uniqueness of the solution of an inverse problem for determining the right-hand side of an elliptic equation *Diff. Eq.* **18**(8) 1450–3

[4] Samarskii AA and Vabishchevich PN 2008 *Numerical methods for solving inverse problems of mathematical physics* (Walter de Gruyter) **52**

[5] Kabanikhin SI 2011 *Inverse and ill-posed problems: theory and applications* (Walter De Gruyter) **55**

[6] Isakov V 2017 *Inverse problems for partial differential equations* (Springer) 3rd ed. **127**

[7] Vabishchevich PN 1985 An inverse problem of reconstruction of the right-hand side of an elliptic equation and its numerical solution *Diff. Eq.* **21**(2) 277–84

[8] Isakov V, Leung S and Qian J 2011 A fast local level set method for inverse gravimetry *Communications in computational physics* **10**(4) 1044–70

[9] Ito K, Kunisch K and Li Z 2001 Level-set function approach to an inverse interface problem *Inverse problems* **17**(5) 1225–42

[10] Logg A, Mardal K-A and Wells G 2012 *Automated solution of differential equations by the finite element method: The FEniCS book* (Springer Science & Business Media)

[11] Funke SW and Farrell PE 2013 A framework for automated PDE-constrained optimisation *Preprint arXiv:1302.3894*

[12] Zhu C, Byrd RH, Lu P and Nocedal J 1997 Algorithm 778: L-bfgs-b: Fortran subroutines for large-scale bound-constrained optimization. *ACM Trans. Math. Softw.* **23** 550–60

Studying of the stability of the flow rates wells unto errors permeability field identification under conditions of single-phase stationary fluid filtration

A V Elesin, A Sh Kadyrova and A I Nikiforov

Institute of Mechanics and Engineering, FRC Kazan Scientific Center, Russian Academy of Sciences, 2/31, Lobachevsky str., Kazan 420111 Russia

E-mail: elesin@imm.knc.ru

Abstract. The problem of calculating the producing wells flow rates in a two-dimensional reservoir is considered under conditions of stationary single-phase fluid filtration. The permeability field of the reservoir is assumed to be unknown and is determined from the solution of the inverse coefficient problem. The identification of the permeability field is carried by the known values of bottomhole pressure and is reduced to the problem of minimizing the residual function.

1. Introduction

To solve the prediction problems arising in the development of hydrocarbon field requires knowledge of the reservoir parameters, in particular, the permeability field. In this paper the model problems of the permeability field identification of a two-dimensional reservoir by the known values of bottomhole pressure in a single-phase stationary fluid filtration are considered. The permeability field is determined from the solution of the inverse coefficient problem, which is reduced to the problem of the residual function minimization [1]-[4]. The residual function is the sum of the squared differences between the calculated and known bottomhole pressure values. The minimization of residual function is carried by Levenberg-Marquardt method. At solving the inverse problem the permeability field is searched as a spline function. The coordinates of interpolation nodes are coincided with the coordinates wells. The calculated permeability field is used to calculate the producing wells flow rates when the flow rates of injection wells change and when adding new injection wells. A comparison is made with the results obtained using the true permeability field and the stability to identification errors is estimated. The calculation results are also influenced by the bottomhole pressure measurement errors. To study this effect, random errors are introduced into the bottomhole pressure measurements.

2. Problem of the permeability field identification

In a two-dimensional reservoir, a single-phase stationary fluid filtration, obeying the Darcy law, is described by the equation [5]

$$\nabla\left(\frac{kh}{\mu}\nabla p\right)=\sum_{i=1}^{M}Q_i\delta\left(x_i,y_i\right), \tag{1}$$

where $k = k(x, y)$ is the coefficient of permeability, $h = h(x, y)$ is the thickness of reservoir, μ is the coefficient of fluid viscosity, p is the pressure, Q_i, (x_i, y_i) are the flow rate and coordinates of i-th wells, $\delta(x_i, y_i)$ is the delta function, M is the number of wells. For the equation (1) the following initial conditions is assigned

$$p\,|_\Gamma = p_\Gamma, \tag{2}$$

where Γ is the boundary of reservoir Ω.

To determine the pressure field from the system (1)-(2), the coefficients of equation (1) must be known. In this paper $h = h(x, y)$ and μ are considered known, and the permeability coefficient $k(x, y)$ is considered unknown. To determine it, the inverse coefficient problem is solved, which is reduced to the problem of the residual function minimization. The residual function is constructed on a known bottomhole pressure measurements in the wells $p^*_{wi}(i = \overline{1, M})$ and has the form

$$J(k) = \frac{1}{2}\sum_{i=1}^{M}\left(p_{wi} - p^*_{wi}\right)^2, \tag{3}$$

where p_{wi} are the bottomhole pressure values obtained as a result of the solution equation (1). The Levenberg-Marquardt method is used for the residual function minimization [6]. Stopping of the minimization process is carried out according to two criteria: achieving of the given accuracy of bottomhole pressure measurements $\max\limits_{j=\overline{1,M}} \left|p_{wj} - p^*_{wj}\right| < 0.01$ MPa or slow convergence of minimization process $J^n - J^{n+1} < 0.01J^n$ during 3 iterations, where n is the iteration number.

In this paper the permeability field parameterization is used in the form of a spline function [7]

$$k(x, y) = \sum_{i=1}^{N} c_i r_i^2 \ln r_i^2 + c_{N+1} + c_{N+2}x + c_{N+3}y,$$

where $r_i^2 = (x - x_i)^2 + (y - y_i)^2$, (x_i, y_i) are the coordinates of interpolation nodes. The number of interpolation nodes N is corresponded with number of identified parameters. The coefficients c_i, $i = \overline{1, N+3}$, are determined from the solution of the equations system

$$k(x_i, y_i) = k_i, i = \overline{1, N},$$

$$\sum_{i=1}^{N} c_i = 0, \sum_{i=1}^{N} c_i x_i = 0, \sum_{i=1}^{N} c_i y_i = 0,$$

where k_i are the permeability values in the interpolation nodes. The using of the spline function avoids the construction of an additional grid for calculating permeability values. Interpolation nodes can be placed arbitrarily and for a single solution to exist, it is sufficient that at least three interpolation nodes do not lie on the same line. In [8, 9] two approaches to the choice of interpolation nodes were used. In this paper the coordinates of interpolation nodes are coincided with the wells coordinates.

3. Model tasks

The reservoir Ω (5000 m $\times$ 8000 m, thickness of reservoir is 10 m) has the impermeable faults (figure 1). The reservoir Ω opens up 12 injection and 123 production wells (figure 1). The flow rate of 100 m^3/day is assigned at injection wells, at production wells flow rate varies from 1.7 m^3/day up to 350 m^3/day. The radius of wells is 0.1 m. At the boundary of Ω is assigned pressure of 20 MPa. The coefficient of fluid viscosity is 10 mPa·c. The model problem of the permeability field identification is constructed as follows. The reservoir Ω is covered with a square grid (5 $\times$ 9). In the nodes of this grid the permeability values are assigned from 0.1 mkm^2 up to 10 mkm^2. The true permeability field $k^{tr}(x, y)$ is constructed by kriging method using these values (figure 2). Then the pressure field p^{tr} and the bottomhole pressure values p^*_{wi}, $i = 1, M$ are calculated from the solution of the equations system (1)-(2) for the field k^{tr}.

The equations system (1)-(2) is solved numerically. For approximation by spatial variables of the equation (1) the method of control volumes is used. The reservoir Ω is covered a square grid with a step of 50 m ($L = 16000$ is number of control volumes). The resulting system of linear algebraic equations is solved by the method of conjugate gradients in the form approximate Cholesky factorization [10, 11]. The values of bottomhole pressure are calculated by using Peaceman formula [12].

Further the permeability field $k(x, y)$ is considered unknown and is restored as a spline function by known bottomhole pressure values. In real bottomhole pressure measurements there is always some error, so the bottomhole pressure measurements p^*_{wi} are randomly entered errors, and their influence on the results of calculations are investigated. For comparison of the calculated permeability fields $k^c(x, y)$ and pressure $p(x, y)$ with true fields are given values of the deviations $\Delta K_{av} = \sqrt{\sum_{i=1}^{L} (k_i^{tr} - k_i^c)^2 / L}$ and $\Delta r_{av} = \sqrt{\sum_{i=1}^{L} (p_i^{tr} - p_i)^2 / L}$, the maximum deviations $\Delta K_{max} = \max_{i=1,L} |k_i^{tr} - k_i^c|$ and $r_{max} = \max_{i=1,L} |p_i - p_i^*|$ in all control volumes of the grid, the maximum residuals for wells $r_{max}^{well} = \max_{j=1,M} \left| p_{wj} - p^*_{wj} \right|$. The results obtained without errors in bottomhole pressure measurements and with six variants errors are shown in table 1. The solutions of all tasks are obtained with the given accuracy on the pressure. The permeability fields obtained without errors in bottomhole pressure measurements and with -0.1 MPa$\leq \varepsilon_i \leq$ 0.1 MPa, 0 MPa$\leq \varepsilon_i \leq$0.1 MPa, -0.1 MPa$\leq \varepsilon_i \leq$ 0 MPa errors are shown in figures 3-6.

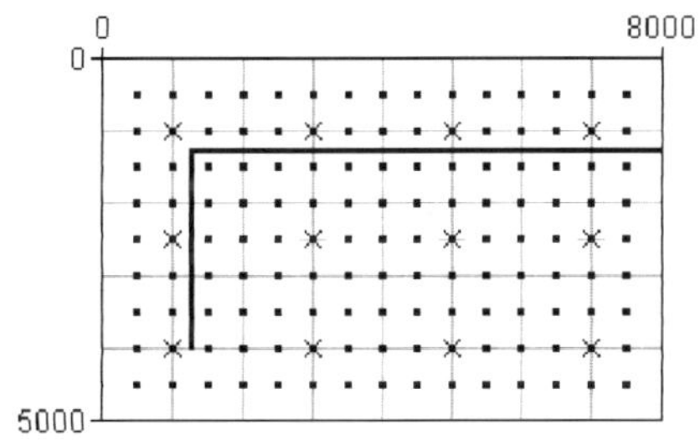

Figure 1. The location of wells and faults, (∎ - production wells, × - injection wells).

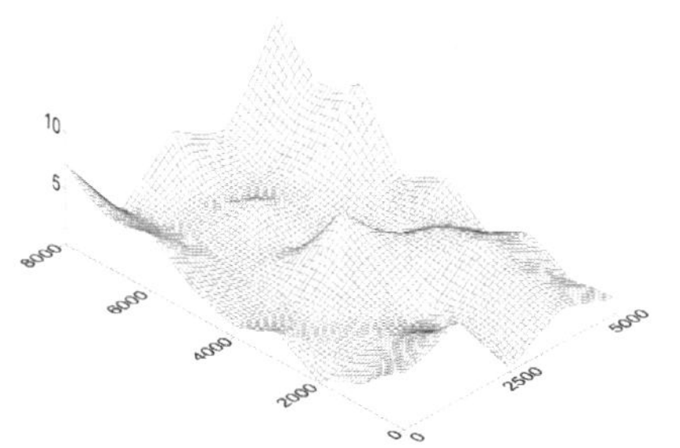

Figure 2. The true permeability field k^{tr}.

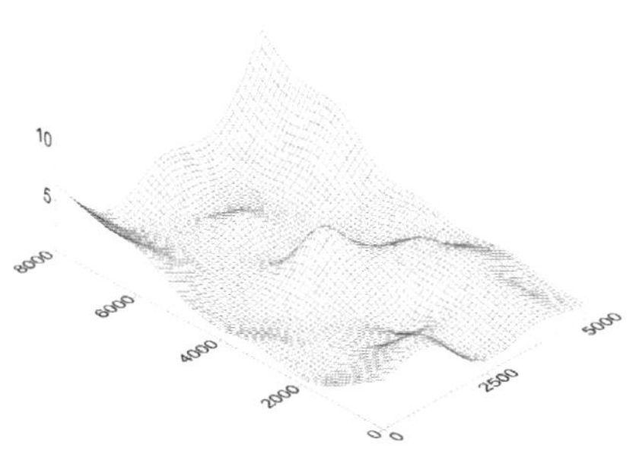

Figure 3. The permeability field k^c, without errors.

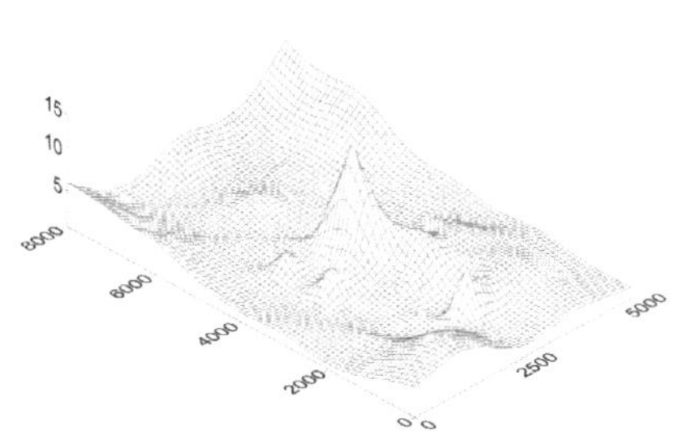

Figure 4. The permeability field k^c, with errors -0.1 MPa$\leq \varepsilon_i \leq$ 0.1 MPa.

Table 1. Comparison the calculated permeability fields with the true permeability field.

	ΔK_{av}, [mkm^2]	ΔK_{max}, [mkm^2]	r_{av}, [MPa]	r_{max}, [MPa]	r^{well}_{max}, [MPa]
without errors	0.1932	1.7412	0.0379	0.3501	0.0047
-0.1 MPa$\leq \varepsilon_i \leq$ 0.1 MPa	0.8391	10.262	0.04364	0.3165	0.0096
-0.01 MPa$\leq \varepsilon_i \leq$ 0.01 MPa	0.1998	1.7228	0.0376	0.3425	0.0048
0 MPa$\leq \varepsilon_i \leq$ 0.1 MPa	0.2150	3.3940	0.0611	0.3291	0.0087
0 MPa$\leq \varepsilon_i \leq$ 0.01 MPa	0.1952	1.7652	0.0389	0.3478	0.0048
-0.1 MPa$\leq \varepsilon_i \leq$ 0 MPa	0.3556	2.8233	0.0426	0.3021	0.0086
-0.01 MPa$\leq \varepsilon_i \leq$ 0 MPa	0.1945	1.6989	0.03684	0.3448	0.0047

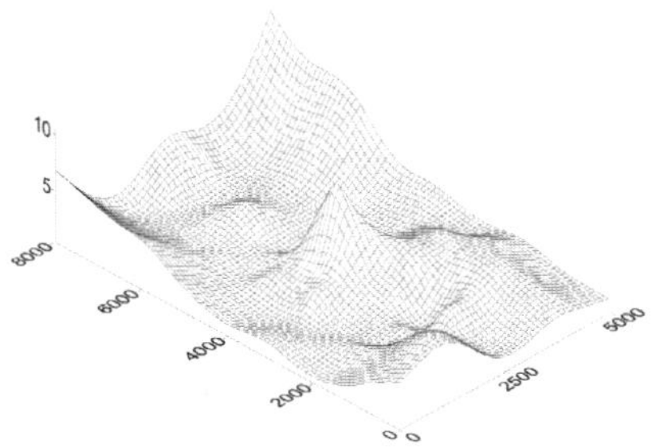

Figure 5. The permeability field k^c, with errors 0 MPa$\leq \varepsilon_i \leq$ 0.1 MPa.

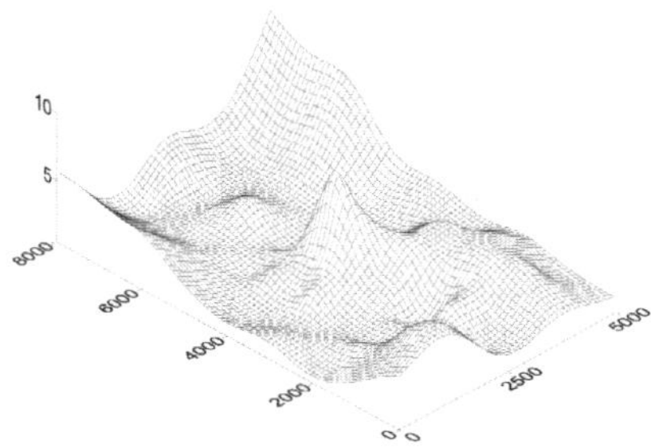

Figure 6. The permeability field k^c, with errors -0.1 MPa$\leq \varepsilon_i \leq$ 0 MPa.

Further, the calculated permeability fields are used to calculate the flow rates of production wells when changing the flow rates of injection wells and when adding new injection wells. At solution of these tasks are assigned the flow rates for the injection wells and the bottomhole pressures in production wells. The bottomhole pressures of production wells were taken to be the same in all variants of the task. Three variants of the debits calculation task are considered.

(i) On all injection wells the flow rate is increased to 150 m^3/day (instead of 100 m^3/day, as it was in the initial task).

(ii) The nine production wells were transferred to injection wells, all injection wells have a flow rate of 100 m^3/day, a total of 21 injection wells (figure 7).

(iii) One production well is transferred to the injection well, on all injection wells the flow rate is 100 m^3/day, the location of the new injection well is determined from the condition of the maximum total flow rate of all production wells.

All tasks are solved with a true permeability field and with calculated permeability fields. The found location of the new injection well for the third variant of the task is shown in figure 8 (□ - for the permeability field in case -0.1 MPa$\leq \varepsilon_i \leq$ 0.1 MPa measurements error, ○ - for the true permeability field and for other variants of the calculated permeability fields). To compare the results the total flow rate of all production wells is given in table 2. The flow rates of production wells for first variant of task are shown in figure 9.

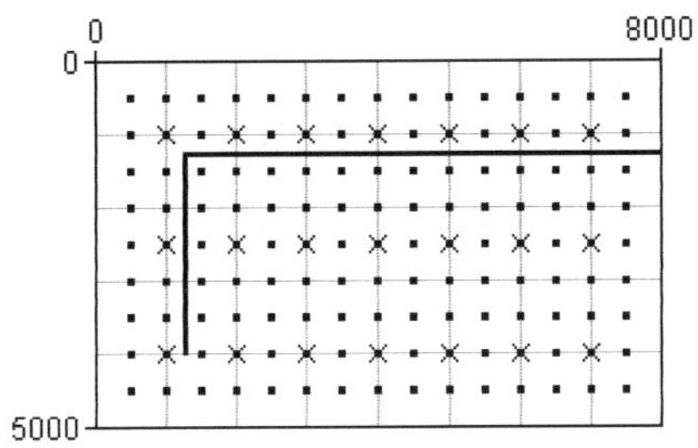

Figure 7. The location of wells and faults, (■ - production wells, × - injection wells).

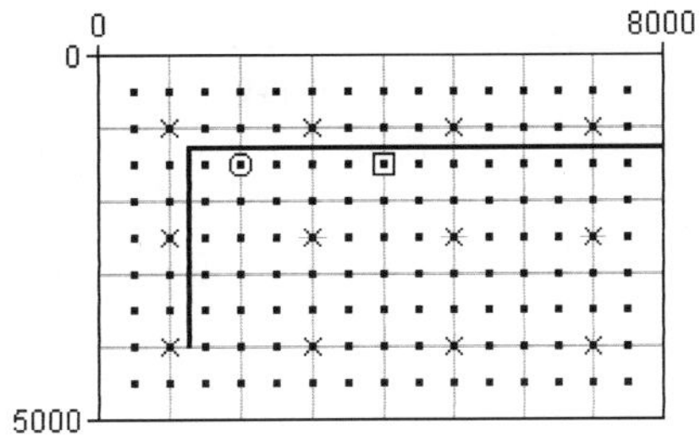

Figure 8. The location of wells and faults, (■ - production wells, × - injection wells).

Table 2. The total flow rate, m^3/day.

	variant 1	variant 2	variant 3
true	8653.23	8815.53	8330.20
without errors	8650.65	8810.87	8330.01
-0.1 MPa$\leq \varepsilon_i \leq$ 0.1 MPa	8653.51	8814.42	8331.89
-0.01 MPa$\leq \varepsilon_i \leq$ 0.01 MPa	8649.92	8810.20	8329.21
0 MPa$\leq \varepsilon_i \leq$ 0.1 MPa	8801.40	8960.38	8481.06
0 MPa$\leq \varepsilon_i \leq$ 0.01 MPa	8665.18	8825.28	8344.58
-0.1 MPa$\leq \varepsilon_i \leq$ 0 MPa	8504.26	8666.24	8182.50
-0.01 MPa$\leq \varepsilon_i \leq$ 0 MPa	8635.40	8795.81	8314.66

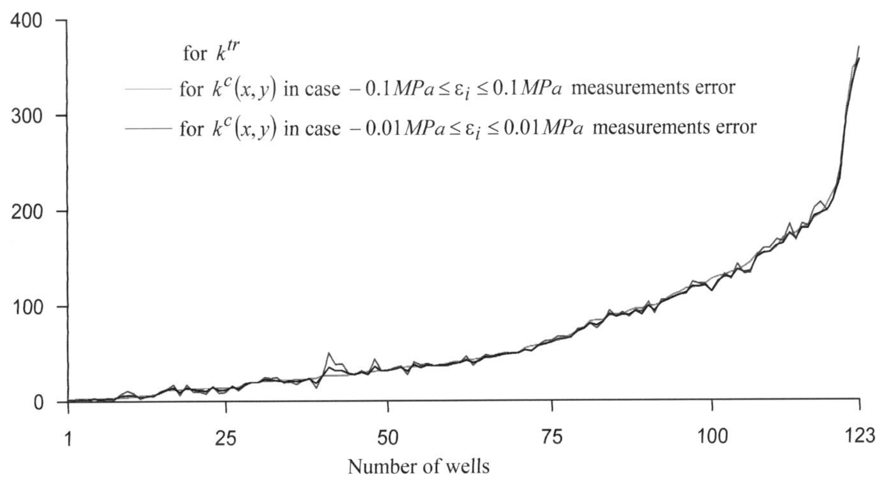

Figure 9. Projected flow rates of production wells are in m^3/day

4. Conclusions

On model tasks the problem stability of calculation of producing wells flow rates is studied in the presence of error permeability field identification. The permeability field is determined in the form of a spline function from the solution of the inverse coefficient problem by known bottomhole pressure measurements. In the identification of permeability field in the bottomhole pressure measurements is entered different error. The results show a good agreement between the predicted production wells flow rates calculated from the true permeability field and the identified permeability fields.

References

[1] Sun N-Z 1994 *Inverse Problems in Groundwater Modeling* (Dordrecht, Kluwer Academic) p 337
[2] Yeh W W-G 1986 *Water Resour. Res.* **22** 95-108
[3] Carrera J, Newman Sh P 1986 *Water Resour. Res.* **22** 228-42
[4] Elesin A V, Kadyrova A Sh 2017 *J. of Engineering Physics and Thermophysics* **90** 1362-68
[5] Aziz Kh, Settari A 1982 *Petroleum Reservoir Simulation* (Moscow: Nedra) p 407
[6] Dennis J E, Schanabel R B 1988 *Numerical Methods for Uncostrained Optimization and Nonlinear Equations* (Moscow: Mir) p 440
[7] Ashkenazy V O 2003 *Splajn-poverhnosti: Osnovy teorii i vychislitel'nye algoritmy: Uchebnoe posobie* (Tver': Tver' State University) p 82
[8] Elesin A V et al 2019 *J. Phys.: Conf. Ser.* **1158** 022039
[9] Elesin A V, Kadyrova A Sh and Nikiforov A I 2018 *Georesources* **2**, 102-107
[10] Golub G H, Van Loan C F 1999 *Matrix Computations* (Moscow: Mir) p 548
[11] Larabi A, De Smedt F 1994 *Water Resour. Res.* **30** 509-21
[12] Peaceman D W 1978 *Soc. Petrol. Eng. J.* **18** 183-94

Journal of Physics: Conference Series **1392** (2019) 012083 doi:10.1088/1742-6596/1392/1/012083

A computational algorithm for constructing a two-dimensional heat wave generated by a non-stationary boundary condition

A L Kazakov[1,2], L F Spevak[2], A A Lempert[1,3] and O A Nefedova[2]

[1] ISDCT SB RAS, 134 Lermontova St., Irkutsk, 664033, Russia
[2] IES UB RAS, 34 Komsomolskaya St., Ekaterinburg, 620049, Russia
[3] INRTU, 83 Lermontova St., Irkutsk, 664074, Russia

E-mail: kazakov@icc.ru; lfs@imach.uran.ru; lempert@icc.ru; nefedova@imach.uran.ru

Abstract. The paper discusses solutions of the nonlinear heat equation, which have the form of a heat wave propagating on a zero background with a finite velocity. Such solutions are not typical for parabolic equations, and their existence is associated with the degeneration of the problem at the wave (zero) front. We propose a numerical algorithm for constructing a two-dimensional heat wave, symmetrical with respect to the origin, with a non-zero boundary condition defined on the moving boundary. The main difficulty of the new task is that at each time point a heat wave front (a domain boundary) is unknown. The solution is carried out in two stages. At first, we change the roles of unknown function and radial polar coordinate. For a new unknown function at each time point, we obtain a boundary value problem for the Poisson equation in a known region. The step-by-step solving of this problem by the method of boundary elements at a given time interval allows us to determine the law of the zero front moving. At second, we approximate the found zero front by an analytical function and construct a generalized self-similar solution. The developed algorithm is implemented and tested on a task set.

1. Introduction

We consider a nonlinear second order parabolic equation, which describes the process of heat propagation in the case of a power dependence of the thermal conductivity coefficient on temperature [1]. It is also reffered as the porous medium equation [2]

$$T_t = \operatorname{div}\left(K(T)\nabla T\right), \quad K(T) = \alpha T^{\sigma}. \tag{1}$$

Here T is an unknown function (temperature), t is time, $K(T)$ is the thermal conductivity coefficient, div, ∇ are the divergence and gradient operators on spatial coordinates α, σ are positive constants.

In addition to modeling heat conductivity [1], Eq. (1) is used to describe the filtration processes of ideal polytropic gas in porous media [2], as well as the migration of biological populations [3].

A lot of publications are devoted to the study of mathematical models of continuum mechanics based on Eq. (1). Among them, we especially note those of Zeldovich [4], Barenblatt et al. [5], Samarsky et al. [1], Sidorov [6], Kudryashov [7], Antontsev and Shmarev [8]. The difference in the approaches to the study is due to a wide range of considered boundary value problems, as well as a large range of possible research objectives.

Our previous studies were devoted to the boundary-value problems with a moving boundary, the solutions of which have the form of a heat wave propagating on a cold (zero) background with a finite velocity. The primary directions of research were to prove existence and uniqueness theorems with the construction of analytical solutions [9, 10], as well as to find numerical solutions for a finite time interval [11, 12, 13, 14].

Here we consider a mathematical model of a two-dimensional heat wave generated by a nonzero condition defined on a moving boundary. The case of circular symmetry is investigated. Equation (1) is reduced to the following equation in polar coordinates by standard substitution $u = T^\sigma$, $t = \alpha t$

$$u_t = uu_{\rho\rho} + \frac{{u_\rho}^2}{\sigma} + \frac{uu_\rho}{\rho}. \tag{2}$$

We previously studied Eq. (2) under the following types of boundary conditions:

$$u|_{\rho=a(t)} = 0, \tag{3}$$

$$u|_{\rho=R} = f(t), \tag{4}$$

$$u|_{\rho=b(t)} = g(t). \tag{5}$$

We proved the existence and uniqueness theorems for an analytical solution [10] and presented step-by-step algorithms for solving by the boundary element method (BEM) at a given time interval [12].

In this study, we deal with problem (2), (5). Equation (5) sets a nonzero boundary condition on the moving boundary $\rho = b(t)$. For an approximate solution of problem (2), (5) we propose to use generalized self-similar solutions, which are presented in Sections 2 and 3. Section 4 proposes a numerical algorithm for the approximate finding of the law of moving of the zero front $\rho = a(t)$ under the given condition (5). In Section 5, for a special approximated function $\rho = a(t)$, we construct a generalized self-similar solution, which is taken as an approximate solution of the problem (2), (5).

2. Exact solutions

Exact solutions of nonlinear equations of mathematical physics are of interest both for the development of the theory of differential equations and for verifying numerical methods.

To construct exact solutions of Eq. (2) with a heat wave type, we use the Clarkson-Kruskal Direct Method [15]. According to this method, the solution is sought as

$$u = \lambda(t)w(z), \quad z = a_0(t) + \frac{\rho}{a_1(t)}. \tag{6}$$

After substituting of Eq. (6) in Eq. (2) and reduction of similar terms, we obtain the following equation:

$$ww'' + \frac{1}{\sigma}(w')^2 + \left[\frac{w}{z - a_0} + \frac{a_1({a_1}'z + {a_0}'a_1 - {a_1}'a_0)}{\lambda(t)}\right]w' - \frac{{a_1}^2(t)\lambda'(t)}{\lambda^2(t)}w = 0. \tag{7}$$

It is easy to see that the necessary condition for separation of variables t and z and Eq. (7) becomes an ordinary differential equation (ODE) for the function $a_0 = \text{const}$. Such solutions are sometimes called *generalized self-similar solutions* [15].

Proposition 1. *Equation (2) has solutions of the form (6) in the following cases:*

1. $a_0 = \text{const}$, $a_1(t) = C_1 e^{C_2 t}$.

2. $a_0 = \text{const}$, $a_1(t) = C_3(t + C_4)^{\omega}$.
Moreover, $\lambda(t) = a_1 a_1{}'$, *and* $w(z)$ *satisfy the following ODE:*

$$ww'' + \frac{1}{\sigma}(w')^2 + \left(\frac{w}{z - a_0} + z - a_0\right) w' - \eta w = 0, \tag{8}$$

where $\eta = 2$ for the first case and $\eta = (2\omega - 1)/\omega$ for the second one, $C_i = \text{const}$, $i = 1, \ldots, 4$.

The proof of the proposition is carried out by integrating a system of ordinary differential equations. It is obtained if $a_0 = \text{const}$ (see above), and the factors in front of w and w' is equated to constants in the left-hand side of Eq. (7). The correctness of the proposition can also be verified by direct substitution in Eq. (7) of the corresponding expressions for λ, a_0, a_1.

In order for the obtained generalized self-similar solutions to have the type of heat wave, it is necessary to supplement Eq. (8) with Cauchy conditions. Let $a(t) = -a_1(t)$. Then the condition given on the heat wave front (3) takes the form $w(a_0 - 1) = 0$. We require that the heat flow at the heat wave front be non-zero. Let us put $z = a_0 - 1$, $w = 0$ in Eq. (8), then $w'|_{z=a_0-1} = -\sigma$.

Thus, we have for Eq. (8) the following Cauchy conditions:

$$w(a_0 - 1) = 0, \quad w'(a_0 - 1) = -\sigma, \tag{9}$$

From the above reasoning, the following statement holds:

Proposition 2. *Problem (2), (3) has generalized self-similar solutions (6), where* a_0, a_1, λ *are found from the condition of Proposition 1, and* w *satisfies Cauchy problem (8), (9).*

Note that for problem (8), (9) classical theorems of existence and uniqueness of the solution are not satisfied, since for $z = a_0 - 1$ the factor in front of the highest derivative vanishes. Nevertheless, the existence and uniqueness of a solution in the class of analytical functions follows from the theorem previously proved by the authors [10]. Find a solution to problem (8), (9) in the form of an analytical formula is possible only in some cases. In the general case, a continuous solution can be obtained by iteration using the boundary element method. An extensive computational experiment shows a fairly high accuracy of such solutions. The resulting generalized self-similar solutions to problem (2), (3) having form (6) are not exact in the full sense; however, they can certainly be used as a reference for testing numerical algorithms.

3. Approximate constructing of an invariant heat wave

The construction of a generalized self-similar solution (6) is possible only for certain types of functions defining the law of motion of the zero front. However, let us consider the possibility of constructing such a solution in the case when the function $\rho = a(t)$ is close to a power or exponential law corresponding to the conditions of Proposition 1. Suppose that the function $\rho = a(t)$ is determined experimentally at some points in time, and the experimental results are approximated with sufficient accuracy by a function that satisfies the conditions of Proposition 1. It can be assumed that solution (6) to Cauchy problem (8), (9) will be an approximate solution to problem (2), (3).

As a typical example, we consider problem (2), (3) when the zero-front moving is given by a segment of the Taylor series of the function $\rho = e^t$:

$$\rho = a_N(t) = \sum_{n=0}^{N} \frac{t^n}{n!}. \tag{10}$$

For the values of this function on the segment $t \in [0, 1]$, at $t_k = kh$, $k = 0, 1, \ldots, K$, $h = 1/K$, the least squares method gives us an approximation function $\rho = C_1 e^{C_2 t}$. Using this approximation as the law of motion of the zero front, we construct a generalized self-similar solution (6), which will be an approximate solution of problem (2), (3), (10). The accuracy of the solution obtained is evaluated by comparing it with a step-by-step solution based on the BEM. Figure 1 shows the solutions for $N = 3$, $K = 10,\ 20$. Similar results for $N = 4$ are shown in Figure 2. In all calculations, hereinafter, for simplicity, it is assumed that $a_0 = 1$. From the figures it can be seen that with an increase in the degree of a segment of a series (10), the time interval increases, where the generalized self-similar solution is close to the BEM one.

Computational results indicate the stability of generalized self-similar solutions of the form (6) with respect to small deviations of the law of motion of the zero front. Consequently, a generalized self-similar approach can be applied to approximate solving boundary value problems for Eq. (2).

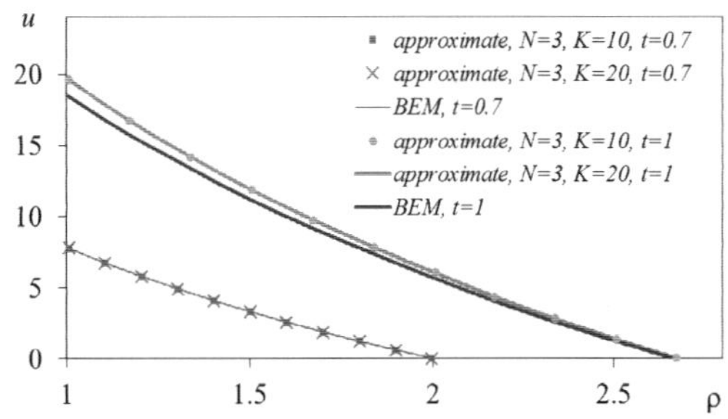

Figure 1. The comparison of numerical solutions for $N = 3$.

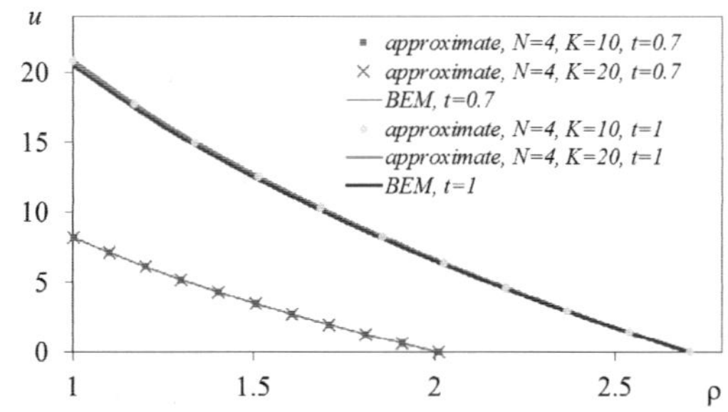

Figure 2. The comparison of numerical solutions for $N = 4$.

4. Approximate constructing of the zero front

We now turn to problem (2), (5). To find the unknown law of moving of the zero front $\rho = a(t)$, we change the role of the unknown function u and spatial variable ρ. Such a replacement was previously used in the papers of Sidorov [6], as well as in the works of the authors. This change of variables is correct in the case of a monotonic function $f(t)$, then at the current time $t = t_k$ the function $u(t_k, \rho)$ is reversible. For the inverse function $\rho = \rho(t_k, u)$ Eq. (2) and the boundary condition (5) take the form:

$$\rho_t {\rho_u}^2 = u\rho_{uu} - \frac{\rho_u}{\sigma} - \frac{u{\rho_u}^2}{\rho}, \tag{11}$$

$$\rho|_{u=g(t)} = b(t). \tag{12}$$

According to the approach proposed in [11, 12, 14], we solve problem (11), (12) in time steps. At the time moment $t_k = kh$, where h is the step size, we consider the boundary value problem in the domain $u \in [L_0, L_k]$, $L_0 = g(0)$, $L_k = g(t_k)$, for the Poisson equation

$$\rho_{uu} = \frac{1}{u}\left(\rho_t {\rho_u}^2 + \frac{\rho_u}{\sigma}\right) + \frac{{\rho_u}^2}{\rho}, \tag{13}$$

$$\rho|_{u=L_k} = g(t_k). \tag{14}$$

The unknown zero front for the original problem corresponds to the values of the unknown function $\rho(t_k, u)$ with $u = L_0$:

$$\rho|_{u=L_0} = a(t_k). \tag{15}$$

Condition (15) arrive us at

$$q^{(\rho)}\Big|_{u=0} = \frac{\partial \rho}{\partial n}\Big|_{u=0} = \frac{1}{\sigma a'(t_k)}, \tag{16}$$

where $q^{(\rho)}$ is the flow for the function $\rho(t, u)$.

An iterative algorithm for solving the problem (13), (14), (16) by the boundary element method allow us to find at each time step a continuous function $\rho = \rho(t_k, u)$, which is inverse to the solution of the original problem. At the same time, we found the zero front $\rho = a(t_k)$ at each moment $t = t_k$ on a given time interval. For further research, the table function obtained in this way can be approximated by a continuous function $\rho = a(t)$.

5. An invariant heat wave construction in the case of a nonzero boundary regime defined on a moving manifold

Based on the results of the computational experiment described in Section 3, we now construct an approximate solution of problem (2), (5), as a generalized self-similar solution (6). Here we use the approximation of the law of zero front moving $\rho = a(t)$, built in steps in time in Section 4.

Let us illustrate the proposed approach with examples. As a first example, consider problem (2), (5) for

$$b(t) = 0.5(1 + e^t), \quad g(t) = U(t, \rho)|_{\rho=0.5(1+e^t)}, \tag{17}$$

where $U(t, \rho)$ is generalized self-similar solution (6) for $a(t) = e^t$, found as a solution of Cauchy problem (8), (9) by BEM. Stepwise solving of problem (13), (14), (16), (17) on the interval $t \in [0, 1]$ allows us to determine the position of the zero front at time points $t_k = kh$, $k = 0, 1, \ldots, K$, h is time step, $h = 1/K$. The obtained data are approximated by the function $\rho = C_1 e^{C_2 t}$ using the least squares method. As an approximate solution of problem (2), (5), (17) we use a solution of the form (6) for $a(t) = C_1 e^{C_2 t}$ for the found values of the parameters of the approximating function. Table 1 shows the values of the parameters C_1, C_2 and standard deviation ε for different numbers of steps at the first stage of the solving. Figure 3 illustrates comparison of approximate solutions with the reference solution $U(t, \rho)$. With an increase in the number of steps, the approximate solution tends to the reference one.

Table 1. Results of approximation of the zero-front motion law.

	$a(t) = C_1 e^{C_2 t}$			$a(t) = C_3(t + C_4)^{\omega}$			
K	C_1	C_2	ε	C_3	C_4	ω	ε
10	1.0004227	1.005316	0.003468	1.019098	0.9796	0.9871	0.002984
20	1.0003852	1.002483	0.001267	1.006263	0.9934	0.9958	0.001501

As a second example, we consider problem (2), (5) for

$$b(t) = 1 + 0.5t, \quad g(t) = V(t, \rho)|_{\rho=1+0.5t}, \tag{18}$$

where $V(t, \rho)$ is generalized self-similar solution (6) for the power function $a(t) = 1 + t$. The values of $\rho = a(t_k)$ obtained by the step-by-step solution are approximated by the power function $\rho = C_3(t + C_4)^{\omega}$. As an approximate solution of problem (2), (5), (18) we use a solution of the form (6) for the found parameters of the approximating function.

Table 1 shows the values of the parameters C_3, C_4, ω and standard deviation ε for different numbers of steps. Figure 4 illustrates comparison of approximate solutions with the reference solution $V(t, \rho)$.

The calculations show that the generalized self-similar approach is applicable for constructing a heat wave generated by a non-zero boundary regime defined on a moving manifold. The proposed algorithm can be used if the zero-front motion found by BEM is approximated with sufficient accuracy by a power or exponential function.

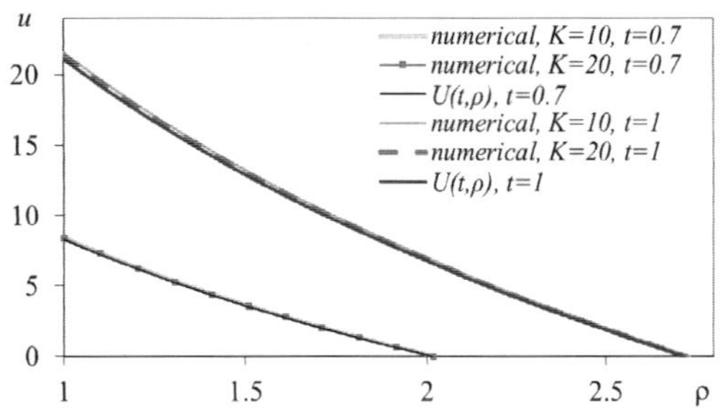

Figure 3. Comparison of approximate solutions of the problem (2), (5), (17) with the reference one.

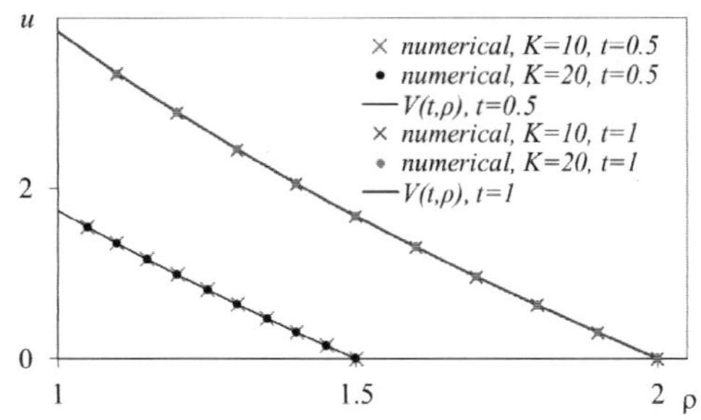

Figure 4. Comparison of approximate solutions of the problem (2), (5), (18) with the reference one.

6. Conclusion

The study shows that the generalized self-similar solutions of the considered equation, having the type of a heat wave, are stable with respect to small perturbations of the law of a heat wave moving. In the case of an arbitrary non-zero boundary condition defined on a moving manifold, this property allows one to construct approximations of heat wave fronts based on generalized self-similar solutions. The performed computational experiment shows a fairly high accuracy of such solutions.

A thought-provoking direction for further research is to obtain qualitative and quantitative estimates of the stability domains of generalized self-similar solutions. These assessments allow us to determine the scope of applicability of the proposed approach.

References

[1] Samarskii A A, Galaktionov V A, Kurdyumov S P and Mikhailov A P 1995 *Blow-up in Quasilinear Parabolic Equations* (Berlin/New York: Walter de Gruyter) p 560
[2] Vazquez J L 2007 *The Porous Medium Equation: Mathematical Theory* (Oxford: Oxford University Press) p 624
[3] Murray J D 2002 *Mathematical Biology: I. An Introduction, Third Edition* Interdisciplinary Applied Mathematics vol 17 (New York: Springer) p 551
[4] Zeldovich Ya B and Raizer Yu P 1966/1967 *Physics of Shock Waves and High-Temperature Hydrodynamic Phenomena* vol 2, ed W D Hayes and R F Probstein (New York: Academic Press) p 478
[5] Barenblatt G I, Entov V M and Ryzhik V M 1990 *Theory of Fluid Flows Through Natural Rocks* (Dordrecht/Boston: Kluwer Academic Publishers) p 396
[6] Sidorov A F 1985 *Dokl. Akad. Nauk SSSR* **280** 47–51 [1985 *Sov. Math. Dokl.* **31** 40–4]
[7] Kudryashov N A 2005 *Computational Mathematics and Mathematical Physics* **45** 1965–72
[8] Antontsev S N and Shmarev S I 2015 *Evolution PDEs with Nonstandard Growth Conditions* Atlantis Studies in Differential Equations vol 4 (Amsterdam: Atlantis Press) p 409
[9] Kazakov A L and Lempert A A 2013 *J. Appl. Mech. Tech. Phys.* **54** 251–58
[10] Kazakov A L and Kuznetsov P A 2018 *J. Appl. Ind. Math.* **12** 255–63
[11] Kazakov A L and Spevak L F 2013 *Applied Mathematical Modelling* **37** 6918–28
[12] Kazakov A L and Spevak L F 2016 *Applied Mathematical Modelling* **40** 1333–43
[13] Spevak L F and Nefedova O A 2016 *AIP Conference Proceedings* **1785** 040077
[14] Spevak L F and Kazakov A L 2016 *AIP Conference Proceedings* **1785** 040076
[15] Polyanin A D and Zaitsev V F 2012 *Handbook of nonlinear partial differential equations, Second Edition* (Boca Raton/London/New York: Chapman & Hall/CRC) p 1912

Recovery of a piecewise constant lower coefficient of an elliptic equation

A E Kolesov[1] , D Kh Ivanov[1], and P N Vabishchevich[1,2]

[1]North-Eastern Federal University, 48 Kulakovskogo str., Yakutsk 677000, Russia
[2]Nuclear Safety Institute of RAS, 52 B. Tulskaya str., Moscow 115191, Russia

E-mail: kolesov.svfu@gmail.com

Аннотация. We propose a new algorithm for the recovery of a piecewise constant lower coefficient of an elliptic problem. The inverse problem is reduced to a shape reconstruction problem. The proposed algorithm is based on the minimization of a cost functional where a control function is the right-hand side of an auxiliary elliptic equation for a level set representation of unknown shape. Numerical implementation is based on the finite element method and the open-source computing platform `FEniCS` and `dolfin-adjoint`. The performance of the algorithm is demonstrated on computationally simulated data.

Keywords: coefficient inverse problem, elliptic equation, finite element method

1. Introduction

Inverse problems of identifying unknown coefficients and the right-hand side of elliptic equations has been studied extensively [1, 2, 3, 4, 5]. Such problems arise when determining the internal structure of objects using measurements made on their boundaries. Coefficient inverse problems are often not well-posed and special algorithms are needed to numerically solve such problems. Also, the stability and uniqueness of their solution are crucial for development of stable numerical algorithms. The main results on these fundamental questions for multi-dimensional elliptic equations can be found in the review paper [6].

In this work, we focus on the recovery of a lower coefficient of two-dimensional elliptic equation. The uniqueness for this problem has been considered in many works [7, 8, 9, 10]. In addition, we assume that the desired coefficient is piecewise-constant and has only one of two possible values, which are known a priori. Therefore, our inverse problem is reduced to a shape reconstruction problem [11].

The general approach of solution of coefficient inverse problems involves minimization of an objective functional with or without some regularization term using iterative iterative algorithms [12, 13]. In [10] the lower coefficient of elliptic equations is recovered by minimizing a discrepancy functional using gradient methods. The identification of lower coefficient in elliptic equation as a shape reconstruction problem was solved using Newton and Landweber algorithms [14]. In recent years, the level set methods were extensively employed to solve shape recovery problems [15, 16]. In [17] the level set approach is applied to identify a piecewise constant conductivity coefficient in electric impedance tomography [17].

In this work, a new algorithm for identification of piecewise-constant lower coefficient of elliptic equation is proposed. The main idea is similar to the level set method and consists in solving an auxiliary elliptic equation to find a representation of unkown shape. The algorithm is based on minimization of a cost functional with a control function taken as the right-hand side of the auxiliary equation. The paper is organised as follows. In the next section, we introduce the inverse problem. Details of reconstruction algorithm and finite element approximation of equations are discussed in section 3. The numerical verification of the proposed algorithm is presented in section 4.

2. Problem statement

Let $\Omega \subset \mathbb{R}^2$ be the bounded domain with sufficiently smooth boundary $\partial\Omega$, and $\boldsymbol{x} = (x_1, x_2) \in \Omega$. We consider the following boundary value problem for second order elliptic equation

$$-\Delta u + c(\boldsymbol{x})u = 0, \quad \boldsymbol{x} \in \Omega, \tag{1}$$

$$u(\boldsymbol{x}) = \phi(\boldsymbol{x}), \quad \boldsymbol{x} \in \partial\Omega. \tag{2}$$

The lower coefficient $c(\boldsymbol{x})$ is defined as

$$c(\boldsymbol{x}) = c_0 + c_1\chi_D(\boldsymbol{x}), \tag{3}$$

where c_0, $c_1 \geq 0$ are given constants and $\chi_D(\boldsymbol{x})$ is the characteristic function of the subdomain $D \subset \Omega$:

$$\chi_D(\boldsymbol{x}) = \begin{cases} 1, & \boldsymbol{x} \in D, \\ 0, & \boldsymbol{x} \in \Omega\backslash D. \end{cases} \tag{4}$$

In this case, the inverse problem of the recovery of the coefficient $c(\boldsymbol{x})$ reduces to the reconstuction of the shape of the subdomain D using only single measurement of the Dirichlet boundary values $\phi(\boldsymbol{x})$ (2) and the additional Neumann boundary values

$$\frac{\partial u}{\partial n} = g(\boldsymbol{x}), \quad \boldsymbol{x} \in \partial\Omega, \tag{5}$$

where $\boldsymbol{n}$ denotes the unit outer normal to $\partial\Omega$. Inverse problems under such conditions were investigated, for example, in [14].

3. Reconstruction algorithm

First, we introduce an auxiliary function $q(\boldsymbol{x})$, which describes D as follows

$$\begin{cases} q(\boldsymbol{x}) \geq 0, & \boldsymbol{x} \in D, \\ q(\boldsymbol{x}) < 0, & \boldsymbol{x} \in \Omega\backslash D. \end{cases} \tag{6}$$

The function $q(\boldsymbol{x})$ is similar to a level set function in the well-known level set methods [16]. To find the function $q(\boldsymbol{x})$ we solve the following elliptic equation

$$\begin{aligned} -\gamma\Delta q + q = f(\boldsymbol{x}), \quad & \boldsymbol{x} \in \Omega, \\ q = 0, \quad & \boldsymbol{x} \in \partial\Omega, \end{aligned} \tag{7}$$

where $\gamma = \text{const} > 0$ is the parameter. When we have the distribution of the function $q(\boldsymbol{x})$ we can define the coefficient $c(\boldsymbol{x})$

$$c(\boldsymbol{x}) = c_0 + c_1 H(q) \tag{8}$$

with the Heaviside function $H(q)$ defined as

$$H(q) = \begin{cases} 1, & q(\boldsymbol{x}) \geq 0, \\ 0, & q(\boldsymbol{x}) < 0. \end{cases} \tag{9}$$

Our reconstruction algorithm is based on the classical approach with the minimization of the cost functional J defined as

$$J(f) = \int_{\partial\Omega} \left(\frac{\partial u}{\partial n}(\boldsymbol{x}, f) - g(\boldsymbol{x}) \right)^2 \mathrm{d}s. \tag{10}$$

Note that the value of functional J is controlled by the right-hand side $f(\boldsymbol{x})$ of equation (7) rather than the desired coefficient $c(\boldsymbol{x})$.

Before delving into steps of reconstruction algorithm, we discuss some details of the numerical implementation. For discretization of our equations the finite element method is used. First, we define the following functional spaces:

$$V = \left\{ v \in H^1(\Omega) : v(\boldsymbol{x}) = 0,\ \boldsymbol{x} \in \partial\Omega \right\}, \quad Q = \left\{ v \in H^1(\Omega) : v(\boldsymbol{x}) = \phi(\boldsymbol{x}),\ \boldsymbol{x} \in \partial\Omega \right\}$$

where $H^1(\Omega)$ is Sobolev space. We obtain a variational problem for (1), (2) by multiplying equation (1) by a test function $v \in V$ and integrating by parts: find $u \in Q$ such that

$$\int_\Omega \nabla u \, \nabla v \, d\boldsymbol{x} + \int_\Omega c \, u \, v \, d\boldsymbol{x} = 0, \quad \forall v \in V. \tag{11}$$

Similarly, we have a variational problem for the auxiliary equation (7): find $q \in V$ such that

$$\int_\Omega \gamma \nabla q \, \nabla v \, d\boldsymbol{x} + \int_\Omega q \, v \, d\boldsymbol{x} = \int_\Omega f \, v \, d\boldsymbol{x}, \quad \forall v \in V. \tag{12}$$

Next, we construct a computational mesh $\Omega_h = \{\omega_1, \omega_2, \ldots, \omega_N\}$ of domain Ω. Here, N is the number of triangular cells, $h = \max_{\omega \in \Omega_h} h_\omega$, where h_ω is the diameter of circle inscribed in a cell ω. Then, in this mesh we define discrete spaces $V_h \subset V$, $Q_h \subset Q$ and restrict the variational problems (11), (12) to these spaces: find $u_h \in Q_h$ and find $q_h \in V_h$ such that

$$\int_{\Omega_h} \nabla u_h \, \nabla v_h \, d\boldsymbol{x} + \int_{\Omega_h} c_h \, u_h \, v_h \, d\boldsymbol{x} = 0, \quad \forall v_h \in V_h. \tag{13}$$

$$\int_\Omega \gamma \nabla q_h \, \nabla v_h \, d\boldsymbol{x} + \int_\Omega q_h \, v_h \, d\boldsymbol{x} = \int_\Omega f_h \, v_h \, d\boldsymbol{x}, \quad \forall v_h \in V_h. \tag{14}$$

In addition, the Heaviside function $H(q)$ in (9) is numerically approximated by the following function $H_\alpha(q)$

$$H_\alpha(q) = \begin{cases} 0, & q < 0, \\ 0.5 - 0.5 \cos\left(\pi \dfrac{q}{\alpha}\right), & 0 \le q < \alpha \\ 1, & q \ge \alpha, \end{cases} \tag{15}$$

where $\alpha > 0$ is smoothing parameter.

Now, we summarize our algorithm for reconstruction of the coefficient $c(\boldsymbol{x})$. The minimization is performed by a gradient based iterative procedure:

1. Define the initial guess for the control function $f(\boldsymbol{x})$;
2. Obtain the function $q(x)$ by solving (14);
3. Modify the coefficient $c(x)$ according to (8);
4. Solve the forward problem (13);
5. Update the control function $f(\boldsymbol{x})$ by a gradient method.
6. Repeat steps 2 – 5 until convergence.

Numerical implementation of the reconstruction is performed in the Python. For solving partial differential equations (13), (14) we use an open-source computing platform FEniCS [18]. The differentiation of the functional J and minimization are implemented using an open source library dolfin-adjoint based on FEniCS [19, 20].

4. Numerical experiments

In this section, to verify the proposed algorithm we will reconstruct the coefficient $c(\boldsymbol{x})$ from computationally simulated data obtained by solving the forward problem (1), (2). We consider a unit square domain Ω. For simplicity, the constants c_0 and c_1 are taken to be equal to 0 and 1, respectively. Also, we use the boundary function $\phi(\boldsymbol{x}) = 1.0$. To test stability of the algorithm we use the function $g(\boldsymbol{x})$ (5) with different levels of noise: 0% noise and 5% noise.

The initial guess for the control function is chosen as follows

$$f_0(\boldsymbol{x}) = \begin{cases} 1.0, & \boldsymbol{x} \in D_0, \\ 0.0, & \boldsymbol{x} \in \Omega \backslash D_0, \end{cases}$$

where D_0 is the initial guess for the subdomain D. It is taken as a circle $D_0 = B_r(\boldsymbol{x}_0)$ with the radius $r = 0.2$ and the center $\boldsymbol{x}_0 = (0.5, 0.5)$. This initial control function $f_0(\boldsymbol{x})$ is shown in Figure 1a. Then, the initial value of the function $q_0(\boldsymbol{x})$ obtained by solving (14) with $\gamma = 0.005$ is depicted in Figure 1b. The corresponding coefficients $c_0(\boldsymbol{x})$ calculated using (8) and the Heaviside functions (15) with $\alpha = 0.1$ and $\alpha = 0.01$ are displayed in Figures 1c and 1d, respectively.

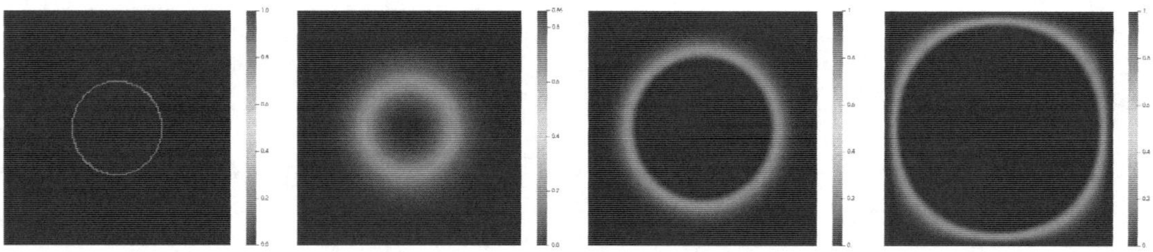

(a) (b) (c) (d)

Рис. 1: The initial guesses: a) the control function $f_0(\boldsymbol{x})$, b) the auxiliary function $q_0(\boldsymbol{x})$, c) the coefficient $c_0(\boldsymbol{x})$ for $\alpha = 0.1$, d) the coefficient $c_0(\boldsymbol{x})$ for $\alpha = 0.01$

Two shapes of the subdomain D are reconstructed: a single rectangle and two circles with different sizes. The rectangle is 0.2×0.4 and is located in the center of domain Ω. The radii of circles are equal to 0.1 and 0.15. The center of the smaller circle is located at $(0.3, 0.3)$ and the center of the other one is $(0.7, 0.7)$.

First, we present reconstruction from noiseless data. The reconstructed coefficient $c(\boldsymbol{x})$ for the rectangle after 10, 15, 20, and 34 iterations of minimization process are depicted in Figure 2. The true shape is shown in black line. These images are obtained using $\gamma = 0.005$ and $\alpha = 0.01$. The minimizion of the functional J is performed using the limited-memory Broyden-Fletcher-Goldfard-Shanno (L-BFGS) algorithm implemented in SciPy [21, 22]. This algorithm showed the fastest convergence rate compared to other techniques. Note that Figure 2d is the final image after the minimization algorithm converged in 34 iterations.

The reconstruction of the two circles after 15, 30, 60, and 121 iterations are presented in Figure 3. The results are obtained using the same coefficient $\gamma = 0.005$. However, we used the larger smoothing parameter $\alpha = 0.1$. We have observed that for more complicated shapes of subdomain D with several objects the larger parameter α leads to much better reconstruction results. We note that the influence of coefficient γ on the performance of our algorithm is subject to more thorough investigation.

Figure 4 shows the reconstruction results for both rectangle and circles in the case of boundary data $g(\boldsymbol{x})$ with 5% noise. We can see that the true shapes are recovered reasonably well. In the first case, L-BFGS algorithm converged in 55 iterations, for the case of circles it congerved in 183 iterations.

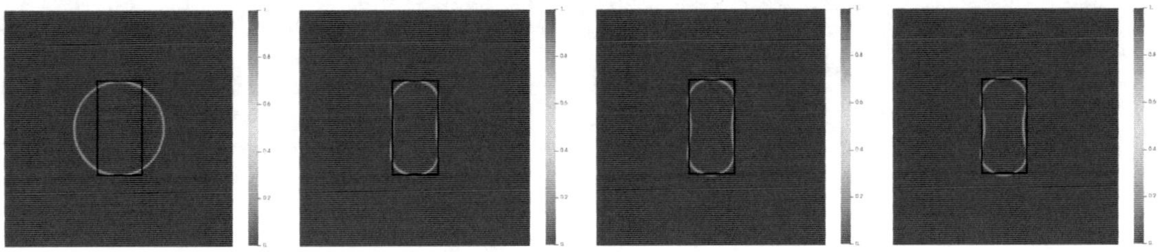

(a) (b) (c) (d)

Рис. 2: Reconstruction results for rectangle after: a) 10, b) 15, c) 20 and d) 34 iterations

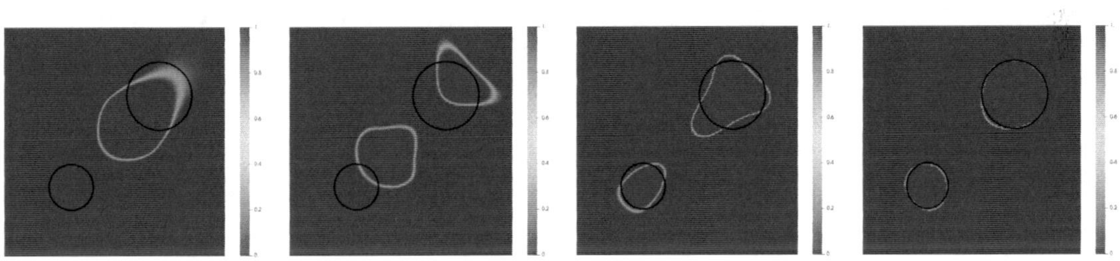

(a) (b) (c) (d)

Рис. 3: Reconstruction results for circles after: a) 15, b) 30, c) 60 and d) 121 iterations

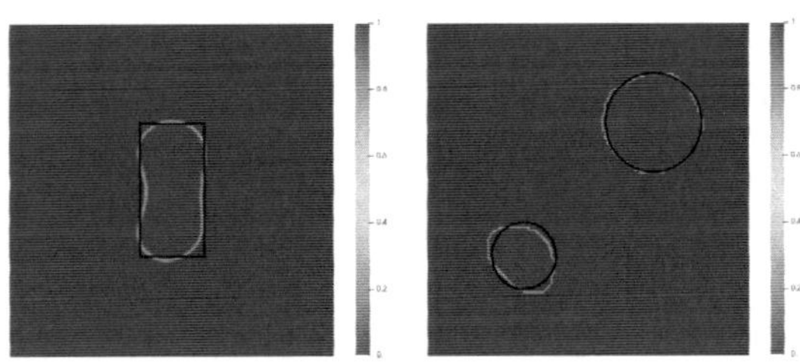

(a) (b)

Рис. 4: Reconstruction results for noisy data: a) rectangle, b) circles

Acknowledgments
The work was supported by Mega grant of Russian Government N14.Y26.31.0013 and RFBR grant N17-01-00689.

References

[1] R. V. Kohn and M. Vogelius. Determining conductivity by boundary measurements ii. interior results. *Comm. Pure Appl. Math*, 38(5):643–667, 1985.

[2] B. D. Lowe and W. Rundell. The determination of a coefficient in an elliptic equation from average flux data. *Journal of Computational and Applied Mathematics*, 70:173–187, 1996.

[3] J. Sylvester and G. Uhlmann. A global uniquness theorem for an inverse boundary value problem. *Ann. Math*, 125(1):153–169, 1987.

[4] M. V. Klibanov, A. E. Kolesov, A. Sullivan, and L. Nguyen. A new version of the convexification method for a 1D coefficient inverse problem with experimental data. *Inverse Problems*, 34(11):115014, 2018.

[5] M. V. Klibanov, A. E. Kolesov, and D.-L. Nguyen. Convexification Method for an Inverse Scattering Problem and Its Performance for Experimental Backscatter Data for Buried Targets. *SIAM Journal on Imaging Sciences*, 12(1):576–603, 2019.

[6] V. Isakov. Uniqueness and stability in multi-dimensional inverse problems. *Inverse problems*, 9:579–621, 1993.

[7] J. R. Cannon and W. Rundell. An inverse problem for an elliptic partial differential equation. *J. Math.Anal. Appl*, 126:329–340, 1987.

[8] L. F. Caudill. A convergent reconstruction method for an elliptic operator in potential form. *Journal of mathematical analysis and applications*, 195(1):44–70, 1995.

[9] A. Nachman, J. Sylvester, and G. Uhlmann. An n-dimensional borg-levinson theorem. *Commun. Math. Phys.*, 115(4):595–605, 1988.

[10] P. N. Vabishchevich. Numerical solution of the identification problem for the lower coefficient of an elliptic equation. *Differential Equations*, 38(7):1000–1006, 2002.

[11] O. Dorn, E. . Miller, and C. M. Rappaport. A shape reconstruction method for electromagnetic tomography using adjoint fields and level sets. *Inverse Problems*, 16(5):1119–1156, 2000.

[12] A. A. Samarskii and P. N. Vabishchevich. *Numerical methods for solving inverse problems of mathematical physics*. De Gruyter, 2008.

[13] S. I. Kabanikhin. *Inverse and ill-posed problems: theory and applications*. De Gruyter, 2011.

[14] F. Hettlich and W. Rundell. Recovery of the support of a source term in an elliptic equation. *Inverse Problems*, 13:959–976, 1997.

[15] F. Santosa. A Level-set Approach for Inverse Problems Involving Obstacles. *Esaim : Control, Optimisation and Calculus of Variations*, 1:17–33, 1996.

[16] O. Dorn and D. Lesselier. Level set methods for inverse scattering. *Inverse Problems*, 22(4):R67–R131, 2006.

[17] E. T. Chung, T. F. Chan, and X. C. Tai. Electrical impedance tomography using level set representation and total variational regularization. *Journal of Computational Physics*, 205(1):357–372, 2005.

[18] M. S. Alnæs, J. Blechta, J. Hake, A. Johansson, B. Kehlet, A. Logg, C. Richardson, J. Ring, M. E. Rognes, and G. N. Wells. The fenics project version 1.5. *Archive of Numerical Software*, 3(100), 2015.

[19] P. E. Farrell, D. A. Ham, S. F. Funke, and M. E. Rognes. Automated derivation of the adjoint of high-level transient finite element programs. *SIAM Journal on Scientific Computing*, 35(4):369–393, 2012.

[20] S. F. Funke and P. E. Farrell. A framework for automated PDE-constrained optimisation. 2013.

[21] E Jones, T Oliphant, P Peterson, and et al. SciPy: Open source scientific tools for Python, 2001.

[22] R. H. Byrd, P. Lu, J. Nocedal, and C. Zhu. A Limited Memory Algorithm for Bound Constrained Optimization. *SIAM Journal on Scientific Computing*, 16(5):1190–1208, 2005.

Journal of Physics: Conference Series **1392** (2019) 012085 doi:10.1088/1742-6596/1392/1/012085

Identification of parameters of adaptive time series models

V. D. Polezhaev[1], K. O. Yusupova[2]
[1]*Plekhanov Russian University of Economics*, Stremyanny per, 36, 117997, Moscow, Russia, e-mail: vpolej@gmail.com;
[2]*Bauman Moscow State Technical University*, ul. Baumanskaya 2-ya, 5/1, 105005, Moscow, Russia, e-mail: uko2304@mail.ru

Abstract. The work is devoted to the study of adaptive models of time series and the development of methods for their construction. The most problematic stage in the implementation of algorithms for time series forecasting methods is the identification of unknown parameters, on which the adequacy of the prediction depends. On the basis of economic data, a number of adaptive models have been studied: using exponential smoothing, Brown, Holt-Winters and Theil-Wage models, which take into account seasonality. It was found that the main difficulties arise in the selection of constant smoothing and the choice of coefficients, since no universal algorithm for their task exists.
The paper proposes a method for integrating the adaptive method and the method of numerical optimization. As an optimization method, the non-gradient method was chosen - the Particle Swarm Optimization (PSO), the use of which provided less error in the final forecast and convergence in all applications. Algorithms for finding the parameters to be identified, and a predictive model based on real data are presented. It is shown that in some cases the proposed model has an advantage over other models of time series.
Keyword: forecasting, adaptive model, time series, particle swarm optimization, identification, exponential smoothing.

1. Introduction

Prediction of time series values based on its previous values is the basis for planning, managing and optimizing production volumes, etc. The task of forecasting a time series is solved on the basis of creating a forecasting model that adequately describes the process under study. To date, there are many models for analyzing and forecasting time series. The most common because of their simplicity and clarity is the additive model of exponential smoothing [1]. A characteristic feature of adaptive forecasting methods is their ability to continuously take into account the evolution of the dynamic characteristics of the studied processes. Very often there is a need to predict indicators subject to seasonal variations [2]. In these cases, Holt-Winters and Theil-Wage models are also used, which include both a trend and a seasonal component.

Objective: Analysis of the main approaches and the development of methods for improving the methodological tools for short-term forecasting of the series, identifying the possibilities of using adaptive forecasting methods using the methods of identifying model parameters to compile the most accurate forecast.

Modification of a method of exponential smoothing for seasonal ranks are the Holt-Winters and Theil-Wage methods. As a model of a series, its representation is used as a combination of a linear trend with a seasonal component superimposed either multiplicatively (Holt-Winters model) or additively (Theil-Wage model. It is assumed that the trend coefficients and the seasonal component may vary slowly over time. In accordance with this the computational process is arranged as an adaptive procedure, controlled by three adaptation parameters (one parameter is level adaptation, the second is the angle of inclination, and the third is seasonality coefficients). In the course of the calculations, a smoothed series is constructed, which at each moment of time t is the forecast according to the data up to (t -1) inclusive.

One of the problems associated with these models is that to start the calculations using the model, you need to somehow set the initial values of the coefficients, but there is no universal principle for specifying these coefficients, and their initial value naturally affects predictive properties of the model. That is, when building a model, it is necessary to select not only the optimal constant smoothing, but also to choose the principle of specifying the model coefficients, thanks to which one could get the most accurate forecast.

2. Theil-Wage model

The Theil-Wage model (H.Theil, S.Wage) is Holt's sophisticated model that takes into account seasonality and an additive trend, unlike the Holt-Winters model, it additively includes a linear trend, which is justified in solving some problems.

In economic practice, exponential trends with multiplicatively imposed seasonality are more common. The additive model, which has an independent value in economic research, is also interesting because it allows you to build a model with multiplicative seasonality and an exponential trend. To do this, it is necessary to replace the values of the original time series by their logarithms, which transforms the exponential trend into a linear and simultaneously multiplicative seasonal model into an additive one. Therefore, before using the additive model the members of the analyzed time series are usually replaced by logarithms, transforming the exponential trend into linear, and the multiplicative seasonality into additive. The advantage of the additive model is the relative simplicity of its computational implementation.

The Theil-Wage model is described by an expression that takes into account linear trend and additive seasonal effects: $y_\tau(t) = a_{1,t} + g_t + u_t$, where row level without seasonal fluctuations is $a_{1,t} = a_{1,t-1} + a_{2,t}$, $a_{2,t}$ – coefficient of additive growth, g_t – seasonality coefficient, u_t – white noise. To fulfill the forecast for τ steps forward, it is necessary to determine the estimate of a number of coefficients and substitute it in: $\hat{y}_\tau(t) = \hat{a}_{1,t} + \hat{a}_{2,t}\tau + \hat{g}_{t-l+\tau}$. The coefficients are estimated using the following formulas:

$$\hat{a}_{1,t} = \alpha_1(y_t - \hat{g}_{t-l}) + (1-\alpha_1)(\hat{a}_{1,t-1} + \hat{a}_{2,t-1});$$
$$\hat{g}_t = \alpha_2(y_t - \hat{a}_{1,t}) + (1-\alpha_2)\hat{g}_{t-l};$$
$$\hat{a}_{2,t} = \alpha_3(\hat{a}_{1,t} - \hat{a}_{1,t-1}) + (1-\alpha_3)\hat{a}_{2,t-1},$$

where $0 < \alpha_1, \alpha_2, \alpha_3 < 1$ – three parameters which are subject to definition, l – number of steps in one seasonal cycle (for monthly observations of l = 12, for quarter l = 4), $\hat{a}_{1,t}, \hat{a}_{2,t}$ – characteristics of development of the studied process, $\hat{g}_1, \hat{g}_{t-1}, \ldots, \hat{g}_{t-l+1}$ – parameters of additive seasonality.

In order to determine the optimal values of the adaptation parameters model assume searching through various sets of their values and comparing the mean-square errors of the predictions resulting from this.

3. Creation of the forecast by Theil-Wage method

The data on income from radio communications, radio broadcasting, television and satellite communications in the Moscow region were obtained from the Unified Interdepartmental Statistical Information System (UniSIS). It is necessary to make a quarterly forecast for 2018 according to data for the 28 quarters of 2011-2017. Actual data for 2018 are known; an estimate of the prediction error was made using their values.

During the preliminary processing of the model range of 28 points (Fig. 1), it was noted that the time series under study is characterized by the presence of a linear trend and an additive component of seasonality. Therefore, for the compilation of the forecast, the Theil-Wage method was chosen [3]. This model contains three unknown parameters, the selection of which used the Particle Swarm Optimization.

Using the standard tools, we find the equation of the linear trend of the original series (Fig. 1):

$$\hat{y}(t) = 110345.553t + 1057876.3246,$$

where t – month number, $\hat{y}(t)$ – budget volume, thousand rubles (fig. 1). The coefficient of determination this case $R^2 = 0.432$.

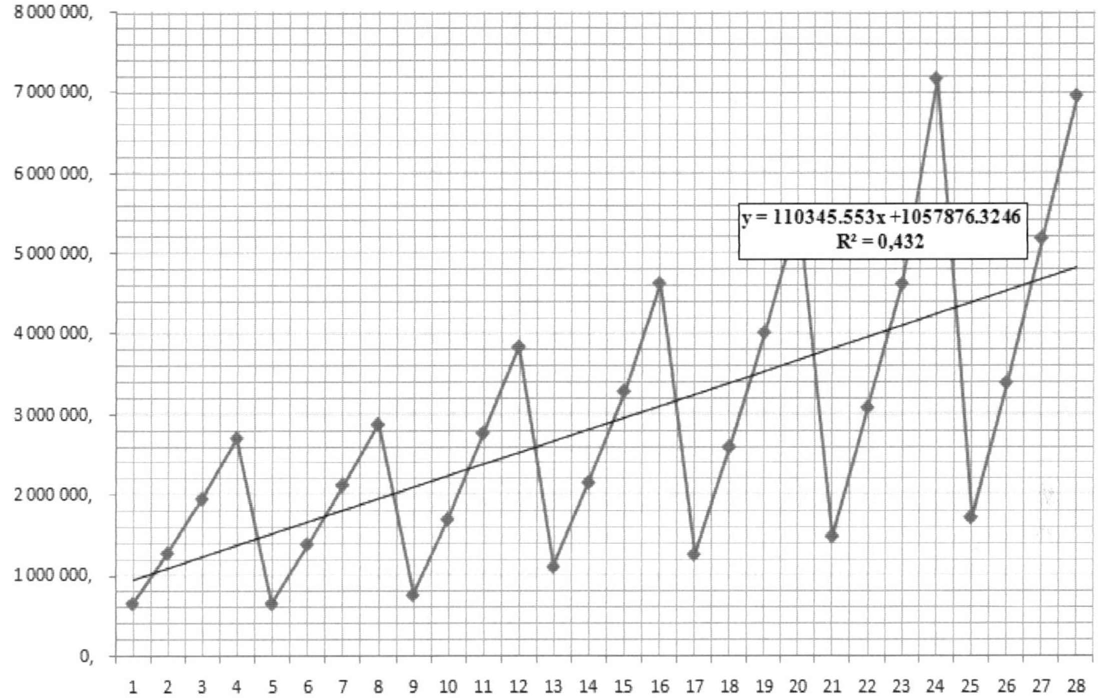

Figure 1. Display of basic data and linear trend

Let us show a detailed description of the calculations using the Theil-Wage method formulas. For empirical reasons we accept model parameters equal $\alpha_1 = 0.4, \alpha_2 = 0.8, \alpha_3 = 0.8$, forecast period τ = 1. Let us show the compilation of the forecast for the next quarter 2018 following the processed data. The initial value of the seasonality coefficients is

$$\hat{g}_t^0 = \frac{1}{2}[(y_t - \hat{y}(t)) + (y_{t+l} - \hat{y}(t+l))].$$

Model values will then be obtained by the following formulas, $t = 1$:

$$\hat{y}_1^* = \hat{a}_{1,0} + \hat{a}_{2,0} + \hat{g}_1^0; \quad \hat{a}_{1,1} = \alpha_1(y_t - \hat{g}_1^0) + (1-\alpha_1)(\hat{a}_{1,0} + \hat{a}_{2,0});$$

$$\hat{g}_1 = \alpha_2(y_t - \hat{a}_{1,1}) + (1-\alpha_2)\hat{g}_1^0; \quad \hat{a}_{2,1} = \alpha_3(\hat{a}_{1,1} - \hat{a}_{1,0}) + (1-\alpha_3)\hat{a}_{2,0},$$

As $\hat{a}_{1,0}$ and $\hat{a}_{2,0}$ we take the coefficients of the linear regression equation without taking into account the seasonality: $\hat{a}_{1,0} = 105787.3246, \hat{a}_{2,0} = 110345.553$.

For t =2 model values are found by the formulas:

$$\hat{y}_2^* = \hat{a}_{1,1} + \hat{a}_{2,1} + \hat{g}_2^0; \quad \hat{a}_{1,2} = \alpha_1(y_t - \hat{g}_2^0) + (1-\alpha_1)(\hat{a}_{1,1} + \hat{a}_{2,1});$$

$$\hat{g}_2 = \alpha_2(y_t - \hat{a}_{1,2}) + (1-\alpha_2)\hat{g}_2^0; \quad \hat{a}_{2,2} = \alpha_3(\hat{a}_{1,2} - \hat{a}_{1,1}) + (1-\alpha_3)\hat{a}_{2,1}.$$

We carry out similar calculations successively up to t = 29. Thus, we obtain a forecast for one period ahead. For the forecast for 2-4 periods we assume τ equal to 2, 3 and 4 respectively. For the case under consideration, the relative error of the Theil-Wage model is determined by the formula $D_{отн} = \frac{|y_t - \hat{y}_t^*|}{y_t} \cdot 100\%$, it is equal to 3.3%.

If we take a different set of parameters of the Theil-Wage model, then the result of the forecast, even with a small change in these parameters, can greatly change. Thus, a change in the two parameters of the model $\alpha_1 = 0.2, \alpha_2 = 0.7, \alpha_3 = 0.8$ entailed increase in an error of the forecast up to 24.4%.

We introduce the automatic determination of the optimal set $\alpha_1, \alpha_2, \alpha_3$. into the forecast generation algorithm. To do this, we require that the parameters be defined so that the model's error at all points is minimal over the entire time interval. The prediction algorithm code was implemented using the Matlab application package.

4. Particle Swarm Optimization for identification of model parameters

The Particle Swarm Optimization is a method of numerical optimization, which we will consider in the context of the problems of identifying unknown parameters of the objective function $f : \mathbb{R}^n \to \mathbb{R}$. The essence of the method lies in the search for optimal solutions at points that are called particles: the coordinates of particles of the swarm are those parameters, the selection of which will solve the problem [4]. The selection is carried out by the movement of these particles according to the principle of the best positions found in space, which is constantly changing when the particles find more advantageous positions. PSO is a non-gradient method: in its implementation algorithm, the exact gradient of the function being optimized is not required.

PSO has a particle search area, which is a n-dimensional space. At the beginning of the algorithm (at the first iteration), the particles are scattered randomly throughout the search area, and each particle has its own random initial velocity vector. At each point the particle has visited, the value of the objective function is calculated. Moreover, each particle remembers what and where the optimal value of the objective function it found, and also each particle knows where the point is located, which is the best among all the points that the particles memorized. At each iteration, the particles adjust the module and the direction of their speed in order, on the one hand, to be closer to the best point that the particle found itself, and, at the same time, move closer to the point that is globally better at the moment [5]. After a certain number of iterations, the particles should gather near the most suitable point.

The most interesting thing in the algorithm is the speed correction, the convergence of the algorithm depends on this step. The work uses a modification of the algorithm, which consists in adding a weight coefficient ω before the current speed, due to which the speed changes more smoothly:

$$v_{i,t+1} = \omega v_{i,t} + \varphi_p r_p (p_i - x_{i,t}) + \varphi_g r_g (g_i - x_{i,t}).$$

Here $v_{i,t}$ – speed component at t iteration of an algorithm, $x_{i,t}$ – i particle coordinate at t iteration of an algorithm, p_i – i coordinate of the best solution found by particle, g_i – i coordinate of the best solution found by all particles, r_p, r_g – random numbers in the range of (0, 1), φ_p, φ_g – weighting coefficients which should be selected under a specific objective.

Let's explain the meaning of these coefficients. ω – the inertial coefficient (determining the tendency of a particle to maintain direction), φ_p – the cognitive coefficient (characterizing the tendency of a particle to its own best position), φ_g – the social coefficient (characterizing the tendency of a particle to the best position in the entire swarm). Each of these coefficients should be selected from certain considerations based on the logic of the method.

The correction of the current coordinates of each particle is carried out according to the formula

$$x_{i,t+1} = x_{i,t} + v_{i,t+1}.$$

After that, the value of the objective function at each new point is calculated, each particle checks whether the new coordinate has become the best among all the points it visited. Then, among all the new points, we check whether we have found a globally better point, and, if we have found, we remember its coordinates and the value of the objective function in it.

When implementing the task, the following parameter values were used:

$$\omega = (0.4-0.9)\frac{k}{S}+0.9, \quad \varphi_p = (0.5-1.5)\frac{k}{S}+1.5, \quad \varphi_g = (1.5-0.5)\frac{k}{S}+0.5,$$

where 0.9 and 0.4 are the largest and smallest values of the parameter, respectively, S is the number of particles in the swarm. At the same time, at the beginning of work, the process tends to a larger value of the parameter, and over time it begins to study the neighborhoods of permissible values ω. Such choice of components allows you to first look for the best position of each particle, and with time the search shifts to determine the best among neighboring particles.

We select the criterion by which the coefficients of the Theil-Wage model will be determined. That is, we make a function, That is, we compose a function, the minimum of which we need to find to determine the parameters α_1, α_2. We make the assumption in the problem that the parameter α_3 is known. Let such a function be the average approximation error (also when selecting parameters, the condition of minimum forecast error will be required separately):

$$A = \frac{1}{n}\sqrt{\frac{\sum_{i=1}^{n} | y_t - \hat{y}_t^* |}{y_t}}.$$

So, the parameter α_3 was taken equal to 0.8. The other two parameters were determined programmatically by Particle Swarm Optimization: $\alpha_1 = 0.2367, \alpha_2 = 0.9991$. In Figure 2, a graph of the process model is plotted against the initial time series.

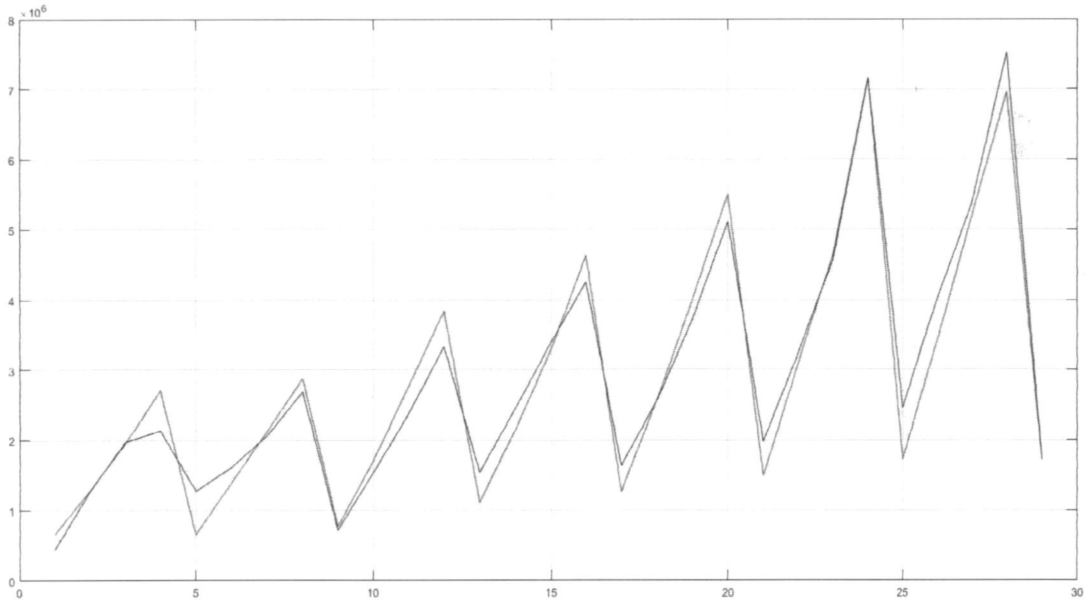

Figure 2. The red graph is the original time series, the blue graph is the forecast time series with a set of parameters $\alpha_1 = 0.2367, \alpha_2 = 0.9991, \alpha_3 = 0.8$.

Using Particle Swarm Optimization an optimal set of parameters was selected. Let's check the quality of the constructed model: the constructed model is considered good if the value of the average approximation error value $A=\frac{1}{n}\sqrt{\frac{\sum_{i=1}^{n}|y_t-\hat{y}_t^*|}{y_t}}\cdot 100\%$ is less than 7%. For a set of parameters obtained using PSO, an indicator $A=0.09\%$ that indicates a satisfactory degree of adequacy of the proposed approach.

Table 1 provides for comparison estimates of the forecast error for this and other sets of values for the Theil-Wage model parameters considered in the paper.

Table 1. Forecast error estimates

№	Set of parameters	A	Forecast error
1	$\alpha_1=0.2367, \alpha_2=0.9991, \alpha_3=0.8$	0.09%	0.25%
2	$\alpha_1=0.2, \alpha_2=0.7, \alpha_3=0.8$	1.15%	24.4%
3	$\alpha_1=0.4, \alpha_2=0.8, \alpha_3=0.8$	1.17%	3.2%

5. Conclusion

In this paper, we considered methods for constructing a forecast in the case of source data in the form of time series. As it turned out, most of the methods in this category have one common problem - unknown parameters. The paper describes in detail the essence of the problem and its solution using the example of economic data. It was proposed to search for the parameters using the integration method of two methods: Theil-Wage (chosen for the specifics of the data) and Particle Swarm Optimization. With this approach, the simulation results gave a set of model parameters in which the average approximation error and the forecast error were minimal. Thus, as a result of solving the problem of short-term forecasting of the process with seasonal dynamics, the following results were obtained:

- identified the possibility of using adaptive forecasting methods using the methods of identifying model parameters to compile the most accurate forecast;
- a method of integrating the adaptive method and numerical optimization of the search for the parameters of the model of the exponential smoothing by the gradient-free method of a particle swarm;
- it is shown the possibility to achieve greater accuracy of the forecast, using the search adjusting procedures of the PSO;
- the code of the forecasting algorithm was developed and implemented using the Matlab application software package;
- it is noted that in some cases, the proposed model has an advantage over other time series models.

References

[1] Lukashin Yu.P. Adaptive methods of short-term forecasting of time series / Yu.P. Lukashin. - Moscow: Finance and statistics, 2003. 413 pages.A reference

[2] Polezhaev V. D., Yusupova K.O. Forecasting of rating of television channels on the basis of these media researches//Basic researches. – 2019. – No. 5.

[3] Svetunkov I.S. Methods of social and economic forecasting in 2 t. T. 2 models and methods: the textbook and a practical work for the academic bachelor degree / I.S. Svetunkov, S.G. Svetunkov. Moscow: Yurayt publishing house, 2018. 447 pages.

[4] Kennedy J. and Eberhart R. Particle Swarm Optimization. In Proceedings of the Fourth IEEE International Conference on Neural Networks, Perth, Australia. IEEE Service Center(1995) 1942-1948.

[5] Use of information and communication technologies for teaching physics at the technical university / Polezhaev V.D., Polezhaeva L.N., Kamenev V.V.//AIP Conference Proceedings "Information Technologies in Education of the XXI Century, ITE-XXI 2015: Proceedings of the International Scientific-Practical Conference "Information Technologies in Education of the XXI Century" 2017. Page 030013. DOI: 10.1063/1.4972452

Management of thermal process for polyethylene gas pipes welding with built-in heater

N P Starostin, M A Vasileva, O A Ammosova

Institute of Oil and Gas problems of the Siberian Branch of the Russian Academy of Science, Avtodorojnaya St. 20, Yakutsk, 677000, Russia

Abstract. Based on the thermal process control, rapid welding technology for polyethylene pipes with built-in heater in the open air at temperatures below standard is proposed. Technological parameters of welding are determined by calculation from the condition of temperature field dynamics in heat-affected zone according to regularities, that are peculiar to welding at permissible temperatures. The results of determining the welding parameters during electrofusion welding and welding of the saddle branch to the polyethylene pipe of the gas pipeline, as well as the results of tests of the joints obtained by the proposed welding technology are presented.

1. Introduction

According to regulatory documents, welding of polyethylene gas pipes using fittings (couplings, saddles, tees, etc.) is recommended to be performed at ambient air temperature from minus 15 ° C to plus 45 ° C. With a wider temperature range, welding is recommended to be performed in rooms (shelters) that ensure compliance with the specified temperature range. Welding in shelters is associated with high energy costs and long preparatory work, which is unacceptable in emergency situations. Even a rather short interruption of gas supply in winter conditions in regions of cold climate can lead to disastrous consequences due to an accident of heating systems. A pressing problem is the development of technology for the rapid welding of polyethylene pipes at low temperatures without using shelters.

In this paper, we consider thermal processes of coupling welding of polyethylene pipes and welding of saddle branches to polyethylene pipes at air temperatures below standard. In the considered methods of welding of polymeric pipes, reflowing is performed by a heating element in the form of a metal spiral integrated into the connecting piece. At ambient air temperatures below the normative, it is proposed to control heat welding process by preheating and using thermal insulation at the cooling stage. Preheating is performed by heating element when voltage is applied that prevents the material from melting. Temperature equalization to temperature acceptable for welding is performed with the power supply turned off. Melting is performed by welding machine automatically in normal mode, simulating the average design temperature in the heat-affected zone on the ambient air temperature sensor. Preheating, temperature equalization, reflow and cooling of the welded joint are performed under a layer of thermal insulation, thickness of which for a certain air temperature is preliminarily calculated from the condition of the joint cooling according to the pattern characteristic of welding under the conditions of acceptable air temperature.

Control of thermal process of welding at low temperatures is reduced to provision of technological parameters (voltage for heating, duration of heating, duration of temperature equalization, thickness of the insulation layer) determined by calculation. Determination of welding parameters will be illustrated with examples.

2. Formulation of the problem

It is obvious that effectiveness of heat welding process controlling depends on adequacy of mathematical model to the real process. In polymeric materials, phase transformations occur in temperature ranges [1]. The temperature distribution during welding of polyethylene pipes is described by equation of thermal conductivity considering two-phase zone during melting and crystallization of polymer material [2-4]:

$$\left(c(T)+\rho^{-}L^{100\%}\frac{dX_C}{dT}\right)\frac{\partial T}{\partial t}=\frac{1}{r}\frac{\partial}{\partial r}\left(r\lambda(T)\frac{\partial T}{\partial r}\right)+\frac{\partial}{\partial z}\left(\lambda(T)\frac{\partial T}{\partial z}\right)+$$
$$+\frac{Q(T)}{n\cdot 2\pi r}\delta(r-R_c)\sum_{i=1}^{n}\delta(z-z_i),\ r,z\in\Omega,\ 0<t\le t_m, \tag{1}$$

where Q is power of the heat source; $L^{100\%}$ is specific heat of phase transformation of completely crystalline polymer; X_C – degree of crystallinity;

$$c(T)=\rho^{+}c^{+}+X_C(T)\cdot\left(\rho^{-}c^{-}-\rho^{+}c^{+}\right),\quad \lambda(T)=\lambda^{+}+X_C(T)\cdot\left(\lambda^{-}-\lambda^{+}\right),$$

$c^{-},\rho^{-},\lambda^{-}$ and $c^{+},\rho^{+},\lambda^{+}$ – specific heat capacity, density and thermal conductivity for solid and liquid phases of pipe material, respectively; n – amount of concentrated heat sources; r_i, z_i – heat source location; δ - Dirac delta function.

Degree of crystallinity X_C is determined by formula:

$$X_C(T)=\begin{cases} X_C^{\infty}, & T\le T_1,\\ X_C^{\infty}-\dfrac{\int_{T_1}^{T}q(u)du}{L^{100\%}v_T}, & T_1<T<T_2,\\ 0, & T\ge T_2.\end{cases} \tag{2}$$

Here $q(u)$ – dependence of heat flux on temperature, referred to unit mass of a substance, recorded by a differential scanning calorimeter (DSC); X_C^{∞} – maximum degree of crystallinity of the studied polymeric material; v_T - temperature change rate.

After differentiating integral in function (2) over variable upper limit, we have:

$$\chi(T)=L^{100\%}\frac{dX_C}{dT}=\begin{cases} 0, & T\le T_1,\\ \dfrac{-q(T)}{v_T}, & T_1<T<T_2,\\ 0, & T\ge T_2.\end{cases} \tag{3}$$

We obtain following equation for temperature in entire computational domain Ω:

$$\left(c(T)-\rho^{-}\chi(T)\right)\frac{\partial T}{\partial t}=\frac{1}{r}\frac{\partial}{\partial r}\left(r\lambda(T)\frac{\partial T}{\partial r}\right)+\frac{\partial}{\partial z}\left(\lambda(T)\frac{\partial T}{\partial z}\right)+$$
$$+\frac{Q(T)}{n\cdot 2\pi r}\delta(r-R_c)\sum_{i=1}^{n}\delta(z-z_i),\ (r,z)\in\Omega,\ 0<t\le t_m, \tag{4}$$

Equation (4) is supplemented by initial

$$T(x,0) = T_0,\ (r, z) \in \Omega, \tag{5}$$

and boundary conditions, where T_0 is ambient temperature. Power of heat source is calculated by formula:

$$Q(t) = \frac{U^2}{R \cdot (1 + \beta(T - 20))}, \tag{6}$$

where R is resistance of electric heating coil at temperature 20 °C; β – temperature coefficient of resistance.

Convective heat exchange α at the contact of the pipe and the coupling (saddle branch) with thermal insulation is determined from the expression [5]:

$$\frac{1}{\alpha} = \frac{1}{\alpha_0} + \frac{h_{iz}}{\lambda_{iz}}, \tag{7}$$

where h_{iz} is the thickness of the thermal insulation layer, λ_{iz} is the thermal conductivity of the insulation material, α_0 is the heat exchange with the environment.

The computational domain Ω triangulation was performed using Gmsh mesh generator. The task (4- (8) was solved by finite element method using the free-access program Dolfin/FEniCS. The visualization of the results obtained was implemented using Paraview package.

3. Calculation of electrofusion welding parameters

In axisymmetric formulation, problem of determining a non-stationary temperature field during coupling welding of polyethylene pipes PE100 SDR 11 63×5.8 at an ambient air temperature of -47 ºC was considered. Figure 1 displays the design scheme of a half of pipe with a coupling. Figure 2 shows temperature distribution after preheating by heating element for 540 seconds until the temperature near the spiral reaches 50 °C and equalizing the temperatures for 2 minutes at ambient temperature minus 47 ° C. The voltage for preheating was 9 V, thickness of thermal insulation layer was 2 cm. Calculations show that the average temperature is about 15 ° C in the field of welding. Heating (reflow) can be made according to the regulated mode.

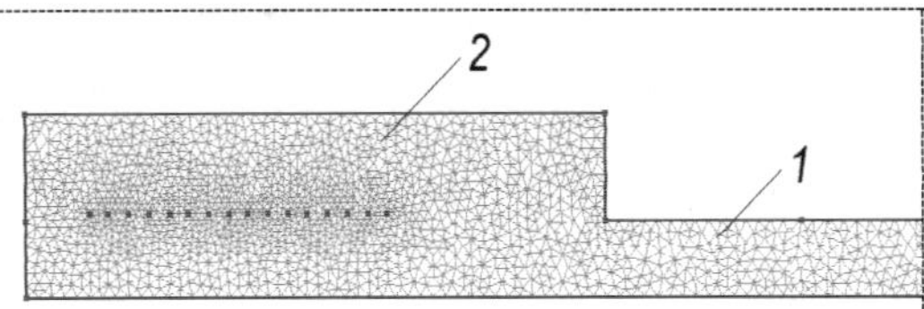

Figure 1. Design scheme: 1 – pipe; 2 – coupling

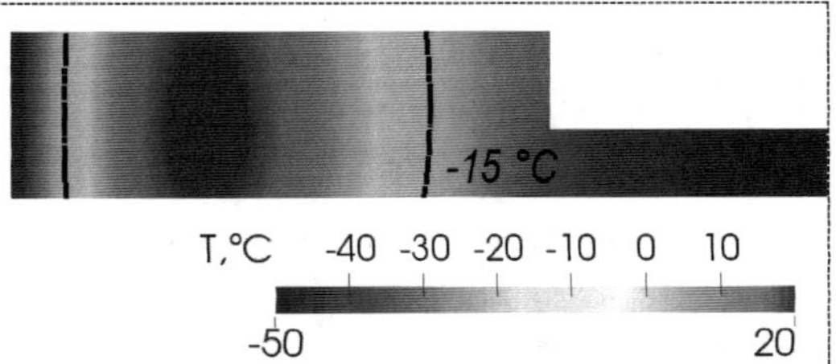

Figure 2. Temperature distribution in cross section of coupling and pipe with heat insulation at ambient air temperature of minus 47 ° C after preheating for 540 s and equalizing for 120 s

Figure 3 shows a comparison of the dimensions of the fused zone in the vertical section when welding at different ambient temperatures at the cooling time for 1 minute after completion of heating. According to calculations the size of the fused zone decreases at low temperatures below minus 15 °C that indicate a smaller volume of the resulting melt. Consequently it will result in insufficient pressure of welded surfaces, more intensive crystallization and formation of a fine structure of the weld material, etc., which lead to a low strength of the joint. Figure 2 (b) shows the cooling time for 1 minute at ambient temperature of minus 47 °C with preheating and thermal insulation and temperature of 5 °C (standard welding). The temperature distributions in the heat-affected zone in both cases are close after the same cooling times that demonstrates formation of identical structures causing the same strength of the joints.

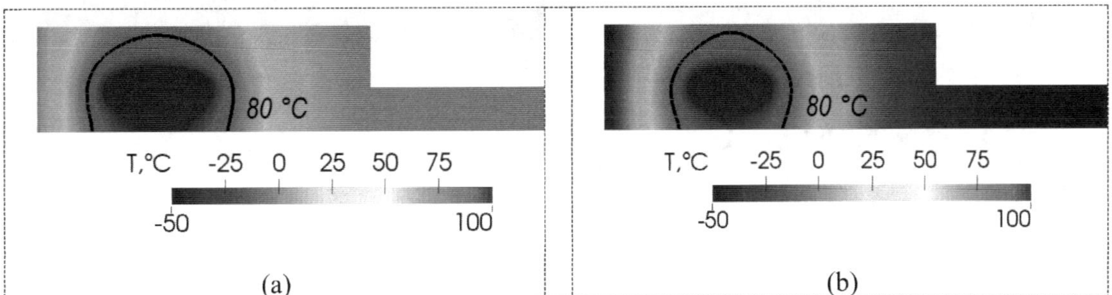

Figure 3. Heat-affected zones in cross section of coupling and pipe during cooling for 1 minute after heating at the ambient temperatures 5 °C (a) and minus 47 ° C (b) with preheating and equalizing under thermal insulation

Figure 4 shows the results of tests on the exfoliation of welded joints obtained at low temperatures. According to the maximum breaking load, the quality of the joints is higher than the quality of the joint obtained at the permissible temperature. The highest values are obtained for the compound obtained for preheating to 50 °C.

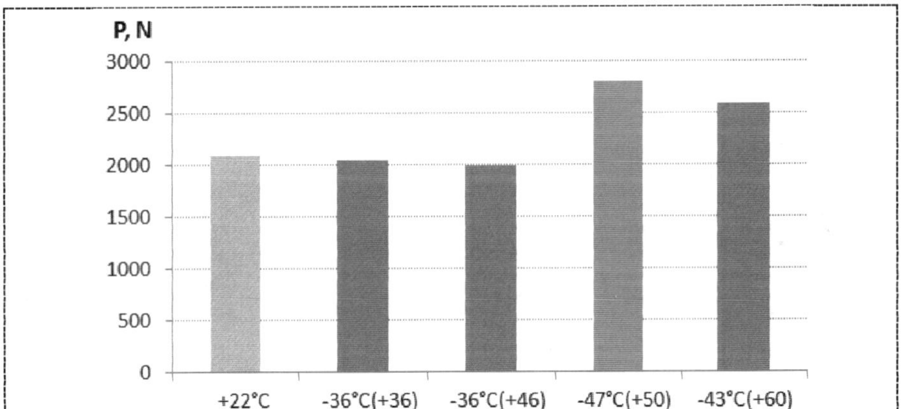

Figure 4. The values of the maximum breaking load of coupling welded joints, obtained at different ambient temperatures. (In brackets the maximum temperatures reached during heating)

4. Calculation of welding saddle parameters.

Welding of saddle 110/63 to polyethylene pipe PE 100 SDR 11 110 × 10.0 was considered at an ambient temperature of -45 °C. (Fig. 4, A). . Consider the design model (Fig. 5 (a)): 1 – saddle branch 1, 2 – polyethylene pipe, 3 – heating element. Figure 5 (b) shows the temperature distribution after preheating for 13.5 minutes and equalizing for 4 minutes. The voltage for preheating was 17 V and the thickness of the thermal insulation layer was 2 cm. The results of preheating and temperature equalizing calculations show that in the area of welding the average temperature is about 15 °C.

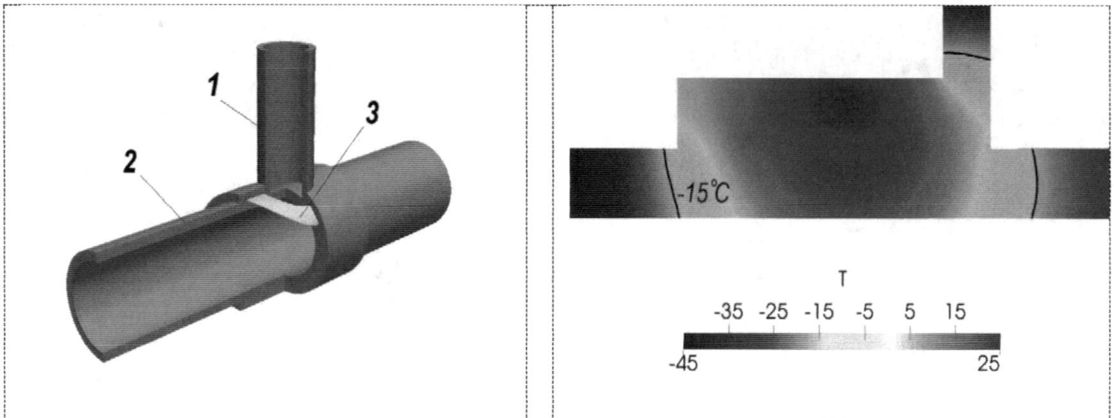

(a) (b)

Figure 5. General view of the saddle branch and pipe (a) and temperature distribution in the joint after preheating for 810 seconds with voltage of 17 V and equalizing for 120 seconds (b)

Calculations of the temperature distribution after heating at different cooling moments were also considered. Figure 6 shows results during cooling for 8 minutes at temperatures of -45 ° C with preheating and thermal insulation and 0 ° C (standard welding). The temperature distributions in the heat-affected zone at low and permissible temperatures are close after the same cooling time that indicates the identity of the crystallization process, which leads to an equivalent strength of the joint. The crystallization process is completed on the ten minute of cooling in both cases.

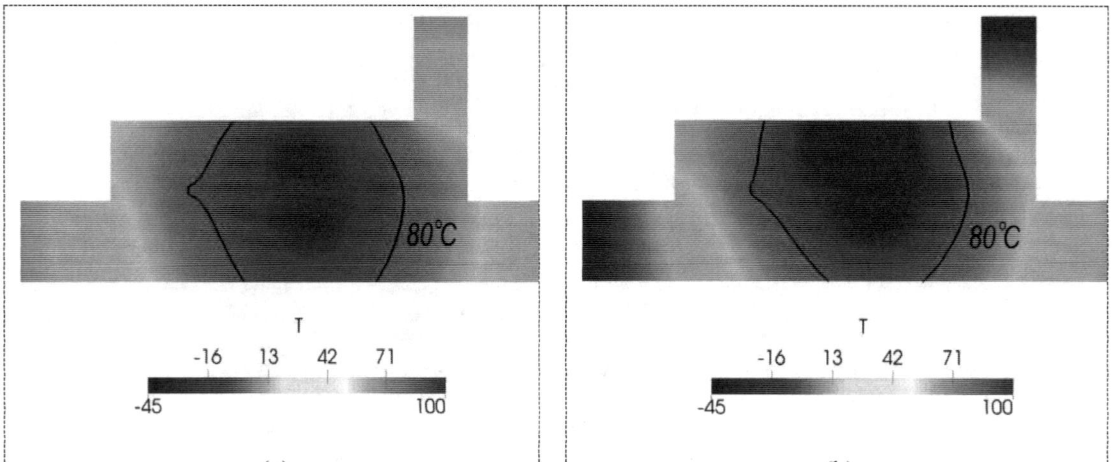

(a) (b)

Figure 6. Heat-affected zones in cross section of saddle branch and pipe during cooling for 8 minute after heating at the ambient temperatures 0 °C (a) and minus 45 ° C (b) with preheating and equalizing under thermal insulation

Breaking loads of tested saddle branches are almost the same (Fig. 7), which testifies to the effectiveness of the proposed technology for welding of saddle branch to polyethylene pipe at ambient temperatures below standard.

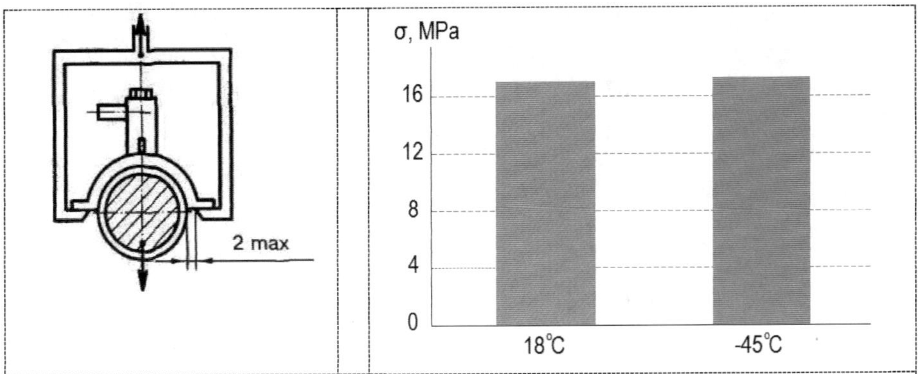

Figure 7. Breaking load of saddle branch welded at ambient air temperatures of 18 °C (standard welding) and minus 45 ° C (according to the proposed technology)

5. Conclusion.

Testing the quality of welded joints has determined the effectiveness of proposed method for calculating the technological parameters of polyethylene pipes welding with built-in heaters at ambient temperatures below the normative ones based on mathematical modeling of the thermal process, considering phase transformation in the temperature range.

References

[1] Gorilovskii M.I., Kalugina E.V., Ivanov A.N., Satdinova F.K., 2005 “Crystal Structure and Thermostability of Pipe Made from Different Types of Polyethylene”, Plast. Massy, **4**, pp. 9-12

[2] Avdonin N A 1980 Mathematical description of crystallization processes (Riga: Zinatnie) p 180

[3] Chebbo Z, Vincent M, Boujlal A, Gueugnaut D, Tillier Y, 2015 “Numerical and Experimental Study of the Electrofusion Welding Process of Polyethylene Pipes”, Polym. Eng. Sci., **55**, pp. 123-31

[4] Spina R, Spekowius M, Hopmann C, 2018 ‘Simulation of crystallization of osotactic polypropylene with different shear”, Thermochimica Acta, **659**, pp. 44-54

[5] Isachenko V P, Osipova V A, Sukomel A S 1969 Heat Transfer (Moscow: Energija) p 440

Acknowledgments

The work has been accomplished within the framework of the State Order of the Federal Agency for Scientific Organizations of the Russian Federation (project No. AAAA-A17-117040710038-8 from 07.04.2017)

Journal of Physics: Conference Series **1392** (2019) 012087 doi:10.1088/1742-6596/1392/1/012087

Identification of the spacewise dependent right-hand side in one-dimensional parabolic equation

LingDe Su, V. I. Vasil′ev

Institute of Mathematics and Information Science, North-Eastern Federal University, Russia

E-mail: sulingde@gmail.com, vasvasil@mail.ru

Abstract. In this paper we consider numerical solution of the inverse problem of determining a spacewise dependent right-hand side function, which is also called source function, in parabolic equation from measured data of the solution at the final time point. Such inverse problem occurs in mathematical modeling of various physics or engineering areas. Special iterative method is constructed for the approximate solution of the inverse problem, based on the source function at the new iterative step can be identified by the time derivative and space derivative at final time moment. The capabilities of the proposed computational algorithm is confirmed by the results of several numerical experiments.

1. Introduction

Inverse problem arises in many mathematical modelings of applied problems in science and engineering applications. It has a long history, its discovery and introduction needs to be traced back to the year 1929, when Ambarzumian studied the problem of determining the equation of a vibrating string [1]. After inverse problem in the CT technology made a major breakthrough in 1979, with the rapid development of calculation methods and the extensive application of computer technology, inverse problem has attracted more and more attention and caused widespread concern. Inverse problems are often considered non-classical issues and they are classified as ill-posed or conditionally well-posed problems [2], so, stable and efficient numerical algorithms need to be developed to obtain numerical solutions of the inverse problem.

Many researchers have made a significant contribution to inverse problem research and contributed to the development of ill-posed problems. In the 1960's Tikhonov produced an important series of papers on ill-posed problems [3–5], the most commonly used method of regularization of ill-posed problems, Tikhonov regularization method was named after Andrey Tikhonov. In 1994, Per Christian Hansen and many other collaborators developed the package of Matlab routines for easy doing experiments with analysis and solution of discrete ill-posed problems by means of regularization methods [6]. In addition, V. Isakov [7], P. N. Vabishchevich [8, 9], V. L. Kamynin [10], M. Li and T. S. Jiang [11], etc., also carried out theoretical or algorithmic research on inverse problems.

Based on the unknowns in the inverse problem, it can be classified geometric inverse problems, in which the domain or a part of it's boundary is unknown; coefficient inverse problems, in which some coefficients in the master equation are unknown; boundary inverse problems, in which boundary conditions are unknown and evolutionary inverse problems (or time inverse problem), in which initial conditions are unknown. The problem in our work, which related to identifying the right-hand side of an equation, belongs to coefficient inverse problem. A lot of works have been done for studying such inverse problem, the existence and uniqueness of the solution and the well-posedness in various functional class were considered in [7, 12–14], also many numerical techniques for solving the inverse source problem were presented in [15–17].

In this paper we construct a special iterative method to recover the right-hand side function, which depends on spatial variables only, in one-dimensional parabolic equation. The new iterative method based on the right-hand side function can be identified with the derivative of time and space at the final time point. The paper is organized as follows, in section 2, we briefly introduce the formulation of the problem. In section 3, we introduce the algorithm and apply on the inverse problems. The results of numerical experiments are presented in section 4. Section 5 is dedicated to a brief conclusion. Finally, some references are introduced at the end.

2. The Problem Formulation

We consider the inverse problem to determine the spacewise dependent right-hand function $f(x)$ in the following parabolic equation,

$$\frac{\partial u}{\partial t} - \frac{\partial^2 u}{\partial x^2} = f(x)g(t), \qquad x \in (0,l), \quad t \in (0,T], \tag{1}$$

where $g(t)$ is a given function and l, T are given positive constants. Also the initial condition is given as

$$u(x,0) = u^0(x), \qquad x \in [0,l], \tag{2}$$

the Dirichlet boundary conditions

$$u(0,t) = u_0(t), \qquad u(l,t) = u_l(t), \qquad t \in [0,T] \tag{3}$$

for solving an inverse problem, some additional conditions are also needed, in this kind of problem, the additional conditions are usually given as the solution at the final time moment, which is also called final overdetermination

$$u(x,T) = u^T(x), \qquad x \in [0,l]. \tag{4}$$

In our work, we focus on numerically solving the inverse problem (1)–(4) of determining a pair of $u(x,t)$ and $f(x)$, we assume that the inverse problem (1)–(4) is well-posed. In [18, 19] provided the existence, uniqueness and stability of such source inverse problem. Also, many numerical works have been proposed for solving the inverse problem (1)–(4), in [15] boundary element method was used to determine the heat source dependent on one variable, B. T. Johansson and D. Lesnic in their work [17] provided a variational method for identifying spacewise dependent heat source, the conjugate gradient method was considered in [20], in the work of Yan with his collaborator [21] a meshless method was used to solve the inverse source problem, P. N. Vabishchevish considered iterative computational identification of spacewise dependent source in [22], and V. I. Vasil'ev and his collaborators also have given a study on numerical solution of the right hand problems in their works [23, 24], etc.

3. The Computational Algorithm

To numerically solve the source inverse problem (1)–(4) iterative algorithm is used, in the new iterative step the right-hand side function $f^{(k+1)}$ is identified for $t = T$,

$$f^{(k+1)}(x) = \Big(\frac{\partial u^{(k)}}{\partial t}(x,T) - \frac{\partial^2 u^T}{\partial x^2}(x)\Big)/g(T), \qquad k = 0,\ 1,\ 2,\ \cdots \tag{5}$$

with some given assumption initial $f^{(0)}$, the symbol $f^{(k)}$ represents the value of $f(x)$ at the k-th iteration step. Then, the parabolic problem (1)–(4) can be solved by the following forms combined with (5) and the evaluation of convergence can be seen in Vabishchevish's work [22].

$$\frac{\partial u^{(k)}}{\partial t} - \frac{\partial^2 u^{(k)}}{\partial x^2} = f^{(k)}(x)g(t), \quad t \in (0, T]. \tag{6}$$

In the calculation of solving equation (6), sweep method is used, we introduce the uniform grid in the computational domain Ω,

$$\omega_h = \{x_i | x_i = ih, \ i = \ 1, \ 2, \ \dots, \ M-1, \quad Mh = l\},$$

where M is a positive integer and $x_0 = 0$, $x_M = l$. For the time we have,

$$\overline{\omega_\tau} = \{t^n | t^n = n\tau, \ n = \ 0, \ 1, \ \dots, \ N, \quad N\tau = T\}.$$

Using the notation $u_i^{n,(k)} = u^{(k)}(x_i, t^n)$ and the implicit scheme, from (6) we get discrete forms,

$$\frac{u_i^{n+1,(k)} - u_i^{n,(k)}}{\tau} - \frac{u_{i+1}^{n+1,(k)} - 2u_i^{n+1,(k)} + u_{i-1}^{n+1,(k)}}{h^2} = f_i^{(k)} g^{n+1}, \tag{7}$$

$$u_i^{0,(k)} = u^{(k)}(x_i, 0) = u^0(x_i), \qquad u_i^{N,(k)} = u^{(k)}(x_i, T) = u^T(x_i), \tag{8}$$

$$u_0^{n,(k)} = u^{(k)}(0, t^n) = u_0(t^n), \qquad u_M^{n,(k)} = u^{(k)}(l, t^n) = u_l(t^n). \tag{9}$$

where the superscript (k) represents the k-th iteration step. We convert (7) into the following form,

$$u_{i+1}^{n+1,(k)} - (2 + \frac{h^2}{\tau})u_i^{n+1,(k)} + u_{i-1}^{n+1,(k)} + \frac{h^2}{\tau}u_i^{n,(k)} + h^2 f_i^{(k)} g^{n+1} = 0, \tag{10}$$

with the conditions (8) and (9). Using the following formula for $u_i^{n+1,(k)}$,

$$u_i^{n+1,(k)} = \alpha_i^{n+1,(k)} u_{i+1}^{n+1,(k)} + \beta_i^{n+1,(k)}, \tag{11}$$

with $\alpha_0 = 0$ and $\beta_0 = u_0$. Bringing (11) into equation (10), we get,

$$\alpha_i^{n+1,(k)} = \frac{1}{2 + \kappa - \alpha_{i-1}^{n+1,(k)}}, \qquad \beta_i^{n+1,(k)} = \frac{\kappa u_i^{n,(k)} + h^2 f_i^{(k)} g^{n+1} + \beta_{i-1}^{n+1,(k)}}{2 + \kappa - \alpha_{i-1}^{n+1,(k)}}, \tag{12}$$

where $\kappa = \frac{h^2}{\tau}$, combining (5) and some given assumption initial $f^{(0)}$, we can find the right-hand side function and the solution for parabolic equation (1). In our computation, we take the initial approximation $f^{(0)}(x)$ in the form as follows,

$$f^{(0)}(x) = -\frac{\partial^2 u^T}{\partial x^2}(x)/g(T) = -\frac{u^T(x + h_0) - 2u^T(x) + u^T(x - h_0)}{h_0^2 g(T)}, \tag{13}$$

and $h_0 = 10^{-3}$.

4. Numerical Examples

To test the effectiveness of the iterative computational algorithm, which mentioned in the previous section, for solving the inverse problem to determine the spacewise dependent right-hand side function in one-dimensional parabolic equation, we present a numerical result with a discontinuous function $f(x)$, $x \in [0, l]$, which is given as,

$$f(x) = \begin{cases} 1, & x \in (0.5, 0.8) \\ \exp(-100(x - 0.25l)^2), & \text{otherwise} \end{cases}$$

the known conditions are also given in the following,

$$g(t) = 1, \quad u(x, 0) = 0, \quad u_0 = u_l = 0, \quad l = 1.$$

In the computation, we use $T = 0.2$, the numerical and exact solution of $f(x)$ with the assumption initial $f^{(0)}$ are presented in Figure 1. In this computation, we take $M = N = 100$ and the total iterative number $K = 10$. As can be seen from the figure, the numerical solution is a good coincidence of the exact solution.

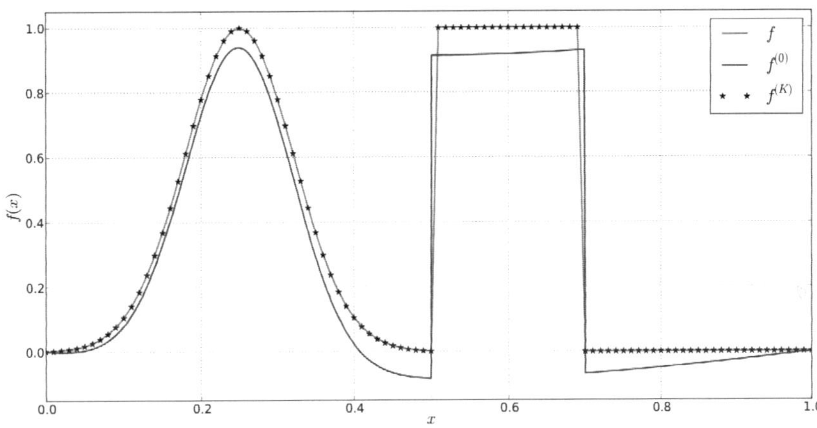

Figure 1. numerical and exact solution of $f(x)$ and the assumption initial $f^{(0)}$.

To evaluate the error of the approximate solution on every iterations, we adopt two errors, which are defined as follows,

$$\epsilon = |f^{(k)}(x) - f(x)|, \qquad \text{RMSE} = \sqrt{\frac{1}{M}\sum_{i=1}^{M}(f^{(k)}(x_i) - f(x_i))^2},$$

where $f^{(k)}(x)$ is the numerical solution of the right-hand side function on the k-th iterative step, while $f(x)$ is the exact solution.

We consider the the errors with different M on every iterative steps, the RMSE and Max ϵ can be obtained from Figure 2 with $N = 100$. In Figure 2(a) the RMSE error with different M is presented, in Figure 2(b) the figure of Max ϵ on every iterative steps can be seen. The figures of RMSE and Max ϵ with different time step are presented in Figure 3, in this computation we use $M = 200$.

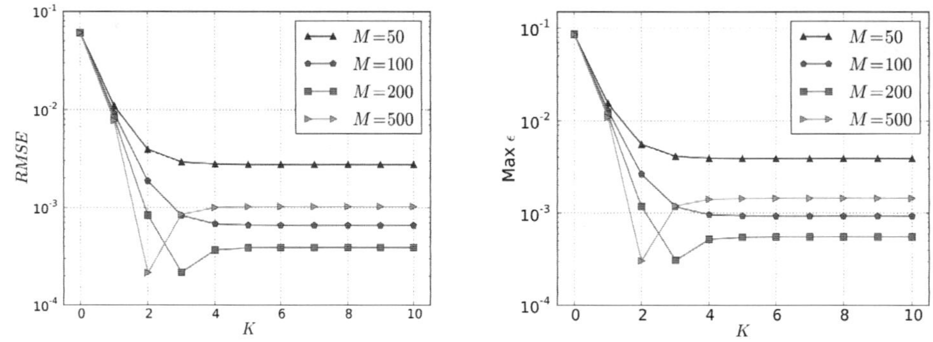

(a) The RMSE errors with M. (b) The Max ϵ errors with M.

Figure 2. The figures of two errors with different M on iterative steps.

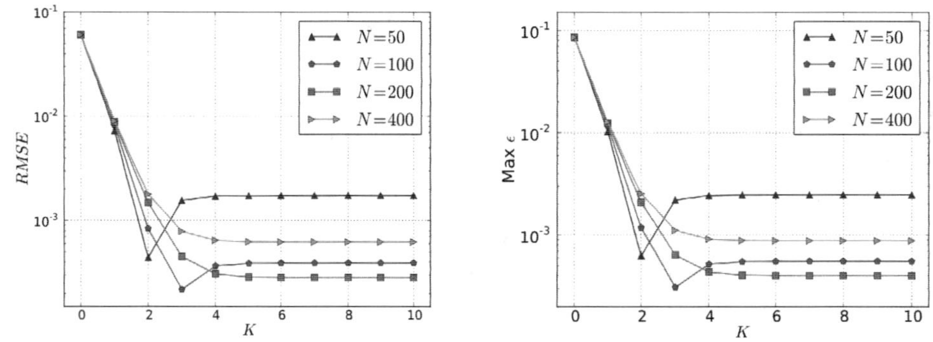

(a) The RMSE errors with N. (b) The Max ϵ errors with N.

Figure 3. The figures of two errors with different N on iterative steps.

Because the error is always accompanied by the actual measurement, we also consider the inverse problem with the noisy measured additional data, which is given by,

$$u_\delta^T = u^T + \delta randn \max(uT)/2,$$

where δ is the tolerated noise level and $randn$ is a random variable uniformly distributed on the interval $[-0.5, 0.5]$. In order to identify the smooth solution, we using a smoothing method, which was given in [25] to get the smooth final solution y,

$$u_\delta^T = -\frac{\tau}{\lambda} y_{xx} + y,$$

where λ is a constant parameter. The figures of the results with different noisy level δ can be seen in Figure 4 with $\lambda = 25$.

5. Conclusions

In this paper we consider the inverse problem of identifying right-hand side function. The numerical method based on the function is identified at the final time point. With some assumption initial source function, iterative method is used combined with sweep method for solving the parabolic equation on each iterative steps. Numerical example of the model problem demonstrates the efficiency of the iterative method for solving the inverse problem.

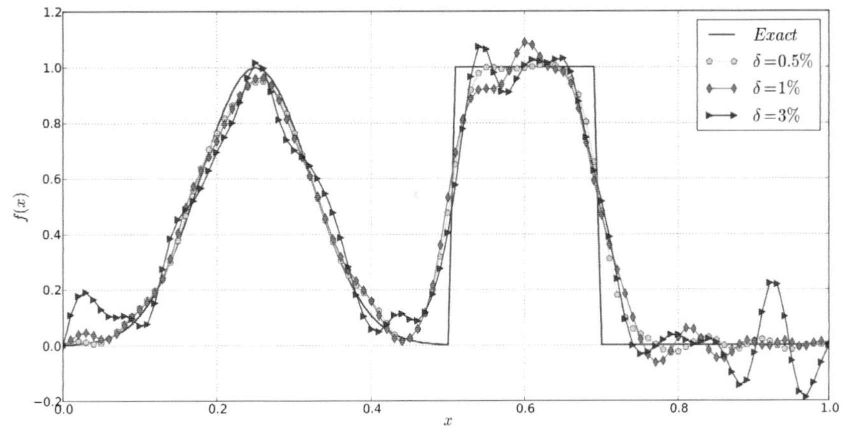

Figure 4. The figure of $f(x)$ with different δ.

Acknowledgments

The authors would like to express their gratitude to professor P. N. Vabishchevich for his helpful and constructive remarks. This work was supported by RFBR (project 17-01-00689).

References

[1] Ambartsumyan V A 1929 *Zeitschrift fúr Physik* **60** 690–695
[2] Tikhonov A N, Goncharsky A, Stepanov V V and Yagola A G 1995 *Numerical Methods for the Solution of Ill-Posed Problems* 1st ed Mathematics and Its Applications 328 (Springer Netherlands)
[3] Tikhonov A N 1963 *Soviet Math. Dokl.* **4** 1624–1627
[4] Tikhonov A N 1963 *Soviet Math. Dokl.* **5** 1035–1038
[5] Tikhonov A N 1963 *Dokl. Akad. Nauk SSSR* **151** 501–504
[6] Hansen P C 1994 *Numer. Algorithms* **6** 1–35
[7] Isakov V 2017 *Inverse Problems for Partial Differential Equations* 3rd ed Applied Mathematical Sciences 127 (Springer International Publishing)
[8] Vabishchevich P N and Vasil'ev V I 2016 *Inverse Probl. Sci. Eng.* **24** 42–59
[9] Samarskii A A and Vabishchevich P N 2007 *Numerical Methods for Solving Inverse Problems of Mathematical Physics* 1st ed Inverse and Ill-Posed Problems Series 52 (De Gruyter)
[10] Kamynin V L 2015 *Math. Notes* **97** 349–361
[11] Jiang T S, Li M and Chen C S 2012 *Numer. Heat Transfer Part A* **61** 338–352
[12] Isakov V 1991 *Comm. Pure Appl. Math.* **44** 185–209
[13] Prilepko A I and Kostin A B 1993 *Math. Notes* **53** 63–66
[14] Gol'dman N L 2008 *Dokl. Math.* **77** 350–355
[15] Farcas A and Lesnic D 2006 *J. Engrg. Math.* **54** 375–388
[16] Xiong X T, Yan Y M and Wang J X 2011 *J. Phys. Conf. Ser.* **290** 12–17
[17] Johansson B T and Lesnic D 2007 *IMA J. Appl. Math.* **72** 748–760
[18] Solov'ev V V 1990 *Differ. Equ.* **25** 1114–1119
[19] Kamynin V L 2005 *Math. Notes* **77** 482–493
[20] Erdem A, Lesnic D and Hasanov A 2013 *Appl. Math. Model.* **37** 10231–10244
[21] Yan L, Yang F L and Fu C L 2009 *J. Comput. Phys.* **228** 123–136
[22] Vabishchevich P N 2017 *Inverse Probl. Sci. Eng.* **25** 1168–1190
[23] Vasil'ev V I, Popov V V and Kardashevsky A M 2018 Conjugate gradient method for identification of a spacewise heat source *Large-Scale Scientific Computing* ed Lirkov I and Margenov S (Springer International Publishing) pp 600–607
[24] Vasil'ev V I, Kardashevsky A M and Popov V V 2018 *AIP Conf. Proc.* **2025** 100011
[25] Samarskii A A, Vabishchevich P N and Vasil'ev V I 1997 *Matematicheskoe Modelirovanie* **9** 119–127

Journal of Physics: Conference Series **1392** (2019) 012088 doi:10.1088/1742-6596/1392/1/012088

Modeling of burden at automated strength calculation of mobile vehicles supporting structures

M Tokareva, T Zubkova

Faculty of mathematics and information technology, *Orenburg State University,* 13 Pobeda ave., Orenburg, 460018, Russian Federation

E-mail: tokareva@mail.osu.ru

Abstract. A method, its algorithmic and software implementation, based on the use of the Ritz variational principle, which allows to calculate the values of stresses, displacements, reactions at the points of interest to the researcher, has been developed to simulate static loading when calculating the strength of structures consisting of flat elements This provides an opportunity to reduce the complexity and timeline of final testing of the supporting structures of mobile vehicles.

1. Introduction

Meeting the needs of society in the effective functioning of vehicles requires the solution of a complex of scientific and technical problems, in particular, the use of high-tech information technologies in the production and restoration of the supporting structures of vehicles. A promising direction of increasing their reliability is the automation of calculation using the methods of the theory of elasticity and structural mechanics of vehicles and structures, which allows to choose designs that are rational in terms of strength and manufacturability at the design stage, which significantly reduces development time and developmental tests [2,3,4,5] .

Depending on the objectives, the following objects may be subjected to testing: the supporting structures in the vehicle; separate complete support systems (frames, dump frames, the foundations of cargo platforms or bodies, etc.); separate closed contours of the supporting structures; open nodal connections and individual elements (spars, cross-pieces, racks, etc.) or their sections.

Testing of objects from the first three groups is most often carried out with the purpose of comparative assessment of the design options durability or individual assessment of the experimental structures durability. Objects that are preferable for conducting final testing are nodal joints or separate parts of structural elements, depending on which local area limits the durability of the entire supporting structure [6].

When repairing the supporting structures, it is necessary to take into account their condition at the time of repair, the possibility of replacing the elements, taking into account the variation of parameters such as geometric shapes and sizes, material properties. With the possible change of these parameters, the construction that has been repaired should have sufficient, as a rule, regulated, reliability and durability with a minimum amount of various types of tests of modified or modernized variants.

The authors have developed a technique enabling rapid assessment of stationary loads, which allows you to choose from a variety of design options for supporting structures assemblies an acceptable design and technological option that meets the requirements for strength and material intensity. The technique uses a modification of the well-known Ritz method, and due to it, it is possible to avoid the numerical integration used in the finite element method [1], which makes it possible to reduce the design calculation time without loss of accuracy.

2. A description of the methodology

Spatial structures consisting of flat elements can be divided into their constituent parts with the introduction of unknown force bonds. Depending on the location of forces relative to the flat element, the latter may be of the following types:

A - with forces acting in the plane of the element;

B - with forces acting perpendicular to the plane of the element;

C - with forces acting both in the plane of the element and perpendicular to it.

In solving the problem, the finite-difference scheme obtained from the Ritz variational principle is used. To this end, the elements are covered with rectangular grids so that they are combined at the joints. Forces can only be applied at grid points. If it is inconvenient to combine an arbitrary force with a node, it should be decomposed into two components applied at the nearest nodes.

The elements of type A with the forces acting along the middle surface of the element can be applied to the ratio of the plane problem of the theory of elasticity.

In particular, the Ritz method and the formula for the specific potential energy, which has the following form for bulk parts, can be applied:

$$W = G(\varepsilon_x^2 + \varepsilon_y^2 + \varepsilon_z^2 + \frac{\nu\Delta^2}{1-2\nu} + \frac{1}{2}(\gamma_{xy}^2 + \gamma_{yz}^2 + \gamma_{zx}^2)), \tag{1}$$

where $\Delta = \varepsilon_x + \varepsilon_y + \varepsilon_z$; ν is Poisson ratio.

In the case of a plane-stressed problem, the condition $\sigma_z=0$ in the formulas of the generalized Hooke's law gives

$$\varepsilon_z = -\frac{\nu}{1-\nu}(\varepsilon_x + \varepsilon_Y).$$

Substituting the values into the formula for the specific potential energy and assuming that $\gamma_{yz}=\gamma_{zx}=0$ we get:

$$W = G(\varepsilon_x^2 + \varepsilon_y^2 + \frac{\nu}{1-\nu}(\varepsilon_x + \varepsilon_y)^2 + \frac{1}{2}\gamma_{xy}^2) \tag{2}$$

It is obvious that the total potential energy in a certain volume V with a constant value of stresses σx, σy, τxy it is equal to $Wn=WV$. In the case of a plane-deformed state, the denominator of the third term instead of $1-\nu$ should be $1-2\nu$.

Talking the coordinate system, as shown in Figure 1, we write the relative deformation:

- for section KI $\varepsilon_x = \frac{\partial U}{\partial x} = \frac{U_x - U_k}{\lambda_i}$;
- for section QK $\varepsilon_y = \frac{\partial V}{\partial y} = \frac{V_q - V_k}{s_i}$;
- for section QKI $\gamma_{xy} = \frac{\partial V}{\partial x} + \frac{\partial U}{\partial y} = \frac{U_q - U_k}{s_i} + \frac{V_i - V_k}{\lambda_i}$.

The total potential energy of the system W_n, divided by a rectangular grid into a series of elements and containing n points can be written as a function of $2n$ unknown displacements:

$$W_n=f(x_1,x_2,x_3,x_4,\dots,x_{2n-1},x_{2n}).$$

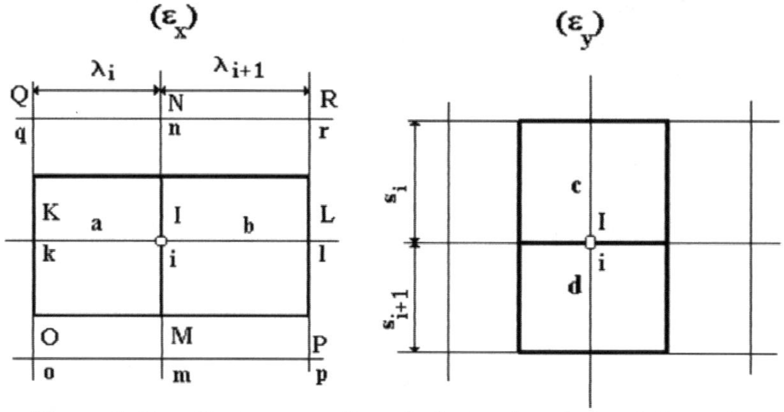

Figure 1. Coordinate system for calculating the relative deformation.

The task is to determine the values of the unknowns x_1, ... ,x_{2n} with a sufficient approximation. The Ritz method gives for this condition the minimum potential energy:

$$\frac{\partial W_n}{\partial x_1} = \frac{\partial W_n}{\partial x_2} = \ldots = \frac{\partial W_n}{\partial x_i} = \ldots = \frac{\partial W_n}{\partial x_1} = 0.$$

You can make sure that for the derivatives $\partial W_n / \partial x_i$ *и* $\partial W_i / \partial x_{i-1}$ there are only unknowns that are located not more than one step away from point *i,* that is, only the unknown displacements of the points indicated in Figure 1. Therefore, we write down the potential values energy for only four plots containing point i. Figure 2 shows a diagram of the regions defining the values of the relative deformations ε_x, ε_y *u* γ_{xy},, as well as 12 regions, in each of which the values of ε_x и ε_y do not change.

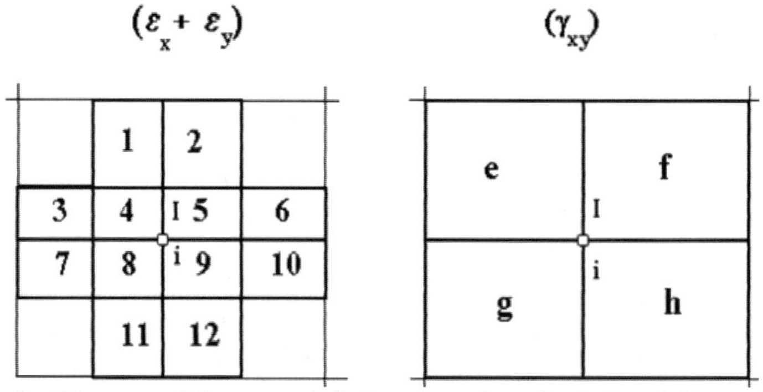

Figure 2 - Diagram of the areas defining the value of relative deformations.

Writing the values of all four terms of the potential energy and taking the derivatives with respect to x_I and x_i,, we obtain 2 equations from the general system consisting of *2n* equations.

The equations are written as follows:

$$A_Q x_Q + A_N x_N + \ldots + A_P x_P + A_q x_q + \ldots + A_p x_p = \frac{P_I}{2G}; \tag{3}$$

$$B_Q x_Q + \ldots + B_P x_P + B_q x_q + \ldots + B_p x_p = \frac{P_i}{2G}, \tag{4}$$

where P_I and P_i are the horizontal and vertical forces applied at point *i*, which are positive if directed along the axes Ox and Oy, respectively; G is the shear modulus equal to $E/2/(1+\nu)$, where *E* is the elastic modulus of the 1st kind, ν is the Poisson ratio.

The coefficients A and B in equations (3), (4) are determined on the basis of the weighting factors, some of which are recorded in the information files of the task, and some - programmatically determined for areas A and B.

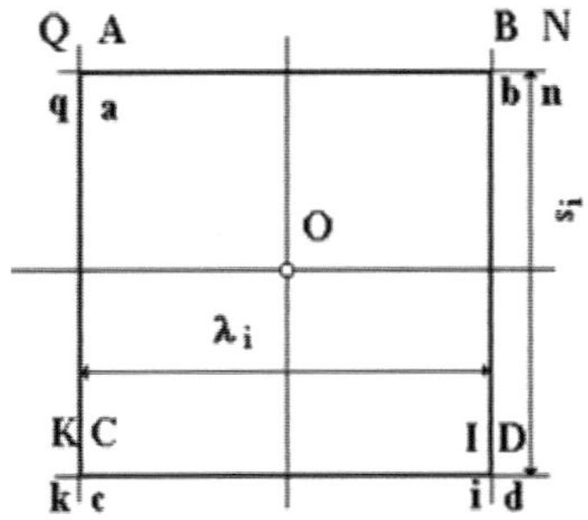

Figure 3. To the calculation of stresses in area A.

For the point O, which lies in the center of the element type A, (Figure 3):

- relative displacements:

$$\varepsilon_x = \frac{x_B - x_A + x_D - x_C}{2\lambda_i}; \varepsilon_y = \frac{x_a - x_c + x_b - x_d}{2s_i}; \tag{5}$$

$$\tau_{xy} = \frac{1}{2}(\frac{x_A - x_C + x_B - x_D}{s_i} + \frac{x_b - x_a + x_d - x_c}{\lambda_i});$$

- tension along the axes *Ox* and *Oy:*

$$\sigma_x = E\frac{\varepsilon_x - \nu\varepsilon_y}{1-\nu^2}; \sigma_y = E\frac{\varepsilon_y + \nu\varepsilon_x}{1-\nu^2}; \tau_{xy} = \frac{E}{2(1-\nu)}\gamma_{xy}; \tag{6}$$

- main stresses:

$$\sigma_{1,2} = \frac{1}{2}((\sigma_x + \sigma_y) \pm \sqrt{(\sigma_x - \sigma_y)^2 - 4\tau_{xy}^{\ 2}}), \tag{7}$$

where σ_1 is large and, σ_2 is smaller in absolute value.

The tangential stresses are equal

$$\tau_{\max} = \frac{\sigma_1 - \sigma_2}{2}.$$

The given stresses are calculated according to the theory of strength formula:

$$\sigma_{np} = \sqrt{\sigma_1^{\ 2} + \sigma_2^{\ 2} - \sigma_1\sigma_2}.$$

The angle of inclination of the main site α_1 is obtained from the dependency:

$$tg\,2\alpha_1 = \frac{2\tau_{xy}}{\sigma_x - \sigma_y}.$$

The positive angle α_1 is deposited from the positive direction of the axis *Ox* clockwise.

For elements of type B, the voltage is determined by the formulas for the plates (Figure 4).

Usually, to solve the problem of stress-strain state of spatial systems, taking into account their plane deformation and bending, the structural components are covered with a grid of a regular structure. However, when solving some problems, it is more convenient to have a grid of irregular structure, since this facilitates the determination of stresses in the places of their concentration. The use of an irregular grid complicates the solution of the problem, therefore, for composite structural elements that carry the load in their plane, an irregular grid is adopted, and for flexural elements - a regular one.

Figure 5 shows one of the structural elements in a plane-stressed state with an irregular grid applied.

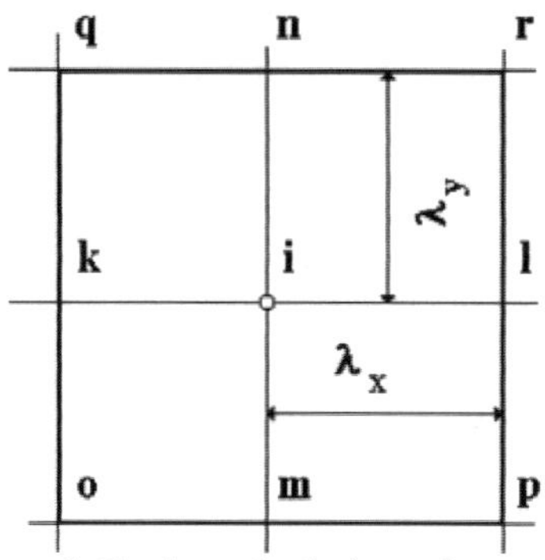

Figure 4. To the calculation of stresses in area B.

At the edges, unknown reactions of bonds R_1, R_2, etc., are applied, which are taken into account in equations (3) and (4) by introducing additional terms like $-R_1/2GB$, where B is the thickness of the element. Missing equations are obtained from the compatibility condition.

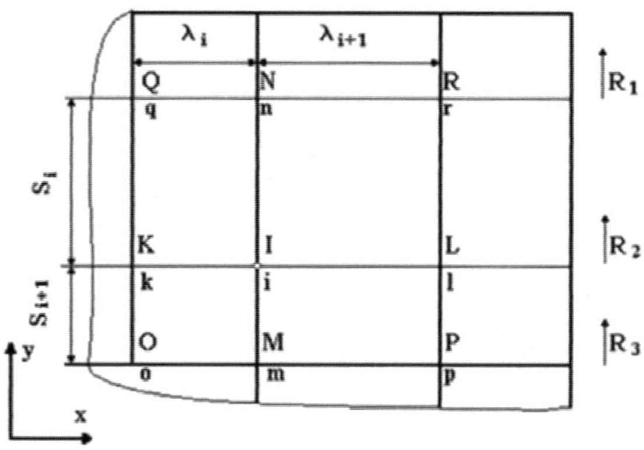

Figure 5. Irregular grid for a flat task.

The method of strength calculation is embodied in the "Plosk" software package, which implements:

1. the algorithm to automate the compilation of the system, taking into account the different configuration of the boundaries of the flat element;
2. obtaining the values of stresses, main stresses, as well as the values of tilt angles and reduced stresses in a user-friendly form;
3. the packaging algorithm and the algorithm for solving the system in a packaged form by the Gauss method with the choice of the main element;
4. the algorithm for organizing software checkpoints, which allows you to take breaks in the decision and resume the decision from the place where the stop occurred;
5. the algorithm for taking into account the input of additional equations or the replacement of already formed ones, which include the equations of support or forced displacements and the compatibility equation;
6. the possibility of correction of weighting coefficients, which allows introducing additional terms with new unknowns into the already formed equations, which may occur when inserting bonds at the joints or when entering bearing links;
7. the algorithm for selective calculation of reactions at points of interest to the researcher;
8. the ability to enter up to three groups of free members corresponding to three types of loading of the same design, which allows one to obtain stresses, displacements and reactions for the specified loading options.

The validity and reliability of the proposed methodology is confirmed by the correct use of the analytical apparatus of the study and the adequacy of the calculated results to the experimental data obtained from the bench tests of local models of tractor semi-trailers and typical operating conditions.

3. Conclusion

Using the software package that implements the modification of the Ritz method, the strength of the frame assembly of the heavy-duty tractor dumping semi-trailer OZTP-9554, which has a wide range of operating conditions and is able to adequately represent other transport vehicles, was calculated. The results of stress calculations are on average 8-10% higher than the experimental ones, i.e. sent to the margin of safety.

Using the results of testing large-scale models with a limited range of modeling and their subsequent approximation to full-scale structures will allow the use of universal testing equipment and significantly reduce energy and material costs in tests related to fatigue life testing of large-sized bearing systems of transport, road-building, metallurgical and other machines working under conditions of periodic loading.

It is possible to use the results in the following main areas:

- studies on the static strength of these structures for various types of loading;
- fine-tuning of new and modified structures and the development of technologies for their production and repair;
- the creation of a data bank on the stress-strain state of various versions of standard design nodes.

Using the results obtained when using the “Plosk” software package, the authors propose a technique and its software implementation for building models of displacements and reactions using correlation and factor analyzes.

The proposed statistical methodology and its software implementation was applied by the authors to determine the conditionality of the sizes of the elements of the chain-drive in the PR-15.875 chain. For the given nominal values of dimensions and their deviations, under the condition of normal distribution, a study matrix was constructed, the column parameters of which were the main technical characteristics with normal distributions of their values, the observations — generation lines.

Further work of the authors in this direction is planned in carrying out calculations of the dynamic characteristics of more complex load-bearing structures and checking the adequacy of the models constructed to the experimental data.

References

[1] Finite-Element Method for Three-Dimensional Problems of Elastic Structures Buckling Theory. *Dimitrienko Yu.I., Bogdanov I.O.* Herald of the Bauman Moscow State Tech. Univ., Nat. Sci. 2016, no. 6, pp. 73–92. DOI: 10.18698/1812-3368-2016-6-73-92

[2] More on the Development of Rigidity Evaluation Techniques of Freight Car Structures and Their Components in the Design and on Trials. Adjustment of Design Loads. *Plotkin V.S., Krasnobaev A.M., Konkova T.Y., Krasnobaev O.A.* JSC Railway Research Institute, 2015, no. 6, pp.50-57.

[3] Increasing the Reliability the Combined Criteria of the Static Strength of a Material of Complexly Loaded Deformable Structures *Zenkov E.V., Tsvik L.B.* Materials physics and mechanics. 2018. т. 40. № 1. pp.124-132.

[4] Numerical Methods for Calculating the Strength and Stability of Stiffened Orthotropic Shell S*karpov V.V., Semenov A.A.* Materials physics and mechanics. 2017. т. 31. № 1-2. с. 16-19.

[5] Application of Unstructured Approximating meshes in discrete-continual finite element method of structural analysis *Akimov P.A., Negrozov O.A.* Materials Physics and Mechanics. 2016. T. 26. № 1. pp. 5-8.

[6] On the issue of assessing the stress-strain state of frame metal structures of mobile vehicles. *Tokareva M.A, Rassokha V.I., Filippov V.Yu.* Control. Diagnostics, 1999. №11.- pp. 7-11.

Numerical identification of the initial condition for parabolic equation

V.I. Vasil'ev, .M. Kardashevsky

677000 Yakutsk, 58, Belinsky str., North-Eastern Federal University

E-mail: vasvasil@mail.ru; kardam123@gmail.com

Abstract. The authors consider a problem of reconstructing of the initial condition in a parabolic equation through the given value of the unknown function at the final moment of time. To perform a numerical solution of the retrospective inverse problem for a multidimensional parabolic equation it is suggested to use a rapidly converging conjugate gradient method, based on a sequential refinement of the initial condition and a direct problem solution at each iteration. The article provides a discussion on the results of a computational experiment conducted on model problems with exact and quasi-realistic solutions, including random errors for desired overdetermination condition.

1. INTRODUCTION

Widely represented, inverse problems for partial differential equations are the most frequently used in construction of mathematical models for various processes.A problem of mathematical model identification through determination of differential equation coefficients, of the right side, area boundary, or initial conditions based on additional observation data or experiment. These tasks belong to inverse problems of mathematical physics and play nowadays a significant role in natural sciences and their applications all over the world. An important feature of inverse problems arising from results processing of a full-scale experiment is that the initial data in these problems is known approximately. This results from the fact that devices used at measurements have a certain level of error. It follows that methods of solving inverse problems should be resistant to small changes in the input data. Besides, an issue of uniqueness of solution is also relevant, and its study, in fact, is an answer to a question on whether the available experimental data is sufficient for unambiguous definition of the desired characteristic of the object or process studied. Studies on establishment of conditional correctness of inverse problems through construction of numerical methods for their solution are summarized in monographs[1] – [7].

Tasks with inverse time (retrospective inverse problems) with given conditions at the final moment of time belong to a class of incorrect problems in classical sense [7]. However, they are correctly set in the class of bounded solutions. Numerical solutions of a retrospective inverse problem are developed, for example, in the following works: a modified boundary element method based on a minimal energy method [8]; an iterative method of the initial condition refining [9]; an iterative algorithm with the boundary element method [10]; modified Tikhonov regularization method [11]; an iterative method with reverse direction of time [12];conjugate gradient method for initial condition refining[13]; use of the Poisson integral [14]; quasi-inversion method [5] and optimal filtering method [4]. In most works indicated above, a regularization method is used, and in this connection we note that incorrect tasks are very sensitive to regularization parameters.

As indicated in monograph [7] iterative algorithms for solving evolutionary inverse problems have long-term prospects that take into account the specifics of problems under consideration. They iteratively refine an initial condition, i.e., a usual direct correct initial-boundary value problem for a parabolic equation is solved at each iteration [9]. In this paper we will use an iterative method of conjugate gradients, where the number of iterations is used as a regularization parameter, which is in agreement with an error of the input data. We present calculations demonstrating the possibilities of the method.

2. PROBLEM STATEMENT

In the domain $\Omega = \{\mathbf{x} \mid \mathbf{x} = (x_1, \ldots, x_p),\ 0 < x_\alpha < l_\alpha,\ \alpha = 1, \ldots, p\}$ we seek a solution of the multidimensional parabolic equation

$$\frac{\partial u}{\partial t} = \sum_{\alpha=1}^{p} \frac{\partial}{\partial x_\alpha}\left(k_\alpha(\mathbf{x})\frac{\partial u}{\partial x_\alpha}\right), \quad \mathbf{x} \in \Omega, \quad 0 \le t < T \tag{1}$$

with homogeneous Dirichlet boundary conditions.

We formulate a retrospective inverse problem as follows: the value of the unknown function is set at the fixed terminal time $u(\mathbf{x}, T)$, and the initial condition $u(\mathbf{x}, 0)$ is to be determined

$$u(\mathbf{x}, T) = \phi(\mathbf{x}), \quad \mathbf{x}\in\overline{\Omega}. \tag{2}$$

Let the coefficients $k_\alpha(\mathbf{x}),\ \alpha = 1, ..., p$ of the differential equation (1) be sufficiently smooth, positive, and bounded functions, that is, satisfy conditions

$$0 < \kappa_1 \le k_\alpha(\mathbf{x}, t) \le \kappa_2 < \infty. \tag{3}$$

The considered non-classical initial-boundary value problem with inverse time belongs to a class of conditionally correct problems of mathematical physics [7].

It should be noted that equation (1) has an exact analytical solution in the case of constant coefficients

$$u(x_1, x_2, ..., x_p, t) = \frac{1}{\sqrt{(\beta + 4t)^p}} e^{-\frac{\|\mathbf{X}-\mathbf{X}_o\|^2}{\beta+4t}}, \quad \beta > 0, \tag{4}$$

where $\mathbf{x}_o = (l_1/2, ..., l_p/2)$.This function is fairly smooth and limited, and it can be further used to establish an accuracy of the proposed iterative method.

We carry out a numerical solution for the parabolic problem (1) – (3) on a rectangular space-time grid in steps $h_\alpha, \alpha = 1, ..., p,\ \tau = T/M$ using a finite-difference scheme [16]:

$$\frac{y^{n+1} - y^n}{\tau} + A\left(\sigma y^{n+1} + (1-\sigma)y^n\right) = 0, \quad \mathbf{x} \in \omega, \quad n = 0, 1, ..., N-1. \tag{5}$$

$$y_M = \phi(x), \quad \mathbf{x} \in \overline{\omega}, \tag{6}$$

where the weight factor is $\sigma \in [0, 1]$.

Journal of Physics: Conference Series **1392** (2019) 012089 doi:10.1088/1742-6596/1392/1/012089

3. SOLUTION ALGORITHM FOR INVERSE PROBLEM

For numerical implementation of the discrete analogue of retrospective inverse problem (5) – (6) we will use an iterative method of conjugate gradients, based on sequential refinement of the initial condition and on solution of the direct problem at each iteration. Let us give the problem a corresponding operator statement. From (5) – (6) for a given y^0 at the final time instant we get

$$y^N = \mathcal{A}y^0, \quad \mathbf{x} \in \overline{\omega}, \quad \mathcal{A} = S^n, \tag{7}$$

where S – is an operator of transition from one temporary layer to another:

$$S = (E + \sigma\tau A)^{-1}(E + (\sigma - 1)\tau A). \tag{8}$$

Thus, it is natural to put a solution of the following grid operator equation to approximate solution of the inverse problem:

$$\mathcal{A}v = \phi(\mathbf{x}), \quad \mathbf{x} \in \overline{\omega}. \tag{9}$$

Due to self-adjointness of an operator A a transition operator S is self-adjoint and, therefore, the operator $\mathcal{A}$ is also self-adjoint in equation (7), its unique solvability will take place, for example, if the operator $\mathcal{A}$ is positive. This condition will be met for the positive transition operator S. Taking representation (9),into consideration, we obtain $S > 0$ under condition of stability

$$\sigma \geq \frac{1}{2} - \frac{1}{\tau\|A\|}. \tag{10}$$

A numerical implementation of the difference scheme will be carried out through an iterative method of conjugate gradients [15], which essence is to refine an initial condition at each iteration. Moreover, at each iteration, correct problems are solved using standard two-layer difference schemes [9].

4. NUMERICAL EXPERIMENTS

To demonstrate a working efficiency of the proposed computational algorithm for solving a retrospective inverse problem for a parabolic equation, let us consider a two-dimensional model problem

$$\frac{\partial u}{\partial t} - \frac{\partial}{\partial x}\Big(k_1(x,y)\frac{\partial u}{\partial x}\Big) - \frac{\partial}{\partial y}\Big(k_2(x,y)\frac{\partial u}{\partial y}\Big) = 0, \quad x \in (0,l) = \Omega, \quad 0 \leq t < T. \tag{11}$$

Let us suppose that homogeneous first-type boundary conditions are given

$$u(0,y,t) = u(x,0,t) = u(l,y,t) = u(x,l,t) = 0, \quad 0 < t \leq T, \tag{12}$$

and an additional condition in the form of function value at the final time instant $t = T$ is specified:

$$u(x,y,T) = \phi(x,y), \quad 0 < x < l, \quad 0 < y < l. \tag{13}$$

The calculations were carried out on the exact solution (4) with $\beta = 1,\ l = 16,\ T = 1,\ n = 50,\ M = 50,\ K = 20$. The best accuracy in solving the direct problem was shown in a scheme with a weight factor $\sigma = 2/3$.

Journal of Physics: Conference Series **1392** (2019) 012089 doi:10.1088/1742-6596/1392/1/012089

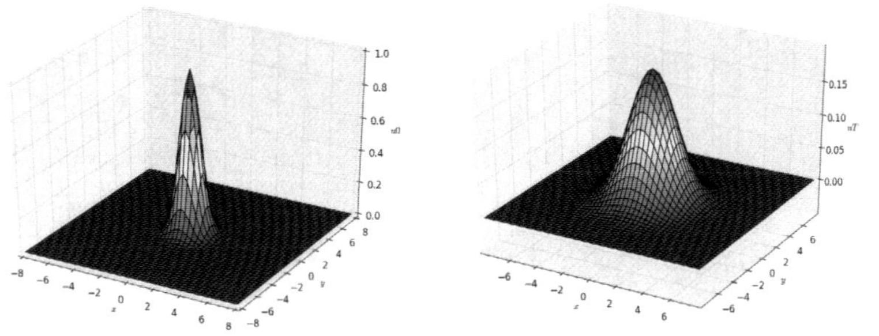

Fig.1. The initial condition (left), the additional condition (right).

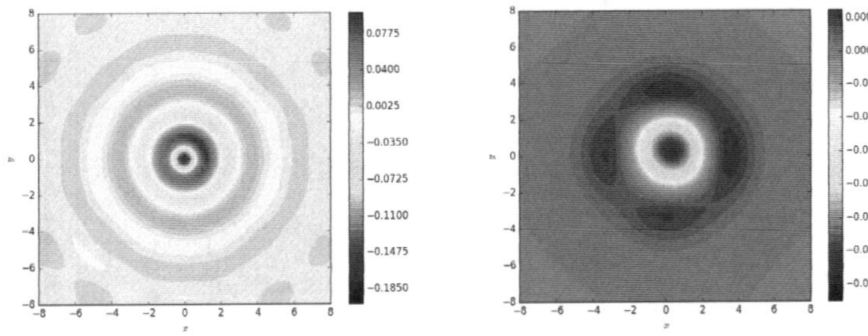

Fig.2. The error in determining the initial condition (left) and a graph of the deviation at the final time from the exact solution (right). Analytical solution.

In Fig.1 on the left there is a graph of the desired function at the initial time, and on the right there is an override condition (4) $u(x,y,T)=\phi(x,y,1)$. In Fig.2 (left) there is an error of the desired function found by the iterative method $u(x,y,1)=\phi(x,y)$. Number of iterations is 20. The right is a graph of the deviation of the found approximate solution at the final time from the exact condition of redefinition. The error is rather large, since at each iteration a difference scheme is implemented, which has a substantial approximation error on this "coarse" grid.

Now let us conduct a computational experiment with an initial condition which is a discontinuous function [10], then, the additional condition will be a quasi-realistic solution obtained from the direct problem (13) – (14) with the initial condition

$$\nu(x)=\begin{cases} 0 & \text{for } x\le 2l/5 \quad \text{or} \quad x\ge 3l/5, \\ 1 & \text{otherwise} \end{cases} \tag{14}$$

As a set condition for redefinition at the final time, we take the value of a direct problem solution using a difference scheme with a weighting factor $\sigma=1/2-h^2/(12\tau)$:

$$\phi(x_i)=y_i^M, \quad i=0,...,n,$$

The calculations were carried out with the following $l=30$, $T=2$, $n=50$, $M=50$, $K=20$.

In Fig.3, on the left there is a graph of the sought discontinuous initial condition, and in the graph on the right there is the overdetermination condition, which is the solution value (**??**) at the final time instant $u(x,y,1)=\phi(x,y)$, calculated using an implicit difference scheme with $\sigma=2/3$. In Fig.4 (left) there is the deviation of the unknown function found by the iteration method $u(x,y,1)=\phi(x,y)$ (left). The number of iterations is 20. On the right there is a graph of the deviation of the found approximate solution at the final time point from the quasi-real override condition. The error is rather large, since at each iteration a difference scheme is implemented, which has a substantial approximation error on this "coarse" grid.

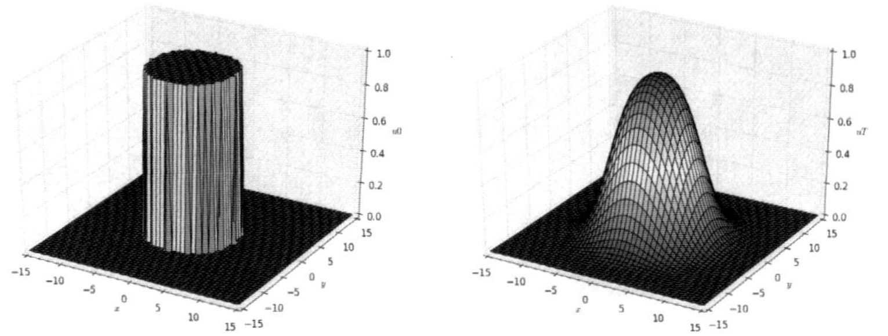

Fig.3. The graph of the initial condition (left), a graph of the additional condition (right).Discontinuous initial condition.

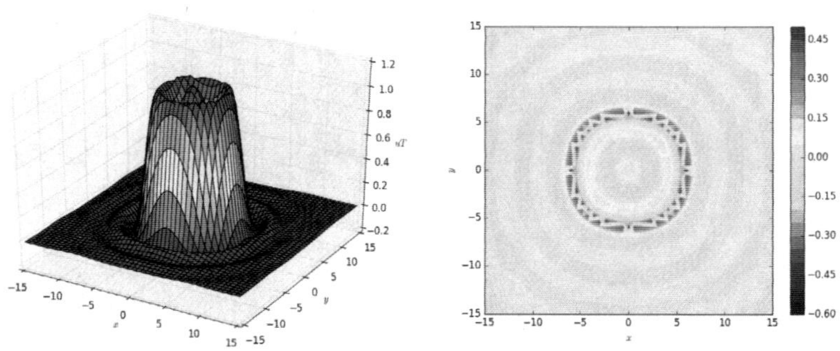

Fig.4. The graph of the restored initial condition (left) and the error in determining the initial condition (right). Discontinuous initial condition.

In inverse problems, the additional conditions are the results of measurements obtained by devices with specified accuracy characteristics, i.e, they should be set with some "noise". It is necessary to find out how small perturbations introduced artificially in an additional condition affect the accuracy of identification of the initial condition. Let us simulate input errors using the following function

$$\varphi_1(x) = \varphi(x_i) + \delta\gamma(-0.5, 0.5),$$

where δ is a coefficient of the defined error, $\gamma(-0.5, 0.5)$ is a generator of random numbers uniformly distributed over the interval $(-0.5, 0.5)$.

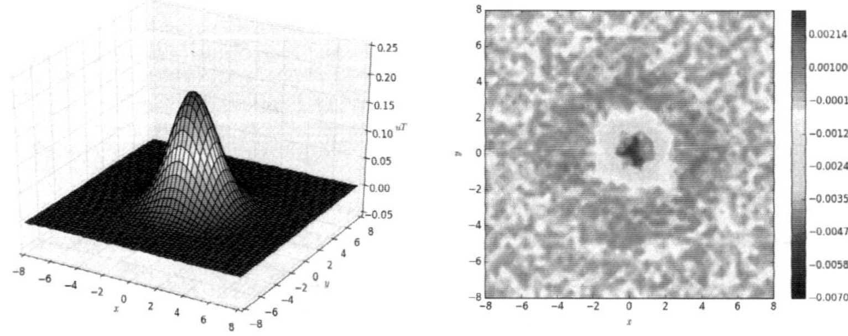

Fig.5. Condition of overdetermination with noise $\delta = 0.007$(left), deviation from the quasi-real overdetermination condition (right).

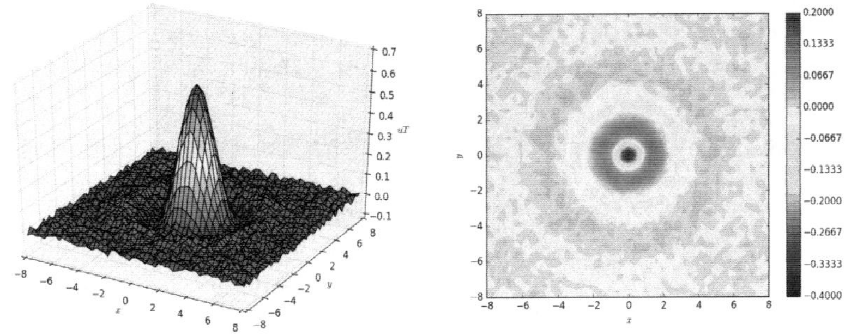

Fig.6. The graph of the restored initial condition (left) and deviation from the exact initial condition. Analytical solution.

The calculations were carried out with the same input data on the same space-time grid as the previous examples. Figure 5 shows the result of calculations for an example with an exact analytical solution (4). The artificially introduced "noise" was of the order of about 1% when specifying the overdetermination condition. Left noisy additional condition. On the right there is a graph of the deviation of the found approximate solution at the final time from the exact condition of redefinition.

The reconstructed initial condition is shown in Fig.6(left), and in the graph on the right there is its deviation from the initial condition. 5 iterations were performed. It should be noted that the accuracy of identification of the initial condition has greatly increased due to conditional correctness of the problem under consideration.

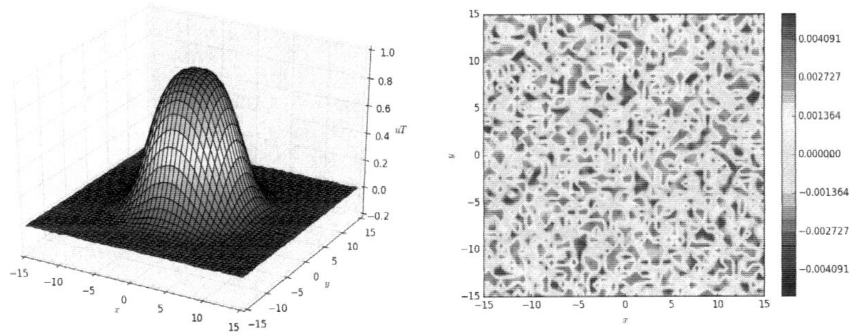

Fig.7. Condition of overdetermination with noise $\delta = 0.005$(left), deviation from the quasi-real overdetermination condition (right).

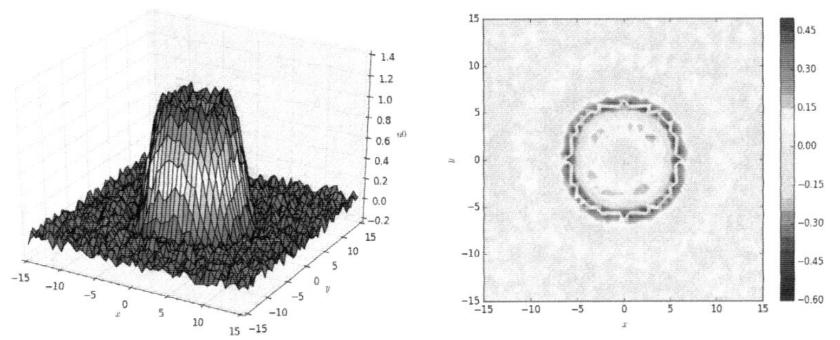

Fig.8. The graph of the restored initial condition (left) and the error in determining the initial condition $\delta = 0.009$ (right).

Fig.7 presents the noisy overdetermination condition $\delta = 0.005$ (1%) (left) and the deviation from the quasi-real overdetermination condition. Fig.8(left) shows the deviation of the unknown function found by the iteration method $u(x, y, 1) = \phi(x, y)$. The number of iterations is 5. On the right there is a graph of the deviation of the found approximate solution at the final moment of time from the quasi-real overdetermination condition. The error is also big enough.

5. CONCLUSION

For numerical solution of the retrospective inverse problem for a parabolic equation, an iterative conjugate gradient method was used. A computational experiment conducted on model two-dimensional problems confirmed the effectiveness of the proposed iterative method for solving a retrospective inverse problem of heat conduction. The method of conjugate gradients can be used to solve other inverse problems and these studies will be further developed in subsequent works.

6. ACKNOWLEDGMENTS

The authors express their sincere gratitude to Professor Petr Nikolaevich Vabishchevich for constructive comments and fruitful discussions. The work was partially supported by the Russian National Science Foundation (project 19-11-00230).

References

[1] *Aster R. C., Borchers B., Thurber C.H.* Parameter estimation and inverse problems. Academic Press. 2004.
[2] *Kabanikhin S. I.* Inverse and Ill-Posed Problems. Theory and Applications. De Gruyter, Germany, 2011.
[3] *Lavrent'ev M. M., Romanov V. G., Shishatskii S. P.* Ill-posed problems of mathematical physics and analysis. American Mathematical Society, 1986.
[4] *Isakov V.* Inverse Problems for Partial Differential Equations, Springer Verlag, New York, 2006.
[5] *Latte's R., Lions J. L.* The Method of Quasi-Reversibility. Applications to Partial Differential Equations. American Elsevier Publishing Company. 1969
[6] *Prilepko A. I., Orlovsky D. G., Vasin I. A.* Methods for solving inverse problems in mathematical physics. Marcel Dekker, 2000.
[7] *Samarskii A. A., Vabishchevich P. N.* Numerical methods for solving inverse problems of mathematical physics. De Gruyter, 2007.
[8] *Han H., Ingham D. B., Yuan Y.* The boundary element method for the solution of the backward heat conduction equation // J. Comput. Phys., 116 (1995), pp. 292–299.
[9] *Samarskii A. A., Vabishchevich P. N., Vasil'ev V. I.* Iterative Solution of a Retrospective Inverse Problem of Heat Conduction // Matematicheskoe Modelirovanie, Volume 9, Number 5, 1997, pp. 119–127.
[10] *Mera N. S., Elliott L., Ingham D. B., Lesnic D.* An iterative boundary element method for solving theone-dimensional backward heat conduction problem // International Journal of Heat and Mass Transfer, Volume 44, Issue 10, May 2001, pp. 1937–1946.
[11] *Zhenyu Zhao, Zehong Meng.* A modified Tikhonov regularization method for a backward heat equation // Inverse Problems in Science and Engineering. Vol. 19, No. 8, December 2011, pp.1175–1182.
[12] *Tsepelev I. A.* Iterative algorithm for solving the retrospective ptoblem of natural thermal convection of a viscous fluid // Computational continuum mechanics 2011. . 4, No 2. pp. 119–127.
[13] *Vasilev V. I., Kardashevsky A. M.* Iterative Solution of the Retrospective Inverse problem for a Parabolic Equation Using the Conjugate Gradient Method / LNCS, 2017, volume 10187. pp. 698–705.
[14] *Vasilev V. I., Kardashevsky A. M.* Numerical solution of the retrospective inverse problem of heat conduction with the help of the Poisson integral // Journal of Applied and Industrial Mathematics. 2018, Volume 12, Issue 3. pp. 577–586.
[15] *Saad U.* Iterative Methods for Sparse Linear Systems. 2-nd Edition. SIAM, 2003.
[16] *Samarskii A. A.* The theory of difference schemes. Marcel Dekker, 2001.
[17] *Savitzky A., Golay M.* Smoothing and differentiation of data by simplified least squares procedures // Anal. Chem. 1964. Vol. 36. pp. 1627–1639.

Journal of Physics: Conference Series **1392** (2019) 012090 doi:10.1088/1742-6596/1392/1/012090

IDENTIFICATION OF HEAT EXCHANGE BOUNDARY CONDITIONS AT VARIOUS NATURAL AND TECHNOGENIC FACTORS

[1]Vinokurova T.A., [2]Permyakov P.P.
[1] P.I. Melnikov Institute of Permafrost of the Russian Academy of Sciences (Siberian Branch)
[2] V.P. Larionov Institute of Physical and Technical Problems of the North of the Russian Academy of Sciences (Siberian Branch)
E-mail: `tatyana_umka91@mail.ru, permyakov2005@mail.ru`

Abstract. The development of permafrost areas requires reliable and long-term projected estimates in a changing climate and man-made impacts. Dangerous natural and man-made factors, such as snowfall, infiltration of precipitation, ice, thermokarst, etc., interact with the ground and determine the stability of the territories of the cryolithozone. In this regard, the work is devoted to the development of new algorithms and software for identifying the boundary conditions on the surface of frozen soil, taking into account the actual process of freezing and thawing of the pore solution of the soil.

Problem Formulation

The temperature field of frozen soils, taking into account the phase transition of pore water, is described by the following heat conduction problem (direct problem) [6] :

$$c^\beta \rho^\beta \frac{\partial T^\beta}{\partial \tau} = \frac{1}{r^n}\frac{\partial}{\partial r}(\lambda^\beta r^n \frac{\partial T^\beta}{\partial r}), \quad R_1 < r < R_N, \quad 0 < \tau <= \tau_m, \quad \beta = \overline{1,N}, \tag{1}$$

$$\lambda^\beta \frac{\partial T^\beta}{\partial r} = 0, \quad r = R_N, \quad 0 < \tau <= \tau_m, \tag{2}$$

$$T^\beta(r,0) = T_0^\beta(r), \quad R_1 <= r <= R_N, \quad \beta = \overline{1,N}, \tau = 0. \tag{3}$$

We see matching conditions at the junction of the two layers:

$$T^\beta|_{R_\beta^-} = T^{\beta+1}|_{R_\beta^+}$$

$$\lambda^\beta \frac{\partial T^\beta}{\partial r}\bigg|_{R_\beta^-} = \lambda^{\beta+1}\frac{\partial T^{\beta+1}}{\partial r}\bigg|_{R_\beta^+}, \quad 0 < \tau <= \tau_m, \quad \beta = \overline{1, n-1},$$

where

$$c^\beta = c^\beta(T) = c_{sk}^\beta + c_i \omega_0^\beta + (c_w - c_i)\omega_{uw}^\beta(T) + L\frac{\partial \omega_{uw}^\beta(T)}{\partial T},$$

$$\lambda^{\beta} = \lambda^{\beta}(T) = \lambda_f^{\beta} + (\lambda_t^{\beta} - \lambda_f^{\beta})\frac{W_{uw}^{\beta}(T) - W_{tb}^{\beta}}{W_0^{\beta} - W_{tb}^{\beta}},$$

It is required to restore the boundary condition on the ground surface ($s = 1$) with $r = R_1$:

$$-\lambda^1 \frac{\partial T^1}{\partial r} = q(\tau), \quad r = R_1, \quad 0 < \tau <= \tau_m, \tag{4}$$

where the desired parameter is the function $u(\tau) = q(\tau)$.

To restore the desired parameter, additional measurements of the temperature inside the sample are required:

$$T(r_i, \tau) = T_i^{э}(\tau), \quad i = \overline{1, n_{\text{т}}}. \tag{5}$$

We formulate this problem as an optimal control problem: find the function $u(\tau)$ from the minimum of the objective functional:

$$J(u) = \sum_{i=1}^{n_{\text{т}}} \int_0^{\tau_m} p_i(\tau)\Big(T(r_i, \tau) - T^{э}(r_i, \tau)\Big)^2 d\tau \tag{6}$$

Here c is effective heat capacity of the soil, taking into account the phase transition in the temperature spectrum, J/(kg K); ρ is soil density, kg/m^3; T is temperature, K; λ is ground thermal conductivity, W/(m K), τ is time, s; r is spatial coordinate, m; n=0,1,2 corresponds to Cartesian, cylindrical and spherical coordinates; β is layer index; R_1, R_2 are area boundaries, m; τ_m is right boundary value of the time interval; L is specific heat of phase transition, J/kg; W_{uw}, W_{tb} is unfrozen and tightly bound water, %; $q(\tau)$ is heat flow, W/m^2; $p_i(\tau)$ is weight factors with dimension K^{-2} c^{-1}; $T(r_i, \tau), T^{э}(r_i, \tau)$ is calculated and measured temperatures in the i-th point of the frozen array.

Solution Algorithm

Boundary inverse problems belong to the class of incorrect ones and are solved by special iterative methods. In this case, the conditions of existence and uniqueness of the solution of the boundary inverse problems of heat conduction [2]. Iterative methods are used with great success in solving ill-posed problems [2], [3], [7]. In this case, the number of iterations acts as a regularization parameter.

In the conjugate gradient method, the iteration sequence is constructed taking into account the previous direction of descent according to the following law:

$$u^{s+1} = u^s - \beta_s S^s, \quad s = 0, 1, ...,$$

$$S^s = J'(u^s) - \gamma_s S^{s-1}, \quad S^0 = J'(u^0), \quad \gamma_0 = 0 \tag{7}$$

The calculation procedure for the conjugate gradient method is as follows. Starting from a certain point u^0 the movement in the direction of the anti-gradient to the minimum is achieved, which is reached at the point u^1. Then, the direction paired with the previous one, in which the transition to the next minimum point occurs, is determined. As a result, a new approximation of u^2 is determined, and so on.

An important feature of conjugate gradients is that the number of iterations to achieve an exact minimum of a quadratic function is theoretically equal to m – the dimension of the desired vector. In practice, due to errors that appear in the course of the calculation, the convergence of the method deteriorates, for this, it is recommended to "stir up" the iterative process, this allows to reduce the accumulation of computational errors.

γ_s parameter is defined in various ways depending on the properties of the target functional. So, for example, if the second Fréchet derivative function changes little, then the parameter is given using the formula:

$$\gamma_s = -\frac{\left(J'(u^s), J'(u^{s-1}) - J'(u^s)\right)_{R^{m_0}}}{||J'(u^s)||^2_{R^{m_0}}}. \tag{8}$$

If the minimized functionality is close to quadratic, then it is determined by the expression:

$$\gamma_s = -\frac{||(J'(u^s)||^2_{R^{m_0}}}{||J'(u^{s-1})||^2_{R^{m_0}}}. \tag{9}$$

Step size β_s of the conjugate gradient method is determined using the formula:

$$\beta_s = \frac{\sum_{i=1}^{n_T} \int_0^{\tau_m} p_i(\tau)\nu_i\left(T_i - T_i^э\right)d\tau}{\sum_{i=1}^{n_T} \int_0^{\tau_m} p_i(\tau)\nu_i^2 d\tau}, \tag{10}$$

where $\nu_i = \nu_i(r, \tau)$ is solution of the boundary value problem for increment.

The gradient of the functional (descent direction) has the form:

$$\frac{\partial J}{\partial b_\alpha} = \int_0^{\tau_m} \psi^1 B_\alpha(\tau)\bigg|_{r=R_1} d\tau, \tag{11}$$

$$\alpha = -1, 0, 1, ..., m_0 + 1,$$

where $\psi^1 = \psi^1(r, \tau)$ is solution of the conjugate problem.

The reliability of the proposed algorithm was tested on a model problem with an exact solution. As a result of a numerical experiment, it was established that the proposed algorithm has a good regularizing property [5].

Numerical Research

Example 1. According to the above proposed algorithm, the heat flux density was restored during the formation of icing in the Ulakhan-Taryn Valley. The length of the icing was about 3 km, width ranged from 120-800 m [1].

The observation was made on four different stationary platforms: 1 – icing; 2 – spruce forest in the icing valley; 3 – pine forest on the watershed, as well as on the "Meadow" Site in Yakutsk on the terrace above the floodplain on the left bank of the Lena River.

Site 3 (pine forest) is 25 m higher from the icing valley at the watershed with total humidity up to 10%, and the soil and vegetation layer is only 0.05 m. Numerical calculations show that the annual course of the heat flux density goes more than intensive compared to other sites (Fig. 1).

At the beginning of the winter period (November), the process of freezing in Icing Valley 1 and Site 2 is almost the same, but in January, ground water arrives, which increases the heat flow density in this massif. The height of the frost reaches up to 2 meters, but in each year its height varies depending on the external temperature and the volume of groundwater. The thawing process depends on the external temperature, the area of ice and sometimes ends at the end of July or does not completely thaw. The depth of thawing of water-saturated ground comes

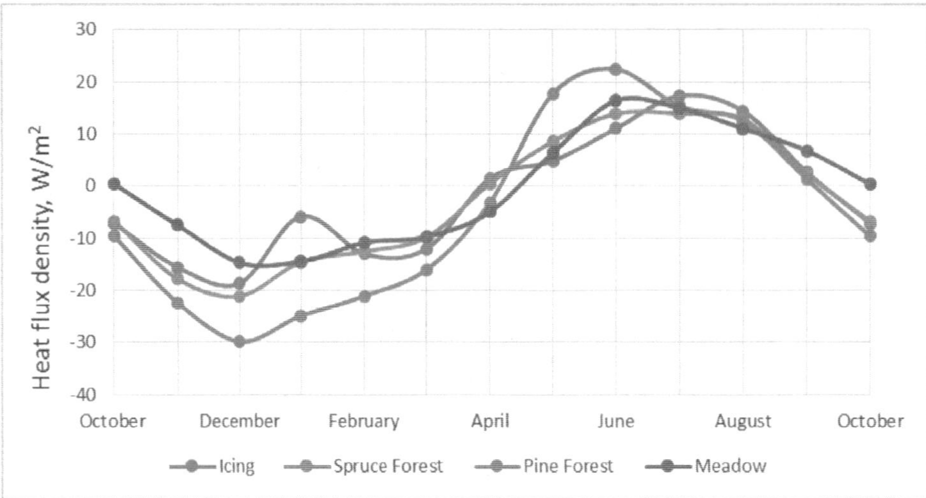

Figure 1. Annual course of the density of the heat flux on the surface of the sites

with some delay, but at the end of the summer period the depth of thawing of natural open ground is reached and due to the large total humidity, full freezing occurs at the most deadline.

The soil section of the "Meadow" Site consists of 0.0-0.3 m of sod-vegetable layer, which is a shielding low heat-conducting layer. Therefore, the heat flux of this area lies in the range from -15 to 18 W/m^2.

Example 2. Considered the restoration of heat flow in "Tuymaada" Station. The initial temperature distribution corresponds to the actual observations at the hospital. As $T^э(r_i, \tau)$ temperature measurements at a depth of 0.2 m are given.

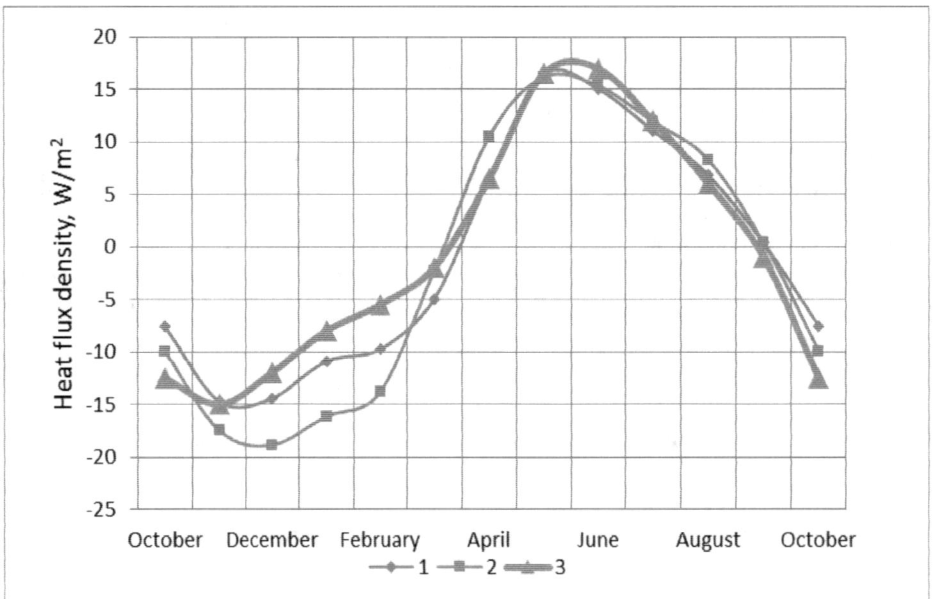

Figure 2. Annual variation of heat flows at the surface of herb meadow (1), bare plot (2) and the data of Pavlov (1979) (3)

From the graph of recovery of heat flux on the surface of the Site is shown (Fig. 2), that at the bare site, the freezing process (curve 1) is more intensive in the wintertime in comparison with the grass of the meadow (curve 2). In March – April, there is an intensive influx of heat flow in the bare area, and in the summertime, the flow intensity is almost the same. Curve 3, which reflects the density of the heat flux, was determined by a numerical method according to the heat balance observation data [4]. The figure shows that the absolute value of the heat flux according to Pavlov A.V. in winter is lower compared with the results of our observations. The intensity of the heat flux depends on the insulating properties of snow, which is determined by the height, density, and structure of the snow cover (density, temperature, water vapor, etc.). In the summer, the heat flux densities do not differ much, but a shading effect of vegetation appears.

Conclusion

With the help of the proposed algorithm, heat fluxes on the frozen soil surface were restored under various exogenous and man-made conditions (ice, snow removal, land cover, etc.).

The results of computational experiments show the high efficiency of the method of iterative regularization and the possibility of solving with it a wide range of nonlinear problems.

The study was carried out with the financial support of the Russian Foundation for Basic Research in the framework of the Research Project No. 18-41-140008.

References

[1] Alekseev V R 1987 *Ice Build-Up* (Novosibirsk: Science) p 160
[2] Alifanov O M, Artyukhin E A and Rumyantsev S V 1998 *Extreme Methods for Solving Ill-Posed Problems* (M.: Science) p 288
[3] Alifanov O M, Budnik S A, Nenarokomov A V, Netelev A V and Titov D M 2013 *Destructive Materials of Thermal Characteristics Definition Using for Space Technology Tests* (Astronautics) vol 85 pp 113–119
[4] Pavlov A V 1979 *Thermal Physics Landscapes* (Novosibirsk: Science) p 284
[5] Permyakov P P, Afanasyeva T A, Varlamov S P and Skryabin P N 2018 *Determination of Boundary Conditions in Modeling the Thermal Regime of Frozen Soils* (Bulletin of the North-Eastern Scientific Center of the Far Eastern Branch of the Russian Academy of Sciences) No. 1 (53) pp 56–62
[6] Permyakov P P, Varlamov S P, Scriabin P N and Skachkov Y B 2016 *Numerical Simulation of the Thermal State of the Cryolithozone in a Changing Climate* (Science, and Education) No. 2 pp 43–48
[7] Danaila S and Chira A 2014 *Mathematical and Numerical Modeling of Inverse Heat Conduction* (Incas Bulletin) vol 6 Issue 4 pp 23–39

AUTHOR INDEX

AUTHOR INDEX